D. Demus, J. Goodby, G. W. Gray,
H.-W. Spiess, V. Vill

Handbook
of Liquid Crystals

 WILEY-VCH

Handbook of Liquid Crystals

D. Demus, J. Goodby,
G. W. Gray,
H.-W. Spiess, V. Vill

Vol. 1:
Fundamentals

Vol. 2 A:
Low Molecular Weight Liquid Crystals I

Vol. 2 B:
Low Molecular Weight Liquid Crystals II

Vol. 3:
High Molecular Weight Liquid Crystals

Further title of interest:

J. L. Serrano:
Metallomesogens

ISBN 3-527-29296-9

D. Demus, J. Goodby, G. W. Gray,
H.-W. Spiess, V. Vill

Handbook of Liquid Crystals

Vol. 3:
**High Molecular
Weight Liquid
Crystals**

WILEY-VCH

Weinheim • New York • Chichester
Brisbane • Singapore • Toronto

Prof. Dietrich Demus
Veilchenweg 23
06118 Halle
Germany

Prof. John W. Goodby
School of Chemistry
University of Hull
Hull, HU6 7RX
U. K.

Prof. George W. Gray
Merck Ltd.
Liquid Crystals
Merck House
Poole BH15 1TD
U.K.

Prof. Hans-Wolfgang Spiess
Max-Planck-Institut für
Polymerforschung
Ackermannweg 10
55128 Mainz
Germany

Dr. Volkmar Vill
Institut für Organische Chemie
Universität Hamburg
Martin-Luther-King-Platz 6
20146 Hamburg
Germany

This book was carefully produced. Nevertheless, authors, editors and publisher do not warrant the information contained therein to be free of errors. Readers are advised to keep in mind that statements, data, illustrations, procedural details or other items may inadvertently be inaccurate.

Library of Congress Card No. applied for.

A catalogue record for this book is available from the British Library.

Deutsche Bibliothek Cataloguing-in-Publication Data:

Handbook of liquid crystals / D. Demus ... – Weinheim ; New York
; Chichester ; Brisbane ; Singapore ; Toronto : Wiley-VCH
    ISBN 3-527-29502-X

    Vol. 3. High molecular weight liquid crystals. – 1998
    ISBN 3-527-29272-1

Composition and Printing: Fa. Konrad Triltsch Druck- und Verlagsanstalt GmbH, D-97070 Würzburg.
Bookbinding: Wilhelm Osswald & Co., D-67433 Neustadt
Printed in the Federal Republic of Germany.

The Editors

D. Demus

studied chemistry at the Martin-Luther-University, Halle, Germany, where he was also awarded his Ph. D. In 1981 he became Professor, and in 1991 Deputy Vice-Chancellor of Halle University. From 1992–1994 he worked as a Special Technical Advisor for the Chisso Petrochemical Corporation in Japan. Throughout the period 1984–1991 he was a member of the International Planning and Steering Commitee of the International Liquid Crystal Conferences, and was a non-executive director of the International Liquid Crystal Society. Since 1994 he is active as an Scientific Consultant in Halle. He has published over 310 scientific papers and 7 books and he holds 170 patients.

J. W. Goodby

studied for his Ph. D. in chemistry under the guidance of G. W. Gray at the University of Hull, UK. After his post-doctoral research he became supervisor of the Liquid Crystal Device Materials Research Group at AT&T Bell Laboratories. In 1988 he returned to the UK to become the Thorn-EMI/STC Reader in Industrial Chemistry and in 1990 he was appointed Professor of Organic Chemistry and Head of the Liquid Crystal Group at the University of Hull. In 1996 he was the first winner of the G. W. Gray Medal of the British Liquid Crystal Society.

G. W. Gray

studied chemistry at the University of Glasgow, UK, and received his Ph. D. from the University of London before moving to the University of Hull. His contributions have been recognised by many awards and distinctions, including the Leverhulme Gold Medal of the Royal Society (1987), Commander of the Most Excellent Order of the British Empire (1991), and Gold Medallist and Kyoto Prize Laureate in Advanced Technology (1995). His work on structure/property relationships has had far reaching influences on the understanding of liquid crystals and on their commercial applications in the field of electro-optical displays. In 1990 he became Research Coordinator for Merck (UK) Ltd, the company which, as BDH Ltd, did so much to commercialise and market the electro-optic materials which he invented at Hull University. He is now active as a Consultant, as Editor of the journal "Liquid Crystals" and as author/editor for a number of texts on Liquid Crystals.

H. W. Spiess
studied chemistry at the University of Frankfurt/Main, Germany, and obtained his Ph. D. in physical chemistry for work on transition metal complexes in the group of H. Hartmann. After professorships at the University of Mainz, Münster and Bayreuth he was appointed Director of the newly founded Max-Planck-Institute for Polymer Research in Mainz in 1984. His main research focuses on the structure and dynamics of synthetic polymers and liquid crystalline polymers by advanced NMR and other spectroscopic techniques.

V. Vill
studied chemistry and physics at the University of Münster, Germany, and acquired his Ph. D. in carbohydrate chemistry in the gorup of J. Thiem in 1990. He is currently appointed at the University of Hamburg, where he focuses his research on the synthesis of chiral liquid crystals from carbohydrates and the phase behavior of glycolipids. He is the founder of the LiqCryst database and author of the Landolt-Börnstein series *Liquid Crystals*.

List of Contributors

Volume 3, Polymeric Liquid Crystals and Lyotropic Crystals

Blunk, D.; Praefcke, K. (**VI**)
Technische Universität Berlin
Institut für Organische Chemie
Straße des 17. Juni 135
10623 Berlin
Germany

Brand, H. R. (**V**)
Theoretische Physik III
Universität Bayreuth
95440 Bayreuth
Germany

Chiellini, E. (**I**:2)
Diparitmento di Chimica e
Chimica Industiale
Università di Pisa
via Risorgimento 35
56126 Pisa
Italy

Dubois, J. C.; Le Barny, P. (**IV**)
Thomson CSF
Laboratoire Central de Recherche
Domaine de Corbeville
91404 Orsay Cedex
France

Fairhurst, C.; Holmes, M. C. (**VII**)
Department of Physics and Astronomy
University of Central Lancashire
Preston, PR1 2HE
U.K.

Finkelmann, H. (**V**)
Albert-Ludwig-Universität
Institut für Makromolekulare Chemie
Stefan-Meier-Straße 31
79104 Freiburg
Germany

Gray, J. (**VII**)
Division of Chemical Sciences
Science Research Institute
University of Salford
Salford, M5 4WT
U.K.

Greiner, A. (**I**:1)
Philipps-Universität Marburg
Fachbereich Physikalische Chemie
35032 Marburg
Germany

Hoffmann, S. (**VIII**)
Martin-Luther-Universität
Halle-Wittenberg
FB Biochemie/Biotechnologie
Kurt-Mothes-Straße 3
06120 Halle (Saale)
Germany

Laus, M. (**I**:2)
Dipartimento di Chimica Industriale
e dei Materiali
Università di Bologna
via Risorgimento 4
40137 Bologna
Italy

Mauzac, M. (**IV**)
CRPP-CNRS
Avenue du Dr. Schweitzer
33600 Pessac
France

Noel, C. (**II** and **IV**)
Laboratoire de Physico-Chimie
Structurale de Macromoleculaire
Laboratoire Associè au CNRS
10, Rue Vauquelin
75231 Paris
France

Ober, C. K.; Mao, G. (**I**:4)
Cornell University
Dept. of Materials Science
and Engineering
322 Bard Hall
Ithaca, NY 14853-1501
USA

Pugh, C.; Kiste, A. (**III**)
The University of Michigan
Dept. of Chemistry
Ann Arbor, MI 4809-1055
USA

Schmidt, H.-W. (**I**:1)
Universität Bayreuth
Makromolekulare Chemie I
Universitätsstraße 30
NW II
95440 Bayreuth
Germany

Tiddy, G. J. T.; Fuller, S. (**VII**)
Department of Chemical Engineering
Institute of Science and Technology
University of Manchester
Manchester, M60 1QD
U.K.

Vill, V. (**VI**)
Universität Hamburg
Institut für Organische Chemie
Martin-Luther-King-Platz 6
20146 Hamburg
Germany

Zentel, R. (**I**:3)
BUGH Wuppertal
Fachbereich Chemie
Gaußstraße 20
42097 Wuppertal
Germany

Zugenmaier, P. (**IX**)
Institut f. Physikalische Chemie
Technische Universität Clausthal
Arnold-Sommerfeld-Straße 4
38678 Clausthal-Zellerfeld
Germany

Outline

Volume 1

Volume 2A

Volume 2B

Volume 3

Contents

Chapter IV: Behavior and Properties of Side Group Thermotropic Liquid Crystal Polymers

Jean-Claude Dubois, Pierre Le Barny, Monique Mauzac, and Claudine Noel

Part III: Amphiphilic Liquid-Crystals

Chapter VI: Amphotropic Liquid Crystals
Dieter Blunk, Klaus Praefcke and Volkmar Vill

Chapter VII: Lyotropic Surfactant Liquid Crystals. 341
C. Fairhurst, S. Fuller, J. Gray, M. C. Holmes, G. J. T. Tiddy

General Introduction

Liquid crystals are now well established in basic research as well as in development for applications and commercial use. Because they represent a state intermediate between ordinary liquids and three-dimensional solids, the investigation of their physical properties is very complex and makes use of many different tools and techniques. Liquid crystals play an important role in materials science, they are model materials for the organic chemist in order to investigate the connection between chemical structure and physical properties, and they provide insight into certain phenomena of biological systems. Since their main application is in displays, some knowledge of the particulars of display technology is necessary for a complete understanding of the matter.

In 1980 VCH published the *Handbook of Liquid Crystals*, written by H. Kelker and R. Hatz, with a contribution by C. Schumann, which had a total of about 900 pages. Even in 1980 it was no easy task for this small number of authors to put together the *Handbook*, which comprised so many specialities; the *Handbook* took about 12 years to complete. In the meantime the amount of information about liquid crystals has grown nearly exponentially. This is reflected in the number of known liquid-crystalline compounds: in 1974 about 5000 (D. Demus, H. Demus, H. Zaschke, *Flüssige Kristalle in Tabellen*) and in 1997 about 70000 (V. Vill, electronic data base LIQCRYST). According to a recent estimate by V. Vill, the current number of publications is about 65000 papers and patents. This development shows that, for a single author or a small group of authors, it may be impossible to produce a representative review of all the topics that are relevant to liquid crystals – on the one hand because of the necessarily high degree of specialization, and on the other because of the factor of time.

Owing to the regrettable early decease of H. Kelker and the poor health of R. Hatz, neither of the former main authors was able to continue their work and to participate in a new edition of the *Handbook*. Therefore, it was decided to appoint five new editors to be responsible for the structure of the book and for the selection of specialized authors for the individual chapters. We are now happy to be able to present the result of the work of more than 80 experienced authors from the international scientific community.

The idea behind the structure of the *Handbook* is to provide in Volume 1 a basic overview of the fundamentals of the science and applications of the entire field of liquid crystals. This volume should be suitable as an introduction to liquid crystals for the nonspecialist, as well as a source of current knowledge about the state-of-the-art for the specialist. It contains chapters about the historical development, theory, synthesis and chemical structure, physical properties, characterization methods, and applications of all kinds of liquid crystals. Two subsequent volumes provide more specialized information.

The two volumes on *Low Molecular Weight Liquid Crystals* are divided into parts dealing with calamitic liquid crystals (containing chapters about phase structures, nematics, cholesterics, and smectics), discotic liquid crystals, and non-conventional liquid crystals.

The last volume is devoted to polymeric liquid crystals (with chapters about main-chain and side-group thermotropic liquid crystal polymers), amphiphilic liquid crystals, and natural polymers with liquid-crystalline properties.

The various chapters of the *Handbook* have been written by single authors, sometimes with one or more coauthors. This provides the advantage that most of the chapters can be read alone, without necessarily having read the preceding chapters. On the other hand, despite great efforts on the part of the editors, the chapters are different in style, and some overlap of several chapters could not be avoided. This sometimes results in the discussion of the same topic from quite different viewpoints by authors who use quite different methods in their research.

The editors express their gratitude to the authors for their efforts to produce, in a relatively short time, overviews of the topics, limited in the number of pages, but representative in the selection of the material and up to date in the cited references.

The editors are indebted to the editorial and production staff of WILEY-VCH for their constantly good and fruitful cooperation, beginning with the idea of producing a completely new edition of the *Handbook of Liquid Crystals* continuing with support for the editors in collecting the manuscripts of so many authors, and finally in transforming a large number of individual chapters into well-presented volumes.

In particular we thank Dr. P. Gregory, Dr. U. Anton, and Dr. J. Ritterbusch of the Materials Science Editorial Department of WILEY-VCH for their advice and support in overcoming all difficulties arising in the partnership between the authors, the editors, and the publishers.

The Editors

Part 1:
Main-Chain Thermotropic Liquid-Crystalline Polymers

Chapter I
Synthesis, Structure and Properties

1 Aromatic Main Chain Liquid Crystalline Polymers

Andreas Greiner and Hans-Werner Schmidt

1.1 Introduction

Aromatic main chain liquid crystalline polymers consist of a sequence of directly connected aromatic moieties, namely, polyarylenes, or a sequence of aromatic moieties linked by an even number of atoms or heterocyclic units. Typical linkage groups in combination with aromatic moieties are, for example, ester and amid groups (Fig. 1). These polymers are semirigid to rigid materials with respect to the conformational freedom along their polymer backbone, and are thermotropic or lyotropic according to the Flory theory, depending mainly on chain extension, melting temperature, and solution behavior [1].

The properties, in particular the solubility and fusibility and as consequence the processability, are controlled by the molecular architecture and chemical structure of the aromatic moieties. Aromatic paralinked polymers with benzene rings as the aromatic moieties, without linking atoms, for example, poly-*p*-phenylene, or with an even number of linking atoms between the benzene rings, for example, poly-*p*-hydroxybenzoic acid, are characterized at relatively low molecular weights by high melting temperatures (often in the range of decomposi-

tion) and by poor solubilities. Consequently, in these systems thermotropic or lyotropic phases can only be observed (if at all) under extreme conditions, often in the range of polymer degradation. Improvement of the solubility and the fusibility are of fundamental interest in order to fully exploit the potential of aromatic liquid crystal polymers (LCPs). Therefore, in the past two decades considerable effort has been made to obtain tractable aromatic LCPs by different structural modification concepts.

It is far beyond the scope of this chapter to present a comprehensive overview on the different structures of aromatic main chain LCPs and take into account the various property and application aspects. The objective of this chapter is to discuss the impact of structural concepts in modifying the properties of aromatic LCPs, focusing here on aromatic thermotropic LC polyesters. This will be discussed for selected examples. Conclusions from the work on polyesters are transferrable to other classes of thermotropic and lyotropic aromatic polymers.

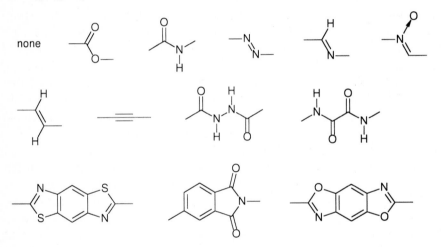

Aromatic units

Linkage group A

Figure 1. Typical aromatic units and linkage groups in aromatic liquid crystalline polymers.

1.2   Structural Modification Concepts of Liquid Crystal Polymers (LCPs)

Under thermodynamic considerations, high melting temperatures are the result of high melting enthalpies and low melting entropies. Consequently, the structural modification concepts are based either on a de-crease of the melting enthalpy and/or an increase of the melting entropy. These can be achieved mainly by a controlled decrease of the symmetry along the polymer backbone, so reducing interchain interactions, or by an increase in the chain flexibility, but without destroying the formation of LC phases. The various structural modifications, which have been extensively investigated, are schematically illustrated in Fig. 2.

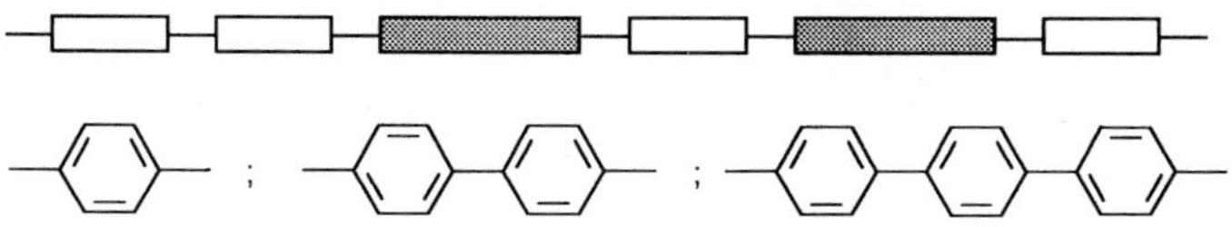

(a)

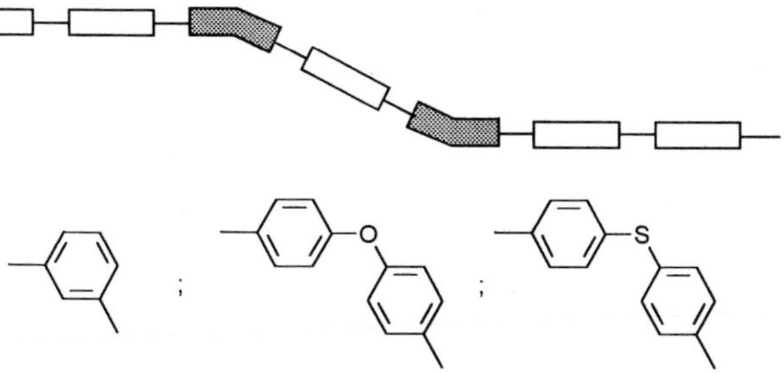

(b)

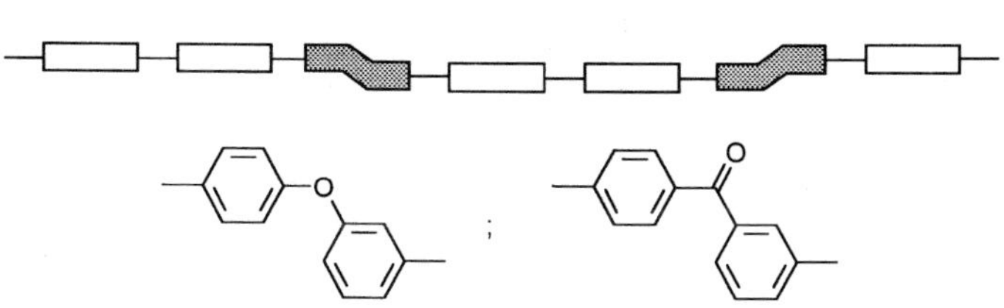

(c)

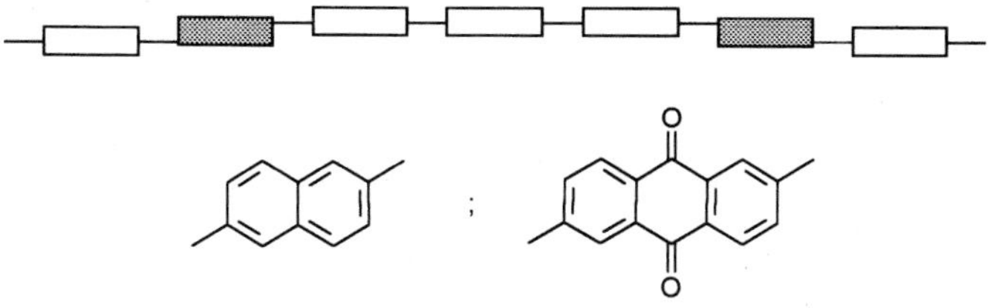

(d)

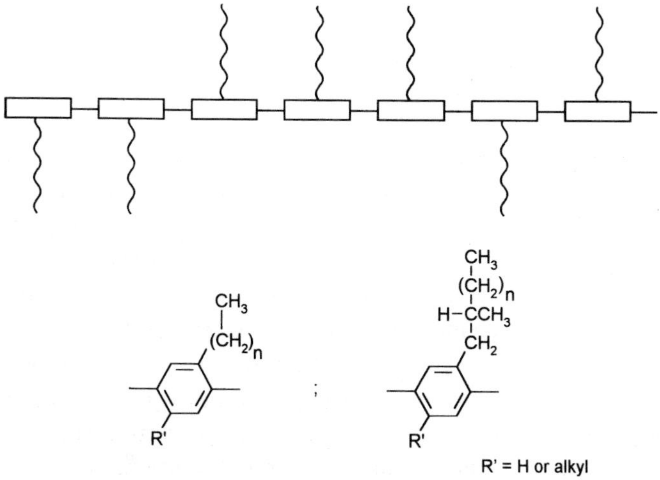

(e)

R' = H or alkyl

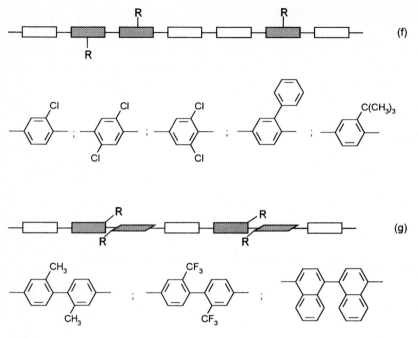

Figure 2. Modification concepts for para-linked aromatic polymers including typical monomer structures: (a) monomer units of different length, (b) kinked comonomers, (c) double kinked comonomers, (d) 'crankshaft' comonomers, (e) flexible lateral substituents, (f) bulky and stiff lateral substituents, and (g) monomers with non-coplanar conformation.

The incorporation of monomer units of different length, such as *p*-phenyl, *p*-biphenyl, or *p*-terphenyl moities (Fig. 2a), causes the linkage groups to be positioned at random distances along the polymer chain. As a consequence, the intermolecular interactions, particularly of the polar linkage groups, are partially excluded. The overall chain stiffness increases with the amount of the longer aromatic unit.

A different structural modification, which is mainly based on lowering the chain stiffness and as a result also on reducing the intermolecular interactions, is the incorporation of kinked and double kinked comonomers (Fig. 2b and c). Typical monomers with kinks are meta-substituted phenylene derivatives or 4,4′-functionalized biphenylethers and 4,4′-functionalized biphenylsulfides. Monomers with double kinks are,

for example, 4,3′-functionalized biphenylethers or 4,3′-substituted functionalized benzophenones. Substitution in the 4,3′ position allows for a linear conformation in which the kink is compensated to some degree. However, in both cases these monomers can only be incorporated as a comonomer in para-linked aromatic polymers up to a critical amount, until the polymer chain becomes too flexible and the LC formation is destroyed. An industrially important concept for the structural modification of LCPs involves using the so-called 'crankshaft' monomers (Fig. 2d), particularly 2,6-substituted naphthalene derivatives. These monomers result in a step-wise shift along the polymer backbone.

The intermolecular interactions can be lowered substantially by lateral substituents, which can be either flexible (Fig. 2e) or bulky and stiff (Fig. 2f). The number of substituents and symmetry of substitution can result in positional isomerism along the polymer backbone and, as a consequence, in a different direction of the substituents with respect to each other. In some cases, the substituent has an influence on the conformation of the linkage group and lowers the chain stiffness. If should be pointed out that these modifications also result in an increase in the chain diameter. The main differences between flexible and bulky substituents are their influence on the thermal stability and the glass transition temperature (T_g). Flexible chains act as an internal plasticizer, whereas bulky substituents increase the T_g.

Monomers with noncoplanar conformation are, for example, 2,2′-substituted biphenylenes or binaphthyl derivatives. Substitution in the 2,2′-position causes the phenyl units to be in a noncoplanar conformation. This reduces the intermolecular interactions between the chains very effectively. These monomers reduce the chain stiffness far less than the bulky substituents. It is obvious that combinations of these different structural modifications can be used and have been utilized in numerous examples to modify LCPs' properties.

An other important concept for the modification of LCPs is the incorporation of flexible spacers between mesogenic units. These semiflexible LCPs have been extensively investigated and have to be distinguished from the above systems. As a consequence, a separate contribution in this handbook is devoted to this class of LCPs (Sect. 2 of this Chapter).

In the following sections, the different concepts of structural modifications of LCPs will be discussed in more detail for thermotropic LC polyesters and briefly for lyotropic polyamides. The same structural modifications have been applied to other classes of para-linked aromatic polymers, e. g., polyimides, poly-(benzobisoxazole)s, poly-(benzobisthiazole)s, poly-p-phenylenes, poly(p-phenylenevinylene)s, and poly(p-phenyleneethinylene)s.

1.3   Aromatic LC (Liquid Crystal) Polyesters

Aromatic para-linked LC polyesters have been investigated extensively and represent the most important class of aromatic LCPs. The historical development, the chemistry, and the physics of aromatic LC polyesters have been summarized in several excellent reviews and book chapters. Some examples are given in the literature [2].

Several synthetic routes are known for the synthesis of aromatic LC polyesters. Melt, solution, and slurry polycondensations are mainly used. Most significant are the polycondensation of terephthalic acid diesters and aromatic diols, the polycondensation of terephthalic acids and acetates of aromatic diols with the addition of transesterification catalysts, and the polycondensation of aromatic diols and aromatic diacid dichlorides [2]. A method successfully utilized for laboratory synhesis is the polycondensation of silated aromatic diols and aromatic diacid dichlorides [3]. Molecular weights depend significantly on the reaction conditions and on the solubility as well as the fusibility of the polyesters, which is relatively poor for para-linked unsubstituted aromatic polyesters.

Polyesters prepared by the polycondensation of monomers based on terephthalic acid (TA) and hydroquinone (HQ) or p-hydroxybenzoic acid (HBA) are insoluble and exhibit very high melting temperatures, which are in the decomposition range (Fig. 3) [4]. Melting temperatures of 600 °C

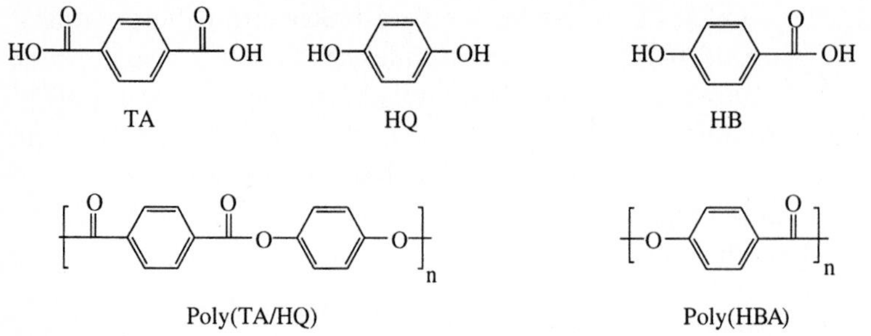

TA   HQ   HB

Poly(TA/HQ)   Poly(HBA)

Figure 3. Monomers and polymer structures of the unsubstituted para-linked aromatic polyesters, poly-(p-phenylene terephthalate) [poly(TA/HQ)] and poly-(p-hydroxybenzoic acid) [poly(HBA)].

for poly(TA/HQ) and 610 °C for poly(HBA) were observed by applying DSC heating rates of 80 K/min in order to minimize decomposition [2 b]. This melting temperature is by far too high to obtain stable liquid crystalline phases on processing from the melt. However, these polymers and also poly(4′-hydroxybiphenyl-4-carboxylic acid) received attention due to the formation of whiskers and the potential applications of such whiskers in composites [5].

General interest in thermally processable LC polymers is driven by a combination of properties that distinguish LC-polymers from random coil polymers. In particular, the formation of thermotropic LC phases reduces the melt viscosities during processing, and the increased chain stiffness reduces the thermal expansion coefficients. Both properties are of technical significance in combination with the high temperature stability. This is the basis for the application of LC polyesters as high precision injection molding materials and high modulus fibers. The different structural modification concepts, which will be discussed below, also have to be looked at in context with this property profile.

Considerable synthetic efforts have been undertaken in order to decrease the melting temperatures of aromatic LC polyesters while retaining LC properties. Melting temperatures below 350 °C are necessary in order to obtain stable, melt-processable poly-

esters. The different structural modifications shown in Fig. 2 were realized for aromatic LC polyesters. Also, combinations of different concepts are used for copolymers in order to modify their thermal properties and solution behavior. In the following sections, selected examples for each structural concept will be discussed focusing mainly on the thermal properties and the formation of thermotropic mesophases. It is important to note that the thermal properties are a function of the molecular weight, and that for a comparison of the structure – property relations the molecular weight has a significant impact. The inherent viscosity dependence of the melting temperature (T_m) and the clearing temperature (T_c), and the molecular weight dependence of the glass transition temperature (T_g) are shown in Fig. 4 for two examples. Nematic poly(2-n-decyl-1,4-phenylene-terephthalate) levels to a plateau for T_m and T_c above an inherent viscosity of 1.0 – 1.5 dl/g [6]. The polyester poly(2,2′-dimethyl-4,4′-biphenylene-phenylterephthalate) forms a nematic melt. Crystallization from the melt is completely suppressed and the polyester only has a T_g. The T_g values reach a plateau at a molecular weight about 8000 g/mol (GPC versus PST standards), which corresponds to an inherent viscosity of 1.2 dl/g [7].

(a)

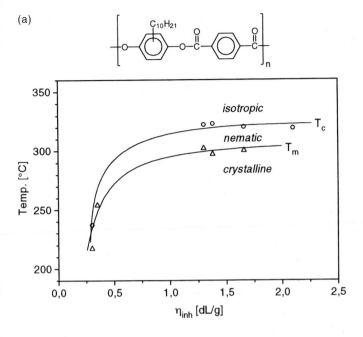

(b)

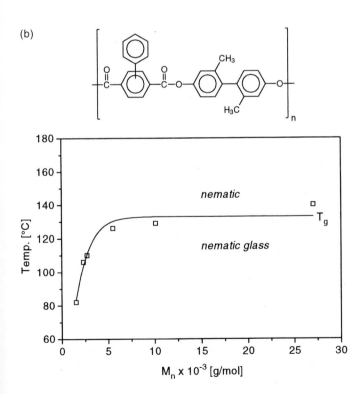

Figure 4. (a) Inherent viscosity dependence of the melting temperature (T_m) and the clearing temperature (T_c) of nematic poly(2-*n*-decyl-1,4-phenylene-terephthalate) [6] and (b) the molecular weight dependence of the glass transition temperature (T_g) of nematic poly(2,2′-dimethyl-4,4′-biphenylene-phenyltereph-thalate)[7].

1.3.1   Polyesters with Moieties of Different Lengths

The combination of monomers such as p-HBA, TA, HQ, and 4,4′-dihydroxybiphenyl (DHB) can be used for the synthesis of copolyesters (Fig. 5). The polycondensation of such monomers causes a structural irregularity in the direction and distance of the ester groups, thus reducing the interchain interactions.

The melting temperatures of polyesters synthesized by the copolymerization of p-HBA, TA, and HQ are still above 500 °C and the polyesters are far from forming stable, processable melts [8]. Depending on the composition, the melting temperatures were decreased by 100–150 °C if HQ was replaced by DHB. These polyesters are injection-moldable and were commercialized under the trade name Ekkcel I-2000 (Carborundum) and currently as Xydar (Amoco). These materials will gain further commercial interest as the cost of DHB decreases. They belong to the class of high temperature LC polyesters. For example, a special Xydar grade (SRT-500) has a melting temperature of 358 °C and is an injection-moldable resin with a tensile modulus of 8.3 GPa and a tensile strength of 126 MPa.

The molecular structure of these copolyesters is complex and is affected significantly by the synthesis conditions, thermal history, and processing conditions. The sequence distribution of the different repeating units is often found to be more or less in blocks, as indicated by cross polarization solid state ^{13}C-NMR and X-ray diffraction [9]. The as-polymerized polyester is highly crystalline, indicating block-like ordered sequences, but becomes less crystalline after processing from the molten state, which is due to further transesterification reactions and the formation of a more random sequence distribution.

1.3.2   Polyesters with Kinked or Double Kinked Moieties

A reduction of the melting temperatures of aromatic LC polyesters can also be accomplished efficiently by the partial incorporation of kinked comonomers. Suitable monomers are meta- or ortho-functionalized benzene rings, meta- or ortho-functionalized six-member heterocycles, and 2,5-functionalized five-member heterocycles. A frequently used alternative is two benzene rings connected by linking groups such as

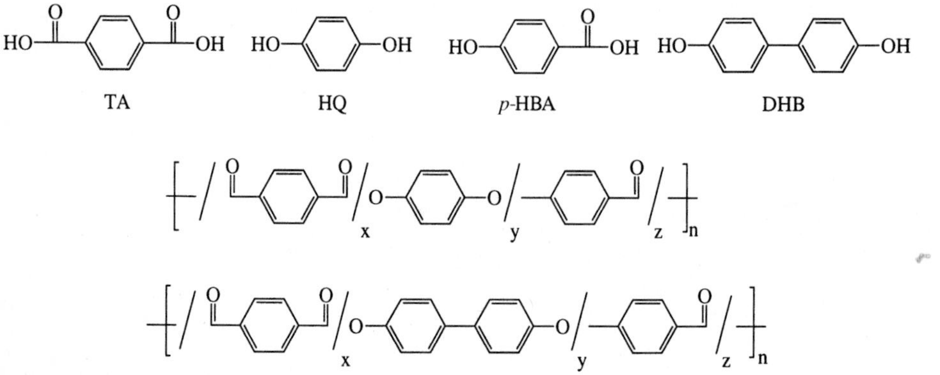

Figure 5. Examples of unsubstituted monomers of different lengths and the corresponding LC copolyesters.

$-CR_2-$, $-O-$, $-S-$, and $-CO-$, which are functionalized in the 4,4′-position or the 3,4′-position. Monomers with functional groups in the 3,4′-position, such as 3,4′-dihydroxy benzophenone (3,4′-DHBP) or 3,4′-dicarboxy-diphenylether (3,4′-CDPE) compensate the kink to a certain extent. These comonomers are also referred to as double kinked comonomers. These types of comonomers can only be incorporated in

para-linked aromatic polyesters up to a critical amount until the polymer backbone becomes too flexible and self-organization into liquid-crystalline order is no longer possible. Examples of structures of kinked and double kinked monomers are given in Fig. 6.

Poly-(p-hydroxybenzoic acid) [poly(HBA)] and poly-(p-phenylene-terephthalate) [poly(TA/HQ)] were modified with monomers

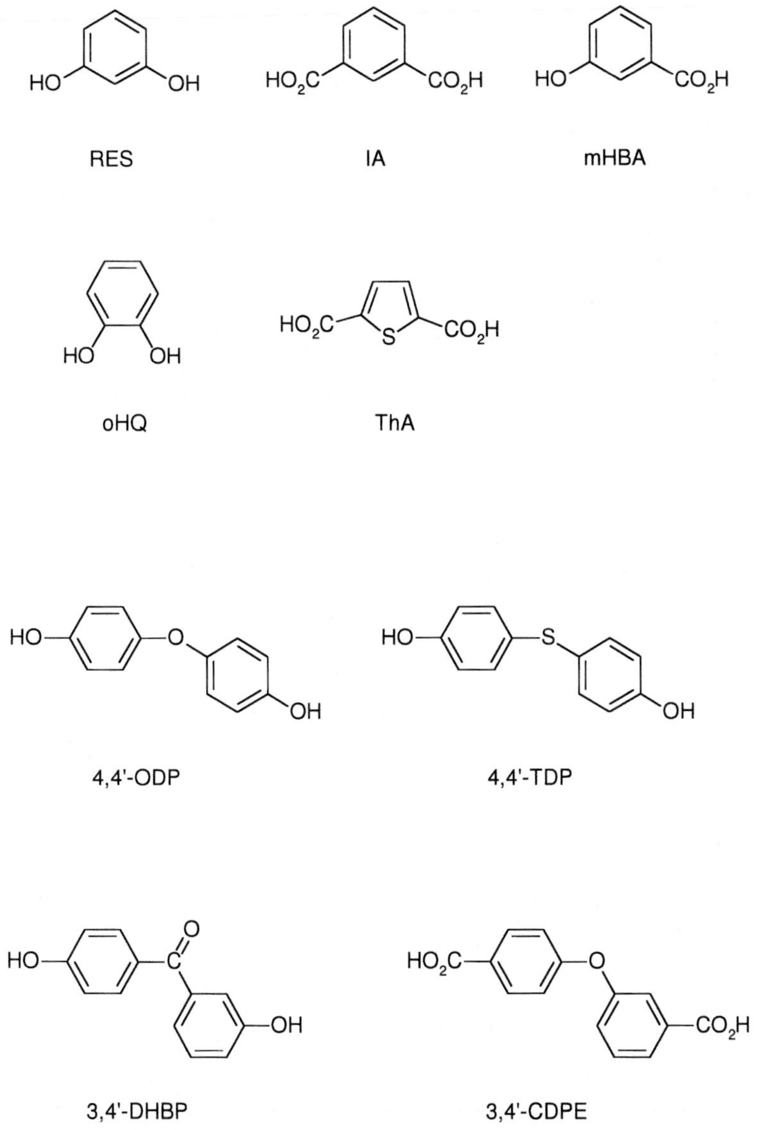

RES   IA   mHBA

oHQ   ThA

4,4'-ODP   4,4'-TDP

3,4'-DHBP   3,4'-CDPE

Figure 6. Examples of kinked and double kinked comonomers for aromatic LC polyesters.

such as isophthalic acid (IA), resorcinol (RES), or *m*-hydroxybenzoic acid (*m*HBA) by several research groups [10]. A liquid-crystalline melt was not found for all compositions. Jackson reported that the content of *m*HBA necessary in order to reduce the melting temperature of poly(TA/HQ) below 400 °C destroys the liquid crystallinity [10 d]. Copolyesters of HQ, TA, and *m*HBA with more than 50% *m*HBA form isotropic melts. At these concentrations, the chain conformation became too flexible due to the angle within the isophthalate moiety. The ability to pack the molecules in an ordered fashion is hindered and, consequently, no liquid-crystalline phases can be formed. Similar results were obtained with copolyesters of HQ, TA, and RES [10 d], HQ/IA/HBA (liquid crystallinity is lost above an HQ–IA content >67%) [10 f], TA/RES/HBA (liquid crystallinity is lost above a TA–RES content of >33%) [10 g], and HBA/*m*HBA (liquid crystallinity is lost above an *m*HBA content of >50%) [10 h]. Lenz et al. reported that the content of *m*-phenylene-terephthalate in poly(chloro-*p*-phenylene-*co*-*m*-phenylene-terephthalate) cannot exceed 60% in order to maintain liquid crystallinity [10 e]. However, poly(*p*-phenylene-terephthalate-*co*-*p*-phenylene isophthalate) containing about 85% of kinked isophthalate units melts at 355 °C, but exhibits a narrow mesophase range of 30 °C [10 i]. Also, the crystallization of para-linked aromatic copolyesters is efficiently suppressed by the incorporation of large amounts of ortho-substituted aromatic units [11]. Compared to meta-substituted monomers, ortho-substituted monomers are more favorable to form an extended chain conformation. For example, a glass transition temperature of 90 °C was reported for a copolyester of TA/methyl substituted HQ/*o*HQ [11 b]. This copolyester, described as amorphous, formed a nematic melt about

19 °C above T_g with a clearing temperature of 380 °C. The broad nematic phase was found to have a content of 50% of *o*HQ and 50% of methyl-substituted HQ.

Kinks in phenylene-based polyesters were also obtained by 2,5-thiophene dicarboxylate (ThA) moieties, which exhibit a core angle of 148° [12]. These copolymers exhibit stable nematic phases, but their transition temperatures are higher than those of the corresponding isophthalates. For example, the polyester composed of chlorohydroquinone and 2,5-thiophene dicarboxylate [poly(ClHQ)/ThA] exhibits a melting transition around 410 °C with a clearing temperature above 500 °C (Fig. 7). In contrast, the corresponding polyester poly(ClHQ/TA) shows only a solid state transition at 375 °C and does not melt below 500 °C. The poly-

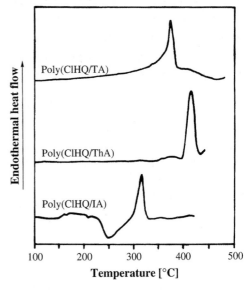

Figure 7. DSC investigations of three different aromatic homopolyesters: (a) LC polyester composed of chlorohydroquinone and terephthalic acid [poly(ClHQ/TA)], (b) LC polyester composed of chlorohydroquinone and 2,5-thiophene dicarboxylic acid [poly(ClHQ)/ThA], (c) polyester composed of chlorohydroquinone and isophthalic acid [poly(ClHQ/IA)] (adapted from [12]).

ester composed of ClHQ/IA has a glass transition endotherm at 146 °C, a recrystallization exotherm around 240 °C, and a melting temperature of 308 °C, but does not exhibit liquid crystallinity.

The incorporation of comonomers based on two benzene rings with single atom linking bridges and functional groups in the 4,4′-position were also investigated for modifying the thermal properties of LCPs [13]. The deviation from the chain linearity of aromatic LC polyesters by such comonomers depends on the type of linking bridge and the amount of incorporated comonomer. Compared to the bent, rigid comonomers discussed above, these comonomers, also referred to as 'swivel'-type monomers, are more flexible.

A comparative study based on poly(chloro-*p*-phenylene-terephthalate), modified by the incorporation of various amounts of 4,4′-functionalized bisphenols of different structures was performed by Lenz and Jin (Fig. 8) [10e]. For example, copolyesters composed of TA/ClHQ/BPA showed melting temperatures of the order of 340 °C with a BPA content of 10–30 mol% with respect to the amount of ClHQ. These copolyesters form anisotropic melts, whereas copolyesters with a BPA content of more than 40% were amorphous, forming isotropic melts. Copolyesters with bis(4-hydroxyphenyl) methane (BPM) or 4,4′-sulfonyldiphenol (SDP) have a stronger tendency to form anisotropic melts. It is possible to replace more of the ClHQ with these comonomers without destroying the liquid crystallinity. Based on these results, different 4,4′-functionalized bisphenols destabilize the liquid-crystalline phase quantitatively in the following order:

$$O < CH_2 \sim S < SO_2 < C(CH_3)_2$$

1.3.3   Polyesters with Crankshaft Moieties

An important structural modification which results in a parallel offset within the polymer backbone is the incorporation of 2,6-difunctionalized naphthalene monomers.

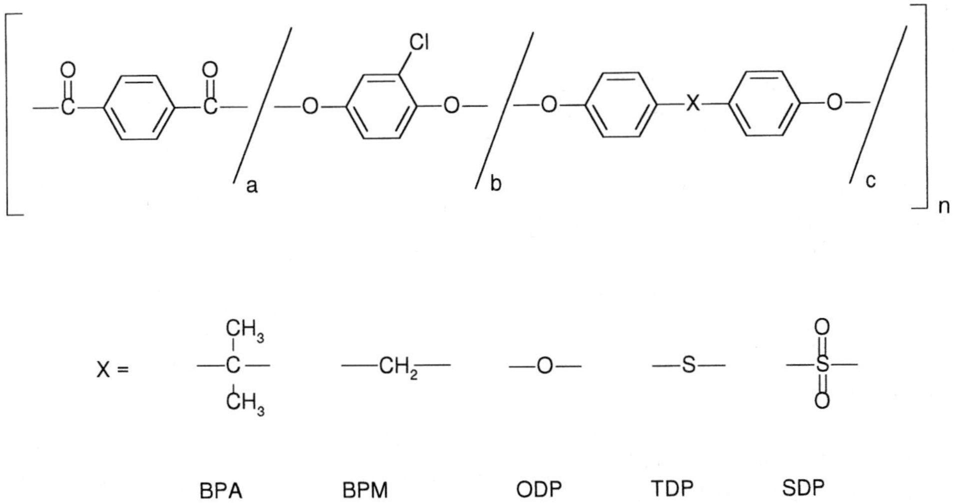

Figure 8. Examples of LC copolyesters derived from terephthalic acid, chlorohydroquinone, and various kinked 4,4′-functionalized bisphenols.

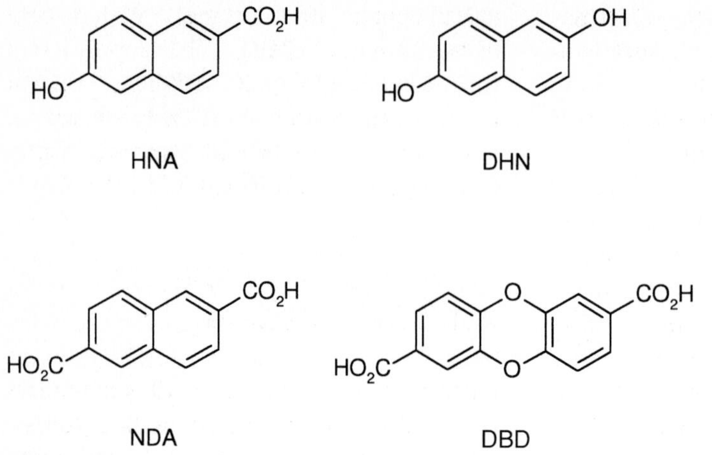

HNA

DHN

NDA

DBD

Figure 9. Examples of different crankshaft monomers.

This class of monomers is also referred to as 'crankshaft' monomers. Examples of such monomers are: 6-hydroxy-2-naphtho-ic acid (HNA), 2,6-dihydroxynaphthalene (DHN), and 2,6-naphthalene dicarboxylic acid (NDA) (Fig. 9) [14].

Figure 10 illustrates the influence of 2,6-naphthalene derivatives in lowering the melting temperature. A copolymer series based on p-HBA/TA/HQ has melting temperatures in the range of 420–500 °C. The melting temperatures can be lowered by about 100 °C by replacing the terephthal-ic acid with 2,6-naphthalene dicarboxylic acid (NDA), but the melting point minimum is still about 323 °C. A further reduction was achieved in the copolymer series p-HBA/TA/DHN. The lowest melting points were observed in a two-component polyester based on 6-hydroxy-2-naphthoic acid (HNA) and p-hydroxybenzoic acid (p-HBA). In this series the melting point could be lowered to a minimum of almost 230 °C.

In particular, the results of the poly (HNA/p-HBA) series are the basis for the commercial development of a family of ther-motropic LCPs by the Hoechst Celanese Cor-

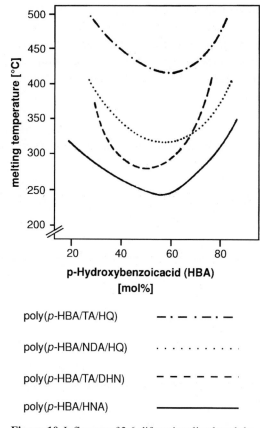

poly(p-HBA/TA/HQ)   — · — · — ·

poly(p-HBA/NDA/HQ)   · · · · · · · · · ·

poly(p-HBA/TA/DHN)   — — — — — —

poly(p-HBA/HNA)   ——————

Figure 10. Influence of 2,6-difunctionalized naphtha-lene comonomers on the melting temperatures of LC polyester based on p-hydroxybenzoic acid (adapted from [14a]).

poration under the trade name of Vectra [14 b]. The Vectra A and Vectra C materials are based on a poly(naphthoate – benzoate) structure. In contrast, the Vectra B series is based on a poly(naphthoate – aminophenol – terephthalate), an aromatic copolyester amide. These LCPs can be processed in a similar manner to conventional thermoplastics using almost all of the common polymer processing techniques. The thermotropic melts have a low viscosity, providing good molding characteristics. Molded parts exhibit low warpage and shrinkage, along with high temperature and dimensional stability, as well as chemical resistance. This property profile opened up commercial applications in areas such as electronics/electrical (connectors, capacitor housing, coil formers), fiber-optics (strength members, compilers, connectors), automotive (full system components, electrical systems), and others (compact disc player

components, watch components, microwave parts, etc.) [14 c].

In addition to the 2,6-naphthalene based LCPs, an interesting aromatic LC polyester with 2,6-substituted DBD moieties has been disclosed in several patents [15, 16].

1.3.4 Polyesters with Lateral Substituents

Different types of lateral substituents can be incorporated by monomers with substituents of different size, flexibility, and polarity. In addition, the number of substituents and the symmetry of the substitution can result in a different positional isomerism along the polymer backbone (Fig. 11). For example, in polyesters derived from monosubstituted monomers, the substituents can point towards each other (*syn*) and away from each other (*anti*), resulting in a random

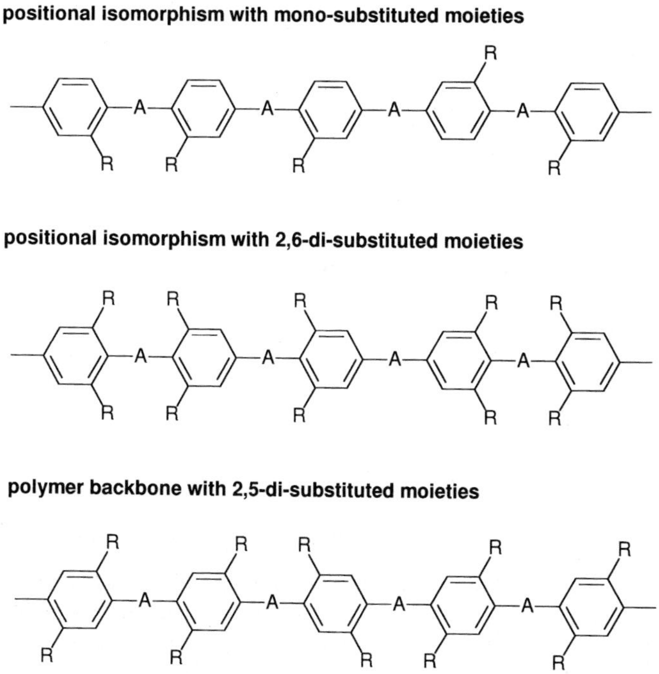

positional isomorphism with mono-substituted moieties

positional isomorphism with 2,6-di-substituted moieties

polymer backbone with 2,5-di-substituted moieties

Figure 11. Structural diversities of aromatic LC polymers with laterally substituted monomers.

sequence along the polymer backbone. A similar, but even more pronounced effect, is found in monomers with 2,6-disubstitution, whereas in a polymer backbone composed of symmetrical 2,5-disubstituted monomers, the substituents are regularly positioned along the backbone and obtain their irregularity by rotational isomerism. A selection of substituted poly(1,4-phenylene terephthalates) with different types of lateral substituents is summarized in Tables 1 and 2.

1.3.4.1 Polymers with Flexible Substituents

As flexible substituents, linear and branched alkyl, alkoxy, or thioalkyl side chains of different lengths have been utilized to modify para-linked aromatic polyesters. The main difference to the systems discussed above is the limited thermal stability caused by the substitution with alkyl chains. Also, the mechanical poperties are substantially lowered with increasing

length and number of the alkyl chains [17]. The dramatic loss of the mechanical properties is obvious due to the reduced number of load-supporting polymer main chains in a given cross-sectional area. In addition, the intermolecular interactions of the alkyl side chains are relatively weak.

Lenz et al. demonstrated for a series of monosubstituted poly(2-n-alkyl-1,4-phenylene-terephthalate)s (Table 1) with the number of C-atoms in the side chain ranging from 6 to 12, that the melting temperatures and the clearing temperatures are lowered, as expected, with increasing length of the alkyl chain [6]. For different samples with inherent viscosities of about 0.5 dl/g (not in the plateau region; see Fig. 4a), the melting temperatures decrease from 277 °C to 228 °C and the clearing temperatures from 323 °C to 291 °C. All the polyesters were found to be nematic. Corresponding polyesters with cyclic and branched alkyl groups (cyclohexyl, cycloheptyl, cyclooctyl, and 3-methylpentyl) [18] showed melting temperatures in the range of 308–

Table 1. Glass transition temperatures, melting temperatures, and clearing temperatures of selected mono- and disubstituted poly-(p-phenylene terephthalate) esters with flexible substituents.

R	R′	T_g (°C)	T_m (°C)	T_c (°C)	Reference
n-hexyl	H	–	277	323	[6]
n-heptyl	H	–	257	302	[6]
n-ocytl	H	–	257	307	[6]
n-nonyl	H	–	237	291	[6]
n-undecyl	H	–	228	292	[6]
cyclohexyl	H	210	363	391	[18]
cycloheptyl	H	–	352	390	[18]
cyclooctyl	H	212	321	385	[18]
3-methylpentyl	H	237	308	377	[18]
–(CH$_2$)$_2$-phenyl	H	–	340	>340	[19]
–(CH$_2$)$_3$-phenyl	H	–	325	>340	[19]
–(CH$_2$)$_2$-phenyl	–(CH$_2$)$_2$-phenyl	60	195	>340	[19]
t-butyl	–(CH$_2$)$_2$-phenyl	110	225	>340	[19]

Table 2. Glass transition temperatures, melting temperatures, and clearing temperatures of selected mono- and disubstitued poly-(p-phenylene terephthalate) esters with bulky and stiff substituents.

R	R'	T_g (°C)	T_m (°C)	T_c (°C)	Reference
Cl	H	220	372	510	[24b]
H	Cl	220	402	475	[24b]
Br	H	–	353	475	[24b]
H	Br	–	405	490	[24b]
phenyl	H	170	346	475	[24b]
H	phenyl	130	265	343	[24b]
phenyl	phenyl	122	–	231	[24b]
phenyl	Cl	108	206	368	[24b]
Cl	phenyl	113	233	360	[24b]
phenyl	Br	120	222	376	[24b]
Cl	Br	120	313	362	[24b]
t-butyl	CF$_3$	141	228	>350	[24c]
phenoxy	Cl	85	241	355	[24c]
phenoxy	Br	90	220	>350	[24c]
phenoxy	CF$_3$	88	–	– [a]	[24c]
phenoxy	phenyl	112	–	– [a]	[24c]
t-butyl	Br	131	–	– [a]	[24c]

[a] No LC phase observed.

363 °C. The clearing temperatures from the nematic melt are between 377 and 391 °C. In some cases glass transitions were found.

As already mentioned, positional isomerism is important for the solubility and fusibility of aromatic LC polyesters. Consequently, polyesters made from symmetrical 2,5-disubstituted or 2,3,4,5-tetrasubstituted monomers should result in polymers that are less soluble and less fusible. This is in general the case with short lateral substituents. Ballauff and others reported that the series of poly(1,4-phenylene-2,5-dialkoxy terephthalate)s with long flexible alkoxy side chains at the terephthalic moiety result in tractable LC polyesters [20] (Fig. 12). These polyesters exhibit a high degree of crystallinity with melting temperatures below 300 °C. Polyesters with short side chains ($2 \leq m \leq 6$) form nematic melts with clearing temperatures $T_c > 350$ °C for $m=2$

and $T_c = 250$ °C for $m=6$. Polyesters with side chains $m \geq 8$ result in the formation of layered mesophases, which can also be found in the solid state. The layer spacing in the mesophase, as revealed by X-ray analysis, increases linearly with increasing length of the alkoxy side chain. Longer alkyl side chains ($m \geq 12$) result in side-chain crystallization with melting transitions around 50 °C.

Tetrasubstituted polyesters with long alkyl side chains on the hydroquinone moiety connected via ether or ester linkages have been reported [20c] (Fig. 12). In these polyesters, the tetrasubstituted moiety can be considered as a discotic mesogen. The polyesters with Z=H are highly soluble and have low melting temperatures ($T_m < 100$ °C) and low clearing temperatures in the range of 123–260 °C. The type of mesophase formed is referred to as "sanid-

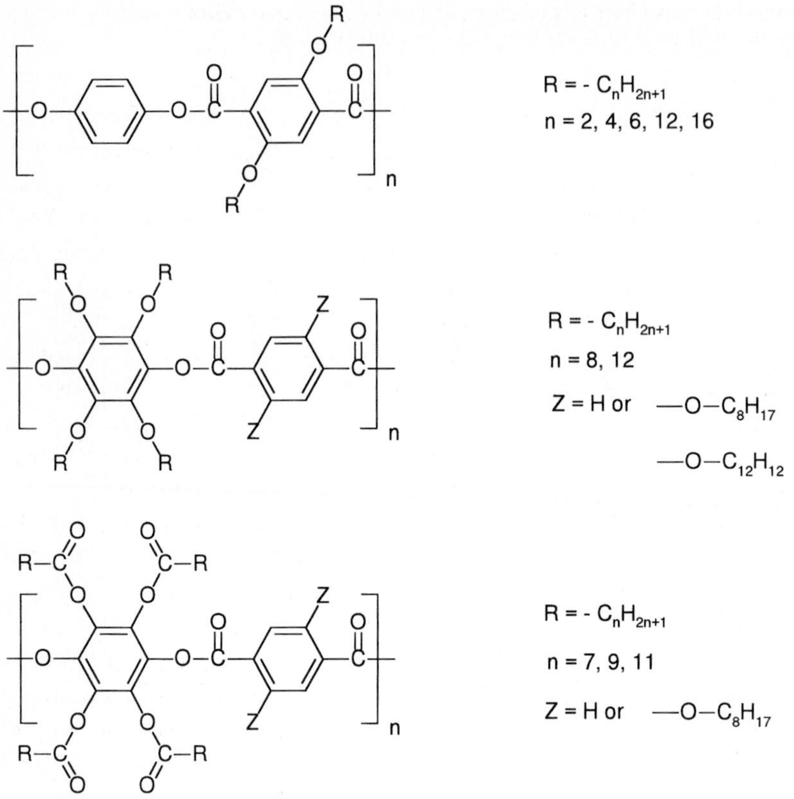

R = - C_nH_{2n+1}

n = 2, 4, 6, 12, 16

R = - C_nH_{2n+1}

n = 8, 12

Z = H or   —O—C_8H_{17}

—O—$C_{12}H_{12}$

R = - C_nH_{2n+1}

n = 7, 9, 11

Z = H or   —O—C_8H_{17}

Figure 12. Structure of 2,5-disubstituted polyesters with flexible alkoxy substituents and of polyesters with tetrasubstituted moieties [20a–c].

ic". If the terephthalic acid moiety is disubstituted with two long alkoxy side chains (Z = –O–C_8H_{17}), the mesophase is destroyed. This observation can be explained by a reduction of the chain stiffness in combination with the effect of the additional two alkoxy side chains. Similar observations have been made with poly(p-phenylene terephthalate)s with one, two, and four thioalkyl side chains of various length [21]. Depending on the degree of substitution and the substitution pattern, thermotropic as well as isotropic polyesters have been obtained.

Chiral sanidic polyesters derived from copolymerizations of (S,S)-2,5-bis(2-methylbutyloxy)terephthalic acid, 2,5-bis(dodecyl)terephthalic acids, and 4,4′-dihydroxybiphenyl have been reported [22]. These copolyesters formed a solid sanidic layer structure with melting temperatures above 200 °C. A broad enantiotropic cholesteric phase is formed above T_m with isotropization temperatures in the range of 275–325 °C.

A different type of flexible substituent are phenylalkyl substituents [19]. Monosubstituted polyesters with phenylalkyl substituents with different lengths of the alkyl linkage are characterized by high melting temperatures (Table 1). The thermal transitions can be lowered significantly if the terephthalic moieties and the phenylene moieties are substituted by phenylalkyl moieties. The replacement of the phenylalkyl substituent at the hydroquinone moiety by t-butylhydroquinone results in an increase of the glass transition temperature. The polycondensation of monosubstituted 1,4-hydroquinone derivatives with monosubstituted terephthalic acid derivatives (AABB monomer system) results in polyesters with substituents arranged randomly along the polymer backbone (Fig. 11). In contrast, the polycondensation of monosubstituted 4-hydroxybenzoic acid derivatives (AB monomer system) leads to polyesters with defined positions of the substituents. A random arrangement of the substituents is also obtained for copolyesters made from 2- and 3-substituted 4-hydroxybenzoic acid derivatives. The role of positional isomerism was clearly demonstrated by a comparison of homo- and copolyesters based on phenethyl substituted poly(4-hydroxybenzoats)s (Fig. 13) [23]. Solubility, melting temperature, and the LC phase were almost identical for the homopolyesters of the 2- and 3-substituted homopolyesters exhibiting melting temperatures of 280–297 °C; these give highly viscous birefringent melts with $T_c > 340$ °C. In contrast, the solubility of the corresponding copolyester is enhanced and the crystallinity is depressed, forming an amorphous solid with T_g at 75 °C and a nematic schlieren texture with T_c at 280 °C. It should be mentioned that the inherent viscosity of this copolyester was 2.80 dl/g and therefore the observed properties cannot be attributed to a low molecular weight.

1.3.4.2   Polyesters with Stiff, Bulky Substituents

The influence of a stiff and bulky substituent on the thermal and the solution properties is governed by the size, polarity, position, and number of the substituents [24]. The main differences between flexible and bulky substituents are their influence on the thermal stability and the glass transition temperature. Whereas flexible chains act as an internal plasticizer and lower the thermal stability, most bulky substituents increase T_g and, depending on the chemical nature, do not lower the thermal stability.

A selection of monosubstituted and disubstituted poly-(p-phenylene-terephthalate)s is compared in Table 2. Poly(p-phenylene-terephthalate)s with methyl, methoxy, chloro, or bromo substituents on either the hydroquinone or the terephthalic acid moiety exhibit melting temperatures of 350 °C or higher. Thermotropic liquid crystalline behavior is observed in these samples, although it is in the range of thermal decomposition. A comparison of the mono- and diphenyl substituted polyesters reveals an important trend. The monosubstituted poly-(p-phenylene-terephthalate) with the phenyl substituent in the hydroquinone moiety melts at 346 °C, also forming a nematic melt up to a clearing temperature of

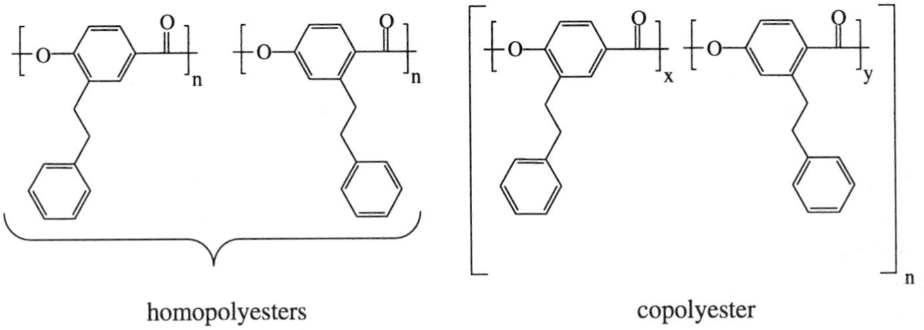

homopolyesters   copolyester

Figure 13. Phenethyl-substituted homopolyesters and copolyesters based on poly(p-hydroxybenzoate) [23].

$T_c = 475\,°C$, which is in the range of decomposition. In contrast, the melting temperature of the polyester with the phenyl substituent on the terephthalic moiety is 265 °C and $T_c = 343\,°C$, which indicates that the chain stiffness is lowered if the bulky substituent is ortho to the carbonyl group of the ester linkage. The substitution of both the hydroquinone and the terephthalic acid moiety with a phenyl substituent suppressed the crystallization and resulted in a polyester glass transition temperature of 122 °C. Below the glass transition temperature, a glass with nematic order is obtained. The clearing temperature of this polyester was detected at 231 °C, which is considerably lower than those of the monosubstituted polyesters. In other disubstituted poly-(p-phenylene terephthalate)s, particularly with polar substituents such as halogen or trifluoromethyl, relatively weak melting transitions in the temperature range of 210–230 °C are usually observed. In all cases, glass transition temperatures ranging from 108 to 141 °C are observable. The glass transition increases as expected with the volume of the bulky substituents. The clearing temperature of the nematic melts is in all cases above 350 °C. However, polyesters with trifluoromethyl-, phenyl-, or bromo-substituents on the terephthalic moieties in combination with a bulky substituent on the hydroquinone moiety were not found to be thermotropic as a result of overly reduced chain stiffness.

Monomers with phenyl substituents based on 4,4'-dihydroxybiphenyl, 4-hydroxy-4'-carboxybiphenyl, and 2,6 functionalized naphthalene have been synthesized and incorporated in various LC copolyester systems. Bhowmik et al. used 3-phenyl-4,4'-dihydroxybiphenyl and 3,3'-dihydroxybiphenyl as comonomers to modify, for example, poly-(p-hydroxybenzoic acid) and poly-(6-hydroxy-2-naphthoic ac-

id) [25]. Jin and co-workers reported on wholly aromatic polyesters derived from 6-hydroxy-5-phenyl-2-naphthoic acid and 4'-hydroxy-3'-phenylbiphenyl-4-carboxylic acid [26]. With these monomers the melting temperatures of copolymers could be lowered substantially without destroying the formation of thermotropic melts, while maintaining thermal stability.

1.3.5 Polyesters with Noncoplanar Aromatic Moieties

Noncoplanar aromatic monomers, such as 2,2'-disubstituted biphenylene derivatives and 4,4'-functionalized 1,1'-binaphthyl derivatives (Fig. 14) have been used as comonomers in para-linked aromatic polyesters with remarkable effects on the phase transition temperatures, the crystallinity, and the solubility. The incorporation of these noncoplanar monomers will not initially reduce the chain stiffness. The phenyl rings are forced by the 2,2'-substitution into a noncoplanar conformation which strongly decreases the intermolecular inter-

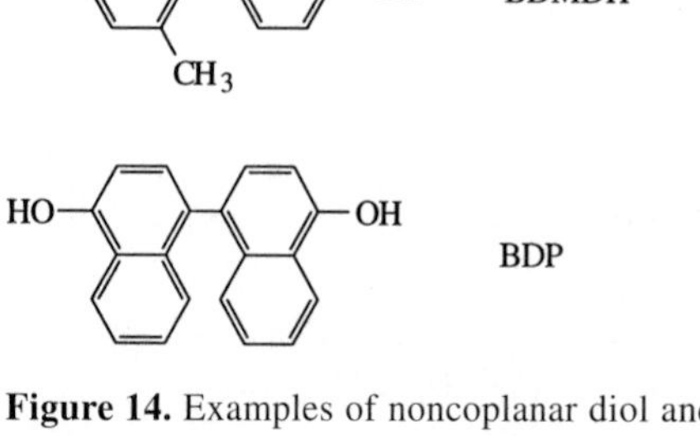

Figure 14. Examples of noncoplanar diol and diacid monomers.

actions. The noncoplanarity correlates with the size of the substituents. As a consequence, the crystallization tendency and transition temperatures are lowered remarkably and the solubilities are significantly enhanced.

Sinta and co-workers [27] demonstrated that in a variety of para-linked aromatic polyesters and copolyesters with 2,2'-di-substituted-4,4'-biphenylene units, mainly based on 2,2'-ditrifluoromethyl-4,4'-biphenyl diacid (BCFDA), the crystallization can be completely suppressed. The glass transition temperatures are generally between 100 and 150 °C with wide nematic phases, and clearing temperatures usually in the range of decomposition. For example, the copolyester based on BCFDA (0.5 mol%), trifluormethyl-substituted terephthalic acid (0.5 mol%), and chlorohydroquinone is amorphous with T_g at 118 °C, and forms an easily shearable nematic melt at 180 °C with clearing temperatures in the decomposition range. Similar results are even obtainable in homopolyesters if the 2,2'-disubstituted biphenylene unit is combined with a monosubstituted phenylene unit [28]. For example, the homopolyester derived from 2,2'-dimethylbiphenyl-4,4'-biphenyl diacid and t-butylhydroquinone has a glass transition temperature of 150 °C and does not crystallize from the nematic melt. LC polyesters with this substitution pattern are highly soluble in common organic solvents such as THF, chloroform, methylene chloride, and tetrachloroethane up to maximum solubilities in the range of 50% (w/v).

Another example of such a type of polyester is shown in Fig. 4. The polyester poly(2,2'-dimethyl-4,4'-biphenylene-phenyl-terephthalate) derived from 2,2'-dimethyl-biphenyl-4,4'-diol (BDMDH) and phenyl-terephthalic acid forms a nematic melt. Crystallization from the melt is completely

suppressed and the polyester displays only one glass transition at around 140 °C. This polyester has been obtained with intrinsic viscosities of up to 9.60 dl/g (CHCl$_3$, 20 °C), corresponding to a weight-average molecular weight (M_w) of about 80 000 g/mol [7a]. The Mark–Houwink–Kuhn parameters of this polyester (CHCl$_3$, 20 °C) were determined as $[n] = 4.93 \times 10^{-4} M_w^{0.877}$. The formation of anisotropic thermoreversible gels was observed with this polyester, for example, in 3-phenoxytoluene as the solvent. These mixtures also retained their anisotropy above the gel melting point at a concentration of 35% or more [29]. In highly aligned fibers spun from the nematic melt and post-drawn, a tensile modulus of 40–45 GPa and a tensile strength of up to 550–650 MPa was reached [28c].

Even higher glass transition temperatures are achievable if 4,4'-functionalized 1,1'-binaphthyl monomers are used [28b, 30]. For example, the homopolyester derived from 4,4'-dihydrox-1,1'-binaphthyl (BDP) and phenylterephthalic acid with an inherent viscosity of 1.90 dl/g has a T_g of 183 °C. The polyester is nematic up to its decomposition temperature and forms a nematic glass below T_g. In a similar homopolyester based on 4,4'-dicarboxy-1,1'-binaphthyl and t-butyl-hydroquinone the T_g was found to be 189 °C. In both cases, good solubility in common chlorinated solvents is maintained. This type of monomers was, for example, also used to modify wholly aromatic copolyesters based on 6-hydroxy-2-naphthoic acid (HNA) and 2,6-naphthalene dicarboxylic acid (NDA) [25a].

In conclusion, the selected examples of this chapter clearly demonstrate that it is possible with the discussed structural modifiations to modify in a systematic manner the melting behavior over a wide temperature range, the degree of crystallization from highly crystalline to amorphous, and the solution behavior from solvent-resistant to highly soluble, while maintaining thermotropic liquid crystalline behavior. In particular, the combination of different structural modifications in copolymers allows the systematic tailoring of the thermal properties, the solution behavior, and the chain stiffness of thermotropic aromatic LC polyesters.

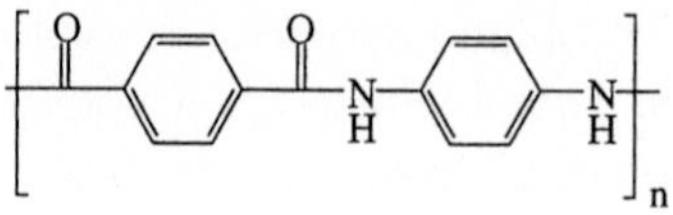

Figure 15. Structure of poly(1,4-phenyleneterephthalamide).

1.4 Para-Linked Aromatic Polyamides

The same structural modification concepts, which were utilized to modify the properties of para-linked aromatic LC polyesters, have also been applied to aromatic polyamides. Para-linked aromatic polyamides are an important class of LC polymers. In contrast to thermotropic LC polyesters, para-linked aromatic polyamides form lyotropic solutions. Due to the formation of intermolecular hydrogen bridges, these polymers are in most cases unable to melt below their thermal decomposition temperature. Infusibility and limited solubility of unsubstituted para-linked aromatic polyamides are characteristic properties which limit synthesis, characterization, processing, and applications.

Already in 1965, the formation of lyotropic liquid crystalline solutions of poly-(p-aminobenzoic acid) in concentrated sulfuric acid was observed. This work was the basis for the commercialization of high strength, high modulus, and heat-resistant poly-(p-phenylene terephthalate) (PPDT) fibers under the trade name Kevlar (Dupont) and Twaron (Akzo) (Fig. 15) [31]. PPDT is typically prepared by the polycondensation of terephthaloyl chloride and 1,4-phenylenediamine in tertiary amidic solvents such as N-methylpyrrolidinone or dimethylacet-

amide containing lithium or calcium chloride. PPDT forms a lyotropic nematic solution in concentrated sulfuric acid solution when a minimum concentration is exceeded. The PPDT fiber is spun from a lyotropic solution in concentrated sulfuric acid.

The solution behavior has been significantly enhanced by the same structural modifications as reported previously for aromatic LC polyesters. For example poly-(p-phenylene terephthalamide) has been modified by bulky, stiff substituents [32], flexible alkyl side chains [33], the incorporation of kinked and double kinked comonomers, and comonomers of different lengths [34], as well as the use of noncoplanar biphenylene monomers [35]. To develop high performance materials, modifications that increase the solubility while maintaining the rod-like character, high glass transition temperatures, and the temperature stability are of particular interest. The solubility and the chain stiffness are critical factors in achieving lyotropic solutions.

Depending on the modifications of poly-(p-phenylene terephthalamide) and analogous aromatic polyamides, it is possible to vary the solubility from concentrated sulfuric acid, to nonpolar aprotic solvents with inorganic salts, or to nonpolar aprotic solvents and to common organic colvents. In particular the incorporation of noncoplanar 2,2'-disubstitution has proven to remarkable enhance the solubility and lower the crystallinity. Gaudiana et al. [35] demonstrated that para-linked aromatic polyamides containing noncoplanar 2,2'-bis(trifluoromethyl)biphenylene units are highly soluble in aprotic polar amide solvents such as DMAc or NMP, and even in common solvents such as THF and acetone. Although high polymer concentrations were achieved, a general lack of lyotropic behavior was observed. In contrast, it was found that para-linked aromatic polyamides containing 2,2'-dimethylbiphenyl units are in some cases capable of forming lyotropic solutions in amide solvents [32 h–j]. The critical concentrations to form a lyotropic phase are strongly influenced by the size and position of the substituents, which govern the overall chain stiffness and capability of chain extension. For some systems, concentrations of up to 40–50 wt.% are required to obtain a lyotropic solution. These results demonstrate the complex interplay of chemical substitution, chain stiffness, choice of solvent, and solvent/polymer chain interaction.

The substituted polyamides with long alkyl- and alkoxychains [33] are highly soluble and form anisotropic melts above their melting temperature. These polyamides are typical examples of sanidic liquid crystalline polymers. Generally, no lyotropic behavior is observed. The temperature stability is obviously substantially lowered due to the substitution with alkyl chains.

It should also be mentioned that a combination of ester and amide linkages was frequently used to prepare thermotropic LC polyesteramides. Frequently, the inexpensive p-aminophenol is incorporated. For example, the Vectra B series is a poly(naphthoate-aminophenolterephthalate) derived from 6-hydroxy-2-naphthoic acid (HNA), p-aminophenol, and terephthalic acid [14]. It is obvious that only a limited amount of p-aminophenol can be incorporated in order to maintain thermotropic liquid crystalline behavior. The amide linkage enhances the intermolecular interactions via hydrogen bridges and improves the solid state properties of the material.

1.5 References

[1] P. J. Flory, *Proc. R. Soc. London* **1956**, *73*, A234.
[2] See, for example: a) M. G. Dobb, J. E. McIntyre, *Adv. Polym. Sci.* **1984**, *60/61*, 61; b) T.-S. Chung, *Polym. Eng. Sci.* **1986**, *26*, 901; c) S. L. Kwolek, P. W. Morgan, J. R. Schaefgen in *Encyclopedia of Polymer Science*, Vol 9 (Eds.: H. F. Mark, N. M. Bikules, C. G. Overberger, G. Menges, J. I. Kroschwitz) Wiley, New York **1987**, 1; d) M. Ballauff, *Angew. Chem. (Int. Ed.)* **1989**, *28*, 253; e) C. M. McCullagh, J. Blackwell in *Comprehensive Polymer Science,* 2nd Suppl. (Eds.: S. Aggarwal, S. Russo), Elsevier, Oxford, U. K. **1996**; f) H. H. Yang, *Aromatic High-Strength Fibers*, Wiley-Interscience, New York **1989**.
[3] H. R. Kricheldorf, *Silicon in Polymer Synthesis*, Springer, Berlin **1996**.
[4] a) W. M. Eareckson, *J. Polym. Sci.* **1959**, *40*, 399; b) W. J. Jackson, Jr., *Br. Polym. J.* **1980**, *12*, 154; c) J. Economy, R. S. Storm, V. I. Matkovich, S. G. Cottis, *J. Polym. Sci., Polym. Chem. Ed.* **1976**, *14*, 2207; d) G. Lieser, G. Schwarz, H. R. Kricheldorf, *J. Polym. Sci., Polym. Phys. Ed.* **1983**, *21*, 1599; e) H. R. Kricheldorf, G. Schwarz, *Makromol. Chem.* **1983**, *184*, 475; f) G. Schwarz, H. R. Kricheldorf, *Makromol. Chem., Rapid Commun.* **1988**, *9*, 717; g) J. Economy, W. Volksen, C. Viney, R. Geiss, R. Siemens, T. Karris, *Macromolecules* **1988**, *21*, 2777.
[5] a) G. Schwarz, H. R. Kricheldorf, *Macromolecules* **1995**, *28*, 3911; b) H. R. Kricheldorf, F. Ruhser, G. Schwarz, T. Adebahr, *Makromol. Chem.* **1991**, *192*, 2371; c) C. Taesler, J. Petermann, *Makromol. Chem.* **1991**, *192*, 2255.
[6] J. Majnusz, J. M. Catala, R. W. Lenz, *Eur. Polym. J.* **1983**, *19*, 1043.
[7] a) R. Schmitt, A. Greiner, W. Heitz, *Makromol. Chem. Symp.* **1992**, *61*, 297; b) W. Heitz, H.-W. Schmidt, *Makromol. Chem. Symp.* **1990**, *38*, 149.
[8] J. Economy, S. G. Cottis, B. E. Nowak, German Patent 2,025,971, **1972**.
[9] W. Volksen, J. R. Lyerla, Jr., J. Economy, B. Dawson, *J. Polym. Sci. Polym. Chem. Ed.* **1983**, *21*, 2249.
[10] a) S. G. Cottis, J. Economy, B. E. Novak, U. S. Patent 3,637,595, **1972**; b) S. W. Kantor, F. F. Holub, U. S. Patent 3 036 990, **1962**; c) I. J. Goldfarb, D. R. Bain, R. McGuchan, A. C. Meeks, *Org. Coat. Appl. Polym. Sci. Proc.* **1971**, *31*, 130; d) W. J. Jackson, Jr., *Br. Polym. J.* **1980**, *12*, 154; e) R. W. Lenz, J.-I. Jin, *Macromolecules* **1981**, *14*, 1405; f) A. B. Eredemir, D. J. Johnson, J. G. Tomka, *Polymer* **1986**, *27*, 441; g) J.-I. Jin, S. H. Lee, H. J. Park, *Polym. Prepr. (Am. Chem. Soc. Polym. Div.)* **1987**, *28(1)*, 122; h) R. Rosenau-Eichin, M. Ballauff, J. Grebo-

wicz, E. W. Fischer, *Polymer* **1988**, *29*, 518; i) R. Cai, E. T. Samulski, *Macromolecules* **1994**, *27*, 135.

[11] a) B. P. Griffin, R. G. Fessey, EU Patent 0 008 855, **1980**; b) C. Noël, C. Friedrich, F. Lauprête, J. Billard, L. Bosio, C. Strazielle, *Polymer* **1984**, *25*, 263; c) J.-I. Jung, J.-H. Chang, *Macromolecules* **1989**, *22*, 93; d) J.-I. Jung, E.-J. Choi, C.-S. Kang, J.-H. Chang, *J. Polym. Sci. Chem. Ed.* **1989**, *27*, 2291; e) F. Navarro, J. L. Serrano, *J. Polym. Sci.: Part A: Polym. Chem.* **1992**, *30*, 1789.

[12] R. Cai, J. Preston, E. T. Samulski, *Macromolecules* **1992**, *25*, 563.

[13] a) B. P. Griffin, M. K. Cox, *Br. Polym. J.* **1980**, *12*, 147; b) H. R. Kricheldorf, V. Döring, *Polymer* **1992**, *33*, 5321; c) R. O. Garray, P. K. Bhowmik, R. W. Lenz, *J. Polym. Sci.: Part A: Polym. Chem.* **1993**, *31*, 1001; d) see also [4 b] and [10 e].

[14] a) G. W. Calundann in *High Performance Polymers: Their Origin and Development* (Eds.: R. B. Seymour, G. S. Kirshenbaum) Elsevier, Amsterdam **1986**, p. 235; b) selected patents: G. W. Calundann (Celanese), U. S. Patents 4,067,852, **1978**; 4,185,996, **1980**; 4,161,470, **1979**; and 4,256,624, **1981**; G. W. Calundann, H. L. Davis, F. J. Gorman, R. M. Mininni (Celanese), U. S. Patent 4,083,829, **1978**; A. J. East, L. F. Charbonneau, G. W. Calundann (Celanese), U. S. Patent 4,330,457, **1982**; c) information package on Vectra products, Hoechst Celanese Corporation.

[15] R. S. Irwin, EU Patent 26,991, **1981**.

[16] J. F. Harris, Jr., US Patent 4,447,592, **1984**.

[17] A. R. Postema, K. Liou, F. Wudl, P. Smith, *Macromolecules* **1990**, *23*, 1842.

[18] H.-R. Dicke, R. W. Lenz, *J. Polym. Sci.: Polym. Chem. Ed.* **1983**, *21*, 2581.

[19] W. Brügging, U. Kampschulte, H.-W. Schmidt, W. Heitz, *Makromol. Chem.* **1988**, *189*, 2755.

[20] a) M. Ballauff, *Makromol. Chem., Rapid Commun.* **1986**, *7*, 407; b) M. Ballauff, G. F. Schmidt, *Mol. Cryst. Liq. Cryst.* **1987**, *147*, 163; c) H. Ringsdorf, P. Tschirner, O. Hermann-Schönherr, J. H. Wendorff, *Makromol. Chem.* **1987**, *188*, 1431.

[21] H. R. Kricheldorf, A. Domschke, *Macromolecules* **1996**, *29*, 1337.

[22] H. R. Kricheldorf, D. F. Wulff, *J. Polym. Sci. Part A – Polym. Chem.* **1997**, *35*, 947.

[23] W. Vogel, W. Heitz, *Makromol. Chem.* **1990**, *191*, 829.

[24] a) I. Goodman, J. E. McIntyre, D. H. Aldred, BP 993 272, **1975**; b) W. R. Krigbaum, H. Hakemi, R. Kotek, *Macromolecules* **1985**, *18*, 965 (1985); c) W. Heitz, N. Nießner, *Makromol. Chem.* **1990**, *191*, 225.

[25] a) P. K. Bhowmik, E. D. T. Atkins, R. W. Lenz, H. Han, *Macromolecules* **1996**, *20*, 1919;

b) P. K. Bhowmik, E. D. T. Atkins, R. W. Lenz, *Macromolecules* **1993**, *26*, 447; c) P. K. Bhowmik, E. D. T. Atkins, R. W. Lenz, *Macromolecules* **1993**, *26*, 447; c) P. K. Bhowmik, E. D. T. Atkins, R. W. Lenz, H. Han, *Macromolecules* **1996**, *29*, 3778.

[26] a) J. I. Jin, S.-M. Huh, *Macromol. Symp.* **1995**, *96*, 125; b) S.-M. Huh, J. I. Jin, *Macromolecules* **1997**, *30*, 3005.

[27] a) R. Sinta, R. A. Gaudiana, R. A. Minns, H. G. Rogers, *Macromolecules* **1987**, *20*, 2374; b) R. A. Gaudiana, R. A. Minns, R. Sinta, N. Weeks, H. Rogers, *Progr. Polym. Sci.* **1989**, *14*, 47.

[28] a) H. W. Schmidt, D. Guo, *Makromol. Chem.* **1988**, *189*, 2029; b) W. Heitz, H.-W. Schmidt, *Macromol. Chem. Macromol. Symp.* **1990**, *38*, 149; c) F. Montamedi, U. Jonas, A. Greiner, H.-W. Schmidt, *Liq. Cryst.* **1993**, *14*, 959.

[29] a) A. Greiner, W. E. Rochefort, K. Greiner, G. W. Heffner, D. S. Pearson, H.-W. Schmidt in *Integration of Fundamental Polymer Science – 5* (Ed.: P. J. Lemstra, L. A. Kleintjens), Elsevier, New York **1991**; b) A. Greiner, W. E. Rochefort, K. Greiner, H.-W. Schmidt, D. S. Pearson, *Makromol. Chem. Rapid Commun.* **1992**, *13*, 25; c) A. Greiner, W. E. Rochefort in *Polymer Liquid Crystals – Mechanical and Thermophysical Properties* (Ed.: W. Brostow), Chapman & Hall, London **1997**, 431.

[30] M. Hohlweg, H.-W. Schmidt, *Macromol. Chem.* **1989**, *190*, 1587.

[31] a) D. Tanner, J. A. Fitzgerald, B. R. Phillips, *Angew. Chem. Adv. Mater.* **1989**, *101*, 665; b) S. L. Kwolek, P. W. Morgan, J. R. Schaefgen, L. W. Gulrich, *Macromolecules* **1977**, *10*, 1390; c) T. I. Blair, P. W. Morgan, F. L. Killian, *Macromolecules* **1977**, *10*, 1396; d) see also [2 c] and [2 f].

[32] See, for example: a) J. Y. Jadhav, W. R. Krigbaum, J. Preston, *J. Polym. Sci., Part A: Polym. Chem.* **1989**, *27*, 1175; b) N. Nagata, N. Tsutsumi, T. Kyotsukuri, *J. Polym. Sci., Part A: Polym. Chem.* **1988**, *26*, 235; c) W. R. Meyer, F. T. Gentile, U. W. Suter, *Macromolecules* **1991**, *24*, 633; d) W. R. Meyer, F. T. Gentile, U. W. Suter, *Macromolecules* **1991**, *24*, 642; e) B. R. Glomm, G. C. Rutledge, F. Küchenmeister, P. Neuenschwander, U. W. Suter, *Macromol. Chem. Phys.* **1994**, *195*, 475; f) H. R. Kricheldorf, B. Schmidt, R. Bürger, *Macromolecules* **1992**, *25*, 5465; g) H. R. Kricheldorf, B. Schmidt, R. Bürger, *Macromolecules* **1992**, *25*, 5471; h) W. Hatke, H.-W. Schmidt, *Makromol. Chem., Macromol. Symp.* **1991**, *50*, 41; i) W. Hatke, H.-W. Schmidt, W. Heitz, *J. Polym. Sci., Part A: Polym. Chem.* **1991**, *29*, 1387; j) W. Hatke, H.-W. Schmidt, *Makromol. Chem. Phys.* **1994**, *195*, 3579.

[33] See, for example: a) M. Ballauff, G. F. Schmidt, *Makromol. Chem., Rapid Commun.* **1987**, *8*, 93; b) M. Ballauff, G. F. Schmidt, *Mol. Cryst. Liq. Cryst.* **1989**, *147*, 163; c) M. Ebert, O. Hermann-Schönherr, J. H. Wendorff, H. Ringsdorf, P. Tschirner, *Makromol. Chem., Rapid Commun.* **1988**, *9*, 445; d) H. Ringsdorf, P. Tschirner, O. Hermann-Schönherr, J. H. Wendorff, *Makromol. Chem.* **1987**, *188*, 1431; e) J. M. Rodriguez-Parada, R. Duran, G. Wegner, *Macromolecules* **1989**, *22*, 2507; f) A. Adam, H. W. Spiess, *Makromol. Chem. Rapid Commun.* **1990**, *11*, 249; g) L. Yu, Z. Bao, R. Cai, *Angew. Chem.* **1993**, *105*, 1392; h) P. Galda, D. Kistner, A. Martin, M. Ballauff, *Macromolecules* **1993**, *26*, 1595; i) M. Sone, I. B. R. Harkness, J. Watanabe, T. Yamashita, T. Torii, K. Horie, *Polym. J.* **1993**, *25*, 997; j) H. R. Kricheldorf, A. Domschke, *Macromolecules* **1994**, *27*, 1509; k) I. G. Voigt-Martin, P.

Simon, S. Bauer, H. Ringsdorf, *Macromolecules* **1995**, *28*, 236; l) M. Steuer, M. Hörth, M. Ballauff, *J. Polym. Sci., Part A: Polym. Chem.* **1993**, *31*, 1609.

[34] See, for example: a) J. Preston, *Polym. Eng. Sci.* **1975**, *15*, 199; b) J. Preston, R. W. Smith, S. M. Sun, *J. Appl. Polym. Sci.* **1972**, *16*, 3237; c) J. Blackwell, R. A. Cageao, A. Biswas, *Macromolecules* **1982**, *20*, 667; d) see also [2c] and [2f].

[35] See, for example: a) R. A. Gaudiana, R. A. Minns, H. G. Rogers, P. S. Kalyanaraman, C. McGowan, W. C. Hollinsed, J. S. Manello, R. Sahatjian, *Macromolecules* **1985**, *18*, 1085; b) R. A. Gaudiana, R. A. Minns, H. G. Rogers, R. Sinta, L. D. Taylor, P. S. Kalyanaraman, C. McGowan, *J. Polym. Sci., Part A: Polym. Chem.* **1987**, *25*, 1249; c) see also [27b], [32h], [32i], and [32j].

2 Main Chain Liquid Crystalline Semiflexible Polymers

Emo Chiellini and Michele Laus

2.1 Introduction

Since the discovery [1] that high performance polymeric fibers could be obtained from rigid-rod polymers forming liquid crystalline mesophases in solution, an enormous research effort has been addressed to the preparation and study of main chain polymers containing rigid anisometric segments. This has given rise to such commercial polymers as poly(*p*-phenylene terephthalamide) from which Kevlar (Du Pont) and Twaron (Akzo) fibers are spun, most of which are constituted of aromatic sequences. As all the poly(aromatic amide)s (aramid) are unmeltable and can only be spun from very aggressive solvents such as sulfuric acid, the macromolecular design was successively directed to the development of melt-processable liquid crystalline polymers [2]. Melt processability can be very effectively obtained by disrupting the regular structure of the main chain, while not substantially reducing the chain stiffness, by random copolymerization, and/or the introduction of rigid nonlinear comonomeric units or of lateral substituents in the chain. This "frustrated chain packing" concept was successfully applied to the design

of the copolyesters consisting of 2-oxy-6-naphthoyl and *p*-benzoyl units, from which Vectra fibers [3] (Hoechst-Celanese) are obtained, and in the preparation of poly(ethylene terephthalate) (PET) modified with ripid *p*-hydroxybenzoic acid units, from which X7G fibers [4] (Eastman-Kodak) are prepared.

The depressed efficiency of chain packing in the solid state results in relatively low melting temperatures and unusual ease in processing. The mechanical properties of these polyesters, including the ultrahigh modulus and strength, high thermal stability, and exceptionally low thermal expansion coefficient, were the subject of many investigations and several reviews [5–8].

However, the synthetic procedures leading to the above copolymers do not guarantee the achievement of the same microstructure. The as-obtained sequence distributions are probably unstable because the presence of transesterification reactions leads to highly complicated solid state behavior, which typically depends on the thermal history and processing conditions of the sample. Definite relations between the phase behavior, the molar mass, and the molar mass distribution are not easily available because of the poor solubility of these polymers. Finally, on account of the relative incompatibility of the different copolymer sequences, driven by their distinct inherent rigidities, multiphase domain structures in the submicrometer scale are obtained whose size and shape affect the physicochemical properties substantially.

Considering the above-mentioned problems, the rigid rod polymers are very interestig for numerous technological opportunities, but they are not suitable candidates for the assessment of reliable correlations between the molecular and structural characteristics and the mechanical and physiochemical properties, including the tendencies to self-assemble and form mesophases.

As an alternative way to reduce the phase transition temperatures, the main chain structure can also be disrupted by the insertion of flexible segments, i. e., molecular segments capable of assuming a variety of different configurations at accessible energy levels. The regularly alternating arrangement of flexible spacers and rigid anisometric groups gives rise to thermotropic main chain semiflexible liquid crystalline polymers (MCLCP). In the great majority of these polymers, rodlike units were employed with their main direction lying along the polymer chain, according to the basic structure illustrated in Fig. 1 a.

However, a few examples of polmers in which the rodlike units are placed with their long axes lying perpendicular to the polymer chain [9], according to the structure illustrated in Fig. 1 b, and in which the anisometric units present were of a disk-like shape (Fig. 1 c), were also reported [10]. A collection of papers and rather comprehensive reviews were published on these polymers [11 – 22].

In general, for these semiflexible polymers, the molecular packing features responsible for their unique thermodynamic

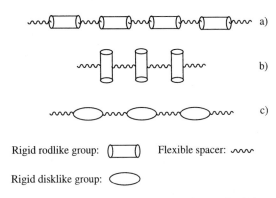

Rigid rodlike group: ▭   Flexible spacer: ∿

Rigid disklike group: ⬭

Figure 1. Schematic representation of the main chain polymers obtained from the combination of rigid anisometric and flexible elements.

and mechanical properties arise from a crucial interplay between the ordering propensity, imparted by the interactions among the rigid groups that are orientation-dependent, and the intra and/or interchain constraints dictated by the flexible spacers connecting the mesogenic groups, including the specific effects exerted by the linking groups. Accordingly, the overall polymer molecule must adopt a spatial configuration whose features define the structure, symmetry, and stability of the mesophase. In addition to these structural modeling features, the polymeric nature influences the liquid-crystalline behavior of main-chain polymers through the molar mass and its distribution characteristics.

2.2   Preparation of Main Chain Liquid Crystalline Polymers (MCLCPs)

2.2.1   Generalities

Semiflexible MCLCPs were prepared following two basic conceptual procedures. The first one (melt transesterification) involves the chemical modification of pre-

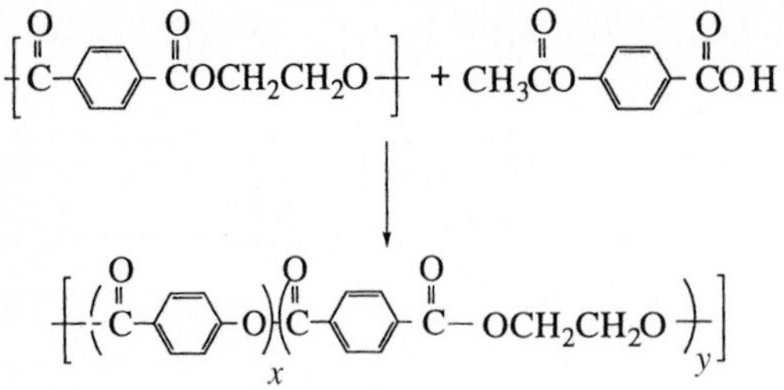

Figure 2. Chemical modification of PET via a melt transesterification reaction with *p*-acetoxybenzoic acid, to obtain a liquid-crystalline copolyester.

formed nonliquid crystalline polymers. By this approach, the poly(ethylene terephthalate) (PET) was reacted [23] with *p*-acetoxybenzoic acid, via a melt transesterification reaction performed at temperatures up to 270 °C and catalyzed by zinc or antimony acetates, to provide a high molar mass liquid crystalline copolyester, according to Fig. 2.

The structure of this copolyester consists of a sequence of ethylene, terephthalic, and oxybenzoic units. Depending on the overall composition and on the reaction conditions, which strongly affect the polymer microstructure, a variety of mesogenic aromatic dyads, triads, and even longer units can be obtained, interconnected by ethylene spacers. Accordingly, on one hand, the overall chain stiffness increases with respect to the starting PET, thus increasing the tendency of the polymer chains to pack into ordered structures. On the other hand, the chain periodicity is destroyed, thus somewhat inhibiting crystallization. The net result of these two conflicting tendencies is the reduction of the melting point and the increase of the isotropization temperature, thus leading to the formation of a liquid-crystalline melt which extends over a wide temperature range.

The second synthetic approach (monomeric components step-growth) is based on more conventional step-growth poly-merization reactions between appropriate monomers. In this case, the mesogenic units can be prebuilt in one precursor species, or they are formed during the polymerization itself. These reactions can be realized through high temperature transesterifications in the melt, for example, of diacetylated diphenols and diacids, or by a variety of conventional polymerization reactions carried out in solution at low or medium temperature, with or without catalysts, in homogeneous or heterogeneous conditions. These latter reactions allow for better control of the polymer microstructure, as long as they are performed under relatively mild conditions, thus preventing molecular scrambling and structure rerrangements. In addition, it is possible to predetermine the end group nature and, to some extent, the molar mass values, by performing the reactions under controlled nonstoichiometric conditions.

Following this second approach, a wide variety of MCLCPs with highly differentiated structures were prepared, including polyesters, polyethers, polyurethanes, polyamides, polysiloxanes, poly(*β*-aminoester)s, and poly(*β*-thioester)s. It should be observed that most of the above polymers contain a variety of different groups in the repeating unit. For example, most of the described polyurethanes contain an ester function, in addition to the urethane functions. Accordingly, this classification is not unequivocal but is made on the basis of the chemical group or functionality, which is particularly important in determining the thermal and mesophasic behavior of the corresponding polymer.

2.2.2   Polyesters

Polyesters represent the majority of the prepared liquid-crystalline polymers. The

Table 1. Representation of some typical structures of polyesters.

Structure[a]	Ref.
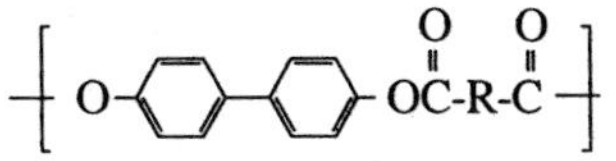	[24, 25]
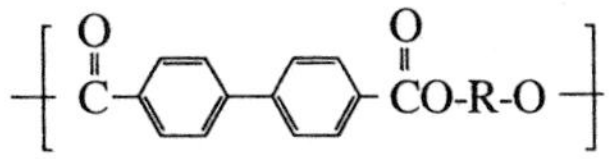	[26, 27]
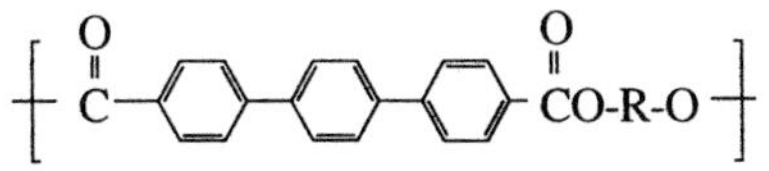	[28, 29]
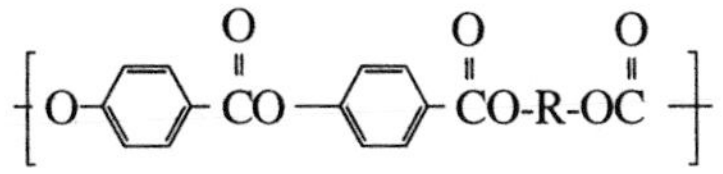	[25, 30, 31]
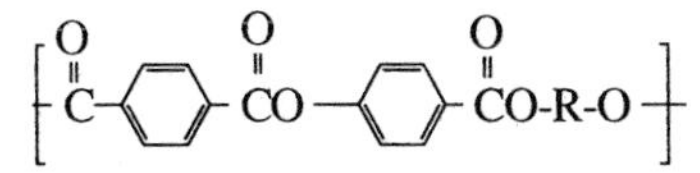	[30, 32]
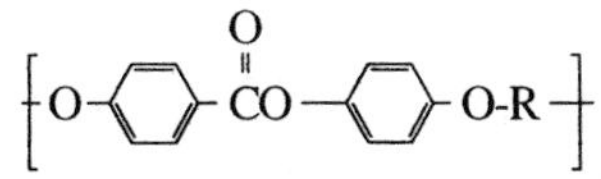	[25]
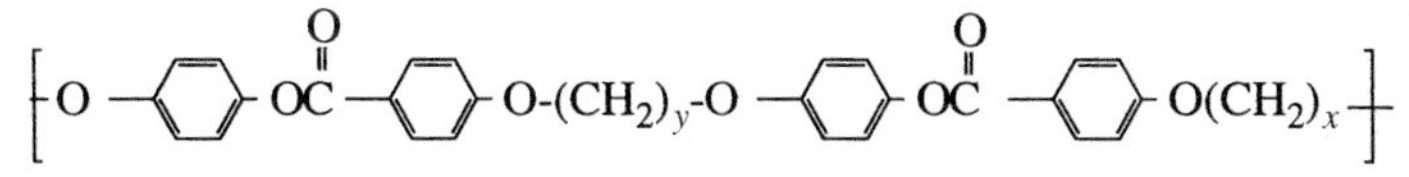	[33, 34]

[a] R = polymethylene segments, segments containing etheroatoms, unsaturation, aliphatic, or aromatic rings.

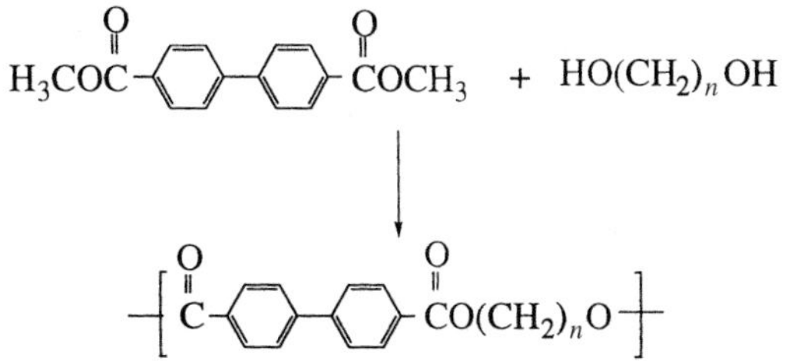

Figure 3. Synthesis of an LC polyester through a melt transesterification reaction between the mesogenic 4,4′-diacetoxybiphenyl and a diol.

structures of representative examples of these polymers are reported in Tables 1–7. LC polyesters are generally prepared by following the monomeric component step-growth procedure involving condensation in the melt or in solution of stoichiometrically defined mixtures of the monomeric components. A typical example of condensation in the melt can be represented by the reaction of transesterification [24, 25] of the mesogenic 4,4′-diacetoxybiphenyl and a polymethylene diol, as illustrated in Fig. 3.

These reactions are usually carried out at temperatures in the 200–280 °C range under an argon or nitrogen atmosphere and at a reduced pressure. Catalysts such as zinc, titanium, or antimonium salts are often employed to speed the reaction. This synthetic approach presents great advantages in that it is possible to obtain high molar mass polymers in quantitative yields without the necessity of solvent removal, and can be recommended when the polymer does not contain chemically different ester functions. However, if distinct ester groups are present within the polymer chain, the above transes-

Table 2. Representation of some typical structures of polyesters.

Structure[a]	Ref.
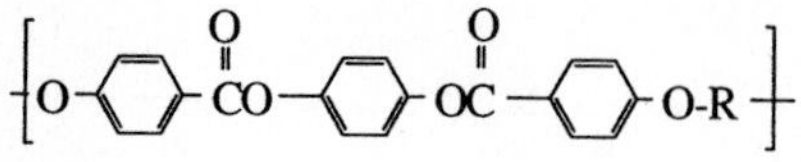	[35]
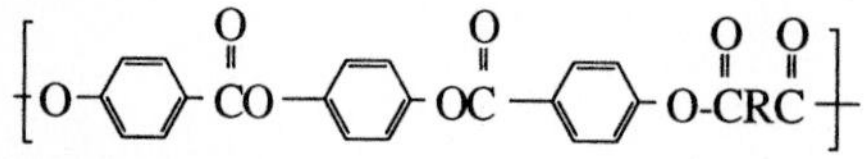	[36]
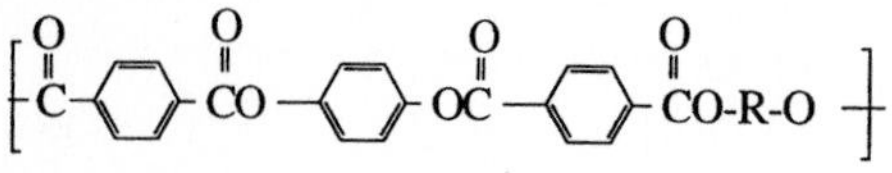	[37, 38]
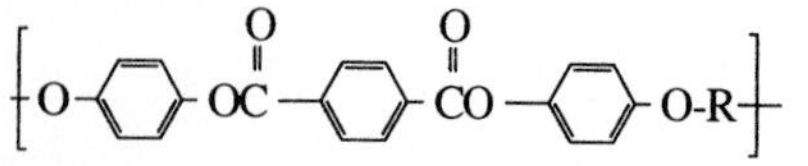	[35, 37]
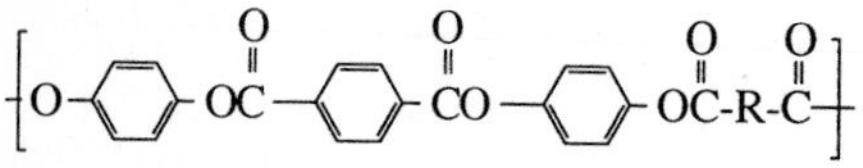	[38, 39]
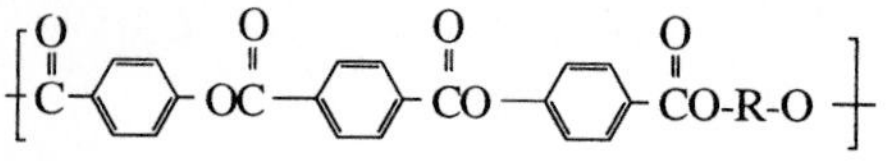	[32, 40, 41 – 46]
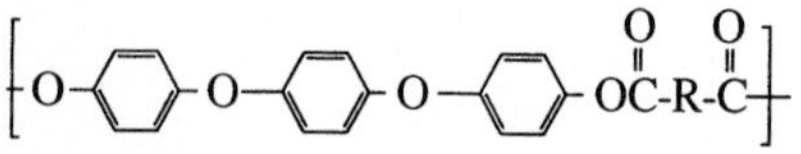	[47]

[a] R = polymethylene segments, segments containing etheroatoms, unsaturation, aliphatic, or aromatic rings.

terification reactions lead to polymers with not well defined structures. In this case, to obtain polymers with controlled structures it is necessary to run the reaction under milder conditions in the presence of solvent, employing either solution or interfacial polycondensation methods. The former reaction was successfully employed in the preparation of polyesters by reacting the mesogenic dihydroxy-4,4′-phenyl benzoate with several α,ω-alkanediacid chlorides in α-chloronaphthalene as the solvent (Fig. 4).

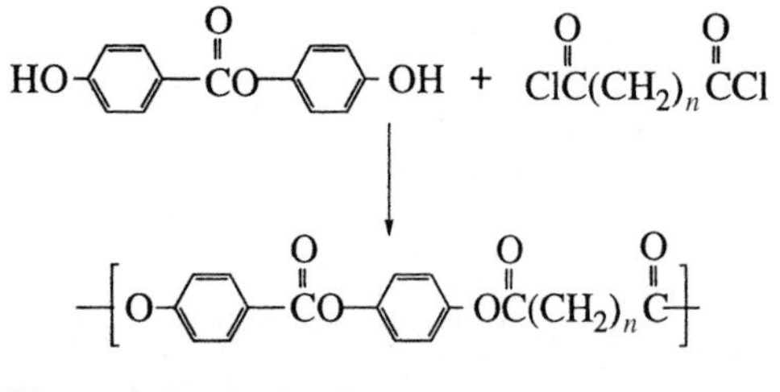

Alternatively, the same polyester can be prepared by reacting polymethylene bis(p-hydroxybenzoates) and polymethylene bis(p-chloroformylbenzoates) in tetrachloroethane using pyridine as the hydrochloric acid acceptor, according to the reaction scheme illustrated in Fig. 5.

Figure 4. Synthesis of LC polyesters by reacting the mesogenic dihydroxy-4,4′-phenyl benzoate with α,ω-alkanediacid chlorides.

However, the majority of polyesters was prepared using the latter reaction, that is, by interfacial polycondensation. Usually, this reaction is carried out at room temperature in a solvent mixture consisting of water and an immiscible organic solvent such as, for example, tetrachloroethane, chloroform, or

Table 3. Representation of some typical structures of polyesters.

Structure		Ref.
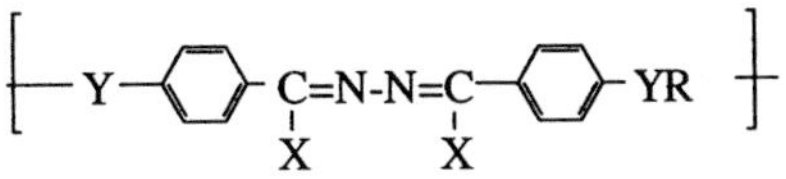 ⌐Y─◯─C=N-N=C─◯─YR⌐ (X, X)	Y=OCO, OCOO X=CH₃, CH₂CH₃ R=(CH₂)ₙ, (CH₂CH₂O)ₙCH₂CH₂	[38, 49–51]
⌐OC─◯─C(H)=C(H)─◯─COR⌐	R=(CH₂)ₙ, (CH₂CH₂O)ₙCH₂CH₂	[52–54]
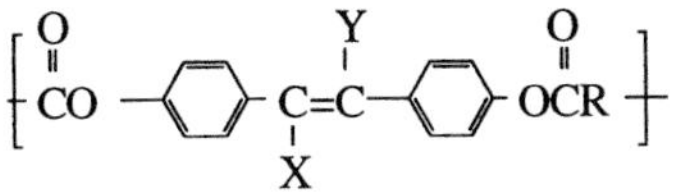 ⌐CO─◯─C(Y)=C(X)─◯─OCR⌐	R=(CH₂)ₙ X=H, Y=H; X=CH₃, Y=H; X=CH₃, Y=CH₃; X=CH₂CH₃, Y=CH₂CH₃	[55–59]
⌐OC─◯─N=N─◯─COR⌐	R=(CH₂)ₙ, (CH₂CH₂O)ₙCH₂CH₂	[60]
⌐OC─◯─N=N(O)─◯─COR⌐	R=(CH₂CH₂O)ₙCH₂CH₂	[61, 62]
⌐CO─◯─N=N(O)─◯─OCR⌐	R=(CH₂)ₙ, (CH₂)₂ₙCH(CH₃)CH₂	[62]
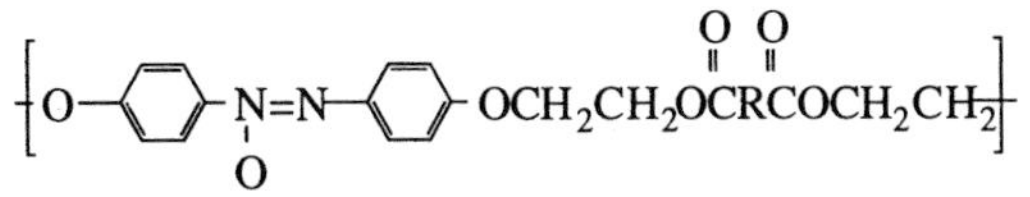 ⌐O─◯─N=N(O)─◯─OCH₂CH₂OCRCOCH₂CH₂⌐	R=(CH₂)ₙ, (CH₂)₂ₙCHCH₃CH₂	[62]
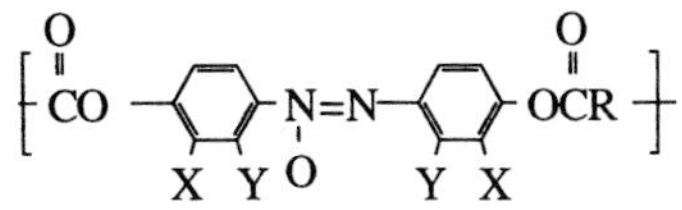 ⌐CO─◯(X Y)─N=N(O)─◯(Y X)─OCR⌐	R=(CH₂)ₙ X=CH₃, Y=H; X=H, Y=CH₅	[63]

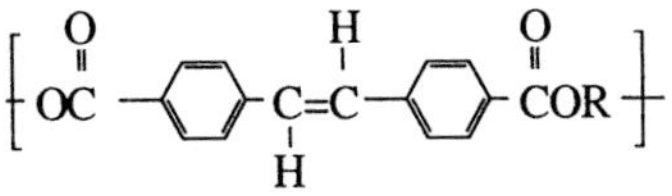

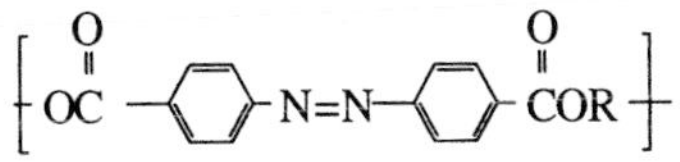

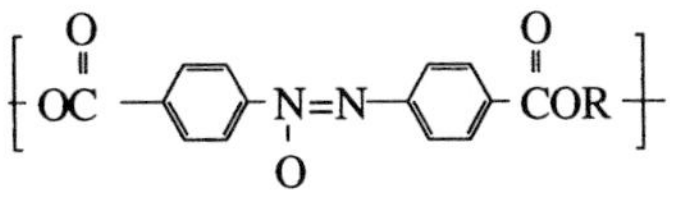

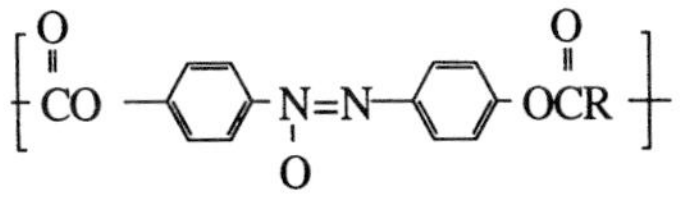

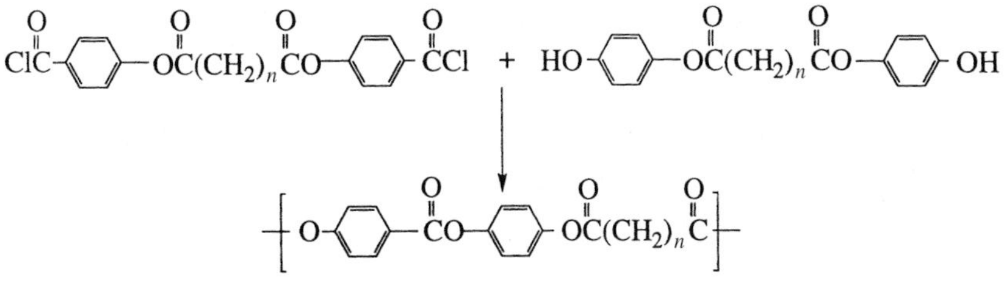

Figure 5. Synthesis of LC polyesters by reacting polymethylene bis(*p*-hydroxybenzoates) and polymethylene bis(*p*-chloroformylbenzoates).

dichloroethane. A monomer, usually a diacid chloride, is dissolved in the organic solvent whereas the other monomer, usually a diphenol, is dissolved in water after reaction with sodium or potassium hydroxide. The two solutions are then mixed together in the presence of a tertiary amine, such as pyridine or triethylamine, or in the presence of

Table 4. Representation of some typical structures of polyesters.

Structure	Ref.
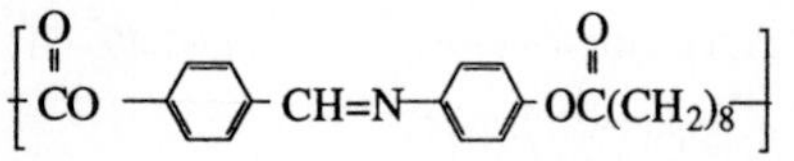	[56]
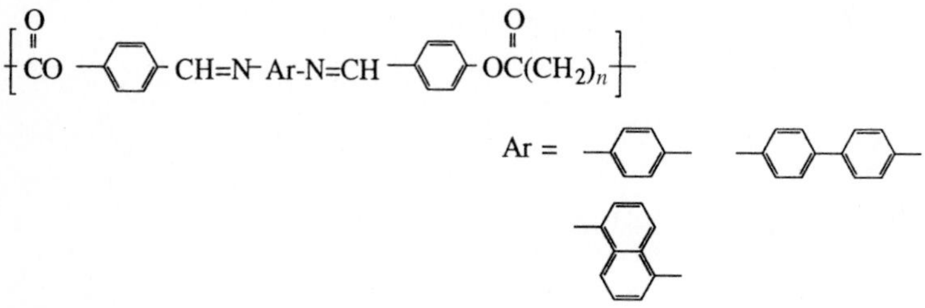	[64]
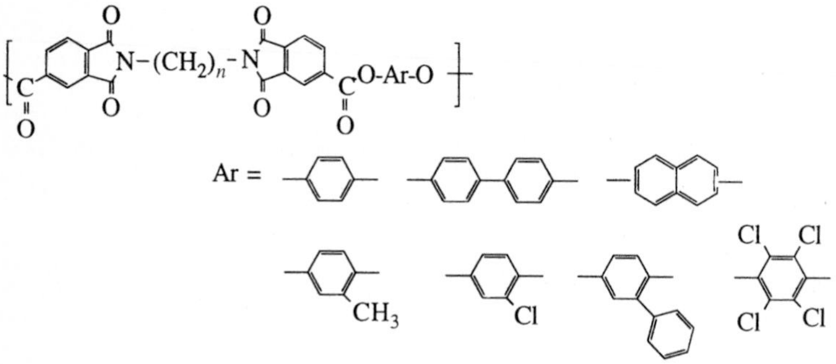	[65, 66]
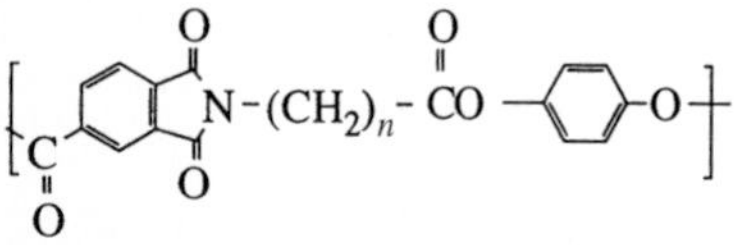	[67]
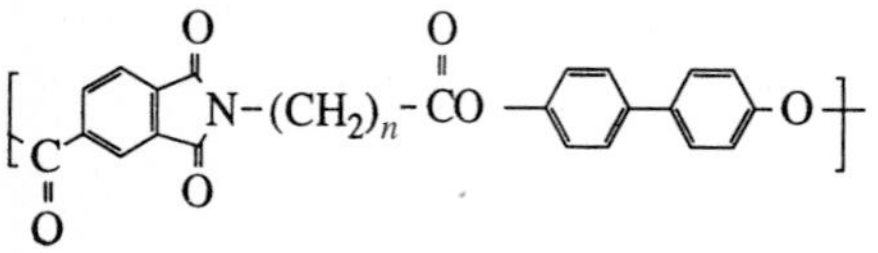	[67]

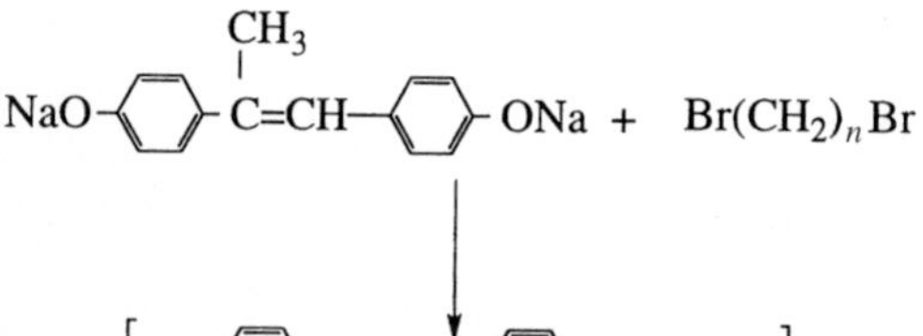

Figure 6. Synthesis of LC polyethers by a phase transfer catalyzed polyetherification reaction.

a phase transfer catalyst such as the benzyltriethylammonium chloride.

2.2.3   Polyethers

Polyethers are usually prepared by a phase transfer catalyzed polyetherification reaction between the sodium salt of the mesogenic diphenol dissolved in water and an α,ω-dibromoalkane dissolved in o-dichlorobenzene or nitrobenzene in the presence of

Table 5. Representation of some typical structures of polyesters.

Structure	Ref.
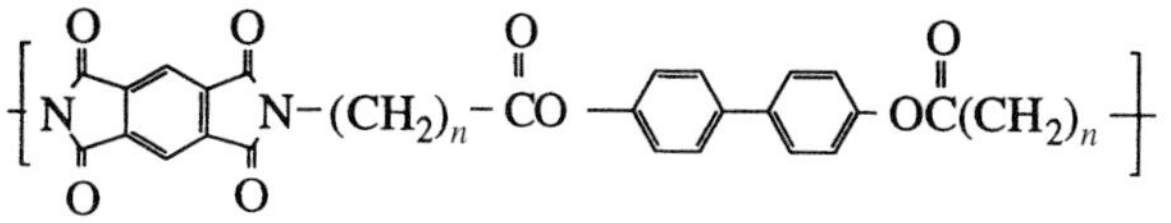	[68]
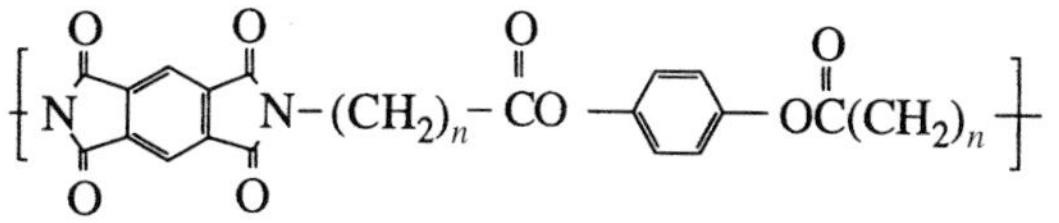	[68]
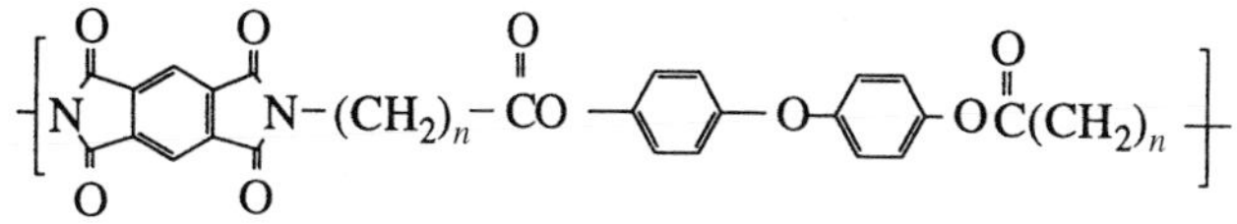	[68]
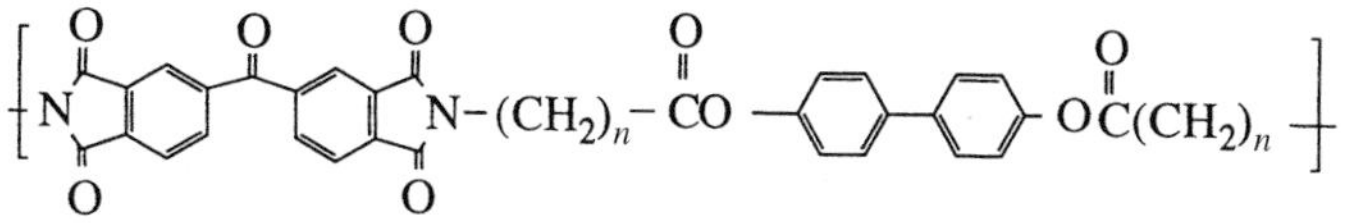	[69]
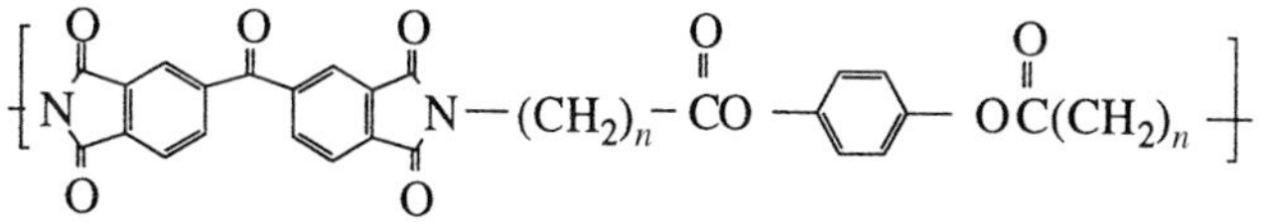	[69]
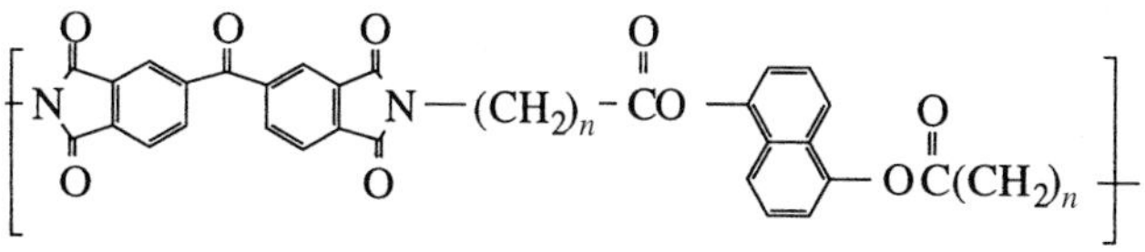	[69]

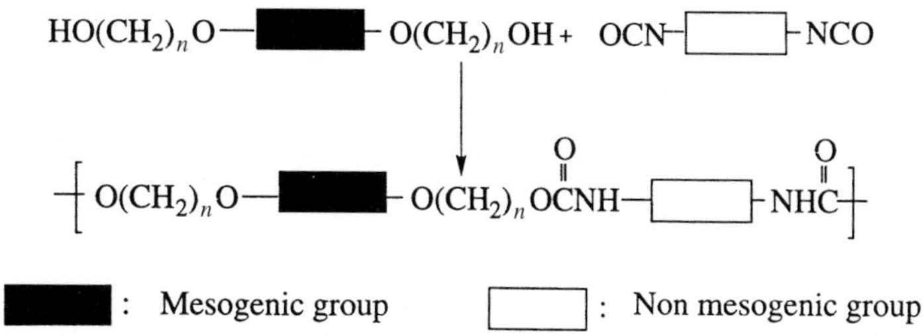

Figure 7. Synthesis of LC polyurethanes by a stepwise polyaddition reaction of a mesogenic diol with an aliphatic or aromatic diisocyanate.

tetrabuthylammonium hydrogen sulfate or tetrabutyl ammonium bromide as the phase transfer catalyst, according to Fig. 6.

The structures of some prominent examples of LC polyethers are collected in Table 8.

2.2.4   Polyurethanes

The preparation of thermotropic LC polyurethanes is performed essentially according to two synthetic pathways involving the stepwise polyaddition reaction of a meso-

Table 6. Representation of some typical structures of polyesters.

Structure		Ref.
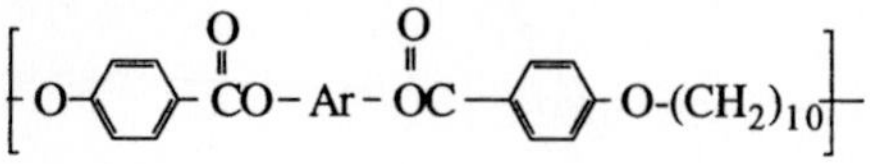 [O—⟨⟩—CO—Ar—OC—⟨⟩—O-(CH₂)₁₀]	 Ar = (ring with X); X=CH₃, Cl, Br, O(CH₂CH₂O)ₙCH₃, O(CH₂CH₂O)ₙCH₂CH₃, n=0, 1, 2	[70–72]
	Ar 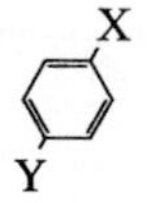 = (ring with X, Y)	
	X, Y= Cl, 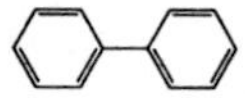 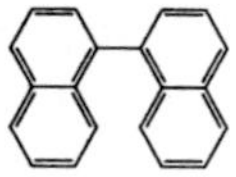 (biphenyl), (binaphthyl)	
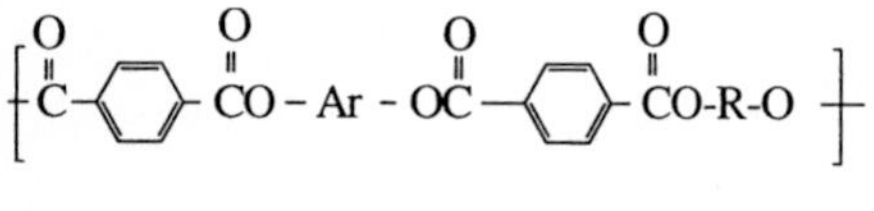 [C—⟨⟩—CO—Ar—OC—⟨⟩—CO-R-O]	 Ar = (ring with X); X=Br, CₙH₂ₙ₊₁	[73, 74]

Table 7. Representation of some typical structures of polyesters and copolyesters.

Structure	Ref.
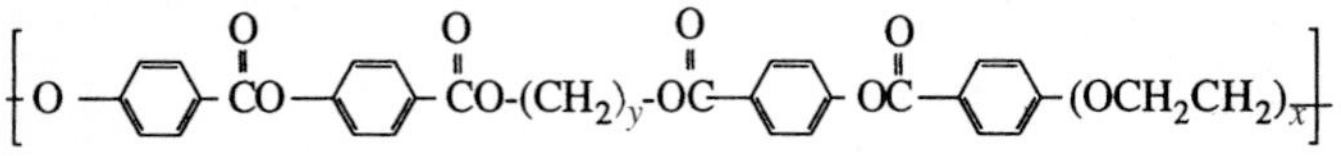 [O—⟨⟩—CO—⟨⟩—CO-(CH₂)ᵧ-OC—⟨⟩—OC—⟨⟩—(OCH₂CH₂)ₓ]	[75]
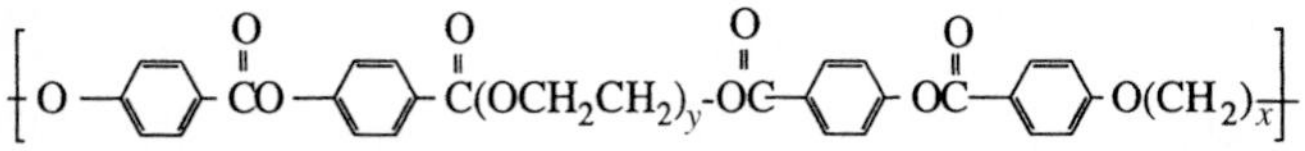 [O—⟨⟩—CO—⟨⟩—C(OCH₂CH₂)ᵧ-OC—⟨⟩—OC—⟨⟩—O(CH₂)ₓ]	[76]
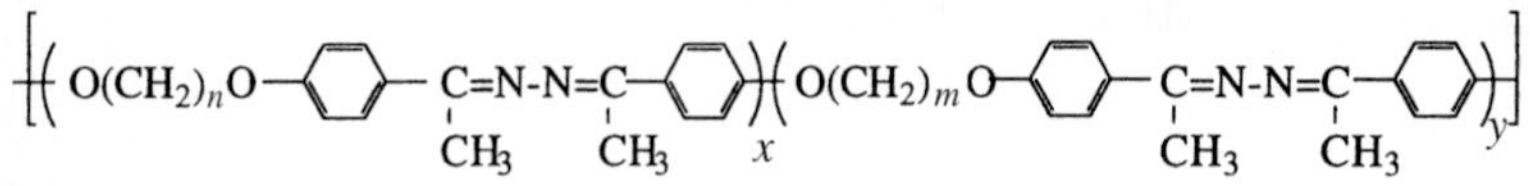 [(O(CH₂)ₙO—⟨⟩—C=N-N=C(CH₃)(CH₃)—⟨⟩)(O(CH₂)ₘO—⟨⟩—C=N-N=C(CH₃)(CH₃)—⟨⟩)]	[77]
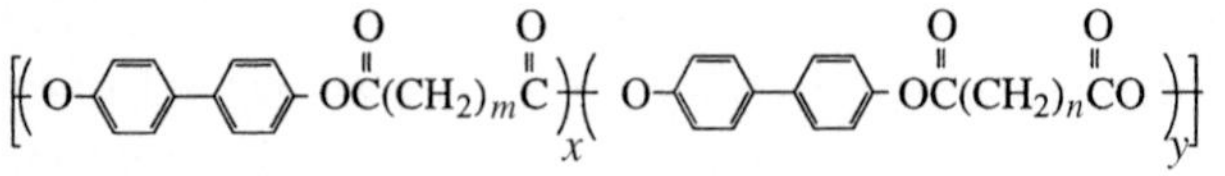 [(O—⟨⟩⟨⟩—OC(CH₂)ₘC)ₓ(O—⟨⟩⟨⟩—OC(CH₂)ₙCO)ᵧ]	[78]
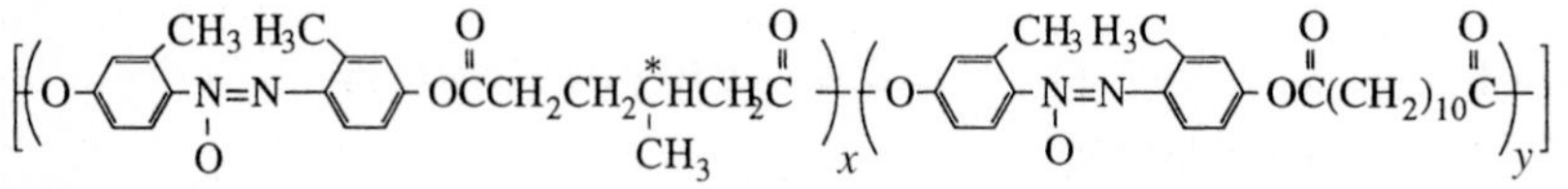 [(O—⟨⟩(CH₃)(H₃C)—N=N—⟨⟩—OCCH₂CH₂ĊHCH₂C(CH₃))ₓ(O—⟨⟩(CH₃)(H₃C)—N=N—⟨⟩—OC(CH₂)₁₀C)ᵧ]	[79]
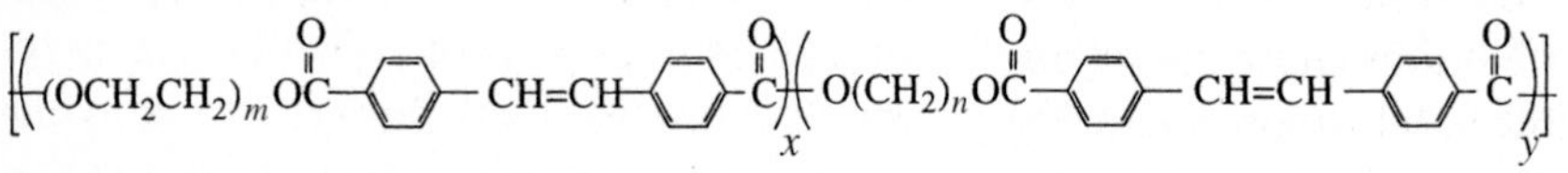 [((OCH₂CH₂)ₘOC—⟨⟩—CH=CH—⟨⟩—C)ₓ(O(CH₂)ₙOC—⟨⟩—CH=CH—⟨⟩—C)ᵧ]	[52]

Table 8. Representation of some typical structures of polyesters and copolyesters.

Structure	Ref.
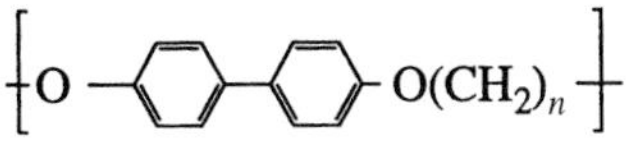 $\left[O-\bigcirc-\bigcirc-O(CH_2)_n\right]$	[80]
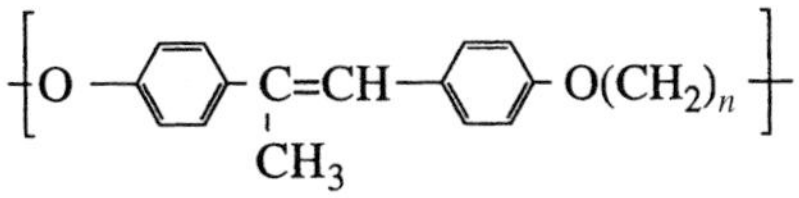 $\left[O-\bigcirc-\underset{CH_3}{C{=}CH}-\bigcirc-O(CH_2)_n\right]$	[81]
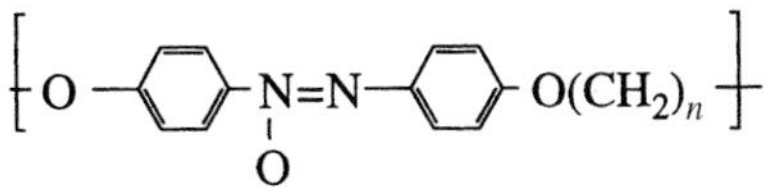 $\left[O-\bigcirc-\underset{O}{N{=}N}-\bigcirc-O(CH_2)_n\right]$	[82]
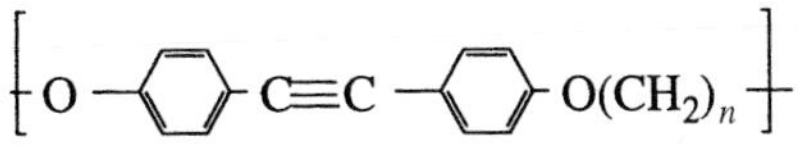 $\left[O-\bigcirc-C{\equiv}C-\bigcirc-O(CH_2)_n\right]$	[83]
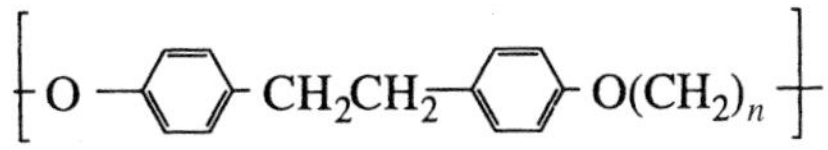 $\left[O-\bigcirc-CH_2CH_2-\bigcirc-O(CH_2)_n\right]$	[84]
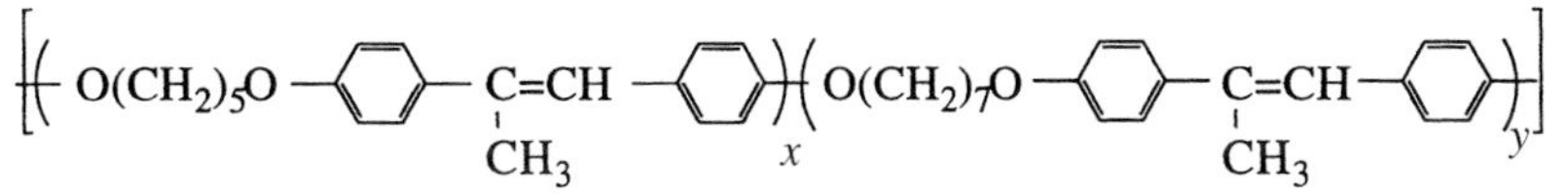 $\left[O(CH_2)_5O-\bigcirc-\underset{CH_3}{C{=}CH}-\bigcirc\right]_x\left[O(CH_2)_7O-\bigcirc-\underset{CH_3}{C{=}CH}-\bigcirc\right]_y$	[85]

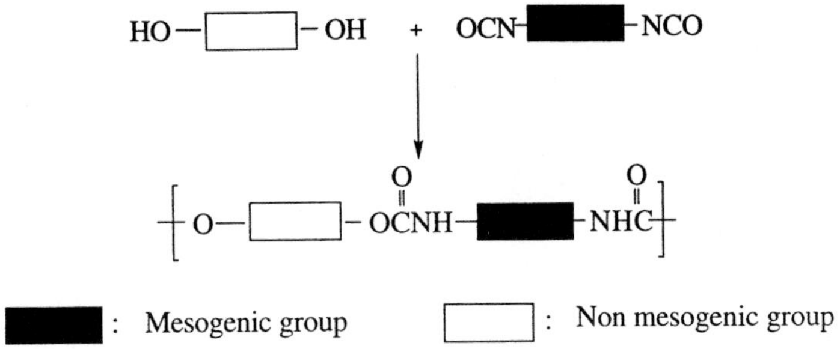

Figure 8. Synthesis of LC polyurethanes by a stepwise polyaddition reaction of a mesogenic diisocyanate with a diol.

genic diol with an aliphatic or aromatic diisocyanate (Fig. 7), or alternatively of a mesogenic diisocyanate with an aliphatic diol (Fig. 8).

Some examples of LC polyurethanes are collected in Table 9. The polymerization reactions were carried out in DMSO or DMF at 120 °C for 24 h without any added catalyst. However, substantial irreproducibility of the molecular characteristics of the poly-

mers was observed in successive preparations, and sometimes an induction period was detected. These effects are probably due to the presence of adventitious water in these solvents, which is able to react with the diisocyanate monomer leading to reactive and unreactive by-products. The addition of catalytic quantities of 1,4-diazabicyclo[2.2.2]octane considerably increases the reaction rate, but only oligomeric products

Table 9. Representation of some typical structures of polyurethanes.

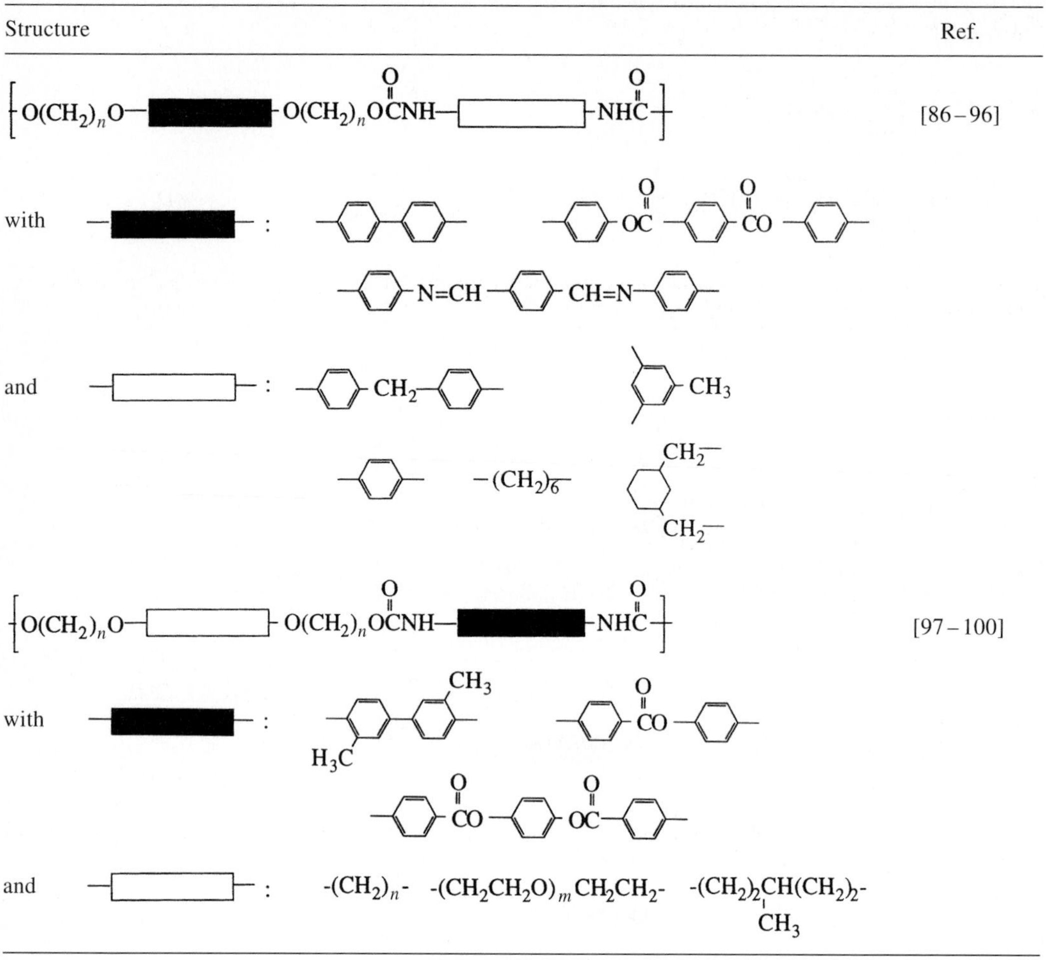

Structure	Ref.

were obtained. Polymeric products with reproducible and high molar masses were obtained by running the reactions at 90 °C in dry chloroform in a nitrogen atmosphere. Alternatively, LC polyurethanes were prepared [101, 102] by reacting dichloroformates and diamines.

2.2.5   Polyamides

Liquid crystalline main chain polyamides with a strictly alternating structure were prepared [103] by reacting aliphatic diacids and bis(aminobenzoates) in the presence of pyridine and triphenylphosphyne according to the Yamazaki procedure at moderate temperatures of 100–115 °C (Fig. 9).

2.2.6   Polysiloxanes

Polysiloxanes are generally prepared, as illustrated in Fig. 10, by reacting equimolar amounts of a mesogenic monomer containing allylic functions with α-dimethylsilanyl-α-hydrogen-oligodimethylsiloxanes in the presence of a chloroplatinic catalyst or

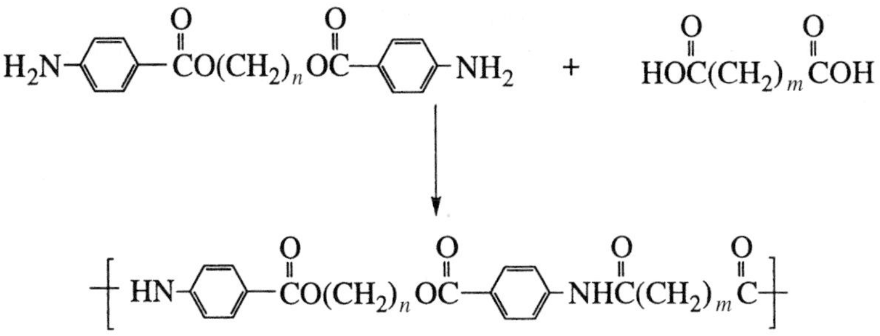

Figure 9. Synthesis of LC polyamides by the polycondensation reaction between aliphatic diacids and bis(aminobenzoate)s.

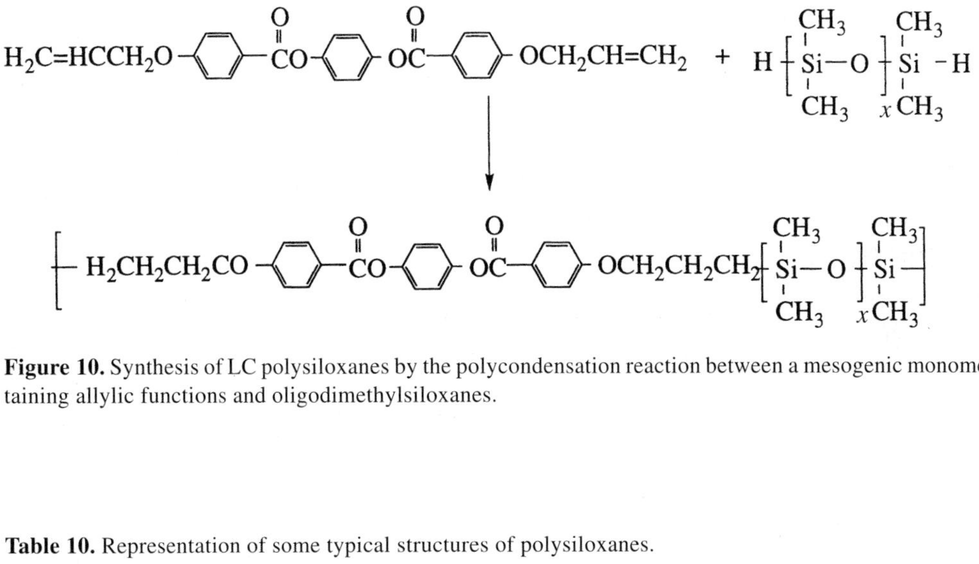

Figure 10. Synthesis of LC polysiloxanes by the polycondensation reaction between a mesogenic monomer containing allylic functions and oligodimethylsiloxanes.

Table 10. Representation of some typical structures of polysiloxanes.

Structure[a]	Ref.
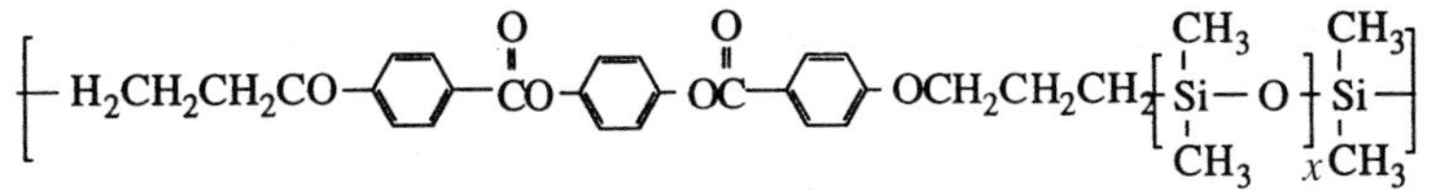	[104]
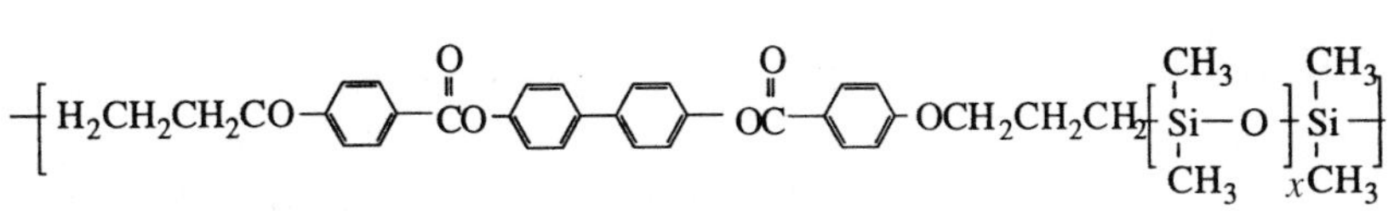	[104]
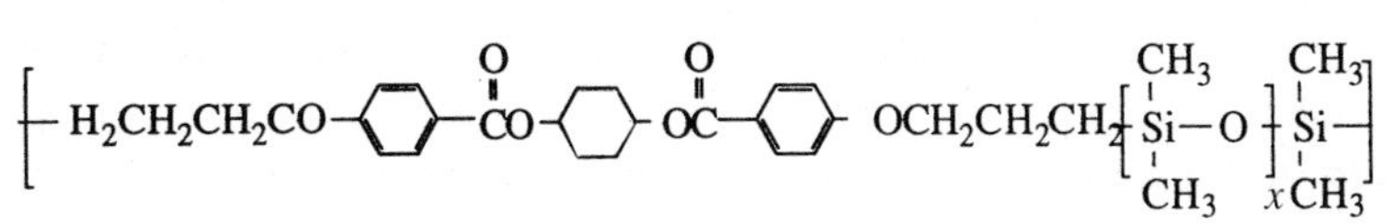	[104]

[a] $x = 2-5$, 13, 38, respectively.

Table 11. Representation of some typical structures of poly(β-aminoester)s and poly(β-thioester)s.

Structure	Ref.

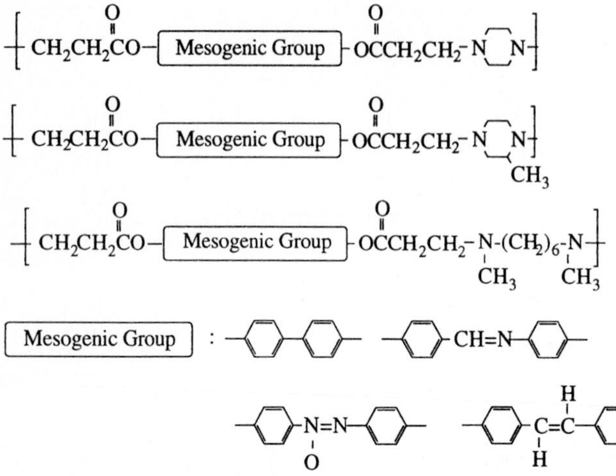

of the platinum divinyltetramethyldisiloxane. The structures of some representative examples of LC polysiloxanes are collected in Table 10.

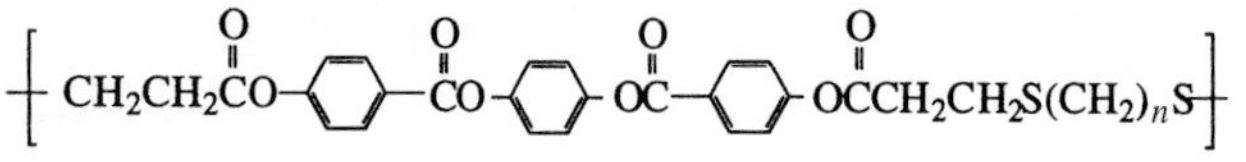

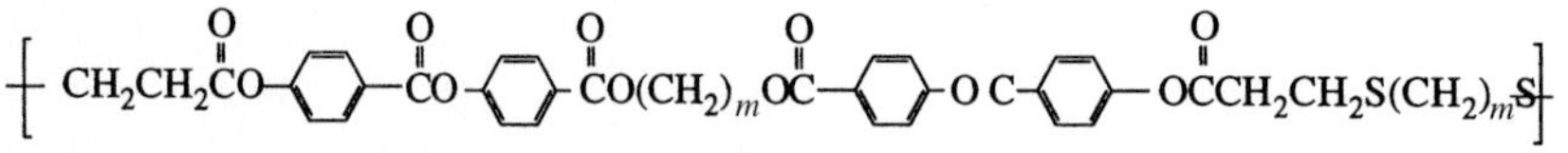

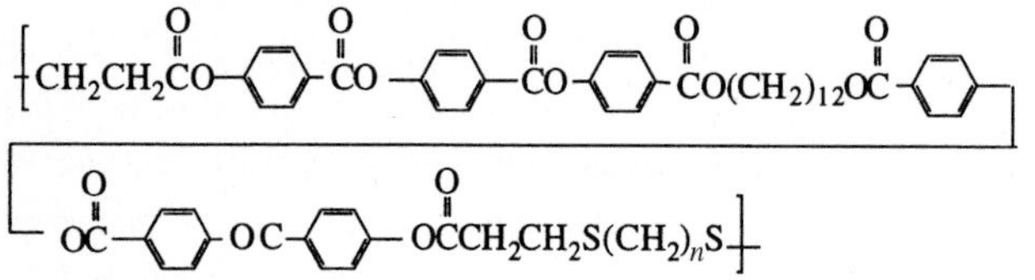

2.2.7 Poly(β-aminoester)s and Poly(β-thioester)s

These functional polymers were prepared, according to Fig. 11, by reacting α,ω-diacrylates containing different mesogenic groups with secondary bisamines, such as piperidine, 2-methylpiperidine, 1,6-dime-

thyl-hexamethylenediamine or α,ω-bisthiols at room temperature in 1,4-dioxane according to a Michael-type reaction. When bisamines are employed no catalyst is necessary, whereas in the case of bisthiols the use of a catalytic amount of triethylamine is necessary to bring the reaction to completion. The structures of some representative examples of LC poly(β-aminoester)s and poly(β-thioester)s are collected in Table 11.

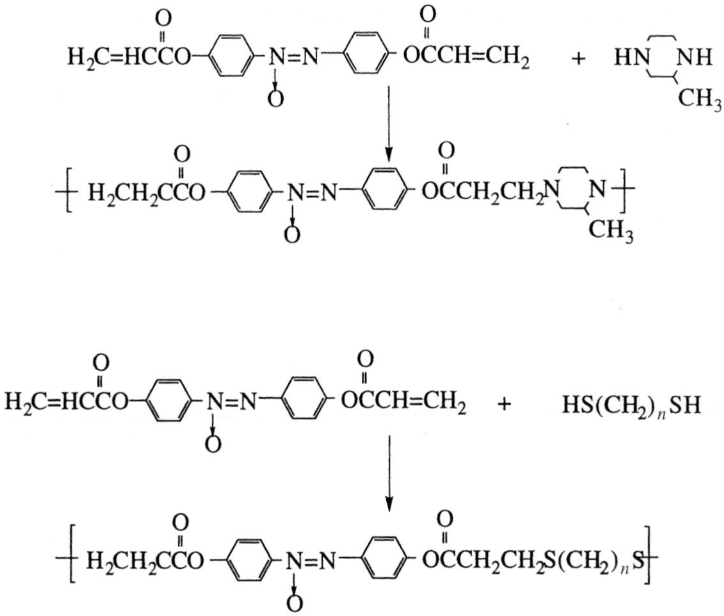

$$H_2C=HC\overset{\overset{\displaystyle O}{\|}}{C}O-\!\!\!\!\bigcirc\!\!\!\!-N=N-\!\!\!\!\bigcirc\!\!\!\!-O\overset{\overset{\displaystyle O}{\|}}{C}CH=CH_2 \quad + \quad HN\quad NH$$

with CH_3 substituent

$$\left[\!\!+ H_2CH_2C\overset{\overset{\displaystyle O}{\|}}{C}O-\!\!\!\!\bigcirc\!\!\!\!-N=N-\!\!\!\!\bigcirc\!\!\!\!-O\overset{\overset{\displaystyle O}{\|}}{C}CH_2CH_2N\quad N\!+\right]$$

with CH_3 substituent

$$H_2C=HC\overset{\overset{\displaystyle O}{\|}}{C}O-\!\!\!\!\bigcirc\!\!\!\!-N=N-\!\!\!\!\bigcirc\!\!\!\!-O\overset{\overset{\displaystyle O}{\|}}{C}CH=CH_2 \quad + \quad HS(CH_2)_n SH$$

$$\left[\!\!+ H_2CH_2C\overset{\overset{\displaystyle O}{\|}}{C}O-\!\!\!\!\bigcirc\!\!\!\!-N=N-\!\!\!\!\bigcirc\!\!\!\!-O\overset{\overset{\displaystyle O}{\|}}{C}CH_2CH_2S(CH_2)_n S\!+\right]$$

Figure 11. Synthesis of LC poly(β-aminoester)s and poly(β-thioester)s by the polycondensation reaction between a mesogenic diacrylic monomer and bissecondary amines and bisthiols, respectively.

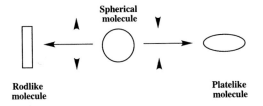

Figure 12. Schematic representation of different molecular shapes.

2.3   Structure – Liquid Crystalline Property Correlations

Although liquid crystallinity is a phase characteristic and not a molecular property, some structural features are well assessed as being conductive to liquid crystal phase formation. Common to all these substances is the asymmetry of molecular shape that produces orientation-dependent interactions both of a hard (self-exclusion) or soft (anisotropy of polarizability) nature. This deviance from spherical symmetry results in two main classes of structurally different liquid crystals, that is, rodlike and platelike molecules (Fig. 12).

General concepts on structure and liquid-crystalline properties in mesomorphic materials based on platelike mesogens are out of the scope of the present chapter and will be treated in other sections of this handbook. The most common liquid crystal substances are those with an elongated, cylinder-like shape. These liquid crystals are referred to as calamitic. Their anisometric features can be measured by the aspect ratio $x = L/d$ where L and d are the length and the diameter of the molecule, respectively. The higher the aspect ratio, the higher the tendency of the molecules to arrange in an ordered assembly at temperatures higher than the melting point, because of the establishment of orientation-dependent interactions of a hard (self-exclusion) nature.

In the case of low molar mass liquid crystals, the overall phase transition behavior depends on a critical balance between the soft and hard interactions. In contrast, for liquid-crystalline polymers, the orientation-dependent interactions of a hard nature are the prevailing ones [112, 113] and the soft interactions can only affect fine details of the molecular and structural organization within the liquid-crystalline mesophase. Depending upon the specific molecular structure, several different types of liquid-crystalline phases or mesophases may occur. The mesophases generated by rodlike molecules usually feature long range uniaxial ordering, which includes both nematic and twisted nematic or cholesteric, or a two-dimensional layered ordering, which includes the low order smectics (smectic A and smectic C) or a three-dimensional crystal-like ordering for the high order smectics (smectics B, E, F, etc.).

From a purely structural viewpoint, a few basic elements can be identified as the fundamental constituents of main chain liquid crystalline polymers, as illustrated in Fig. 1. The essential building blocks which constitute the molecular architecture are the rigid mesogenic groups and the spacers, including also the relevant linking groups. The effects of each individual constituent of the polymer unit, including the molar mass and the molar mass distribution, will be described briefly in the following sections.

2.3.1   Mesogenic Group Effects

The structural features of the mesogenic groups affect the onset, nature, and stability of the liquid-crystalline mesophases through their rigidity, their polarizability along either the main or minor axis direc-

tions, and their aspect ratio. Indeed, the rod-like or cylinderlike part of the molecule must be substantially rigid, even at high temperature, and accordingly it is usually constituted of a sequence of para-linked aromatic (or cycloaliphatic) rings joined together by a rigid link which maintains the linear alignment of the aromatic groups. Linking groups such as imino, azo, azoxy, ester, and *trans*-vinylene groups, etc., as well as a direct link, such as a biphenyl and terphenyl units, are frequently employed since they restrict the freedom of rotation. In addition, linking groups containing multiple bonds can conjugate with phenylene rings, enhancing the anisotropic polarizability and introducing dipole moments which impart orientation-dependent interactions of a soft nature. The ester and amide groups are also effective, since resonance interactions confer a double bond character at the link, thereby restricting rotation from the *trans* to *cis* conformer. In all the above examples, the extended nature of the mesogenic group is maintained through the linearity and rigidity of its constituents. Flexible rodlike mesogens are also introduced when the linking unit consists of ethane or methyleneoxy units. Although these groups are quite flexible, they are able to preserve the colinearity of the aromatic groups through the adoption of the *anti* conformation (1), which is in dynamic equilibrium with the less preferred *gauche* conformation.

Anti

Gauche

X= CH$_2$, O

1

It is practically impossible to draw definite correlations between these structural

Table 12. Representation of some typical structures of polyesters containing the flexible spacer $-(CH_2)_{10}-$ and their melting and isotropization temperatures.

Structure	T_m (°C)	T_i (°C)
—OC—⟨benzene⟩—⟨benzene⟩—CO—	154	160
—OC—⟨benzene⟩—⟨benzene⟩—⟨benzene⟩—CO—	256	311
—OC—⟨benzene⟩—N=N—⟨benzene⟩—CO—	210	–
—CO—⟨benzene⟩—N=N—⟨benzene⟩—OC—	225	245
—OC—⟨benzene⟩—N=N(O)—⟨benzene⟩—CO—	200	–
—CO—⟨benzene⟩—N=N(O)—⟨benzene⟩—OC—	216	265
—CO—⟨benzene, CH$_3$⟩—N=N(O)—⟨benzene, H$_3$C⟩—OC—	118	163
—CO—⟨benzene, CH$_3$⟩—N=N(O)—⟨benzene, CH$_3$⟩—OC—	196	218

features and their specific effects. However, some interesting correlations can be observed from the comparison of the mesophase behavior of various polymers in which different mesogenic units are connected by the same flexible spacer, that is, a polymethylene chain consisting of ten methylene groups. The structures of the mesogenic groups and the relevant phase transition data are collected in Table 12. The aspect ratio of the mesogenic group has a strong influence on the both the melting and isotropization temperatures. In particular, both transition temperatures increase as the axial ratio increases, as clearly observed for the biphenyl and terphenyl units, in which the introduction of a phenylene unit increases the mesophase stability by 150 °C, whereas the melting temperature increases by about 100 °C. This effect produces an increase in the mesophase existence range on increasing the aspect ratio of the mesogenic unit. The effect of reversing the ester group can be evidenced by comparing the polymers containing the azo and the azoxy groups as the bridging group, respectively. In the case of carboxy-terminated flexible spacers, polymers with stable and persistent mesophases are obtained. In contrast, no mesophase can be seen in polymer samples with the reverse ester orientation. This effect can derive from both a polarizability decrease of the mesogenic group and the introduction of strong dipoles perpendicular to the major axis of the mesogenic groups. The introduction of methyl substituents on the mesogenic group decreases both the melting and the isotropization temperatures, because they prevent the mesogenic groups from close packing by steric hindrance. In addition, the position of the methyl groups on the mesogen plays an important role in determining the mesophase stability. The depression in LC mesophase forming tendency is maximum when the methyl groups are placed in adjacent positions on the mesogen, because they can cause a steric effect that results in a twisting of the molecular structure and in a reduction of the collinear disposition of the phenyl groups, thus substantially decreasing the inherent propensity of the mesogenic groups to liquid crystal phase formation.

The effect of substituents other than the methyl group on the mesophase behavior was investigated in very few studies. In general, it appears that small substituents such as Br or Cl do not substantially affect the isotropization temperature, but reduce the melting temperature, thus increasing the mesophase's existence range. This effect was believed to derive from two conflicting tendencies, that is, from the decrease in the mesophase stability due to the steric interactions that reduce the packing efficiency, and from the increased polar interactions which should provide an additional stabilization to the mesophase. Indeed, the introduction of very polar substituents such as NO_2- or $-CN$ decreases both the melting and isotropization temperatures, but the decrease in the latter is less pronounced, as in the case of the methyl group whose size is similar to that of the nitro or cyano groups. Also, in this case, the occurrence of polar interactions was claimed to be responsible for the partial mesophase stabilization.

The effect of the introduction of lateral alkyl chains of variable length on the mesogenic group was systematically studied for a polyester [70] with structure **2**.

$$(CH_2)_n\,CH_3$$

2

With respect to the unsubstituted polymer, the introduction of a methyl group reduces both the melting and isotropization temperatures by about 80 °C. An increase in the alkyl chain further reduces the transition temperatures, which gradually level off and reach their saturation values corresponding to the polymer homolog with $n=5$.

2.3.2   Flexible Spacer Effects

The regular insertion of segments of flexible molecules to separate the mesogenic units along the polymer chain preserves the chemical periodicity of the molecule, although the repeat distance is increased. The flexible spacers provide extra flexibility to the polymer backbone; however, they also impart an effective decrease of the overall aspect ratio of the chain. There is nowadays a huge amount of evidence that the spacer segments do not merely dilute the mesogens, but rather they play a key role in affecting the global thermotropic behavior as far as the stability and order of the mesophases are concerned. Due to the commercial availability of a great variety of aliphatic diols, dichlorides, and diacids with different chain lengths, most typical spacer segments consist of flexible polymethylene $-(CH_2)_n-$ sequences of varying length (n). Spacer segments other than polymethylene have also been introduced into liquid-crystalline polymers. Polyethylene oxide segments extend the spacer length greatly, up to as much as $-(CH_2CH_2O)_8-$ without completely destroying the mesophase [40]. Polysiloxane spacers appear to be more effective per unit length in reducing the transition temperatures, on account of their greater flexibility and larger diameter.

Less frequently, functional spacers such as diamino-alkylene [106] or disulfide-alkylene segments [109] have been used. Specific reactions performed on the reactive sites of both spacers allow for the chemical modification of liquid-crystalline polymers [114]. The possibility of establishing correlations between the liquid-crystalline behavior, as indicated by the transition temperatures, and the nature and relative properties of the mesogenic and flexible sequences is, therefore, a central issue in designing and developing liquid-crystalline polymers of this type.

2.3.2.1   Polymethylene Spacers

The effect of the polymethylene flexible spacer on the liquid-crystalline behavior of

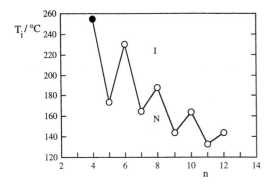

Figure 13. Variation of the isotropization temperature T_i with the flexible spacer length n ($\bullet$ with thermal decomposition).

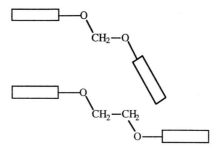

Figure 14. Limit conformations of odd-numbered and even-numbered flexible spacers.

main chain polymers is well assessed in a number of publications. Both T_m and T_i decrease significantly with increasing length of the flexible spacer, and T_i normally decreases in a zig-zag fashion in homologous series in which the spacer length is regularly increased [63]. By this so-called odd–even effect, T_i tends to be higher when there is an even number n of methylene groups in the spacer, but this oscillation is attenuated on ascending the series. This effect is exemplified in Fig. 13 for nematic polymers of structures **3**.

$$\left[\!\!-O\!\!-\!\!\bigcirc\!\!-\!\!\underset{\underset{O}{\overset{CH_3}{|}}}{N}\!\!=\!\!N\!\!-\!\!\bigcirc\!\!-\!\!\overset{H_3C}{}\!\!OC(CH_2)_nC\!\!-\!\!\right]\quad \textbf{3}$$

Evidently, the even members allow for a more favorable packing of the polymer chains in the mesophase; this results in higher transition temperatures. This effect is best understood by assuming the conformation of the methylene spacer to be all *trans*. This conformation has the lowest energy, and there is considerable evidence that it is in fact a very likely conformation of these spacers in both liquid-crystalline and crystalline phases. The conformational distribution of a linear polymethylene chain, when

connected to a rigid group at both ends, is strongly dependent on its number (n) of methylenes. In a nematic phase, the more extended conformers are selectively preferred for steric packing reasons. This preference results in a change of their statistical weight in the anisotropic phase with respect to the isotropic one [115]. An even-numbered polymethylene spacer possesses a set of low energy *trans* conformers that force the rigid units to adopt a collinear disposition. In contrast, an odd-numbered spacer displays a relatively small fraction of *trans* conformers, which places the two mesogenic groups in angled orientations disfavoring the ordering of a nematic phase (Fig. 14).

This basic difference between the configurational [116] characters of the extended conformers of the flexible spacer determines the largely oscillating transition temperatures and can also explain the higher order assumed by the even members. Indeed, the clearing enthalpy and entropy (ΔH_i and ΔS_i) follow a zig-zag rising trend with increasing n, the even members exhibiting consistently greater values (Fig. 15). The increase in ΔS_i is a manifestation of the greater flexibility and conformational freedom of the polymer chain in the isotropic phase than in the anisotropic one, and confirms the participation of the flexible spacer in the ordering process in such a way that even members are endowed with great-

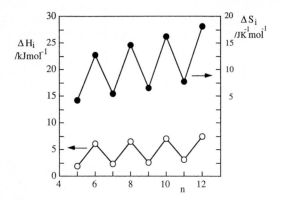

Figure 15. Dependence of the clearing enthalpy ΔH_i and the entropy ΔS_i on the flexible spacer length n.

er stability (T_i) and order (ΔS_i). Order parameter measurements have detailed these general conclusions for particular classes of main-chain polymers [17]. Whereas the above trend is well documented for the nematic–isotropic transition, unequivocal experimental evidence is not available for the smectic–nematic transition of semi-rigid polymers [117]. Correlations of the thermodynamic parameters $(T_{Sm-N}, \Delta H_{Sm-N})$ with the structure of the polymer repeat unit must take into account very fine structural details (**4**).

portant to recognize that the oxyethylene unit comprises three atoms and, accordingly, comparisons are only possible within polymers with the same spacer length. Comparing polymers containing poly(ethylene oxide) spacers with analogous polymers containing polymethylene spacers with the same number of atoms in the spacers, lower isotropization temperatures and enthalpy and entropy changes are observed, unless the mesogenic group is so effective at dictating the phase transition behavior that the flexible spacers play a very minor role in affecting the mesophase onset and stability. This result clearly indicates that the poly(ethylene oxide) spacers have an inherently low mesogenic propensity relative to the polymethylene segments, which is in agreement with the high energy and low probability of existence associated with the extended conformers of poly(ethylene oxide) due to the preference of the *gauche* over the *trans* state relative to the rotational state around the C–C bond [118]. The effect of the "parity" of the poly(ethylene oxide) spacer was studied for a polymer system with structure **5**, in which the two mesogenic *p*-oxybenzoate dyads are interconnected by both polymethylene and poly(ethylene oxide) spacers in a regularly alternating fashion. This polymer system is particularly interesting because independent varia-

2.3.2.2   Poly(ethylene oxide) Spacers

Various mesogenic groups were connected through oligomers of poly(ethylene oxide), even with a polydisperse distribution of chain lengths. However, to understand the individual effect exerted by the poly(ethylene oxide) spacers it is necessary to employ monodisperse molecular species. It is im-

tions are allowed in both the chemically different spacers, with n between 2 and 4 and m between 5 and 10, thus elicidating the distinct effects of each spacer. Within each series, keeping n constant, the isotropization temperatures varied with a typical even–odd effect in which the even member possessed the higher temperature. Interestingly, the odd members are characterized by an

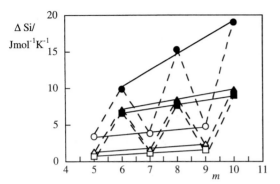

Figure 16. Dependence of the nematic isotropic transition entropy ΔS_i on the flexible spacer length m: $\square$, $\blacksquare$ $n=2$; $\circ$, $\bullet$ $n=3$; $\triangle$, $\blacktriangle$ $n=4$.

increase in the isotropization temperature as the series is ascended, whereas the isotropization temperature decreases for the even members. This dual behavior derives from two conflicting tendencies, that is, the usual decrease due to the increased conformational flexibility of the macromolecular chain on increasing the length of the flexible spacers and the increase in the transition temperature due to a better packing possibility of the mesogenic groups through the establishment of favorable conformational states in the flexible spacer, which enhances the anisotropy in the molecular interactions. Figure 16 shows collectively the trends of the phase transition entropies of all the polymer samples as a function of the number of methylene units in the polymethylene chain.

Within each series, the isotropization entropies for homologs with the same parity lie on smooth curves which increase with increasing spacer length, even members giving the upper curve. The isotropization entropies of the members of the series with $n=2$ and 4 containing the same number m of methylene units are very similar to each other, but significantly lower than those of the corresponding members of the series with $n=3$. In addition, the slope of the iso-

tropization entropy of the even members of the series with $n=3$ is substantially higher than that of the ones of the other series, irrespective of the parity of the polymethylene spacer. This indicates that in the case of the even members of the series with $n=3$ the most favourable combination of the conformational properties of both flexible spacers occurs. In fact both spacers are of "even" parity and accordingly both can propagate the orientation correlation from a rigid group to the successive one. As a consequence, better collinearity and packing of the mesogens in the nematic state is realized. In contrast, the presence of an "odd" element in the macromolecular chain seems to attenuate the cooperativity between adjacent units.

2.3.2.3   Polysiloxane Spacers

Polysiloxane spacers are much more flexible than polymethylene and poly(ethylene oxide) spacers. In addition, the bulky structure and irregular conformation of the siloxane unit reduce the effective chain packing efficiency and interchain forces. Accordingly, lower transition temperatures and enthalpies are usually observed in polymers with polysiloxane spacers with respect to analogous polymers based on the more conventional polymethylene and poly(ethylene oxide) spacers. However, the introduction of polysiloxane spacers decreases the melting temperature and strongly inhibits the crystallization propensity of the polymeric materials, thus producing polymers with liquid-crystalline phases extending over a wide temperature region from the glass transition temperature, sometimes located below room temperature, to the isotropization temperature.

2.3.3   Spacer Substituent Effects

The introduction of substituents such as methyl or ethyl groups in the flexible spacers decreases the steric packing efficiency of the chain molecules in the liquid-crystalline state and produces a perturbation of the conformational characteristics of the flexible spacer which partly loses its ability of propagating the orientation correlation from a rigid group to the successive one. As a result, a decrease of the isotropization temperature with respect to the unsubstituted polymer sample is usually observed, accompanied by a decrease in the propensity of giving rise to smectic phases. However, the steric packing efficiency in the solid state is even more reduced and accordingly, polymers with a very low, if any, melting temperature are observed, thus leading to mesophases extending over wide temperature regions.

Another important effect associated with the introduction of lateral substituents in the spacer is the creation of chiral groups along

nize that within these two polymer systems, the basic mesogenic behavior is independent of the enantiomeric excess which, however, affects fine details of the mesophase structure and physical properties.

2.3.4   Copolymerization Effects

Copolymerization is a very effective way of tailoring phase transitions and tuning the properties of desired levels. In principle two or more flexible spacers can be inserted along the polymer chain, keeping the mesogenic group constant or, alternatively, two or more mesogenic groups can be inserted keeping the flexible spacer constant. Concerning the former case, a complete study was performed on the statistical copolymer samples given by **7**.

the polymer chain. When a prevalent chirality is present, the nematic mesophase becomes cholesteric and the smectic mesophases are turned into their chiral counterparts.

The influence on the mesomorphic properties of polymer structural parameters, for example, the length and nature of the polymer, the substituent's structure, the structural isomerism enantiomeric excess, e. g., of the chiral spacer segment, has been elucidated for the polymer systems (**6**) [42–46] based on the 4,4'-terephthaloyldioxydibenzoyl mesogen. It is important to recog-

It was found that copolymerization lowers the crystal–mesophase transition temperature and increases the mesophase temperature range. This effect becomes progressively more pronounced as the difference in length of the flexible spacers increases. In addition, spacers of comparable length but different parity are not compliant with the steric-packing requirements of the crystalline solid state, thus decreasing the crystallization propensity of the copolymer with respect to the corresponding homopolymers. However, the nematic and smectic mesophases can accommodate different units with minimal disruption of the meso-phase structure provided that the two flexible spacers are not totally different. Accordingly, for both mesophases the transition temperatures and corresponding enthalpy changes are the same as the weight-averaged values of the similar parameters of the parent homopolymers. In this context, it is worth mentioning that by appropriate adjustment of the gross composition of random copolymers comprising a chiral spacer with prevalent chirality with an achiral one, such as, for example, in copolymer system **8**, it is possible to produce chi-

these polymers [119–125]. One reason for this is certainly related to the intrinsic experimental difficulty of obtaining thermotropic main chain polymers with predetermined molar mass and narrow molar mass distribution. This task has mainly been pursued through a time-consuming fractionation technique using conventional chromatographic techniques, e.g., for polyesters with structure **9**.

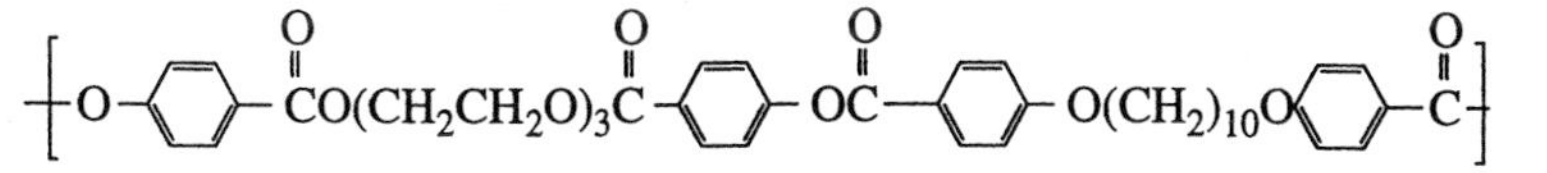

Such polymer samples ($M_n = 2800 - 37200$; $M_w/M_n = 1.27 - 1.84$) form a nematic phase, and both T_m and T_i increase, steeply at first

ral liquid crystalline materials with precisely predetermined mesophase sequence and tunable properties.

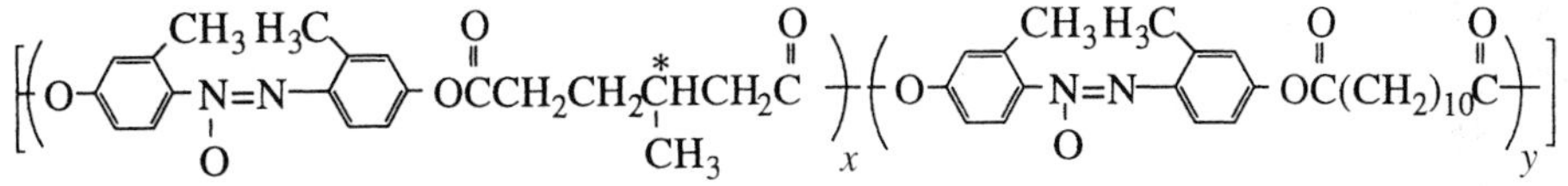

Also, for copolymers consisting of either two different mesogenic units, or a mesogenic unit and a nonmesogenic rigid one, interconnected by a flexible spacer, a depression of the crystallization tendency is observed, whereas the liquid-crystalline transitions are affected very little, in agreement with substantial thermodynamic control of the transitions involving liquid-crystalline mesophases.

2.3.5   Molar Mass and Molar Mass Distribution Effects

Comparatively little attention has been devoted to the elucidation of the dependence of the phase transition parameters on the molar mass and molar mass distribution of

and then more gradually, with increasing molar mass up to a saturation value for $M_n = 10000$. Beyond this value, the transition temperatures remain essentially unaffected (Fig. 17).

The increase in T_i with molar mass is more pronounced than for T_m, and the mesophasic range is consistently wide for samples with greater M_n. A quite similar dependence on M_n is observed for the ΔS_i, which increases from 11.7 J mol^{-1} K^{-1} to a saturation value of 17.5 J mol^{-1} K^{-1}. Such an increase in the thermodynamic phase transition parameters with molar mass is ascribed to an increased cooperativity between distant units belonging to the same macromolecular chain, as mediated by the orientational field of the nematic phase [124, 125]. This cooperative ordering of the rigid groups enhances the overall rigidity of the polymer chain and consequently imparts an increasing first-order character to the

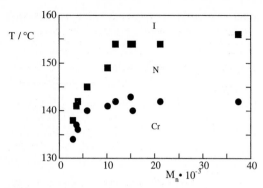

Figure 17. The melting (●) and nematic–isotropic (■) temperatures as functions of the molar mass.

isotropization transition. While an M_n of 10000 seems to be a very general value for main chain nematic polymers at which both T_i and ΔS_i saturate, the specific responses of these transition parameters in the region of lower M_n values cannot be anticipated. Therefore, samples of sufficiently high M_n (and low M_w/M_n), beyond the saturation limit, must be considered when making reliable comparisons of the transitional behaviors of different polymers.

The above-described behavior is well accepted and is considered a general one, although there are some interesting exceptions. For example, within the homologous series of polyesters with the general structure [126] given by **10**, and marked $CmCn$,

where m and n identify the number of methylene units in both the polymethylene spacers; the phase transition parameters relevant to both the smectic C–nematic and the nematic–isotropic transitions of polymer samples C_7C_7 and C_7C_{10} increase with the number average molar mass up to the limiting value of $M_r = 6000-8000$ g mol^{-1} and then remain constant. Accordingly, the molar mass dependence of the phase transition parameters of both the smectic C–nematic

and the nematic–isotropic transitions of these polymers is in agreement with the above-described general trend. In complete contrast, the stability of the smectic mesophase of polymers C_5C_5 and C_5C_7 increases, whereas that of their nematic mesophase decreases with increasing number-average molar mass. This is particularly evident for the former polymer, for which the nematic mesophase changes from enantiotropic to monotropic with increasing number-average molar mass, and then disappears on reaching molar mass values of about $M_n = 7000$ g mol^{-1}. In addition, a wide molar mass distribution tends to increase the stability of the nematic mesophase. These effects clearly indicate that the increase in the molar mass tends to stabilize specific conformations of the flexible spacers which create an overall polymer chain configuration that is no longer compatible with the onset of a nematic mesophase, but is highly compatible with a smectic mesophase. This unconventional behavior of polymers C_5C_5 and C_5C_7 was supposed to be mainly related to the short length and odd parity of both the flexible spacers, which should place the mesogenic groups in angled orientations unfavorable to the onset of nematic ordering. The conformational selection appears to be more and more effective as the molar mass increases and the molar mass distribution narrows.

Biphasic and segregation phenomena [127–131] at the nematic-to-isotropic transition reveal the multicomponent nature of the main-chain polymers as due to the wide molar mass distribution. The thermodynamic width of the biphasic gap was delineated by annealing the samples inside the apparent biphasic gap, giving a result of about $7-15\,°C$. The use of polymer "blends" with a specifically designed molar mass and degree of polydispersity was found to be very effective in elucidating this behavior. Samples with high molar mass ($M_n > 20\,000$ g mol^{-1}) and narrow molar mass distribution ($M_w/M_n < 1.5$) and samples with low molar mass ($M_n < 6000$ g mol^{-1}) and narrow or relatively wide molar mass distribution ($M_w/M_n < 1.6$) do not show nematic to isotropic biphase separation. In contrast, samples with intermediate molar mass ($M_n = 8000 - 15\,000$ g mol^{-1}) and relatively narrow or wide molar mass distributions ($M_w/M_n = 1.7-2.4$) show nematic–isotropic biphase separation. Therefore, the phase segregation behavior appears to be due to the effects of the molar mass and the molar mass distribution of the polymer. The location of the biphasic region depends on the number-average molar mass, whereas the width of the biphasic region and the separation between the isotropization peak temperatures of the biphase demixed components depend on the width of the molar mass distribution.

2.4 References

[1] S. L. Kwolek (DuPont), U. S. Patent 3,600,350, **1971**.
[2] S. G. Cottis, J. Economy, B. E. Nowak (Carborandum), U. S. Patent 3,637,595, **1972**.
[3] G. W. Calundann (Celanese), U. S. Patent 4,161,470, **1979**.
[4] W. J. Jackson, Jr., F. Kuhfuss, *J. Polym. Sci., Polym. Chem. Ed.* **1976**, *14*, 2043.
[5] G. W. Calundann, M. Jaffe in *Proc. of the Robert A. Welch Conference on Chemical Research XXVI. Synthetic Polymers* (Houston, Texas, November 1982) **1982**, p. 247.
[6] M. Ballauff, *Angew. Chem.* **1988**, *28*, 253.
[7] M. G. Northolt, D. J. Sikkema, *Adv. Polym. Sci.* **1990**, *98*, 118.
[8] C. L. Jackson, M. T. Shaw, *Int. Mater. Rev.* **1991**, *36*, 165.
[9] V. Krone, B. Reck, H. Ringsdorf, presented at 16. Freiburger Arbeitstagung Flüssigkristalle, Freiburg **1986**.
[10] W. Kreuder, H. Ringsdorf, P. Tschirner, *Makromol. Chem., Rapid. Commun.* **1985**, *6*, 367.
[11] A. Blumstein (Ed.), *Liquid Crystalline Order in Polymers,* Academic, New York, **1978**.
[12] A. Ciferri, W. R. Krigbaum, R. B. Mayer, *Polymer Liquid Crystals,* Academic, New York **1982**.
[13] C. K. Ober, J. I. Jin, R. W. Lenz, Adv. Polym. Sci. **1984**, *59*, 104.
[14] M. Gordon, N. A. Plate (Eds.) *Advances in Polymer Science*, Vols. 59–61, Springer, Berlin **1984**.
[15] A. Blumstein (Ed.), *Polymeric Liquid Crystals,* Plenum, New York **1985**.
[16] L. L. Chapoy (Ed.), *Recent Advances in Liquid Crystalline Polymers,* Elsevier, London **1986**.
[17] M. Jaffe in *Encyclopedia of Polymer Science and Engineering*, Vol. 7 (Eds.: J. I. Kroschwitz, A. Salvatore, A. Klingsberg, R. Piccininni), Wiley, New York, **1987**, p. 699.
[18] C. Noël, *Makromol. Chem. Macromol. Symp.* **1988**, *22*, 95.
[19] R. A. Weiss, C. K. Ober (Eds.) *Liquid Crystalline Polymers*, American Chemical Society, Washington **1990**.
[20] W. Brostow, *Polymer* **1990**, *31*, 979.
[21] J. Franek, Z. J. Jedlinski, J. Majnusz in *Handbook of Polymer Synthesis* (Ed.: H. R. Kricheldorf), Marcel Dekker, New York **1992**.
[22] P. Magagnini in *Thermotropic Liquid Crystal Polymer Blends* (Ed.: F. P. la Mantia), Technomic, Basel **1993**, p. 1.
[23] W. J. Jackson, H. F. Kuhfuss, *J. Polym. Sci., Polym. Chem. Ed.* **1976**, *14*, 2043.
[24] D. Van Luten, L. Strzelecki, *Eur. Polym. J.* **1980**, *16*, 303.
[25] J. Asrar, H. Toriumi, J. Watanabe, W. R. Krigbaum, A. Ciferri, *J. Polym. Sci., Polym. Phys. Ed.* **1983**, *21*, 1119.
[26] W. R. Krigbaum, J. Asrar, H. Toriumi, A. Ciferri, J. Preston, *J. Polym. Sci., Polym. Lett. Ed.* **1982**, *20*, 109.
[27] W. R. Krigbaum, J. Watanabe, *Polymer* **1983**, *24*, 1299.
[28] P. Meurisse, C. Noël, L. Monneri, B. Fayolle, *Br. Polym. J.* **1981**, *13*, 55.
[29] L. Bosio, B. Fayolle, C. Friedrich, F. Lauprétre, P. Meurisse, C. Noël, J. Virlet in *Liquid Crystals and Ordered Fluids* (Ed.: A. C. Griffin, J. F. Johnson), Plenum, New York **1984**.
[30] J. S. Moore, S. I. Stupp, *Macromolecules* **1988**, *21*, 1217.
[31] W. R. Krigbaum, R. Kotek, T. Ishikawa, H. Hakemi, *Eur. Polym. J.* **1984**, *20*, 225.
[32] C. K. Ober, J. I. Jin, R. W. Lenz, *Polym. J. (Jpn.)* **1982**, *14*, 9.
[33] A. C. Griffin, S. J. Havens, *Mol. Cryst. Liq. Cryst. Lett.* **1979**, *49*, 239.
[34] A. C. Griffin, S. J. Havens, *J. Polym. Sci., Polym. Phys. Ed.* **1981**, *19*, 951.

[35] J. I. Jin, S. Antoun, C. K. Ober, R. W. Lenz, *Br. Polym. J.* **1980**, *12*, 132.
[36] A. Y. Bilibin, A. A. Shepelevskii, S. Y. Frenkel, S. S. Shorokhodov, *Vysokomol. Soed.* **1980**, *B22*, 739.
[37] Q. F. Zhou, R. W. Lenz, *J. Polym. Sci., Polym. Chem. Ed.* **1983**, *21*, 3313.
[38] Z. Jedlinksi, J. Franek, P. Kuziw, *Makromol. Chem.* **1986**, *187*, 2317.
[39] P. Kuziw, J. Franek, *Chem. Stos.* **1986**, *2*, 273.
[40] G. Galli, E. Chiellini, C. K. Ober, R. W. Lenz, *Makromol. Chem.* **1982**, *183*, 2693.
[41] A. Y. Bilibin, A. V. Tenkovtsen, O. N. Piraner, S. S. Skorokhodov, *Vysokomol. Soed.* **1984**, *A26*, 2570.
[42] E. Chiellini, G. Galli, C. Malanga, N. Spassky, *Polym. Bull. (Berlin)* **1983**, *9*, 336.
[43] E. Chiellini, P. Nieri, G. Galli, *Mol. Cryst. Liq. Cryst.* **1984**, *113*, 213.
[44] E. Chiellini, G. Galli, S. Carrozzino, S. Melone, G. Torquati, *Mol. Cryst. Liq. Cryst.* **1987**, *146*, 385.
[45] E. Chiellini, R. Po, S. Carrozzino, G. Galli, B. Gallot, *Mol. Cryst. Liq. Cryst.* **1990**, *179*, 405.
[46] E. Chiellini, G. Galli, S. Carrozzino, B. Gallot, *Macromolecules* **1990**, *23*, 2106.
[47] A. H. Al-Dujaili, A. D. Jenkins, D. R. M. Walton, *Makromol. Chem., Rapid Commun.* **1984**, *5*, 33.
[48] A. Roviello, A. Sirigu, *J. Polym. Sci., Polym. Lett. Ed.* **1975**, *13*, 455.
[49] A. Roviello, A. Sirigu, *Eur. Polym. J.* **1979**, *15*, 423.
[50] A. Roviello, A. Sirigu, *Gazz. Chim. It.* **1980**, *110*, 403.
[51] P. Iannelli, A. Roviello, A. Sirigu, *Eur. Polym. J.* **1982**, *18*, 753.
[52] C. Carfagna, A. Roviello, S. Santagata, A. Sirigu, *Makromol. Chem.* **1986**, *187*, 2123.
[53] A. Roviello, S. Santagata, A. Sirigu, *Makromol. Chem., Rapid Commun.* **1984**, *5*, 141.
[54] W. J. Jackson, J. C. Morris, *J. Appl. Polym. Sci., Appl. Polym. Symp.* **1985**, *41*, 307.
[55] A. Roviello, A. Sirigu, *Makromol. Chem.* **1984**, *180*, 2543.
[56] K. N. Sivaramakrishnan, A. Blumstein, S. B. Clough, R. B. Blumstein, *Polym. Prep.* **1979**, *19(2)*, 190.
[57] A. Blumstein, K. N. Sivaramakrishnan, R. B. Blumstein, S. B. Clough, *Polymer* **1982**, *23*, 47.
[58] A. Roviello, S. Santagata, A. Sirigu, *Makromol. Chem., Rapid Commun.* **1983**, *4*, 281.
[59] M. Sato, *J. Polym. Sci., Polym. Chem. Ed.* **1988**, *26*, 2613.
[60] K. Iimura, N. Koide, R. Ohta, M. Takeda, *Makromol. Chem.* **1981**, *182*, 2563.
[61] S. Vilasagar, A. Blumstein, *Mol. Cryst. Liq. Cryst. Lett.* **1980**, *56*, 263.

[62] A. Blumstein, S. Vilasagar, S. Ponrathnam, S. B. Clough, R. B. Blumstein, *J. Polym. Sci., Polym. Phys. Ed.* **1982**, *20*, 877.
[63] A. Blumstein, O. Thomas, *Macromolecules* **1982**, *15*, 1264.
[64] D. Sek, *Polym. J. (Jpn.)* **1985**, *17*, 427.
[65] H. R. Kricheldorf, R. Pakull, *Macromolecules* **1988**, *21*, 551.
[66] H. R. Kricheldorf, R. Pakull, *Polymer* **1987**, *28*, 1772.
[67] H. R. Kricheldorf, R. Pakull, *Mol. Cryst. Liq. Cryst.* **1988**, *13*, 157.
[68] H. R. Kricheldorf, R. Pakull, *Polymer* **1989**, *30*, 659.
[69] H. R. Kricheldorf, R. Pakull, S. Bruckner, *Macromolecules* **1988**, *21*, 1929.
[70] S. Antoun, R. W. Lenz, J. I. Jin, *J. Polym. Sci., Polym. Chem. Ed.* **1981**, *19*, 1901.
[71] R. W. Lenz, A. Furukawa, C. N. Wu, E. D. Atkins, *Polym. Prep.* **1988**, *29(1)*, 480.
[72] B. W. Jo, R. W. Lenz, J. I. Jin, *Makromol. Chem. Rapid. Commun.* **1982**, *3*, 23.
[73] Q. F. Zhou, R. W. Lenz, *J. Polym. Sci., Polym. Chem. Ed.* **1983**, *21*, 3313.
[74] J. I. Jin, B. W. Jo, R. W. Lenz, *Polymer (Korea)* **1982**, *6*, 136.
[75] E. Chiellini, G. Galli, A. S. Angeloni, D. Caretti, M. Laus, *Liq. Cryst.* **1989**, *5*, 1593.
[76] D. Caretti, A. S. Angeloni, M. Laus, E. Chiellini, G. Galli, *Makromol. Chem.* **1989**, *190*, 1655.
[77] A. Roviello, A. Sirigu, *J. Polym. Sci., Polym. Lett. Ed.* **1979**, *15*, 61.
[78] J. Watanabe, W. R. Krigbaum, *Macromolecules* **1984**, *17*, 2288.
[79] S. Vilasagar, A. Blumstein, *Mol. Cryst. Liq. Cryst. Lett.* **1980**, *56*, 263.
[80] V. Percec, T. D. Shaffer, H. Nava, *J. Polym. Sci., Polym. Lett. Ed.* **1984**, *22*, 637.
[81] V. Percec, T. D. Shaffer in *Advances in Polymer Synthesis* (Eds.: B. M. Culbertson, J. M. McGrath), Plenum, New York **1984**, p. 133.
[82] P. Keller, *Makromol. Chem., Rapid Commun.* **1985**, *6*, 255.
[83] V. Percec, R. Yourd, *Macromolecules* **1989**, *22*, 3229.
[84] V. Percec, R. Yourd, *Makromol. Chem.* **1990**, *191*, 49.
[85] V. Percec, H. Navara, H. Jonsson, *J. Polym. Sci., Polym. Chem. Ed.* **1987**, *25*, 1943.
[86] M. Tanaka, T. Nakaya, *Makromol. Chem.* **1986**, *187*, 2345.
[87] M. Tanaka, T. Nakaya, *J. Macromol. Sci., Chem.* **1987**, *24*, 777.
[88] M. Tanaka, T. Nakaya, *Makromol. Chem.* **1989**, *190*, 3067.
[89] C. M. Brunette, S. L. Hsu, W. J. McKnight, *Macromolecules* **1982**, *15*, 71.
[90] P. J. Stenhouse, E. M. Valles, S. W. Kantor, W. J. McKnight, *Macromolecules* **1989**, *22*, 1467.

[91] S. K. Pollack, D. Y. Shen, S. L. Hsu, Q. Wang, H. D. Stadham, *Macromolecules* **1989**, *22*, 551.

[92] G. Smyth, E. M. Valles, S. K. Pollack, J. Grebowicz, P. J. Stenhouse, S. L. Hsu, W. J. McKnight, *Macromolecules* **1990**, *23*, 3389.

[93] R. Lorenz, M. Els, F. Haulena, A. Schmitz, O. Lorenz, *Angew. Makromol. Chem.* **1990**, *180*, 51.

[94] E. Chiellini, G. Galli, S. Trusendi, A. S. Angeloni, M. Laus, O. Francescangeli, *Mol. Cryst. Liq. Cryst.* **1994**, *243*, 135.

[95] A. S. Angeloni, M. Laus, E. Chiellini, G. Galli, O. Francescangeli, B. Yang, *Eur. Polym. J.* **1995**, *31*, 253.

[96] O. Francescangeli, B. Yang, M. Laus, A. S. Angeloni, G. Galli, E. Chiellini, *J. Polym. Sci., Polym. Phys.* **1995**, *33*, 699.

[97] W. Mormann, *Makromol. Chem., Rapid Commun.* **1988**, *9*, 175.

[98] W. Mormann, M. Brahm, *Makromol. Chem.* **1989**, *190*, 631.

[99] W. Mormann, S. Benadda, *Makromol. Chem., Macromol. Symp.* **1991**, *50*, 229.

[100] W. Mormann, M. Brahm, *Macromolecules* **1991**, *24*, 1096.

[101] H. R. Kricheldorf, J. Awe, *Makromol. Chem.* **1989**, *190*, 2579 and 2597.

[102] H. R. Kricheldorf, J. Awe, *Makromol. Chem., Rapid Commun* **1988**, *9*, 681.

[103] S. M. Aharoni, *Macromolecules* **1988**, *21*, 1941.

[104] C. Aquilera, J. Bartulin, B. Hisgen, H. Ringsdorf, *Makromol. Chem.* **1983**, *184*, 253.

[105] G. Galli, M. Laus, A. S. Angeloni, P. Ferruti, E. Chiellini, *Makromol. Chem., Rapid Commun.* **1983**, *4*, 681.

[106] A. S. Angeloni, M. Laus, C. Castellari, G. Galli, P. Ferruti, E. Chiellini, *Makromol. Chem.* **1985**, *186*, 977.

[107] G. Galli, E. Chiellini, M. Laus, A. S. Angeloni, *Polym. Bull.* **1989**, *21*, 563.

[108] A. S. Angeloni, M. Laus, E. Burgin, G. Galli, E. Chiellini, *Polym. Bull.* **1985**, *13*, 131.

[109] G. Galli, E. Chiellini, A. S. Angeloni, M. Laus, *Macromolecules* **1989**, *22*, 1120.

[110] A. S. Angeloni, E. Chiellini, G. Galli, M. Laus, G. Torquati, *Makromol. Chem., Macromol. Symp.* **1989**, *24*, 311.

[111] M. Laus, A. S. Angeloni, V. Milanesi, G. Galli, E. Chiellini, *Polym. Rep.* **1987**, *28*, 82.

[112] L. Onsager, *Ann. N. Y. Acad. Sci.* **1949**, *51*, 627.

[113] P. J. Flory, *Proc. R. Soc.* **1956**, *A234*, 73.

[114] E. Chiellini, G. Galli, A. S. Angeloni, M. Laus in *Liquid Crystalline Polymers, ACS Symposium Series* (Eds.: R. A. Weiss, C. K. Ober), **1990**, p. 435.

[115] D. Y. Yoon, S. Bruckner, W. Volksen, J. C. Scott, A. C. Griffin, *Faraday Discuss. Chem. Soc.* **1985**, *79*, 41.

[116] D. Y. Yoon, S. Bruckner, *Macromolecules* **1982**, *18*, 651.

[117] P. Esnault, M. M. Gauthier, F. Volino, J. F. D'Allest, R. B. Blumstein, *Mol. Cryst. Liq. Cryst.* **1988**, *157*, 273.

[118] P. J. Flory, *Statistical Mechanics of Chain Molecules,* Interscience, New York **1969**.

[119] W. R. Krigbaum, T. Ishikawa, J. Watanabe, H. Toriumi, K. Kubota, J. Preston, *J. Polym. Sci., Polym. Phys. Ed.* **1983**, *21*, 1851.

[120] R. B. Blumstein, E. M. Stickles, M. M. Gauthier, A. Blumstein, F. Volino, *Macromolecules* **1984**, *17*, 177.

[121] G. Galli, E. Chiellini, A. S. Angeloni, M. Laus, *Macromolecules* **1989**, *22*, 1120.

[122] M. Laus, A. S. Angeloni, G. Galli, E. Chiellini, *Makromol. Chem.* **1990**, *191*, 147.

[123] V. Percec, A. Keller, *Macromolecules* **1990**, *23*, 4347.

[124] G. Sigaud, D. Y. Yoon, A. C. Griffin, *Macromolecules* **1983**, *16*, 875.

[125] P. G. de Gennes, *Mol. Cryst. Liq. Cryst. Lett.* **1984**, *102*, 95.

[126] M. Laus, A. S. Angeloni, A. Spagna, G. Galli, E. Chiellini, *J. Mater. Chem.* **1994**, *4*, 437.

[127] J. F. D'Allest, P. P. Wu, A. Blumstein, R. B. Blumstein, *Mol. Cryst. Liq. Cryst. Lett.* **1986**, *3*, 103.

[128] J. F. D'Allest, P. Sixou, A. Blumstein, R. B. Blumstein, *Mol. Cryst. Liq. Cryst.* **1988**, *157*, 229.

[129] D. Y. Kim, J. F. D'Allest, A. Blumstein, R. B. Blumstein, *Mol. Cryst. Liq. Cryst.* **1988**, *157*, 253.

[130] M. Laus, D. Caretti, A. S. Angeloni, G. Galli, E. Chiellini, *Macromolecules* **1991**, *24*, 1459.

[131] M. Laus, A. S. Angeloni, G. Galli, E. Chiellini, *Macromolecules* **1992**, *25*, 5901.

3 Combined Liquid Crystalline Main-Chain/Side-Chain Polymers

Rudolf Zentel

3.1 Introduction

The starting point for the preparation of combined liquid crystalline (LC) main-chain/side-chain polymers ('combined' LC polymers) was merely a curiosity about what would happen if the structural properties of main-chain and side-chain polymers were combined (see Figure 1) [1]. As a result of these syntheses a variety of new LC polymers were obtained which showed a surprisingly broad range of LC phases. Further interest in combining main-chain and side-chain structures arose from the search for LC polymers with properties, such as the orientation in drawn fibers [1, 2], intermediate between those of these two classes of LC polymers. As the mesogens are oriented parallel to the fibre axis in main-chain polymers, but perpendicular to it in most side-chain polymers [1, 2], it seemed possible that the combined LC polymers would show a biaxial orientation and thus good mechanical properties in all directions. The initial results [1] seemed to indicate that such an orientation did indeed occur, with the side-group mesogens oriented perpendicular to the main-chain mesogens.

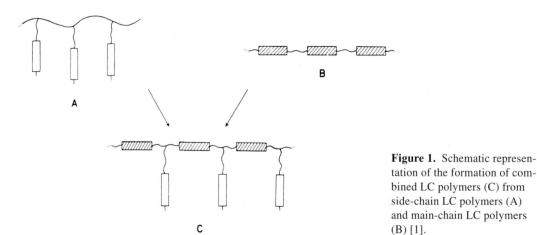

Figure 1. Schematic representation of the formation of combined LC polymers (C) from side-chain LC polymers (A) and main-chain LC polymers (B) [1].

However, this conclusion later proved to be false.

Due to the interesting LC properties of combined LC polymers (i.e. broad LC phases, and the occurrence of different smectic phases and a nematic phases at different temperatures) and their intermediate nature between that of side-chain and that of main-chain polymers, a lot of research has been undertaken on these materials. Most of the research has been directed towards the preparation of cross-linkable polymers and LC elastomers [3–11] and of chiral combined LC polymers [4, 6, 7, 9, 12–16].

Quite early on, structural investigations showed [17] that the mesogens in the main chain and the side chain orient parallel to one another. This was confirmed by electron microscopy [18], X-ray diffraction [6, 7, 13, 14, 17, 19, 20] and ^{2}H NMR spectroscopy [21]. Following the general classification of LC side-chain polymers (Figure 2) [22], these polymers should be classified as type III. In these combined LC polymers, the polymer chains and mesogens orient parallel to one another to define the LC director.

3.2   Molecular Structure of Combined LC Polymers

3.2.1   Polymers with Side-Chain Mesogens Linked at the Main Chain Spacer

3.2.1.1   Achiral Combined LC Polymers

The first combined LC polymers prepared by Reck and Ringsdorf [1] were obtained by means of a melt polycondensation, as shown in Scheme 1. The phase transitions of some of the polymers prepared in this way are collected in Figure 3. It can be seen that the LC phases observed exist mostly over a broad temperature range. Different smectic phases and a nematic phase can often exist in one material at different temperatures. The preference for smectic or nematic phases depends on whether the main chain and side chain spacers are of comparable (smectic) or strongly different lengths. The synergistic stabilization of the LC phase due to the presence of mesogens in the main chain and as side groups is best demonstrated by comparing the corresponding polymers shown in Figure 4. The clearing temperature of the

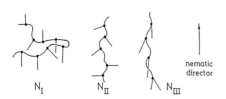

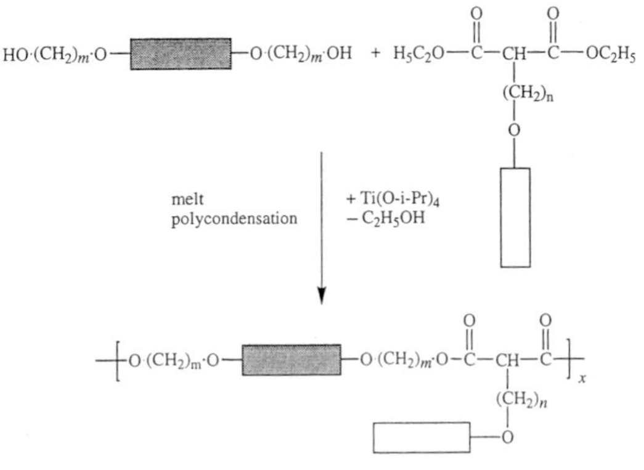

Figure 2. Schematic representation of different types of polymer with mesogenic side groups that form LC phases. In case I (N_I), only the mesogenic side groups give rise to LC ordering. In case II (N_{II}), only the polymer chain orients, whereas in case III (N_{III}) both the main chain and the side groups orient parallel to one another [22].

Scheme 1. Melt polycondensation for the preparation of combined LC polymers.

Phase assigment of combined main chain / side chain polyesters and
visualisation of the orientation of both types of mesogens.

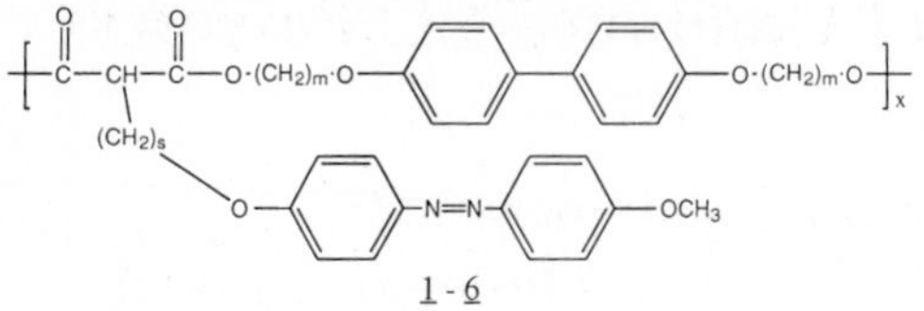

1 - 6

No.	m	s	Phase Transitions [a] /°C
1	2	2	k 148 s176 i
2	2	6	k 124/129 s$_C$ 153 s$_A$ 162 n 181 i
3	2	10	k 162 (n 159) i
4	6	2	k 109/120 n 153 i
5	6	6	k 108/112 s$_C$ 131 s$_A$ 136 n 155 i
6	6	10	g 72 s$_X$ 120 s$_C$ 139 s$_A$ 154 i

a) k: Crystalline or higher ordered smectic phase; s: smectic A or C; s$_X$: unidentified ordered
smectic phase; s$_C$: smectic C; s$_A$: smectic A; n: nematic; i: isotropic.

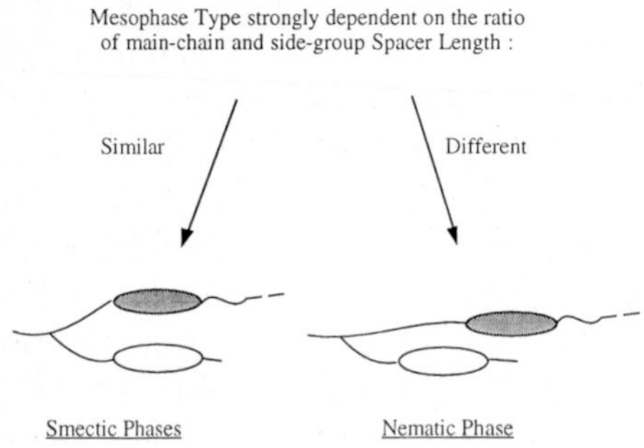

Mesophase Type strongly dependent on the ratio
of main-chain and side-group Spacer Length :

Similar   Different

Smectic Phases   Nematic Phase

Figure 3. Phase assignment of combined main-chain/side-chain polyesters, and visualization of the orientation of the two types of mesogen.

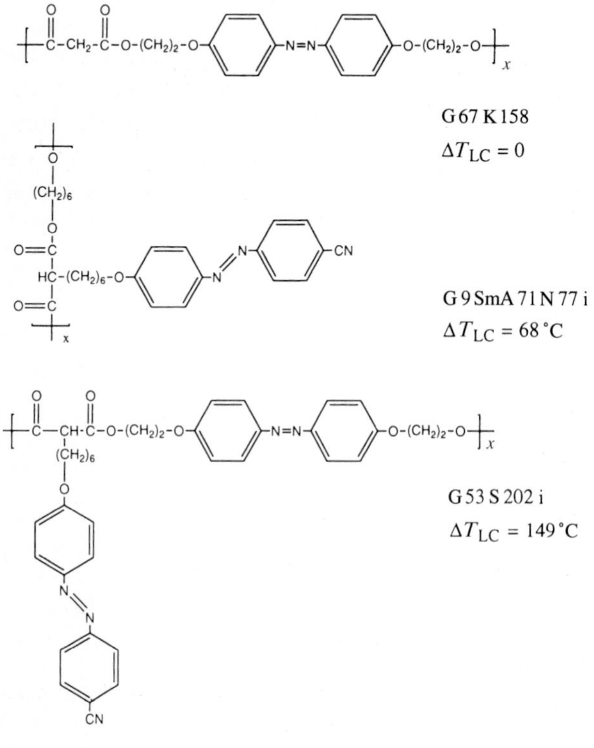

G 67 K 158
$\Delta T_{LC} = 0$

G 9 SmA 71 N 77 i
$\Delta T_{LC} = 68\,°C$

G 53 S 202 i
$\Delta T_{LC} = 149\,°C$

Figure 4. Comparison of a combined main-chain/side-chain polyester with the corresponding main-chain and side-group polymalonates. ΔT_{LC}, the temperature range of the LC phase.

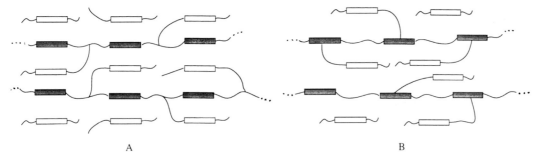

Figure 5. Schematic representation of the arrangement of the main-chain (■) and side-chain (▭) mesogens of combined LC polymers of type A (3.2.1) and B (3.2.2).

combined LC polymers is much higher than that of the corresponding main-chain and side-chain polymers. Obviously the temperature range of the LC phase is strongly broadened in the combined LC polymers due to the synergistic effect. For a detailed discussion of the structure–property relationships see Endres et al. [19] and Diele et al. [20]. The structural model derived from these observations is presented in Figure 5 (type A).

3.2.1.2   Chiral Combined LC Polymers

Because of the interest in chiral LC phases in general, which give rise to selective reflection of light (cholesteric phase) or ferroelectricity (chiral smectic C* phase), chiral combined LC polymers were prepared quite early on [4]. Polymers with cholesteric and chiral smectic C* phases could be prepared easily. As these polymers were synthesized using to the polycondensation process shown in Scheme 1, the chiral groups had to be selected carefully in order to prevent racemization during polycondensation [4, 7, 12, 13].

In order to make more labile (referring to racemization) chiral groups accessible, a new synthetic route to combined LC polymers was developed (Scheme 2), which involves the esterification of chiral acids with

a preformed polyphenolic polymer [9, 15, 16, 23]. The phase transitions of some of the polymers prepared in this way are summarized in Table 1.

From Table 1 [16] the following structure–property relations can be extracted. Polymers without lateral substituents tend to be crystalline at room temperature. However, this crystallization can be suppressed

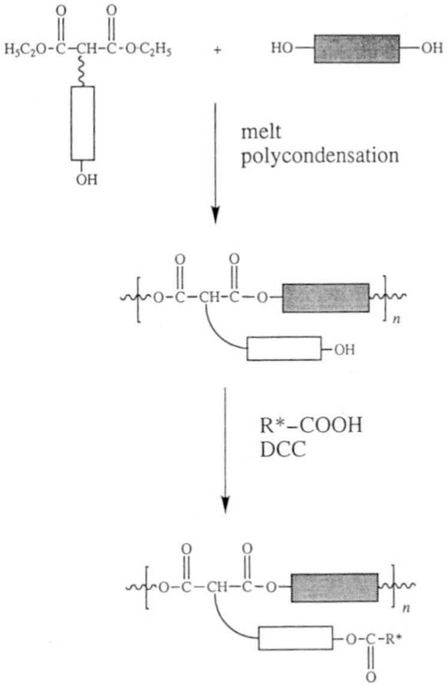

Scheme 2. Synthesis of chiral combined LC polymers by esterification of chiral acids with a preformed polyphenolic polymer. DCC, dicyclohexylcarbodiimide.

Table 1. Phase assignment of chiral combined main-chain/side-chain polymers.

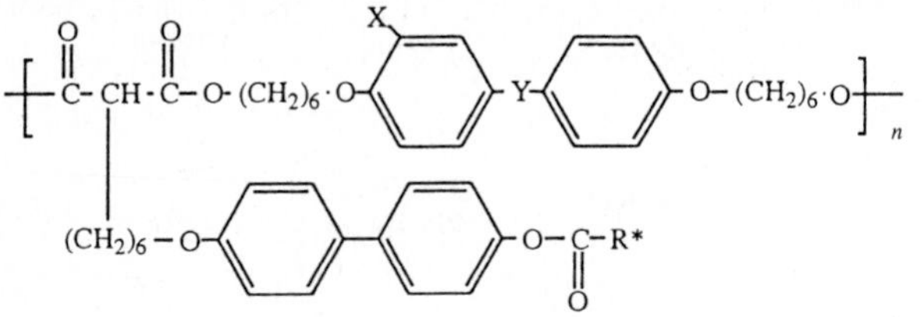

	R*	X	Y	Phase transition temperature (°C)
1.	$-C^{*}H \cdot C^{*}H \cdot C_2H_5$ with Cl, CH_3	H	–	K 115 SmX 124 SmC* 135 I
2.	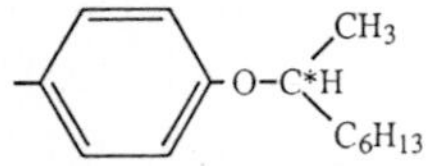 (phenyl)$-O-C^{*}H$ with CH_3, C_6H_{13}	H	–	K 134 N* 145 I
3.	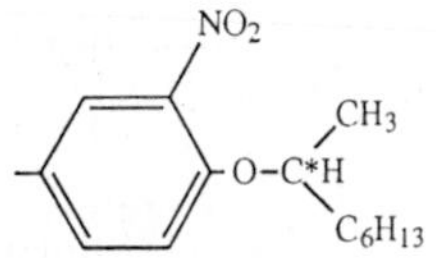 (phenyl, NO_2)$-O-C^{*}H$ with CH_3, C_6H_{13}	H	–	G 21 K 19 N* 122 I
4.	$-C^{*}H-C^{*}H-C_2H_5$ with Cl, CH_3	Br	–	G 13 S 91 I
5.	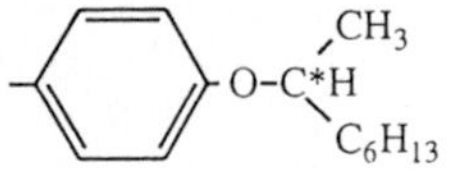 (phenyl)$-O-C^{*}H$ with CH_3, C_6H_{13}	Br	–	G 17 SmC* 55 SmA 89 N* 113 I
6.	$-C^{*}H-C^{*}H-C_2H_5$ with Cl, CH_3	H	$-N(O)=N-$	G 12 S 108 N* 116 I
7.	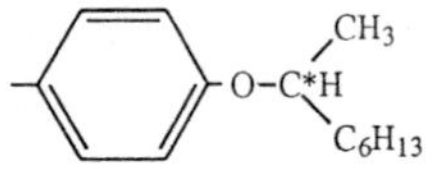 (phenyl)$-O-C^{*}H$ with CH_3, C_6H_{13}	H	$-N(O)=N-$	G 11 S 106 N* 149 I
8.	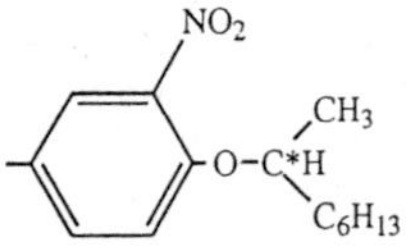 (phenyl, NO_2)$-O-C^{*}H$ with CH_3, C_6H_{13}	H	$-N(O)=N-$	G 19 SmC* 47 SmA 103 N* 125 I

to give glassy freezing materials (with a T_g of about room temperature) by the introduction of lateral substituents. This can be either a nitro group at the chiral acid (polymers 3 and 8), a bromine atom at the main chain mesogen (polymers 4 and 5) or the oxygen atom in the azoxybenzene unit (polymers 6–8). Such glassy freezing materials are of interest for use in the preparation of cross-linked LC elastomers (see below).

3.2.1.3   Cross-Linked LC Elastomers

In order to combine LC properties and rubber elasticity a variety of LC elastomers were prepared [8]. In these systems a reor-

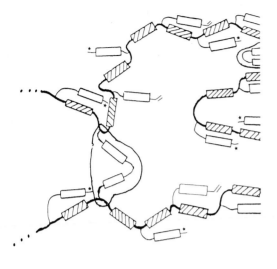

Figure 6. Schematic representation of LC elastomers prepared from chiral combined main-chain/side-chain copolymers. ▭ main-chain mesogens; ▭ side-chain mesogens; * chiral groups [7].

ientation of the director is coupled with a deformation of the polymer network, and vice versa. This makes, for example, a reversible mechanical orientation of the LC phase possible. The LC elastomers prepared from combined LC polymers (Figure 6) are interesting in this respect, because here the coupling between a deformation of the polymer network and the orientation of the mesogenic groups should be stronger than in LC side-chain polymers (see Chapter V of this volume).

Cross-linked LC elastomers from combined LC polymers were mostly obtained by means of the hydrosililation reaction shown in Figure 7 [3–7, 10]. Later, a thermal or photochemical polymerization of acrylates was used [9, 11]. As can be seen from Figure 7, this cross-linking process, which

transforms a soluble polymer into an insoluble gel, does not change the LC phase sequence. It leads rather to a small decrease in the phase transition temperatures due to the incorporation of the non-mesogenic siloxane cross-linker. In this way elastomers with broad chiral smectic C* and cholesteric phases are accessible; these are of interest because of their piezoelectric properties (see Sec. 3.3.3 of this Chapter, Figure 15). All elastomers prepared in this way are soft materials, and can be deformed greatly.

3.2.2 Polymers with Side-Chain Mesogens Linked at the Main Chain Mesogen

As an alternative concept to combined LC polymers, polymers with the side-chain mesogens linked at the main-chain mesogen were prepared [24] (see Figure 5, type B). These were prepared by a solution polycondensation of a *bis*(phenol) and a *bis*(acid chloride) as shown in Scheme 3 [24, 25]. The resulting polymers (Table 2) again show broad LC phases due to the synergistic interaction of the main-chain and side-chain mesogenic groups. However, compared to the other types of combined LC

Table 2. Structure, molecular weight and phase behaviour of the combined polymers in which the side-chain mesogen is linked to the main-chain mesogen.

R*	m	Phase transitions** (°C)
–OCH$_3$	2	G 90 N 310 I$_{\text{deg}}$ [a)]
–OCH$_3$	6	G 51 SmX 158 N 257 I
–OCH$_3$	9	G 41 SmX 68 N 216 I
–OCH$_3$	10	G 37 SmX 129 N 214 I
–CN	10	G 43 N 209 I

  * See Scheme 3.
 ** See Figure 3.
[a)] I_{deg}: clearing under degradation.

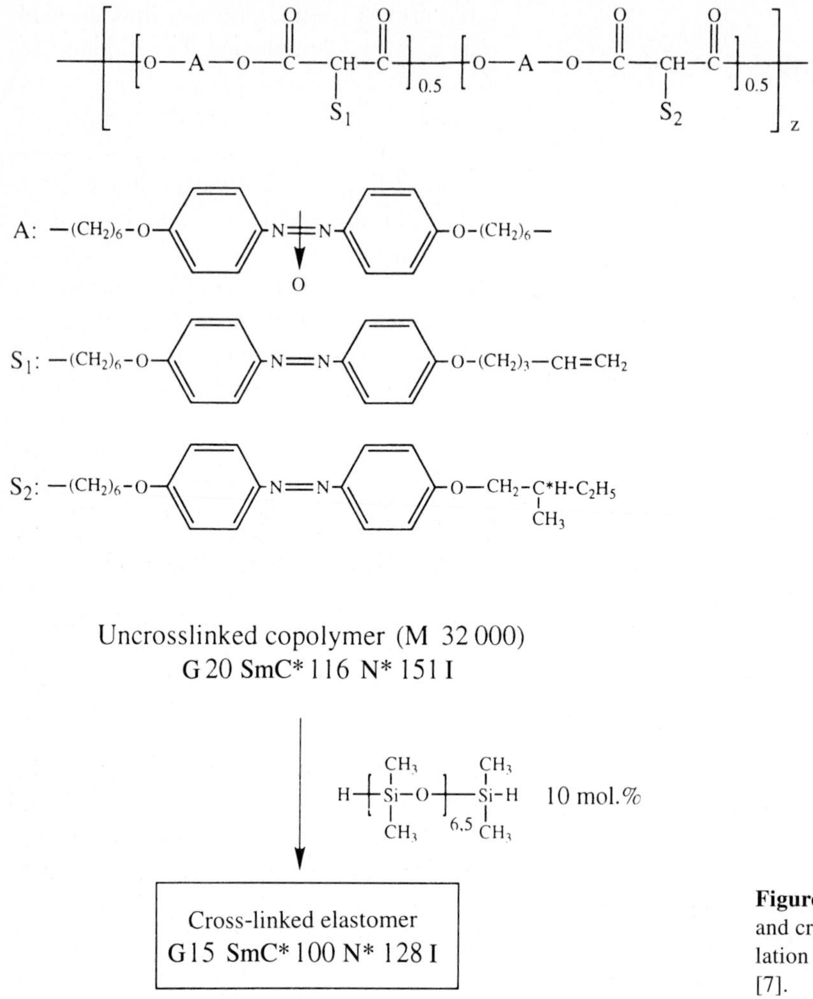

A: —(CH₂)₆-O— ... —O-(CH₂)₆—

$$\text{A: } -(CH_2)_6-O-\bigcirc-\bigcirc-N=N-\bigcirc-\bigcirc-O-(CH_2)_6-$$

$$\text{S}_1: -(CH_2)_6-O-\bigcirc-\bigcirc-N=N-\bigcirc-\bigcirc-O-(CH_2)_3-CH=CH_2$$

$$\text{S}_2: -(CH_2)_6-O-\bigcirc-\bigcirc-N=N-\bigcirc-\bigcirc-O-CH_2-C^*H\cdot C_2H_5$$
$$\qquad\qquad\qquad\qquad\qquad\qquad\qquad\qquad\qquad CH_3$$

Uncrosslinked copolymer (M 32 000)
G 20 SmC* 116 N* 151 I

$$H-\left[Si-O\right]_{6.5}\!-Si-H \quad 10 \text{ mol.\%}$$

Cross-linked elastomer
G 15 SmC* 100 N* 128 I

Figure 7. Chemical structure and cross-linking by hydrosililation of a chiral copolymer [7].

polymers (see Figure 5, type A), nematic phases are strongly favoured. The preference for nematic phases can be rationalized by considering their structure (see Figure 5). For the combined LC polymers of type A a smectic structure may already be formed locally. For the combined LC polymers of type B, however, the side-chain mesogens will typically be located at about the position of the main-chain spacer. This arrangement disfavours the formation of smectic phases.

3.3 Properties of Combined LC Polymers

3.3.1 Structure–Property Relationships and Types of LC Phase

The structural model given in Figure 5 was obtained from comparing the temperature ranges of the LC phases of combined LC polymers and side-chain or main-chain

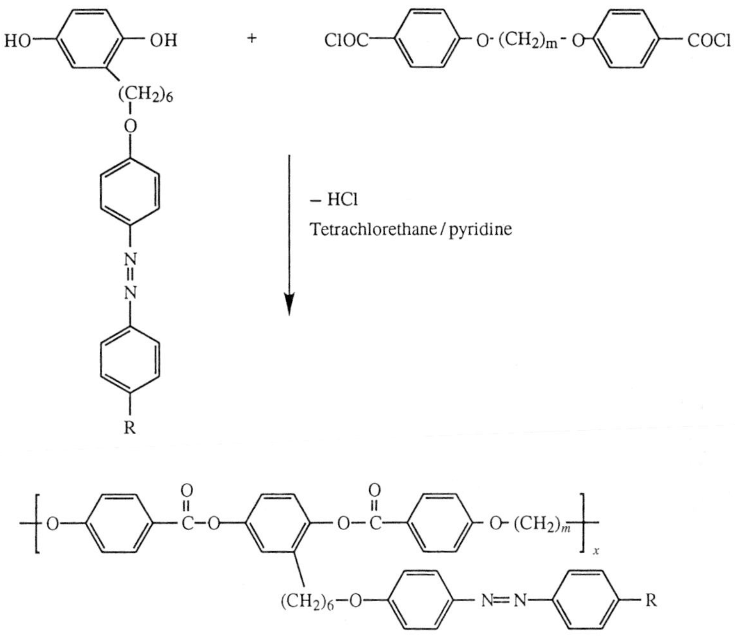

Scheme 3. Preparation of combined LC polymers by a solution condensation of a *bis*(phenol) and a *bis*(acid chloride).

polymers (Figure 4). The model assumes that a synergistic interaction of the main-chain and side-chain mesogens stabilizes the LC phases. In addition, it has been found [19] that a decrease in the spacer length leads to an increase in the glass transition temperature. The length of the spacer also strongly influences the clearing entropy and enthalpy [19]. These facts indicate that strong conformational changes occur at the clearing point. The clearing temperature, however, is not strongly dependent on the spacer length.

X-ray measurements [7, 11, 13–15, 17, 19, 20] performed on combined LC polymers show that the LC phases are analogous to low molar mass liquid crystals, which also show nematic, smectic A, smectic C and higher ordered smectic phases at different temperatures (see Figure 8).

The results of these X-ray investigations have been used to obtain the following structural models. For polymers of type A for which the repeating distance along the main chain (L_1) is larger than the length of the side chain (L_2) the model shown on the left-hand side in Figure 9(a) is proposed [20, 25]. In this case the main-chain spacer is partly bent in order to adjust for the shorter side chain. For polymers in which L_1 is smaller than L_2, the model on the right-hand side in Figure 9(a) is proposed. This model assumes that there are separate layers for main-chain and side-chain mesogens and a sharp bending of the main spacer. For the smectic phase of polymers of type B the model shown in Figure 9(b) has been proposed [19].

Using results obtained by dielectric spectroscopy, four relaxation processes have been assigned (Figure 10): a reorientation of the long axis of the side-chain mesogen (δ relaxation) [19], a glass process (α relaxation) [26] and two separate processes (β_m and β_s) [26] of the main-chain and side-chain mesogens around their long axes.

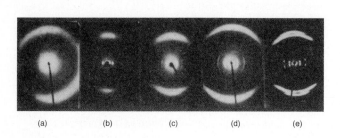

Sm$_1$ 108 Sm$_2$ 112 SmC 131 SmA 136 N155 I

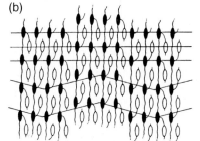

Figure 8. Temperature-dependent X-ray measurements performed within the nematic (a), smectic A (b), smectic C (c), smectic 2 (d) and smectic 1 (e) phases [20, 25]. The arrow indicates the direction of orientation.

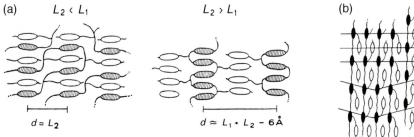

Figure 9. (a) Structural models for the smectic phases of LC polymers of type A (see Figure 5), for which the length of repeating unit along the polymer chain (L_1) is longer or shorter than the length of the mesogenic side groups (L_2) [20, 25]. (b) Structural model for the smectic phase of combined LC polymers of type B (see Figure 5) [19, 25].

These two β relaxations are observed above and below the glass transition temperature, respectively.

Combined LC polymers have been investigated intensively in a search for biaxial nematic phases [21]. However, it has never been possible to prove the existence of such phases.

3.3.2   Interaction of Main-Chain and Side-Chain Mesogens

In order to understand the synergistic interaction of the main-chain and side-chain mesogens, ^{2}H NMR [21] and neutron scattering [27] experiments have been performed on selectively deuterated samples. ^{2}H NMR [21] (see Volume I, Chapter VIII, Sect. 1 of this Handbook) shows clearly that the main-

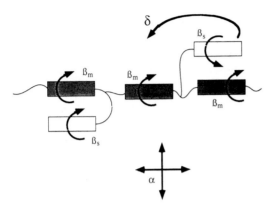

Figure 10. Schematic representation of the relaxation processes (α, β_m, β_s and δ) in combined LC polymers [19, 26].

chain and side-chain mesogens orient parallel to one another (Figure 11). The order parameter of the main-chain mesogens is, however, somewhat higher. This corresponds to the situation in main-chain polymers, for which rather high-order parameters can be found.

Small angle neutron scattering (SANS) [27] shows that the main chain orients parallel to the director. The detailed behaviour of the combined LC polymers is intermediate between that of LC main-chain and LC side-chain polymers. The polymer chain adopts an anisotropy of about 30% in the ne-

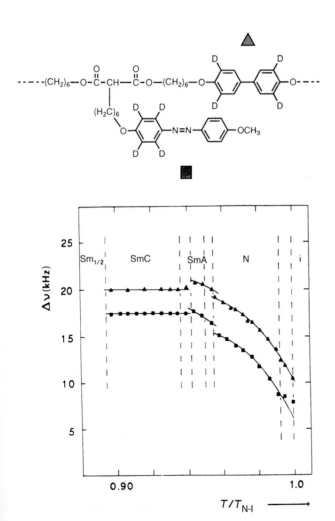

Figure 11. The temperature dependence of the quadrupole splitting Δv (^{2}H NMR) of combined LC polymers selectively deuterated at either the main-chain (▲) or the side-chain (■) mesogen [21, 25].

matic phase. This is far more than for side-chain polymers, but less than for main-chain polymers. In the smectic A phase the polymer chain is highly stretched, but the effect is still smaller than for main-chain polymers.

3.3.3 Rheology and LC Elastomers

Rheological investigations on combined LC polymers [10] have shown that the complex viscosity η^*, behaves as shown in Figure 12. The viscosity decreases strongly at the glass transition temperature, but levels off at a rather high value in a smectic A or C phase. At the transition into the nematic phase there is a second decrease in viscosity, to values that are lower than for the isotropic phase at higher temperatures. This is presumably due to a strong shear thinning effect in the nematic phase.

Cross-linking of these polymers, which was first [10] done by a hydrosilylation reaction according (Figure 7), did not change this general behaviour. It did, of course, lead to a strong increase in the viscosity and the moduli above the glass transition temperature. The influence of the amount of cross-linking on the rheological properties was studied in more detail for LC elastomers cross-linked by means of a radical polymerization of pendant acrylate groups (Scheme 4) [11]. The resulting change in the shear modulus, G', is shown in Figure 13 [11]. It can be seen that a continuous increase in the cross-linking leads to a continuous increase in G'. In this case the LC phase did not appear in the most strongly cross-linked sample. This is presumably due to the fact that the new polyacrylate chain formed during polymerization is linked without a spacer to the mesogenic groups.

All the LC elastomers prepared from combined LC elastomers were soft materi-

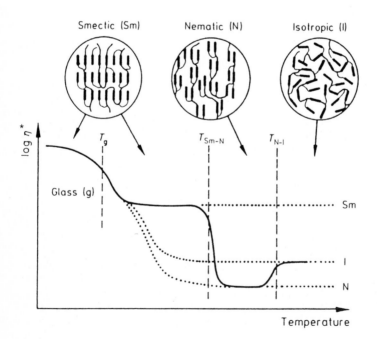

Figure 12. Schematic representation of the mechanical behaviour of an LC sample in relation to its structure and phase transitions [10]. The dotted lines represent the hypothetical behaviour of smectic, nematic and isotropic phases if they were to extend over the entire temperature range above the glass transition.

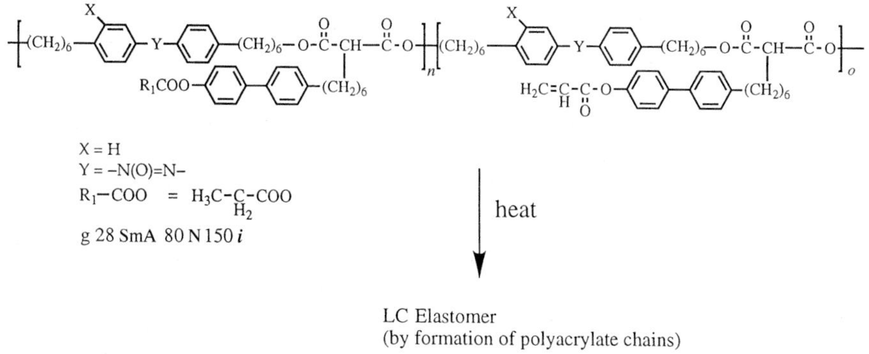

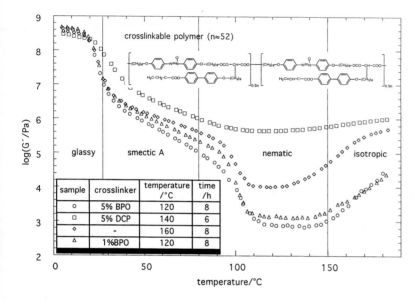

X = H
Y = −N(O)=N−
R_1−COO = H_3C-$\underset{H_2}{C}$-COO

g 28 SmA 80 N 150 *i*

heat

LC Elastomer
(by formation of polyacrylate chains)

Scheme 4. Cross-linking of LC elastomers by radical polymerization of pendant acrylate groups.

sample	crosslinker	temperature /°C	time /h
○	5% BPO	120	8
□	5% DCP	140	6
◇	-	160	8
△	1%BPO	120	8

Figure 13. Shear storage modulus of the combined LC polymer specified in Scheme 4 as a function of the extent of cross-linking [11].

als, which could be deformed strongly [7] and swelled strongly if brought into contact with a solvent. For such elastomers it is interesting to investigate the reversible mechanical orientation of the LC phase [8] (see Sec. 3.2.1.3). This mechanical orientability is especially interesting with regard to ferroelectric LC phases (chiral smectic C* phase) (Figure 14). As long as the helical superstructure is undeformed, the electric polarizations from different smectic layers cancel each other. As soon as the helix becomes deformed (or for the extreme case of a complete helix unwinding), a macroscopic polarization occurs.

X-ray measurements on LC elastomers have shown [6–8] that the reversible transition between a chiral smectic C* phase with and without a helical superstructure can be induced mechanically. The helix untwisted state corresponds in this case to a polar ferroelectric monodomain. The piezoelectricity arising from this deformation of the helical superstructure (which does not require a complete untwisting) has been demonstrated [9] for polymers cross-linked by polymerization of pendant acrylate groups (Figure 15).

Further research on this topic requires either the preparation of samples with an unpartial unwinding   partial unwinding

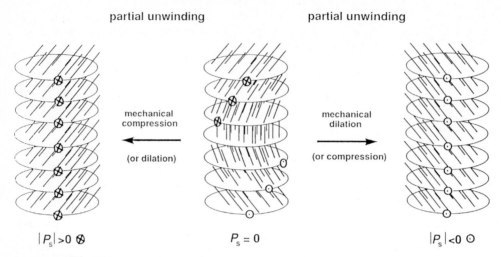

$|P_s|>0$ ⊗ $P_s = 0$ $|P_s|<0$ ⊙

Figure 14. Idealized presentation of the orientation process, which leads to piezoelectricity in chiral smectic C* elastomers (only the mesogens are shown) [28], P_s: macroscopic polarization). The deformed states with a partially unwound helix (left and right) are prepared from the ground state with a helical superstructure (middle) by mechanical forces. ⊗ and ●: direction of the spontaneous polarization in and out of the plane of drawing, respectively.

disturbed helix and a very good understanding of how the helix is deformed mechanically, or the preparation of polar monodomains and the mechanical disturbance of their orientation. In this respect, the use of ferroelectric LC side-chain polymers, which can be easily oriented in external electric fields, has proven to be advantages. For such systems it is relativly easy to prepare large ferroelectric monodomains and to cross-link them photochemically [23].

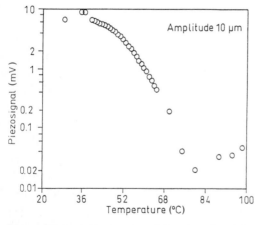

Figure 15. Logarithmic plot of the piezosignal for an LC elastomer with the phase sequence g 1 SmC* 81 I [19]. The cross-linking was done according to Scheme 4.

3.4   References

[1] B. Reck, H. Ringsdorf, *Makromol. Chem., Rapid Commun.* **1985**, *6*, 291.
[2] G. Canessa, B. Reck, G. Reckert, R. Zentel, *Makromol. Chem., Macromol. Symp.* **1986**, *4*, 91.
[3] R. Zentel, G. Reckert, *Makromol. Chem.* **1986**, *187*, 1915.
[4] R. Zentel, G. Reckert, B. Reck, *Liq. Cryst.* **1987**, *2*, 83.
[5] S. Bualek, R. Zentel, *Makromol. Chem.* **1988**, *189*, 791.
[6] R. Zentel, *Liq. Cryst.* **1988**, *3*, 531.
[7] R. Zentel, G. Reckert, S. Bualek, H. Kapitza, *Makromol. Chem.* **1989**, *190*, 2869.
[8] R. Zentel, *Angew. Chem. Adv. Mater.* **1989**, *101*, 1437.
[9] S. U. Vallerien, F. Kremer, E. W. Fischer, H. Kapitza, R. Zentel, H. Poths, *Makromol. Chem., Rapid Commun.* **1990**, *11*, 593.
[10] T. Pakula, R. Zentel, *Makromol. Chem.* **1991**, *192*, 2401.

[11] M. Brehmer, R. Zentel, *Mol. Cryst. Liq. Cryst.* **1994**, *243*, 353.

[12] S. Bualek, R. Zentel, *Makromol. Chem.* **1988**, *189*, 797.

[13] H. Kapitza, R. Zentel, *Makromol. Chem.* **1988**, *189*, 1793.

[14] S. U. Vallerien, R. Zentel, F. Kremer, H. Kapitza, E. W. Fischer, *Makromol. Chem., Rapid Commun.* **1989**, *10*, 333.

[15] H. Poths, R. Zentel, S. U. Vallerien, F. Kremer, *Mol. Cryst. Liq. Cryst.* **1991**, *203*, 101.

[16] H. Kapitza, R. Zentel, *Makromol. Chem.* **1991**, *192*, 1859.

[17] R. Zentel, G. F. Schmidt, J. Meyer, M. Benalia, *Liq. Cryst.* **1987**, *2*, 651.

[18] I. G. Voigt-Martin, H. Durst, B. Reck, H. Ringsdorf, *Macromolecules* **1988**, *21*, 1620.

[19] B. W. Endres, M. Ebert, J. H. Wendorff, B. Reck, H. Ringsdorf, *Liq. Cryst.* **1990**, *7*, 217.

[20] S. Diele, M. Naumann, F. Kuschel, B. Reck, H. Ringsdorf, *Liq. Cryst.* **1990**, *7*, 721.

[21] K. Kohlhammer, G. Kothe, B. Reck, H. Ringsdorf, *Ber. Bunsenges. Phys. Chem.* **1989**, *93*, 1323.

[22] M. Warner in *Side Chain Liquid Crystal Polymers* (Ed.: C. B. McArdle), Blackie, Glasgow, **1989**.

[23] M. Brehmer, R. Zentel, G. Wagenblast, K. Siemensmeyer, *Macromol. Chem. Phys.* **1994**, *195*, 1891.

[24] B. Reck, H. Ringsdorf, *Makromol. Chem., Rapid Commun.* **1986**, *7*, 389.

[25] B. Reck, *Synthesen und Struktur-Eigenschaftsbeziehungen von flüssigkristallinen Polyestern*, Jo. Gu. Universität Mainz, Germany, **1988**.

[26] F. Kremer, S. U. Vallerien, R. Zentel, H. Kapitza, *Macromolecules* **1989**, *22*, 4040.

[27] L. Noirez, H. Poths, R. Zentel, C. Strazielles, *Liq. Cryst.* **1995**, *18*, 123.

[28] H. Kapitza, H. Poths, R. Zentel, *Makromol. Chem., Macromol. Symp.* **1991**, *44*, 117.

4 Block Copolymers Containing Liquid Crystalline Segments

Guoping Mao and Christopher K. Ober

4.1 Introduction

Self-organization is a common phenomenon in nature [1, 2]. Both liquid crystalline polymers (LCPs) [3–5] and block copolymers (BCPs) [6, 7] are two classes of synthetic materials which can readily undergo self-organization. By combining both of these components in a single molecular system, the competition between their self-organizing behavior offers opportunities for simultaneously creating ordered structures at many length scales in polymer systems [8]. Combination of microphase separation and liquid crystallinity may result in new materials with superior and unanticipated properties. Such polymers may also serve as models providing insight to the ordering of more complicated biological systems in which multiple ordering processes are present [9, 10]. Several review papers have been published so far on this topic [11–13]. By definition, LC block copolymers contain at least one LC segment and may have structures which include rod–rod, rod–coil, side group LC (SGLC)–coil and other block combinations. In this review, we would like to summarize the most recent advances in the field of LC block copolymers.

4.1.1 Block Copolymers: a Brief Review

Block copolymers are a unique class of materials. Covalently connected dissimilar polymer chains undergo microphase separation because an unfavorable mixing enthalpy and a very small mixing entropy drive the system into phase separation while the covalent bond between these two polymer chains prevents the system from undergoing macrophase separation. By minimizing the interfacial area (e. g. via creating curvature and chain stretching normal to the interface), conventional coil–coil diblock copolymers form morphologies including spheres, cylinders, bicontinuous phases, and lamellae. The important parameters which govern microphase separation are the total degree of polymerization $N(=N_A+N_B)$, the Flory–Huggins χ parameter, and the volume fraction of the A block, f_A. The structure–property relationship for coil–coil diblock copolymer systems is relatively well understood both experimentally [14–19] and theoretically [20–23]. Chain stretching normal to the interface leads to a positive deviation of the scaling exponent [14]. For example, scaling behavior for a conventional styrene–isoprene (SI) diblock copolymer in the lamellar regime is $D \propto M_n^{2/3}$ in which D is the lamellar domain size and M_n is the total molecular weight [14, 23], while for a random coil, theory predicts that $D \propto M_n^{1/2}$.

Block copolymers can be regarded as composite materials [6, 24] with domain sizes on submicron scales. For example, conventional poly(styrene-*b*-isoprene) block copolymers have properties ranging from rubbery (isoprene matrix) to plastic (styrene matrix). Block copolymers also offer an opportunity for engineering new, tailored materials using self-organizing liquid crystal components. For example, high strength LC cylinders in a ductile continuous phase may form a strong, tough polymer. Switchable LC groups in a block copolymer environment can be processed to provide prealigned structures.

4.1.2 Liquid Crystalline Polymers: Architecture

The field of liquid crystal polymers (LCP) has undergone tremendous development since the discovery of Kevlar™, an extraordinarily strong fiber spun from a lyotropic liquid crystal solution [25]. As shown in Fig. 1, the principle structures of LC polymers include main chain LCPs [26], side groups LCPs [27], combined main chain/side group LCPs [28], mesogen-jacketed polymers [29], well-defined LC stars [30], hyperbranched LCPs [31, 32], and (more recently) LC dendrimers [33]. The structure–property relationships of LCPs are quite well documented in the literature. The mesogenic structures can be calamitic or disc-like groups [34, 35] and can also be formed via molecular recognition using H-bonding [36]. Essentially all known LC phases have now been reproduced in polymeric systems.

4.1.3 Architecture of Liquid Crystalline Block Copolymers

Liquid crystal block copolymers are a recently explored group of polymers which combine microphase separation and liquid crystallinity. Yet, as early as 1963, Gratzer and Doty [37] reported the first block copolymers containing two polypeptide blocks in which one of the blocks, poly(γ-benzyl-L-glutamate) (PBLG), was already well known to be liquid crystalline. In principle, all the LCP structures shown in Fig. 1 can be incorporated into block copolymer structures with a second flexible coil or LC block. Some of the possible LC–BCP structures are shown in Fig. 2. Due to the limited nature of this review, structures such as the grafted LC copolymers [38] and various multiblock LC copolymers [39] will not be covered.

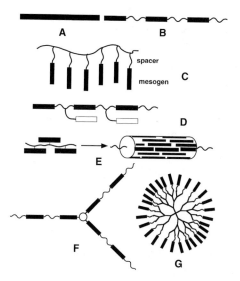

Figure 1. Structures of liquid crystalline polymers (LCPs): (A) rigid rod LCP [25, 65]; (B) main chain LCP with flexible spacer [26]; (C) side group LCP with flexible spacer [27]; (D) combined main/side group LCP [28]; (E) side group LCP without flexible spacer or 'mesogen jacketed LCP' [29]; (F) well-defined three-arm star [30]; (G) LC dendrimer [33].

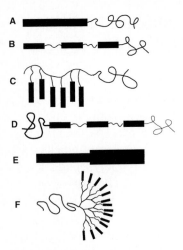

Figure 2. Structure of liquid crystalline block copolymers (LC-BCPs): (A) rod–coil diblock copolymer;
(B) rod–coil diblock copolymer with flexible spacer
in the rod block; (C) side group liquid crystal–coil
(SGLC–coil) diblock copolymers; (D) coil–rod–coil
ABC triblock copolymers (predicted to be novel ferroelectric fluid by R. G. Petschek and K. M. Wiefling,
Phys. Rev. Lett., **1987**, *59(3)*, 343–346); (E) rod–rod
diblock copolymer (one example of well-defined po-
ly(*n*-hexyl isocyanate-*b*-*n*-butyl isocyanate) rod–rod
diblock copolymer was given by Novak et al. [68],
however, no morphology studies were reported); (F)
dendritic liquid crystal–coil (DLC–coil) diblock copolymer (not reported).

4.1.4   General Features of Liquid Crystalline Block Copolymers

Liquid crystallinity and block microphase
separation both compete during the minimization of free energy of the system. As
we will show later in this review, in the case
of a rod–coil diblock copolymer, liquid
crystallinity plays a very important role in
the microphase separation process and leads
to morphologies distinctly different from
the conventional spheres, cylinders and
lamellar microstructures and include the
arrow head, zig-zag, and wavy lamellae
phases [40, 41]. In the case of SGLC–coil

diblock copolymers, microphase separation
dominates the microstructure in which only
conventional morphologies have been observed [42, 43] so far and the LC microphase is formed within the conventional microdomains.

Important transitions observed after introducing LC segments into block copolymers include: a glass transition temperature
for each distinct block, phase transition temperatures for the LC block, and the block
microdomain order–disorder transition temperature (ODT). It is generally expected that
the χ parameters in LC–BCP systems are
much larger than for conventional coil–coil
BCPs due to mesophase formation and such
a large χ parameter leads to a much higher
ODT [44]. For example, the ODT of a
poly(styrene-*b*-isoprene) (PSI) diblock copolymer with M_n of 12 000 and 9 000 (SI-
12/9) for each block is 152 °C [45]. Due to
the potentially high ODT in LC–BCP
systems, it may be impossible to determine
the ODT because the polymer may decompose before reaching its order–disorder
transition. Another very important feature
of LC–BCP systems is that the large χ parameter and high ODT in LC–BCP systems
also guarantees that a microphase separated
structure will be maintained at temperatures
above the liquid crystalline clearing transition temperature.

LC–BCPs provide theorists with novel
model systems for developing new theories
of polymer behavior [41]. Consider the process of microphase separation in LC–BCPs
starting from the disordered BCP structure
(or isotropic solution) (Scheme 1). As the
temperature decreases, the system will
undergo microphase separation to form ordered microdomain structures. As temperature falls past the LC clearing transition temperature (T_i) the LC block will shift from
the isotropic phase to a LC phase. If this temperature is lower than the T_g of the other

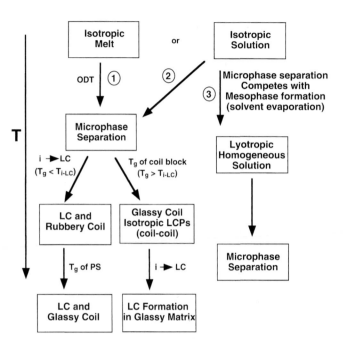

Scheme 1. Microphase separation process for LC–BCPs. Solvents have the same effect as raising temperature due to a reduced $\chi = \chi_\phi$. Path 1 and 2: for SGLC–coil systems (e. g. SICN5 systems, see Scheme 9 for structures); path 3: for rod–coil systems (e. g. poly(styrene-*b*-*n*-hexyl isocyanate)).

flexible coil block, the LC transition will be in a confined microdomain. Finally, the LC block may undergo further LC transitions. Since the ODT for LC–BCPs is high, at temperature above the T_i, the LC–BCP can be regarded as coil–coil BCPs. In solvent casting, this situation may be different. If the solvent is nonpreferential to each block, as the solvent evaporates, the formation of liquid crystalline mesophase may compete with microphase separation until an ordered microdomain structure forms as a result of this competition.

In order to study the detailed behavior of LC block copolymers, it would be ideal to create monodispersed LC–BCP samples with well defined architecture for each block over a wide range of molecular weights. LC–BCP systems with narrow polydispersity should form well-ordered microdomain structures while LC–BCPs with broad molecular weight distributions would probably not. Synthesis of such materials still remains a challenge.

In this review, we will concentrate on rod–coil and side group LC–coil diblock copolymers. Here we use the term rod–coil to emphasize the rigid rod nature of the main chain LC polymer in which the mesogens (or rigid rods) are located along the polymer chain, that is main chain LC–coil (MCLC–coil). Diblock systems have been chosen since they are the simplest block copolymer system and can serve as excellent model materials. It is beyond the scope of this review to cover all the aspects in the field of LC–BCPs, but rather to point out the main principles behind their molecular design and to summarize the most recent results. It is our hope that this review will generate interest and provide greater insight into these novel systems.

4.2   Rod – Coil Diblock Copolymer Systems

4.2.1   Polypeptides as the Rod Block

Block copolymers containing a biological-ly active, polypeptide block were original-ly studied as models for biological mem-branes. The variety of conformations of polypeptides (α-helix, β-sheet, and random coil) were expected to produce new copol-ymers of technical interest [10].

Gallot and coworkers [9, 10, 46, 47] played a leading role in the synthesis and characterization of AB diblock copolymers consisting of a flexible coil A block (poly-butadiene or polystyrene) and a rod B block (polypeptide). The synthetic methods used to produce these polymers are shown in Scheme 2 [9]. The coil block was synthe-sized by conventional living anionic poly-merization initiated by sec-BuLi followed by end-capping with a primary amine group. Three methods were employed for the end-capping as shown in Scheme 2 A. In the first method, the living polymer (e. g. polybuta-diene) was end-capped with ethylene oxide to produce a hydroxyl end group, followed by reaction with phosgene to convert the hy-droxyl group into the corresponding chloro-formate. The addition of 20 equivalents of diamine (e. g. hexamethylenediamine) af-forded the primary amine end-capped poly-mer with 75% conversion. Due to the pres-ence of a weak urethane linkage, further modification of the peptide block via hy-drolysis is not possible. The second method gives poor yields ($\approx 5\%$) of the amine group. In the third method, the living polymer chain end was converted to a carboxylic acid group with CO_2 followed by coupling with a diamine (such as hexamethylenediamine) using DCC (dicyclohexylcarbodiimide).

This method produced amine end-capped polymer in more than 80% yield. The pri-mary amine group containing flexible poly-mer was then used as macroinitiator for the polymerization of the N-carboxyl anhydride (NCA) to produce the polypeptide block. For example, PBLG was chosen as homo-polypeptide because of its lyotropic liquid crystalline mesophase (Scheme 2 B). It is stable and soluble in many common organ-ic solvents and water in its α-helical con-formation [48]. It is also worth noting that Deming [49] has summarized the recent de-velopment in the area of controlled poly-merization of NCAs using transition metal catalysts [50] including the synthesis of block copolymers.

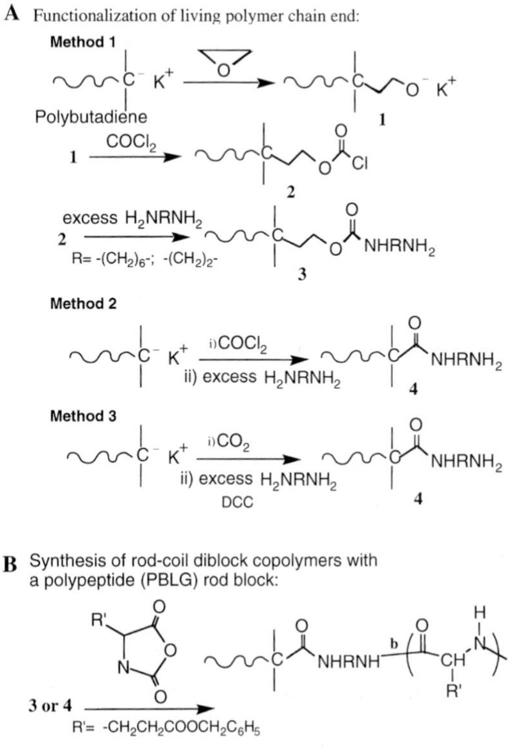

A Functionalization of living polymer chain end:

B Synthesis of rod-coil diblock copolymers with a polypeptide (PBLG) rod block:

Scheme 2. Functionalization of living chain end (A) and synthesis of rod – coil diblock copolymer with a polypeptide rod block (B) [9].

The final block copolymers were carefully purified and fractionated. The polybutadiene homopolymer without an amine end-group (because it cannot initiate the polymerization of NCA) was extracted with a mixture of ethyl acetate/methanol (85/15, v/v), while trace amounts of a polypeptide homopolymer (PBLG) were extracted in DMF. Block copolymers containing 15–75 wt% PBLG with polybutadiene molecular weights ranging from 16 000 to 26 000 g/mol were synthesized and all of them exhibited lamellar morphology [9, 46]. The PBLG segments in these block copolymers also have α-helix conformations as confirmed by circular dichroism. They assembled into hexagonal arrays in the polypeptide domain with chain folding [9] as established by X-ray studies and TEM (transmission electron microscopy). The lamellar morphology packing model is shown in Fig. 3 A. These block copolymers also form lyotropic mesophases in dioxane and various chlorinated solvents such as 1,2-dichloroethane and 2,3-dichloro-1-propene [9].

Another type of polypeptide-containing block copolymer, amphiliphilic rod–coil diblock copolymers such as poly{(N-trifluoroacetyl-L-lysine)-b-sarcosine} (Kt–Sa), were also synthesized and characterized by Gallot and coworkers [47]. The hydrophobic rod block poly(N-trifluoroacetyl-L-lysine) (Kt) was prepared by polymerization of Kt–NCA using N-hexylamine as the initiator. After fractionation using DMF (good solvent)/water (nonsolvent), the narrowly dispersed polymer (Kt) was then used as macroinitiator to initiate polymerization of the second monomer (Sa–NCA) to afford the hydrophilic block. Final elimination of Kt and Sa homopolymers were performed by precipitation with acetone and water respectively. The synthesis of Kt–Sa diblock copolymer is shown in Scheme 3.

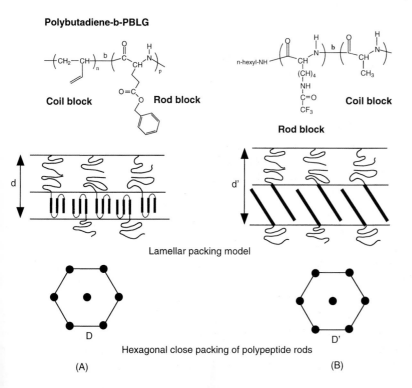

Polybutadiene-b-PBLG

Coil block   Rod block

Rod block   Coil block

Lamellar packing model

Hexagonal close packing of polypeptide rods

(A)   (B)

Figure 3. Lamellar packing model for polypeptide containing rod–coil diblock copolymers: (A) Model for polybutadiene-*b*-PBLG diblock copolymer in which the PBLG chains fold in the lamellar layer and adopt hexagonal packing. This corresponds to a SmB mesophase [9]. (B) Model for amphiphilic polypeptide diblock copolymer. Note that the rods are tilted relative the lamella layer normal while maintaining hexagonal close packing with a constant domain D′ independent of tilt angle. This corresponds to a SmF mesophase [47].

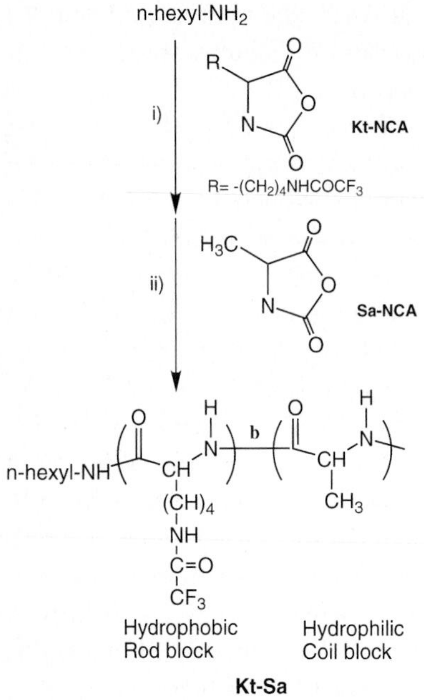

Scheme 3. Synthesis of amphiphilic diblock copolymers with polypeptide blocks [47].

X-ray diffraction studies showed lamellae formation in Kt–Sa rod–coil diblock copolymers containing 25–73 wt% Sa block [47]. The hydrophilic polysarcosine block (Sa) has a coil conformation while the hydrophobic poly((N-trifluoroacetyl-L-lysine) (Kt) block contains an α-helix conformation as indicated by IR spectroscopy with absorption peaks at 1655 and 1545 cm^{-1} (for amide I and amide II respectively) which is assembled in a hexagonal array in a plane normal to the helix axis. Interestingly, the rods (Kt block) are tilted relative to the lamellar layer as indicated in Fig. 3 B. The tilt angle increases as the water content increases, while the D spacing (=1.44 nm) of the hexagonally packed polypeptide array remains constant [47] at the same value as in the solid state. Therefore, water only swells the hydrophilic Sa block. This swelling increases the area per chain of the Sa block at the polymer–polymer inter-

face and forces the rod of the hydrophobic block to tilt in order to increase the interfacial chain density for efficient chain packing. For the same reason, increasing the coil molecular weight also results in an increase of rod (Kt block) tilt angle. Thin films of Kt–Sa block copolymers were studied by XPS and the surface was found to be rich in the hydrophobic domain (Kt block) at the polymer–air interface [51]. This may be the first report of a low surface energy component segregating to the surface in a block copolymer system.

4.2.2 Rod–Coil Block Copolymers with Short Rod Blocks

Stupp and coworkers [52] presented an elegant approach to produce well-defined rod–coil block copolymers using DiPC (diisopropylcarbodiimide) coupling of a rod block (containing eight phenyl groups) and a polyisoprene block ($M_n \approx 3000-8000$) with a carboxylic acid end group (Scheme 4). The rod was synthesized by conventional organic synthetic methods and has a well-defined structure with a fully extended rod length of 6 nm. The final polymers could be purified by flash column chromatography using silica gel because the total molecular weights were less than 10000 g/mol. This method produced well-defined rod–rod block copolymers with narrow molecular weight distribution. The rod volume fractions range from 0.19 to 0.36 as listed in Table 1. Even though the degree of polymerization (DP) was small, microphase separated structures can still be observed because the Flory–Huggins χ parameter between the highly immiscible rod and coil blocks is large. The small DP leads to domain sizes which were much smaller than conventional coil–coil diblock copolymers with much higher DP. This was termed 'nanophase

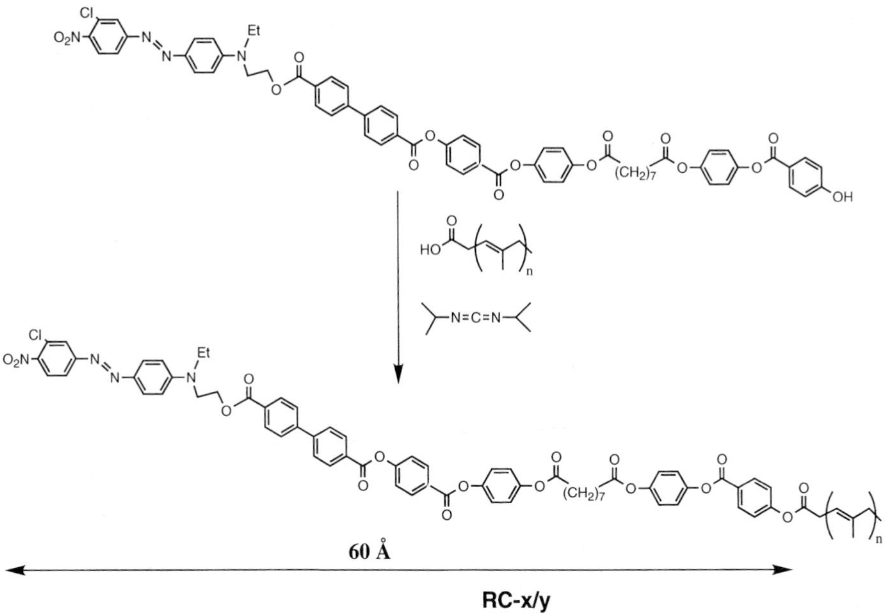

60 Å

RC-x/y

Scheme 4. Stupp's approach for synthesizing well defined rod–coil polymers [52].

Table 1. Rod–coil diblock copolymers synthesized by Stupp and coworkers (adapted from [52]).

M_n (Rod–coil)[a]	M_w/M_n	Rod volume fraction, f_{rod}	Morphology[b]
RC-1.16/3.2	1.08	0.36	Alternating strips
RC-1.16/4.2	1.04	0.30	Coexistence of strip and aggregates
RC-1.16/5.4	1.05	0.25	Discrete aggregates
RC-1.16/7.6	1.05	0.19	No nanophase separation

[a] For structure, see Scheme 4. The coil block is polyisoprene.
[b] In as-cast film.

separation' [53] by Stupp and coworkers in order to emphasize the small in-plane dimensions (less than 10 nm) in their system. Interesting morphologies ranging from alternating strips and discrete aggregates [53] were observed with their rod–coil samples. The results are summarized in Table 1.

Quite recently, Lee and coworkers [54] reported the first observation of a series of rod–coil molecules in which a hexagonal columnar liquid crystalline mesophase was first observed (Scheme 5 A). Interestingly, the coil block they chose contains either po-

ly(ethylene oxide) or poly(propylene oxide) which forms complexes with Li^+ salts [55, 56]. The rod–coil **16-4** (see Scheme 5 A for structure) has a SmB mesophase while its Li^+ complex displayed a cylindrical micellar mesophase as observed by polarized optical microscopy observation [55, 56]. This behavior is interesting because columnar mesophases are generally observed only with disc-like mesogenic groups [35].

Another interesting rod–coil diblock copolymer is poly(styrene-b-polyphenylene) [57] (Scheme 5 B) reported by Zhong and

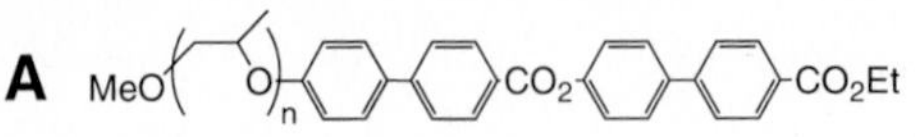

A

P-12-4: n=12; K 21.5°C Col 33.5°C I

P-7-4: n=7; K 61.5°C SmC 119.9°C SmA 139°C I

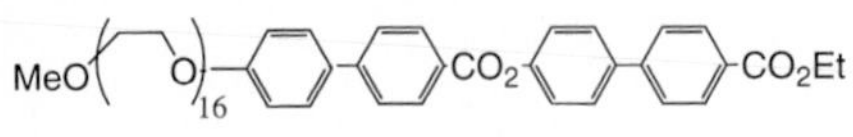

16-4: g32°C K56°C K120°C SmB 123°C I

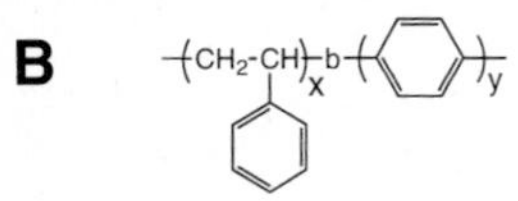

B

Scheme 5. Lee's rod–coil molecules [54–56]; (B) poly(styrene-*b-p*-phenylene) [57].

Francois. This material was synthesized from styrene and 1,3-cyclohexadiene via anionic polymerization. The poly(1,3-cyclohexadiene) block was then converted to poly(para-phenylene) by aromatization using *p*-chloranil. Through the use of UV – vis spectroscopy, the researchers concluded that the polyphenylene block was constructed of short para-phenylene sequences with an average of seven units of phenyl rings separated by defects. Unfortunately its LC properties were not investigated [57]. These interesting diblock copolymers have very unique nonequilibrium morphologies, termed the 'honeycomb morphology' [58] in which essentially monodispersed pores are observed to arrange in hexagonal arrays. This novel morphology [58] was believed to be due to micelle formation.

4.2.3 Rod–Coil Block Copolymers Based on Polyisocyanates

Although polypeptide rod–coil block copolymers [9, 10, 59] are very interesting materials, the study of a true rod–coil system requires a simpler model material because polypeptides have so many possible

configurations [60]. Because polypeptide chains can fold rather than tilt, these two issues may also compete making interpretation difficult. Other rod–coil systems known in the literature only have low molecular weight rod blocks [52–56].

Ober, Thomas, and coworkers [40, 41, 61, 62] chose poly(*n*-hexyl isocyanate) as their rigid rod building block and poly(styrene) (PS) as the flexible coil block. Poly(*n*-hexyl isocyanate) (PHIC) is well known to have an 8_3 helical (8 units with 3 turns) conformation [40] with a translation of 0.195 nm and rotation of 135° per monomer unit [40] and a persistent chain length of 50–60 nm [63, 64]. Its lyotropic [65] and thermotropic [66, 67] LC behaviors were thoroughly studied by Aharoni's pioneering work in the late seventies and early eighties. By sequential living anionic polymerization, the rod–coil diblock copolymers, poly(styrene-*b-n*-hexyl isocyanate)s (SHIC), were synthesized [40, 62] as shown in Scheme 6. Fractionation was generally necessary to remove PS homopolymer. The synthesized rod–coil diblock copolymers are listed in Table 2

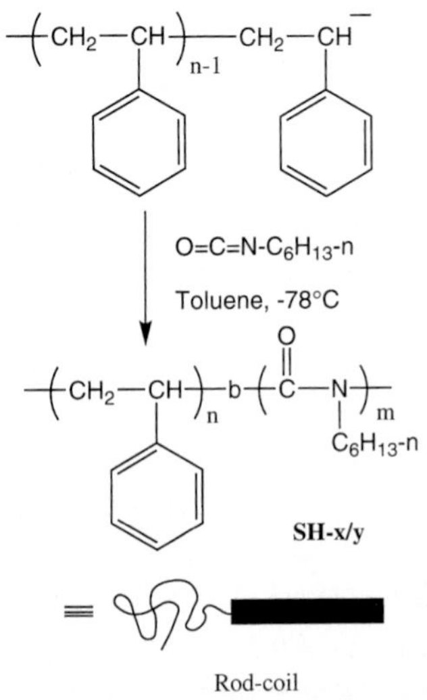

Scheme 6. Synthesis of poly(styrene-*n*-hexyl isocyanate) rod–coil diblock copolymers [40, 62].

Table 2. Summary of poly(styrene-*b*-*n*-hexyl isocyanate) rod–coil diblock copolymers (adapted from [41, 61, 62]).

Sample[a]	M_w (GPC)[b]	M_w/M_n	DP_{PS}[c] (M_w)	DP_{PHIC} (M_w)	f_{PHIC}[d]
SH-104/73	240,000	1.11	1000 (104 K)	575 (73 K)	0.42
SH-14/36	68,000	1.40	135 (14 K)	283 (36 K)	0.73
SH-7/58	222,000	1.86	63 (7 K)	457 (58 K)	0.90
SH-9/245	1,820,000	2.52	89 (9 k)	1930 (245 K)	0.96
SH-7/386	1,430,000	3.11	68 (7 k)	3040 (386 K)	0.98

[a] SH-*x/y*: S stands for styrene block, H for poly(*n*-hexyl isocyanate) block, *x* for the real molecular weight of styrene block, and *y* for the polyisocyanate block in kg/mol.
[b] Relative molecular weight from GPC measurements. Polystyrene standards were used as calibration. The GPC gives M_w values *much larger* than the real molecular weight due to the rigid nature of the molecule which results in a shorter retention time in the GPC column.
[c] DP_{PS}: Degree of polymerization for PS block. All samples of the PS block have polydispersity between 1.04–1.08. The M_n of the rod block were calculated from molar ratio obtained from ^{1}H NMR and the M_n of styrene block were measured from GPC.
[d] Volume fraction of the rod block.

[40, 62]. It is worth noting that one must be very careful in the determination of the molecular weight for this kind of rod–coil diblock copolymers because GPC (gel permeation chromatography) gives much larger M_n values than the real molecular weight. Novak and coworkers reported that GPC gave a M_n of PHIC about five times larger than its absolute molecular weight determined by viscosity and end-group analysis [68].

Novel morphologies including lenticular aggregates [61] (previously reported as wavy lamellae [41]), zig-zag lamellae [40, 41] to arrow-head [41] microdomain structures were observed by TEM as shown in Fig. 4. These strikingly new morphologies are believed to be a result of the interplay of block microphase separation and liquid crystallinity. As a result of solvent evaporation, this system first forms a homogeneous nematic mesophase with an orientational order of the rods followed by microphase separation to form ordered smectic-like lamellar morphologies (see Scheme 1, path 3). The structural packing model is shown in Fig. 5.

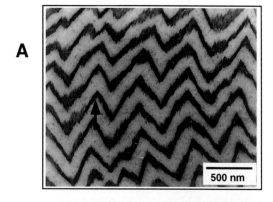

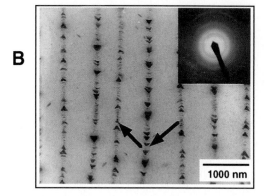

Figure 4. TEM of poly(styrene-*n*-hexyl isocyanate) rod–coil diblock copolymers. The orientation of rods in LC domain is indicated by arrows. (Reprinted from [41], with permission).

A

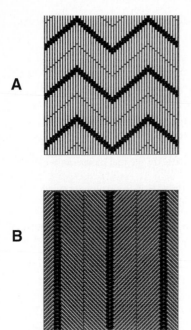

B

Figure 5. Packing model for poly(styrene-*n*-hexyl isocyanate) rod–coil diblock copolymers. (Reprinted from [41], with permission).

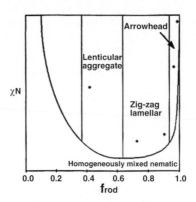

Figure 6. Preliminary phase diagram of poly(styrene-*n*-hexyl isocyanate) rod–coil diblock copolymers [44].

These materials represent the first observation of the SmC (zig-zag) and SmO (arrow head) structure in rod–coil diblock copolymers [41] in contrast to the homopolymer of poly(*n*-hexyl isocyanate) which only form a nematic mesophase (both lyotropic [65] and thermotropic [66]). This confirms the idea by Halperin [60, 69] that rod–coil systems are a microscopic model for smectic liquid crystals in general. Although the SHIC rod–coil system has a relatively broad polydispersity, a smectic mesophase over a size scale of as much as 10 μm has been observed (Fig. 4 B). This indicates that microphase separation plays a very important role in determining the self-assembly of the liquid crystalline process of these blocks. The existence of only a nematic phase in the rod homopolymer system is probably due to its broad polydispersity in contrast to the fact that a smectic meso-

phase was observed in the monodispersed ($M_w/M_n = 1.0$) PBLG homopolymer reported by Tirrell and coworkers [70]. From the TEM pictures, one can easily see that each rod domain is quite uniform but has a different size (Fig. 4 B, the arrow-head phase). On the other hand, due to the formation of lyotropic mesophase prior to microphase separation, the microdomain structure was greatly influenced which leads to novel morphologies. A tentative phase diagram for the poly(styrene-*n*-hexyl isocyanate) rod–coil system [44] is shown in Fig. 6.

4.2.4   Discussion of Rod–Coil Systems

Due to the conformation asymmetry in rod–coil diblock copolymer systems, the packing is expected to be totally different from conformationally symmetric coil–coil block copolymers. Semenov and Vasilenko [71] have predicted that a N–SmA transition can be either a first-order transition (in the case of large coil fraction) or a second-order transition (in the case of small coil fraction) and a SmC phase in a rod–coil system is also expected for $f_{rod} < 0.36$.

Later Halperin [60, 69] predicted that the lamellae are expected to exhibit a tilting, SmA – SmC first-order phase transition which is driven by the competition between the surface and deformation free energies.

It is this competition which determines the tilt angle in the smectic phase [60] for rod–coil diblock copolymers. Rod tilt increases the area per chain, and lowers the free energy of the system because flexible coil deformation is reduced thus gaining conformational entropy for the coil block. On the other hand, the interface area also increases which results in an increase in surface tension. When the deformation energy of the flexible coil is large enough, tilted rods representing a SmC phase are expected. Halperin's theory [60] offers a simple physical origin for the explanation of rod tilt in lamellae system. In the case of a rod–coil system with a large coil fraction, Williams and Fredrickson [72] predicted the existence of 'hockey puck' micelles, where the rods are packed axially into cylinders. For ABA coil–rod–coil systems, Raphael and de Gennes [73] predicted 'needles' and 'fence' morphologies. The fence morphology can only be stabilized by anisotropic bonding between adjacent rods (such as H-bonding).

All the above theories are based on geometric considerations. By applying self-consistent field theory, Müller and Schick [74] have predicted that the only thermodynamically stable morphologies for rod–coil systems are those with the coils on the convex side of the interface. Very recently Gurovich [75] developed a statistical theory which treats the microphase separation in LC block copolymer melts near the spinodal and predicts orientational and reorientational phase transitions driven by the configurational separation and four different phases.

Halperin's theory [60, 69] was essential for understanding the rod tilt in the SHIC rod–coil system (see Fig. 4). However, his prediction that a SmC phase is favored with high coil volume fraction is contrary to the experimental results found in the SHIC system. The SmA phase predicted by Semenov and Vasilenko [71] and the 'hockey-puck' phase predicted by Williams and Fredrickson [72] were not observed in the SHIC rod–coil systems. Furthermore, none of the observed morpologies were predicted by previous theory. Recent theory by Gurovich [75] shows some promising improvements. The LC nature of the A block is treated in his theory through an orientational interaction which is approximated by a self-consistent molecular field. The prediction that a homogeneously mixed nematic mesophase forms before microphase separation as solvent evaporates for $f_{rod} > 0.27$ is consistent with experimental findings (Fig. 6). The predicted phase diagram is in qualitative agreement with experiment, especially the prediction of smectic morphologies with the rods tilted relative to the IMDS (intermaterial dividing surface) over a wide composition range. The observed zigzag and arrowhead morphologies are both tilted smectic phases (SmC and SmO respectively) with different orientation of the rods (see Fig. 5).

The study of the SHIC rod–coil system offers an excellent model system for testing existing theories and opens routes to a new world of materials which combine aspects of liquid crystals, statistical physics, and solid-state physics.

4.3 Side Group Liquid Crystal – Coil Diblock Copolymer Systems

4.3.1 Synthesis and Characterization of the Side Group Liquid Crystal – Coil Systems

The first SGLC–coil system was reported in 1989 by Gronski and coworkers [8] utilizing polymer analogous chemistry (Scheme 7 A). Since then, SGLC–coil diblock copolymers have drawn great attention. Almost all kinds of living polymerization methods have been employed to synthesize well-defined SGLC–coil diblock copolymers such as group transfer polymerization (GTP), living anionic, living cationic, living ring opening metathesis polymerization (ROMP) and living radical polymerization. In this section we would like to briefly summarize the recent developments in this field (Schemes 8 and 9).

In 1986, Ringsdorf and coworkers [76] first successfully employed living group transfer polymerization (GTP) to polymerize a mesogenic methacrylate to afford LC methacrylate homopolymers and random copolymers. Block copolymers were not reported in this publication. Interestingly, the random copolymer containing 50 wt% (19 mol%) mesogenic monomer did not show a LC mesophase. In 1990, Hefft and Springer [77] reported the first AB block copolymer produced with GTP (Scheme 7 B) with a polydispersity of 1.07–1.09 and with a DP_{LC} of 16–22, although one of the monomers containing cyanophenyl benzoate failed with GTP. The failure was proposed [77] to be due to the back biting of the silyl ketene acetal functional group to the center ester group containing an electron withdrawing group (–CN).

Percec and Lee [78] employed living cationic polymerization (Scheme 7 C) in 1992 to create SGLC–coil diblock copolymers with a cyanobiphenyl mesogenic block and a non-LC fluorinated block. The degree of polymerization (DP) was very low (around 7–12) for each block. An interesting feature for their block copolymers was the persistence of birefringence at temperatures well above the isotropic transition temperature (T_i) of the LC block. This was explained by microphase separation above T_i and the big difference of refractive indices of these two blocks [78]. Omenat et al. [79] at Philips also employed living cationic polymerization to synthesize the first chiral SGLC–coil diblock copolymers with SmC* mesophases. In contrast to the bistable switching behavior of the homopolymer, their block copolymer showed only monostable switching behavior.

Living ring opening metathesis polymerization methods (ROMP) were first employed to synthesize LC–coil diblock copolymers by Komiya and Shrock [80] in 1993. The structure of their polymer system is shown in Scheme 7 D. Recent work from Grubbs group also used a novel ruthenium catalyst which can tolerate more functional groups [81] to synthesize well-defined LC–coil block copolymers [82]. The ROMP polymer backbone can be hydrogenated to create saturated structure to improve its stability.

In 1994, Finkelmann and coworkers [83] first reported LC–coil block copolymers with anionic polymerization of mesogenic methacrylates (see Scheme 7 E). However, monomer purification was a problem due to the crystalline nature of the mesogenic monomer. Finkelmann et al. used triethyl aluminum as an in-situ drying agent for successful anionic polymerization to obtain block copolymers with high molecular weights and narrow polydispersity

(1.08–1.3). (The effect of $AlEt_3$ on anionic polymerization has been studied by Müller et al. [84].) Watanabe et al. [85, 86] also reported using anionic polymerization of mesogenic methacrylates to produce SGLC–coil diblock copolymers with well-controlled molecular weight and narrow molecular weight distribution (≈1.05). However, their system also suffers from the difficulties of monomer purification. As a result, the M_n of the LC block can only reach ≈15 000 g/mol. Chiral SmC* SGLC–coil

diblock copolymers were reported by Hammond and coworkers [87] with anionic polymerization of chiral mesogenic methacrylates. Interestingly, the LC homopolymers produced from anionic polymerization and radical polymerization of mesogenic methacrylates have totally different LC properties. For example, Finkelmann et al. reported SmA and N mesophases for the anionically polymerized mesogenic methacrylate in contrast to a single N mesophase for the radically polymerized polymer sam-

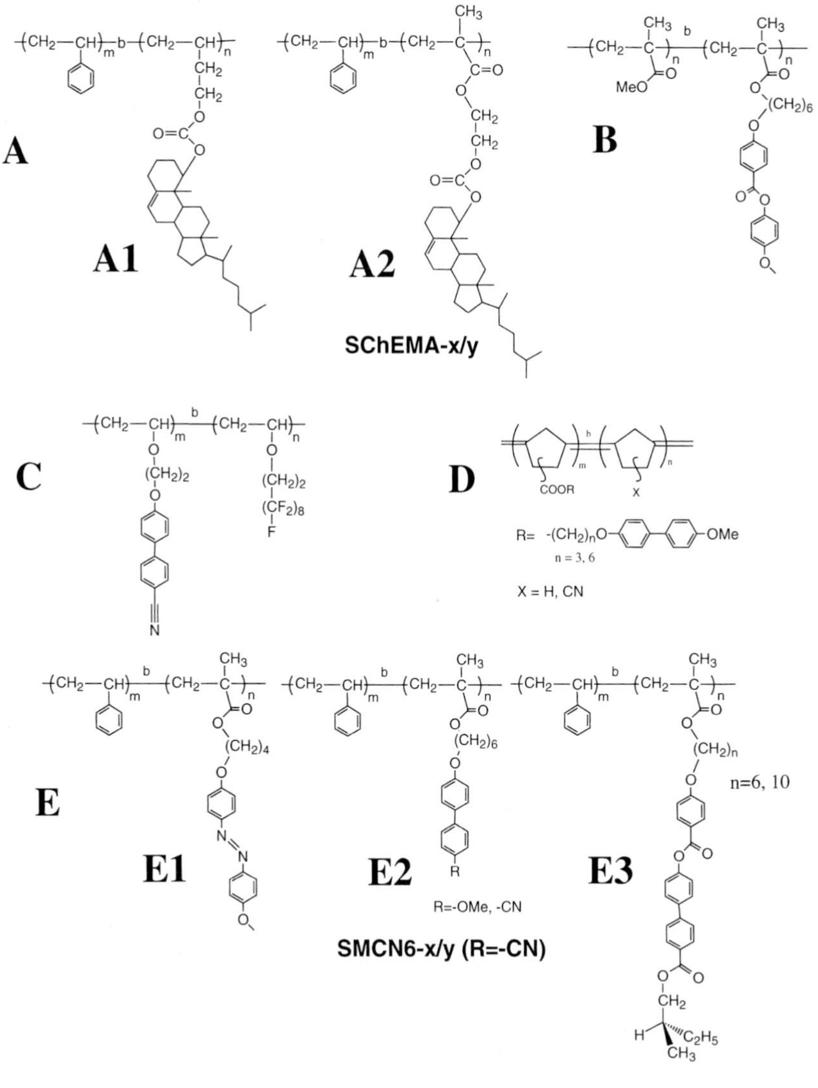

Scheme 7. Structures of well-defined SGLC–coil systems synthesized via: (A) polymer analogous reaction: A1 [8]; A2 [42]; (B) GTP [77]; (C) living cationic polymerization [78, 79]; (D) living ROMP [80]; (E) anionic polymerization [83, 85–87].

ples [83]. It is worth noting that polymers synthesized by GTP of the same mesogenic methacrylate [76] displayed LC properties similar to anionically produced LC polymers [83]. This major difference in LC properties obtained from different polymerization methods may be attributed to tacticity. Anionic polymerization of mesogenic monomer can produce LCPs with high syndiotactic (rr) diad content which favors a smectic mesophase compared with the atactic from radical polymerization. It is worth noting here that Watanabe et al. [85] first employed additives to promote the living anionic polymerization of mesogenic methacrylates (i.e. they added LiCl to their system to obtain narrow molecular weight distribution for the anionic polymerization of mesogenic methacrylate). We believe

that recent advances in anionic polymerization [88, 89] will have a strong impact on the synthesis of well-defined LC–BCPs via direct anionic polymerization.

Since Xerox researchers [90] reported the stable free radical polymerization (SFRP), controlled 'living' radical polymerization has drawn great attention [91]. Recently atom transfer radical polymerization (ATRP) was developed in several groups [92–96] as another type of 'controlled'/'living' radical polymerization method. Bignozzi and Ober [97] have employed this SFRP method to successfully synthesize liquid crystalline block copolymers. Their synthetic approach is shown in Scheme 8. The GPC traces shown in Fig. 7 clearly indicate the block copolymer formation occurs in a controlled/'living' fashion. We believe that controlled/'living' radical polymerization [both SFRP and recent work by Matyjaszewski's group [94] on atom transfer radical polymerization (ATRP)] will have great

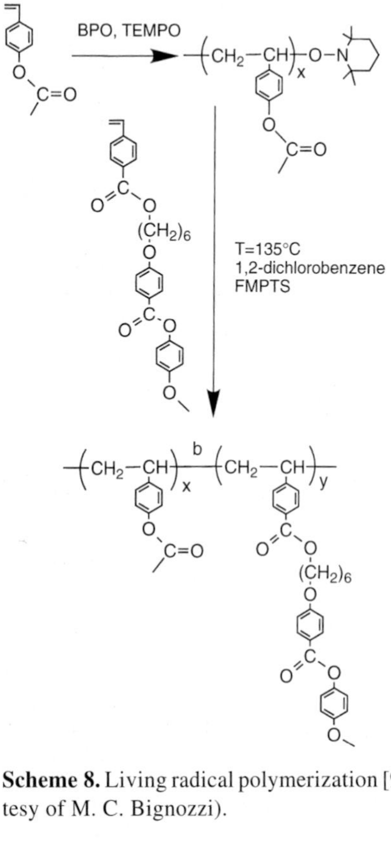

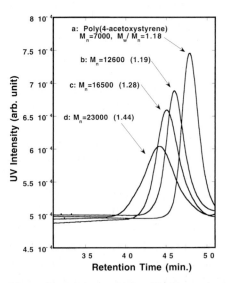

Figure 7. GPC traces of LC-coil BCPs synthesized by controlled/'living' radical polymerization (for structure, see Scheme 8): a: Coil precursor; b–d; LC–coil diblock copolymers with the same M_n of coil block. (Courtesy of M. C. Bignozzi).

Scheme 8. Living radical polymerization [97]. (Courtesy of M. C. Bignozzi).

Table 3. Comparison of different synthetic methods for liquid crystalline block copolymers.

Synthetic method for SGLC–coil BCP	Advantages	Disadvantages
Polymer analogous reaction	Wide range of MW, narrow MWD Wide range of functional group	Small defect in LC block
Living Anionic	Good control of MW, narrow MWD	Moderate MW, limited functional groups Difficult monomer purification
GTP, ROMP Living cationic	Good control of MW, narrow MWD	Moderate MW, limited functional groups Special monomer
Living radical	Good control of MW, rather narrow MWD, tolerates functional groups	Needs more study
Direct coupling (rod–coil)	Good control of MW, narrow MWD	Limited MW, to purify

impact on the synthesis of liquid crystalline block copolymers since radical polymerization is relatively simple and can tolerate water and many functional groups. In fact, we note that ATRP has been used in the synthesis of well controlled LC homopolymers by Pugh and coworkers [98]. Controlled/ 'living' radical polymerization will play an important role in meeting the challenge of synthesizing well-defined liquid crystalline block copolymers. A comparison of the methods of producing SGLC–coil BCPs is listed in Table 3.

Besides the work by Gronski and coworkers [8, 99–102], polymer-analogous reactions have also been employed by Fischer et al. [42, 103–107] (see Scheme 7 A2) and Ober et al. [43, 44, 62, 108–111] (Scheme 9). Systematic morphology studies were only carried out in these two SGLC–coil systems. Table 4 lists some of the LC–BCPs synthesized and characterized by Ober et al. Well-controlled SGLC–coil structures with a wide range of molecular weight and a narrow molecular weight distribution (MWD) have been reported. Typical GPC traces are shown in Fig. 8. In our system, purification of the final block copolymers was successfully conducted with Soxhlet extraction us-

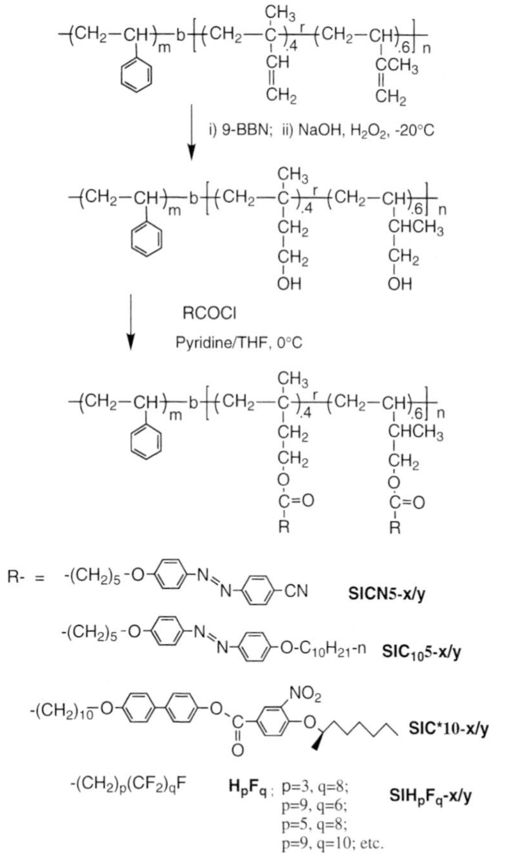

Scheme 9. Synthetic route adopted for producing well-defined LC–coil BCPs over a wide range of molecular weight with narrow molecular weight distribution [43, 44, 108–111].

Table 4. Summary of typical SGLC–coil diblock copolymers synthesized by Ober et al.

Sample[a]	Theoretical M_n (PS/LC)[b]	M_w/M_n	f_{LC}[c]	Thermal transitions (°C)[d]	Bulk Morphology	Ref.
SICN5-176/55	176K/55K	1.13	0.22	g45 g102 S_A 179 I	LC cylinder	[43]
SICN5-59/62	59K/62K	1.10	0.49	g45 g101 S_A 158 I	Lamella	[43]
SICN5-176/78	176K/78K	1.15	0.28	g45 g101 S_A 185 I	Bicontinuous	[43]
SIC$_{10}$5-13/56	13K/56K	1.09	0.79	g~50 g97 S_F110 S_C 170 I	Coil cylinder	[62]
SIC$_{10}$5-41/53	41K/53K	1.08	0.53	g~50 g102 S_F93 S_C 169 I	Coil cylinder	[62]
SIC*10-41/63	41K/63K	1.13	0.61	g55 g101 S_C^* 118 S_A^* 130 I	Coil cylinder	[109]
SIH$_5$F$_8$-41/64	41K/64K	1.05	0.53	S_B48 S_A67 I g101	Lamella	[111]
SIH$_{10}$F$_{10}$-41/85	41K/85K	–[e]	0.61	S_B97 g101 S_A 114 I	Coil cylinder	[110]

[a] SI**–x/y: S stands for styrene block, I** for LC block (see scheme 9 for structures), x for the real molecular weight of styrene block, and y for M_n of the LC block in kg/mol.
[b] Absolute molecular weight of SGLC–coil block copolymers. The M_n for PS block was measured by GPC calibrated with polystyrene standards. M_n for LC block was calculated from ^{1}H NMR.
[c] The volume fraction of LC block.
[d] Data from DSC second heating run.
[e] Not determined due to insolubility in tetrahydrofuran (THF).

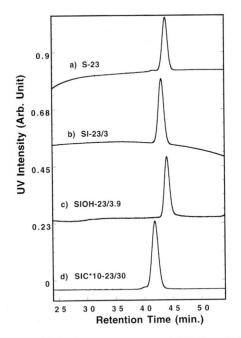

Figure 8. Typical GPC traces of SGLC–coil BCPs via polymer analogous reaction [62] (see Scheme 9 for structures): (a) polystyrene precursor; (b) styrene–isoprene diblock copolymer; (c) hydroxylated styrene–isoprene diblock copolymer (unimolecular micelle in tetrahydrofuran); (d) final FLC–coil diblock copolymer.

ing 95% ethanol to remove small molecule mesogens [43] as indicated by GPC and TLC (thin layer chromatography). However, in Fischer's system, HPLC was used to remove the small molecular mesogens [106]. This is due to a much higher T_g (>120 °C) for the LC block in their system while the T_g for our system is ≈50 °C. The low T_g of the LC block in our system offered chain mobility for the successful removal of any small molecular mesogenic groups via simple extraction.

4.3.2   Properties of Side Group Liquid Crystal–Coil Systems

4.3.2.1   Liquid Crystal Properties in Side Group Liquid Crystal–Coil Systems

In general, SGLC-coil diblock copolymers have mesophases similar to those of the parent LC homopolymer [80, 83, 85] in contrast to the fact that a 50 wt% random copolymer of mesogenic monomer with a non-mesogenic monomer does not show any LC mesophase [76]. This is due to the unique microphase separation of SGLC–coil systems. Another interesting feature is the clearing entropy loss of SGLC–coil block copolymers [8] compared to its LC homopolymer. The entropy loss was found to be highly dependent on the composition [43, 80, 83]. This observation was originally [8, 80, 83] believed to be due to the much smaller LC domain size and a rather disordered interfacial region between the LC and the coil components. Later work by Gronski et al. [101] using ^{2}H NMR showed no evidence of an isotropic region at the LC–coil interface. This unusual phenomenon needs further investigation.

4.3.2.2 Phase Diagram for Side Group Liquid Crystal–Coil Systems

It was Fischer and coworkers who first reported a phase diagram [42] (strictly speaking, a morphology diagram) for LC–coil diblock copolymers (Fig. 9). Similar phase diagrams were also found in another LC–coil system with a lower T_g coil block [106] and even LC triblock copolymer systems [107]. In their phase diagram, PS spheres, PS cylinder, and lamellar morphology were observed. The LC cylinder morphology, however, was not observed. Although LC

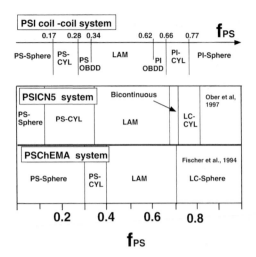

Figure 9. Phase diagram (strictly speaking, morphology diagram) of SGLC–coil diblock copolymer systems [42, 43, 62] and conventional coil–coil diblock copolymers [23].

spheres were observed, the LC mesophase in the sphere microdomain only showed a nematic phase rather than the smectic mesophase which appeared in other morphologies and the LC homopolymer. This interesting phenomenon was explained by arguing that a thermodynamically stable SmA layered structure cannot be formed within a cylindrical or a spherical microdomain.

However, from simple packing considerations of layered mesogens in a cylindrical microdomain, we can form two kinds of structures as indicated in Fig. 10. One is ho-

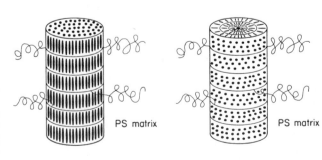

Figure 10. Packing consideration for a LC cylinder embedded on a hexagonal lattice. Mesogen ∥ cylinder IMDS, homogenous packing; mesogen ⊥ cylinder IMDS, homeotropic packing. (Reprinted from [43], with permission).

mogeneous packing in which all the mesogens are aligned parallel to the intermaterial dividing surface (IMDS) and in the other
they are aligned perpendicular to the IMDS
and have homeotropic packing. It can be
understood easily that homogeneous packing is a more energetically favorable state
than the homeotropic form due to the +1 disclination which would exist in every cylinder. Similar arguments were also proposed
by Walther and Finkelmann [13] in their review paper. Based on Fischer's results, they
pointed out that it would be interesting to see
whether a nematic mesophase can be observed in an LC cylinder morphology if not
a smectic mesophase. Thomas and coworkers employed symmetry arguments [44, 61]
for the efficient packing in SGLC–coil
systems to provide a rationale for predicting
morphologies. Their work suggested that not
only a smectic mesophase but also a columnar mesophase could exist in a cylinder microdomain because of a match of symmetry.

Ober and coworkers [43] have first observed unambiguously a LC cylinder morphology with SICN5-175/55 (see Scheme 9
for structure) in their block copolymers by
both TEM (Fig. 11 D) and SAXS (small angle X-ray scattering) [43]. WAXD (wide angle X-ray diffraction) clearly showed a
smectic mesophase [43]. This is the first observation of a LC cylinder morphology with
a SmA mesophase packed homogeneously
in the cylinder microdomain. Other morphologies such as coil cylinders, lamellae,
and a bicontinuous structure were also observed with TEM (Fig. 11) [43]. The packing models derived from SAXS were shown
in Fig. 12. We have observed that the structure of the famous leaning Tower of Pisa
(Fig. 13) is similar to that found in the LC
cylinder morphology with a smectic mesophase in the LC domain.

Interestingly, the clearing transition temperature for the LC cylinder (SICN5-

176/55, $T_i = 178\,°C$) was found to be 22 °C
higher than that of the LC lamellar structure
(SICN5-66/60, $T_i = 156\,°C$; SICN5-59/62,
$T_i = 155\,°C$) [43]. This is believed to be due
to the confining effect of the cylinder morphology which stabilizes the mesophase
within it, since the ODT of this system is
much higher than T_i. Similar confining effects were reported by Dadmum and Muthukumar in their small molecular nematic liquid crystal embedded in porous glasses
[112] in which an increase of the N–I transition temperature was observed and attributed to the surface induced ordering. Clearly, a smectic mesophase can exist within the
cylinder microdomain, therefore, it would
be interesting to explore the structure of a
nematic SGLC block packed within the cylinder microdomain to see whether it can be
stabilized or even turned into a smectic mesophase.

A bicontinuous morphology was also
observed in Ober's LC–coil systems with
SICN5-176/78 [43]. This is interesting
because it was previously suggested by
Thomas et al. [44] that this morphology
might not be able to form due to the periodic LC defect presented in the microphase
separated structure. Ober et al. also reported a preliminary phase diagram [43] based
on their LC–coil system as shown in Fig. 9.
Besides the two new morphologies observed in their system, another interesting
feature is that coil cylinder (PS) morphology was observed at much smaller volume
fraction of coil (PS) (see Fig. 9). It is believed that a cylinder morphology can offer
a larger area per chain for effective packing
of mesogenic groups.

Interestingly, the LC cylinder morphology was also observed by Watanabe et al.
[86] with SMCN6-17/20 (for structure, see
Scheme 7 E2) since the LC weight fraction of this sample is very close to 50% for
which a lamellar morphology was expect-

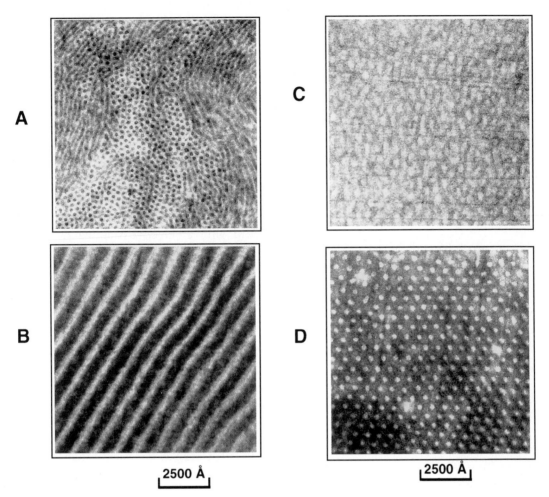

Figure 11. TEM of SGLC–coil system by Ober et al. (Reprinted from [43], with permission). A. SICN5-41/53 (Coil Cylinder); B. SICN5-59/26 (Lamellar); C. SICN5-176/78 (Bicontinuous); D. SICN5-176/55 (LC Cylinder).

ed. Other samples in the same series, such as SMCN6-5.6/5.5, SMCN6-11.7/9.6, and SMCN6-31/28 all showed a lamellar morphology with a *nematic* phase while the mesophase in the LC cylinder was *smectic*. The mesogen packing was also observed to be homogeneous as indicated by SAXS which means the mesogens are packed parallel to the IMDS (Fig. 10). Tacticity would not be responsible for the phenomenon because all the polymers were prepared under the same experimental conditions.

Another interesting observation by Fischer [113] is the existence of a tetragonal structure in the coil cylinder morphology in their SGLC–coil system (PS–ChEMA) rather than the hexagonal close packing in coil–coil cylinders. This is in contrast to our observation that hexagonal packing was observed for both coil cylinders and LC cylinders (Fig. 11 A, D) [43, 61].

So far, no reported novel morphology has been observed in any SGLC–coil diblock copolymer systems [8, 42–44, 83, 105, 106]

a)

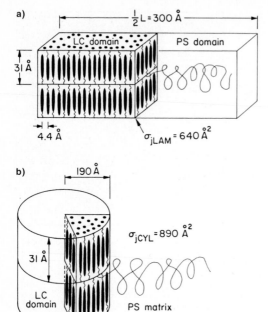

b)

Figure 12. Packing models for lamellar and LC cylinder morphology. (Reprinted in part from [43], with permission).

Figure 13. Structure of the famous leaning Tower of Pisa. Note that the packing is also identical to that found in LC cylinder morphology with a smectic mesophase in the LC domain. The tower represents the cylinder microdomain and the columns represent mesogens which are packed in a layer structure like the SmA phase.

in contrast to the rod–coil systems [40, 41, 44, 53, 61] in which several novel morphologies were discovered. This is mainly due

to the rather flexible nature of the main chain and the spacer in side group LC polymers. This flexibility offers the SGLC polymer chain mobility in many ways similar to conventional coil–coil BCPs. We feel that novel morphologies would most probably be observed in block copolymers with increased chain rigidity.

4.3.2.3   Interface Thickness

Fischer directly measured the SGLC–coil interface thickness ($t=2.1\pm1$ nm) using TEM with a selective staining technique [114]. This interface thickness value was also found to be independent of morphology and molecular weight which was similar to a coil–coil system. Quantitative estimation of interface thickness was given by Ober et al. using the SAXS technique. For SICN5 system, values of 1.7 nm were reported [43] which is quite similar to that of coil–coil BCPs (1.8–2.4 nm) reported by Hashimoto et al. [14]. A much sharper interface thickness of 1.1 nm was estimated for the SIH_5F_8–41/64 polymer (semifluorinated system, see Scheme 9 and Table 4) [111]. This is due to a much larger χ parameter for the semifluorinated block copolymer system since semifluorinated compounds are very immiscible with hydrocarbons.

4.3.2.4   Interplay of Liquid Crystallinity and Microphase Separation

In SGLC–coil systems, the LC mesophase must form within the block microdomain and adapt to the domain boundary conditions [44, 111]. One needs to realize that the ODT in this system is much higher than the isotropic transition temperature. This domain boundary condition can act to stabilize the orientation of the mesogen in one of the domains [111] as indicated in Fig. 14. In Fig. 14, WAXD patterns are shown for sample SIH_5F_8-41/64 (phase transition: SmB

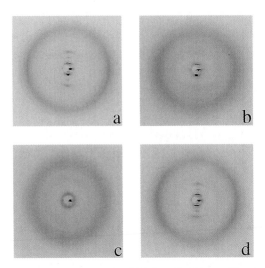

Figure 14. WAXD of a SGLC–coil with semifluorinated groups: SIH$_5$F$_8$-41/64. The sample was simply cast from dilute solution with trifluoromethyl-benzene over a period of one week coupled with drying for 2 days at room temperature and annealing at 140 °C for 4 days in a vacuum oven: (a) at 30 °C SmB phase; (b) at 60 °C, SmA phase; (c) at 80 °C, isotropic phase; (d) cooled to 30 °C from isotopic phase. (Courtesy of J. Wang).

48 °C SmA 67 °C I g 101 °C). As can be seen from Fig. 14, the smectic layer is oriented as a result of its well oriented lamellar structure. The mesophase was identified as SmB at room temperature [110, 111] due to the sharp arcs in the wide angle of the WAXD pattern (Fig. 14a). At 60 °C in the SmA phase, the LC block underwent a SmB–SmA transition so that the arcs at wide angle become very diffuse (Fig. 14b). At 80 °C the isotropic transition happened in which the sharp smectic layer diffraction disappeared (Fig. 14c). A small degree of mesogen orientation can still be seen even in the isotropic state. After cooling from the isotropic state, the orientation of the mesogen was fully recovered (Fig. 14d). This is believed to be due to the block domain boundary which stabilizes LC orientation [111].

If the block microstructure is disordered, it is impossible to generate a LC monodomain in the SGLC–coil system at equilibrium since the LC domain must adapt to the block microdomain boundary condition [43]. However, once the block microdomain has monodomain character (e. g. via single liquid crystal processing methods such as roll-casting [24, 115, 116], shearing [117–120], E-field [121–124], parallel plate [125, 126], etc.), the LC mesophase will also form a monodomain in the SGLC–coil system. This LC monodomain structure will be stabilized by the block microdomain structure since the ODT is higher than its T_i. This could be very useful for processing monodomain smectic structures for reducing light scattering. Called a 'stabilized monodomain structure', it has been taken advantage of for constructing FLC displays [109, 130] (see next section for details). It is worth pointing out that the defect structure of both the block microdomain and the LC subphase and the interaction between these two kinds of structures will be very interesting. We believe this area will draw great attention in the near future.

4.4   Applications of Liquid Crystal–Block Copolymers

It is our belief that block copolymers containing LC segments are materials with novel and unencountered properties which will offer great opportunities for developing high performance materials. Here we would like to give two examples. One example is a microphase stabilized ferroelectric liquid crystal (MSFLC) [109] for potential flat panel display applications, while the other is a material for stable, low surface energy [110] application.

When chiral mesogenic groups [127] are introduced into polymer systems, chiral mesophases such as the SmC* [128] or cholesteric mesophase (i. e. chiral nematic) [129] will form. For LC–BCPs containing chiral mesogenic groups, the question is what the effect microdomain structure will have on these chiral mesophases. One can imagine that in the LC–BCP system, microphase separated structures will form with one LC domain and a second amorphous domain. Since the pitch of the SmC* LCPs (in the range of 1–2 µm) is much larger than the domain size which is generally on the submicron scale (10–60 nm), the pitch will then be unwound by the microdomain structure. This feature would make the SmC* mesophase in the block copolymer different from that of the homopolymer in solid bulk state. This feature was realized by Ober and coworkers and used for the production of a microphase stabilized FLC (MSFLC) which exhibited bistable switching [109]. The cholesteric mesophase also has a characteristic pitch for selective light reflection if the wavelength (and polarization) of light matches the pitch of the twisted chiral nematic mesophase [129]. The pitch value is usually in the 400–10000 nm range which is much larger than the typical block microdomain size. Therefore, the cholesteric pitch could also be unwound by the block microdomain. It is then expected that cholesteric mesophases may not exist in SGLC–coil block copolymers because selective light reflection may not be observed unless the domain size is bigger than the pitch. However, it can still be called a chiral nematic mesophase since it is nematic and also optically chiral. It is worth noting that Hammond et al. [87] have reported that cholesteric mesophases were observed with POM (polarized optical microscopy) in their chiral SGLC–coil diblock copolymers. Based on the above argument, this iden-

tification has to be confirmed by light reflection studies. One also needs to realize that the unwinding power in the LC–BCP block system may depend on the composition.

4.4.1   Microphase Stabilized Ferroelectric Liquid Crystal Displays

Ferroelectric liquid crystals (FLC) are of great interest due to their fast electro-optical response which is about 1,000 times faster than conventional twisted nematic cells [131]. The geometry used is called a 'surface stabilized FLC' cell which utilizes a very thin gap (≈ 2 µm) to unwind the FLC supramolecular pitch ($\approx 1–2$ µm) since the bulk FLC materials do not show macroscopic polarization. This very thin gap, however, leads to difficulties in manufacturing large panels and very poor shock resistance. Researchers have proposed the concept of 'microphase stabilized FLC' [79, 109, 130] using FLC–coil diblock copolymers for electro-optical applications as shown in Fig. 15. This concept takes advantage of ferroelectric liquid crystallinity and block copolymer microphase separation since the block

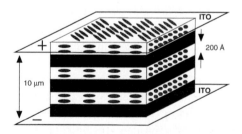

Figure 15. Concept of microphase stabilized ferroelectric liquid crystal (MSFLC). The black domain represents the coil block. Only lamellar microdomain morphology is shown in the figure. The FLC supramolecular pitch is unwound by the block microdomain [109, 130].

microdomain is in the submicron length scale which is much smaller than the FLC pitch. If the FLC can be oriented with a bookshelf structure, bistable switching should be possible.

However, the bistable switching was not reported in Omenat et al.'s FLC–coil block copolymers [79]. We speculate that it may be due to a relatively small LC segments and small domain sizes in their system. Ober et al. have prepared a FLC–coil diblock copolymer and demonstrated its bistable switching behavior using the concept of MSFLC [109, 130] although their system is far from practical application due to the slow response time (≈ 0.1 s) and high voltage ($V_{pp} = 700$ V/10 μm cell) needed for the bistable switching.

It can be seen clearly from Fig. 15 that MSFLC cells offer many advantages. For example, a large gap of 10 μm can be used [109, 130] instead of 2 μm used in SSFLC cells [131]. The polymeric nature of the LC materials offers excellent shock resistance and mechanical stability. No rubbing of the substrate is needed. We believe that practically useful electro-optical cells can be made utilizing this concept if the FLC–coil diblock copolymers are properly designed, synthesized, and processed.

Since the ODT of LC block copolymers is in general much higher than the LC isotropic transition temperature T_i (unless the total molecular weight is too small, i.e. a small χN) [43, 62], all the LC mesophase transitions lie below the block microdomain boundary conditions. These microdomain boundary conditions were shown [111] to stabilize the LC mesophase. The orientation of the mesogen within the block microdomain can be recovered when cooling from the isotropic state as already shown in Fig. 14. On the other hand, the texture of the LC block copolymer is very homogeneous and almost defect-free compared to that of the LC homopolymer which has many defects due to a polydomain structure for the LC subdomain. The orientation of block microdomains makes it possible for the mesogens to be oriented in the LC domain parallel to the IMDS. The stabilized smectic monodomain structure, we believe, will have great potential for flat panel display application.

4.4.2   Self-healing, Stable, Low Surface Energy Materials Based on Liquid Crystal–Block Copolymers

The idea of using block copolymers for low surface energy materials is to take advantage of surface segregation [132, 133]. A semifluorinated block segment offers a low surface energy material. By introducing a LC forming semifluorinated group (Scheme 9, SIH$_p$F$_q$), the liquid crystallinity enables the fluorinated groups to self-assemble into surface mesophases with almost all CF$_3$ groups at the air–polymer interface and it offers extra stability for the surface due to the formation of the SmB mesophase [110]. Extra energy is needed to destroy the LC phase in order to reconstruct the surface. Contact angle measurements showed an advancing contact angle (with water) as high as 123°, which is much higher than that of Teflon® ($\approx 110°$). The calculated surface tension is ≈ 8 mN m^{-1} for SmB mesophase using Zisman analysis [110]. In this process, the introduction of a flexible spacer on the LC side group is extremely important. It provides enough mobility for the polymer to self-organize into well-ordered systems with a hierarchy of ordered structures to produce stable, low energy surfaces. Block copolymers also offer processing advantages due to micelle formation since fluorinated materials are famous for poor solubility in organic solvents. These micelle forming abilities in semifluorinated block copolymers enable the fluorinated materials to be synthesized and processed easily in organic solvents. The polymer analogous reaction approach also offers opportunity for the rationale design of semifluorinated block copolymers to generate materials with various properties [110].

4.5   Future Work

The field of liquid crystalline block copolymers has undergone great development in the last 10 years showing increasing attraction for both polymer chemists and polymer physicists. More work needs to be carried out in order to fully explore this class of materials. Novel architectures such as LC–BCP with a dendritic LC block, rod–rod block with different rod diameters, and other novel structures will prove to be very interesting.

Acknowledgments

Funding by NSF grants DMR94-01845 and DMR92-01845 is gratefully acknowledged. We would like to thank Professor E. L. Thomas, Dr John T. Chen, Mary Jane O'Rourke, and Dr Jianguo Wang for their collaboration and enlightening discussions. The authors also thank Dr M. C. Bignozzi for providing unpublished data.

4.6   References

[1] A. Ciferri, *Prog. Polym. Sci.* **1995**, *20*, 1081–1120.

[2] C. K. Ober, G. Wegner, *Adv. Mater.* **1997**, *9*, 17–31.

[3] H. Ringsdorf, B. Schlarb, J. Venzmer, *Angew. Chem. Int. Ed. Engl.* **1988**, *27*, 113–158.

[4] R. A. Weiss, C. K. Ober (Eds), *Liquid-Crystalline Polymers, ACS Symposium Series 435.* ACS, Washington DC, **1996**.

[5] V. V. Tsukruk, J. H. Wendorff, *Trends Polym. Sci.* **1995**, *3*, 81–89.

[6] F. S. Bates, *Science* **1991**, *251*, 898–905.

[7] E. L. Thomas, R. L. Lescanec, *Phil. Trans. R. Soc. London, Ser. A* **1994**, *348*, 149–166.

[8] J. Adams, W. Gronski, *Makromol. Chem., Rapid Commun.* **1989**, *10*, 553–557.

[9] B. Perly, A. Douy, B. Gallot, *Makromol. Chem.* **1976**, *177*, 2569–2589.

[10] B. Gallot, *Prog. Polym. Sci.* **1996**, *21*, 1035–1088.

[11] E. Chiellini, G. Galli, A. S. Angeloni, M. Laus, *Trends Polym. Sci.* **1994**, *2*, 244–250.

[12] H. Fischer, S. Poster, *Acta Polymer.* **1996**, *47*, 413–428.

[13] M. Walther, H. Finkelmann, *Prog. Polym. Sci.* **1996**, *21*, 951–979.

[14] T. Hashimoto, M. Shibayama, H. Kawai, *Macromolecules* **1980**, *13*, 1237–1247.

[15] T. Hashimoto, N. Nakamura, M. Shibayama, A. Izumi, H. Kawai, *J. Macromol. Sci., Polym. Phys.* **1980**, *B17*, 389–406.

[16] E. L. Thomas, D. B. Alward, D. J. Kinning, D. C. Martin, D. L. Handlin, Jr., L. J. Fetters, *Macromolecules* **1986**, *19*, 2197–2202.

[17] H. Hasegawa, K. Tanaka, K. Yamasaki, T. Hashimoto, *Macromolecules* **1987**, *20*, 1651–1662.

[18] K. Almdal, K. A. Koppi, F. S. Bates, K. Mortensen, *Macromolecules* **1992**, *25*, 1743–1751.

[19] D. A. Hajduk, P. E. Harper, S. M. Gruner, C. C. Honeker, G. Kim, E. L. Thomas, L. J. Fetters, *Macromolecules* **1994**, *27*, 4063–4075.

[20] D. J. Meier, *J. Polym. Sci., Part C* **1969**, *26*, 81–98.

[21] E. Helfand, Z. R. Wasserman, *Macromolecules* **1978**, *11*, 961–966.

[22] L. Leibler, *Macromolecules* **1980**, *13*, 1602–1617.

[23] F. S. Bates, G. H. Fredrickson, *Ann. Rev. Phys. Chem.* **1990**, *41*, 525–557.

[24] R. J. Albalak, *Polymer* **1994**, 4115–4119.

[25] S. L. Kwolek, P. W. Morgan, J. R. Schaefgen, L. W. Gulrich, *Macromolecules* **1977**, *10*, 1390–1396.

[26] C. K. Ober, J. I. Jin, Q.-F. Zhou, R. W. Lenz, *Adv. Polym. Sci.* **1984**, *59*, 102–146.

[27] C. B. McArdle (Ed.), *Side Chain Liquid Crystal Polymers,* Chapman & Hall, New York **1989**.

[28] B. Reck, H. Ringsdorf, *Macromol. Chem. Rapid Commun.* **1985**, *6*, 291–299.

[29] Q.-F. Zhou, H.-M. Li, X.-D. Feng, *Macromolecules* **1987**, *20*, 233–234.

[30] S. R. Clingman, C. K. Ober, *PMSE (Polym. Mater. Sci. Eng.)* **1995**, *72*, 238–239.

[31] Y. H. Kim, *J. Am. Chem. Soc.* **1992**, *114*, 4947–4948.

[32] V. Percec, P. Chu, M. Kawasumi, *Macromolecules* **1994**, *27*, 4441–4453.

[33] S. A. Ponomarenko, E. A. Rebrov, A. Y. Bobrovsky, N. I. Boiko, A. M. Muzafarov, V. P. Shibaev, *Liq. Cryst.* **1996**, *21*, 1–12.

[34] D. Adam, D. Haarer, F. Closs, T. Frey, D. Fun-
hoff, K. Siemensmeyer, P. Schuhmacher, H.
Ringsdorf, *Ber. Bunsenges. Phys. Chem.* **1993**,
97, 1366–1370.

[35] H. Ringsdorf, R. Wüstefeld, E. Zerta, M. Ebert,
J. H. Wendorff, *Angew. Chem. Int. Ed. Engl.*
1989, *28*, 914–918.

[36] T. Kato, N. Hirota, A. Fujishima, J. M. J. Fre-
chet, *J. Polym. Sci., Part A. Polym. Chem.* **1996**,
34, 57–62.

[37] W. B. Gratzer, P. Doty, *J. Am. Chem. Soc.* **1963**,
85, 1193–1197.

[38] A. Gottschalk, H.-W. Schmidt, *Liq. Cryst.* **1989**,
5, 1619–1627.

[39] D. Pospiech, L. Häubler, H. Komber, D. Voigt,
D. Jehnichen, A. Janke, A. Baier, K. Eckstein, F.
Böhme, *J. Appl. Polym. Sci.* **1996**, *62*, 1819–
1833.

[40] J. T. Chen, E. L. Thomas, C. K. Ober, S. S.
Hwang, *Macromolecules* **1995**, *28*, 1688–1697.

[41] J. T. Chen, E. L. Thomas, C. K. Ober, G. Mao,
Science **1996**, *273*, 343–346.

[42] H. Fischer, S. Poser, M. Arnold, W. Frank, *Macromolecules* **1994**, *27*, 7133–7138.

[43] G. Mao, J. Wang, S. R. Clingman, C. K. Ober,
J. T. Chen, E. L. Thomas, *Macromolecules* **1997**,
30, 2556–2567.

[44] E. L. Thomas, J. T. Chen, M. J. E. O'Rourke,
C. K. Ober, G. Mao, *Macromol. Symp.* **1997**,
117, 241.

[45] K. I. Winey, S. S. Patel, R. G. Larson, H. Watanabe, *Macromolecules* **1993**, *26*, 2542–2549.

[46] J.-P. Billot, A. Douy, B. Gallot, *Makromol.
Chem.* **1977**, *178*, 1641–1650.

[47] A. Douy, B. Gallot, *Polymer* **1987**, *28*, 147–154.

[48] V. S. Ananthanayanan, G. Davenport, E. R. Stim-
son, H. A. Scheraga, *Macromolecules* **1973**, *6*,
559–563.

[49] T. J. Deming, *Adv. Mater.* **1997**, *9*, 299–311.

[50] T. J. Deming, *J. Am. Chem. Soc.* **1997**, *119*,
2759–2760.

[51] M. Gervais, A. Douy, R. Erre, B. Gallot, *Polymer* **1987**, *27*, 1513–1526.

[52] L. H. Radzilowski, J. L. Wu, S. I. Stupp, *Macromolecules* **1993**, *26*, 879–882.

[53] L. H. Radzilowski, S. I. Stupp, *Macromolecules*
1994, *27*, 7747–7753.

[54] M. Lee, N.-K. Oh, W.-C. Zin, *Chem. Commun.*
1996, 1787–1788.

[55] M. Lee, N.-K. Oh, *J. Mater. Chem.* **1996**, *6*,
1079–1086.

[56] M. Lee, N.-K. Oh, H.-K. Lee, W.-C. Zin, *Macromolecules* **1996**, *29*, 5567–5573.

[57] X. F. Zhong, B. Francois, *Makromol. Chem.
Rapid Commun.* **1988**, *9*, 411–416.

[58] G. Widawski, M. Rawiso, B. Francois, *Nature*
1994, *369*, 387–389.

[59] B. Gallot, A. Douy, H. H. Hassan, *Mol. Cryst.
Liq. Cryst.* **1987**, *153*, 347–356.

[60] A. Halperin, *Macromolecules* **1990**, *23*, 2724–
2731.

[61] J. T. Chen, *PhD Thesis* **1997**, MIT.

[62] G. Mao, *PhD Thesis* **1997**, Cornell University.

[63] M. N. Berger, *J. Macromol. Sci.-Rev. Macromol.
Chem.* **1973**, *c9*, 269–303.

[64] A. J. Bur, L. J. Fetters, *Chem. Rev.* **1976**, *76*,
727–746.

[65] S. M. Aharoni, *Macromolecules* **1979**, *12*, 94–
103.

[66] S. M. Aharoni, *J. Polym. Sci., Polym. Phys. Ed.*
1980, *18*, 1303–1310.

[67] S. M. Aharoni, *Macromolecules* **1981**, *14*, 222–
224.

[68] T. E. Pattern, B. M. Novak, *J. Amer. Chem. Soc.*
1991, *113*, 5065–5066.

[69] A. Halperin, *Europhys. Lett.* **1989**, *10*, 549–553.

[70] G. Zhang, M. J. Fournier, T. L. Mason, D. A.
Tirrell, *Macromolecules* **1992**, *25*, 3601–3063.

[71] A. N. Semenov, S. V. Vasilenko, *Sov. Phys.
JETP* **1986**, *63*, 70–79.

[72] D. R. M. Williams, G. H. Fredrickson, *Macromolecules* **1992**, *25*, 3561–3568.

[73] E. Raphael, P. G. de Gennes, *Makromol. Chem.,
Macromol. Symp.* **1992**, *62*, 1–17.

[74] M. Müller, M. Schick, *Macromolecules* **1996**,
29, 8900–8903.

[75] E. Gurovich, *Macromolecules*, in press.

[76] W. Kreuder, O. W. Webster, H. Ringsdorf, *Makromol. Chem., Rapid Commun.* **1986**, *7*, 5–13.

[77] M. Hefft, J. Springer, *Makromol. Chem., Rapid
Commun.* **1990**, *11*, 397–401.

[78] V. Percec, M. Lee, *J. Macromol. Sci.-Pure Appl.
Chem.* **1992**, *A29*, 723–740.

[79] A. Omenat, R. A. M. Hikmet, J. Lub, P. van der
Sluis, *Macromolecules* **1996**, *29*, 6730–6736.

[80] Z. Komiya, R. R. Schrock, *Macromolecules*
1993, *26*, 1387–1392.

[81] R. H. Grubbs, *J. Macromol. Sci., Pure and Applied Chemistry* **1994**, 1829–1833.

[82] R. H. Grubbs, Lectures at Cornell.

[83] R. Bohnert, H. Finkelmann, *Macromol. Chem.
Phys.* **1994**, *195*, 689–700.

[84] H. Schlaad, A. H. E. Müller, *Macromol. Rapid
Commun.* **1995**, *16*, 399–406.

[85] M. Yamada, T. Iguchi, A. Hirao, S. Nakahama,
J. Watanabe, *Macromolecules* **1995**, *28*, 50–58.

[86] M. Yamada, T. Iguchi, A. Hirao, S. Nakahama,
J. Watanabe, *PMSE (Polym. Mater. Sci. Eng.)*
1997, *76*, 306–307.

[87] W. Y. Zheng, P. T. Hammond, *Macromol. Rapid
Commun.* **1996**, *17*, 813–824.

[88] J.-S. Wang, H. Zhang, R. Jerome, P. Teyssie,
Macromolecules **1995**, *28*, 1758–1764.

[89] S. Antoun, J.-S. Wang, R. Jerome, P. Teyssie,
Polymer **1996**, *37*, 5755–5759.

[90] M. K. Georges, R. P. N. Veregin, P. M. Kaz-
maier, G. K. Hamer, *Macromolecules* **1993**, *26*,
2987–2988.

[91] C. J. Hawker, *J. Am. Chem. Soc.* **1994**, *116*, 1185–1186.

[92] M. Kato, M. Kamigaito, M. Sawamoto, T. Higashimura, *Macromolecules* **1995**, *28*, 1721–1723.

[93] J.-S. Wang, K. Matyjaszewski, *J. Am. Chem. Soc.* **1995**, *117*, 5614–5615.

[94] K. Matyjaszewski, *Polym. Prep.* **1997**, *38*, 683–711.

[95] V. Percec, B. Barboiu, A. Neumann, J. C. Ronda, M. Zhao, *Macromolecules* **1996**, *29*, 3665–3668.

[96] V. Percec, B. Barboiu, *Macromolecules* **1995**, *28*, 7970–7972.

[97] M. C. Bignozzi, C. K. Ober, **1997**.

[98] A. Kasko, A. M. Heintz, Y. Wang, C. Pugh, *Polym. Prep.* **1997**, *38*, 675–676.

[99] J. Adams, W. Gronski in *Liquid-Crystalline Polymers ACS Symposium Series 435* (Eds.: R. A. Weiss, C. K. Ober) American Chemical Society, Washington DC **1990**, p. 174–184.

[100] J. Adams, J. Sänger, C. Tefehne, W. Gronski, *Macromol. Rapid. Commun.* **1994**, *15*, 879–886.

[101] A. Martin, C. Tefehne, W. Gronski, *Macromol. Rapid Commun.* **1996**, *17*, 305–311.

[102] J. Sänger, W. Gronski, *Macromol. Rapid Commun.* **1997**, *18*, 59–64.

[103] B. Zaschke, W. Frank, H. Fischer, K. Schmutzler, M. Arnold, *Polym. Bull. (Berlin)*, **1991**, *27*, 1–8.

[104] M. Arnold, S. Poser, H. Fischer, W. Frank, H. Utschick, *Macromol. Rapid Commun.* **1994**, *15*, 487–496.

[105] H. Fischer, S. Poser, M. Arnold, *Liq. Cryst.* **1995**, 503–509.

[106] H. Fischer, S. Poser, M. Arnold, *Macromolecules* **1995**, *28*, 6957–6962.

[107] S. Poser, H. Fischer, M. Arnold, *J. Polym. Sci., Part A: Polym. Chem.* **1996**, *34*, 1733–1740.

[108] G. Mao, S. R. Clingman, C. K. Ober, T. E. Long, *Polym. Prepr. (Am. Chem. Soc., Div. Polym. Chem.)*, **1993**, *34*, 710–711.

[109] G. Mao, J. Wang, C. K. Ober, M. J. O'Rourke, E. L. Thomas, M. Brehmer, R. Zentel, *Polym. Prep.* **1997**, *38*, 374–375.

[110] J. Wang, G. Mao, C. K. Ober, E. J. Kramer, *Macromolecules* **1997**, *30*, 1906–1914.

[111] C. K. Ober, J. Wang, G. Mao, E. J. Kramer, J. T. Chen, E. L. Thomas, *Macromol. Symp.* **1997**, *117*, 141.

[112] M. D. Dadmum, M. Muthukumar, *J. Chem. Phys.* **1993**, *98*, 4850–4852.

[113] H. Fischer, *Polymer* **1994**, *35*, 3786–3788.

[114] H. Fischer, *Macromol. Rapid Commun.* **1994**, *15*, 949–953.

[115] R. J. Albalak, E. L. Thomas, *J. Polym. Sci., Part B, Polym. Phys.* **1993**, *31*, 37–46.

[116] R. J. Albalak, E. L. Thomas, *J. Polym. Sci., Part B, Polym. Phys.* **1994**, *32*, 341–350.

[117] G. Hadziioannou, A. Mathis, A. Skoulios, *Colloid Polym. Sci.* **1979**, *257*, 136–139.

[118] F. A. Morrison, H. H. Winter, *Macromolecules* **1989**, *22*, 3533–3540.

[119] K. A. Koppi, M. Tirrell, F. S. Bates, K. Almdal, R. H. Colby, *Journal De Physique II* **1992**, *2*, 1941–1959.

[120] R. M. Kannan, J. A. Kornfield, *Macromolecules* **1994**, *27*, 1177–1186.

[121] K. Amundson, E. Helfand, D. D. Davis, X. Quan, S. S. Patel, S. D. Smith, *Macromolecules* **1991**, *24*, 6546–6548.

[122] K. Amundson, E. Helfand, X. Quan, S. D. Smith, *Macromolecules* **1993**, *26*, 2698–2703.

[123] K. Amundson, E. Helfand, X. N. Quan, S. D. Hudson, S. D. Smith, *Macromolecules* **1994**, *27*, 6559–6570.

[124] E. Gurovich, *Phys. Rev. Lett.* **1995**, *74*, 482–485.

[125] Y. M. Zhang, U. Wiesner, *J. Chem. Phys.* **1995**, *103*, 4784–4793.

[126] Y. M. Zhang, U. Wiesner, H. W. Spiess, *Macromolecules* **1995**, *28*, 778–781.

[127] J. W. Goodby, *J. Mat. Chem.* **1991**, *1*, 307–318.

[128] M. V. Kozlovsky, L. A. Beresnev, *Phase Transition* **1992**, *40*, 129–169.

[129] Ya. S. Freidzon, E. G. Tropsha, V. P. Shibaev, N. A. Plate, *Macromol. Chem., Rapid Commun.* **1985**, *6*, 625–629.

[130] M. Brehmer, R. Zentel, G. Mao, C. K. Ober, *Macromol. Symp.* **1997**, *117*, 175.

[131] N. A. Clark, S. T. Lagerwall, *Appl. Phys. Lett.* **1980**, *36*, 899–901.

[132] S. S. Hwang, C. K. Ober, S. Perutz, D. R. Iyengar, L. A. Schneggenburger, E. J. Kramer, *Polymer* **1995**, *36*, 1321–1325.

[133] D. R. Iyengar, S. M. Perutz, C.-A. Dai, C. K. Ober, E. J. Kramer, *Macromolecules* **1996**, *29*, 1229–1234.

Chapter II
Defects and Textures in Nematic Main-Chain Liquid Crystalline Polymers

Claudine Noël

1 Introduction

Liquid crystalline phases show a variety of beautiful optical patterns when observed with a polarizing microscope, each characteristic of defects (local imperfections that break the order of molecular arrangement) that are peculiar to the state of molecular order prevailing in that mesophase. The first analysis of defects in low molecular weight liquid crystals (LMWLCs) was carried out very early by Lehmann [1], Friedel [2], and Grandjean [3]. The accepted classification and terminology of liquid crystals (LCs) is based almost entirely on Friedel's observations. His pioneering work introduced and defined the terms used today to describe the LC state: nematic (N), cholesteric (N*) and smectic (Sm). His conclusions were supported by citations of work in other fields (e.g., X-ray diffraction, behavior in electric and magnetic fields). Although Friedel noted the possibility of smectic polymorphism, he was sceptical of its actual occurrence. Much work on the characterization and classification of smectic phases has been done in Halle by Demus, Sackmann and associates [4, 5]. The work consisted essentially of the attempt to determine experimentally the number of different smectic phases by miscibility studies and to characterize them by their textures. Two relevant books with photographic illustrations have been published dealing deeply with this subject [6,

7]. Defects and textures in LMWLCs are also presented in certain general articles and reviews [2, 3, 8–13]. Although physicists have learned much about defects and textures from optical studies, from solutions to elastic equations, and from simply drawing pictures, only since 1976 have researchers [14, 15] used the topology to introduce a general scheme for classifying defects into condensed-matter physics. Much progress was made during the 1980s in elucidating the relationships between defects and structure due to the work of Kléman and others [14, 16–18] who elaborated a complete theory relating defects of any dimensionality (point, lines, wall-like defects ...) in ordered systems to the generic singularities of the order parameter, via the topology of an adequate order parameter space. Of late, the properties of defects and textures in LMWLCs have been discussed on the basis of their material-dependent properties including their energetical stability and the nature of the order in the central region or core of the defect and also from the point of view of topology [19–23].

As pointed out by Kléman in early 1985 [24], liquid crystalline polymers (LCPs) and LMWLCs are equivalent from the point of view of topology since they exhibit the same symmetries and the same type of order parameters. Defects are indeed characteristic breaks in local symmetry of the order parameter and can be classified according to their dimensionality and the specific symmetry they break. This implies that defects of a given type have equal topological stability in LCPs and LMWLCs. However, the relative occurrence of defects of different types and the arrangement of defects into various textures depend on energetic considerations. Since these are directly related to the molecular structure, LCPs are not energetically equivalent to LMWLCs.

Despite the fact that main-chain liquid crystalline polymers (MCLCPs) have been studied for two decades, relatively little information is available on the influence of the structure on the properties of defects and very few conclusions of a general nature have been reached, except that the defects observed pertain to the topological classes expected [25]. In this regard, it is important to note that microscopic observations are sometimes misleading owing to the difficulty with which MCLCPs give specific textures in the liquid crystalline state. This might be due to their multiphase nature (existence of polycrystalline and amorphous material), polydispersity and the high viscosities of the liquid crystalline melts. In most cases, samples must be annealed for hours or days at a suitable temperature if equilibrium textures reminiscent of those of LMWLCs are to be seen [26–28]. Also, the high melting points of MCLCPs have limited progress on their study. These temperatures are often above the operating range of the sample heaters, and in any case, the extended heating of polymer samples at such temperatures can cause degradation.

The effect of structural disclinations upon texture and the origins of these in local deformations of or discontinuities in the arrangement of the molecules have been studied only in the case of nematic MCLCPs. The primary aim of this chapter is therefore to present to the reader defects and textures in nematic MCLCPs. Some of consequences of constructing a nematic phase from long polymeric chains will be presented. We will concentrate on understanding the role of properties such as scarcity of chain ends, chain rigidity, high degree of anisotropy in the elasticity and viscosity that clearly distinguish MCLCPs from LMWLC materials. We will consider neutral, rigid or semiflexible polymers. In electing to discuss the geometry and topology of defects in nematic MCLCPs, this chapter does differ somewhat from others [29–32] dealing with textures in both side chain (SC) and MCLCPs, which were intended as practical and useful experimental guides to the textures and classification of nematic cholesteric and smectic LCs of different polymorphic types.

2   Textural Assignments

2.1   The Basic Equations

In ideal nematics, the molecules are aligned along one common direction which is referred to as the director and represented by a dimensionless unit vector n. The system is uniaxial and the states (n) and $(-n)$ are undistinguishable since there is no polarity. In most practical circumstances, however, this ideal configuration will not be compatible with the constraints which are imposed by the limiting surfaces of the sample and/or by external fields acting on the molecules. There will be some deformation of the alignment and the orientation of the director will change continuously and in a systematic manner from point to point in the sample. The distorted state may then be described by a variable director $n(r)$. Three types of distortions are present simultaneously in nematics. As illustrated in Fig. 1, one can distinguish three situations which are pure splay, pure twist and pure bend. The corre-

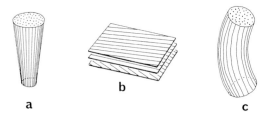

Figure 1. Distortions in nematics. (a) Pure splay, (b) pure twist and (c) pure bend [19].

sponding elastic constants (often referred to as Frank constants) are K_{11}, K_{22} and K_{33}, respectively. The free-energy per unit volume of a deformed specimen relative to the undeformed one is given by [33, 34].

$$F = 1/2\,[K_{11}(\nabla \cdot \mathbf{n})^2 + K_{22}(\mathbf{n} \cdot \nabla \times \mathbf{n})^2$$
$$+ K_{33}(\mathbf{n} \times \nabla \times \mathbf{n})^2] \qquad (1)$$

This is the fundamental formula of the continuum theory for nematics. The total energy of the system is:

$$E = \int F\,d\mathbf{r} \qquad (2)$$

Minimization of the total energy yields the conditions for equilibrium in the bulk:

$$\left(\frac{\partial F}{\partial n_{i,j}}\right)_{,j} - \frac{\partial F}{\partial n_i} = 0 \qquad (3)$$

where we use the Cartesian tensor notation, repeated tensor indices being subject to the usual summation convention, and the comma denotes partial differentiation with respect to spatial coordinates.

Distortions and defects can be interpreted in terms of the continuum theory through equations derived from the expressions of the elastic energy and the imposed boundary conditions. Solutions are known in certain simple situations. Oseen [35] has found configurations, named 'disinclinations' by Frank [33], or 'disclinations' today, which are solutions of this problem for planar samples in which the director $\mathbf{n}$ is confined to

the xy plane parallel to the glass surfaces, and is not a function of z, the normal to the film. Taking the components of the director to be $n_x = \cos\phi$, $n_y = \sin\phi$, $n_z = 0$, and making the simplifying assumption that the medium is elastically isotropic (i.e. $K_{11} = K_{22} = K_{33}$), Eqs. (1) and (3) reduce to:

$$F = \frac{1}{2}\,K(\nabla\phi)^2 \qquad (4)$$

$$\nabla^2\phi = 0 \qquad (5)$$

The solutions of Eq. (5) are $\phi = S\theta + \theta_0$ where $\theta = \tan^{-1}(y/x)$ and θ_0 is a constant $(0 < \theta_0 < \pi)$. This equation describes the director configuration around the disclination, the singular line is along the z-axis and the director orientation changes by $2\pi S$ on going round the line. In the nematic phase, the orientational order is apolar and hence S must be an integer or a half integer. Fig. 2, taken from Frank's article, shows the molecular orientation in the neighborhood of a disclination for a few values of S. The curves represent the projection of the director field in the xy plane. S is called the strength of the disclination.

The elastic energy (per unit length) associated with an isolated disclination is:

$$E = \pi K S^2 \ln(R/r_c) + E_c \qquad (6)$$

where R is a typical dimension of the specimen or the distance to other disclinations and r_c the radius of the inner region or 'core' (of the order of molecular dimensions). E_c is the energy of the core which is hard to calculate with any accuracy. Lines of strength $\pm 1/2$ are expected to be common since the energy scale as S^2. However, in LMWLCs the most frequent disclinations have $S = \pm 1$. The reason for this is that defects with integral values of S can 'escape into the third dimension' [36, 37], that is the director can simply rotate to lie along the line in the central region, obviating the need for a core.

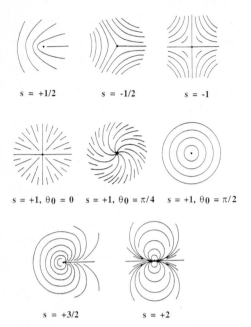

Figure 2. Director field in the neighborhood of the core of a disclination line of strength S: Wedge disclinations (after Frank [33]).

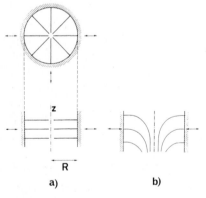

Figure 3. (a) Wedge disclination of strength +1 in a capillary: the arrangement near the center has a discontinuity and must involve a core. (b) 'Escape' of the disclination in the third dimension: the arrangement is continuous with no core [34].

Fig. 3b shows the director 'escape' at the center of a disclination of strength $S=\pm 1$ in a thin capillary of radius R. The arrangement is continuous with no singular line. The deformation involves splay and bend, but no

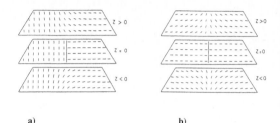

Figure 4. Twist disclinations: the director patterns for (a) $S=1/2$, $\theta_0=0$ and (b) $S=1$, $\theta_0=0$.

twist. Other escaped configurations exist [22]. The energy (per unit length) is:

$$E = \begin{cases} 3\pi K & \text{for } S=+1 \\ \pi K & \text{for } S=-1 \end{cases} \tag{7}$$

On the other hand, for the simplest arrangement where the director is everywhere radial and the deformation is pure splay, the energy is easily calculated to be:

$$E = \pi K \ln(R/r_c) + E_c \tag{8}$$

Then, the escaped configuration is more favorable energetically than the line as soon as R is large enough.

In addition to the above-discussed disclinations, which are referred to as wedge disclinations, there are twist disclinations. The director is always parallel to the xy plane, but the axis of rotation (z-axis) is normal to the singular line (y-axis). Figure 4 shows the director patterns for (a) $S=1/2$, $\theta_0=0$ and (b) $S=1$, $\theta_0=0$. ϕ is a linear function of the angle $\theta=\tan^{-1}(z/x)$.

$$\phi = S \tan^{-1}(z/x) + \theta_0 \tag{9}$$

where S is the strength of the disclination and θ_0 a constant. The deformation involves simultaneously splay, twist and bend. If the medium is elastically isotropic ($K_{11}=K_{22}=K_{33}=K$), then the expressions derived for the wedge disclinations are applicable to the present situation. It should be noted, however, that the energy (per unit length) for the escaped configuration is now $2\pi K|S|$.

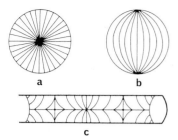

Figure 5. Singular points in droplets: (a) spherically symmetric radial configuration with the director normal to the surface and (b) bipolar structure with the director tangential to the surface; (c) singular points in a capillary [22].

Points defects are another class of defects in nematic LCs. They occur in droplets and can be also seen in thin capillaries (Fig. 5). The total energy stored in the elastic field around a point defect grows linearly with the radius, R, of the volume enclosed:

$$E \approx KR \qquad (10)$$

where K is a suitable Frank constant coefficient. Point of singularities of equal and opposite strengths attract one another and are annihilated.

For LMWLCs, it has long been recognized that degenerate anchoring at the constraining surfaces may allow formation of domains of different molecular orientation, these being separated by diffuse walls or by disclinations. Energy considerations suggest that walls, rather than disclinations, may be observed if the sample thickness, h, obeys the inequality:

$$h < 2K/W_s \qquad (11)$$

where K is the Frank elastic constant and W_s the anchoring energy. For LMWLCs this is hard to achieve because $W_s \approx 10^{-3}$ Jm^{-2} leading to the need for film thickness of less than 10 nm. However, for MCLCPs, although measurements of W_s have not been made, it seems that walls may be somewhat

easier to achieve. If rubbing of the constraining surfaces has been used to induce a homogeneous or planar alignment (i.e., one in which the molecules lie parallel with the interface) two types of wall may be identified. In both cases a reorientation of the molecules by 180° is observed on crossing the wall. These two types of walls are termed inversion walls of the first and second kind [4] depending whether the reorientation is accomplished by rotation about an axis normal to the sample plane so that the molecules remain in that plane at all points, or about an axis within the sample plane, so that the molecules themselves rotate out of the plane and pass through the homeotropic orientation (i.e., normal to the plane) at the center of the wall.

2.2 Some Consequences of Elastic Anisotropy

We have so far assumed $K_{11}=K_{22}=K_{33}=K$. Real nematics are of course, elastically anisotropic. From the values of the energies of the disclinations calculated for the case where $K_{11}=K_{33}\neq K_{22}$ it follows that the wedge disclinations are more stable than the twist disclinations if $K_{22}>K=K_{11}=K_{33}$, and vice versa. Anisimov and Dzyaloshinskii [38] have shown that lines of half-integral strength may be stable against three-dimensional perturbations if the twist elastic constant $K_{22}\neq 1/2(K_{11}+K_{33})$. More precisely,

(1) $S=\pm 1/2$ wedge disclinations are energetically favored when $K_{22}>1/2(K_{11}+K_{33})$ and are stable against any out-of-plane distortion.
(2) $S=\pm 1/2$ twist disclinations are favored when $K_{22}<1/2(K_{11}+K_{33})$, but are unstable against an out-of-plane distortion.

The effects of elastic anisotropy ($K_{11} \neq K_{22} \neq K_{33}$) have been investigated by Dzyaloshinskii [39]. He found that, qualitatively, the molecular orientation close to the line is the same as in the elastically isotropic case, except for $S=1$. The only $S=+1$ wedge lines that are stable versus three-dimensional perturbations have a purely circular ($\phi = \theta + \pi/2$) or radial ($\phi = \theta$) configuration. Cladis and Kléman [36] have calculated the energy of the nonsingular line for the bend–twist and splay–bend cases. Whether the stable geometry is circular or radial with or without a singular core will depend on the nature and the size of the core and on the degree of anisotropy of the elastic constants. It is worth noting, however, that if singular integral lines are favored, the energy $E \approx S^2$ predicts these should dissociate into two lines of half-order strength.

In LMWLCs, where the differences between K_{11}, K_{22} and K_{33} are usually small, nonsingular integral lines are the most commonly observed defects. By contrast, as was first pointed out by Kléman [25], one can infer from the arguments above that in MCLCPs, where the elastic anisotropy is large, the half-integral lines will occur more frequently than the integral ones and that these integral lines will be singular if there exist core arrangements of moderate energy. De Gennes [40] has proposed a model for the splay modulus for semiflexible polymers, based on intermolecular strain free energy, which predicts $K_{11} \approx L^2$ (where L is the contour length of the chains). K_{33} is expected to be normal and K_{22} independent of L. Meyer [41] has estimated the magnitude of K_{11} by an argument based on chain end entropy and has predicted that K_{11} should increase essentially with L. K_{33} strongly reflects the chain rigidity and is limited by the persistence length. K_{22} should still remain small and independent of L. Although some

disagreement appears between the qualitative description of de Gennes's and Meyer's models, it is clear that splay deformation becomes difficult in a nematic composed of long molecules. Only few experimental studies have been carried out on thermotropic MCLCPs [42–45]. Measurements of the elastic constants with conventional methods, particularly with the Freedericksz distortion technique, have been faced with experimental difficulties. This has been due mainly to the lack of strong surface anchoring, large viscosity and thermal degradation of polymers. Zheng-Min and Kléman [42] have evaluated the elastic constants of the polyester:

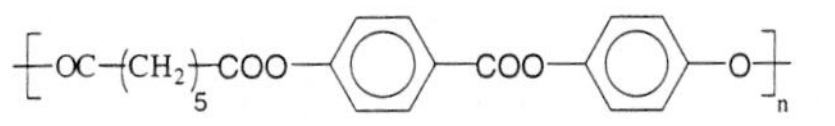

1

from the Freedericksz' transition. The splay elastic constant ($K_{11} \approx 3 \times 10^{-6}$ dyn) was found to be one order of magnitude larger than in LMW nematics while the twist elastic constant ($K_{22} \approx 3 \times 10^{-7}$ dyn) and the bend elastic constant (which was more difficult to measure because of chemical degradation) exhibited more conventional values. Volino et al. [43, 44] have used a simple NMR technique to evaluate the ratio K_{33}/K_{11} for the polyesters:

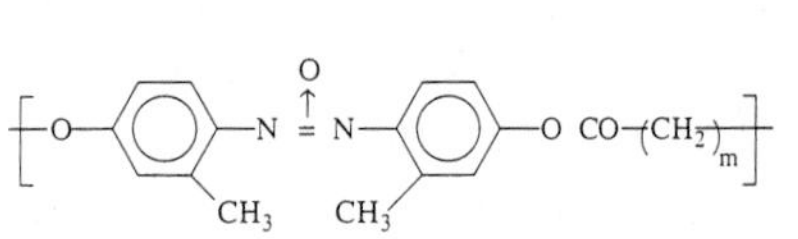

2

where $m=7$, 10. They found $K_{33}/K_{11} \approx 0.3$–0.5, a value close to that of ≈ 0.1 reported by Zheng-Min and Kléman. It should be noted, however, that the three elastic constants of the copolyesteramide Vectra B 950® is about 100 times higher than corresponding values obtained for the above thermotropic MCLCPs [45]. To what extent these large values are related to the chemi-

cal structure of this polymer or to its biaxiality are unclear.

Windle et al. [46, 47] have developed a
model that simulates the evolution of microstructure for different ratios of the elastic
constants and thus for both LMWLCs and
MCLCPs. The results obtained on two dimensional lattices are in good agreement
with those predicted by the continuum theory both when the elastic constants are equal
(LMWLCs) and when they are different
(MCLCPs). For example, the trajectories of
the directors around the $S=-1/2$ disclinations are observed to be insensitive to the
magnitudes of the elastic constants while
those around the $S=+1/2$ disclinations undergo a marked change. In 3-dimensional
models simulated using equal elastic constants, most disclinations are of the type
$S=\pm1$ and have point character. The reason
for this is proposed to lie in the process of
escape into the third dimension. By contrast,
a predominance of $S=\pm1/2$ line disclinations is found when 3-dimensional simulations are performed with free boundary conditions and with the splay energy set much
higher than that of bend or twist. This is due
to the fact that escape in the third dimension
is no longer favored. However, a number of
problems have been encountered with this
approach. In particular, the model needs to
account correctly for both splay–splay compensation and the differentiation between
splay and bend distortions in three dimensions. Quite recently, new developments of
the central algorithm of this model has enabled the splay and bend components to be
successfully partitioned in three dimensions, and splay–splay compensated structures to be handled properly [48]. Modeling
using the new algorithm for the case where
the splay constant is dominant leads to the
generation of structures in which twist
escaped +1 line singularities are a predominant feature. 3-Dimensional simulations

were also performed with planar boundary
conditions at the top and bottom surfaces
and with splay energy one hundred times
that of bend and twist [47]. The structures
were found to form layered morphology
with little matching from layer to layer and
a sinuous trajectory of the directors within
the layers. This undulating structure is very
reminiscent of the banded [49] and tight [50,
51] textures observed in thermotropic copolyesters.

2.3   Topology of Director Fields: Homotopy Groups and Classification of Defects

2.3.1   Uniaxial Nematics

The term 'nematic' is derived from the
Greek word 'nema', a thread, and refers to
the presence of a large number of apparent
threads in relatively thick samples. Stirring
a nematic leads to a pronounced increase in
their number. When the motion stops, many
of them disappear and other progressively
stabilize. In uniaxial nematics, the threads
are of two types: either thin or thick, clearly distinguishable, both directly observable
in light microscopy (Fig. 6).

The thin threads are line defects of half
integral strength $S=\pm1/2, \pm3/2 \dots$ about
which the director rotates by an angle of
$2\pi S$ and which are topologically stable.
They have the same topology as a Möbius
strip. The central region, or core, of the disclination appears as a fine thread running
through the material as it scatters light
strongly. The threads never show any free
extremity in the interior of the liquid crystal. Topological considerations show that
they have to form closed loops or end at the
boundary surfaces or any interface (i.e.,

Figure 6. Thin and thick threads in a typical low molecular weight liquid crystal: threaded texture of *p*-methoxybenzylidene-*p*-*n*-butylaniline (MBBA).

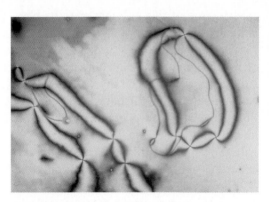

Figure 7. Schlieren texture of terephthalidene-bis-(4-*n*-butylaniline) (TBBA). Integral nuclei (four brushes) and half-integral nucleus (two brushes).

between isotropic liquid and mesophase). In some cases, they are nearly perpendicular to the boundary surfaces, their ends being attached to the glass slide and coverslip, respectively. Often, however, they lie more or less horizontally, their ends being attached to the glass and the other parts moving and floating freely around in the nematic. The threads have elastic properties and may be extended and curved due to flow of the liquid. Some threads may also be stuck to the glass surfaces.

As already discussed, the lines of integral strength $S=\pm 1, \pm 2, \ldots$, about which the director rotates by an angle $2\pi S$ are not topologically stable. The director can escape into the third dimension obviating the need for a core. In this case, the integral line shows up a blurred contrast under the polarizing microscope and thus appears thick.

A half-integral line of any strength can be continuously deformed into the half-integral line of opposite strength as illustrated in [19]. So it is with integral lines.

The threaded texture changes to the Schlieren texture when the lines are all perpendicular to the sample boundaries. This texture observed between crossed polarizers displays dark brushes (also called black

stripes or Schlieren) which have irregular curved shapes and correspond to extinction positions of the nematic (Fig. 7). Accordingly, the director lies either parallel or normal to the polarizer or analyzer planes, respectively. These brushes meet at point singularities on the surface of the preparation or rather, point singularities are the origins of the brushes. For the configurations $S=\pm 1$, the director field forms $+2\pi$ or -2π disclinations and one observes point singularities with four dark brushes ('Maltese crosses'), due to the two orientations of extinction. The configurations $S=\pm 1/2$ lead to singularities with only two brushes, corresponding to $+\pi$ or $-\pi$ disclinations. Neighboring singularities which are connected by brushes have opposite signs and may attract each other. They can vanish, if their S values add up to zero.

It should be noted that twist lines are visible in threaded textures while wedge lines are observed in Schlieren textures. The continuous transformation of a wedge S line into a wedge $-S$ line involves the passage through an intermediate twist $|S|$ line [19].

Point defects are another class of defects in nematics. When a line, such as the wedge disclination of Fig. 3a, with $S=1$, escapes

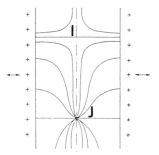

Figure 8. Singular points in a capillary with normal boundary conditions.

in the third dimension, it may do it in two ways: either upwards or downwards. As shown in Fig. 8, an upwards portion and a downwards portion will be linked by a singular point (J).

The classification of defects in nematics represents a straightforward example of the applications of homotopic group theory [14, 15]. The reader is referred to reviews of the subject [19–21, 23, 52]. This topological approach confirms the absence of walls, the existence of Möbius lines, the mutual annihilation of thin threads and the existence of singular points. More importantly, it shows that defects combine and merge according to the rules of multiplication of the two-element Abelian group Z_2.

2.3.2 Biaxial Nematics

The applicability of homotopic theory becomes much less obvious for liquid crystal phases with more complicated order parameters such as biaxial nematics and cholesterics, which are both locally defined by three directors forming a tripod. This gives rise to a description of the line singularities in terms of the quaternion group, Q. This is particularly interesting because the quaternion group Q is non-Abelian, a property that

has important consequences when one considers combining defects or entanglements of defects [52]. In the biaxial nematic, there are three classes of line defects involving π rotations about each of the three inequivalent symmetry axes and a class of line defects of equivalent 2π rotations. Because there is an orthogonal set of axes, one cannot rule out a strength 1 or 2π singular line as one could in the uniaxial nematic. However, the continuous deformations of the director that are equivalent to 'escape' into the third dimension make all 2π lines equivalent. Lines of integral strength 2 or 4π are not topologically stable and must appear as nonsingular in biaxial nematics. It should be noted that two defects of opposite signs can be continuously transformed into one another if they belong to the same class. Similarly, a π defect and the $-\pi$ defect that annihilates it must be of the same class. Since the mobility of line defects can play an important role in determining the macroscopic properties of a medium, it is important to determine whether two given line defects can or cannot cross each other. In media with Abelian fundamental groups all line defects can cross each other without the formation of a connecting line. In the non-Abelian case of the biaxial nematic, an examination of the multiplication table for the quaternion group reveals that two mobile lines can always cross without the production of a connecting line except for the case of two π disclinations of distinct types, which are necessarily joined after crossing by a 2π disclination. Finally, it is worth noting that singular points are not topologically stable in biaxial nematics. All the peculiar features of the defects in biaxial nematics are of importance since the characteristics of a few thermotropic MCLCPs have been related to biaxiality [45, 51, 53–56].

3   Defects in Nematic Main-Chain Liquid Crystalline Polymers

As a rule, thin lines of strength ±1/2 or singular lines of strength ±1 are seen in the threaded textures of nematic thermotropic MCPs [25, 57, 58]. Rare cases of thick lines have been observed. As we have already pointed out, the reason for this is that the elastic anisotropy is large in these systems. The disclinations with $S=\pm 1/2$ were also reported to be the most abundant in the Schlieren textures of nematic copolyesters [59–65].

The first experimental studies of the defects in the nematic phase of thermotropic MCPs were reported by Kléman and coworkers [57, 58]. They examined in detail the threaded texture of polyester **1** for two specific values of the degree of polymerization. The molecular weight was found to affect the threaded texture of the polyester with $n=5$. The thin lines ($S=\pm 1/2$) and thick lines ($S=\pm 1$) observed in samples with lower molecular weight ($\overline{M_w} \approx 1000$) are the familiar defects seen in ordinary LMW nematics. It should be noted that in these samples the threads have a strong tendency to disappear with time and give place to a well-defined texture of Friedel's nuclei with only integral lines and singular points. Integral lines or nuclei are occasionally found in samples of higher molecular weights ($\overline{M_w} > 10\,000$), but half integral lines are the most abundant. This is in agreement with the fact that both the splay elastic constant, K_{11}, and the bend elastic constant, K_{33}, are larger than the twist elastic constant, K_{22} [42]. Under these conditions, any splitting of the core of an $|S|=1$ line tending to remove the core singularity is not favored and since the line energy is proportional to $|S|^2$

when a core singularity is present, $|S|=1/2$ lines are favored. In planar samples, the half integral lines appear as thin loops floating in the bulk. These loops, which have a twist character, become smaller and smaller with time and to a great extent disappear after a few minutes. In samples with strong planar anchoring, surface lines with their ends attached to the same glass plate are observed. Very thin free droplets show disclination lines, the length of which increases with time, thus suggesting that they terminate on the free surface rather than on the glass surface.

In LMWLCs, as one approaches the center of disclination, there is an isotropic core region. Its radius is estimated to be a few molecular lengths far from the isotropization temperature, T_{N-I}, but increases rapidly at T_{N-I}. In MCLCPs, the cores appear quite large. Mazelet and Kléman [57] reported important differences between the cores of the wedge parts of the half integral lines observed in free droplets of polyester **1**. The +1/2 cores are much larger than the −1/2 cores. Positive and negative defects also display different mobilities. While the +1/2 defects are practically insensitive to any gentle flow given in the sample, the −1/2 ones move immediately. Mazelet and Kléman interpreted their observations for $S=+1/2$ by a high concentration of chain ends inside the core, which scatters light. Taking into account that for polyester **1**, $K_{11} \gg K_{33} > K_{22}$, they assumed that there is no splay involved outside the core. For $S=-1/2$, they developed a model which minimizes elastic energy by concentrating it in some angular sectors outside the core. The energy balance between splay and bend, which are the only contributions, leads to a subtle geometry where most of the bend energy is concentrated in a manner that avoids too much splay energy. In central part of the core, the molecular configuration is verti-

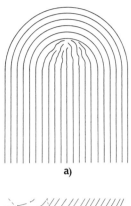

Figure 9. Schematic models for the cores of half integral lines in polyester **1**. Chain ends segregate (a) in the core ($S=+1/2$) or (b) along three quasi walls ($S=-1/2$) [57].

cal. Schematic representations of $S=+1/2$ and $S=-1/2$ wedge disclinations are shown in Fig. 9.

Integral lines have been observed in the threaded texture [25] of a random copolyester prepared by the Eastman Kodak Company from 40 mol% poly(ethylene terephthalate) and 60 mol% p-acetoxybenzoic acid and commercially known as X-7G:

$$\left[\!\!-O-CH_2-CH_2-O-OC-\!\!\left\langle\bigcirc\right\rangle\!\!-CO-\!\!\right]_{0.4}\left[\!\!-O-\!\!\left\langle\bigcirc\right\rangle\!\!-CO-\!\!\right]_{0.6}$$

3

They show a blurred contrast in linearly polarized light, but remain weakly contrasted in unpolarized light which may be taken as an indication of their singular character. Integral lines have also been observed in the Schlieren textures of a number of copoly-

esters [59, 60, 64, 66]. The core of the nuclei is much larger and the contrast is much fuzzier than in the Schlieren textures of ordinary LMW nematics. However, no detailed experimental observations of the disclination structure have been made and it is therefore difficult to ascertain whether these lines are singular or nonsingular.

As a rule, only $S=\pm1/2$ and $S=\pm1$ are seen in nematics. Rare cases of higher strength defects ($|S|$ up to 4) have been observed in lyotropic systems [67] as also in two-component or impure thermotropic systems [68]. In the former cases, the defects were unstable as the microemulsion broke up after several minutes and the high strength singularities divided into conventional disclinations of $S=\pm1$. In the latter cases, however, the texture was found to be stable. More recently, transient disclinations of strength $\pm3/2$ were observed in the nematic phase of a single component, thermotropic, rigid rod polytolane [69]. They were characterized by six extinction brushes. Their formation was promoted by a combination of mechanical agitation and smaller undercooling below the clearing point. They had a lifetime of ≈30 s. Their small number and short lifetime were discussed in terms of their energy per unit line length, and in terms of the force that causes them to be attracted to and annihilated at other disclinations.

For thermotropic MCPs, stable Schlieren texture with high strength defects ($|S|=3/2$ and $|S|=2$) was first reported by Galli et al. [70] for the poly(β-thioester):

$$\left[\!\!-CH_2-CH_2-CO-O-\!\!\left\langle\bigcirc\right\rangle\!\!-N\overset{O}{=}N-\!\!\left\langle\bigcirc\right\rangle\!\!-O-CO-CH_2-CH_2-S-(CH_2)_5-S-\!\!\right]_n$$

4

maintained at a temperature close to the $I-N$ transition. These defects were characterized by a single thin core. However, sometimes they possessed a blurred inner region from which the dark brushes were seen to emerge.

Under these circumstances, it was difficult to distinguish between true high strength singularities and 'apparent' singularities consisting of several points bunched together around an immiscible isotropic droplet.

Quite recently, high strength defects were observed in the nematic polymers of the following structure:

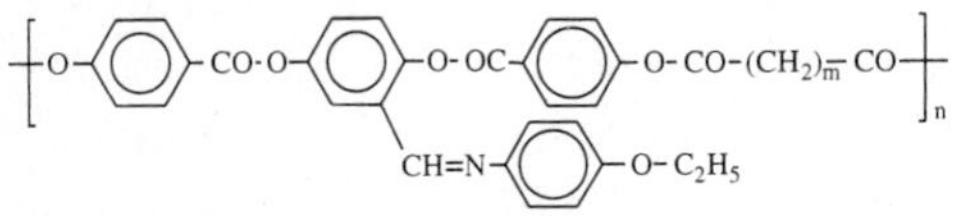

(polymer **5A** m=6, polymer **5B** m=8)

5

For the polymer **5A**, disclinations of strength $S=\pm 3/2$, $+2$ and $+5/2$ were observed [71]. Most of these high strength disclinations were connected with neighboring defects of lower strengths of $S=\pm 1/2$ and $S=\pm 1$. Rare cases of pairs of $+3/2$ and $-3/2$ and of $+3/2$ and -2 disclinations were noted. These disclinations were found to be quite stable. They persisted even after annealing for several hours. For the polymer **5B**, pairs of $+2$ and $-3/2$ and of $+1$ and $-3/2$ disclinations were observed [72]. An examination of the textures showed that the core of these high strength disclinations was wider than that of the ordinary ones with $S=\pm 1/2$ and $S=\pm 1$. However, further studies are required for a better understanding of the core size and topology.

4 Biaxial Nematic Main-Chain Liquid Crystalline Polymers

In 1970 the existence of a biaxial nematic phase for thermotropic LCs was theoretically predicted by Freiser [73]: the first-order transformation from the isotropic state to the uniaxial nematic state may be followed at lower temperature by a second-order transformation to a biaxial nematic state. At this phase transformation, the rotation of the molecules around their long axes becomes hindered and the mesogens achieve planar alignment.

As far as thermotropic polymers are concerned, the presence of a biaxial nematic phase was first suggested by Viney and Windle [59] for copolyester **3**. Between 340 and 350°C, the beginning of the growth of the isotropic phase, this polymer shows a continuous Schlieren texture with both $\pm 1/2$ and ± 1 disclinations, the former being relatively more numerous than the latter. Such a behavior is consistent with a uniaxial nematic phase. Below 340 °C, however, the melt exhibits interrupted Schlieren textures with disclinations of strength ± 1, but absence of any of strength $\pm 1/2$, and occasionally large areas of homogeneous alignment separated by well defined walls [59, 66]. As pointed out by Viney and Windle [59] these features, which are similar to those seen in LMWLCs with SmC structure, may in fact indicate the presence of a biaxial nematic phase.

Windle and coworkers [51, 54, 56] have then interpreted some discrepancies between optical microstructures and X-ray diffraction patterns of nematic domains in copolyesters **3** and **6** in terms of biaxiality.

Thin microtomed sections of these random copolyesters were

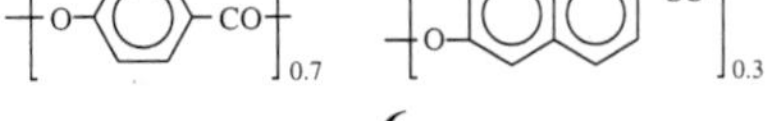

6

examined by X-ray scattering and shown to have a high level of preferred molecular orientation. However, when the same thin slices were viewed in the polarizing microscope, no suggestion of macroscopic orientation was seen. Long range rotational correlations of the molecules about their long axes, giving othorhombic or lower symmetry and thus biaxial optical properties, would account for these observations. Another indication of biaxiality was provided by examination of thin samples of copolyester **7** prepared directly from the melt. The microstructure of such samples showed domains clearly delineated by walls [51, 55].

7

Even if there was some indication that copolyesters **3**, **6** and **7** could present biaxiality, no clear evidence concerning their defects signature was given. Recently, however, De'Nève, Kléman and Navard [45, 53] reported what we believe to be the first unambiguous observations of biaxial defects in a conventional thermotropic polymer commercially known as Vectra B 950® (**8**). These authors showed the existence of three types of half integer disclination lines (called E_x, E_y and E_z), sometimes associated with integer disclination lines. Figure 10 shows a structural model of the chain (assuming that the aromatic planes are all parallel) and Fig. 11 illustrates the three possible half integer twist disclinations in the biaxial nematic. Interestingly, if Vectra B 950® is biaxial at rest, an external mag-

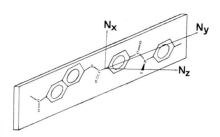

Figure 10. Schematic representation of the chain (Vectra B 950®). N_x, N_y and N_z are the three principal refractive indices [53].

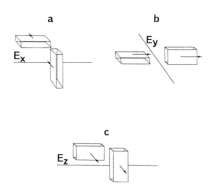

Figure 11. The three possible half integer disclination lines in the biaxial nematic phase of Vectra B 950®. (a) Rotation around the X axis, (b) rotation about the Y axis and (c) rotation around the Z axis [53]. It is worth noting that the E_y line is perpendicular to the main director Y.

netic field or a weak shear flow disrupts the biaxiality and transforms the liquid crystal into a uniaxial nematic, thus indicating that the biaxiality is weak.

8

5 Defect Associations and Textures at Rest

Textures correspond to various arrangements of defects. When the isotropic liquid is cooled, the nematic phase may appear at the deisotropization point in the form of separate small, round objects called droplets (Fig. 12). These can show extinction crosses, spiral structures, bipolar arrangements, or some other topology depending on boundary conditions. Theoretical studies based on a simple model confirm the stability of radial or bipolar orientation (Fig. 5) [22]. Considerations based on improved theoretical models yield stable twisted

(a)

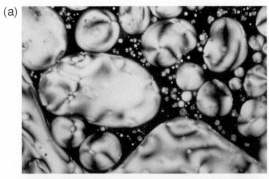

(b)

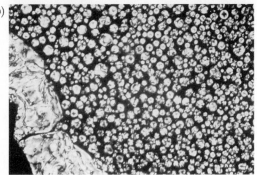

Figure 12. The separation of the nematic phase in the form of droplets from the isotropic liquid of

(a) polyester:

(and (b) polyester:)

and (b) polyester:

Figure 13. Marbled texture of copolyester:

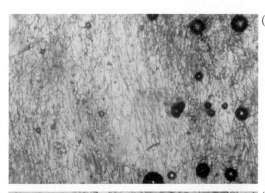

(

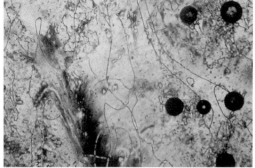

(

Figure 14. Threaded texture of copolyester:

obtained at 255 °C. (a) $t=10$ min and (b) $t=30$ min.

structures. Nematic droplets characterize a type-texture of the nematic phase since they occur nowhere else. Upon cooling, the droplets grow and join together to form larger structures from which stable texture finally forms.

The marbled, threaded and Schlieren textures are the ones most often cited in the literature for thermotropic MCLCPs. Examples of typical polymeric textures are shown in Figs. 13–15.

Nematic marbled texture consists of several areas with different molecular orientation. On observing the preparation between

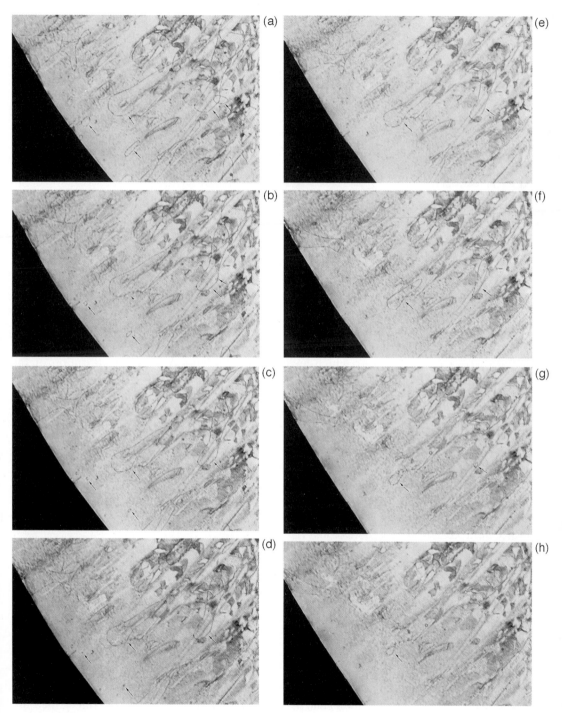

Figure 17. Closed loops viewed between crossed po-larizers at 250°C: (a) $t=0$, (b) $t=2$ min, (c) $t=5$ min, (d) $t=6$ min, (e) $t=8$ min, (f) $t=15$ min, (g) $t=22$ min and (h) $t=30$ min (from [75]).

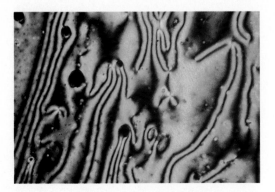

Figure 15. Schlieren texture of copolyester:

Figure 16. Closed loop viewed between crossed polarizers at 250°C. Copolyester **11** (from [75]).

crossed polarizers, one can note that the interference color is nearly constant within the individual areas, indicating quasihomogeneous regions. Marbled textures very often appear with glass surfaces which have not been specially treated.

The threaded (or line) texture is observed in preparations where threads lie more or less horizontally, their ends being attached to the glass surfaces and the other parts floating freely in the liquid crystal. The threads may be extended and curved due to flow of the liquid. They can be closed in the form of loops (Fig. 16) and, in this case, generally shrink and disappear more or less rapidly (Fig. 17, s. page 107) [74, 75].

The Schlieren texture also called nuclei texture ('plage à noyaux') appears when the threads lie vertically (Figs. 18 and 19). When viewed between crossed polarizers this texture shows an irregular network of black brushes branching out from a number of scattered points or 'nuclei' and passing continuously from one nucleus to another. Four brushes meet at integral nuclei and two brushes at half integral nuclei – see Sec. 2.3.1 of this chapter. From the observation of $S=\pm1/2$ lines, the mesophase can be identified as a nematic phase since these singu-

larities occur nowhere else. Smectic C phases, which can exhibit the Schlieren texture, only form singularities with four brushes corresponding to $S=\pm1$.

Transmission electron microscopy (TEM) studies conducted by Thomas et al. [62, 63] have permitted direct visualization of the molecular director distribution in flow-oriented thin films prepared from semiflexible thermotropic LCPs of the following structure:

polymer **9A** $X=$H, polymer **9B** $X=$CH$_3$)

9

These authors have used the technique of 'lamellar decoration' [76] which enables detailed assessment of characteristic mesophase defects and texture on a much finer scale than previously possible with conventional electron-microscopy preparations. The defects and texture existing in the polymer melt state are first retained by thermal quenching of the polymer fluid to room temperature. The glassy LCP film is then annealed above its glass transition, but below the melting point. Crystalline lamellae grow perpendicular to the local chain axis and effectively 'decorate' the molecular director field throughout the sample. The most common types of disclinations seen in these polyesters are those of strength $S=\pm1/2$. Dipoles consisting of a positive and a negative π disclination at a typical separation of micrometers are frequently observed. Apparently, annihilation occurs if pairs of opposite sign disclinations approach each other more closely. These observations are consistent with those reported by Ford et al. [77] and Dong and Deng [78] who have used the same technique to reveal disclinations of strength $S=\pm1/2$ in a MCLCP containing flexible spacers between the mesogenic groups, and in an aromatic copolyester, respectively. Interestingly, in addition to isolated and variously paired disclination dipoles, arrays involving equal numbers of positive and negative disclinations are observed [76]. The molecular director pattern in regions defined by two pairs of oppositely signed half integer disclinations bears a likeness to that of the 'effective domain' structure hypothesized by Marrucci [79] to account for certain rheological aspects of MCLCPs. More recently, Chen et al. [64] have used the combined techniques of 'lamellar decoration' and ruthenium tetraoxide staining to reveal the director trajectories around disclinations of strength $S=\pm1/2$ and ±1 in polymer **9B**. All six topologies represented in Fig. 2 as well as $(1/2, -1/2)$ and $(1, -1)$ defect pairs have been observed, consistent with the fact that disclinations of opposite signs attract. More surprisingly, pairs of two positive $(+1, \theta_0=\pi/4; +1/2)$ or two negative $(-1/2, -1/2)$ disclinations have been seen repeatedly. These defect pairs correspond presumably to a nonequilibrium state and to an unstable texture.

Quite recently, Chen et al. [71b, 80] have developed the technique of banded texture decoration. These authors showed that in thick samples, when the disclination density is small enough due to prior annealing,

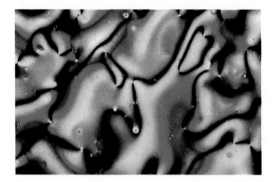

Figure 18. Schlieren texture of copolyester formed by transesterification of poly(ethylene-1,2-diphen-oxyethane-p,p'-dicarboxylate) with p-acetoxybenzoic acid. Recognizable singularities $S=\pm1/2$ (crossed polarizers, 250 °C, ×200) (from [60]).

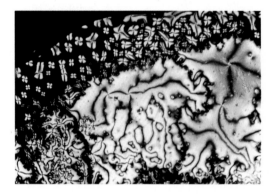

Figure 19. Schlieren texture of the polyester:

at the $I-N$ transition. Recognizable singularities $S=\pm1$ (crossed polarizers, 278 °C, ×200).

quenching from the nematic state can lead to the appearance of a banded texture similar to that seen in flowing or deformed systems after removal of the external orienting influence. The banded texture formation is not crystallization-induced as suggested by Hoff et al. [81], since the nematic polymer of the following structure

10

Figure 20. Grainy texture of polyester **6**.

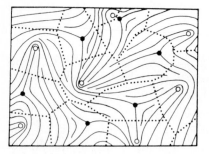

Figure 21. Schematic representation of the polydomain structure produced by spatial distribution of disclination lines of $S=+1/2$ (open circles) and $S=-1/2$ (filled circles) [65].

showed a banded texture in the quenched amorphous glassy state. The bands lie perpendicular to the chain direction, thus offering the possibility of mapping the director orientations [82]. As compared to the 'lamellar decoration', the banded texture technique provides an extremely useful mechanism for direct visualization of defects and textures in both crystalline and non-crystalline materials by polarizing optical microscopy. The detailed observations by Chen et al. [71b, 82] point to a large predominance of disclinations of strength $S=\pm1/2$, some disclinations of strength $S=+1$ and a few disclinations of strength $S=-1$. They also reveal (+1/2, –1/2) and (+1, –1/2) defect pairs and the formation of inversion walls.

In addition to these well characterized textures, an ill defined grainy texture appearing as a fine speckle pattern between crossed polarizers is commonly obtained at the $I-N$ transition (Fig. 20). This texture has been called a tight texture, a dense texture or a polydomain texture. Only a fine scale mottle is apparent (of say $0.1-1$ μm). In general, this texture is observed for high molecular weight specimens, in thick samples and/or at low temperatures [83]. Determination of the molecular organization associated with this microstructure presents a challenge to the microscopist because the domain size approaches the resolution limit of the optical microscope. Transmission electron microscopy studies [63, 65] have shown that this microstructure can be considered as an assembly of 'effective domains' due to a particular arrangement of disclinations. The fine grains seen are believed to correspond to regions of uniform orientation correlation within nematic structures bounded by disclinations as illustrated in Fig. 21. Disclination densities up to $50-100$ μm^{-2} have been observed [62]. However, the number of disclination lines decreases with annealing time [27, 81, 84] leading to a nematic texture that approaches the texture exhibited by LMWLCs although the resulting increased size range still remains with only several microns [85]. These observations seem independent on the specific MCLCP, the only difference being the time scale of this kinetic behavior. The coarsening of this nematic texture seems to be related to the dynamics of disclinations [65, 86]. When a nematic phase is formed from the isotropic phase, its structure is dependent on the kinetics and mechanism of the phase transition. The initial texture is usually not stable under the given boundary conditions. The annealing process at $T>T_g$ can lead to the relaxation of the excess free energy associated with the disclination density. The coarsening of the initially grainy texture, which is an ordering process, seems to involve a decrease of the number density of disclinations, $\rho(t)$, and an increase in the domain size $\xi_d(t)$, by annihilation of the disclination lines, which occurs as a consequence of the diffusion and coalescence of the lines of opposite signs. The two quantities are interrelated:

$$\xi_d(t) \approx \rho(t)^{-1/d} \tag{12}$$

where d is a spatial dimensionality. For copolyester **3** (X-7G), Shiwaku et al. [65, 86] by plotting the average spacing between neighboring disclinations measured by optical microscopy as a function of annealing time found that:

$$\xi_d(t) \approx \rho(t)^{-1/2} \approx t^{0.35} \tag{13}$$

Rojstaczer et al. [28] established through small angle light scattering (SALS) experiments that the size of the domains, characterized by the reciprocal scattering vector, Q_{max}, scales with time as:

$$1/Q_{max} \approx t^{0.37} \tag{14}$$

for a semiflexible aromatic polyester based on triad mesogenic unit containing a methyl arylsulfonyl substituted hydroquinone and a decamethylene spacer. It should be noted that the rate of decrease of the number of disclinations per unit volume is strongly influenced by the molecular weight of the polymer [26] and the rigidity of the polymer backbone [84]. Rieger [87] has proposed a simple mean field theory for the description of the decay of the density of disclinations in nematic films of thermotropic MCLCPs, which is based on results for diffusion reaction systems of the type $A + B \rightarrow 0$, where A and B correspond to two different types of end points of a disclination line, respectively. This approach was found to approximate the functional behavior of the time dependence of the number of disclination points slightly better than the algebraic law. The problem of describing SALS patterns for MCLCPs was first approached by Hashimoto et al. [61]. These authors developed a scattering theory for a model having isolated disclination lines contained in discs. The results poorly described the observed scattering. Later on, Greco [88] proposed a model consisting in a collection of isolated discs, randomly oriented, each containing a disclination dipole together with the corresponding director field. This simple two-body interaction of the $S = \pm 1/2$ disclinations could explain all the essential features of SALS experiments. The only parameter of the model (i.e. the distance between two disclinations) was sufficient to explain the evolution of the patterns as a function of annealing time.

A suitable treatment of the glass surfaces between which the polymer is observed allows one to obtain either planar layers in which the director lies parallel to the surface or homeotropic films in which the molecules are aligned normal to the surface. Whereas, for LMWLCs, several techniques are now available for developing uniform orientation at surfaces, their application to orientation in MCLCPs is proving more difficult to achieve. However, a number of techniques have been used to create reproducible planar alignment. Rubbing of a raw glass plate with a diamond paste leads to uniformly aligned samples. SiOx layers evaporated obliquely (at 60° incidence) [75] and freshly cleaved mica surfaces [60] produce similar effects. Polymer coatings on glass substrates [58] and surface-deposited hexadecyltrimethylammonium bromide [75] can also be used to align MCLCPs homogeneously. However, since a nondegenerate planar alignment is generally required, rubbing of the substrate after deposition of the polymer is used. Viewing planar samples from the top between crossed polariz-

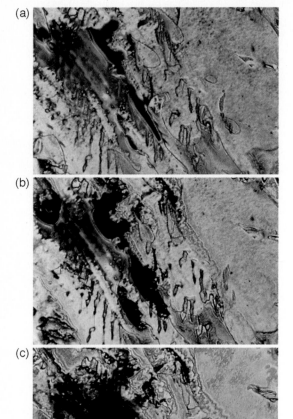

Figure 22. Development of homeotropic texture from copolyester **11** at 330°C (crossed polarizers). Annealing time (a) $t=4$ min, (b) $t=8$ min and (c) $t=28$ min (from [75]).

ers results in the observation of four positions of extinction.

Homeotropic alignment can be achieved by simple treatments of glass slides with boiling chromic–sulphuric acid, acetone and methanol (sequentially interspersed with water rinses) and rinsing with hot distilled water in an ultrasonic cleaner [75]. Films thus prepared appear completely dark

when viewed vertically between crossed polarizers because the director is oriented perpendicular to the glass surfaces and parallel to the light beam. If the cover glass is touched, the originally dark field of view brightens instantly, thus distinguishing between homeotropic and isotropic texture. Homeotropic alignment was also achieved by using a rocksalt substrate [89, 90]. It should be noted that ageing MCLCPs at high temperature seems to encourage the homeotropic orientation [58, 75, 89, 90]. This effect has been observed for polyesters **1**, **6** and **11** (Fig. 22). A factor which seems to control homeotropic orientation at surfaces is the density of chain ends. For polyester **6**, the tendency towards overall homeotropy was found to increase as the average molecular weight of the specimens decreased [90]. For polyesters **1** and **11**, clearly some degradation of the samples occurred at the high temperatures required for homeotropic alignment leading to the formation of shorter chains with their relatively high density of chain ends [42, 75]. This was borne out by the experiments on polyester **1** which showed that samples aged in an atmosphere of argon did not become homeotropic [42]. As has been discussed by Meyer [41], a perfect, ideal polymer nematic composed of infinite chains (i.e. with no chain ends) must lie with the chains parallel to the boundary surfaces for energetic reasons. For finite, but long, macromolecular chains, entropy leads to keep chain ends 'dissolved' in the bulk rather than condensed on the surfaces. On the other hand, short chains (i.e. abundant chain ends) will favor a homeotropic texture: while entropy will tend to distribute chain ends randomly, energy will tend to place them so as to relieve strains.

11

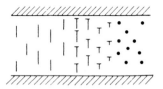

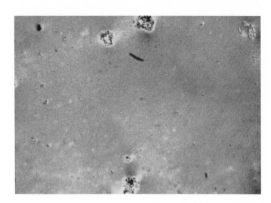

Figure 23. Schematic representation of the geometry at a wall with a 90° rotation of the molecules across the wall [58].

Figure 24. Nematic sample with the director making a given angle with the glass surfaces.

Kléman et al. [58] reported that walls formed during the slow transformation from the planar to the homeotropic texture induced by annealing samples of polyester **1** for long times. The walls separated untransformed and transformed regions, with a consequence of 90° rotation of the molecules across them. The proposed geometry, akin to a 90° Bloch wall in ferromagnets, is shown in Fig. 23. Clearly, the molecules suffer a practically pure twist inside the transition region. Walls have also been observed in thin films of polyester **6** [89, 90]. In these films, the walls also developed as the molecules attempted to transform to a homeotropic alignment, but the geometry was more akin to the 180° Bloch walls of ferromagnets. In this system, the walls do not separate planar and homeotropic regions, but rather regions where the molecules in both domains have some finite misorientation with respect to the plane of the sample, but in opposite senses. The domain structure adopted seems to be an example of splay compensation in operation, with the splay distortions in one plane being compensated by a negative splay component in an orthogonal plane.

Classification of LC alignment into planar or homeotropic is an oversimplification. Under suitable conditions, the nematic director may make an angle with the substrate surface (Fig. 24). The observations between crossed polarizers of such samples are the

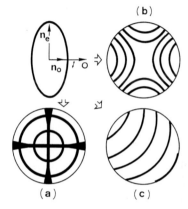

Figure 25. Ellipsoid of indices for a positive ($n_e > n_o$) uniaxial LC. The typical conoscopic images with average beam direction perpendicular to the plates are shown for (a) homeotropic, (b) planar and (c) oblique orientation [31].

same as in the planar case. Such tests are insufficient to check the uniformity of alignment and observations of the conoscopic images formed by illuminating the specimen with highly convergent light and inserting a so-called Bertrand lens below the eyepiece are needed (Fig. 25) [91].

The conoscopic interference patterns are also quite useful to identify uniaxial and biaxial nematics. For a uniaxial specimen, the conoscopic image consists of a dark cross

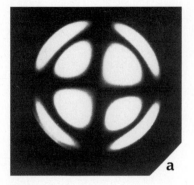
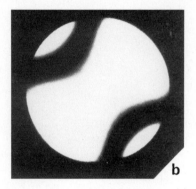

Figure 26. Basal interference figures of biaxial media. Sections perpendicular to the bisectrix of the acute angle. The cases depicted are those for thin sections of material and axial angle of about 10° (a) and 45° (b), respectively [31].

coupled with concentric circles, its position depending on the 'crystal' orientation. The cross will have four-fold symmetry. Figure 25a corresponds to the special case where the unique optic axis of the specimen is parallel to the microscope axis. For a biaxial material, the symmetry of the conoscopic image will be two-fold as illustrated in Fig. 26. Whereas this technique was used successfully for side chain liquid crystalline polymers (SCLCPs) [92–94], its application to MCLCPs appears not so easy to achieve owing to the difficulty with which MCLCPs give 'monodomains' in the liquid crystalline state. This is due mainly to the lack of strong surface anchoring, large viscosity and thermal degradation of polymers. However, with the aid of conoscopic observations, Noël et al. [75] have proved the positive uniaxial character of polyester **11**.

A final remark will concern the 'glassy liquid crystal'. All the textures observed in the nematic state can easily be quenched and supercooled to room temperature. In spite of the thermal shock, both the planar alignment and the homeotropic one can be retained in the glassy state [59, 60, 66, 95].

6   Flow-induced Textures and their Relaxation Behavior

The evolution from the static state to different textures observed as a consequence of controlled shearing experiments and the subsequent relaxation behavior at the cessation of flow were first analyzed by Graziano and Mackley [96] for a number of thermotropic MCLCPs. This pioneering work introduced and defined the terms used today to classify shear-induced textures: 'worm' texture, 'ordered' texture and 'banded' texture. It should be noted that these textures have also been observed in lyotropic nematics.

The shearing behavior of MCLCPs depends on the initial state of the material. When molten MCLCPs exhibiting a threaded texture in the quiescent state are subjected to shear, the threads first extend and curve and disclination loops distort within the shear. However, at a rather well defined shear strain multiplication of the defects occurs [83, 84, 96, 97]. These defects appear as short, dark curled entities which flow along the streamlines whilst continually changing their shapes, and the overall view of them during shear resembles a 'vat of teeming worms'. The resulting disordered polydomain texture at the scale of a few microns (i.e. below the size of the specimen and independent of it) has been named a 'worm' texture by Graziano and Mackley [96]. The density of worms increases with increasing shear rate. At still higher shear rates, a sharp transformation can be induced into a birefringent texture whose optical appearance is dependent on the direction of crossed polarized light, thus indicating that flow alignment of some species is occurring [83, 84, 96, 97]. This texture has been labelled the 'ordered' texture [96]. In general, the ordered texture is favored at low temperatures, high shears and small thicknesses of the samples, whereas the worm texture exists under less severe shear conditions and higher temperatures. On cessation of flow, the 'worms' relax by vanishing or coalescing to form threads with a relaxation time that depends on the polymer, the molecular weight of the specimen, the value of the applied shear and the temperature. Typically this might range from a few seconds to tens of minute. Upon cessation of shear, in the ordered texture, the worm texture develops within less than 1 s and subsequently relaxes as described above.

Starting with a molten polymer exhibiting an ill-defined grainy texture, shearing causes this texture to be modified by the distortion and multiplication of defects [83, 84, 96]. Further shearing at higher rates leads to the apparent disappearance of any defect textures and the emergence of pure birefringence with the optic axis in the plane of shear (ordered texture). Upon cessation of shear, the material will relax back to the original dense defect texture. It should be noted, however, that another relaxation behavior has been observed for high molecular weight specimens [96, 98–100]: a banded texture characterized by bands aligned perpendicular to the prior flow direction appeared after cessation of shear in the ordered texture. In general, it is a low temperature, high shear relaxation phenomenon.

Very little is known about the defects which are present in the worm texture because a direct optical study is hampered by the difficulty of isolating such moving objects [96]. However, the analysis of defects in Vectra B 950® (polymer **8**) carried out by De'Nève et al. [101, 102] suggested that the defects are mostly half-integer twist disclination loops lying in the shearing plane. The chains are highly oriented along the flow direction near the core of the defects, thus indicating that the core is composed of a 'perfect' nematic, an assumption previously made by Mazelet and Kléman [57] for polyester **1**. A peculiar feature of the cores of the defects is their large macroscopic size (of the order of 1 μm), which is probably related to the fact that the lines are in motion. They may move by dragging a large amount of nematic fluid in order to minimize distortion energy. Far from the defects, the directors are more or less aligned in the flow direction. However, in the vicinity of the defects two perpendicular orientations are observed: the director is oriented along the flow direction inside the loops, but along the vorticity axis outside the loops. The nucleation process of the loops is therefore related to a rheological regime in which these two perpendicular orientations are favored. As pointed out by De'Nève et al. [97, 102] the simplest mechanism for this is the occurrence of tumbling.

Concerning the mathematical modeling of LCPs, most attempts have concentrated on lyotropic systems. Doi [103] has developed a defect free model for lyotropic rod-like polymers in the isotropic and anisotropic states and Larson and Mead [104] have proposed a polydomain phenomenological description of lyotropic systems. Marrucci and Maffettone [105] solved a diffusion equation for rodlike lyotropic molecules Their two-dimensional numerical simulation in the shearing plane predicts a periodic rotation of the director at low shear rates, the so-called tumbling regime, and a higher shear alignment regime. This two-dimensional approach was extended in three dimensions by Öttinger and Larson [106]. These authors found that, at low shear rates, the director reaches two attractors. The first one is the shearing-plane and the second one is the vorticity axis. Tumbling is also shown to occur in three dimensions, but in a more complex manner, carrying both in-plane and out-of-plane components. However, these molecular theories do not at this stage predict or include defect effects.

To account for experimentally observed microscopic observations of the worm texture behavior at low shear rates, De'Nève et al. [97, 102] assumed that tumbling generates large distortions that decrease their free energy by producing a large number of defects. An equilibrium density of defects should be obtained when the distance between neighboring defects prevents the director from tumbling freely. At this point defects start to annihilate. After having reached a first minimum, the defect density increases again. Such an explanation is consistent with the model developed by Marrucci and Greco [107] which is based on the hypothesis that defects prevent the free rotation of the director. It also accounts for the damped oscillation phenomenon observed when the intensity of the light transmitted through the sample is plotted as a function of strain. These oscillations, which are related to the light scattered by the core of the defects, would correspond to the alternate action of nucleation and annihilation of defects. In addition, this tumbling mechanism agrees with the fact that the shear strain period of the defect density does not depend on shear rate since the tumbling time peri-

od is inversely proportional to shear rate [108]. The ordered texture does not seem to appear suddenly at a given shear rate, since there is a progressive increase in flow orientation [97]. Experimental data suggest that this texture may contain defects of the type seen in the worm texture, but with dimensions smaller than the resolution of optical microscope [102]. Quite recently, Gervat et al. [84] developed a director simulation that incorporates optical, X-ray and rheological data. Essential elements of the model are concerned with a local molecular anisotropy, a molecular correlation coefficient, and a local defect texture. The main prediction of the simulation is that the presence of defects influences orientation relaxation after shear and that the defects themselves do not appear to have a profound effect on rheology.

Following thin film shear [49, 109–111], uniaxial fiber drawing [112], injection molding [113], or elongational flow [114], nematic polymers typically exhibit a banded microstructure. This is especially evident when thin specimens are observed between crossed polarizers (Fig. 27). Dark bands lie perpendicular to the shear or flow direction Z, and are regularly spaced a few microns apart if Z is aligned parallel to the vibration direction of either polarizer. Detailed analysis of this microstructure [49, 50, 115–118] indicates that the global molecular alignment introduced by processing has partially relaxed, and that the molecular orientation instead varies periodically relative to the macroscopic shear direction. Banded texture is one of the characteristics of oriented specimens of both lyotropic and thermotropic LCPs. This subject has been reviewed in a recent article by Viney and Putnam [110]. The bands can appear during continuous shear [109, 119] or after cessation of shear [83, 120]. In either case, a critical deformation rate must be exceeded [83,

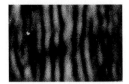

Figure 27. Banded texture of polyester **6**. The shear direction was horizontal.

109, 119–123], which has been predicted theoretically [124, 125]. A finite time must elapse from the onset of shear [109] or the cessation of shear [120, 122] before the bands can form.

The molecular organization in the banded texture is understood in some detail as a result of optical microscopy [49, 50, 83, 110–112, 115–122, 126], transmission electron microscopy [49, 76, 89, 90, 115, 127] and scanning electron microscopy [50]. It has been shown that the orientation variation of the molecules in this texture is near sinusoidal about the shear direction with most of the misorientation within the shear plane. There is a high degree of register of the molecules between each of the layers [50]. The influence of a number of parameters on the formation of bands has been studied: the molecular weight [96], the sample thickness [110, 120–122], the shear rate [109], the time during which the shear stress was applied [122] and the total shear deformation [121].

Various theoretical models have been proposed to account for this texture. Orientation of the directors within alternate planes by creation of alignment-inversion walls has been invoked [109]. Subsequent experiments seemed to indicate rather a smooth transition of the orientation between domains of a given direction [126]. The tumbling instability has been suggested as an explanation [128], but would not fulfill this condition.

The development of novel instabilities of the director pattern was also observed as the response of uniformly aligned nematic LCPs to a suddenly applied reorienting magnetic field [129, 130]. With conventional LCs composed of moderate length-to-width ratio molecules, the spatially periodic response to reorientating fields is not easily observed because it decays quickly to the homogeneous ground state. However, for nematics composed of very long molecules, the large anisotropy in their elasticity and viscosity has two important consequences. First, the periodic instabilities appear under a very broad range of circumstances, and second the stripe patterns produced are very long lived [130–132]. An extension of the theoretical model by Guyon et al. [133] for LMWLCs was used to interpret the data and it was suggested that a similar mechanism could explain the banded textures of LCPs in a shear flow.

Unfortunately, because there are so many variables it is rarely possible to compare the observations made in different studies and the literature contains many conflicting results. For example, there is some contention as to whether the bands form during shear [109, 119] or on stress relaxation afterwards [83, 120]. Ernst and Navard [120] considered a rather pure relaxation process via a periodic distortion after shear flow has ceased. According to this model, the macromolecules are stretched in the flow and then relax back. It is not clear, however, which mechanism in the relaxation process is responsible for the observed periodic structure. There is also disagreement as to whether the shear threshold does [110, 121] or does not [120, 122] depend on sample thickness. A shear threshold inversely proportional to the sample thickness, as reported in [121], would support the instability concept. Zielinska and Ten Bosch [124, 125] argued that, for simple shear flow, the band structures are due to an instability mechanism present in the Leslie–Ericksen equations. They considered the existence of an instability producing a periodic pattern with a distortion perpendicular to the shear plane as observed in the optical experiments. They determined the critical shear rate and the critical wavevector for the instability to occur and showed that these depend strongly on the values of the elastic constants and anisotropic viscosities. They also obtained solutions for the director and velocity field.

As discussed in the previous section, the literature on banded textures contains many attempts to correlate change in a parameter describing a banded texture with change in a molecular characteristic or processing conditions. However, experiments have sometimes given conflicting results. Also, theory is not sufficiently developed to suggest which microstructural parameter will be most sensitive to changes in any given molecular or processing parameter, and anticipated correlations frequently cannot be demonstrated conclusively.

7 References

[1] O. Lehmann, *Flüssige Kristalle, sowie Plastizität von Kristallen im Allgemeinen, Molekulare Umlagerungen und Aggregatzustandsänderungen*, W. Engelmann, Leipzig, **1904**.

[2] G. Friedel, *Ann. Phys. (Paris)* **1922**, *18*, 273.

[3] G. Friedel, F. Grandjean, *Bull. Soc. Fr. Miner.* **1910**, *33*, 192 and 409.

[4] H. Sackmann, D. Demus, *Fortschr. Chem. Forsch.* **1969**, *12*, 349; *Mol. Cryst.* **1966**, *2*, 81; *Mol. Cryst. Liq. Cryst.* **1973**, *21*, 239.

[5] H. Arnold, H. Sackmann, *Z. Physik. Chem. (Leipzig)* **1960**, *213*, 137, 145 and 262.

[6] D. Demus, L. Richter, *Textures of Liquid Crystals*, Verlag Chemie, Weinheim, **1978**.

[7] G. W. Gray, J. W. Goodby, *Smectic Liquid Crystals: Textures and Structures*, Leonard Hill, Glasgow, **1984**.

[8] M. Kléman, *Points, Lignes, Parois dans les Fluides Anisotropes et les Solides Cristallins*, Vol. 1, Les Editions de Physique, Orsay, **1977**.

[9] M. Kléman, J. Friedel, *J. Physique Colloq.* **1969**, *30*, C4–43.

[10] J. Friedel, M. Kléman in *Fundamental Aspects of Dislocation Theory*, Vol. 1, (Eds.: J. A. Simmons, R. de Wit, R. Bullough), National Bureau of Standards USA, Special Publication No. 317, **1970**, p. 607.

[11] J. Friedel in *Dislocations in Solids*, Vol. 1 (Ed.: F. R. N. Nabarro), North-Holland Publishing Company, Amsterdam, **1979**, p. 3.

[12] M. Kléman, ibid, Vol. 5, **1980**, p. 243.

[13] Y. Bouligand, ibid, Vol. 5, **1980**, p. 299.

[14] G. Toulouse, M. Kléman, *J. Phys. (Lett.)* **1976**, *37*, L 149.

[15] G. E. Volovik, V. P. Mineev, *Sov. Phys. J. E. T. P. (Lett.)* **1976**, *24*, 561; *Sov. Phys. J. E. T. P.* **1977**, *45*, 1186.

[16] V. Poenaru, G. Toulouse, *J. Phys.* **1977**, *38*, 887.

[17] M. Kléman, L. Michel, *Phys. Rev. (Lett.)* **1978**, *40*, 1387.

[18] L. Michel, *Rev. Mod. Phys.* **1980**, *52*, 617.

[19] Y. Bouligand in *Physique des Défauts/Physics of Defects*, Les Houches Session XXXV (Eds.: R. Balian et al.), North-Holland Publishing Company, Amsterdam, **1981**, p. 668.

[20] W. F. Brinkman, P. E. Cladis, *Physics Today* May **1982**, 48.

[21] M. Kléman, *Points, Lines and Walls*, Wiley, Chichester, New York, **1983**.

[22] S. Chandrasekhar, G. S. Ranganath, *Adv. Phys.* **1986**, *35*, 507.

[23] M. Kléman, *Rep. Progr. Phys.* **1989**, *52*, 555.

[24] M. Kléman, *Faraday Dis. Chem. Soc.* **1985**, *79*, 215.

[25] M. Kléman in *Liquid Crystallinity in Polymers: Principles and Fundamental Properties* (Ed.: A. Ciferri), VCH, Weinheim, **1991**, p. 365.

[26] V. Percec, D. Tomazos, C. Pugh, *Macromolecules* **1989**, *22*, 3259.

[27] V. Percec, R. Yourd, *Macromolecules* **1989**, *22*, 3229.

[28] S. Rojstaczer, B. S. Hsiao, R. S. Stein, *Div. of Polym. Chem., Am. Chem. Soc. Polymer Preprints* **1988**, *29*, 486.

[29] C. Noël in *Polymer Liquid Crystals* (Ed.: A. Blumstein), Plenum Press, New York, **1985**, p. 21.

[30] C. Noël in *Recent Advances in Liquid Crystalline Polymers* (Ed.: L. Chapoy), Elsevier Applied Science Publishers, London **1985**, p. 135.

[31] C. Noël in *Side Chain Liquid Crystal Polymers* (Ed.: C. B. McArdle), Blackie, Glasgow, **1989**, p. 159.

[32] C. Noël in *Liquid Crystal Polymers: From Structures to Applications* (Ed.: A. A. Collyer), Elsevier Applied Science, London, **1992**, p. 31.

[33] F. C. Frank, *Discuss. Faraday Soc.* **1958**, *25*, 19.

[34] P. G. de Gennes, *The Physics of Liquid Crystals*, Clarendon Press, Oxford, **1974**.

[35] C. W. Oseen, *Trans. Faraday Soc.* **1933**, *29*, 883.
[36] P. E. Cladis, M. Kléman, *J. Phys. (Paris)* **1972**, *33*, 591.
[37] R. B. Meyer, *Phil. Mag.* **1973**, *27*, 405.
[38] S. I. Anisimov, I. E. Dzyaloshinskii, *Sov. Phys. J. E. T. P.* **1972**, *36*, 774.
[39] I. E. Dzyaloshinskii, *Sov. Phys. J. E. T. P.* **1970**, *31*, 773.
[40] P. G. de Gennes in *Polymer Liquid Crystals* (Eds.: A. Ciferri, W. R. Krigbaum, R. B. Meyer), Academic Press, New York, **1982**, pp. 115–131.
[41] R. B. Meyer in *Polymer Liquid Crystals* (Eds.: A. Ciferri, W. R. Krigbaum, R. B. Meyer), Academic Press, New York, **1982**, pp. 133–163.
[42] Sun Zheng-Min, M. Kléman, *Mol. Cryst. Liq. Cryst.* **1984**, *111*, 321.
[43] A. F. Martins, P. Esnault, F. Volino, *Phys. Rev. (Lett.)* **1986**, *57*, 1745.
[44] P. Esnault, J. P. Casquilho, F. Volino, A. F. Martins, A. Blumstein, *Liq. Cryst.* **1990**, *7*, 607.
[45] T. De'Nève, M. Kléman, P. Navard, *Liq. Cryst.* **1994**, *18*, 67.
[46] S. E. Bedford, T. M. Nicholson, A. H. Windle, *Liq. Cryst.* **1991**, *10*, 63.
[47] S. E. Bedford, A. H. Windle, *Liq. Cryst.* **1993**, *15*, 31.
[48] J. R. Hobdell, A. H. Windle, *J. Chem. Soc., Faraday Trans.* **1995**, *91*, 2497.
[49] A. M. Donald, C. Viney, A. H. Windle, *Polymer* **1983**, *24*, 155.
[50] S. E. Bedford, A. H. Windle, *Polymer* **1990**, *31*, 616.
[51] A. H. Windle, C. Viney, R. Golombok, A. M. Donald, G. R. Mitchell, *Faraday Discuss. Chem. Soc.* **1985**, *79*, 55.
[52] N. D. Mermin, *Rev. Mod. Phys.* **1979**, *51*, 591.
[53] T. De'Nève, M. Kléman, P. Navard, *J. Phys. II France* **1992**, *2*, 187.
[54] C. Viney, G. R. Mitchell, A. H. Windle, *Polym. Commun.* **1983**, *24*, 145.
[55] A. M. Donald, C. Viney, A. H. Windle, *Phil. Mag. B* **1985**, *52*, 925.
[56] C. Viney, G. R. Mitchell, A. H. Windle, *Mol. Cryst. Liq. Cryst.* **1985**, *129*, 75.
[57] G. Mazelet, M. Kléman, *Polymer* **1986**, *27*, 714.
[58] M. Kléman, L. Liebert, L. Strzélécki, *Polymer* **1983**, *24*, 295.
[59] C. Viney, A. H. Windle, *J. Mater. Sci.* **1982**, *17*, 2661.
[60] C. Noël, F. Lauprêtre, C. Friedrich, B. Fayolle, L. Bosio, *Polymer* **1984**, *25*, 808.
[61] T. Hashimoto, A. Nakai, T. Shiwaku, H. Hasegawa, S. Rojstaczer, R. S. Stein, *Macromolecules* **1989**, *22*, 422.
[62] S. D. Hudson, E. L. Thomas, R. W. Lenz, *Mol. Cryst. Liq. Cryst.* **1987**, *153*, 63.
[63] B. A. Wood, E. L. Thomas, *Nature* **1986**, *324*, 655.
[64] S. Chen, L. Cai, Y. Wu, Y. Jin, S. Zhang, Z. Qin, W. Song, R. Qian, *Liq. Cryst.* **1993**, *13*, 365.
[65] T. Shiwaku, A. Nakai, H. Hasegawa, T. Hashimoto, *Macromolecules* **1990**, *23*, 1590.
[66] M. R. Mackley, F. Pinaud, G. Siekmann, *Polymer* **1981**, *22*, 437.
[67] H. Lee, M. M. Labes, *Mol. Cryst. Liq. Cryst. (Lett.)* **1982**, *82*, 199.
[68] N. V. Madhusudana, R. Pratibha, *Mol. Cryst. Liq. Cryst.* **1983**, *103*, 31.
[69] C. Viney, D. J. Brown, C. M. Dannels, R. J. Twieg, *Liq. Cryst.* **1993**, *13*, 95.
[70] G. Galli, M. Laus, A. S. Angeloni, P. Ferruti, E. Chiellini, *Eur. Polym. J.* **1985**, *21*, 727.
[71] W. Song, S. Chen, R. Qian, (a) *Makromol. Chem., Rapid Commun.* **1993**, *14*, 605; (b) *Macromol. Symp.* **1995**, *96*, 27.
[72] Qi-Feng Zhou, X.-H. Wan, F. Zhang, D. Zhang, Z. Wu, X. Feng, *Liq. Cryst.* **1993**, *13*, 851.
[73] M. J. Freiser, *Phys. Rev. (Lett.)* **1970**, *24*, 1041.
[74] B. Millaud, A. Thierry, C. Strazielle, A. Skoulios, *Mol. Cryst. Liq. Cryst. (Lett.)* **1979**, *49*, 299.
[75] C. Noël, C. Friedrich, F. Lauprêtre, J. Billard, L. Bosio, C. Strazielle, *Polymer* **1984**, *25*, 263.
[76] E. L. Thomas, B. A. Wood, *Faraday Discuss. Chem. Soc.* **1985**, *79*, 229.
[77] J. R. Ford, D. C. Bassett, G. R. Mitchell, *Mol. Cryst. Liq. Cryst. B* **1990**, *180*, 233.
[78] Y. Dong, H. Deng, *Acta Polym. Sin.* **1991**, 584.
[79] G. Marrucci, *Proc. IX Int. Congr. Rheol. (University of Mexico)*, **1984**.
[80] S. Chen, C. Du, Y. Jin, R. Qian, Q. Zhou, *Mol. Cryst. Liq. Cryst.* **1990**, *188*, 197.
[81] M. Hoff, A. Keller, J. A. Odell, V. Percec, *Polymer* **1993**, *34*, 1800.
[82] S. Chen, W. Song, Y. Jin, R. Qian, *Liq. Cryst.* **1993**, *15*, 247.
[83] N. J. Alderman, M. R. Mackley, *Faraday Discuss. Chem. Soc.* **1985**, *79*, 149.
[84] L. Gervat, M. R. Mackley, T. M. Nicholson, A. H. Windle, *Phil. Trans. R. Soc. Lond. A* **1995**, *350*, 1.
[85] J. L. Feijoo, G. Ungar, A. J. Owen, A. Keller, V. Percec, *Mol. Cryst. Liq. Cryst.* **1988**, *155*, 187.
[86] T. Shiwaku, A. Nakai, H. Hasegawa, T. Hashimoto, *Polym. Commun.* **1987**, *28*, 174.
[87] J. Rieger, *Macromolecules* **1990**, *23*, 1545.
[88] F. Greco, *Macromolecules* **1989**, *22*, 4622.
[89] A. M. Donald, *J. Mat. Sci. (Lett.)* **1984**, *3*, 44.
[90] A. M. Donald, A. H. Windle, *J. Mat. Sci.* **1984**, *19*, 2085.
[91] N. H. Hartshorne, A. Stuart in *Crystals and the Polarising Microscope*, 4th edn., Edward Arnold, London, **1970**.
[92] H. Finkelmann in *Liquid Crystals of One- and Two-Dimensional Order* (Eds.: W. Helfrich, G. Heppke), Springer-Verlag, Berlin, **1980**.
[93] A. I. Hopwood, H. J. Coles, *Polymer* **1985**, *26*, 1312.

[94] F. Hessel, H. Finkelmann, *Polym. Bull.* **1986**, *15*, 349.

[95] R. W. Lenz, J. I. Jin, *Macromolecules* **1981**, *14*, 1405.

[96] D. J. Graziano, M. R. Mackley, *Mol. Cryst. Liq. Cryst.* **1984**, *106*, 73.

[97] T. De'Nève, P. Navard, M. Kléman, *J. Rheol.* **1993**, *37*, 515.

[98] S. Chen, Y. Jin, S. Hu, M. Xu, *Polym. Commun.* **1987**, *28*, 208.

[99] X. Liu, D. Shen, L. Shi, M. Xu, Q. Zhou, X. Duan, *Polymer* **1990**, *31*, 1894.

[100] J. Hou, W. Wu, D. Shen, M. Xu, Z. Li, *Polymer* **1994**, *35*, 699.

[101] T. De'Nève, M. Kléman, P. Navard, *C. R. Acad. Sci. (Paris)* **1993**, *316*, Série II, 1037.

[102] T. De'Nève, P. Navard, M. Kléman, *Macromolecules* **1995**, *28*, 1541.

[103] M. Doi, *J. Polym. Sci.; Polym. Phys. Ed.* **1981**, *19*, 229.

[104] R. G. Larson, D. W. Mead, *J. Rheol.* **1989**, *33*, 1251.

[105] G. Marrucci, P. L. Maffettone, *Macromolecules* **1989**, *22*, 4076.

[106] R. G. Larson, H. C. Öttinger, *Macromolecules* **1991**, *24*, 6270.

[107] G. Marrucci, F. Greco, *J. Non-Newtonian Fluid Mech.* **1992**, *44*, 1.

[108] R. G. Larson, *Macromolecules* **1990**, *23*, 3983.

[109] G. Kiss, R. S. Porter, *Mol. Cryst. Liq. Cryst.* **1980**, *60*, 267.

[110] C. Viney, W. S. Putnam, *Polymer* **1995**, *36*, 1731.

[111] J. Hou, W. Wu, M. Xu, Z. Li, *Polymer* **1996**, *37*, 5205.

[112] S. C. Simmens, J. W. S. Hearle, *J. Polym. Sci.; Polym. Phys. Ed.* **1980**, *18*, 871.

[113] H. Thapar, M. Bevis, *J. Mater. Sci. (Lett.)* **1983**, *2*, 733.

[114] E. Peuvrel, P. Navard, *Macromolecules* **1991**, *24*, 5683.

[115] C. Viney, A. M. Donald, A. H. Windle, *Polymer* **1985**, *26*, 870.

[116] C. Viney, A. H. Windle, *Polymer* **1986**, *27*, 1325.

[117] M. Horio, S. Ishikawa, K. Oda, *J. Appl. Polym. Sci.; Appl. Polym. Symp.* **1985**, *41*, 269.

[118] P. Navard, A. E. Zachariades, *J. Polym. Sci. Part B; Polym. Phys.* **1987**, *25*, 1089.

[119] P. Navard, *J. Polym. Sci.; Polym. Phys. Ed.* **1986**, *24*, 435.

[120] B. Ernst, P. Navard, *Macromolecules* **1989**, *22*, 1419.

[121] G. Marrucci, N. Grizzuti, A. Buonaurio, *Mol. Cryst. Liq. Cryst.* **1987**, *153*, 263.

[122] E. Marsano, L. Carpaneto, A. Ciferri, *Mol. Cryst. Liq. Cryst.* **1988**, *158B*, 267.

[123] T. Takebe, T. Hashimoto, B. Ernst, P. Navard, R. S. Stein, *J. Chem. Phys.* **1990**, *92*, 1386.

[124] B. J. A. Zielinska, A. Ten Bosch, *Phys. Rev. A* **1988**, *38*, 5465.

[125] B. J. A. Zielinska, A. Ten Bosch, *Liq. Cryst.* **1989**, *6*, 553.

[126] C. Viney, A. M. Donald, A. H. Windle, *J. Mater. Sci.* **1983**, *18*, 1136.

[127] A. M. Donald, A. H. Windle, *J. Mater. Sci.* **1983**, *18*, 1143.

[128] G. Marrucci, *Pure Appl. Chem.* **1985**, *57*, 1545.

[129] C. R. Fincher, *Macromolecules* **1986**, *19*, 2431.

[130] A. J. Hurd, S. Fraden, F. Lonberg, R. B. Meyer, *J. Phys.* **1985**, *46*, 905.

[131] F. Lonberg, S. Fraden, A. J. Hurd, R. B. Meyer, *Phys. Rev. (Lett.)* **1984**, *52*, 1903.

[132] F. Lonberg, R. B. Meyer, *Phys. Rev. (Lett.)* **1985**, *55*, 718.

[133] E. Guyon, R. B. Meyer, J. Salan, *Mol. Cryst. Liq. Cryst.* **1979**, *54*, 261.

Part 2:
Side-Group Thermotropic Liquid-Crystalline Polymers

Chapter III
Molecular Engineering of Side Chain Liquid Crystalline Polymers by Living Polymerizations *

Coleen Pugh and Alan L. Kiste

1 Introduction

1.1 Chain Polymerizations

Unil recently, most side chain liquid crystalline polymers (SCLCPs) were prepared by either hydrosilations of mesogenic olefins with poly(methylsiloxane)s or by free radical polymerizations of acrylates, methacrylates and chloroacrylates [1, 2]. Both routes involve chain polymerizations, either directly or prior to a polymer analogous reaction. Chain polymerizations involve the four elementary reactions shown in Scheme 1 [3]. In contrast to step polymerizations,

1. **Initiation** $\qquad$ $I + M \xrightarrow{k_i} IM*$

2. **Propagation** $\quad IM* + n\,M \xrightarrow{k_p} I(M)_n M*$

3. **Transfer** $\qquad I(M)_n M* + A \xrightarrow{k_{tr}} P + A*$

4. **Termination** $\quad I(M)_n M* + A \xrightarrow{k_t} P$

Scheme 1. Four elementary reactions of a chain polymerization.

chain polymerizations require an initiator (I) to produce reactive centers. In order for a vinyl monomer to be polymerizable, it must contain a substituent capable of stabilizing the resulting active site. Therefore,

most 1-substituted alkenes undergo facile radical polymerizations (although α-olefins do not), whereas an electron-donating substituent is required to activate olefins to cationic initiation and propagation, and an electron-withdrawing substituent is required to activate olefins to anionic polymerizations.

In addition to initiation and propagation, chain polymerizations may also undergo transfer and termination reactions (with reagent A for example), in which inactive chains (P) are formed. Chain transfer may occur to monomer, solvent, polymer, or a chain transfer agent. In order for this to be a transfer reaction rather than termination, the molecule (A*) generated must be reactive enough to reinitiate polymerization. Solvent is not required in radical polymerizations, although it retards autoacceleration and helps to diffuse heat; solvent is required in ionic polymerizations to control the heat generated. Since solvent is the component present in the highest concentration in most ionic polymerizations, solvents which either terminate the active sites or which act as chain transfer agents must be avoided. In contrast to chain transfer to solvent, which would be prevalent from the initial stages of a polymerization due to the solvent's high concentration, chain transfer to polymer often does not compete noticeably with propagation until the end of the polymerization when monomer is depleted.

Termination can be avoided or suppressed in ionic and metathesis polymeriza-

* This article is reprinted with permission from C. Pugh, A. Kiste, *Progress in Polymer Science*, Vol. 22: Molecular Engineering of Side-Chain Liquid Crystalline Polymers by Living Polymerization, (**1997**), Elsevier Science Ltd., Oxford, England.

tions under appropriate conditions. In contrast to ionic polymerizations, in which the growing species can not react with each other due to their like charge, termination does occur in radical polymerizations by combination or disproportionation of two growing polymeric radicals. Therefore, termination is unavoidable in radical polymerizations.

In contrast to the synthesis of small molecules, it is impossible to obtain from a polymerization a completely pure product that has a unique chain length and therefore a unique molecular weight. Chains of vastly different lengths will form if there are multiple propagating species with varying activities which do not exchange rapidly. This distribution of chain lengths and molecular weights is described by the moments of a statistical distribution of molecular weights (Eq. 1). The first moment of the distribution is the number average molecular weight (M_n), which is the total weight of the sample $(\sum w_i)$ divided by the number of chains in that sample $(\sum n_i)$. The weight (M_w) and z-average (M_z) molecular weights are the second and third moments of the distribution, respectively. That is, $M_z > M_w > M_n$.

$$\overline{M}_n = \frac{\sum n_i M_i}{\sum n_i} = \frac{\sum w_i}{\sum n_i}$$

$$\overline{M}_w = \frac{\sum n_i M_i^2}{\sum n_i M_i} = \frac{\sum w_i M_i}{\sum w_i}$$

$$\overline{M}_z = \frac{\sum n_i M_i^3}{\sum n_i M_i^2} = \frac{\sum w_i M_i^2}{\sum w_i M_i} \qquad (1)$$

n_i = number of chains of length i,
M_i = molecular weight of chains of length i.

However, the thermotropic behavior of SCLCPs is generally most accurately described as a function of the number average degree of polymerization (DP_n), rather than the average molecular weight. As shown in Eq. (2), the number (DP_n), weight (DP_w) and z-average (DP_z) degrees of polymerization correspond to the number, weight and z-average molecular weights, respectively.

$$\overline{M}_n = \overline{DP}_n M_0$$
$$\overline{M}_w = \overline{DP}_w M_0$$
$$\overline{M}_z = \overline{DP}_z M_0$$

M_0 = molecular weight of repeat unit and/or monomer. (2)

Since the number of polymer chains resulting from a chain polymerization is proportional to the amount of initiator used, the degree of polymerization is inversely proportional to initiator concentration. The ratio of the concentration of reacted monomer and polymer chains is determined by the ratio of the rates of propagation (R_p) and all chain forming reactions, including transfer (R_{tr}) (Eq. 3).

$$\overline{DP}_n = \frac{\Delta[M]}{[Polymer]} = \frac{R_p}{R_i + R_{tr} + \dots} \qquad (3)$$

As shown in Fig. 1, high molecular weight polymer forms rapidly in a chain polymerization if initiation is slow and propagation is relatively fast, as in classic radical polymerizations. The molecular weight is limited by one of the chain breaking reactions (transfer or termination). In this case, the polymerization system consists of monomer and high molecular weight polymer at all stages of the polymerization, with only the yield of polymer increasing with increasing monomer conversion. In a radical polymerization which terminates by coupling and which has a constant rate of initiation, high molecular weight polymer forms at low (<1%) conversion. However, the average molecular weight decreases with further conversion since the number average degree of polymerization (DP_n) is determined by the ratio of the rates of propagation and initiation.

The breadth of the molecular weight distribution is described by the ratio of the weight and number average molecular weights or degrees of polymerization, and is referred to as the polydispersity index

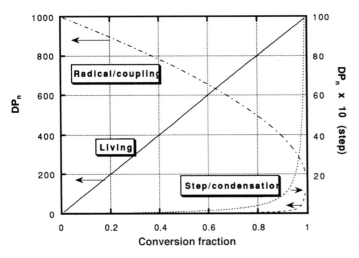

Figure 1. Polymer molecular weight as a function of monomer conversion in chain-growth, step-growing and living polymerizations [4].

(pdi) or molecular weight distribution (mwd) (Eq. 4).

$$\text{pdi} = \frac{\overline{M}_w}{\overline{M}_n} = \frac{\overline{DP}_w}{\overline{DP}_n} \tag{4}$$

Chain polymerizations which do not terminate by bimolecular chain coupling generally result in polymers with the most probable molecular weight distribution of 2.0; free radical polymerizations which terminate by bimolecular chain coupling result in pdi = 1.5. However, chain transfer to polymer, autoacceleration, slow initiation, and slow exchange between active species of different reactivities result in much higher polydispersities.

1.2 Living Polymerizations

Since the 1960s, much of the research in polymer synthesis has been directed at establishing living conditions for chain polymerizations [5–8]. The only requirement for a polymerization to be considered truly living is that all chain breaking reactions such as chain transfer and termination are absent. That is, the rate constants of both chain transfer and termination should be equal to zero ($k_{tr} = 0$, $k_t = 0$). As shown in Fig. 2 for a polymerization which is first order in monomer, a change in the number of active species due to termination is detected by a nonlinear dependence of monomer conversion as a function of time. (Such plots have never been used to detect termination in polymerizations of mesogenic monomers.)

In contrast, transfer reactions are not detected by following the monomer conversion if the rate of reinitiation is comparable to that of propagation. In this case, transfer is detected by a nonlinear dependence of the polymer molecular weight or degree of polymerization as a function of monomer conversion, polymer yield or the ratio of monomer to initiator concentrations, $[M]_0/[I]_0$ (Fig. 2). However, a linear dependence of molecular weight on conversion is often used erroneously to demonstrate that a polymerization is living [9–11]. That is, termination does not affect the number of chains in the system, and therefore does not affect M_n or DP_n if initiation is fast and monomer conversion goes to completion (i.e. if all of the chains are not terminated).

In practice, linear semilogarithmic kinetic plots and linear dependencies of molecular weight on monomer conversion require only that the rate constants of chain transfer

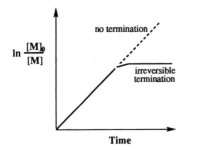

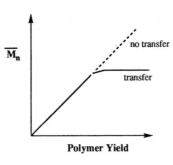

Figure 2. The effect of termination and transfer reactions on monomer conversion and polymer molecular weight.

and termination are much less than that of propagation ($k_p/k_t \geq 10 \times [M]_0/[I]_0 \approx 10 \times DP_n$; $k_p/k_{tr} \geq 10 \times [M]_0/[I]_0 \approx 10 \times DP_n$) [10, 11]. Termination is therefore less important at lower ratios of monomer to initiator (Eq. 5).

$$\ln([M]_0/[M]_\infty) = [I]_0 k_p/k_t \qquad (5)$$

This is therefore the practical requirement for the synthesis of well-defined polymers, such that complete monomer conversion can be reached and the chain ends can be quantitatively functionalized. However, since chain breaking reactions are actually present, such systems are more appropriately labeled *controlled* polymerizations rather than *living* polymerizations. In fact, conditions have recently been established for controlled radical polymerizations, even though it is impossible to avoid bimolecular termination [12–20]. The extent of the 'livingness' or controllability of a polymerization can be ranked if the individual or relative rate constants of propagation, transfer and termination are known [10, 11].

If initiation is also faster than, or at least comparable to, propagation all chains begin growing simultaneously and the molecular weight of the polymer increases linearly with conversion. In this case, the degree of polymerization is determined by the ratio of the concentrations of reacted monomer and initiator (Eq. 6). Although this formally

$$DP_n = \frac{\Delta[M]}{[I]_0} = \frac{[M]_0 - [M]_t}{[I]_0} \qquad (6)$$

requires that $k_i = \infty$, controllable molecular weight is obtained if the rate constant of initiation is greater than or comparable to that of propagation ($k_i \geq k_p$) [21]. If initiation is slow, the number average molecular weight is initially too high but becomes equal to the theoretical value once initiator is consumed.

In addition to the absence of chain breaking reactions and fast initiation, two additional requirements must be met to obtain polymers with narrow molecular weight distributions: all chains must have equal reactivity ($k_p^x = k_p^y = k_p^z$, etc.) or the active sites must exchange rapidly, and propagation must be irreversible ($k_p > k_d$). In this case, the molecular weight distribution depends on only the degree of polymerization as defined by the Poisson distribution (Eq. 7).

$$\frac{DP_w}{DP_n} = 1 + \frac{DP_w}{(1 + DP_n)^2} = 1 + \frac{1}{DP_n} \cong 1 \qquad (7)$$

Living polymerizations in which initiation is fast and quantitative and which have irreversible growth offer several advantages over conventional polymerizations. In addition to the ability to obtain polymers with controlled molecular weights and narrow molecular weight distributions, it is also possible to control the polymer architecture and chain end functionality. For example, diblock and triblock copolymers containing liquid crystalline blocks have been prepared by living polymerizations.

2 'Living' Polymerizations used to Synthesize Side Chain Liquid Crystalline Polymers

The elementary synthetic and structural principles of SCLCPs are now being elucidated using living polymerizations in which the effect of a single structural feature can be isolated, while other structural variables remain constant. In particular, living polymerizations are used to determine the molecular weight at which the thermotropic be-

2.1 Anionic and Group Transfer Polymerizations of Olefins

Alkenes polymerize anionically by nucleophilic attack of a growing carbanion at the least hindered position of the double bond of the monomer. Therefore, the monomer must be electrophilic and capable of stabilizing the resulting negative charge. In addition, the double bond must be the most electrophilic functionality in the monomer. Some vinyl monomers which can polymerize anionically are listed in Eq. (8) in their

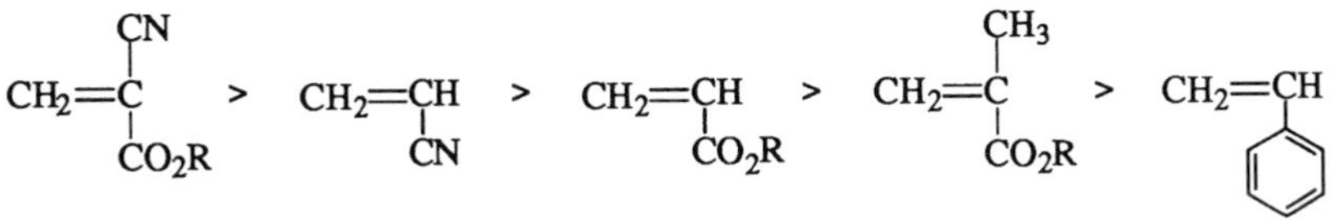

$$
\underset{\underset{CO_2R}{|}}{\overset{\overset{CN}{|}}{CH_2{=}C}} \quad > \quad \underset{\underset{CN}{|}}{CH_2{=}CH} \quad > \quad \underset{\underset{CO_2R}{|}}{CH_2{=}CH} \quad > \quad \underset{\underset{CO_2R}{|}}{\overset{\overset{CH_3}{|}}{CH_2{=}C}} \quad > \quad CH_2{=}CH \bigcirc \tag{8}
$$

havior of SCLCPs no longer depends on the addition or subtraction of another repeat unit. Additional comparisons should then be made only between samples whose thermotropic behavior is independent of molecular weight. In addition, living polymerizations with fast and quantitative initiation can be used to minimize the effect of polydispersity, although it is questionable as to whether or not polydispersity has any effect on the thermotropic behavior of SCLCPs [22, 23]. Living chain ends are also required for preparing block copolymers by sequential monomer addition, and are convenient for preparing statistical copolymers whose composition depends on conversion. If the copolymerization is living, copolymers with compositions equal to that of the initial monomer feed are guaranteed by complete comonomer conversion. Thus far, anionic, cationic, metalloporphyrin, and ring-opening metathesis polymerizations have been used to prepare SCLCPs in a controlled manner.

order of reactivity, which corresponds to the electron-withdrawing ability of their substituents. However, the reactivity of the growing carbanions (and enolate anions) follows the opposite order shown above, which the most stable carbanions being the least reactive [5, 6, 24].

The choice of initiator is very important for a controlled anionic polymerization. The initiator will be nucleophilic enough to initiate polymerization if the pK_a of the conjugate acid is similar to or greater than that of the propagating anion [25]. However, the initiator should not be so reactive that it causes side reactions. Common solvents for anionic polymerizations include hydrocarbons and ethers. Protonic solvents, dissolved oxygen, and carbon dioxide result in either transfer or termination in most anionic polymerizations, depending on the nucleophilicity of the anion generated, and are therefore not used. Halogenated solvents also terminate anionic polymerizations and should be avoided.

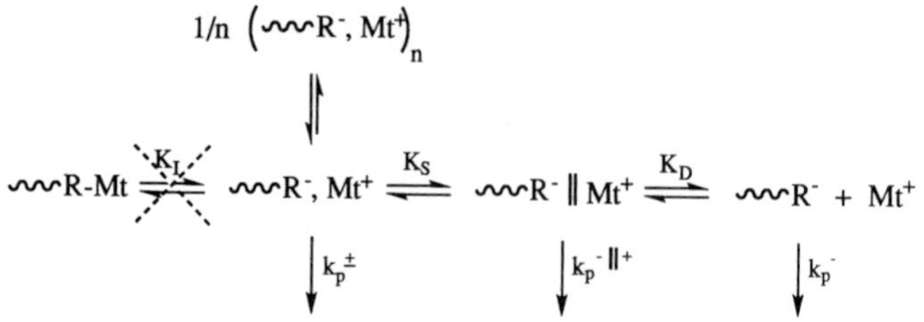

$$1/n \left(\text{\large\raisebox{2pt}{$\sim\!\!\!\sim\!\!\!\sim$}} R^-, Mt^+ \right)_n$$

$$\sim\!\!\!\sim\!\!\!\sim R\text{-}Mt \underset{\substack{K_L}}{\overset{}{\rightleftharpoons}} \sim\!\!\!\sim\!\!\!\sim R^-, Mt^+ \overset{K_S}{\rightleftharpoons} \sim\!\!\!\sim\!\!\!\sim R^- \parallel Mt^+ \overset{K_D}{\rightleftharpoons} \sim\!\!\!\sim\!\!\!\sim R^- + Mt^+$$

$$\Big\downarrow k_p^{\pm} \qquad\qquad \Big\downarrow k_p^- {}^{\parallel +} \qquad\qquad \Big\downarrow k_p^-$$

Scheme 2

In contrast to radical polymerizations in which there is only one type of propagating species, ionic polymerizations may involve several active species, each with different reactivities and/or lifetimes. As outlined in Scheme 2, ionic polymerizations may potentially involve equilibria between covalent dormant species, contact ion pairs, aggregates, solvent separated ion pairs, and free ions. Although ion pairs involving alkali metal countercations can not collapse to form covalent species, group transfer polymerization apparently operates by this mechanism. In anionic polymerizations, free ions are much more reactive than ion pairs ($k_p^- \approx 10^5 k_p^{\pm}$), although the dissociation constants are quite small ($K_D \approx 10^{-7}$) [5]. The equilibria between these different species in anionic polymerizations of nonpolar alkenes are dynamic enough relative to the rate of propagation to generate monomodal and narrow molecular weight distributions [26, 27].

Although anionic polymerizations of nonpolar monomers such as styrene and budadiene are the most inherently living polymerizations available [5, 6], only one mesogenic styrene has been polymerized by an anionic mechanism [28]. As shown

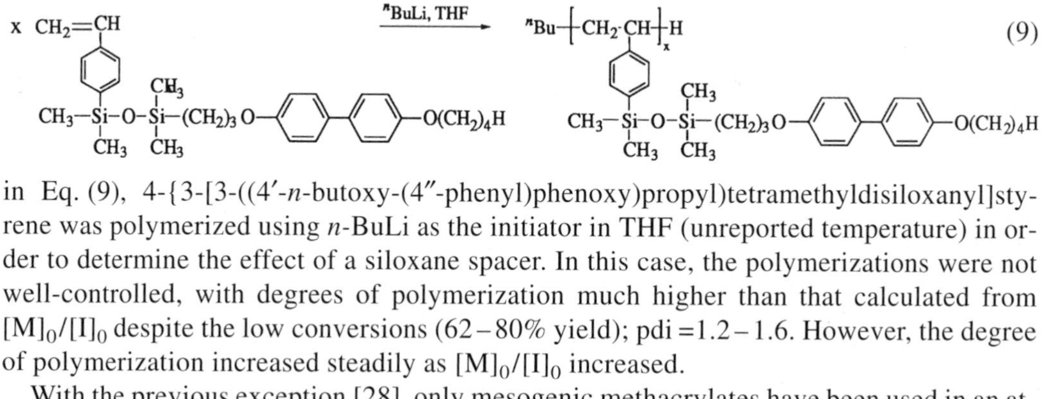

$$x \ CH_2{=}CH \qquad \xrightarrow{\text{\textit{n}BuLi, THF}} \qquad {}^nBu{-}\!\!\left[CH_2{-}CH\right]{-}\!H \qquad\qquad (9)$$

in Eq. (9), 4-{3-[3-((4′-n-butoxy-(4″-phenyl)phenoxy)propyl)tetramethyldisiloxanyl]styrene was polymerized using n-BuLi as the initiator in THF (unreported temperature) in order to determine the effect of a siloxane spacer. In this case, the polymerizations were not well-controlled, with degrees of polymerization much higher than that calculated from $[M]_0/[I]_0$ despite the low conversions (62–80% yield); pdi =1.2–1.6. However, the degree of polymerization increased steadily as $[M]_0/[I]_0$ increased.

With the previous exception [28], only mesogenic methacrylates have been used in an attempt to prepare well-defined SCLCPs directly by an anionic addition mechanism. In this case, the most troublesome side reactions to avoid are intermolecular (Eq. 10), and intra-

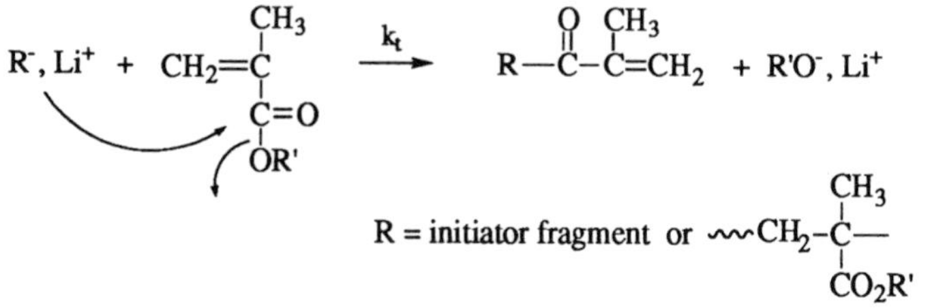

$$R^-, Li^+ + CH_2{=}\underset{\substack{|\\C=O\\|\\OR'}}{\overset{\substack{CH_3\\|}}{C}} \xrightarrow{k_t} R{-}\overset{\substack{O\ CH_3\\\parallel\ |}}{C}{-}C{=}CH_2 + R'O^-, Li^+$$

$$(10)$$

$$R = \text{initiator fragment or} \ \sim\!\!\!\sim CH_2{-}\underset{\substack{|\\CO_2R'}}{\overset{\substack{CH_3\\|}}{C}}{-}$$

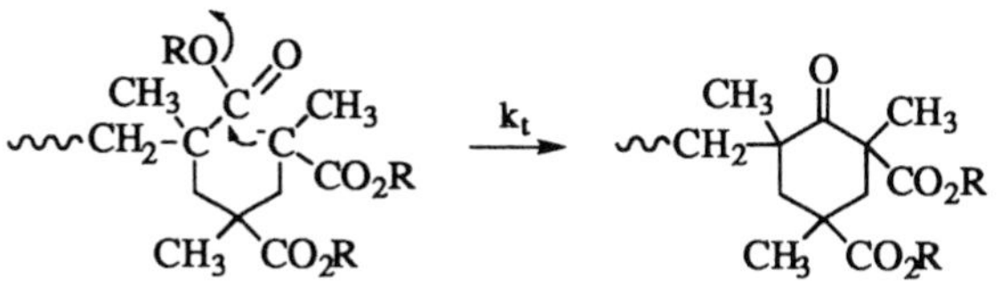

$$\text{\tiny (structure)} \xrightarrow{k_t} \text{\tiny (structure)} + RO^- \qquad (11)$$

molecular (Eq. 11), termination by nucleophilic attack at the carbonyl group of the monomer or polymeric repeat unit to produce unreactive alkoxide anions [29]; other electrophilic functionalities within the mesogen are also susceptible to nucleophilic attack. This is minimized in the initiation step by using a bulky initiator such as 1,1-diphenylhexyl lithium at low temperature (−78 °C). Low temperature and more polar solvents such as THF also minimize nucleophilic attack by the relatively hindered enolate anion during propagation [30].

In addition, the propagating ion pairs and free ions of these polar monomers may exchange slowly relative to propagation to produce polymers with multimodal molecular weight distributions [31–34]. For example, anionic polymerization of t-butyl-methacrylate results in broad (pdi = 2.00 – 2.72) and multimodal molecular weight distributions when performed in a mixture of toluene and THF at −78 °C using 1,1-diphenylhexyl lithium as the initiator [35]. However, narrow and unimodal molecular weight distributions are produced in either pure THF or pure toluene, in which propagation apparently occurs exclusively by free ions or ion pairs, respectively. Although 1,1-diphenyl-3-methylpentyl lithium and 1,1-diphenylhexyl lithium initiated polymerizations of methyl methacrylate in THF at −78 °C effectively yield 'living' polymers

with narrow molecular weight distributions (pdi ≈ 1.15), addition of lithium chloride decreases the rate of polymerization and narrows the polydispersity to 1.09 [36]. These conditions are also effective at controlling the anionic polymerization of t-butyl acrylate [37]. As outlined in Scheme 3, lithium chloride narrows the polydispersity by shifting the equilibrium to LiCl adducts, thereby breaking up less reactive aggregates of the propagating polymer and shifting the equilibrium from free ions to ion pairs [38].

In addition to determining the effect of molecular weight on the thermotropic behavior of SCLCPs, 'living' anionic polymerizations have been used extensively to study the effect of tacticity. Although toluene is generally used as the nonpolar solvent in anionic polymerizations of methacrylates to promote association and isotactic placement [39, 40] it is a poor solvent for mesogenic methacrylates and/or the corresponding polymers. This results in only low yields of low molecular weight polymers with insufficient levels of isotacticity. Highly isotactic poly{6-[{4′-[4″-[(S)-2‴-methylbutoxy]phenyl]phenoxy}hexyl methacrylate} and poly{[4-(4′-methoxyphenyl)-phenoxy]-n-alkyl methacrylate}s (n=2–6) were instead obtained in chloroform at −20 to −30 °C using t-BuMgBr as the initiator (Table 1), although termination evidently broadened the polydispersity to 1.54–3.06

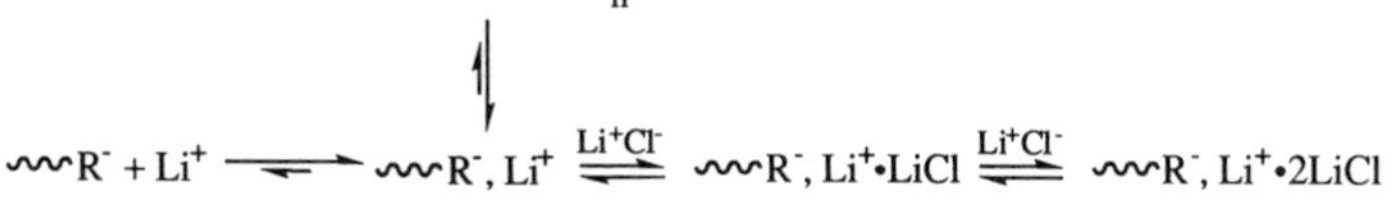

$$\tfrac{1}{n}\left(\text{\tiny ⟂⟂R}^-, \text{Li}^+\right)_n$$

$$\text{\tiny ⟂⟂R}^- + \text{Li}^+ \rightleftharpoons \text{\tiny ⟂⟂R}^-, \text{Li}^+ \underset{}{\overset{\text{Li}^+\text{Cl}^-}{\rightleftharpoons}} \text{\tiny ⟂⟂R}^-, \text{Li}^+\!\cdot\!\text{LiCl} \underset{}{\overset{\text{Li}^+\text{Cl}^-}{\rightleftharpoons}} \text{\tiny ⟂⟂R}^-, \text{Li}^+\!\cdot\!2\text{LiCl}$$

Scheme 3

Table 1. Anionic polymerizations of n-[4′-(4″-methoxyphenyl)phenoxy]alkyl methacrylates ($n = 2 - 6$).

Initiating system	Solvent	Temp. (°C)	$\dfrac{[M]_0}{[I]_0}$	GPC[a]		Tacticity	Ref.
				DP_n	pdi		
t-BuMgBr/Et$_2$O	CHCl$_3$	−30 to −20	10	13 – 39	1.54 – 3.06	(mm) = 0.90 – 0.93	[42]
			20	24 – 68	2.35 – 14.0	(mm) = 0.90 – 0.96	[42]
t-BuMgBr/Et$_2$O	THF	−78 to −60	10	14 – 21	1.23 – 2.27	(rr) = 0.79 – 0.81	[42]
			20	19 – 33	1.23 – 1.57	(rr) = 0.79 – 0.85	[42]

CH$_3$ Ph
  | |
CH$_3$CH$_2$CHCH$_2$C⁻, Li⁺
  |
3 LiCl   Ph
	THF	−40	11	13[b]	1.13	(rr) = 0.80	[44]

			22	23[b]	1.26	(rr) = 0.80	[44]

[a] Relative to polystyrene standards.
[b] Relative to PMMA standards.

at $[M]_0/[I]_0 = 10$, and to pdi = 2.35 – 14.0 at $[M]_0/[I]_0 = 20$ [41, 42]. Termination of the propagating chains is also demonstrated by the lower conversions at higher attempted degrees of polymerization, whereas partial consumption of the initiator is demonstrated by the much higher degrees of polymerization obtained relative to the theoretical values, even at these low polymer yields.

Table 1 demonstrates that t-BuMgBr initiated polymerizations in THF at −60 to −78 °C are more controlled, yielding polymers with narrower polydispersities [41, 42]. These conditions involve freely propagating chains and therefore result in syndiotactic polymers at low temperature [43]. Although 'living' polymethacrylates with the highest syndiotacticities [(rr) = 0.92 – 0.96] have been achieved using t-BuMgBr/(n-C$_8$H$_{17}$)$_3$Al (1:3) in toluene at −78 °C [40], these conditions have not been applied to mesogenic methacrylates, probably because of the corresponding polymers limited solubility in toluene. Instead, the initiating system developed by Teyssié [36] of

1,1-diphenyl-3-methylpentyl lithium in THF in the presence of at least three equivalents of LiCl has been used most successfully for preparing well-defined syndiotactic poly{n-[4′-(4″-methoxyphenyl)phenoxyl]alkyl methacrylate}s with narrow molecular weight distributions (pdi = 1.05 – 1.46) (Table 1) [44]. Although the polymers precipitated during the polymerization at −78 °C, Fig. 3 demonstrates that the polymerization is controlled and transfer is not detectable at −40 °C. However, the polymers precipitate during the polymerization when $[M]_0/[I]_0 > 40$, even at −40 °C, resulting in bimodal molecular weight distributions.

In the absence of LiCl, 1,1-diphenyl-3-methylpentyl lithium initiated polymerizations of 6-[4′-(4″-methoxyphenoxycarbonyl)phenoxy]hexyl methacrylate [45] and 4-[4′-(4″-methoxyphenylazo)phenoxy]butyl methacrylate [46] in THF at −70 °C yielded polymers (Scheme 4) with pdi = 1.09 – 1.31 and molecular weights close to the theoretical values up to $[M]_0/[I]_0 = 50$.

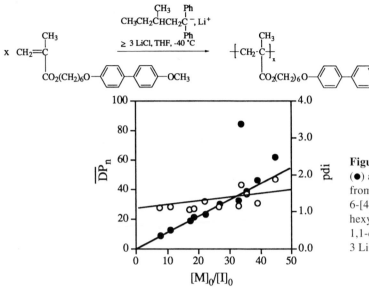

Figure 3. Degree of polymerization (●) and polydispersity (○) resulting from anionic polymerizations of 6-[4′-(4″-methoxyphenyl)phenoxy]-hexyl methacrylate initiated by 1,1-diphenyl-3-methylpentyl lithium/ 3 LiCl in THF at $-40\,°C$ [44].

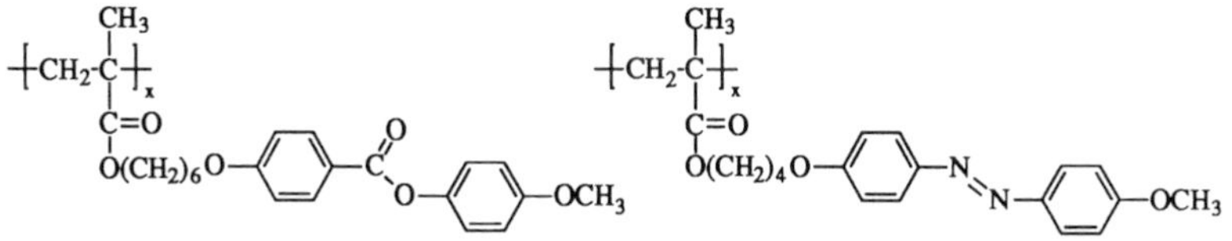

Scheme 4

In the latter case, the (co)polymerization was improved by adding a trace amount of triisobutylaluminium to the monomer solution in order to remove any residual water [46].

SCLCPs have also been prepared by group transfer polymerization (GTP) of mesogenic methacrylates are room temperature [47, 48]. The livingness of the nucleophilic catalyzed GTP was originally attributed to a new mechanism (Eq. 12), in-

volving a (symmetry-forbidden [49, 50]) eight-member cyclic transition state in which the trialkylsilyl group is transferred from the growing chain end to the incoming monomer [51].

Alternatively, GTP may be an anionic polymerization in which a small concentration of enolate anions are in dynamic equilibrium with dormant silyl ketene acetal chain ends (Scheme 5). This dynamic equi-

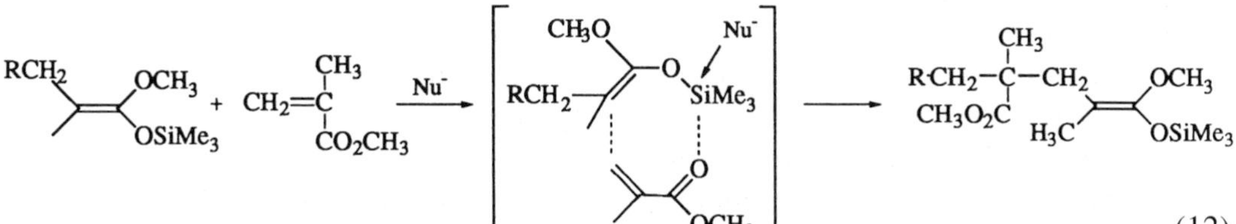

$$(12)$$

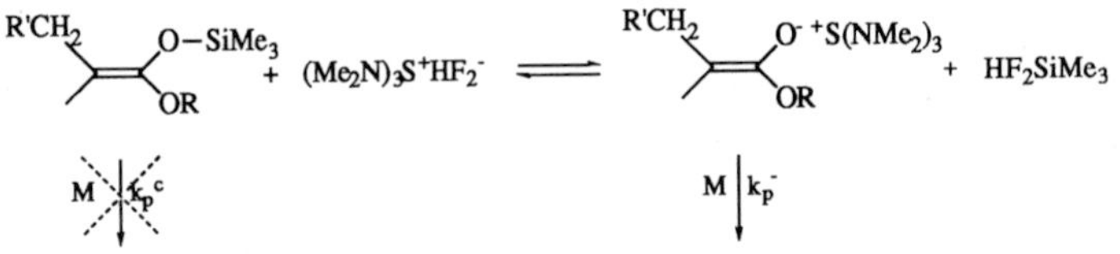

Scheme 5

librium between a small concentration of enolate anions and a high concentration of dormant chain ends would reduce the lifetime of the active enolate anions, thereby accounting for the reduced rate [52, 53] of polymerization compared to classic anionic polymerizations of methacrylates. This shift in the equilibrium to only one type of active species would also account for the narrower polydispersities. Since the countercations are bulky tris(dimethylamino)sulfonium [51] or tetrabutylammonium [54] ions, the active species would be free ions stabilized by solvent, which are less reactive towards intramolecular backbiting (Eq. 11).

Most experimental results indicate that GTP is a classic anionic polymerization operating by reversible termination (dissociative mechanism) [55, 56], and/or degenerative transfer [57, 58]. For example, termination occurs once monomer is consumed by backbiting at the pen-penultimate carbonyl to generate cyclic β-ketoester endgroups as in classic anionic polymerizations (Eq. 11) [59]. Acids with $pK_a < 18$ also terminate 'group transfer' polymerizations, whereas

so similar to those in anionic copolymerizations [55, 61, 62] rather than $r_1 \approx r_2 \approx 1$ as expected for an associative mechanism. However, the most compelling evidence supporting an anionic mechanism is that the stereochemistry resulting from group transfer polymerizations is nearly identical to that of an anionic polymerization involving freely propagating enolate anions [55, 63, 64]. That is, the polymers are not highly isotactic as expected for the coordinated chain ends of an associative mechanism. Instead, freely propagating chains favor syndiotactic placement, especially at low temperature. In addition, ^{1}H-NMR experiments recently demonstrated that the silyl end groups of two different living polymethacrylates exchange in the presence of tris(dimethylamino)sulfonium bifluoride [65].

6-[4'-(4''-Methoxyphenoxycarbonyl)-phenoxy]hexyl methacrylate was first polymerized by GTP using 1-methoxy-1-(trimethylsiloxy)-2-methyl-1-propene as the initiator and tris(dimethylamino)sulfonium bifluoride as the catalyst in THF at room temperature (Eq. 13) [47].

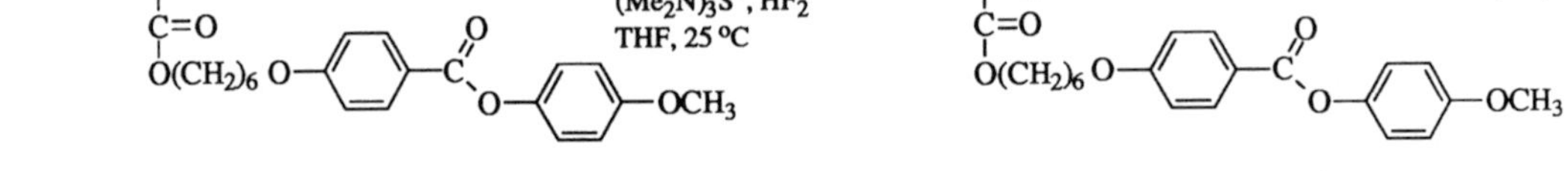

$$(13)$$

acids with $pK_a = 18-25$ act as chain transfer agents [60]. The reactivity ratios in GTP copolymerizations of methacrylates are al-

Although the polymer yield was high, the polymerization was not well-controlled, apparently due to termination at the phenyl benzoate carbonyl group of the mesogen, resulting in broader molecular weight distributions (pdi $= 1.64 - 1.77$) and molecular weights substantially different than that calculated by $[M]_0/[I]_0$. As stated previously, molecular weights closer to the theoretical values were obtained by anionic polymerization of this monomer up to $[M]_0/[I]_0 = 50$ using the bulky anionic initiator, 1,1-diphenyl-3-methylpentyl lithium, at low temperature ($-70\,°C$) in THF [45]. However, the molecular weight distribution was still higher than desired (pdi $= 1.19 - 1.31$).

Relatively controlled 'group transfer' polymerizations are possible when the monomer lacks a phenylbenzoate group (Eq. 14). Poly[6-(4'-methoxy-4"-α-methylstilbeneoxy)hexyl

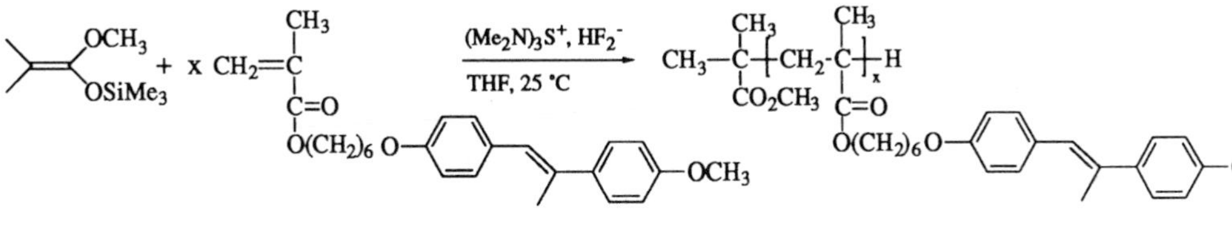

$$\tag{14}$$

methacrylate] was prepared using 1-methoxy-1-(trimethylsiloxy)-2-methyl-1-propene as the initiator and tris(dimethylamino)sulfonium bifluoride as the catalyst in THF at room temperature ($DP_n = 2.8 - 28$, pdi $= 1.09 - 1.36$) [48].

2.2 Polymerizations with Metalloporphyrins

Aluminum porphyrins initiate controlled ring-opening polymerizations of oxiranes [67 – 69] β-lactones [70 – 72], δ-valerolactone [74], ε-caprolactone [74] and D-lactide [75], as well as controlled addition polymerizations of methacrylates [76] and methacrylonitrile [77] (Eq. 15). As shown in Eq. (16), propagation occurs by a coordinative anionic mechanism

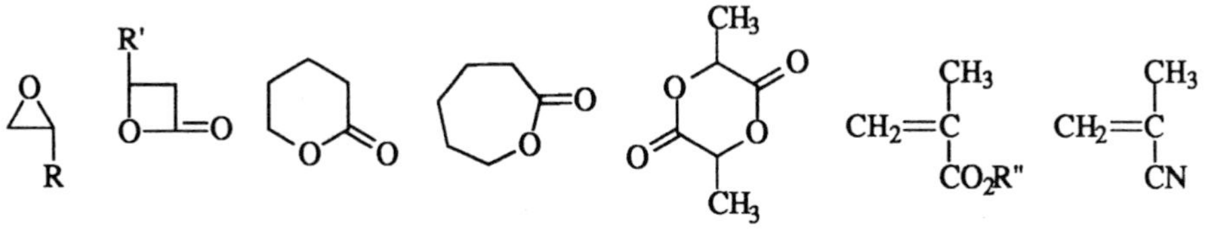

$$\tag{15}$$

R = -H, -CH$_3$, -CH$_2$CH$_3$, -CH$_2$Cl, -CH$_2$OCH$_3$
R' = -H, -CH$_3$
R" = -CH$_3$, -Et, -^{i}Pr, -^{n}Bu, -^{i}Bu, -CH$_2$Ph, -(CH$_2$)$_{12}$H

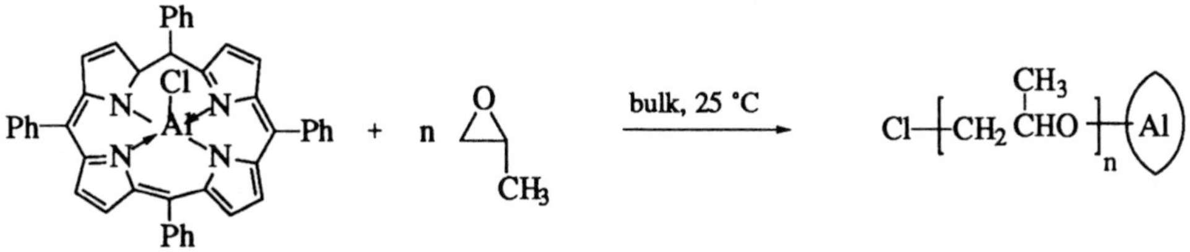

$$\tag{16a}$$

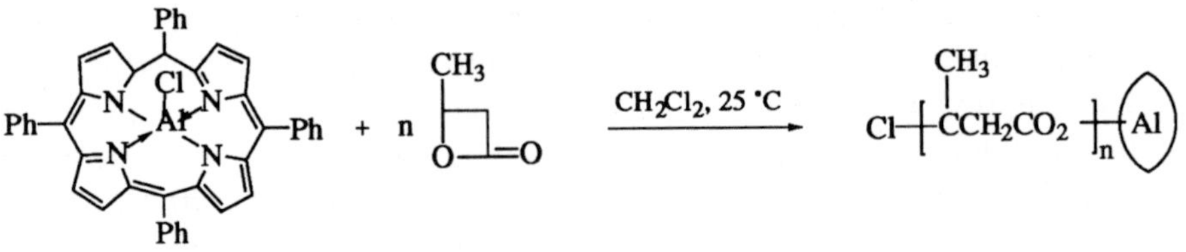

$$(16b)$$

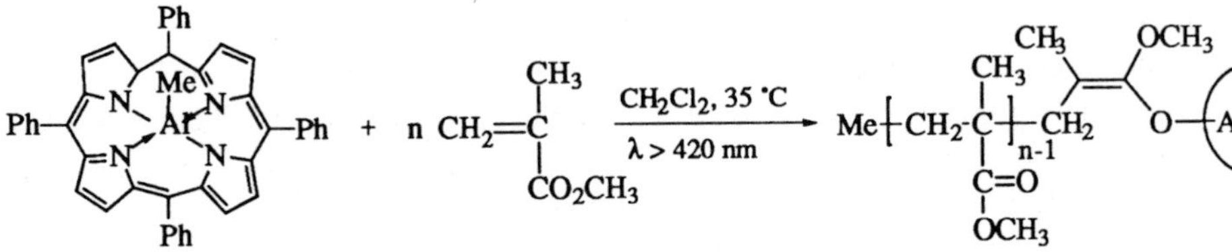

$$(16c)$$

with insertion of the monomer into the aluminum-axial ligand bond of (5,10,15,20-tetra-phenylporphinato)aluminum derivatives [(TPP)AlCl, (TPP)AlMe, (TPP)AlOMe] to generate an aluminum alkoxide (oxiranes [69], δ-valerolactone [73], ε-caprolactone [78], D-lactide [75]), aluminum carboxylate (β-lactones) [71] or aluminum enolate (methacrylates) [76] as the growing species. Inoue coined the term 'immortal' [79] to describe these polymerizations because the active chain ends are not 'killed' by proton sources. For example, the oxirane polymerizations undergo rapid and reversible exchange with alcohols such as methanol, hydroxyethyl methacrylate and polyethylene glycol (Eq. 17) [78–81]; lactone

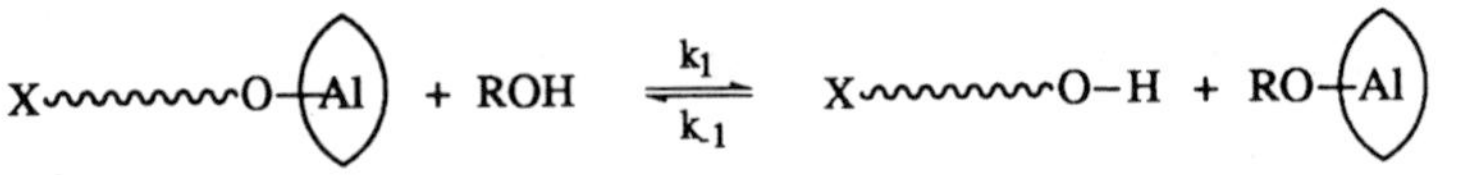

$$(17a)$$

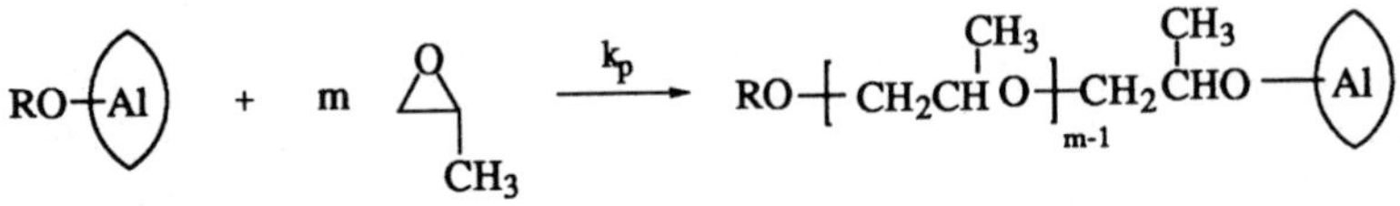

$$(17b)$$

chain ends exchange rapidly with carboxylic acids [81]. In this case, the degree of polymerization corresponds to the molar ratio of reacted monomer to the sum of the initial concentrations of initiator and proton source ($\Delta[M]/[TPP\text{-}AlX]+[ROH]_0$). Since the reversible termination is rapid compared to propagation ($k_1/k_p = 10$) [81], the polydispersity of the resulting polymers remains narrow at pdi ≈ 1.1.

Although the aluminum porphyrin initiated polymerizations take days or weeks to go to complete monomer conversion in the presence of a proton source, their rate is increased substantially by adding a sterically hindered Lewis acid such as methylalu-

minum bis(2,6-di-t-butyl-4-methylphenolate) [82]. In contrast to the oxirane and β-lactone polymerizations, ε-caprolactone and 2,2-dimethyltrimethylene carbonate are polymerized by methylaluminum bis(2,6-diphenylphenolate) in the absence of an aluminum porphyrin; these polymerizations are relatively rapid and controlled in the presence of i-propanol or methanol [83].

In contrast to the heterocyclic polymerizations, (TPP)AlCl does not initiate polymerization of methacrylates [76] and methacrylonitrile [77], whereas (TPP)AlMe initiation requires irradiation by visible light. These extremely slow polymerizations are also accelerated by addition of bulky Lewis acids such as methylaluminum bis(2,6-di-*t*-butylphenolate) [77, 84–86] triphenylphosphine [86], or organoboron [87] compounds such as triphenylboron or tris(pentafluorophenyl)boron [88]. These Lewis acids accelerate the polymerization by coordinating to the monomer's carbonyl group, thereby increasing the olefin's electrophilicity. In addition, light is not needed to polymerize methacrylates if more nucleophilic thiolatealuminum porphyrins such as propylthio- and phenylthio(5,10,15,20-tetraphenylporphinato)aluminum [(TPP)-AlSPr, (TPP)AlSPh] are used, preferably in the presence of methylaluminum bis(2,6-di-*t*-butyl-4-methylphenolate) [89]. In all cases, the molecular weight is determined by the ratio of the concentrations of reacted

monomer to initiator up to at least $[M]_0/[I]_0 = 300$, and the polydispersity is narrow at pdi $= 1.05-1.2$.

Photoirradiated (TPP)AlMe was used to polymerize 6-[4'-(4''-methoxyphenoxycarbonyl)phenoxy]hexyl methacrylate [90], 6-[4'-(4''-*n*-butoxyphenoxycarbonyl)phenoxy]hexyl methacrylate [91] and 6-[4'-(4''-cyanophenoxycarbonyl)phenoxy]hexyl methacrylate [92] in order to determine the effect of molecular weight and tacticity [(*rr*) ≈ 0.75]. Although 'high speed' conditions [84–89] were not used, the polymerizations reached 76–92% conversion in 10 h to produce polymers with the expected molecular weight ($DP_n < 35$) and pdi $= 1.09-1.35$. In contrast, that of 6-[4'-(4''-cyanophenyla-zo)phenoxy]hexyl methacrylate (Eq. 18)

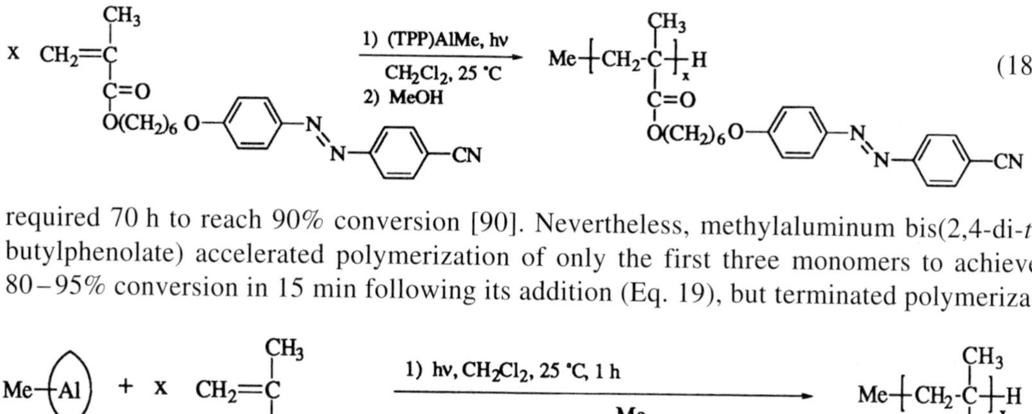

(18)

required 70 h to reach 90% conversion [90]. Nevertheless, methylaluminum bis(2,4-di-*t*-butylphenolate) accelerated polymerization of only the first three monomers to achieve 80–95% conversion in 15 min following its addition (Eq. 19), but terminated polymeriza-

(19)

tion of 6-[4'-(4''-cyanophenylazo)phenoxy]hexyl methacrylate [90].

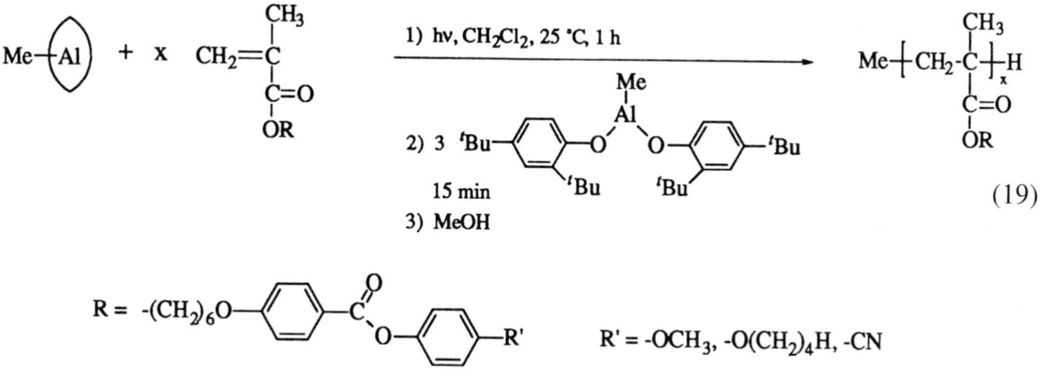

2.3   Cationic Polymerizations of Olefins

Alkenes polymerize cationically by electrophilic addition of the monomer to a growing carbenium ion [8]. Therefore, the monomer must be nucleophilic and capable of stabilizing the resulting positive charge. In addition, the double bond must be the most nucleophilic functionality in the monomer. Some vinyl monomers which polymerize cationically are listed in Eq. (20) in their order of reactivity, which corresponds to the electron-do-

$$\text{(20)}$$

nating ability of their substituents. Sufficiently nucleophilic alkenes such as vinyl ethers, styrenes and isobutylene polymerize cationically to generate stabilized carbenium ions as the propagating species. However, the reactivity of the growing carbenium ions follows the opposite order shown above, with the most stable carbenium ions being the least reactive.

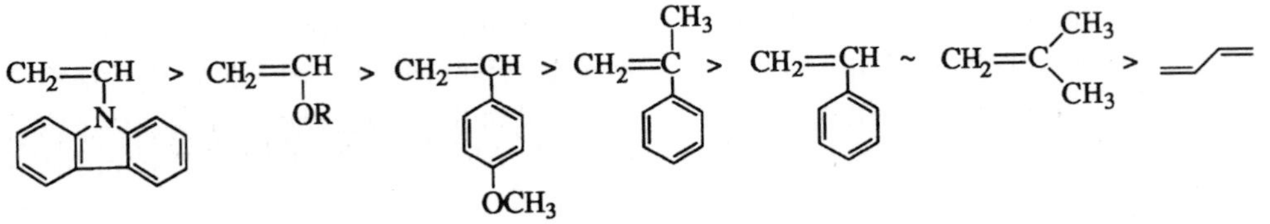

Initiators include strong protonic acids and the electrophilic species generated by reaction of a Lewis acid with water, an alcohol, ester or alkyl halide (Scheme 6). In this case, initiation occurs by electrophilic addition of the vinyl monomer to the proton or carbenium ion, which may be accompanied by reversible or irreversible collapse of the ion pair (Eq. 21).

$$\text{H}^+, \text{MtX}_n^- + \text{CH}_2\!\!=\!\!\underset{\text{R}}{\text{CH}} \;\rightleftharpoons\; \text{CH}_3\!\!-\!\!\underset{\text{R}}{\text{CH}}^+, \text{MtX}_n^- \;\rightleftharpoons\; \text{CH}_3\!\!-\!\!\underset{\text{R}}{\text{CH}}\!\!-\!\!\text{X} + \text{MtX}_{n\text{-}1} \qquad \text{(21)}$$

As demonstrated by the styrene polymerization in Scheme 7, propagation occurs in carbocationic polymerizations by electrophilic addition of the vinyl monomer to a growing carbenium ion. The resulting carbenium ions are very reactive and therefore difficult to control, with rate constants of propagation $k_p = 10^4 - 10^6 \, 1 \, (\text{mol s})^{-1}$. In addition, a significant amount of the positive charge is distributed over the β-H atoms, making them prone to abstraction by either monomer ($k_{tr,M}$) or counteranion (k_{tr}). Chain transfer by β-proton elimination from the propagating carbenium ions is the most common and detrimental side reaction in cationic polymerizations of alkenes. The first requirement for a controlled cationic polymerization is therefore to use only components which are nonbasic in order to reduce the possiblity of transfer by β-proton elimination. In the case of styrene polymerizations,

$$
\begin{aligned}
\text{H}_2\text{O} + \text{MtX}_n &\rightleftharpoons \text{H}^+, (\text{MtX}_n\text{OH})^- \\
\text{HX} + \text{MtX}_{n\text{-}1} &\rightleftharpoons \text{H}^+, \text{MtX}_n^- \\
\text{HY} + \text{MtX}_n &\rightleftharpoons \text{H}^+, \text{MtX}_n\text{Y}^- \\
\text{RX} + \text{MtX}_{n\text{-}1} &\rightleftharpoons \text{R}^+, \text{MtX}_n^- \\
\text{RY} + \text{MtX}_n &\rightleftharpoons \text{R}^+, \text{MtX}_n\text{Y}^-
\end{aligned}
$$

$$
\begin{aligned}
&\text{Y} \\
&\text{R'O-} \\
&\text{R'CO}_2\text{-} \\
&\text{R'CONH-} \\
&\text{R'SO}_3\text{-}
\end{aligned}
$$

Scheme 6

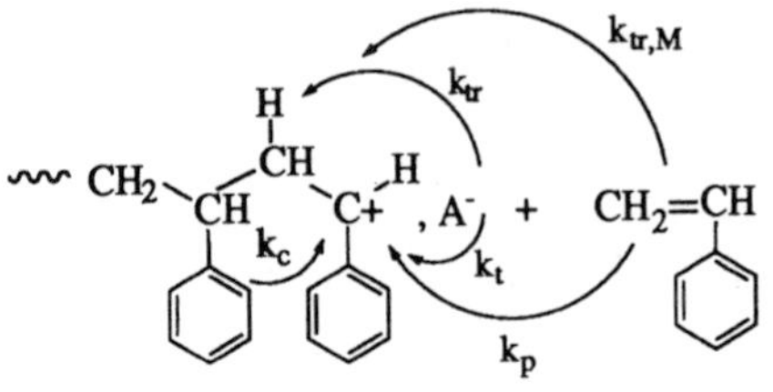

Scheme 7

transfer also occurs by intramolecular Friedel–Crafts cyclization (k_c) to form polymers with indanyl end groups, especially at high conversion.

However, due to their higher activation energies compared to propagation, these chain transfer reactions can be suppressed by polymerizing at lower temperatures in hydrocarbon and chlorinated solvents [93]; nitro solvents are also used. Protonic solvents and amines which result in either transfer or termination should be avoided. In the absence of impurities and/or terminating agents, termination generally occurs in cationic polymerizations by unimolecular collapse of the ion pair (k_t). However, termination may be reversible depending on the nature of the ligand. For example, anions containing chloride and bromide ligands such as $SnCl_5^-$, $SbCl_6^-$, $SnBr_5^-$, BCl_4^- and BBr_4^- usually decompose reversibly. Reversible systems which exchange fast in comparison to propagation provide 'living' systems with narrow molecular weight distributions.

Cationic polymerizations often involve both contact ion pairs and free ions as the propagating species (Scheme 8); solvent-separated ion pairs have not been identified

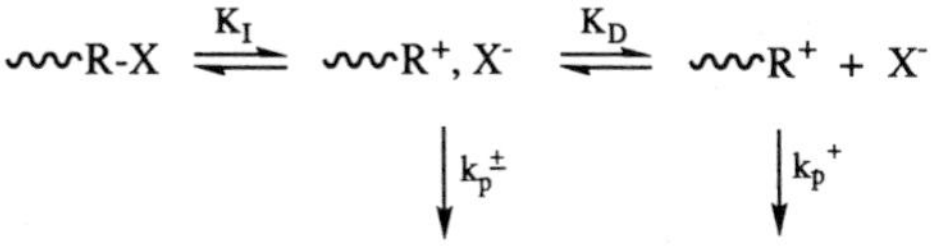

spectroscopically. Propagation via covalent species has also been proposed for many of the new living carbocationic systems [94]. However, ions have been detected indirectly using optically active compounds and/or by various salt, substituent and solvent effects [8], and more directly by ^{1}H-NMR of exchange reactions in polymerizations and model systems [95–99]. Although the reactions of these ion pairs and free ions are similar ($k_p^{\pm} \approx k_p^+$) [100–102], their different lifetimes lead to high polydispersities [10, 103].

In order to achieve more controlled cationic polymerization conditions, the overall rate of polymerization must be decreased, thereby increasing the time available for functionalizing the chain ends. This has been accomplished by decreasing the stationary concentrations of active carbenium ions by shifting the equilibrium to dormant species. Since most cationic polymerizations involve an equilibrium between covalent adducts and ionic species, the equilibrium can be shifted from free ions to dormant covalent species by adding a salt with a common counteranion [104]. Alternatively, the carbenium ions can be deactivated with nucleophiles, such as phosphines [105], hindered amines [106] dialkyl sulfides [107, 108], ethers [109], sulfoxides [110], esters [111], and amides [112] to generate dormant onium ions (Scheme 9). The rapid dynamic exchange between active and dormant species, and the trapping of free carbenium ions with salts or nucleophiles also decreases the polydispersities (pdi < 1.2).

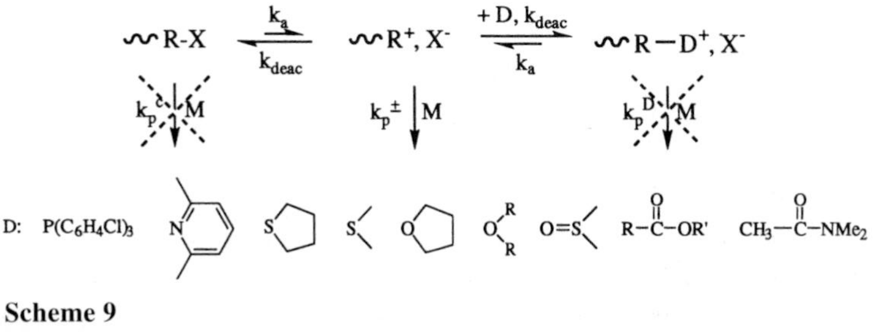

$$\text{\textasciitilde\textasciitilde\textasciitilde R-X} \underset{}{\overset{K_I}{\rightleftharpoons}} \text{\textasciitilde\textasciitilde\textasciitilde R}^+, X^- \underset{}{\overset{K_D}{\rightleftharpoons}} \text{\textasciitilde\textasciitilde\textasciitilde R}^+ + X^-$$

$$\downarrow k_p^{\pm} \qquad\qquad \downarrow k_p^+$$

Scheme 8

D: $P(C_6H_4Cl)_3$

Scheme 9

Table 2. Relative rates of initiation in cationic polymerizations initiated by covalent initiators in the presence of a Lewis acid catalyst [113].

Initiator	Fast initiation	Similar initiation and propagation rates	Slow initiation
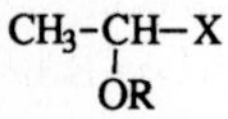 CH₃–CH–X OR	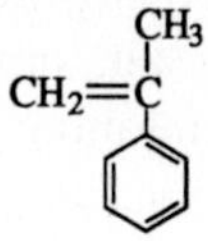 CH₂=C(CH₃) (phenyl)	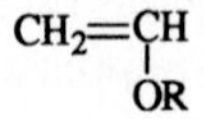 CH₂=CH–OR	
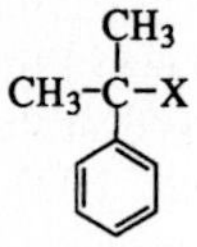 CH₃–C(CH₃)–X (phenyl)	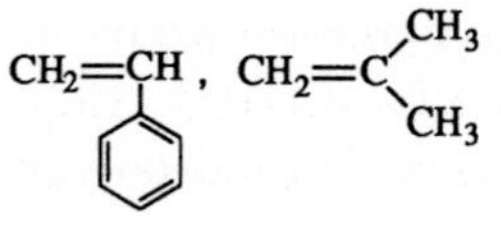 CH₂=CH (phenyl) , CH₂=C(CH₃)(CH₃)	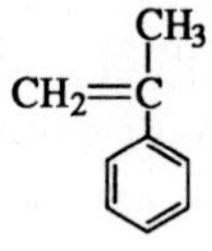 CH₂=C(CH₃) (phenyl)	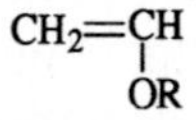 CH₂=CH–OR
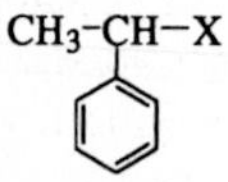 CH₃–CH–X (phenyl)		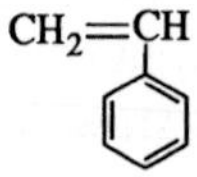 CH₂=CH (phenyl)	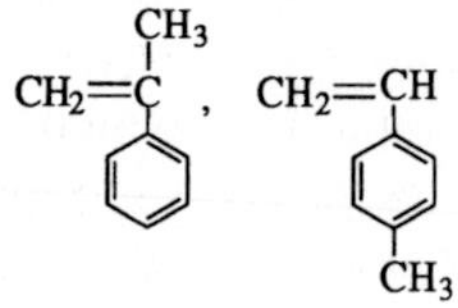 CH₂=C(CH₃)(phenyl) , CH₂=CH (phenyl–CH₃)
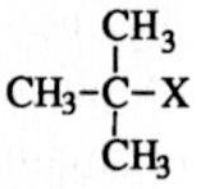 CH₃–C(CH₃)(CH₃)–X		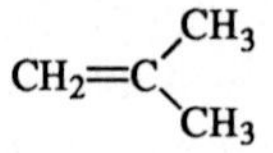 CH₂=C(CH₃)(CH₃)	

Table 3. Examples of 'living' cationic polymerizations of alkenes.

Type	Polymerization system	Solvent	Temperature (°C)	Ref.
HA	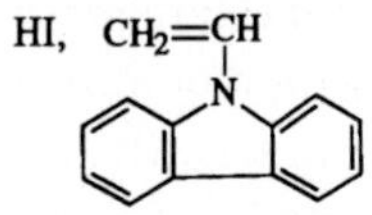 HI, CH₂=CH (N-carbazole)	CH₂Cl₂	−78	[115]
HA or RX/LA	HI, I₂, CH₂=CH–O-i-Bu	n-hexane	−15	[116]
	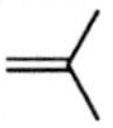 CumOAc, BCl₃, (isobutylene)	CH₂Cl₂	−30	[117]
HA or RX LA/N⁺X⁻	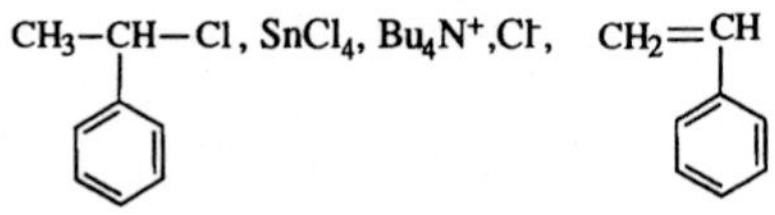 CH₃–CH–Cl (phenyl), SnCl₄, Bu₄N⁺,Cl⁻, CH₂=CH (phenyl)	CH₂Cl₂	−15	[119]
HA/Nu	CF₃SO₃H, SMe₂, CH₂=CH–O-i-Bu	CH₂Cl₂	−15	[107]

The choice of initiator is also very important for a controlled cationic polymerization since initiation is often not quantitative in classic systems. Covalent initiators will be reactive enough to initiate polymerization if the initiator ionizes faster than the (dormant) propagating chain ends. Many of the new 'living' systems therefore use initiators with better leaving groups and tertiary versus secondary active centers. The emerging relative rates of initiation that are summarized in Table 2 correlate with the order of monomer reactivity listed in Eq. (20); completion of the table should therefore be obvious. Nevertheless, the new 'living' systems also use a much higher concentration of these well-defined initiators. This results in lower molecular weight polymers than in classic systems, which makes transfer less easily detected [114], and decreases the probability of spontaneous coinitiation with adventitious moisture.

Examples of the different types of 'living' cationic polymerization systems are listed in Table 3. All involve relatively fast initiation and optimal equilibria between a low concentration of active carbenium ions and a high concentration of dormant species (Scheme 9). Only hydroiodic acid initiated polymerizations of N-vinyl carbazol are controlled in the absence of a Lewis acid activator or a nulceophilic deactivator [115].

polymerization of isobutylene [117] to produce polymers with $-Cl$ terminated chains since boron has a greater affinity for acetate ligands than for chlorine [118]. The BCl_3 system therefore has the added advantage that the initiator with an acetate leaving group is more easily ionized than the propagating chain with a chlorine leaving group. [118] Table 3 also provides two examples in which it is necessary to add a common ion salt [119] or a nucleophile [107] to decrease the concentration of carbenium ions.

Although many controlled cationic polymerizations that have been developed [120], only mesogenic vinyl ethers have been used in an attempt to prepare well-defined SCLCPs by a cationic addition mechanism [121]. Nevertheless, these polymerizations (and copolymerizations) provide the most complete series of SCLCPs with the widest range of structural variables. As shown in Eq. (22), most of these monomers, includ-

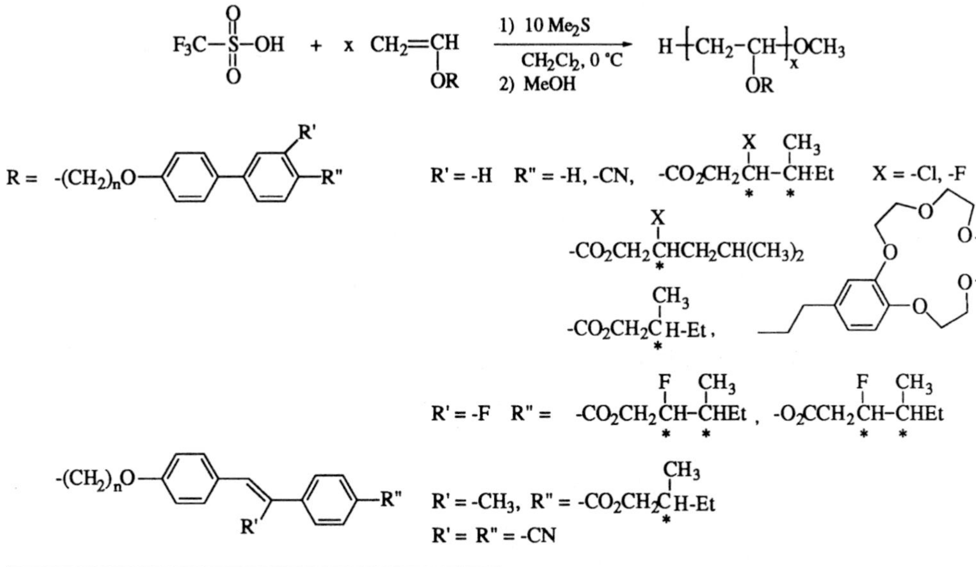

In contrast, all other adducts generated by addition of HI to the monomer's double bond require activation with a Lewis acid such as I_2 [116]. Similarly, cumyl acetate requires activation by a Lewis acid to initiate

ing n-[(4'-(4''-cyanophenyl)phenoxy)alkyl]-vinyl ethers $(DP_n = 2.1 - 32, pdi = 1.02 - 1.54)$ [122–127], 2-[(4'-biphenyloxy)-ethyl]vinyl ether $(DP_n = 3.8 - 22, pdi = 1.07 - 1.11)$ [128], n-[(4'-(S(-)-2-methyl-

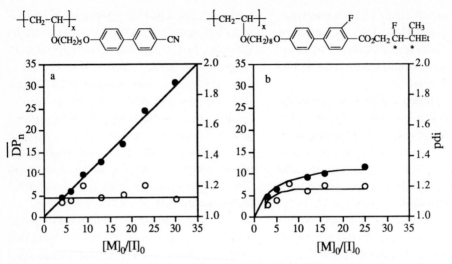

Figure 4. Degree of polymerization (●) and polydispersity (○) resulting from cationic polymerizations of (a) 5-[(4′-(4″-cyanophenyl)phenoxy)pentyl]vinyl ether [125, 126] and (b) 8-[(4′-(2R,3S)-2-fluoro-3-methyl-pentyloxycarbonyl)-3′-fluorophenyl-4″-phenoxy)octyl]vinyl ether [139] initiated by triflic acid in CH_2Cl_2 at $0\,°C$ in the presence of dimethyl sulfide.

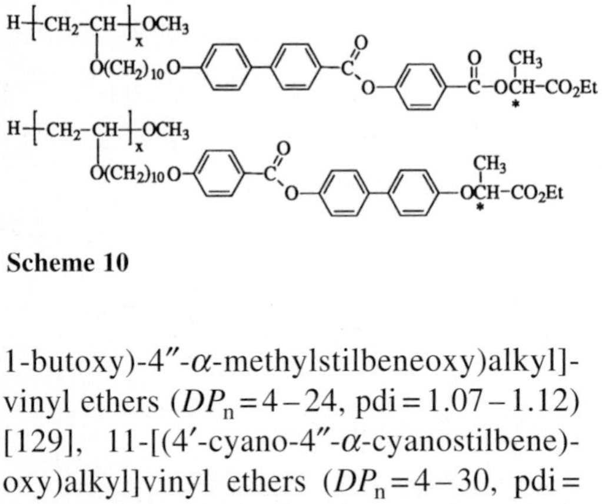

Scheme 10

1-butoxy)-4″-α-methylstilbeneoxy)alkyl]-vinyl ethers ($DP_n = 4-24$, pdi $= 1.07-1.12$) [129], 11-[(4′-cyano-4″-α-cyanostilbene)-oxy)alkyl]vinyl ethers ($DP_n = 4-30$, pdi $= 1.05-1.09$) [130], n-[(4′-(S-(−)-2-methyl-1-butoxy)phenyl-4″-phenoxy)alkyl]vinyl ethers ($DP_n = 4-26$, pdi $= 1.04-1.10$) [131, 132], n-[(4′-(2S,3S)-(+)-2-chloro-3-methylpentyloxycarbonyl)phenyl-4″-phenoxy)-alkyl]vinyl ethers ($DP_n = 4-21$, pdi $= 1.07-1.16$) [133, 134], n-[(4′-(2R(2S),3S)-2-fluoro-3-methylpentyloxycarbonyl)phenyl-4″-phenoxy)alkyl]vinyl ethers ($DP_n = 3.4-16$, pdi $= 1.05-1.20$) [135, 136], n-[(4′-(2R(2S),3S),3S)-2-fluoro-3-methylpenta-noate)phenyl-4″-phenoxy)alkyl]vinyl ethers ($DP_n = 3.5-15$, pdi $= 1.04-1.20$)

[137], n-[(4′-(2R,3S)-2-fluoro-3-methyl-pentanoate)-3′-fluorophenyl-4″-phenoxy)-alkyl]vinyl ethers ($DP_n = 4.0-16$, pdi $= 1.06-1.13$) [138], n-[(4′-(2R,3S)-2-fluoro-3-methylpentyloxycarbonyl)-3′-fluorophe-nyl-4″-phenoxy)alkyl]vinyl ethers ($DP_n = 4.5-27$, pdi $= 1.08-1.22$) [139], 8-[(4′-(2R(2S))-2-chloro-4-methylpentyloxycar-bonyl)phenyl-4″-phenoxy)octyl]vinyl ethers ($DP_n = 3.9-15$, pdi $= 1.12-1.22$) [140], 8-[(4′-(2R(2S))-2-fluoro-4-methyl-pentyloxycarbonyl)phenyl-4″-phenoxy)oc-tyl]vinyl ethers ($DP_n = 4.1-16$, pdi $= 1.11-1.33$) [141], and 4-{2-[4′-(11-vinyloxyun-decyloxy)biphenyl-4-yl]ethyl}benzo-15-crown-5 ($DP_n = 6-19$) [142], were polymer-ized using triflic acid as the initiator and dimethyl sulfide as a deactivator in CH_2Cl_2 at $0\,°C$.

Although Webster et al. found that triflic acid/dimethyl sulfide initiated polymeriza-tions of i-butylvinyl ether are not living at $0\,°C$ in CH_2Cl_2 [107], plots of the degree of polymerization against $[M]_0/[I]_0$ demon-strate that transfer is usually not detectable at the molecular weights attempted in these polymerizations (e. g. Fig. 4 a); the polydispersities are generally narrow at pdi ≈ 1.1. Exceptions (e. g. Fig. 4 b) occur when the monomers are not sufficiently pure [135–141].

The CF_3SO_3H/SMe_2 initiating system has also been used to cyclopolymerize 11-[(4'-cyanophenyl-4"-phenoxy)alkyl]undecanyl-3,4-bis(ethenyloxyethoxy)benzoate to form SCLCPs with crown ethers in the polymer backbone, although the polymerization is accompanied by termination [143]. Polymerization of the chiral vinyl ethers shown in Scheme 10 were reported to give only oligomers with $DP_n = 5$ under the same condi-

tions, except with an unreported $[M]_0/[I]_0$ [144]. However, this is apparently due to the use of impure monomers, reagents and/or reaction conditions, since similar functional groups were tolerated in previously polymerized monomers (Eq. 22). That is, cationic polymerizations tolerate cyano groups, phenyl benzoates, olefins, crown ethers and chiral centers with alkyl, chloro or fluoro substituents. However, they do not tolerate azobenzene groups [144].

Percec et al. have demonstrated by 1D and 2D ^{1}H-NMR (COSY) that chain ends due to transfer by β-H elimination (Scheme 7) or termination by alkyl abstraction (Eq. 23) are absent or barely detectable,

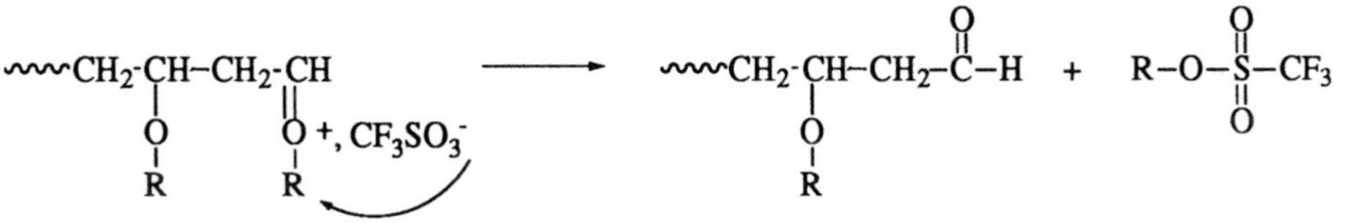

$$\text{(23)}$$

respectively, for poly{11-[(4'-cyanophenyl-4"-phenoxy)undecyl]vinyl ether}s with $DP_n \leq 30$ [145]. The ability to endcap growing polymers of 3-[(4'-(4"-cyanophenyl)phenoxy)propyl]vinyl ether with 2-hydroxyethyl methacrylate (Eq. 24), 10-undecene-1-ol and

$$\text{(24)}$$

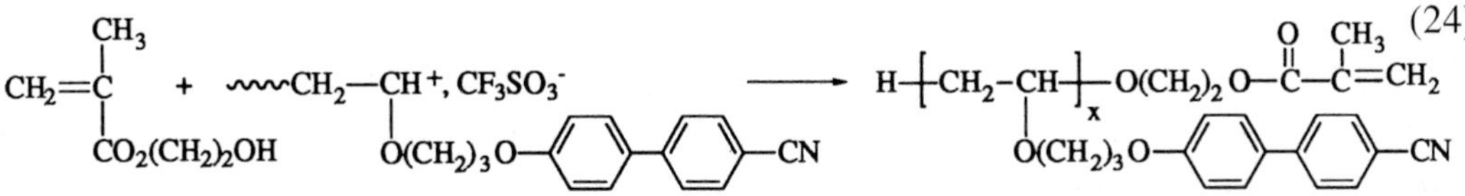

2-[2'-(2"-allyloxyethoxy)ethoxy]ethanol also demonstrates that transfer and termination are negligible at low degrees of polymerization ($DP_n = 5-7$) [146].

As shown in Eq. (25), well-defined poly{2-[(4'-biphenyloxy]ethyl]vinyl ether}s ($DP_n = 22-50$, pdi = 1.2) [147] poly{2-[(4'-n-alkoxyphenyl-4"-phenoxy)ethyl]vinyl ether}s

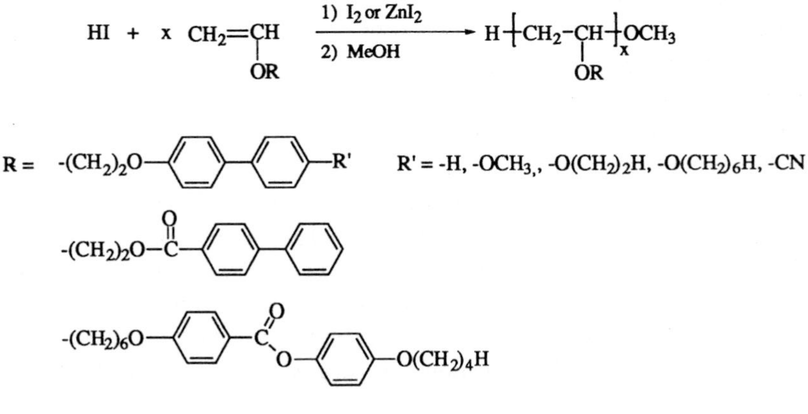

$$\text{(25)}$$

(DP_n = 7.0 – 24, pdi = 1.02 – 1.4) [147, 148] poly{2-[(4'-cyanophenyl-4''-phenoxy)-ethyl]vinyl ether}s (DP_n = 4.9 – 28, pdi = 1.04 – 1.2) [149], and poly{4-(2'-vinyloxyethyl)-4''-phenylbenzoate}s (DP_n = 18 – 32, pdi = 1.2 – 1.3) [147] were prepared using HI as the initiator and either I_2 or ZnI_2 as the catalyst in CH_2Cl_2 at –5 °C. {6-[(4'-n-Butoxyphenyl-4''-benzoate)hexyl]vinyl ether (DP_n = 7.6 – 38, pdi = 1.10 – 1.15) was polymerized similarly using HI/I_2, except that toluene was used as the solvent at –40 °C [150]. Low molecular weight oligo(vinyl ether)s (DP_n = 10 – 19, pdi = 1.07 – 1.22) containing olefinic, nitro and cyano groups were also prepared using residual H_2O in CH_2Cl_2 as the initiator and $AlCl_2Et$ as catalyst in presence of dimethyl sulfide at 0 °C [151, 152].

2.4   Ring-Opening Metathesis Polymerizations

As outlined in Scheme 11, cycloolefins polymerize by ring-opening metathesis polymerization (ROMP) via a [2 + 2] cycloaddition with a propagating metallaolefin to form a metallocyclobutane intermediate, which then ring opens to regenerate a metallaolefin at the terminus of the growing chain [7, 153, 154]. The overall ring-opening of the cyclic monomer therefore occurs by cleavage of the double bond. Since polymers with narrow molecular weight distributions are only obtained if propagation is irreversible ($k_p > k_d$), only highly strained cyclobutene derivatives and derivatives of the bicyclo[2.2.1]hept-2-ene ring system, including norbornenes, norbornadienes, 7-oxanorbornenes and 7-oxanorbornadienes, are candidates for controlled ROMP (Eq. 26).

$$(26)$$

Classic initiating systems for ROMP involve a transition metal such as WCl_6, $WOCl_4$, or $MoCl_5$, combined with an alkyl metal such as n-BuLi or a Lewis acid such as $AlEtCl_2$, $SnMe_4$, or $SnPh_4$. These systems generate low concentrations of unstable metallaolefin complexes which decompose during the polymerization [7]. In contrast, polymers with stable propagating chain ends are generated using well-defined initiators containing either a metallacyclobutane or a metallaolefin stabilizing by bulky ligands such as tricyclohexylphosphine, tertiary alkoxide groups and arylimido ligands with bulky isopropyl substituents at the ortho positions [7, 55 – 163]. Most of the initiators which produce well-defined polymers are either ruthenium(II) complexes [161 – 163], or high oxidation state complexes of titanium, tungsten or molybdenum (Table 4) [7, 155 – 160]. However, the active chain ends and initiators containing titana- [155] and tantallacyclobutanes [156] rapidly decompose once monomer is consumed at the temperatures (50 – 75 °C) required to open the metallocyclobutane in chain growth. In contrast, the coordinatively unsaturated molybdenum and tungsten alkylidene initiators are stable for weeks in an inert atmosphere [7, 164 – 167], whereas the $RuCl_2(CHCH = CPh_2)(PR_3)_2$ complexes are moderately stable in air and are even tolerant of water in an inert atmosphere [161, 162]. However, $RuCl_2(CHAr)(PPh_3)_2$ complexes decompose in solution via bimolecular reactions [168].

In addition to preventing spontaneous termination, increased stability of the metal center also minimizes chain transfer and other termination reactions. The elementary reactions outlined in Scheme 11 demonstrate that each repeat unit of the polymer contains a double bond which may participate in secondary metathesis reactions. Therefore, the most troublesome chain

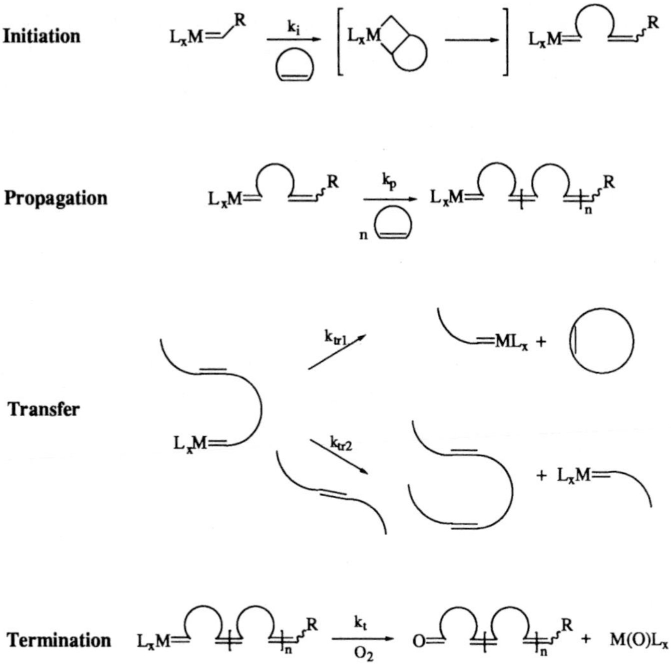

Initiation $L_xM{=}\!\!{\diagup}^{R}$ $\xrightarrow{k_i}$ $\left[L_xM \diagup \bigcirc \right]$ $\longrightarrow$ $L_xM{=}\!\!\curvearrowright_{R}$

Propagation $L_xM{=}\!\!\curvearrowright_{R}$ $\xrightarrow[n\,\bigcirc]{k_p}$ $L_xM{=}\!\!\curvearrowright\!\!\curvearrowright_{\overline{n}}{}^{R}$

Transfer

$\xrightarrow{k_{tr1}}$ ${=}ML_x + \bigcirc$

$L_xM{=}$ $\xrightarrow{k_{tr2}}$ $+ L_xM{=}$

Termination $L_xM{=}\!\!\curvearrowright\!\!\curvearrowright_{\overline{n}}{}^{R}$ $\xrightarrow[O_2]{k_t}$ $O{=}\!\!\curvearrowright\!\!\curvearrowright_{\overline{n}}{}^{R} + M(O)L_x$

Scheme 11

Table 4. Metathesis of *cis*-pentene in toluene or benzene at 25 °C.

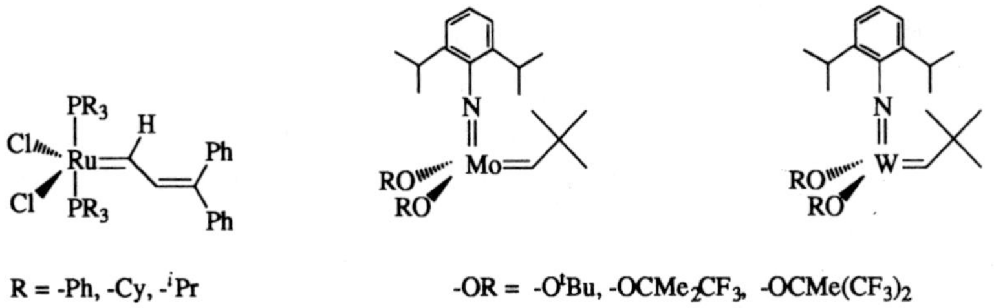

$R = $ -Ph, -Cy, -^{i}Pr   -OR = -O^{t}Bu, -OCMe$_2$CF$_3$, -OCMe(CF$_3$)$_2$

Catalyst	Turnovers per second	Ref.
W(CH-*t*-Bu)(N-2,6-C$_6$H$_3$-*i*-Pr$_2$)[OCMe(CF$_3$)$_2$]$_2$	1.7×10^1	[164]
Mo(CH-*t*-Bu)(N-2,6-C$_6$H$_3$-*i*-Pr$_2$)[OCMe(CF$_3$)$_2$]$_2$	4.1×10^0	[166]
RuCl$_2$(CHCH=CPh$_2$)(PCy$_3$)$_2$	7.2×10^{-3}	[162]
RuCl$_2$(CHCH=CPh$_2$)(P-*i*-Pr$_3$)$_2$	6.1×10^{-3}	[162]
Mo(CH-*t*-Bu)(N-2,6-C$_6$H$_3$-*i*-Pr$_2$)[OCMe$_2$CF$_3$]$_2$	1.4×10^{-3}	[166]
W(CH-*t*-Bu)(N-2,6-C$_6$H$_3$-*i*-Pr$_2$)(O-*t*-Bu)$_2$	5.5×10^{-4}	[7]
Mo(CH-*t*-Bu)(N-2,6-C$_6$H$_3$-*i*-Pr$_2$)(O-*t*-Bu)$_2$	$\approx2\times10^{-5}$	[166]
RuCl$_2$(CHCH=CPh$_2$)(PPh$_3$)$_2$	0	[161]

transfer reaction to eliminate from ring-opening metathesis polymerizations is chain transfer to polymer. Intramolecular chain transfer to polymer results in macro-cycle formation with elimination of a short-er linear chain endcapped with the metalla-olefin (or metallacyclobutane). Intermolec-ular chain transfer to polymer results in randomization of the chain lengths and broadening of the molecular weight distri-bution to the most probable distribution (pdi = 2).

Secondary metathesis is prevented by using initiators which do not metathesize internal olefins. That is, if the metallaolefin or metallacyclobutane complex is reactive enough to metathesize an internal olefin, it is generally too reactive to obtain a living polymerization. As shown in Table 4, the reactions and selectivities of the commonly used initiators correlate with their rates of metathesis of cis-2-pentene. In general, the reactivity increases and selectivity decreases with increasing electrophilicity of the metal center. This is determined by both the metal itself and the electronegativity of the ligands. For example, W(CH-t-Bu)(N-2,6-C$_6$H$_3$-i-Pr$_2$)[OCMe(CF$_3$)$_2$]$_2$ metathesizes cis-2-pentene at over 17 turnovers per second in toluene at 25 °C, and therefore does not produce a well-defined polynorbornene [164]. In contrast, Mo(CH-t-Bu)(N-2,6-C$_6$H$_3$-i-Pr$_2$)(Ot-Bu)$_2$ metathesizes cis-2-pentene at only $\approx 2 \times 10^{-5}$ turnovers per second at 25 °C and is therefore much more selective at producing well-defined polynorbornene [166].

These transfer reactions can also be minimized by working with bicyclic monomers such as norbornene derivatives, which generate polymers whose double bonds are sterically hindered and therefore resistant to secondary metathesis [153]. Polymerizations of cyclobutene monomers are less controlled than norbornene polymerizations because the double bonds in the resulting polymers are highly susceptible to secondary metathesis reactions. However, well-defined polymers of cyclobutene [169, 170] and methylcyclobutene [171, 172] are obtained by adding a Lewis base (PR$_3$) to W(CH-t-Bu)(N-2,6-C$_6$H$_3$-i-Pr$_2$)(O-t-Bu$_2$) initiated polymerizations, thereby decreasing the reactivity of the initiator and growing chain ends and increasing the rate of initiation relative to that of propagation. In addition, 3,3- [173] and 3,4-disubstituted [174, 175] cyclobutenes generate polymers with sterically hindered double bonds, and are therefore polymerized in a controlled manner using Mo(CH-t-Bu)(N-2,6-C$_6$H$_3$-i-Pr$_2$)(O-t-Bu)$_2$ without an additive.

Molecular oxygen and sometimes the carbonyl groups of aldehydes, ketones, esters, and amides terminate the active metallaolefin chain ends of all but the new ruthenium initiators. For example, reaction of the growing chain with molecular oxygen generators a metal oxide and an aldehyde terminated polymer (Scheme 11). The resulting aldehyde chain end can then react with the active end of another polymer chain to generate a second molecule of the metal oxide and a polymer of higher molecular weight [176, 177]. Since the rate of reaction with molecular oxygen is slower than the rate of propagation in polymerizations using the molybdenum and tungsten alkylidene initiators, it occurs at the end of the polymerization, resulting in a double molecular weight fraction [176, 177].

Of the molybdenum and tungsten alkylidene initiators shown in Table 4, Mo(CH-t-Bu)(N-2,6-C$_6$H$_3$-i-Pr$_2$)(O-t-Bu)$_2$ is the least reactive and therefore the most selective initiator for ROMP. Its polymerizations are endcapped with benzaldehyde, pivaldehyde and acetone (benzophenone is less efficient) [7], although it tolerates a variety of nonprotonic functional groups, including acetate, cyano, ethers, esters, phenyl benzoates, N-substituted imides and trifluoromethyl groups in both the monomer [175, 177–182] and endcapping [183] agent. With the exception of cyano groups [184], ruthenium(II) complexes tolerate the same groups, as well as alcohols, aldehydes, and ketones [162]. These polymerizations are terminated with ethyl vinyl ether.

The final requirement for a controlled ring-opening metathesis polymerization is that the rate of initiation be greater than or

Table 5. Relative rates of initiation and propagation in ring-opening metathesis polymerizations of norbornene in C_6D_6 at $17-22\,°C$.

Initiator	k_i/k_p	Ref.
$RuCl_2(CHCH=CPh_2)(PPh_3)_2$	0.006	[185]
$Mo(CH$-t-$Bu)(N$-$2,6$-C_6H_3-i-$Pr_2)(O$-t-$Bu)_2$	0.083	[186]
$RuCl_2(CH$-p-$C_6H_4X)(PPh_3)_2$		
X=−Cl	1.2	[185]
X=−NO_2	2.3	[185]
X=−NMe_2	2.6	[185]
X=−OCH_3	2.6	[185]
X=−CH_3	2.9	[185]
X=−F	4.8	[185]
X=−H	9.0	[185]

equal to that of propagation. As shown in Table 5 for the most selective initiators, only the $RuCl_2(CHAr)(PPh_3)_2$ complexes initiate norbornene polymerizations faster than propagation occurs, thereby producing polymers with very narrow molecular weight distributions (pdi = 1.04−1.10) [185]. Although one would expect the rate of initiation to be highest with benzylidenes substituted with electron-withdrawing groups (most electrophilic metal center), the electronic effect of the alkylidene substituents is obviously minor. The aryl alkylidene is therefore evidently less sterically demanding than the insertion product (Eq. 27). In contrast, the low initiation rates of the ruthenium diphenylvinyl alkylidenes produce poly-

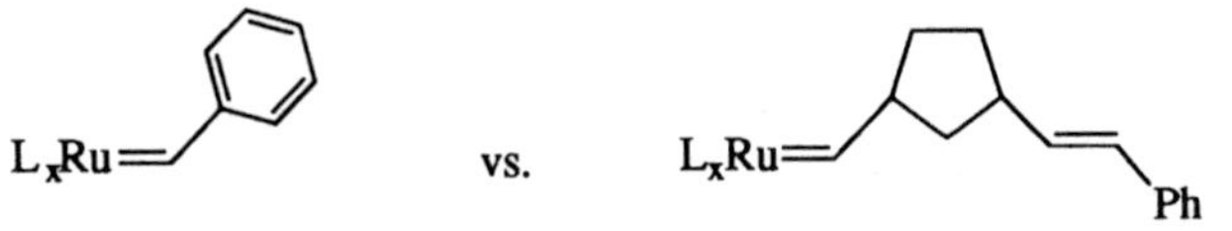

$$L_xRu= \qquad vs. \qquad L_xRu= \qquad \qquad (27)$$

norbornenes with polydispersities of approximately 1.25 [185], although the diphenylvinyl alkylidene appears to be less sterically demanding than the cyclopentylidene insertion product (Eq. 28). Its lower reactivity was attributed to conjugation [185]. The rate constant of

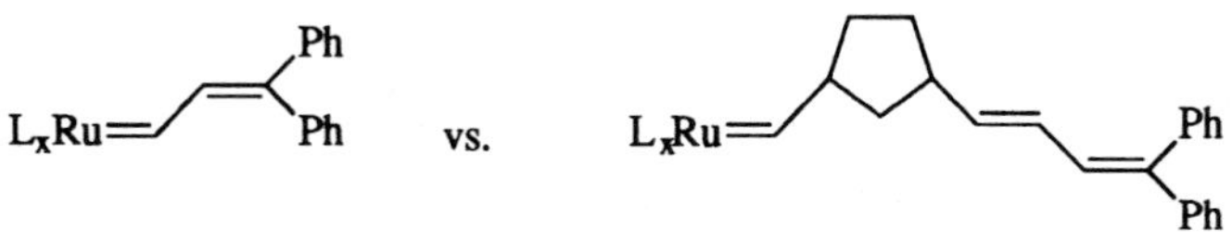

$$L_xRu= \qquad vs. \qquad L_xRu= \qquad \qquad (28)$$

initiation using $Mo(CH$-t-$Bu)(N$-$2,6$-C_6H_3-i-$Pr_2)(O$-t-$Bu)_2$ is also slightly less than that of propagation, producing polynorbornenes with pdi <1.10 [186]. This is apparently because the neopentylidene ligand with a tertiary carbon β to the metal is more sterically demanding than the insertion products, which contain secondary carbons β to the metal (Eq. 29).

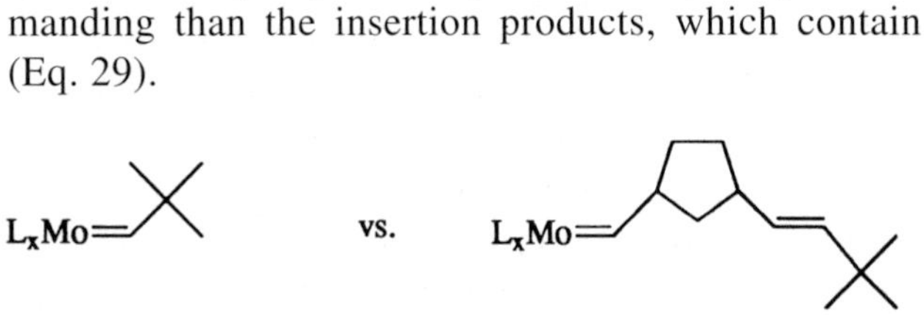

$$L_xMo= \qquad vs. \qquad L_xMo= \qquad \qquad (29)$$

With one exception [187], only mesogen-ic norbornenes have been used to prepare well-defined SCLCPs by a ring-opening metathesis mechanism. Norbornene mono-mers containing terminally attached p-me-thoxybiphenyl [22, 188–190] and p-cyano-biphenyl [189–191] mesogens were poly-merized using Mo(CH-t-Bu)(N-2,6-C$_6$H$_3$-i-Pr$_2$)(O-t-Bu)$_2$ as the initiator in order to determine the effect of molecular weight, polydispersity, flexible spacer length and mesogen density on the thermotropic behav-ior or SCLCPs with polynorbornene back-bones (Eqs. 30 and 31). As shown in Fig.

tion varied from 6 to 151, with polydisper-sities of 1.05–1.28 [22, 188]. Polymeriza-tions of 5-{[n-[4′-(4″-methoxyphenyl)phen-oxy]alkyl]methyleneoxy}bicyclo[2.2.1]-hept-2-enes (DP$_n$ = 9–342, pdi = 1.15–1.19) [189] and 5-{[n-[4′-(4″-cyanophe-nyl)phenoxy]alkyl]carbonyl}bicyclo[2.2.1]-hept-2-ene (DP$_n$ = 7–290, pdi = 1.08–1.27) [189] yielded similar results. The higher polydispersities were usually the result of a minor amount of a double molecular weight fraction in addition to the expected molec-ular weight due to insufficient degassing of molecular oxygen.

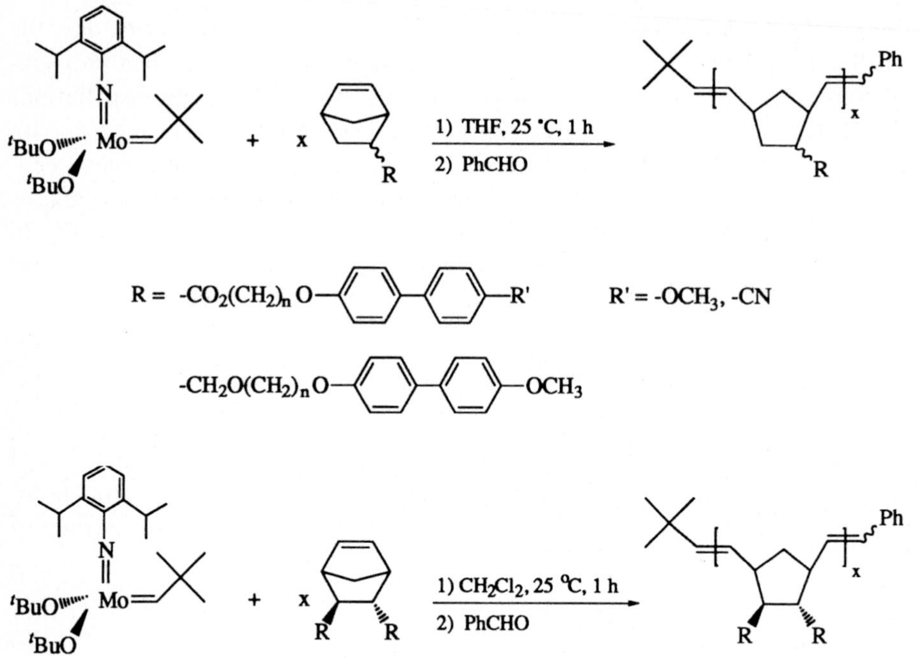

$$R = -CO_2(CH_2)_n O\!\!-\!\!\bigcirc\!\!-\!\!\bigcirc\!\!-R' \qquad R' = -OCH_3, -CN$$

$$-CH_2O(CH_2)_n O\!\!-\!\!\bigcirc\!\!-\!\!\bigcirc\!\!-OCH_3$$

$$R = -CO_2(CH_2)_n O\!\!-\!\!\bigcirc\!\!-\!\!\bigcirc\!\!-R \qquad R = -OCH_3, -CN$$

5 (a) using a 5-{[n[4′-(4″-methoxyphenyl)-phenoxy]alkyl]carbonyl}bicyclo[2.2.1]-hept-2-ene as an example, the GPC-deter-mined number average degree of polymer-ization (relative to polystyrene) agrees well with the initial ratio of monomer to initia-tor. In this case, the degree of polymeriza-

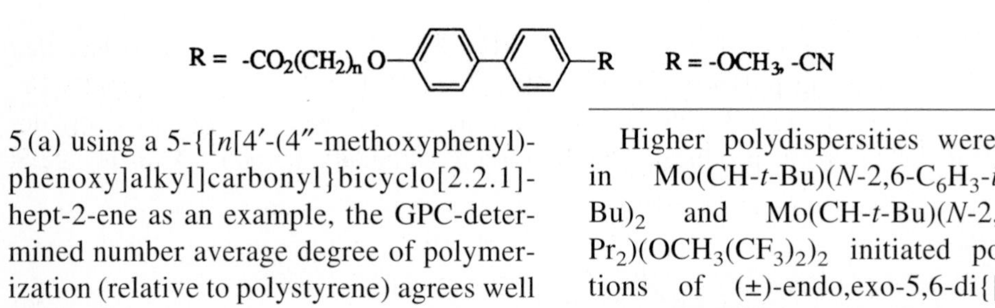

Higher polydispersities were obtained in   Mo(CH-t-Bu)(N-2,6-C$_6$H$_3$-i-Pr$_2$)(O-t-Bu)$_2$ and   Mo(CH-t-Bu)(N-2,6-C$_6$H$_3$-i-Pr$_2$)(OCH$_3$(CF$_3$)$_2$)$_2$ initiated polymeriza-tions of (±)-endo,exo-5,6-di{[n-[4′-(4″-cyanophenyl)phenoxy]alkyl]carbonyl}bi-cyclo[2.2.1]hept-2-ene (pdi = 1.21–1.60)

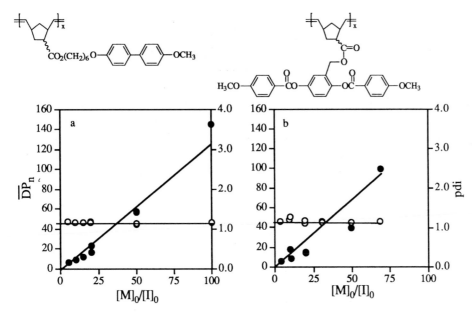

Figure 5. Degree of polymerization (●) and polydispersity (○) resulting from ring-opening metathesis polymerizations of (a) 5-{[6'-[4"-(4'''-methoxyphenyl)phenoxy]hexyl]carbonyl}bicyclo[2.2.1]hept-2-ene [22] and (b) 5-[[[2',5'-bis[(4"-methoxybenzoyl)oxy]benzyl]oxy]carbonyl]bicyclo[2.2.1]hept-2-ene [182] initiated by Mo(CH-*t*-Bu)(N-2,6-C$_6$H$_3$-*i*-Pr$_2$)(O-*t*-Bu)$_2$ or Mo(CHCMe$_2$Ph)(N-2,6-C$_6$H$_3$-*i*-Pr$_2$)(O-*t*-Bu)$_2$, respectively, in THF at 25 °C.

[191] and (±)-endo,exo-5,6-di{[*n*-[4'-(4"-methoxyphenyl)phenoxy]alkyl]carbonyl}-bicyclo[2.2.1]hept-2-ene (pdi = 1.20–2.75) [190], respectively, due to oxygen and monomer impurities (Eq. 31). The degrees of polymerization (DP_n = 41 – 266) were substantially different from the attempted ratio of monomer to iniator ([M]$_0$/[I]$_0$ = 100), which were attributed to difficulties in actually controlling this ratio. Surprisingly, the 'absolute' molecular weights determined using a GPC with a viscometry detector and a universal calibration curve were even higher, with degrees of polymerization (DP_n = 87 – 372) approximately twice those calculated relative to polystyrene [191].

Several polynorborenes with laterally attached mesogens have also been prepared using Mo(CHCMe$_2$Ph)(N-2,6-C$_6$H$_3$-*i*-Pr$_2$)(O-*t*-Bu)$_2$ as the initiator in order to determine the effect of molecular weight

[182], the effect of the length of the *n*-alkoxy substitutent [182, 192–197] as well as to design polymers with SmC mesophases by proper choice of the mesogen [192, 193, 196, 197] and to test various concepts for inducing smectic layering in nematic liquid crystals (Scheme 12) [192–195]. In particular, polymerizations of both 5-[[[2',5'-bis[(4"-*n*-alkoxybenzoyl)oxy]benzyl]-oxy]carbonyl]bicyclo[2.2.1]hept-2-enes (DP_n = 5.1 – 168, pdi = 1.12–1.27) and 5,8-bis[(4'-*n*-alkoxybenzoyl)oxy]-1,2,3,4-tetra-hydronaphthalene (DP_n = 40 – 101, pdi = 1.06–1.21) are fairly well controlled as demonstrated by the similar degrees of polymerization and [M]$_0$/[I]$_0$ (Fig. 5 b), and by the polymers' narrow polydispersities [182]. Polymerization of 5-{[[2',5',-bis[2-(3"-fluoro-4"-*n*-alkoxyphenyl)ethynyl]-benzyl]oxy]carbonyl}bicyclo[2.2.1]hept-2-enes also results in the expected molecu-

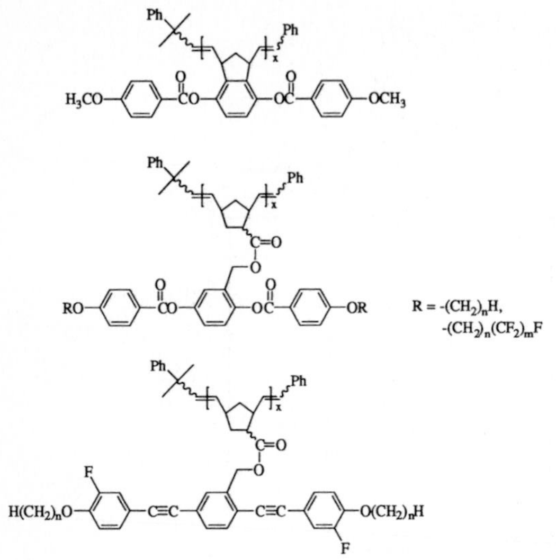

Scheme 12

lar weights ($DP_n = 39 - 76$, pdi = 1.09 – 1.29) at $[M]_0/[I]_0 \approx 50$) [196, 197]. In this case, higher polydispersities result when the monomers are viscous oils which can not be purified by recrystallization. 5-{[[2′,5′-bis[(4″-n-((perfluoroalkyl)alkoxy)ben-zoyl)oxy]benzyl]oxy]carbonyl}bicy-clo[2.2.1]hept-2-enes were also polymer-ized by ROMP, although it was necessary to use higher temperatures (40 °C) in order to prevent the polymer from precipitating out of solution during the polymerization; this

broadens the polydispersity to pdi ≈ 1.5 [195].

Both norbornene and butene monomers containing terminally attached p-nitrostil-bene mesogens were recently polymerized using $RuCl_2(CHPh)(PPh_3)_2$ as the initiator in order to determine the effect of molecu-lar weight and flexibility of the polymer backbone on their thermotropic behaviour (Scheme 13) [187]. Although the polydis-persities are quite narrow (pdi = 1.07–1.11), Fig. 6 a shows that the molecular weight re-sulting from polymerizations of 5-[n-[4′-(4″-nitrostilbeneoxy)alkyl]carbonyl}bicy-clo[2.2.1]hept-2-enes ($DP_n = 15 - 47$) using $RuCl_2(CHPh)(PPh_3)_2$ as the initiator may be slightly less controlled than those using $Mo(CHCMe_2R)(N-2,6-C_6H_3-i-Pr_2)(O-t-Bu)_2$ to polymerize norbornene monomers, although nitro groups do react slowly with the latter initiator [183]. That is, Fig. 5 dem-onstrates that the number average degree of polymerization of mesogenic polynorbor-nenes determined relative to polystyrene standards generally match $[M]_0/[I]_0$ quite closely. In contrast to the 5-{[n-[4′-(4″-ni-trostilbeneoxy)octyl]alkyl]carbonyl}bicy-clo[2.2.1]hept-2-enes, Fig. 6 b shows that polymerization of the corresponding cyclo-butene monomers yield polymers with de-

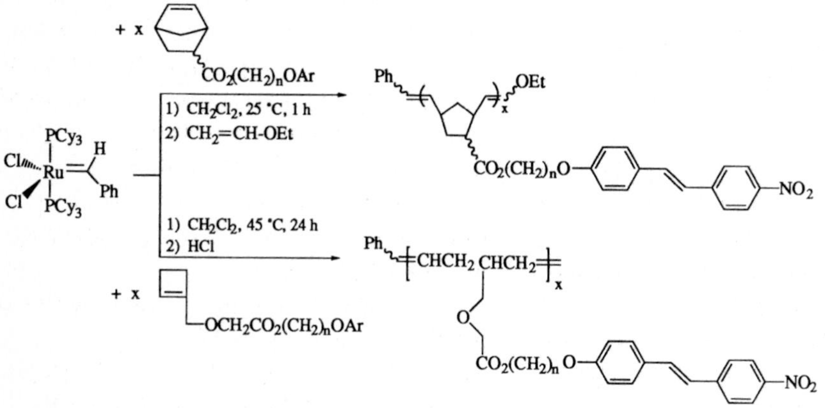

Scheme 13

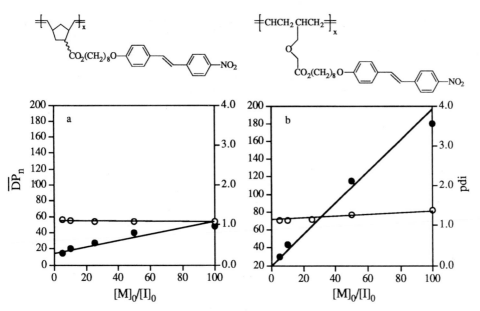

Figure 6. Degree of polymerization (●) and polydispersity (○) resulting from ring-opening metathesis polymerizations of (a) 5-{[8'-[4"-(4'''-nitrostilbeneoxy)octyl]carbonyl}bicyclo[2.2.1]hept-2-ene and (b) poly[3-((8'-(4"-(4'''-nitrostilbeneoxy)octyl)carbonyl)methyleneoxy)methyl]cyclobutene} initiated by $RuCl_2(CHPh)(PPh_3)_2$ in CH_2Cl_2 at 25 and 45 °C, respectively [187].

grees of polymerizations which are two to six times higher than $[M]_0/[I]_0$ ($DP_n = 30 - 72$, pdi $= 1.11 - 1.38$).

$Mo(CHCMe_2Ph)(N-2,6-C_6H_3-iPr_2)(O-t-Bu_2)$ has also been used to prepare low molar mass liquid crystalline polyenes [198].

2.5 Polymer Analogous Reactions on Well-Defined Precursor Polymers

In order to synthesize homopolymers by polymer analogous reactions, the reaction must go to 100% conversion. Hydrosilations of mesogenic olefins are the most common polymer analogous reactions used to prepare SCLCPs [199]. However, the precursor poly(methyl siloxane)s are generally not prepared by living polymerizations, nor do the hydrosilations readily go to comple-

tion. Only recently has a communication appeared describing a well-defined side-chain liquid crystalline polysiloxane [200]. As shown in Scheme 14, the precursor polysiloxane ($DP_n \sim [M]_0/[I]_0$, pdi < 1.2) was prepared by anionic ring-opening polymerization of pentamethylvinylcyclotrisiloxane using lithium trimethylsilanolate as the initiator in THF at 0 °C, followed by termination with t-butyldimethylsilyl chloride. The resulting poly(dimethylsiloxane-co-methylvinylsiloxane) was functionalized by hydrosilation of the vinyl groups with the monoadduct of a mesogenic olefin and 1,1,3,3-tetramethylsiloxane. The thermotropic behavior of these (co)polymers has not been reported yet.

Polybutadiene was the first well-defined polymer used in a polymer analogous reaction to synthesize SCLCPs [201–203], primarily for comparison to the corresponding block copolymers discussed in Sec. 4.1.1 of this chapter. As shown in Scheme 15, butadiene was polymerized by a living anionic mechanism using *s*-butyl lithium as the initiator in THF at −78 °C to produce poly(1,2-butadiene) ($DP_n = 1.1 \times 10^3$, pdi = 1.10). Poly[((2-cholesteryloxycarbonyloxy)ethyl)-ethylene] [201] (pdi = 1.13), poly{[4-(4'-methoxyphenylazo)(2",3",4",6"-D$_4$)phenyl glutarate]ethylene} [202] (pdi = 1.15) and poly{[4-(4'-methoxyphenylcarbonyl)phenyl glutarate]ethylene} [203] were then prepared by hydroboration of the remaining double bonds with 9-borabicyclo[3.3.1]-nonane (9-BBN), followed by oxidation and formylation or esterification with the corresponding mesogenic chloroformate or acid chloride. Another cholesterol-containing polymer, poly[(2-cholesteryloxycarbonyl-oxy)ethyl methacrylate], was prepared by anionic polymerization of a protected hydroxyethyl methacrylate for comparison to the corresponding diblock and triblock copolymers (Sec. 4.1.1) [204–210]. As shown in Scheme 16, 2-(trimethylsiloxy)-ethyl methacrylate was polymerized in THF

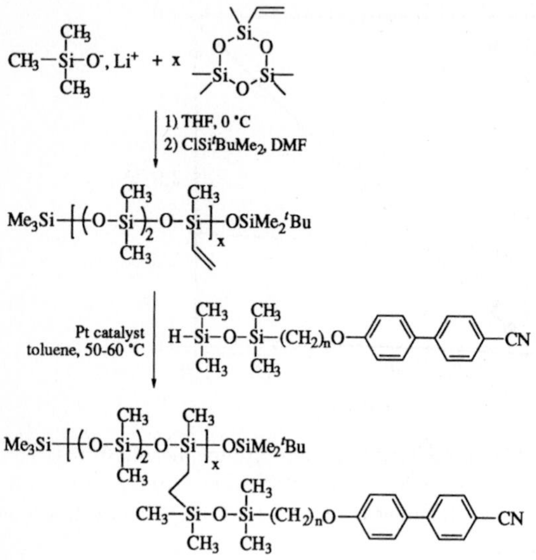

Scheme 14

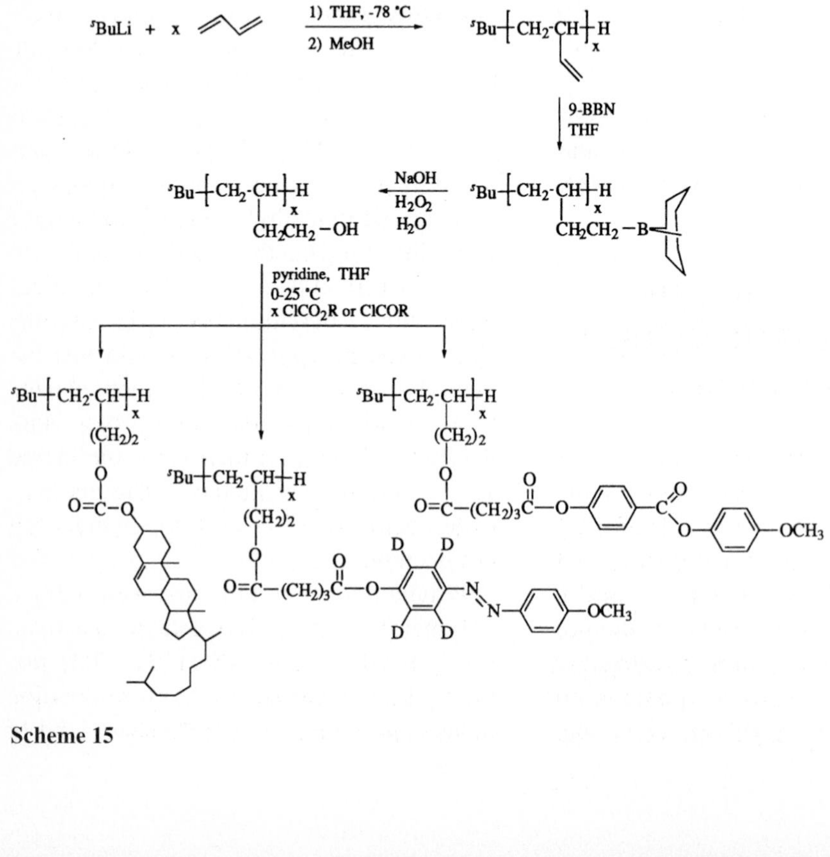

Scheme 15

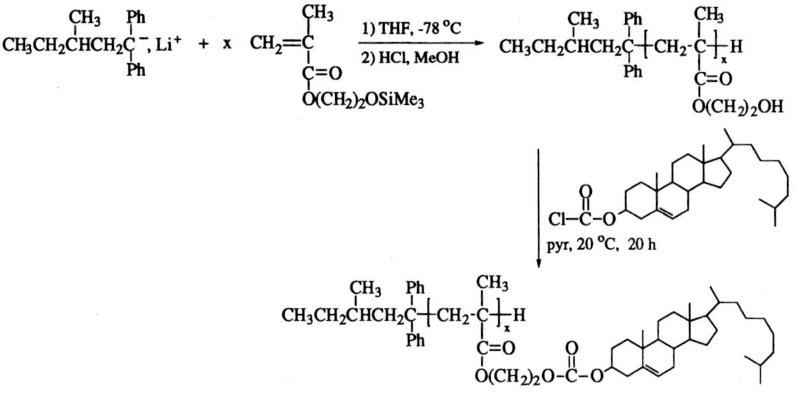

Scheme 16

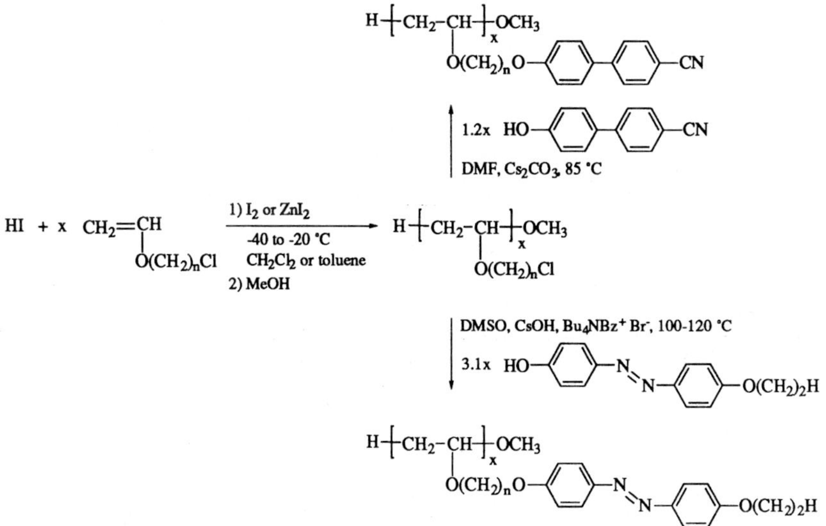

Scheme 17

at −78 °C using 1,1-diphenyl-3-methylpen-tyl lithium as the initiator, which was generated in situ by reacting s-butyl lithium with 1,1-diphenylethylene. The trimethylsilyl protecting group was removed when the polymer was quenched and precipitated in acidic methanol. The resulting poly(2-hydroxyethyl methacrylate) block was then quantitatively functionalized with cholesterylchloroformate. Although the molecular weight distribution of poly[(2-trimethylsilyloxy)ethyl methacrylate] is reported to be broad (pdi = 1.30) when polymerized under

these conditions and narrow (pdi = 1.08) when polymerized in the presence of 10 equivalents of LiCl [211], the molecular weight distribution of the final poly[(2-cholesteryloxycarbonyloxy)ethyl methacrylate]s prepared in these studies varied from pdi = 1.0 − 1.2 when $DP_n < 159$ [204, 208], and pdi = 1.90 when $DP_n = 461$ [208].

In contrast to the direct cationic polymerizations of mesogenic vinyl ethers discussed in Sec. 2.3 of this chapter, poly{n-[(4'-cyanophenyl-4″-phenoxy)alkyl]vinyl ether}s

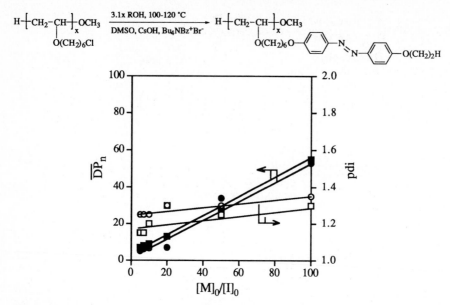

Figure 7. Degree of polymerization (●, ■) and polydispersity (○, □) of the precursor polymer resulting from the cationic polymerization of 6-chlorohexylvinyl ether (●, ○) initiated by HI/I_2 in toluene at $-40\,°C$, and of the final poly{6-[(4'-n-ethoxy-4''-azobenzene)hexyl]vinyl ether}s prepared by polymer analogous reactions (■, □) [214].

can also be prepared by (quantitative) etherification of poly[n-(chloroalkyl)vinyl ether]s as shown in Scheme 17 [212, 213], with reportedly no change in the polydispersity (pdi = 1.1–1.3) and only the expected increase in molecular weight. However, the more complete set of data plotted in Fig. 7 shows that the degrees of polymerization of poly{6-[4'-(4''-ethoxyphenylazo)phenoxyhexyl]vinyl ether}s and their precursor polymers are lower than expected from $[M]_0/[I]_0$ [214]. In this case, the extent of grafting was less than quantitative of 80–97%.

SCLCPs have also been prepared by alkylation and/or quaternization of linear poly(ethylene imine)s [215, 216] and poly(4-vinylpyridine)s [217–223] with mesogenic alkyl halides or carboxylic acids. Although these precursor polymers can be prepared by controlled cationic [224] and anionic [225, 226] polymerizations, respectively, 'living' polymerizations were either not used to prepare the SCLCPs, or the mo-

lecular weight and polydispersity data were not reported for the precursor polymers and the resulting SCLCPs. These systems are therefore not discussed here.

3   Structure/Property Correlations Determined using 'Living' Polymerizations

3.1   The Effect of Molecular Weight

The most basic and fundamental structure/property question that must be answered with any new system, especially one prepared by a living polymerization, is the molecular weight or degree of polymerization at which its thermotropic behavior no longer depends on the addition or subtraction of another repeat unit. (However, the isotropization enthalpy of SCLCPs is essentially molecular weight independent.) Comparisons of additional structural variables can then be made between samples whose thermotropic behavior is independent of molecular weight.

The thermal transitions of liquid crystalline polymethacrylates reach their limiting values at less than 50 repeat units. For example, Fig. 8 shows that the glass and nematic–isotropic transition temperatures of syndiotactic $[(rr) = 0.70 - 0.77]$ poly{6-[4'-(4"-methoxyphenoxycarbonyl)phenoxy]hexyl methacrylate} prepared by controlled anionic polymerization levels off at approximately 35 repeat units [45]. This is in contrast to the corresponding polymers prepared by radical polymerizations, which have transition temperatures that are substantially higher and level off at approximately 25 repeat units [45], in addition to having different end groups and perhaps branched architectures, the radically prepared polymers are also less syndiotactic $[(rr) = 0.40 - 0.66]$. However, Fig. 8 also plots the transition temperature(s) as a function of the inverse degree of polymerization in order to calculate the transition temperatures of an infinite molecular weight polymer from the intercepts. Although the transition temperatures plateau as a function of the degree of polymerization, they do not correspond to the transitions of an infinite molecular weight polymer, which is similar to that of the radically prepared polymers. Nevertheless, the rapid increase in transition temperatures upon going from $DP_n = 20$ to $DP_n = 30$ indicate that this discrepancy may be due simply to errors in the calculated molecular weights of the low molecular weight polymers, and therefore to an inaccurately high (absolute) slope used for the extrapolation.

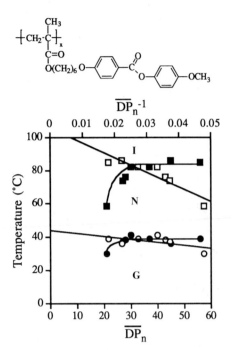

Figure 8. Dependence of the glass (●, ○) and nematic–isotropic (■, □) phase transition temperatures of syndiotactic $[(rr) = 0.70 - 0.77]$ poly{6-[4'-(4"-methoxyphenoxycarbonyl)phenoxy]hexyl methacrylate} as a function of the number average degree of polymerization (●, ■) and the inverse number average degree of polymerization (○, □) [45]. Infinite molecular weight transitions: G 44 N 105 I.

The G–SmA–N–I transition temperatures of syndiotactic poly{6-[4'-(4"-n-butoxyphenoxycarbonyl)phenoxyl)phenoxy]-hexyl methacrylate} prepared by aluminum porphyrin initiated polymerizations also level off at approximately 25 repeat units [91]. Similarly, the glass and nematic–isotropic transition temperatures of poly[6-(4'-methoxy-4"-α-methylstilbeneoxy)hexyl methacrylate] prepared by group transfer polymerization become independent of molecular weight at approximately 20 repeat units [48]. Both polymethacrylates reach the same transition temperatures as the corresponding polymers prepared by radical polymerizations, which have nearly identical tacticities.

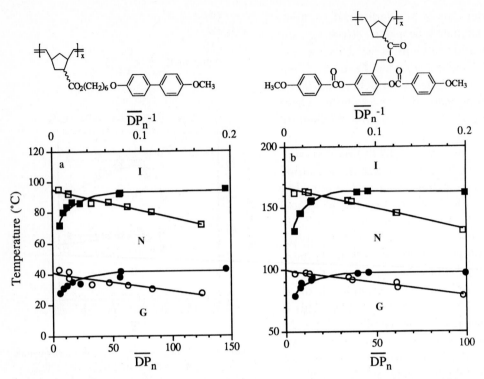

Figure 9. Dependence of the glass (●, ○) and nematic–isotropic (■, □) phase transition temperatures of (a) poly{5-{[6'-[4"-(4'''-methoxyphenyl)phenoxy]hexyl]carbonyl}bicyclo[2.2.1]hept-2-ene} [22] and (b) poly{5-[[[2',5'-bis[(4"-methoxybenzoyl)oxy]benzyl]oxy]carbonyl]bicyclo[2.2.1]hept-2-ene} [182] as a function of the number average degree of polymerization (●, ■) and the inverse number average degree of polymerization (○, □). Infinite molecular weight transitions: (a) G 41 N 95 I; (b) G 99 N 167 I.

The thermotropic behavior of liquid crystalline polynorbornenes also reach their limiting values at 50 repeat units or less [22, 182, 188–190]. For example, Fig. 9 demonstrates that the glass and nematic–isotropic transitions of both terminally and laterally attached systems level off at 25–50 repeat units, and correspond to the transition temperatures of the infinite molecular weight polymers. The same is true of the crystalline melting and smectic-isotropic transition temperatures of poly{(±)-endo, exo-5,6-di{[n-[4'-(4"-methoxyphenyl)-phenoxy]hexyl]carbonyl}bicyclo[2.2.1]-hept-2-ene} [190].

Although most well-defined SCLCPs have been prepared by cationic polymeriza-

tions of mesogenic vinyl ethers, most of these studies did not prepare polymers of high enough molecular weight to prove that the transition temperatures had leveled off. For example, the highest molecular weight poly{6-[(4'-cyanophenyl-4"-phenoxy)hexyl]vinyl ether} prepared by direct cationic polymerization contains approximately 30 repeat units [124]. Figure 10 compares the thermotropic behavior of these poly(vinyl ether)s to the same polymers prepared by polymer analogous [212, 213] reactions. The higher molecular weight polymers prepared by polymer analogous reactions confirm that the glass and isotropization transitions saturate at approximately 30 repeat units, although the SmC–SmA transition

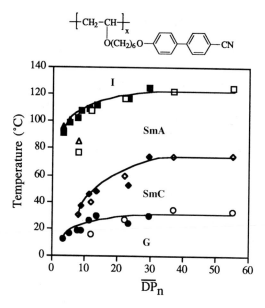

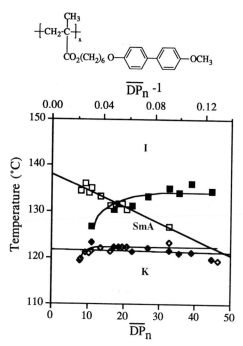

Figure 10. Dependence of the transition temperatures from the glass (●, ○), SmC (◆, ◇), SmA (■, □) and nematic (▲, △) phases of poly{6-[(4'-cyanophenyl-4"-phenoxy)hexyl]vinyl ether} prepared by direct cationic polymerization (●, ◆, ■) [124] and by polymer analogous reactions (○, ◇, □) [212, 213] as a function of the GPC-determined degree of polymerization. Infinite molecular weight transitions: G 31 SmC 81 SmA 122 I.

Figure 11. Dependence of the transition temperatures from the crystalline (◆, ◇) and SmA (■, □) phases of syndiotactic [(rr)=0.80] poly{n-[4-(4'-methoxyphenyl)phenoxy]hexyl methacrylate} as a function of the number average degree of polymerization (◆, ■) and the inverse number average degree of polymerization (◇, □) [44]. Infinite molecular weight transitions: K 122 SmA 138 I.

has not leveled off yet. Similarly, the crystalline–nematic and nematic–isotropic transitions of poly{6-[4'-(4"-ethoxyphenylazo)phenoxyhexyl]vinyl ether} prepared by polymer analogous reactions reach their limiting values at 27 repeat units [211].

The systems shown in Fig. 8 and 9 display very simple thermotropic behavior, with the same phase sequence observed at all molecular weights. However, Fig. 10 exemplifies systems in which the number and type of mesophases vary as a function of molecular weight. This is simply due to the phases having different dependencies on molecular weight. That is, the slope of the glass transition versus degree of polymerization of poly{6-[(4'-cyanophenyl-4"-

phenoxy)hexyl]vinyl ether} is less than that of the SmC–SmA transition; the SmC mesophase is therefore observed only at higher molecular weights, whereas a nematic mesophase is observed only at very low molecular weights. Figure 11 shows a second example in which the SmA–I transition temperature of syndiotactic poly{n-[4-(4'-methoxyphenyl)phenoxy]hexyl methacrylate} increases and levels off an approximately 30 repeat units, whereas the crystalline melting temperature is relatively constant [44]. In this case, the lowest molecular weight oligomer melts directly to the isotropic state. The crystalline melting of isotactic poly{n-[4'-(4"-methoxyphenyl)phenoxy}hexyloxy methacrylate} fractionated into samples with $DP_n = 16-133$ (pdi $= 1.16-1.24$) is also essentially constant (3.4 °C variation) [42].

As explained by Percec and Keller [227], the temperature range of at least the highest temperature mesophase therefore increases with increasing polymer molecular weight due to the greater dependence of isotropization on the degree of polymerization compared to crystalline melting and the glass transition. This is due to a greater decrease in entropy and increase in free energy of the isotropic liquid with increasing molecular weight compared to that of the more ordered phases.

3.2   The Effect of the Mesogen

The extensive literature on low molar mass liquid crystals demonstrates that specific mesogens (specific chemical structures) tend to form specific mesophases, which vary somewhat with the length of the flexible substituent. We therefore expect that the type of mesogen, including the terminal substituent(s) and the length of the spacer should be the primary factors determining the specific mesophase(s) exhibited by a given SCLCP. The nature of the polymer

Table 6. Thermotropic behavior of 4-n-alkoxy-4'-cyanobiphenyls [228], poly{n-[(4'-(4''-cyanophenyl)phenoxy)alkyl]vinyl ether}s ($DP_n = 17-32$, pdi $= 1.09-1.21$) [122–127, 212, 213], n-[(4'-(4''-cyanophenyl)phenoxy)alkyl]vinyl ethers [122–127] and α-ethoxy-ω-(4-n-alkoxy-4'-cyanobiphenyl)s [122–127].

n	H(CH₂)ₙO—⬡—⬡—CN		{CH₂–CH}ₓ, O(CH₂)ₙO—⬡—⬡—CN		
		Thermotropic behavior (°C)			
1	K 104	(N 86) I	G 81×86		I
2	K 102	(N 91) I	G 64×68		N 104 I
3	K 71	(N 64) I	G 47×58		N 88 I
4	K 78	(N 76) I	G 38	N_re 69 SmA 104	N 115 I
5	K 53	N 68 I	G 29	SmC 67 SmA 125	I
6	K 58	N 76 I	G 21	SmA 139	I
7	K 54	N 75 I	G 23	SmC 67 SmA 155	I
8	K 54	SmA 67 N 80 I	G 23	SmA 152	I
9	K 65	SmA 76 N 80 I	G 14	K 56 (SmC 49) SmA 156	I
10	K 61	N 84 I	G 17	K 66 (SmC 52) SmA 165	I
11	K 71	N 87 I	G 21		
12	K 69	SmA 89 I			

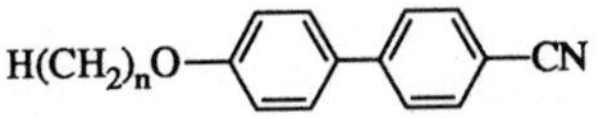

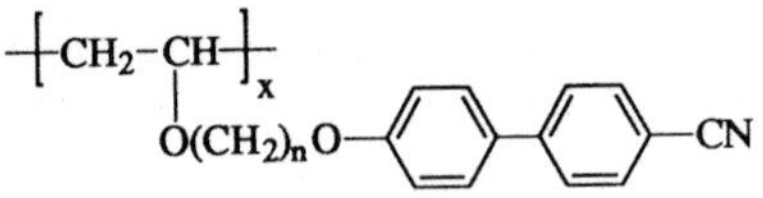

n	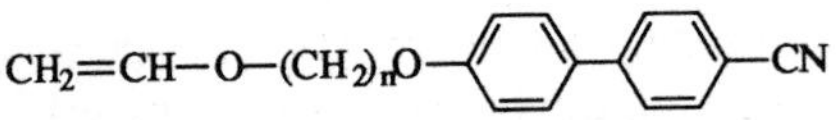 CH₂=CH—O—(CH₂)ₙO—⬡—⬡—CN		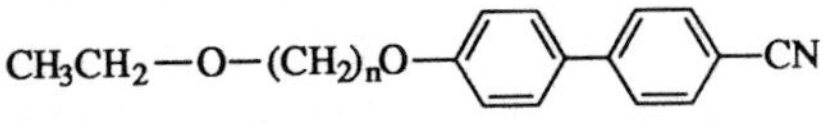 CH₃CH₂—O—(CH₂)ₙO—⬡—⬡—CN		
2	K 118	I	K 69		I
3	K 79	I	K 65		I
4	K 73	N 77 I	K 65		(N 33) I
5	K 54	(N 39) I	K 54		(N 35) I
6	K 65	(N 60) I	K 65		(N 60) I
7	K 59	(N 54) I	K 56		(N 55) I
8	K 54	N 71 I	K 63	(SmA 58	N 61) I
9	K 63	(N 59) I	K 75	(SmA 41	N 50) I
10	K 65	N 70 I	K 69	(SmA 65)	I
11	K 71	(SmA 61 N 71) I	K 69	SmA 60	I

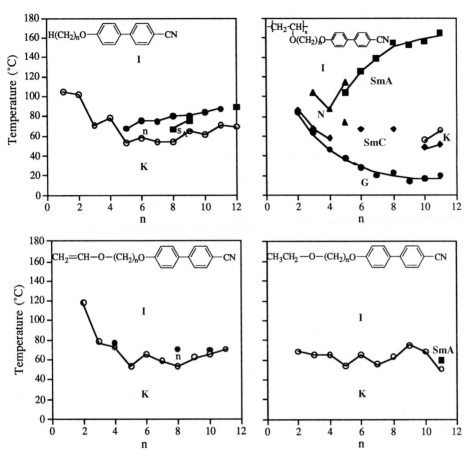

Figure 12. Transition temperatures observed on heating 4-*n*-alkoxy-4'-cyanobiphenyls [228], poly{*n*-[(4'-(4''-cyanophenyl)phenoxy)alkyl]vinyl ether}s ($DP_n = 17 - 32$, pdi = 1.09 – 1.21) [122 – 127, 212, 213], *n*-[(4'-(4''-cyanophenyl)phenoxy)alkyl]vinyl ethers [122 – 127] and α-ethoxy-ω-(4-*n*-alkoxy-4'-cyanobiphenyl)s [122 – 127] from the glassy (●), crystalline (○), SmC (◆), SmA (■) and nematic (▲) states.

backbone, tacticity, polydispersity, etc. should then be secondary factors in enhancing, altering, or disrupting the natural ordering of the mesogenic side chains.

If the mesogen and the length of the spacer are the primary factors determining the specific mesophase(s) exhibited by a given SCLCP, then its thermotropic behavior should be predictable by comparison to an appropriate model compound. Three possibilities for an appropriate model compound are shown in Table 6 and Fig. 12 using poly{*n*-[(4'-(4''-cyanophenyl)phenoxy)-

alkyl]vinyl ether}s [122 – 127, 212, 213] as an example. Not surprisingly, the monomers themselves, which contain olefin endgroups that are very different from the chemical structure of the polymer backbone, are the least appropriate model compounds for poly{*n*-[(4'-(4''-cyanophenyl)phenoxy)alkyl]vinyl ether}s. That is, the polymer exhibits enantiotropic nematic mesophases at spacer lengths of *n* = 3 – 5 and enantiotropic SmA mesophases at spacer lengths of *n* = 5 – 12. In addition, enantiotropic and monotropic SmC mesophases are sometimes observed, as well as a higher ordered phase. In contrast, the monomer forms a SmA mesophase only at $n=11$, which is monotropic; the nematic mesophases observed at $n=4-11$ are also primarily monotropic.

The more obvious choice for a model compound corresponds to exactly one repeat unit of the polymer. The α-ethoxy-ω-(4-n-alkoxy-4'-cyanobiphenyl)s exhibit nematic mesophases at $n=4-9$ and SmA mesophases at $n=8-11$, and therefore match the thermotropic behavior of the polymer better than the vinyl ether monomer. However the SmA mesophase is enantiotropic only at $n=11$, and the nematic mesophase is monotropic at all of these spacer lengths. Compounds which take into account only the mesogen and spacer are also good, if not better, models of the polymers. In contrast to the ethyl ethers, all of the SmA and most of the nematic mesophases are enantiotropic, which means that the melting temperature mimics the relative temperature of the glass transition of the polymer backbone better. However, the nematic mesophase still appears at $n=6-11$, and the SmA mesophase doesn't appear at $n=5-7$, 10, 11.

Nevertheless, the poly(vinyl ether)s and all of these model compounds demonstrate that liquid crystals based on cyanobiphenyl mesogens tend to form nematic and SmA mesophases.

Model compounds which take into account only the mesogen and spacer also mimic the thermotropic behavior of laterally attached SCLCPs well. For example, Table 7 demonstrates that both 2,5-bis[(4'-n-alkoxybenzoyl)oxy]toluenes [228] and all corresponding SCLCPs, such as polynorbornenes [182], laterally attached through a one carbon spacer with 2,5-bis[(4'-n-alkoxybenzyoyl)oxy]benzene mesogens exhibit only nematic mesophases. (However, Sec. 5 of this chapter will demonstrate that smectic layering can be induced in these SCLCPs and their low molar mass analogs by terminating the mesogen's n-alkoxy substituents with immiscible fluorocarbon segments [193–195]).

In order to determine whether or not the alkyl substituent is necessary for low molar mass liquid crystals to mimic the thermotropic behavior of laterally attached SCLCPs, both 1,4-bis[(3'-fluoro-4'-n-alkoxyphenyl)ethynyl]benzenes   [193, 196]

Table 7. Thermotropic behavior of 2,5-bis[(4'-n-alkoxybenzoyl)oxy]toluenes [228] and poly{5-[[[2',5'-bis[(4''-methoxybenzoyl)oxy]benzyl]oxy]carbonyl]bicyclo[2.2.1]hept-2-ene}s ($DP_n=23-100$, pdi$=1.12-1.24$) [182].

n	Thermotropic behavior (°C)	
1	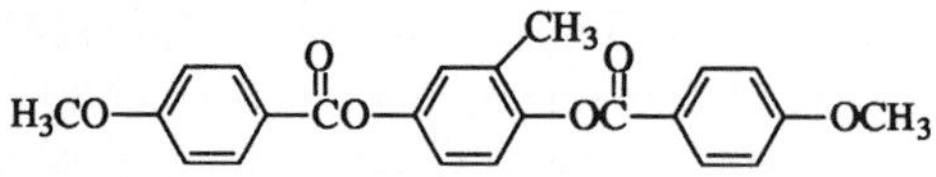 K 166 N 252 I	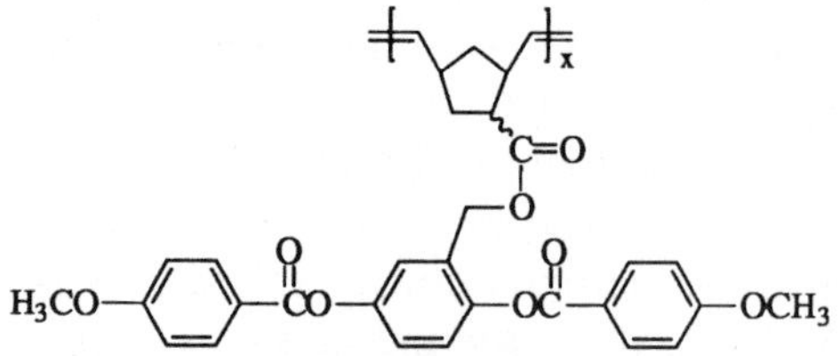 G 98 N 164 I
2	K 187 N 248 I	G 92 N 172 I
3	K 138 N 209 I	G 83 N 140 I
4	K 115 N 206 I	G 73 N 138 I
5	K 90 N 178 I	G 60 N 123 I
6	K 88 N 173 I	G 56 N 126 I

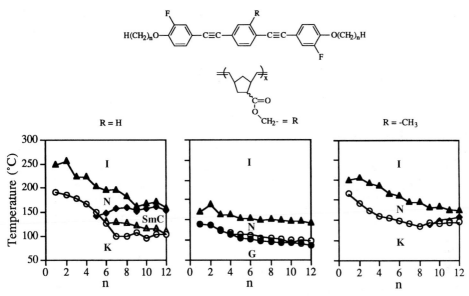

Figure 13. Transition temperatures of 1,4-bis[(3'-fluoro-4'-*n*-alkoxyphenyl)ethynyl]benzenes [193, 196], 1,4-bis[(3'-fluoro-4'-*n*-alkoxyphenyl)ethynyl]toluenes [196] and poly{5-{[[2',5'-bis[2-(3"-fluoro-4"-*n*-alkoxyphenyl)ethynyl]benzyl]oxy]carbonyl}bicyclo[2.2.1]hept-2-ene}s ($DP_n = 31 - 59$, pdi = 1.09 – 1.60) [197]; from the glassy (●), crystalline (○), SmE (△), SmC (◆) and nematic (▲) states as a function of the length of the *n*-alkoxy substituents.

and 1,4-bis[(3'-fluoro-4'-*n*-alkoxyphenyl)-ethylnyl]toluenes [196] were synthesized and their thermotropic behavior compared to that of poly{5-{[[2',5',-bis[2-(3"-fluoro-4"-*n*-alkoxyphenyl)ethynyl]benzyl]-oxy]carbonyl}-bicyclo[2.2.1]hept-2-ene}s [197]. As shown in Fig. 13, the 1,4-bis[(3'-fluoro-4'-*n*-alkoxyphenyl)ethynyl]ben-zenes exhibit a SmE – SmC – N – I phase se-quence when *n* = 6 – 12. In contrast, the poly-mers exhibit a nematic mesophase, and those with *n* = 3 – 12 slowly organize into a crystalline phase which occurs over an ex-tremely narrow temperature range. The 1,4-bis[(3'-fluoro-4'-*n*-alkoxyphenyl)ethynyl]-benzenes are therefore not appropriate mod-el compounds for these polynorbornenes, evidently because they do not take the benzylic spacer into account. In contrast, the corresponding 1,4-bis[(3'-fluoro-4'-*n*-alk-oxyphenyl)ethynyl]toluenes mimic the polynorborne's thermotropic behavior quite

well. For example, Fig. 13 demonstrates that only a nematic mesophase is observed at most substituent lengths, in addition to a monotropic or an enantiotropic SmC meso-phase when *n* = 9 – 12.

Further comparisons to appropriate mod-el compounds and the effect of increased mesogen density are discussed in the next sections.

3.3 The Effect of the Spacer

Increasing the spacer length has much the same effect on the thermotropic behavior of SCLCPs as increasing the length of the flex-ible substituent has on that of low molar mass liquid crystals. That is, it destabilizes some phases and stabilizes others. For ex-ample, just as increasing the length of the flexible substituent depresses the melting

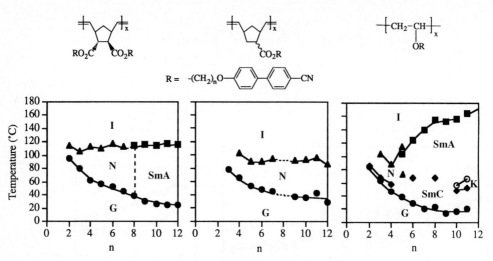

Figure 14. Transition temperatures of poly{(±)-endo,exo-5,6-di{[n-[4′-(4″-cyanophenyl)phenoxy]alkyl]carbonyl}bicyclo[2.2.1]hept-2-ene}s (*DP*ₙ=41–266, pdi=1.21–1.60) [191], poly{5-{[n-[4′-(4″-cyanophenyl)phenoxy]alkyl]carbonyl}bicyclo[2.2.1]hept-2-ene}s (*DP*ₙ=43–290, pdi=1.08–1.27) [189] and poly{n-[(4′-(4″-cyanophenyl)phenoxy)alkyl]vinyl ethers (*DP*ₙ=17–32, pdi=1.09–1.21) [122–127, 212, 213] as a function of the number of methylenic units in their *n*-alkyl spacers; from the glassy (●), crystalline (○), SmC (◆), SmA (■) and nematic (▲) states.

point of low molar mass liquid crystals, increasing the spacer length depresses the glass transition of SCLCPs, and consequently often uncovers mesophase(s) that are not observed without a spacer. Therefore, many of the first SCLCPs synthesized with the mesogen directly attached to the polymer backbone did not exhibit liquid crystallinity [2]. It was only after Finkelmann, Ringsdorf and Wendorff introduced the spacer concept in 1978 [229], that SCLCPs were synthesized routinely. Rather than simply suppressing the glass transition and/or crystalline melting, they attributed the spacer concept to a decoupling of the motions of the mesogens, which want to order anisotropically, from that of the polymer backbone, which tends to adopt a random coil conformation. However, neutron scattering experiments have demonstrated that the polymer backbone is deformed in both nematic and smectic mesophases, and that it is therefore impossible to completely decouple its motions from those of the

mesogens to which it is chemically linked [230].

In spite of the number of well-defined SCLCPs described in Sec. 2 of this chapter, only a few poly(vinyl ether)s and polynorbornenes have been synthesized with a complete and homologous set of spacer lengths. Figure 14 plots the phase diagrams of three polymers with terminally attached 4-(4′-cyanophenyl)phenoxy mesogens as a function of the spacer length: Fig. 15 plots the corresponding phase diagrams of three polymers terminally attached with 4-(4′-methoxyphenyl)phenoxy mesogens. Both figures demonstrate that the glass transition decreases as the length of the spacer increases. This is due to decreased packing density, and is often referred to as internal plasticization.

The temperature of crystalline melting of tactic SCLCPs also decreases as the spacer length increases (Fig. 15), albeit in an odd–even alternation as in low molar mass liquid crystals as a function of the length of the

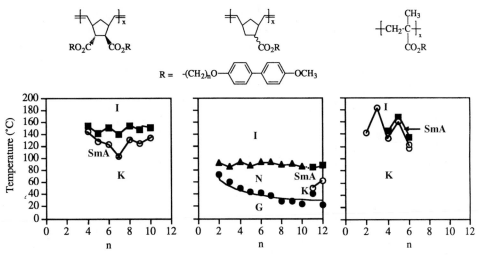

Figure 15. Transition temperatures of (presumably) isotactic poly{(±)-endo,exo-5,6-di{[n-[4'-(4"-methoxyphenyl)phenoxy]alkyl]carbonyl}bicyclo[2.2.1]hept-2-ene}s ($DP_n=30-173$, pdi = 1.20 – 2.75) [190], poly{5-{[n-[4'-(4"-methoxyphenyl)phenoxy]alkyl]carbonyl}bicyclo[2.2.1]hept-2-ene}s ($DP_n=72-151$, pdi = 1.05 – 1.28) [22, 188] and syndiotactic [(rr) = 0.79 – 0.86] poly{n-[4'-(4"-methoxyphenyl)phenoxy]alkyl methacrylate}s ($DP_n=19-33$, pdi = 1.14 – 2.08) [42, 44] as a function of the number of methylenic units in their n-alkyl spacer; from the glassy (●), crystalline (○), SmA (■) and nematic (▲) states.

flexible substituent. However, without additional order within the polymer backbone itself due to high tacticity (Sec. 3.4 and 3.5 of this chapter), the mesogenic side chains are generally not able to crystallize until the spacer is sufficiently long. Therefore, both poly[n-[(4'-(4"-cyanophenyl)phenoxy)alkyl]vinyl ethers [122 – 127] (Fig. 14) and poly{5-{[n-[4'-(4"-methoxyphenyl)phenoxy]alkyl]carbonyl}bicyclo[2.2.1]-hept-2-ene}s [188] (Fig. 15) undergo side chain crystallization only when $n \geq 10$, or when polymers with singly shorter spacers are of low molecular weight.

Depending on the specific mesogen, the nematic mesophase of low molar mass liquid crystals is generally destabilized by increasing substituent length, whereas smectic mesophases tend to be stabilized [228]. That is, long alkyl or alkoxy substituents favor smectic mesophases, whereas short substituents favor nematic mesophases. Figures 14 and 15 confirm that long spacers also favor smectic mesophases, and that short

spacers favor nematic mesophases. These plots also seem to confirm that the transition temperature of nematic disordering decreases slightly with increasing spacer length in SCLCPs, whereas that of the SmA mesophase tends to increase, with at least a slight odd – even alternation. Nevertheless, since the glass transition temperature decreases more rapidly than isotropization with increasing spacer length, the temperature window over which these mesophase(s) are observed increases.

Figure 16 shows the change in enthalpy and entropy as a function of the length of the spacer for the three polymer series plotted in Fig. 14. It demonstrates that both the change in enthalpy and entropy of isotropization from the nematic and s_A mesophases increase linearly with increasing spacer length. Although such increases in ΔH_i and ΔS_i have been attributed to more efficient decoupling of the mesogen from the polymer backbone and therefore to increased order [231], it simply corresponds to a con-

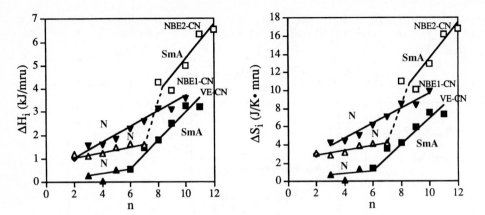

Figure 16. Change in enthalpy and entropy of isotropization from the nematic ($\triangle$) and SmA ($\square$) mesophases of poly{($\pm$)-endo,exo-5,6-di{[n-[4'-(4''-cyanophenyl)phenoxy]alkyl]carbonyl}bicyclo-[2.2.1]hept-2-ene}s (NBE2-CN, DP_n=41–266, pdi=1.21–1.60) [191]; from the nematic ($\blacktriangledown$) mesophase of poly{4-{[n-[4'-(4''-cyanophenyl)phenoxy]alkyl]carbonyl}bicyclo[2.2.1]hept-2-ene}s (NBE1-CN, DP_n=43–290, pdi=1.08–1.27) [189]; and from the nematic ($\blacktriangle$) and SmA ($\blacksquare$) mesophases of poly{n-[(4'-(4''-cyanophenyl)henoxy)alkyl]vinyl ethers (VE–CN, DP_n=17–32, pdi=1.09–1.21) [122–127] as a function of the number of methylenic units in their n-alkyl spacers.

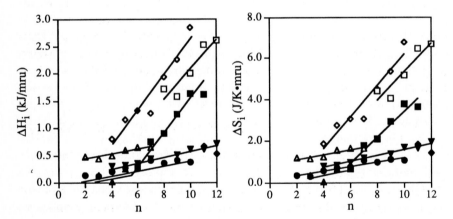

Figure 17. Normalized [232] change in enthalpy and entropy of isotropization from the nematic ($\triangle$) and SmA ($\square$) mesophases of poly{($\pm$)-endo,exo-5,6-di{[n-[4'-(4''-cyanophenyl)phenoxy]alkyl]carbonyl}-bicyclo[2.2.1]-hept-2-ene}s (NBE2-CN, DP_n=41–266, pdi=1.21–1.60) [191]; from the nematic ($\blacktriangledown$) mesophase of poly{5-{[n-[4'-(4''-cyanophenyl)phenoxy]alkyl]carbonyl}bicyclo[2.2.1]hept-2-ene}s (NBE1-CN, DP_n=43–290, pdi=1.08–1.27) [189]; from the nematic ($\blacktriangle$) and SmA ($\blacksquare$) mesophases of poly{n-[(4'-(4''-cyanophenyl)phenoxy)alkyl]vinyl ethers (VE–CN, DP_n=17–32, pdi=1.09–1.21) [122–127]; from the SmA ($\diamond$) mesophase of poly{($\pm$)-endo,exo-5,6-di{[n-[4'-(4''-methoxyphenyl)phenoxy]alkyl]carbonyl}bicyclo[2.2.1]hept-2-ene}s (NBE2-OMe, DP_n=30–173, pdi=1.20–2.75) [190]; and from the nematic ($\bullet$) and SmA ($\blacklozenge$) mesophases of poly{5-{[n-[4'-(4''-methoxyphenyl)phenoxy]alkyl]carbonyl}bicyclo[2.2.1]hept-2-ene}s (NBE1-OMe, DP_n=72–151, pdi=1.05–1.28) [22, 188] as a function of the number of methylenic units in their n-alkyl spacers.

stant increase in ΔH_i and ΔS_i per methylenic unit, and therefore to the disordering of an additional $-CH_2-$ unit of the spacer. The slope of these lines ($\Delta\Delta H_i/-CH_2-$ and $-\Delta\Delta S_i/-CH_2-$) are nearly equivalent for both nematic and smectic disordering, re-

Table 8. Normalized [232] changes in enthalpy and entropy of isotropization per methylenic unit in the spacer of poly{(±)-endo,exo-5,6-di{[n-[4'-(4"-cyanophenyl)phenoxy]alkyl]carbonyl}bicyclo[2.2.1]hept-2-ene}s (NBE2-CN) [191], poly{5-{[n-[4'-(4"-cyanophenyl)phenoxy]alkyl]carbonyl}bicyclo[2.2.1]hept-2-ene}s (NBE1-CN) [189], poly{n-[(4'-(4"-cyanophenyl)phenoxy)alkyl]vinyl ethers (VE-CN) [122–127], poly{(±)-endo,exo-5,6-di{[n-[4'-(4"-methoxyphenyl)phenoxy]alkyl]carbonyl}bicyclo[2.2.1]hept-2-ene}s (NBE2-MeO) [190], and poly{5-{[n-[4'-(4"-methoxyphenyl)phenoxy]alkyl]carbonyl}bicyclo[2.2.1]hept-2-ene}s (NBE1–MeO) [22, 188].

SCLCP	$\Delta\Delta H_i/-CH_2-$ (kJ/mru)		$\Delta\Delta S_i/-CH_2-$ (J/K mru)	
	Nematic	Smectic A	Nematic	Smectic A
NBE2–CN	0.044	0.28	0.12	0.70
NBE1–CN	0.051	–	0.14	–
VE–CN	0.058	0.28	0.14	0.62
NBE2–MeO	–	0.32	–	0.76
NBE1–MeO	0.038	–	0.11	–
Mean	0.048±0.008	0.29±0.02	0.13±0.02	0.69±0.06

spectively, of poly{(±)-endo,exo-5,6-di{[n-[4'-(4"-cyanophenyl)phenoxy]alkyl]carbonyl}bicyclo[2.2.1]hept-2-ene} and poly{n-[(4'-(4"-cyanophenyl)phenoxy)alkyl]vinyl ether, and are half those values of poly{5-{[n-[4'-(4"-cyanophenyl)phenoxy]alkyl]carbonyl}bicyclo[2.2.1]hept-2-ene} since the latter has half as many mesogenic side chains. Therefore, if the mesogen density is taken into account for the four amorphous polymer series plotted in Figs. 14 and 15 [232], $\Delta\Delta H_i/-CH_2-$ and $\Delta\Delta S_i/-CH_2-$ are equivalent for the same mesophase (Fig. 17 and Table 8), regardless of the type of polymer [poly(vinyl ether), mono- and di-substituted polynorbornene] and regardless of the mesogen [4-(4'-cyanophenyl)phenoxy and 4-(4'-methoxyphenyl)phenoxy mesogens].

As expected, Figs. 16 and 17 and Table 8 also demonstrate that the enthalpy and entropy of isotropization from the more ordered smectic mesophase is higher than those from the nematic phase. This discontinuity and/or change in the slope with a change in the type of mesophase can therefore be used as additional confirmation that a phase change has occurred with the addition or subtraction of one methylenic unit in the spacer of a homologous series.

The effect of varying the nature of the spacer from hydrocarbon to siloxane is starting to be investigated using living polymerizations [28]. However, these polymers have not been characterized yet.

3.4   The Effect of the Nature of the Polymer Backbone

Since the effect of the nature of the polymer backbone has not been studied extensively using well-defined polymers, it is worth summarizing the observations from less controlled systems based on comparisons of poly(phosphazene), poly(methylsiloxane), poly(vinyl ether), poly(acrylate), poly(methacrylate), poly(chloroacrylate), and polystyrene backbones [233–241]. In general, the mesogen is more 'decoupled' from the polymer backbone as the latter's flexibility increases. This is because the dynamics of ordering increase with increasing flexibility, which increases the ability of the mesogenic side-chains to form more ordered mesophases and/or crystallize. (The relative dynamics of these SCLCPs correlate directly with the speed with which they form identifiable microscopic textures between crossed polarizers [235]). In addition, the side chains are able to crystallize at shorter spacer lengths as the polymer flexibility increases [236–238]. Therefore, SCLCPs with extremely flexible backbones may form only a monotropic or virtual mesophase, whereas an analogous polymer with a more rigid backbone is more likely to form an enantiotropic mesophase. However, as the flexibility of the backbone increases, the glass transition temperature usually decreases, whereas that of (melting and) isotropization typically increase. Additional mesophases may therefore be revealed if the glass transition is low enough, and if the mesophase is thermodynamically stable relative to the crystalline phase.

Secondary structural variables, including the nature of the polymer backbone, will only be elucidated by synthesizing complete and homologous series of well-defined polymers. As discussed previously, only a few of the poly(vinyl ether)s, polynorbornenes and polymethacrylates contain both identical mesogens and spacers. These systems confirm the above trends. For example, when (4′-n-butoxyphenoxycarbonyl)phenoxy mesogens are attached to polymethacrylate and poly(vinyl ether) backbones through an n-hexyl spacer, the more flexible poly(vinyl ether) forms a highly ordered SmF or SmI mesophase with no evidence of a glass transition (Table 9). In contrast, the less flexible polymethacrylate is glassy. Otherwise, both have nearly identical mesophases and transition temperatures, except poly{6-[(4′-n-butoxyphenyl-4″-benzoate)hexyl]vinyl ether} forms a SmC mesophase (X-ray evidence) in addition to the nematic mesophase, whereas poly{6-[4′-(4″-n-butoxyphenoxycarbonyl)phenoxy]hexyl methacrylate} reportedly forms a SmA mesophase in the same temperature range.

Comparison of the phase diagrams plotted in Fig. 14 of poly{5-{[n-[4′-4″-cyanophenyl)phenoxy]alkyl]carbonyl}bicyclo[2.2.1]hept-2-ene}s [189] and poly{n-[(4′-(4″-cyanophenyl)phenoxy)alkyl]vinyl ethers [122–127, 212, 213] which contain a single mesogen per repeat unit demonstrates that the glass transition temperature decreases as the flexibility of the polymer backbone increases from polynorbornene to poly(vinyl ether), whereas the isotropization temperature increases. In addition to revealing additional mesophases at lower temperatures, this increase in polymer flexibility enables the poly(vinyl ether)s to form more ordered mesophases. That is, poly{5-{[n-[4′-(4″-cyanophenyl)phenoxy]alkyl]carbonyl}bicyclo[2.2.1]-hept-2-ene}

Table 9. Comparison of the thermotropic behavior of poly{6-[4′-(4″-n-butoxyphenoxycarbonyl)phenoxy]hexyl methacrylate} and poly{6-[(4′-n-butoxyphenyl-4″-benzoate)hexyl]vinyl ether}.

Polymer backbone	DP_n	pdi	Thermotropic behavior (°C)	Ref.
Methacrylate	19	1.13	G 38 SmA 104 N 110 I	[91]
Vinyl ether	38	1.15	S 47 SmC 106 N 111 I	[150]

exhibits only a nematic mesophase at all spacer lengths, whereas the corresponding poly(vinyl ether) exhibits SmA and SmC mesophases with increasing spacer lengths, and also undergoes side-chain crystallization at spacer lengths of 10–11 carbons, (at which point the SmC mesophases becomes monotropic).

When the polynorbornene is substituted with two mesogens rather than one, the internal mobility of the chain should not change, although its overall flexibility and free volume should decrease. As shown in Fig. 14, doubling the mesogen density has very little effect on the glass transition, but causes an increase in the temperature of isotropization, especially at longer spacer lengths. Increasing the mesogen density also enables the mesogens to form smectic layers when the spacer length reaches eight carbon atoms, which is typical of SCLCPs terminally attached with cyano-substituted mesogens [2].

Since the glass transition of a polymer is determined not only by the polymer flexibility, but also by the packing density (free volume), the interactions between chains and the chain length, it is difficult to assess the relative flexibility of two polymers from the glass transition alone. Therefore, it is not immediately obvious from the glass transitions of syndiotactic poly(methyl methacrylate) (105 °C) and polynorbornene (43 °C) which of the two backbones are more flexible. That is, although polynorbornene has the lower T_g, the double bonds and cyclopentane ring along the backbone should be more constrained than the single bonds along a polymethacrylate backbone.

The phase diagrams of syndiotactic [(rr)=0.79–0.86] poly{n-[4'-(4″-methoxyphenyl)phenoxy]phenoxy]alkyl methacrylate [42, 44], poly{5-{[4'-(4″-methoxyphenyl)phenoxy]-alkyl]carbonyl}bicyclo[2.2.1]hept-2-ene [22, 188] and poly

{(±)-endo,exo-5,6-di{[n-[4'-(4″-methoxyphenyl)phenoxy]alkyl]carbonyl}bicyclo[2.2.1]hept-2-ene} [190] were plotted in Fig. 15. Assuming that the polynorbornene backbone is more rigid than the polymethacrylate backbone, increasing the backbone flexibility increases the ability of the side chains to form more ordered mesophases and crystallize at shorter spacer lengths as expected. However, the more ordered side-chains of the poly{n-[4'-(4″-methoxyphenyl)phenoxy]alkyl methacrylates may also be due to their high tacticity as discussed in Sec. 3.5 of this chapter.

Increasing the mesogen density from one to two mesogenic groups per repeat unit of the polynorbornene backbone again allows the side chains to organize into smectic layers. However, in this case, Mo(CH-t-Bu)(N-2,6-C$_6$H$_3$-i-Pr$_2$)(OCH$_3$(CF$_3$)$_2$)$_2$ was used as the initiator to polymerize the (±)-endo,exo-5,6-di{[n-[4'-(4″-methoxyphenyl)phenoxy]alkyl]carbonyl}bicyclo[2.2.1]hept-2-enes instead of Mo(CH-t-Bu)(N-2,6-C$_6$H$_3$-i-Pr$_2$)(O-t-Bu)$_2$ [190]; Mo(CH-t-Bu)(N-2,6-C$_6$H$_3$-i-Pr$_2$)(OCH$_3$(CF$_3$)$_2$)$_2$ polymerizes (±)-endo,exo-5,6-[disubstituted]bicyclo[2.2.1]hept-2-enes to produce isotactic polymers with approximately 85% cis double bonds, whereas the corresponding Mo(CH-t-Bu)(N-2,6-C$_6$H$_3$-i-Pr$_2$)(O-t-Bu)$_2$ polymerizations produce atactic polymers with approximately 95% trans double bonds [242]. Therefore, the poly{(±)-endo,exo-5,6-di{[n-[4'-(4″-methoxyphenyl)-phenoxy]alkyl]carbonyl}bicyclo[2.2.1]-hept-2-ene}s shown in Fig. 15 are presumably highly isotactic and able to undergo side-chain crystallization, in contrast to the atactic poly{(±)-endo,exo-5,6-di{[n-[4'-(4″-cyanophenyl)phenoxy]alkyl]carbonyl}bicyclo[2.2.1]hept-2-ene}s presented in Fig. 14. Although a more ordered mesophase is formed by the more tactic polymers, the increased ability of the side chains to

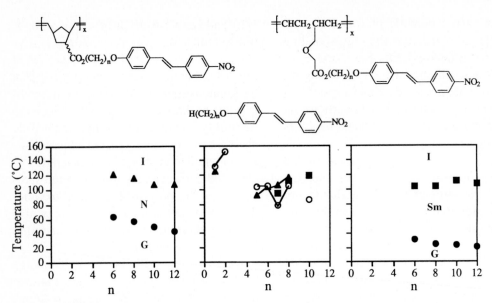

Figure 18. Transition temperatures from the glass (●), crystalline (○), nematic (▲) and smectic (■) phases of poly{5-{[n-[4''-(4'''-nitrostilbeneoxy)alkyl]carbonyl}bicyclo[2.2.1]hept-2-ene}s ($DP_n = 27-42$, pdi = 1.08–1.11) [187], 4-n-alkoxy-4'-nitro(stilbene) [228] and poly[3-[((n-(4'-(4''-nitrostilbeneoxy)alkyl)carbonyl)methyleneoxy)methyl]cyclobutene}s ($DP_n = 61-72$, pdi = 1.14–1.16) [187] as a function of the number of methylenic carbons in their n-alkyl spacers.

crystallize also results in much narrower temperature ranges over which the liquid crystalline phase is observed.

A final example demonstrating that more ordered phases are possible when more flexible polymer backbones are used is shown in Fig. 18. Although this is not a complete series with consecutive spacer lengths, it does demonstrate that the more flexible polybutadiene backbone allows n-[4'-(4''-nitrostilbeneoxy)alkyl] side chains with 6–12 carbons in the spacer to organize into smectic layers, whereas the corresponding polynorbornenes exhibit only nematic mesophases [187]. This is especially interesting since the corresponding low molar mass liquid crystals transform from a monotropic nematic phase when the alkoxy substituent contains six carbons, to an enantiotropic smectic and nematic sequence at 7–8 carbons, and finally to only enantiotropic smectic at ten carbons [228]. The corre-

sponding polymers with less than six carbons in the spacer (as well as the rest of the series) should therefore by synthesized and characterized to determine to what extent the polymer backbone overrides the natural ordering tendency of the side chains.

As discussed in Sec. 3.3 of this chapter, the change in enthalpy and entropy of isotropization increases linearly with increasing spacer length. The slopes of these lines ($\Delta\Delta H_i/-CH_2-$ and $\Delta\Delta S_i/-CH_2-$) therefore correspond to the disordering of one methylenic unit of the spacer, and are evidently equivalent for a given mesophase if they are normalized [232] over the mesogen density. This is true regardless of the type of polymer [polynorbornene or poly(vinyl ether)] and regardless of the mesogen (p-cyanobiphenyl or p-methoxybiphenyl). Since the slope of the normalized changes in enthalpy and entropy of isotropization are equivalent for a specific mesophase, the degree

Table 10. Changes in enthalpy and entropy of isotropization of poly{(±)-endo-exo-5,6-di{[n-[4'-(4''-cyanophenyl)phenoxy]alkyl]carbonyl}bicyclo[2.2.1]hept-2-ene}s (NBE2–CN) [191], poly{5-{[n-[4'-(4''-cyanophenyl)-phenoxy]alkyl]carbonyl}bicyclo[2.2.1]hetp-2-ene}s (NBE1–CN) [189], poly{n-[(4'-(4''-cyanophenyl)phenoxy)alkyl]vinyl ethers (VE–CN) [122–127].

SCLCP	n	ΔH_i (kJ/mru)	ΔS_i (J/K mru)	T_i (°C)	Mesophase
NBE2–CN	4	0.492	1.27	114	Nematic
NBE1–CN	4	0.312	0.830	103	
VE–CN	4	0.0314	0.0870	104	
NBE2–CN	5	0.596	1.56	110	
NBE1–CN	5	0.322	0.884	91	
VE–CN	5	0.262	0.675	115	
NBE2–CN	9	1.58	4.05	116	Smectic A
VE–CN	9	1.26	2.96	152	
NBE2–CN	10	2.00	5.17	115	
VE–CN	10	1.63	1.70	156	
NBE2–CN	11	2.53	6.46	118	
VE–CN	11	1.61	1.68	165	

of order of at least the spacer, and presumably the mesogen, within a specific mesophase must be equivalent for the five polymer series summarized in Fig. 17 and Table 8. However, the different absolute values (e. g. the intercepts) of these lines indicate that the disordering of the spacer is not independent of the rest of the polymer, and instead varies with the polymer backbone to which it is chemically linked.

The data in Table 10 demonstrates that for a constant spacer length and mesophase, both the change in enthalpy and entropy of isotropization decrease as the flexibility of the polymer backbone increases from poly-norbornene to poly(vinyl ether). However, the change in entropy decreases more rapidly than the change in enthalpy, and the isotropization temperature ($T_i = \Delta H_i / \Delta S_i$) therefore increases with increasing flexibility. Since lower entropies of fusion are associated with more rigid structures, the lower entropy of isotropization of poly(vinyl ether)s is obviously not due to a lack of inherent flexibility of its polymer backbone, but rather to the more flexible backbone being more ordered and therefore more

constrained (anisotropic) within the mesophase, as was recently demonstrated by neutron scattering experiments [230].

3.5 The Effect of Tacticity

The effects of tacticity were first observed and discussed in 1981 [243–245]. However, relatively little effort has been made in examining the effect of tacticity on well-defined SCLCPs prepared through living polymerizations. Although these limited results are contradictory, we believe that trends are emerging which indicate that high tacticity in SCLCPs will cause either of two effects. In most systems, the high tacticity evidently places the mesogenic side groups in the proper configuration to crystallize and/or form more orderes mesophases. However, if the mesogenic groups are not in the proper configuration to form ordered phases, the primary effect of increasing the tacticity is to decrease the flexibility of the polymer backbone relative to that of the atactic polymer.

Table 11. Comparison of the thermotropic behavior of isotactic and syndiotactic poly{n-[4'-(4"-methoxyphenyl)phenoxy]alkyl methacrylate}s.

$$\left[CH_2-\underset{\underset{CO_2(CH_2)_nO-\bigcirc-\bigcirc-OCH_3}{|}}{\overset{\overset{CH_3}{|}}{C}}\right]_x$$

n	GPC		%			Phase transitions (°C)	Ref.
	DP_n	pdi	(mm)	(mr)	(rr)		
2	37	2.35	94	6	0	K 193 I	[42]
2	23	1.43	2	19	79	K 142 I	[42]
3	68	2.80	97	3	0	K 147 I	[42]
3	33	2.08	5	12	83	K 183 I	[42]
4	24	3.22	95	4	1	K 141 I	[42]
4	26	1.23	2	12	86	K 133 SmA 145 I	[42]
5	36	14.0	96	3	1	K 145 I	[42]
5	19	1.57	2	18	80	K 160 SmA 167 I	[42]
6	20	2.21	90	9	1	K 136 I	[42]
6	28	1.41	1	14	85	K 116 SmA 131 I	[42]
6	33	1.14	–	–	≈80	K 122 SmA 135 I	[44]

Discussion of the effect of the polymer backbone in Sec. 3.4 of this chapter already provided examples of highly isotactic poly{($\pm$)-endo,exo-5,6-di{[n-[4'-(4"-methoxyphenyl)phenoxy]alkyl]carbonyl}bicyclo[2.2.1]hept-2-ene}s [190] and syndiotactic poly{n-[4'-(4"-methoxyphenyl)phenoxy]alkoxy methacrylate}s [42, 44] which crystallize and form more ordered mesophases than those of the corresponding atactic polymers (Fig. 15). Although more flexible backbones are more able to achieve the conformation necessary to order, the side chains are evidently already attached to the polymer backbone of these tactic polymers with the proper configuration to order, which obviates the need to distort their conformation for such purposes.

As discussed in Sec. 2.1 of this chapter, isotactic poly{n-[4'-(4"-methoxyphenyl)-phenoxy]alkyl methacrylate}s have also been prepared, although the isotactic polymerizations are much less controlled than the syndiotactic polymerizations [42, 44].

The thermotropic behavior of both the isotactic and syndiotactic poly{n-[4'-(4"-methoxyphenyl)phenoxy]alkyl methacrylate}s is summarized in Table 11. All of the tactic polymers crystallize. With the exception of poly{2-[4'-(4"-methoxyphenyl)phenoxy]ethyl methacrylate} (n=2), the melting temperature of the isotactic polymers is almost independent of the spacer length. In contrast, the syndiotactic polymers melt with a large odd–even alternation. However, only the syndiotactic polymers with at least four carbons in the spacer exhibit an enantiotropic smectic mesophase, which occurs over only a very narrow temperature range. The greater order of the isotactic polymers is evidently due to the greater segmental mobility of isotactic versus syndiotactic polymethacrylate backbones [246, 247].

Surprisingly, the mesogenic groups of syndiotactic [(rr)=0.70–0.78] poly{6-[4'-(4"-methoxyphenoxycarbonyl)phenoxy]-hexyl methacrylate}s are not in the proper

Table 12. Comparison of the thermotropic behavior of poly{6-[4′-(4″-methoxyphenoxycarbonyl)phenoxy]hexyl methacrylate} with different tacticities.

Polymerization mechanism	GPC		%			Phase transitions (°C)	Ref.
	DP_n	pdi	(mm)	(mr)	(rr)		
Radical	48	1.51	–	41	59	G 39 SmA 72 N 107 I	[45]
	84	3.65	1	34	65	G 40 N 107 I	[90]
	290	3.10	–	34	66	G 38 N 112 I	[45]
	953	1.23	–	–	–	G 41 SmA 72 N 111 I	[248]
GTP	42	1.71	–	44	50	G 42 N 107 I	[47]
Anionic	37	1.24	–	28	72	G 39 N 82 I	[45]
(TPP)AlMe[a]	21	1.31	–	28	72	G 38 SmA 64 N 105 I	[90]
(TPP)AlMe-ALC[b]	30	1.43	–	22	78	G 40 N 107 I	[90]

[a] Methyl(5,10,15,20-tetraphenylporphinato)aluminum.
[b] Methyl(5,10,15,20-tetraphenylporphinato)aluminum and methylaluminum bis(2,4-di-*t*-butylphenolate).

configuration to crystallize [45, 90], and the primary effect is therefore that of decreased flexibility compared to the atactic polymer. As discussed in subsection 3.4, more rigid backbones suppress side-chain crystallization and the formation of more ordered mesophases. In addition, the glass transition temperature increases whereas that of isotropization tends to decrease.

However, as shown in Table 12, the glass transition temperature of poly{6-[4′-(4″-methoxyphenoxycarbonyl)phenoxy]hexyl methacrylate} is essentially independent of tacticity. With the exception of the polymer prepared by anionic polymerization [45], the nematic–isotropic transition also appears to be independent of tacticity. However, as demonstrated by the data in Fig. 8, the extrapolated transition temperatures (G 44 N 105 I) of an infinite molecular weight polymer prepared by anionic polymerization are nearly identical to those of the rest of the polymers presented in Table 12.

Although the more ordered SmA mesophase is not detected in two of the three more syndiotactic poly{6-[4′-(4″-methoxyphenoxycarbonyl)phenoxy]hexyl methacrylate)s, it is also not detected in some of the less tactic polymers prepared by free radical and group transfer polymerizations. These results are therefore inconclusive as to whether or not increasing syndiotacticity and decreasing flexibility suppresses the formation of the more ordered SmA mesophase in poly{6-[4′-(4″-methoxyphenoxycarbonyl)phenoxy]hexyl methacrylate}s.

3.6 The Effect of Polydispersity

One of the most fundamental questions still open regarding the structure/property relationships of SCLCPs is the effect of polydispersity. In some cases, the temperature and nature of the mesophase(s) depend not only on the degree of polymerization, but also on the molecular weight distribution [147–149, 212]. For example, the s_C mesophase of the broad polydispersity (pdi = 1.9) poly{6-[(4'-cyanophenyl-4''-phenoxy)hexyl]vinyl ether] prepared by direct polymerization was not detected by DSC, although it was readily apparent in samples prepared by polymer analogous reactions (pdi = 1.2) [212].

However, broad polydispersity is generally believed to manifest itself in broad phase transitions. In contrast to low molar mass liquid crystals (LMMLCs) which undergo phase transitions over a few degrees, SCLCPs often exhibit extremely broad transitions. This wide (and even split [249, 250]) biphasic region, in which a mesophase coexists with either the isotropic melt or another mesophase [251], is evidently due to the fact that polymers are polydisperse. For example, fractionated samples of poly{6-[(4'-cyanophenyl-4''-phenoxy)undecyl]acrylate} undergo (SmC – SmA, SmA – I) transition over a narrower temperature range (15–25 °C) than the original polymer (40 °C) prepared by free radical polymerization [252]. (Unfortunately, the actual molecular weight and polydispersity were not determined for any of the samples.)

In contrast, the transitions of most well-defined SCLCPs prepared by controlled polymerizations are relatively narrow. The effect of polydispersity was therefore investigated by blending well-defined (pdi < 1.28) poly{5-{[6'-[4''-(4'''-methoxyphenyl)phenoxy]alkyl]carbonyl}bicyclo[2.2.1]hept-2-ene}s of varying molecular weights (DP_n = 5, 10, 15, 20, 50, 100) to create polydisperse samples (pdi = 2.50–4.78) [22]. In this case, both monodisperse samples and multimodal blends underwent the nematic – isotropic transition over a narrow temperature range. Polydispersity also had no effect on the temperatures of transitions, which were determined simply by the number average degree of polymerization.

Since polydispersity created by blending linear polymers has no effect on the thermotropic behavior of polynorbornenes [22], whereas fractionation of polyacrylates prepared by radical polymerization results in narrower biphasic regions [252], we must consider what the various sources of polydispersity are. Most SCLCPs are still prepared by free radical polymerizations. Since mesogenic monomers are highly functionalized with a number of sites which can cause chain transfer, their free radical polymerizations should result in chain branching and broad polydispersity at high monomer conversion. In addition, entanglements are trapped in high molecular weight polymers.

Therefore, the broad phase transitions of SCLCPs may be caused by the immiscibility of a mixture of molecular architectures resulting from chain branching at the end of the polymerization [23], and/or by entanglements; only chain branching increases the molecular weight distribution. This argument is supported by recent theoretical and experimental studies of polyethylene. That is, while polydisperse blends of linear polyethylene are homogeneous [253], blends of linear and branched polyethylene phase segregate [254–258] with the immiscibility increasing as the branch content of the two components becomes more dissimilar [259, 260]. In addition, entropic corrections to the Flory – Huggins theory predict that mixtures of linear polymer and low concentrations of branched polymer will phase separate [261].

4 Chain Copolymerizations

Polymers which contain more than one type of monomeric repeat unit are called copolymers. By copolymerizing two or more monomers in varying ratios and arrangements, polymeric products with an almost limitless variety of properties can be obtained [3]. As shown in Scheme 18, there are four basic types of copolymers as defined by the distribution of, for example, comonomers A and B. While the properties of a random copolymer are intermediate between those of the two homopolymers, block and graft copolymers exhibit the properties of both homopolymers. The properties of an alternating copolymer are usually unique.

The first three types of copolymers can be prepared by polymerizing a mixture of the two monomers simultaneously. In this case, the distribution of comonomers is determined by their relative concentrations and reactivity ratios. The reactivity ratios (r_1, r_2, etc.) are the ratios of the rate constants of homopropagation and cross-propagation (Eq. 32).

$$r_1 = \frac{k_{11}}{k_{12}}, \quad r_2 = \frac{k_{22}}{k_{21}} \tag{32}$$

Using the terminal model of copolymerization, there are four possibilities for propagation (Scheme 19).

$$\begin{array}{ccc}
\text{\small$\sim\!\!\sim$M}_1\!^{\bullet} + M_1 & \xrightarrow{k_{11}} & \text{\small$\sim\!\!\sim$M}_1\!^{\bullet} \\
\text{\small$\sim\!\!\sim$M}_1\!^{\bullet} + M_2 & \xrightarrow{k_{12}} & \text{\small$\sim\!\!\sim$M}_2\!^{\bullet} \\
\text{\small$\sim\!\!\sim$M}_2\!^{\bullet} + M_2 & \xrightarrow{k_{22}} & \text{\small$\sim\!\!\sim$M}_2\!^{\bullet} \\
\text{\small$\sim\!\!\sim$M}_2\!^{\bullet} + M_1 & \xrightarrow{k_{21}} & \text{\small$\sim\!\!\sim$M}_1\!^{\bullet}
\end{array}$$

Scheme 19

If there are only two active sites (M_1^* and M_2^*) whose concentration are at steady state and if monomer is consumed entirely by propagation, the molar ratio of the two monomer units in the polymer ($d[M_1]/d[M_2]$) is defined by Eq. (33), in which $[M_1]$ and $[M_2]$ are the concentrations of the two monomers in the polymer feed.

$$\frac{d[M_1]}{d[M_2]} = \frac{[M_1]}{[M_2]} \cdot \frac{r_1[M_1]+[M_2]}{r_2[M_2]+[M_1]} \tag{33}$$

Random copolymers in which the comonomer distribution follows Bernoullian or zero-order statistics are formed by ideal copolymerizations in which $r_1 r_2 = 1$. However, a truly random distribution of the two units results only if the Bernoullian distribution is symmetric (i.e. when $r_1 = r_2 = 1$). In this case, the two monomers have equal probability of reacting with a given active center, regardless of the monomer it is derived from, and the copolymer composition equals the comonomer feed composition at all conversions. Random copolymers are generally formed by radical copolymerizations, whereas ionic copolymerizations tend to favor propagation of one of the comono-

RANDOM	poly(A-*ran*-B)	AABAABBBABBAAABABAABB
BLOCK	poly(A-*block*-B)	AAAAAAAAAAAABBBBBBBBBB
ALTERNATING	poly(A-*alt*-B)	ABABABABABABABABABABA
GRAFT	poly(A-*graft*-B)	AAAAAAAAAAAAAAAAAAAAAA

Scheme 18

mers much more than the other, yielding blocky sequences of that comonomer. That is, it is usually difficult to copolymerize monomers by an ionic mechanism due to large differences in their reactivity ratios: $r_1 \gg 1$, $r_2 \ll 1$. Alternating copolymers result when both reactivity ratios are close to zero due to little tendency of either monomer to homopolymerize ($r_1 = r_2 = 0$).

Block copolymers containing long blocks of each homopolymer in a diblock, triblock, or multiblock sequence are formed by simultaneous polymerization of the two comonomers only when $r_1 \gg 1$ and $r_2 \gg 1$. However, block copolymers are more effectively prepared by either sequential monomer addition in living polymerizations, or by coupling two or more telechelic homopolymers subsequent to their homopolymerization. Alternatively, if the two monomers do not polymerize by the same mechanism, a block copolymer can still be formed by sequential monomer addition if the active site of the first block is transformed to a reactive center capable of initiating polymerization of the second monomer.

Living polymerizations have been used to prepare copolymers based on either two mesogenic monomers, or one mesogenic monomer and one nonmesogenic monomer, with the comonomers distributed either statistically or in blocks or grafts.

4.1   Block Copolymers

Block and graft copolymers based on two or more incompatible polymer segments phase separate and self-assemble into spatially periodic structure when the product (χN) of the Flory–Huggins interaction parameter and the total number of statistical segments in the copolymer exceeds a critical value [25, 262, 263]. Since the incompatible

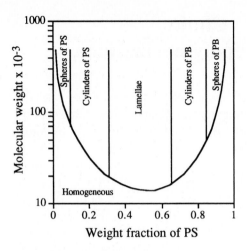

Figure 19. Morphology of poly(styrene-*block*-butadiene) as a function of molecular weight and composition [25].

blocks are chemically linked to each other, macroscopic phase separation is prevented, and phase separation is instead limited to the microscopic level. The resulting morphology minimizes the free energy of the system by minimizing unfavorable contacts and the interfacial curvature energy, while maximizing the conformational entropy and maintaining a uniform segment density throughout the phase [262–266]. The geometry and size of the microdomains are therefore determined not only by the degree of incompatibility χN, but also by the volume fraction (ϕ_A, ϕ_B) of the different blocks, the block length(s), the total molecular weight, the conformational asymmetry and fluctuations [262–271].

An isothermal morphology diagram of poly(styrene-*block*-butadiene) is shown in Fig. 19 as a function of molecular weight and copolymer composition; the classic morphologies include spherical microdomains ($0 \leq \phi_A \leq 0.15$) packed in a body-centered cubic lattice, hexagonally packed cylindrical microdomains ($0.15 \leq \phi_A \leq 0.3$), and alternating lamellae of approximately symmetric diblocks ($0.3 \leq \phi_A \leq 0.5$). Several new morphologies and thermotropic transitions were discovered by blending diblock copolymers with homopolymer [272–278] or diblock [279, 280] copolymer, and by varying temperature [281, 282] or hydrostatic pressure [283] in order to change χ at a fixed composition. Although the first nonclassic morphology was characterized as an ordered bicontinuous double diamond phase [284–289], it has since been identified as a gyroid bicontinuous cubic phase [278, 290], which is apparently unstable [291]. Linear ABA triblock [25, 267–270], diblock star [289, 292] and cyclic diblock [293] copolymers generally form the same morphologies as diblock copolymers of the same composition. However, a new tricontinous cubic structure was recently discovered in miktoarm star copolymers [294] and ABC triblock [295–297] copolymers; ABC triblock copolymers also form spheres and cylinders at lamellae interfaces [298–301].

Block copolymers containing a side-chain liquid crystalline segment are being synthesized in order to answer structure/property questions, and to create new materials that combine the anisotropic ordering of liquid crystals with the mechanical properties of block copolymers. For example, in contrast to block copolymers which have more uniform mechanical properties, even main-chain liquid crystalline polymers are notorious for lacking strength in the direction transverse to chain extension. Self-supported liquid crystalline materials should also be possible by synthesizing liquid crystalline thermoplastic elastomers, in which the ABA triblock copolymer contains a liquid crystalline B block with hard segments as the A blocks.

However, understanding the thermodynamics of phase separation in liquid crystalline block copolymers is in its infancy. The morphology of such block copolymers will be influenced by the competition between microphase separation of the chemically-different blocks and anisotropic ordering of the liquid crystalline block(s). This may result in unique morphologies, such as those recently observed with rod–coil diblock copolymers of poly(styrene-*block*-hexyl isocyanate) [302, 303]. This microphase separation may also influence the equilibrium packing of the mesogens, and therefore the onset, nature and stability of the mesophase(s), especially at the interface and especially at curved surfaces.

The molecular weight, block lengths and volume fraction of each block are three factors controlling the morphology of block copolymers. In addition, phase separation should be most discrete with the narrowest interfacial areas when the block lengths are monodisperse [25]. Block copolymers are therefore generally synthesized using living polymerizations which readily control both the molecular weight and result in copolymers with narrow polydispersity. Conversely, the ability to form well-defined block copolymers by sequential monomer addition provides further evidence that a polymerization and its propagating chain ends are living. All of the 'living' polymerization mechanisms used to prepare well-defined SCLCPs have been used to prepare block copolymers containing a liquid crystalline block, and all of the well-defined side-chain liquid crystalline block copolymers have been prepared by sequential monomer addition. In this case, the liquid crystalline monomer is either polymerized directly or the mesogen is introduced in a polymer analogous reaction. When the liquid crystalline block is generated by a polymer analogous reaction, well-defined blocks require quantitative functionalization.

4.1.1   Anionic and Group Transfer Copolymerizations

In order to prepare block copolymers by sequential monomer addition, the crossover reaction must be efficient. The order of monomer addition is therefore extremely important in anionic polymerizations, since the propagating anion of the first block must be nucleophilic enough to initiate polymerization of the second monomer. Since the reactivity of the propagating anion correlates with the pK_a of its conjugate acid, any monomer listed in Table 13 (such as styrene) will polymerize any monomer listed

Table 13. Monomer and propagating anion's reactivity in anionic polymerizations.

Monomer	pK_a	Ref.
Styrenes	41–42	[25]
Dienes	43	[25]
2-Vinyl pyridine	29	[304]
Methacrylic esters	27–28	[25]
Acrylonitrile	32	[25]
4-Vinyl pyridine	26	[304]
Vinyl ketones	19	[25]
Oxiranes	16–18	[25]
Thiiranes	12–13	[25]
Siloxanes	10–14	[25]
Cyanoacrylates	11–13	[25]
Nitroalkanes	10–14	[25]
Lactones	4–5	[25]

below it (such as methyl methacrylate) with a similar or lower pK_a [25]. As outlined in Scheme 20, this means that the more reactive propagating species is used to polymerize the more reactive monomer, whereas the less reactive propagating species can not initiate polymerization of the less reactive monomer.

Block copolymers of butadiene and styrene are therefore readily synthesized anionically, with either of the two monomers polymerized first. Precursor copolymers of poly(styrene-*block*-butadiene) have been used to prepare well-defined liquid crystalline block copolymers by the same polymer analogous reaction described in Sec. 2.5 of this chapter (Scheme 21) [201–203]. Following anionic polymerization by sequential monomer addition, the polymer analogous reactions of the cholesterol (PS–PBCh) [201] and azobenzene (PS–PBAz) [203] derivatives were essentially quantitative, while that of the phenyl benzoate (PS–PBBz) block went to up to 94% conversion [202]. The polydispersities of the liquid crystalline copolymers (pdi = 1.13–1.23) were nearly as narrow as those of their precursor copolymers (pdi = 1.08–1.21, $M_n = 8.09–9.2 \times 10^4$) [201, 203].

The ability of butadiene anions to initiate styrene polymerization, and vice versa, is

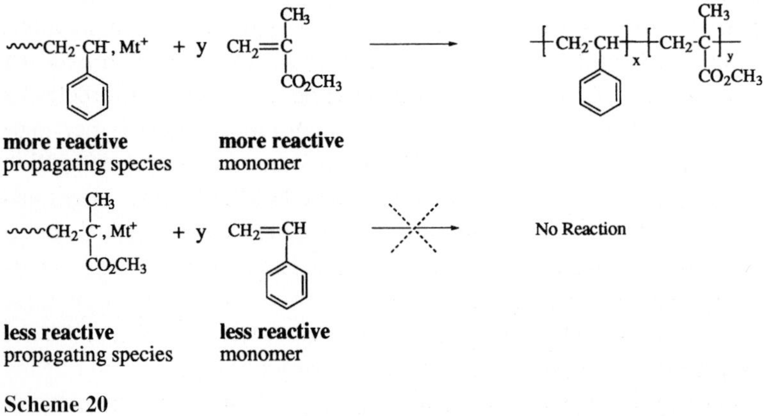

more reactive propagating species   **more reactive monomer**

less reactive propagating species   **less reactive monomer**   No Reaction

Scheme 20

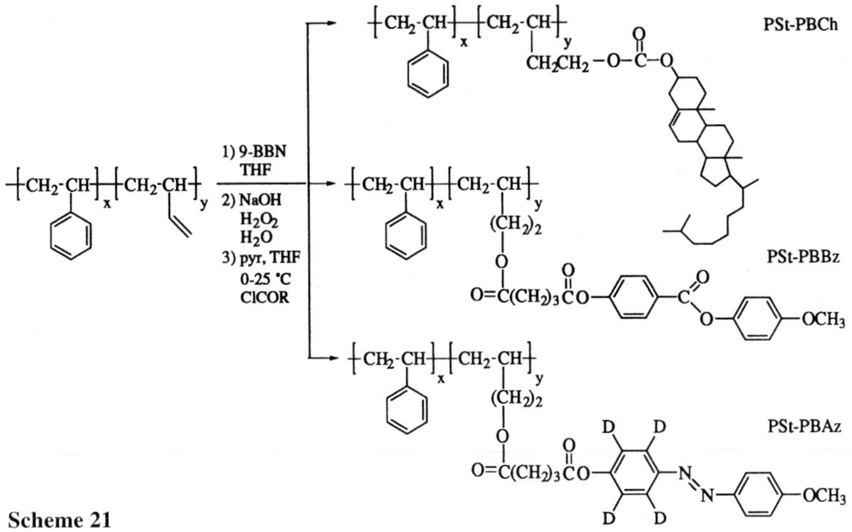

Scheme 21

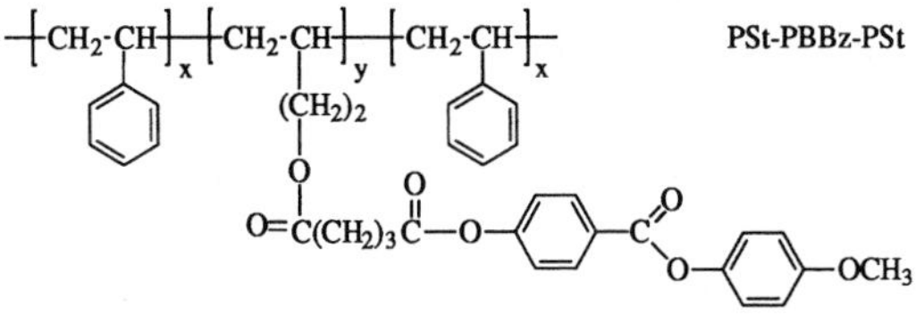

PSt-PBCh

1) 9-BBN
THF
2) NaOH
H$_2$O$_2$
H$_2$O
3) pyr, THF
0-25 °C
ClCOR

PSt-PBBz

PSt-PBAz

Scheme 22

PSt-PBBz-PSt

demonstrated by the synthesis of triblock poly(styrene-*block*-butadiene-*block*-styrene) by three sequential monomer additions, which was then used to synthesize the liquid crystalline triblock copolymer shown in Scheme 22.

Similarly, diblock copolymers involving two (or more) methacrylate monomers can generally by polymerized in the order which is most convenient since the two propagating anions should have similar pK_as. For example, diblock copolymers of methyl methacrylate and 4-[4'-(4"-methoxyphenylazo)phenoxy]butyl methacrylate (PMMA – PMAzM) were synthesized by polymerizing the mesogenic block first (Scheme 23) [46].

Although block copolmers of methyl methacrylate and 6-(4'-(4"-methoxyphenyl)-phenoxy)hexyl methacrylate have been synthesized by group transfer polymerization (Scheme 24) [305], they have not been characterized. This same block copolymers has also been synthesized by coupling acyl chloride endcapped polystyrene with hydrazide endcapped poly{6-[4'-(4"-methoxyphenyl)phenoxy]hexyl methacrylate] [306]. However, the two coupled homopolymers were prepared by radical polymerizations, and the resulting block copolymers are therefore not monodisperse in either block length or molecular weight.

In contrast to block copolymers of two methacrylates, block copolymers of a methacrylate and styrene can only be prepared anionically by polymerizing styrene first. As shown in Scheme 25, well-defined (pdi = 1.03 – 1.11) diblock copolymers of styrene and 6-[4'-(4"-Methoxyphenyl)phenoxy]hexyl methacrylate (PS – PMPPHM) were synthesized directly by sequential anionic polymerization of styrene and then the methacrylate in THF at –78 °C using s-butyl lithium as the initiator [44]. The reactivity of the growing polystyrene anions were reduced by reaction with 1,1-diphenylethylene be-

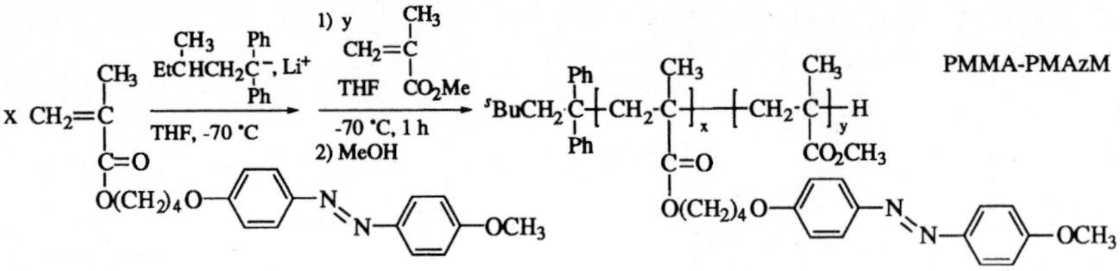

Scheme 23

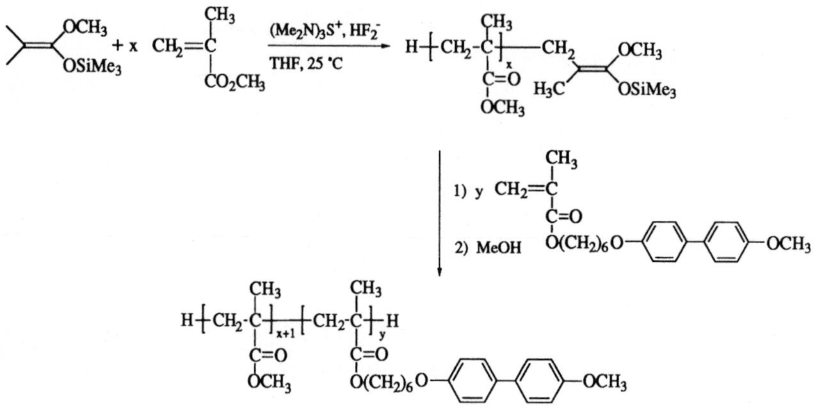

Scheme 24

fore polymerizing (6-[4′-(4″-methoxyphe-nyl)phenoxy]hexyl methacrylate at −40 °C in the presence of lithium chloride. Diblock copolymers of styrene and 4-[4′(4″-meth-oxyphenylazo)phenoxy]butyl methacrylate (PS – PMAzM) were prepared similarly (Scheme 25), except that the methacrylate block was polymerized at −70 °C without adding lithium chloride [46].

Poly[styrene-*block*-2-(cholesteryloxy-carbonyloxy)ethyl methacrylate] (PS – PChEMA) copolymers and similar diblock copolymers with amorphous butadiene (PB – PChEMA) [204] and *n*-butylmeth-acrylate (PBMA – PChEMA) [204, 210] blocks were synthesized by a route analo-gous to that shown in Scheme 16, except that the nonmesogenic block was polymer-ized first. For example, styrene was poly-merized first in THF at −78 °C using *s*-BuLi as the initiator (Scheme 26). Diblock

PS – PChEMA copolymers were generated by reacting the growing polystyrene anions with 1,1-diphenylethylene before polymer-izing 2-(trimethylsiloxy)ethyl methacry-late, following by deprotection and reaction of the poly(2-hydroxyethyl methacrylate) block with cholesterylchloroformate. The molecular weights of the diblock copoly-mers corresponded to $DP_n = [M_1]_0 + [M_2]_0/[I]_0$; the polydispersities of PB-PChEMA (pdi = 1.3) [204], PBMA – PChEMA (pdi = 1.2) [204] and PS – PChEMA (pdi = 1.02 – 1.18 [205, 207, 209]) were generally very narrow.

Since triblock copolymers can not be syn-thesized anionically by a third monomer ad-dition of styrene to the growing methacry-late anion, PS – PChEMA – PS triblock co-polymers were generated by coupling the living diblock copolymers with terepthaloyl chloride before quenching/deprotection

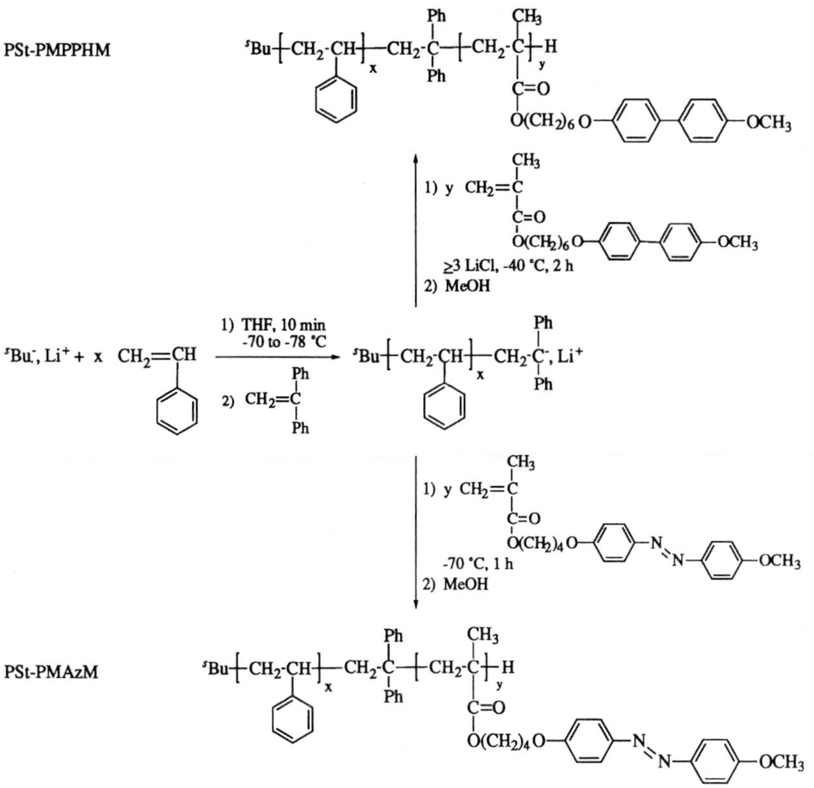

PSt-PMPPHM

$^tBu\left[CH_2\text{-}CH\right]_x\text{-}CH_2\text{-}\underset{Ph}{\overset{Ph}{C}}\text{-}\left[CH_2\text{-}\underset{y}{\overset{CH_3}{C}}\right]H$

with Ph substituent on chain and

$\underset{\overset{\displaystyle |}{C=O}}{}$
$O(CH_2)_6O$—⬡—⬡—OCH_3

1) y $CH_2=\underset{\overset{\displaystyle |}{\underset{C=O}{C}}}{\overset{CH_3}{C}}$
$O(CH_2)_6O$—⬡—⬡—OCH_3

≥3 LiCl, -40 °C, 2 h
2) MeOH

$^tBu^-, Li^+ + x\ CH_2=CH$—⬡

1) THF, 10 min
-70 to -78 °C
⟶
2) $CH_2=\underset{Ph}{\overset{Ph}{C}}$

$^tBu\left[CH_2\text{-}CH\right]_x\text{-}CH_2\text{-}\underset{Ph}{\overset{Ph}{C}}{}^-, Li^+$

1) y $CH_2=\underset{\overset{\displaystyle |}{\underset{C=O}{C}}}{\overset{CH_3}{C}}$
$O(CH_2)_4O$—⬡—N=N—⬡—OCH_3

-70 °C, 1 h
2) MeOH

PSt-PMAzM

$^tBu\left[CH_2\text{-}CH\right]_x\text{-}CH_2\text{-}\underset{Ph}{\overset{Ph}{C}}\text{-}\left[CH_2\text{-}\underset{y}{\overset{CH_3}{C}}\right]H$

$\underset{\overset{\displaystyle |}{C=O}}{}$
$O(CH_2)_4O$—⬡—N=N—⬡—OCH_3

Scheme 25

(Scheme 26). Alternatively, triblock PChEMA – PS – PChEMA copolymers can be prepared using a difunctional initiator as shown in Scheme 27. All of the triblock PS – PChEMA – PS (pdi = 1.07 – 1.25) [209] and PChEMA – PS – PChEMA (pdi = 1.01 – 1.47) [206, 207, 209] copolymers are also well-defined in terms of molecular weight and polydispersity. Although only 70 – 88% of the 2-hydroxyethyl methacrylate groups were initially functionalized with cholesterylcarbonate [204], complete conversion was subsequently achieved using 100% excess cholesterylchloroformate and longer reaction times of up to 4 days [208].

4.1.2   Cationic Copolymerizations

As in anionic copolymerizations, the order of monomer addition is extremely important in achieving efficient crossover reactions in cationic copolymerizations. However, in contrast to anionic polymerizations the most efficient crossover reaction is not achieved by reacting the more reactive monomer with the more reactive propagating species. Instead, initiation of the second monomer is apparently fast and quantitative if the most reactive monomer is polymerized first. As demonstrated by the examples in Scheme 28, this generates the more easily ionized covalent chain end, which therefore corresponds to the efficiency of covalent initiators discussed in Sec. 2.3 of this chapter and

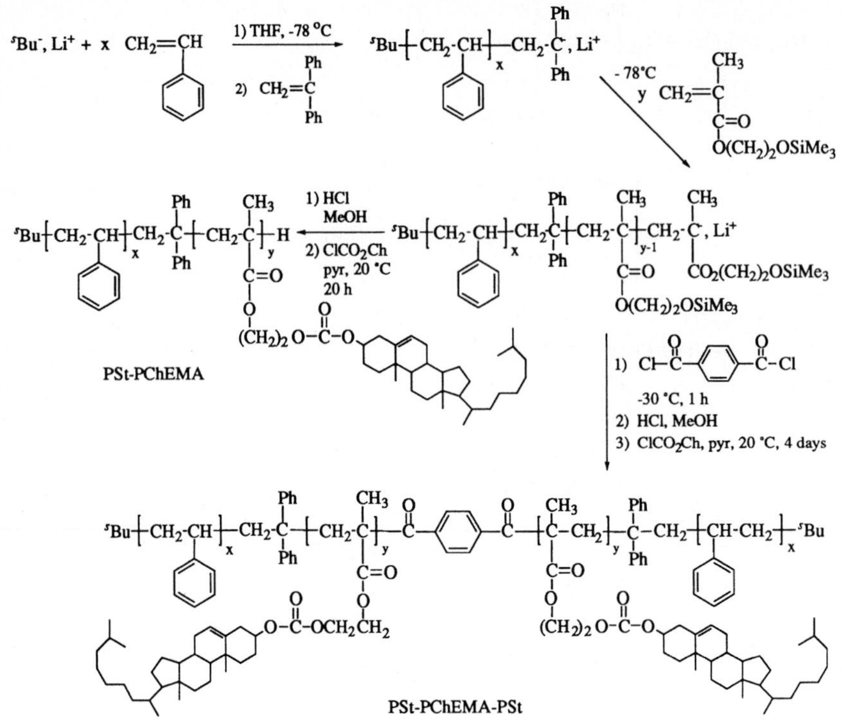

Scheme 26

outlined in Table 2. However, since the optimum polymerization conditions generally vary with each monomer, the temperature and catalyst type and concentration may have to be adjusted accordingly with each monomer addition. For example, Scheme 28 shows that fairly well-defined (pdi = 1.33) diblock copolymers of styrene (45 repeat units) and methyl vinyl ether (20 repeat units) are obtained when the vinyl ether is polymerized first, although the second stage polymerization requires higher temperature and a higher concentration of the Lewis acid in conjunction with a common ion salt [307]. The synthesis of block copolymers of styrene – 2-chloroethyl vinyl ether [307], p-methoxystyrene – i-butyl vinyl ether [308], p-methoxystyrene – methyl vinyl ether [308], p-t-butoxystyrene – i-butyl vinyl ether [309], and p-methylstyrene – i-butyl vinyl ether [309] also require poly-

merization of the most reactive monomer first. The opposite order of first stage polymerization of the less reactive monomer generally produces a mixture of the block copolymer and the homopolymer of the second monomer [307–311].

Monomer pairs of similar reactivity, such as p-t-butoxystyrene – p-methoxystyrene [309], and α-methylstyrene – 2-chloroethyl vinyl ether [310] can be polymerized starting from either monomer. However, even monomer pairs with similar stabilities and reactivities have an optimum order of addition. For example, Scheme 29 shows that the molecular weight distribution of triblock copolymers of styrene (50–56 repeat units) and p-methylstyrene (60–62 repeat units) is narrower when p-methylstyrene is polymerized first (using a difunctional initiator) [312], thereby generating a more easily ionized macroinitiator.

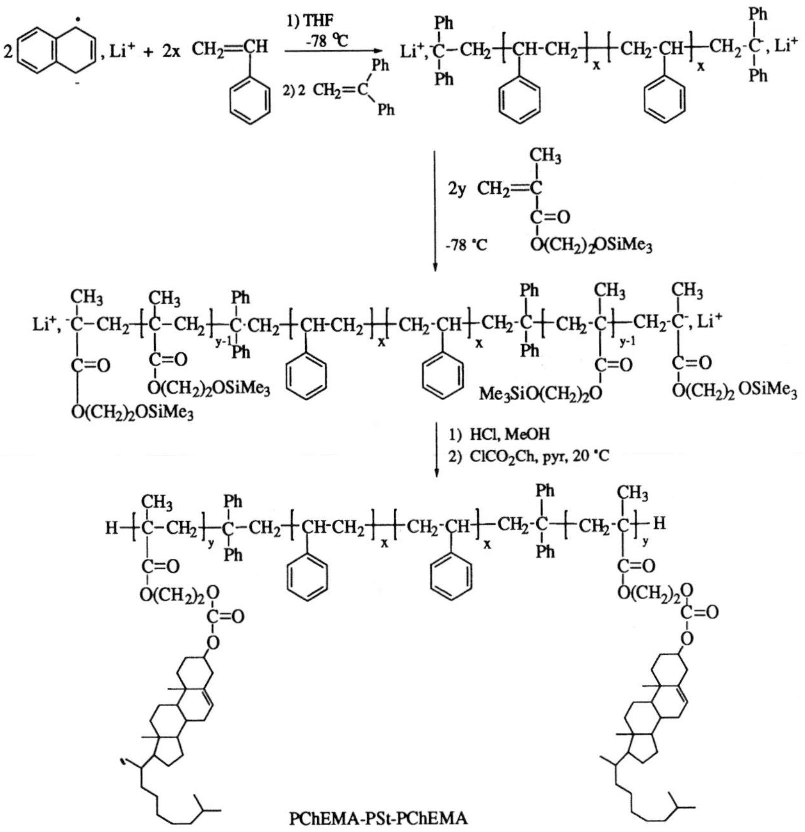

Scheme 27

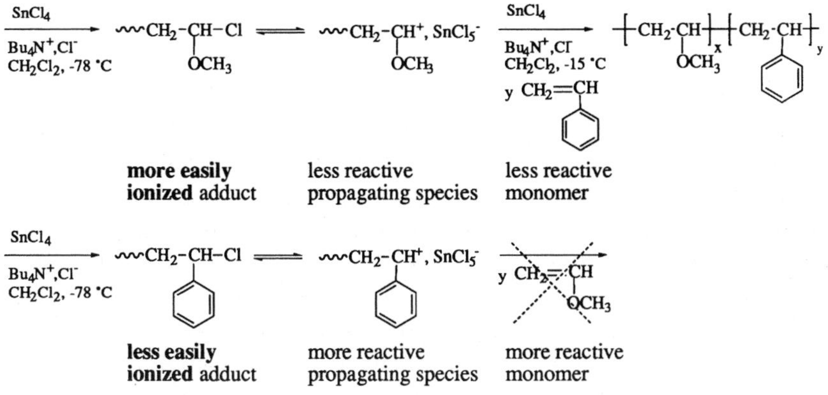

more easily
ionized adduct

less reactive
propagating species

less reactive
monomer

less easily
ionized adduct

more reactive
propagating species

more reactive
monomer

PChEMA-PSt-PChEMA

Scheme 28

Nevertheless, efficient crossover reactions can be achieved even when the second monomer is more reactive than the first monomer by endcapping the first block with a more easily ionized, but nonpolymerizable monomer. As outlined in Scheme 30, Faust et al. prepared well-defined (pdi < 1.1) diblock and triblock copolymers of isobuty-

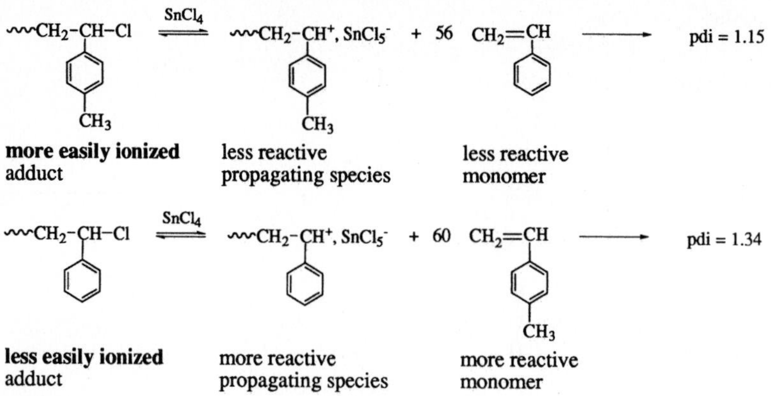

Scheme 29

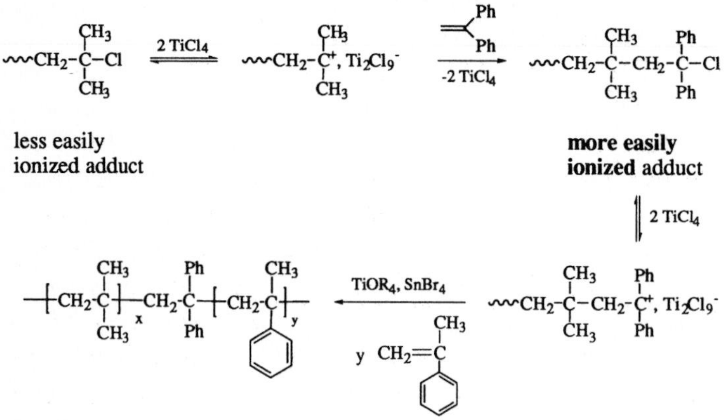

Scheme 30

lene and α-methylstyrene (or other styrene derivatives) with 100% crossover efficiencies by reacting the first block with 1,1-diphenylethylene before adding α-methylstyrene [313–315]. In this case, it is also necessary to deactivate the first Lewis acid (TiCl$_4$) by adding titanium(IV) isopropoxide or butoxide, and to add a second Lewis acid (SnBr$_4$) [315]. Diblock copolymers (pdi < 1.1) of isobutylene and methyl vinyl ether were prepared similarly [316].

Therefore, techniques are emerging for generating well-defined block copolymers by sequential monomer addition in cationic polymerizations of vastly different vinyl monomers. However, side-chain liquid

crystalline block copolymers have only been prepared by sequential monomer addition of two vinyl ethers of approximately the same reactivity. Both diblock copolymers of (perfluorooctyl)ethyl vinyl ether and n-[(4'-(4''-cyanophenyl)phenoxy)alkyl]vinyl ethers (PR$_f$-PCNVEn) or 2-[(4'-biphenyl-oxy)ethyl]vinyl ether (PR$_f$-PbiPHVE2) were prepared by polymerizing the mesogenic block first using triflic acid as the initiator and dimethyl sulfide as a deactivator in CH$_2$Cl$_2$ at 0 °C, followed by addition of the less soluble fluorocarbon monomer (Scheme 31) [317]. Somewhat well-defined (pdi = 1.30–1.40) diblock copolymers of 6-[4'-(4''-n-(S)-2-methylbutyl vinyl ether

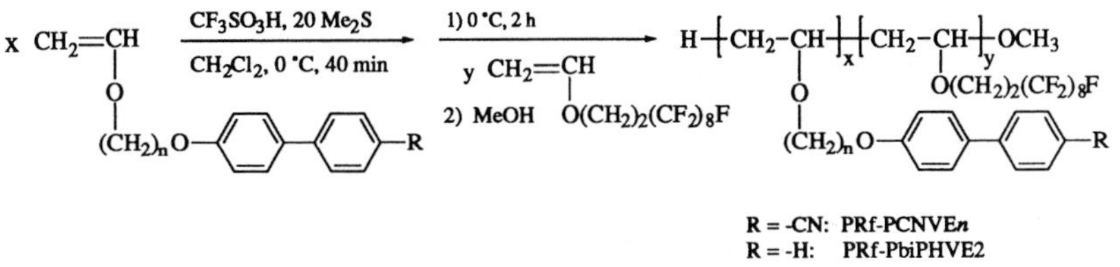

Scheme 31

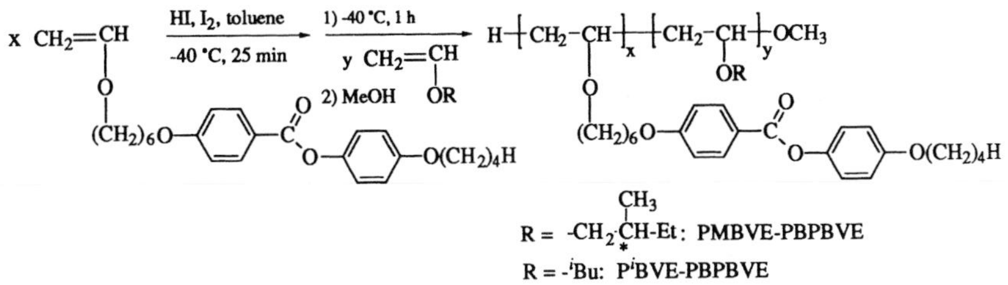

Scheme 32

(PMBVE–PBPBVE) were also prepared by polmerizing the mesogenic block first, except that HI was used as the initiator and I_2 as the catalyst in toluene at $-40\,°C$, followed by addition of the nonmesogenic monomer (Scheme 32) [150]. Although a minor amount of homopolymer was obtained from the second stage polymerization, this was removed by selective extraction.

4.1.3 Copolymerizations with Metalloporphyrins

Aluminum porphyrins have been used to prepare block copolymers by sequential monomer addition of two oxiranes [68, 318], of two β-lactones [319], of a β-lactone and an oxirane [319], of ethylene oxide and ε-caprolactone [74] of propylene oxide and D-lactide [75], and of two methacrylates [76]. In the one example involving monomer pairs (β-lactone and oxirane) which generate two different types of propagating species, the β-lactone had to be polymerized

first since the resulting aluminum carboxylate was able to initiate polymerization of the oxiranes, whereas aluminum alkoxides were unable to initiate polymerization of β-lactone [319]. Metalloporphyrins have also been useful in preparing liquid crystalline block copolymers of methyl methacrylate and either 6-[4′-(4″-n-butoxyphenoxycarbonyl)phenoxy]hexyl methacrylate (PMMA–PMOB) [90, 91] or 6-[4′-(4″-cyanophenylazo)phenoxy]hexyl methacrylate (PMMA–PMAC) [90] with narrow polydispersity (pdi = 1.14 – 1.19) and the expected molecular weight using methyl(5,10,15, 20-tetraphenylporphinato)aluminum as the initiator (Scheme 33).

4.1.4 Ring-Opening Metathesis Copolymerizations

Well-defined diblock and triblock copolymers (pdi = 1.08 – 1.14) were first prepared by ring-opening metathesis polymerization by sequential monomer addition of norbornene, dicyclopentadiene and benzonorbor-

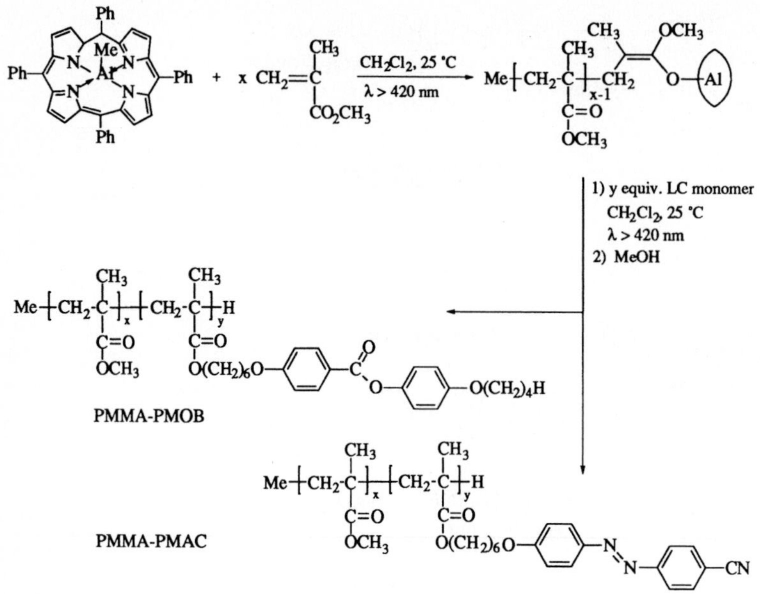

Scheme 33

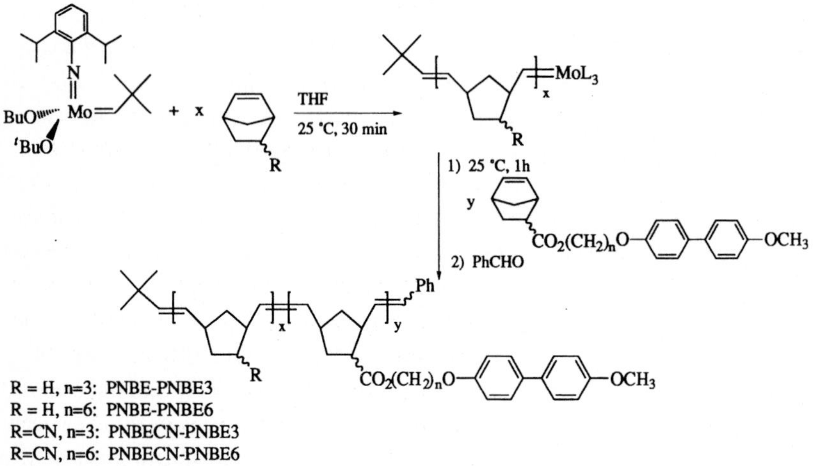

R = H, n=3: PNBE-PNBE3
R = H, n=6: PNBE-PNBE6
R=CN, n=3: PNBECN-PNBE3
R=CN, n=6: PNBECN-PNBE6

Scheme 34

nadienes using the bis(η^5-cyclopentadien-yl)titanacyclobutane derivative of 3,3-di-methylcyclopropene as the initiator [320, 321]. Norbornenes and norbornadienes have also been sequentially polymerized us-ing M(CHCMe$_2$R)(N-2,6-C$_6$H$_3$-i-Pr$_2$)(O-t-Bu)$_2$(M = Mo, W; R = –CH$_3$, –Ph) as the initiator (pdi = 1.05 – 1.12, up to 488 repeat

units per block) [7, 186, 322 – 334], as have norbornenes and cyclobutane derivatives (pdi = 1.04 – 1.23, up to 200 repeat units per block) [335 – 339]. In these systems, revers-ing the order of monomer addition has no effect on the copolymerization. More re-cently, block copolymers of norbornenes and 7-oxanorbornenes (pdi = 1.21 – 1.26, up

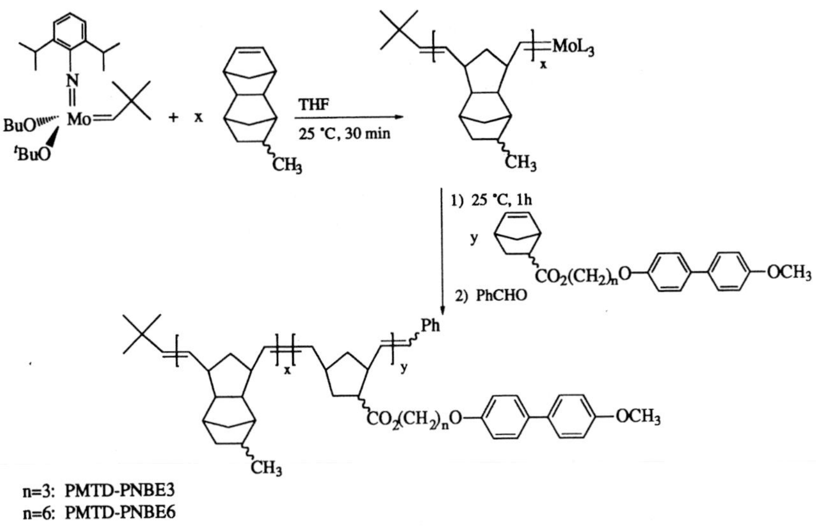

n=3: PMTD-PNBE3
n=6: PMTD-PNBE6

Scheme 35

to 292 repeat units per block) [340], and of an oxanorbornadiene and a functionalized norbornene (pdi = 1.32) [341] were prepared using RuCl$_2$(CHCH=CPh$_2$)(PR$_3$)$_2$ and RuCl$_2$(CHPh)(PPh$_3$)$_2$, respectively, as the initiator. Triblock copolymers of 7-oxa-norbornenes have also been prepared using a difunctional (PCy$_3$)$_2$Cl$_2$RuCH-Ph-CH$_2$RuCl$_2$(PCy$_3$)$_2$ initiator (pdi = 1.11–1.26, up to 200 repeat units per block) [342].

However, only Mo(CH-t-Bu)(N-2,6-C$_6$H$_3$-i-Pr$_2$)(O-t-Bu)$_2$ has been used to prepare well-defined, liquid crystalline block copolymers. As shown in Schemes 34 and 35, the nonmesogenic monomer was polymerized first in THF at 25 °C, followed by addition of 5-{[n-[4′-(4″-methoxyphenyl)phenoxy]alkyl]carbonyl}bicyclo[2.2.1]hept-2-ene (n=3: PNBE3; n=6: PNBE6), to produce diblock copolymers (pdi = 1.06–1.25) with norbornene (PNBE), methyltetracyclodecene (PMTD) and 5-(cyano)bicyclo[2.2.1]hept-2-ene (PNBECN) [343].

4.1.5   Morphology and Thermotropic Behavior of Side-Chain Liquid Crystalline Block Copolymers

The amorphous segment of microphase-separated amorphous/liquid crystalline block copolymers may influence the ordering of the mesogens at the interface, as well as the size and discreteness of that interface. Living copolymerizations are therefore being used to determine the effect of the morphology and domain size on the thermotropic behavior of side-chain liquid crystalline block copolymers.

As summarized in Table 14, block copolymers of methyl methacrylate and 6-[4′-(4″-n-butoxyphenoxycarbonyl)phenoxy]hexyl methacrylate (PMMA–PMOB) [90, 91] or 6-[4′-(4″-cyanophenylazo)phenoxy]hexyl methacrylate (PMMA–PMAC) [90] demonstrate that liquid crystalline block copolymers behave similarly to standard block copolymers. That is, entry 1 shows that short blocks are miscible, such that this PMMA–PMOB copolymer exhibits a

Table 14. Phase behavior of block copolymers of methyl methacrylate and 6-[4'-(4"-n-butoxyphenoxycarbo-nyl)phenoxy]hexyl methacrylate (PMMA–PMOB) [90, 91] and 6-[4'-(4"-cyanophenylazo)phenoxy]hexyl methacrylate (PMMA–PMAC) [90] (Scheme 33).

Entry	Copolymer	Block length		Wt. ratio	Thermotropic behavior (°C)	Ref.
		PMMA	LC	PMMA/LC		
1		56	13	50/50	G 75 I	[90]
2	PMMA – PMOB	109	12	67/33	G 40 G 108 I	[91]
3		109	24	50/50	G 42 G 100 SmA/N 100 I	[91]
4	PMMA – PMAC	28	8	51/49	G 44 G 92 I	[90]

Table 15. Phase behavior of block copolymers of 6-[4'-(4"-n-butoxyphenoxycarbonyl)phenoxy]hexyl vinyl ether with i-butyl vinyl ether (P-i-BVE–PBPBVE) or (S)-2-methylbutyl vinyl ether (PMBVE–PBPBVE) (Scheme 32) [150].

Entry	Copolymer	(Theoretical) block length		(Actual) wt. ratio	Thermotropic behavior (°C)[a]	$\dfrac{\Delta H_{i,co}}{\Delta H_{i,homo}}$ [b]
		non-LC	LC	non-LC/LC		
1		2	10	3/97	s 28 SmC 87 N 97 I	1.1
2	P-i-BVE – PBPVE	3	10	8/92	s 24 SmC 83 N 93 I	1.1
3		10	10	10/90	s 25 SmC 83 N 94 I	0.99
4		25	10	25/75	s 29 K 43 SmC 83 N 93 I	0.85
5	PMBVE – PBPBVE	10	10	21/79	K 27 Sm 33 SmC 60 N 70 I	nr
6		25	10	32/68	K 20 Sm 28 I	nr

[a] PBPBVE homopolymer $DP_n = 10$: Sm 31 SmC 92 N 102 I [150]; PBPBVE homopolymer $DP_n = 38$: Sm 47 SmC 106 N 111 I [150]. P-i-BVE: T_g –20 °C [344]; T_m 110 °C [345]; T_m 165 °C [346].
[b] nr = not reported.

single glass transition at a temperature intermediate between that of poly(methyl methacrylate) (105 °C) and poly{6-[4'-(4"-n-butoxyphenoxycarbonyl)phenoxy]hexyl methacrylate} (G 38 SmA 104 N 110 I); increasing the poly(methyl methacrylate) block length causes it to phase separate (entry 2). In addition, the mesogenic block must be sufficiently long for it to organize anisotropically (entry 3). The final entry demonstrates that poly(methyl methacrylate) and poly{6-[4'-(4"-cyanophenylazo)phenoxy]hexyl methacrylate} (G 47 SmA 154 I) are evidently more incompatible than PMMA–PMOB since phase separation occurs at lower block lengths, although the mesogenic block length is still not long enough to exhibit liquid crystallinity. (The morphology

of these block copolymers was not determined.)

In contrast, block copolymers of 6-[4'-(4"-n-butoxyphenoxycarbonyl)phenoxy]-hexyl vinyl ether and i-butyl vinyl ether (P-i-BVE–PBPBVE) or (S)-2-methylbutyl vinyl ether (PMBVE–PBPBVE) exhibit liquid crystallinity even at very short block lengths (Table 15). However, the weight fraction of the liquid crystalline segment is 68–98%, which should correspond to cylinders or spheres of the nonmesogenic block in a matrix of poly{6-[4'-(4"-n-butoxyphe-noxycarbonyl)phenoxy]hexyl vinyl ether}. Without microscopic or X-ray evidence, it is also not clear from the data presented in Table 15 whether microphase separation occurs, especially at block lengths of two to

Table 16. Phase behavior of block copolymers of (perfluorooctyl)ethyl vinyl ether and n-[(4′-(4″-cyanophenyl)phenoxy)alkyl]vinyl ethers (PR$_f$–PCNVEn) or 2-[(4′-biphenyloxy)ethyl]vinyl ether (PR$_f$–PbiPHVE2) (Scheme 31) [317].

Copolymer	Block length		Wt. ratio	Thermotropic behavior (°C)[a]		$\dfrac{\Delta H_{i,co}}{\Delta H_{i,homo}}$
	PR$_f$	LC	PR$_f$/LC	Copolymer	Poly(LC)[b]	
PR$_f$–PCNVE2	7	14	45/55	K –6 K 22 G 75 X 83 LC 250 dec	G 78 X 86 I	0.61
PR$_f$–PCNVE3	8	15	54/46	K –12 K 15 G 60 X 64 N 97 LC 250 dec	G 60 X 64 N 98 I	0.43
PR$_f$–PCNVE9	5	16	32/68	K –7 G 12 K 23 SmC 41 SmA 140 I	G 14 SmA 142 I	0.81
PR$_f$–PCNVE11	5	14	31/69	G 10 K 54 K 64 SmA 146 I	G 17 K 66 SmA 155 I	0.95
PR$_f$-PbiPHVE2	6	13	44/56	K 30 K 66 LC 250 dec	G 59 K 70 I	0.79

[a] PR$_f$ homopolymer: K 23 Sm 41 I [347]; X = unidentified phase, LC = unidentified liquid crystalline phase, dec = decomposes.
[b] Of comparable molecular weight of the liquid crystalline block length.

three repeat units of the non-mesogenic poly(vinyl ether). That is, transitions due to poly(i-butyl vinyl ether) are not observed in the first three entries, and all of the transition temperatures of the poly{6-[(4′-n-butoxyphenyl-4″-benzoate)hexyl]vinyl ether} segment are depressed relative to those of the homopolymer. The transition temperatures are even further depressed in those block copolymers (entries 4–6) which exhibit a crystalline melting, presumably due to the nonmesogenic, semi-crystalline block. The fact that this transition (43 °C) is substantially depressed relative to that of poly(i-butyl vinyl ether) (110 °C [345] – 165 [346] °C depending on tacticity) further indicates that some mixing of the blocks occurs.

Microphase separation does occur at shorter block lengths when the two blocks are highly immiscible, such as with hydrocarbon and fluorocarbon blocks. For example, all of the diblock copolymers of (perfluorooctyl)ethyl vinyl ether and n-[(4′-cyanophenyl-4″-phenoxy)alkyl]vinyl ethers (PR$_f$-PCNVE$_n$) or 2-[(4′-biphenyloxy)ethyl]vinyl ether (PR$_f$-PbiPHVE2) listed in Table 16 exhibit the phases characteristic of both of the corresponding homo-

polymers. However, the three copolymers with the shortest spacers reportedly remain in an ordered state above temperatures corresponding to isotropization of the homopolymers.

Table 17 presents the thermotropic behavior of diblock copolymers of 5-{[n-[4′-(4″-methoxphenyl)phenoxy]alkyl]carbonyl}bicyclo[2.2.1]hept-2-ene (n=3: PNBE3; n=6: PNBE6) with norbornene (PNBE), methyltetracyclodecene (PMTD) and 5-(cyano)bicyclo[2.2.1]hept-2-ene (PNBECN) with at least 50 wt% liquid crystalline blocks of 20 or 50 repeat units [343]. With the exception of 45/55 wt% PNBECN–PNBE3, all of the diblock copolymers exhibit enantiotropic nematic mesophases and undergo phase transitions at approximately the same temperatures as the two corresponding homopolymers, although both glass transitions are not always detectable when they either occur in the same temperature region, or when the corresponding block is present as a low weight fraction of the copolymer. The ability of the mesogen to pack into microphase separated liquid crystalline regions is therefore independent of the identity of the amorphous block in these systems.

Table 17. Phase behavior of polynorbornene block copolymers (Schemes 34 and 35) [343].

Copolymer	Block length		Wt. ratio	Thermotropic behavior (°C)			$\dfrac{\Delta H_{i,co}}{\Delta H_{i,homo}}$
	Amorph.	LC	Amorph/LC	Copolymer	Poly(amorph)	Poly(LC)	
PNBE–PNBE3	300	20	48/52	G 44 G 58 N 84 I	G 43 I		*
	100	20	22/78	G 39 G 55 N 92 I		G 57 N 79 I	0.65
PNBECN–PNBE3	180	20	45/55	G 112 I	G 116 I		
PNBE–PNBE3	100	50	11/89	G 39 G 59 N 86 I	G 43 I		0.97
	50	50	6/94	G 36 G 60 N 87 I			0.90
PNBECN–PNBE3	180	50	27/73	G 67 N 88 G 119 I	G 116 I	G 61 N 86 I	*
	89	50	13/87	G 63 N 82 I			0.49
	36	50	9/91	G 63 N 82 I			0.40
PNBE–PNBE6	300	20	40/60	G 42 N 97 I	G 43 I		*
	100	20	18/82	G 38 N 94 I			0.75
PMTD–PNBE6	100	20	46/54	G 41 N 85 G 208 I	G 214 I	G 35 N 87 I	*
PNBECN–PNBE6	180	20	40/60	G 40 N 87 G 112 I	G 116 I		*
PNBE–PNBE6	100	50	9/91	G 38 N 91 I	G 43 I		0.97
	56	50	5/95	G 38 N 93 I			0.93
PMTD–PNBE6	100	50	27/73	G 41 N 92 G 207 I	G 214 I	G 42 N 93 I	0.60
PNBECN–PNBE6	180	50	23/77	G 43 N 93 G 114 I	G 116 I		0.60
	89	50	15/85	G 43 N 91 I			0.82
	36	50	7/93	G 45 N 91 I			0.81

* Transition too broad to accurately calculate enthalpy.

Table 18. Phase behavior of polystyrene block copolymers (Schemes 21, 22, 25).

Copolymer	Block length		Wt. ratio	Thermotropic behavior (°C)*		$\dfrac{\Delta H_{i,co}}{\Delta H_{i,homo}}$	Ref.
	St	LC	PS/LC	Copolymer	Poly(LC)		
PS–PBCh	692	165	48/52	G 98 S 203 I	G 75 S 206 I	0.38	[201]
PS–PBBz	725	183	50/50	G 27 SmC 58 N 101 I	G 24 SmC 60 N 101 I	0.30	[202]
PS–PBBt–PS	306	313	11/89	G 20 N 88 I	G 18 N 53 I	1.29	[202]
	180	41	55/45	G 90 K 104 SmA 132 I		0.71	[44]
PS–PMPPHM	144	41	50/50	G 92 K 105 SmA 133 I	K 121 SmA 135 I	0.69	[44]
	103	49	37/63	G 91 K106 SmA 135 I		0.77	[44]

* Amorphous PS: $T_g = 85\,°C$.

Well-defined PS–PBCh [201], PS–PBBz [202] and PS–PBAz [203] (Scheme 21) diblock copolymers with extremely long liquid crystalline block lengths (10^2 repeat units) form distinct lamellae typical of 50:50 wt% block copolymers, although the glass transitions of both blocks are not always detected due to either their similar temperatures or masking by a liquid crystalline transition; the morphology of the PS–PBBz–PS triblock copolymer (Scheme 22) has not been determined. As summarized in Table 18, the diblock copolymers form the same phases at the same temperatures as the corresponding homopolymers.

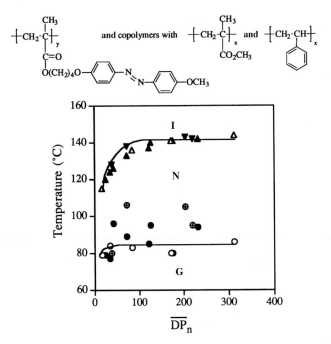

Figure 20. Thermotropic behavior of poly[4-[4′-(4″-methoxyphenylazo)phenoxy]butyl methacrylate} (○, △), poly{(methyl methacrylate)-*block*-[4-[4′-(4″-methoxyphenylazo)phenoxy]butyl methacrylate]} (●, ▲) (Scheme 23) and poly{styrene-*block*-[4-[4′-(4″-methoxyphenylazo)phenoxy]butyl methacrylate]} (⊕, ▼) (Scheme 25) as a function of the number of repeat units in the liquid crystalline block [46].

The same is true of diblock copolymers of styrene and 6-[4′-(4″-methoxyphenyl)phenoxy]hexyl methacrylate (PS–PMPPHM) with at least 45 wt% liquid crystalline blocks of 40–50 repeat units (Table 18) [44]. That is, the glass transition of polystyrene and the SmA to isotropic transition of poly[6-[4′-(4″-methoxyphenyl)phenoxy]hexyl methacrylate} occur at essentially the same temperatures in the copolymers. However, the crystalline melting of the liquid crystalline block is depressed in the block copolymers, which broadens the temperature range of the SmA mesophase. All of the diblock copolymers form well-defined lamellae, although the layer spacings of the mesogens in the diblock copolymers are slightly lower than that in the homopolymer. In addition, the lamellae orient perpendicular to the interface of the two blocks in the copolymer and therefore perpendicular to the fiber axis, in contrast to the orientation in the homopolymers.

Although the temperatures of the transitions of the liquid crystalline block generally match those of the corresponding homopolymer, the transition temperatures may be molecular weight dependent. For example, Figure 20 demonstrates that the glass and nematic–isotropic transition temperatures of poly{4-[4′-(4″-methoxyphenylazo)phenoxy]butyl methacrylate} level off at approximately 100 repeat units [46]. In addition, the isotropization temperature of its diblock copolymers with styrene (PS–PMAzM) and methyl methacrylate (PMMA–PMAzM) follow approximately the same dependence on the block length of the liquid crystalline block, although the polystyrene (19–365) and poly(methyl

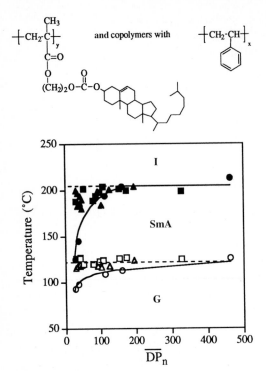

Figure 21. Thermotropic behavior of poly{2-(choles-teryloxycarbonyloxy)ethyl methacrylate} (○, ●), and diblock (PS–PChEMA) [208, 209] (□, ■) and triblock [PChEMA–PS–PChEMA [208] (△, ▲) and PS–PChEMA–PS [209] (▽, ▼)] copolymers of 2-(choles-teryloxycarbonyloxy)ethyl methacrylate and styrene (Schemes 26 and 27) as a function of the number of repeat units in the liquid crystalline block.

methacrylate) (30–529) block lengths vary widely; the glass transition temperatures of the block copolymers are much more scattered.

In contrast, the data in Table 17 demonstrates that the nematic–isotropic temperature of polynorbornene copolymers with either 20 or 50 repeat units in the liquid crystalline block are more similar to the higher molecular weight homopolymer than to that with 20 repeat units. This trend is demonstrated more conclusively by the diblock (PS–PChEMA) [208, 209] and triblock (PChEMA–PS–PChEMA [208] and PS–PChEMA–PS [209]) copolymers of 2-(cholesteryloxycarbonyloxy)ethyl meth-

acrylate and styrene. Although the exact temperatures vary greatly, Fig. 21 shows that the glass and smectic A–isotropic transition temperatures vary much less as a function of the length of the liquid crystalline block, and are essentially independent of the type of block copolymer and the morphology.

Studies of the copolymers discussed so far have therefore demonstrated that the mesogens within the liquid crystalline segment of block copolymers can organize anisotropically, and that the same mesophase is generally formed by both the homopolymer and copolymers containing at least 50 wt% of the liquid crystalline block. However, systematic studies of diblock (PS–PChEMA) and triblock (PS–PChEMA–PS, PChEMA–PS–PChEMA) copolymers of styrene and (2-cholesteryloxycarbonyl-oxy)ethyl methacrylate have demonstrated that this varies with lower weight or volume fractions of the liquid crystalline block [204–209]. Most importantly, the morphology and thermotropic behavior of the diblock (PS–PChEMA) [208, 209] and triblock (PChEMA–PS–PChEMA [208] and PS–PChEMA–PS [209]) copolymers of 2-(cholesteryloxycarbonyloxy)ethyl methacrylate and styrene are identical when the volume fraction of the blocks are equal, which is consistent with the behavior of amorphous AB and ABA block copolymers [25, 267–270]. This is true regardless of the length of the individual blocks. That is, the extent of immiscibility of all of the block copolymers is above the critical value ($\chi N > \chi N_{ODT}$), and the order–disorder transition (ODT) is not observed in any of the copolymers studies. As summarized in Fig. 22, those samples with volume fractions of polystyrene $0 < \phi_{PS} < 0.7$ exhibit the classic morphologies of polystyrene spheres or rods embedded in a smectic A matrix of poly[2-cholesterylformyl)ethyl methacry-

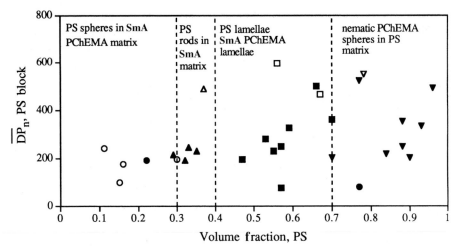

Figure 22. Morphology of AB diblock (○, △, □, ▽) and ABA and BAB triblock (●, ▲, ■, ▼) copolymers of styrene and 2-(cholesteryloxycarbonyloxy)ethyl methacrylate (Schemes 26 and 27): PS spheres in a SmA PChEMA matrix (○, ●); PS rods in a SmA PChEMA matrix (△, ▲); alternating lamellae of PS and smectic A PChEMA (□, ■); nematic PChEMA spheres in a PS matrix (▽, ▼) [208, 209].

late], and alternating lamellae of polystyrene and poly[2-(cholesterylformyl)ethyl methacrylate] (SmA). However, the phase diagram is asymmetric about $\phi_{PS}=0.5$. Such asymmetry is observed in amorphous block copolymers when the conformational and volume-filling characteristics of the two blocks are different [263, 348]. The extreme asymmetry of the phase diagram shown in Fig. 22 is therefore consistent with the conformational asymmetry of amorphous/side-chain liquid crystalline block copolymers.

In contrast to samples with $\phi_{PS}<0.7$, those with $\phi_{PS}>0.7$ exhibit only a spherical morphology of *nematic* poly[2-(cholesterylformyl)ethyl methacrylate] spheres in a polystyrene matrix, with no occurrence of PChEMA rods in a polystyrene matrix. As proposed by the researchers, both the curvature of the interface and the small size of the spheres (17 nm) evidently prevent the liquid crystalline block from organizing into its equilibrium smectic A structure, which has a layer distance of 4.5 nm. That is, the equilibrium packing of the mesogens is constrained in the microphase separated re-

gions, which results in further asymmetry in the phase diagram. However, no new morphologies are observed. Diblock PBMA–PChEMA copolymers exhibit a very similar, asymmetric phase diagram, which lacks the morphology of PChEMA rods in the amorphous matrix; PChEMA is also nematic in the PChEMA sphere morphology [210].

The last structure/property relationship to be studied in side-chain liquid crystalline block copolymers is the change in enthalpy of the liquid crystalline transitions per mol mesogenic repeat unit. Tables 14–18 list the relative changes in enthalpy of isotropization for the copolymers and corresponding liquid crystalline homopolymers whenever possible. As shown, the change in enthalpy is often depressed relative to that of the homopolymer, although higher percentages seem to be obtained when the block copolymer contains a greater fraction of the liquid crystalline block. Nevertheless, it is just as often close to 100%, as in poly{(methyl methacrylate)-*block*-4-[4′-(4″-methoxyphenylazo)phenoxy]butyl methacrylate} (75–120%; Scheme 23) and poly{styrene-*block*-4-[4'-(4"-methoxyphenylazo)phenoxy]butyl methacrylate} (100–110%; Scheme 25) [46].

The relative decrease in enthalpy should represent the extent of disorder at the interface of the two blocks, and has actually been directly correlated to the thickness of the interphase. That is, the ΔH of isotropization of lamellar PS–PChEMA (Scheme 26) containing $\phi_{PChEMA} = 0.56$ is 81.4% of that of the homopolymer, indicating that the interphase should be approximately 18.6% of the liquid crystalline lamellae (12.5 nm). The calculated value of 2.3 nm corresponds very well to that measured by TEM following preferential staining of the interphase [207]. Nevertheless, variations in the extent of disorder at the interface as measured by decreased enthalpies of transitions may be due to variations in sample preparation and thermal history. For example, Gronski et al.'s ^{2}H-NMR experiments on deuterated PS–PBAz (Scheme 21) indicate that the disordered interphase present in powder samples is eliminated when the samples are oriented by shear for extensive time in the nematic mesophase [203].

4.2   Graft Copolymers

Although copolymers with equivalent compositions but different molecular architec-

tures generally form the same morphology, the temperature of the order-disorder transition (T_{ODT}) varies [349]. This is because the critical value of χN, in which χ is inversely proportional to temperature, varies with molecular architecture. For example, when $\phi_A = 0.5$ weak segregation theory calculates (in the absence of fluctuation corrections) that ordering occurs when $\chi N \geq 10.5$ for linear diblock copolymers [264], $\chi N \geq 11$ for graft copolymers (depending on the number of grafts) [350, 351], and $\chi N \geq 18$ for linear triblock copolymers (Eq. 34) [352]. Therefore, graft copolymers

$$(\chi N)_{ODT, blend} < (\chi N)_{ODT, diblock}$$
$$< (\chi N)_{ODT, graft} < (\chi N)_{ODT, triblock} \qquad (34)$$

are more likely to remain homogeneous at a given temperature than the corresponding diblock copolymers.

In addition, both the shape and distribution of the domains of the minor phase in the matrix of the major phase of graft copolymers is more irregular than the spatially periodic structures of block copolymers [353]. This is probably due to the greater limitations to phase separation due to the greater number of connections between the two blocks, as well as to the generally random distribution of grafts along the length of the main-chain block.

As shown in Eq. (35), graft copolymers containing a mesogenic monomer have been synthesized by free radical copolymer-

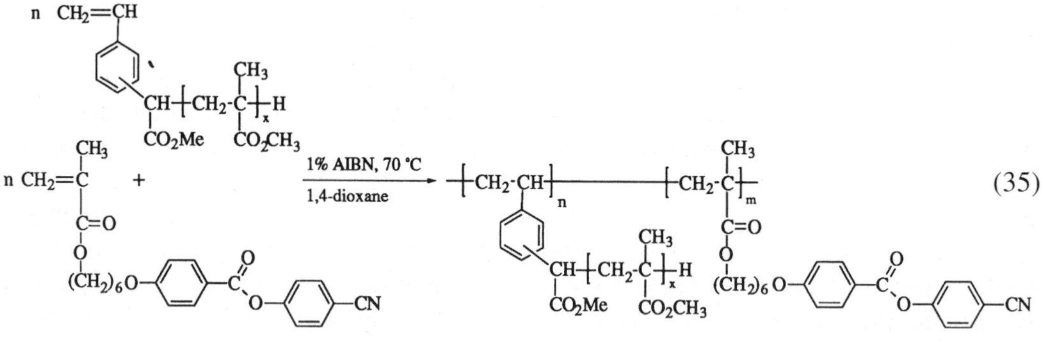

(35)

Table 19. Phase behavior of poly{6-[4′-(4″-cyanophenoxycarbonyl)phenoxy]hexyl methacrylate – *graft* – methyl methacrylate}s, Eq. (35) [354, 355].

Entry	(Number average) block length[a]		Wt. ratio[b]	Thermotropic behavior (°C)[c]	Weighted average[c]
	PMMA	LC	PMMA/LC		T_g (°C)
1	73	35	55/45	G 46 SmA, N 114 I	87
2	73	37	55/45	G 56 SmA, N 112 I	87
3	73	48	59/41	G 58 SmA, N 106 I	89
4	73	88	63/37	G 56 SmA, N 98 I	92
5	73	74	67/33	G 103 I	94
6	73	50	74/26	G 91 I	98
7	28	76	57/43	G 54 I	77
8	59	64	64/36	G 74 I	88
9	85	71	70/30	G 105 I	97
10	127	53	75/25	G 113 I	101

[a] Calculated assuming pdi of LC main chain is equivalent to pdi of copolymer.
[b] Calculated using weight average molecular weight of graft and LC main chain relative to PMMA standards.
[c] PMMA graft: $T_g = 94 - 117$ °C depending on block length; poly{6-[4′-(4″-cyanophenoxycarbonyl)phenoxy]hexyl methacrylate}: G 54 SmA 104 N 110 I [354].

ization of styryl-terminated poly(methyl methacrylate) and 6-[4′-(4″-cyanophenoxy-carbonyl)phenoxyl]hexyl methacrylate [354, 355]. In this case, the block length of the graft is nearly monodisperse ($DP_n = 28 - 127$, pdi ≤ 1.10). However, the copolymers are not well-defined in molecular weight (pdi $= 2.21 - 4.00$) since a living mechanism was not used for the copolymerization. More importantly, the distance between grafts along the backbone is not uniform. Nevertheless, the copolymers contain nearly the same ratio of the comonomers as was present in the feed, with total molecular weights up to $M_w = 2.5 \times 10^5$.

Table 19 presents the thermotropic behavior of these graft copolymers with varying compositions and graft lengths. Only the first four entries are microphase separated with the mesogenic block organized anisotropically. In these cases, the glass transition of the PMMA block overlaps the SmA – N – I transitions. All other entries exhibit a single glass transition. However, only entries 6 and 8 appear to form a homogeneous solution with a glass transition corresponding to the weighted average of the two blocks. The glass transition of the other entries seems to correspond to either the PMMA graft (entries 5, 9, 10) or the mesogenic main-chain (entry 7). The T_g of the mesogenic main-chain is apparently not detected in entries 5, 9, and 10 because of its low weight fraction, whereas the T_g of PMMA is not detected in the sample in which it is present in the lowest weight fraction (entry 7). (The actual morphology and therefore confirmation of phase separation by these graft copolymers has not been determined.)

Comparison of the data in Tables 14 and 19 of graft and block copolymers, respectively, based on methyl methacrylate and a mesogenic methacrylate confirm that block copolymers phase separate more easily than graft copolymers. Although not exactly comparable due to the different mesogenic methacrylates, the block copolymers phase separate at shorter block lengths than the graft copolymers. In addition, the distribution of amorphous grafts along the mesogenic block disrupts the ability of the mesogens to organize anisotropically much more than the single connection in diblock copolymers.

4.3   Statistical Binary Copolymers

As discussed in the beginning of Sec. 4 of this chapter, a truly random distribution of two comonomers is achieved only in an azeotropic copolymerization in which $r_1 = r_2 = 1$. In this case, the two monomers have equal probability of reacting with a given active center, regardless of the monomer it is derived from, and the copolymer composition equals the comonomer feed composition at all conversions. Therefore, if the copolymerization also follows as 'living' mechanism all of the copolymer chains will have approximately the same chain length and the same random distribution of comonomer units. If $r_1 \neq r_2 \neq 1$, not only will the comonomers not be randomly distributed along the copolymer chain, but the comonomer feed and therefore the copolymer composition will also change as a function of conversion, resulting in a mixture of chemically heterogeneous chains.

With one exception [90], side-chain liquid crystalline binary copolymers have only been prepared by a 'living' mechanism using cationic polymerizations of vinyl ether comonomers (Eq. 36). However, these

nonmesogenic vinyl ethers or vinyl ethers containing other biphenyl mesogens. The reactivity ratios of the two comonomers have not been determined in any of these systems. Nevertheless, the reactivity of a given functional group is independent of a second functional group if they are separated by at least three methylenic carbons [356]. Therefore, the reactivity ratio of a given mesogenic monomer should be identical to that of the corresponding n-alkyl substituted ($n \geq 3$) monomer with an identical polymerizable group if the n-alkyl spacer also contains at least three carbons. Similarly, the reactivity ratio of a mesogenic monomer should be independent of the length of the n-alkyl spacer if it contains at least three carbons.

If a mesogenic monomer is copolymerized with a nonmesogenic monomer, the mesophase is observed above a minimum concentration of the mesogenic monomer. For example, neither poly{(methyl methacrylate)-co-[6-[4'-(4"-cyanophenylazo)phenoxy]hexyl methacrylate]} ($DP_n \approx 35$, pdi = 1.35) nor poly{(methyl methacrylate)-co-[6-[4'-(4"-n-butoxyphenoxycarbonyl)phenoxy]hexyl methacrylate]} ($DP_n \approx 69$, pdi = 1.12) (Scheme 36) prepared by metalloporphyrin-initiated copolymerizations and containing 18–20 mol% of the mesogenic monomer exhibit the SmA mesophase of the corresponding SCLCPs [90].

However, the SmA mesophase of poly{{[n-butyl vinyl ether]-co-[11-[(4'-cyanophenyl-4"-phenoxy)undecyl]vinyl ether]} is retained at compositions containing at

$$x\ CH_2{=}CH \quad + \quad y\ CH_2{=}CH \quad \xrightarrow[\substack{CH_2Cl_2,\ 0\ °C \\ 2)\ MeOH}]{1)\ CF_3SO_3H,\ 10\ Me_2S} \quad H{-}{[}CH_2{-}CH{]}_x{[}CH_2{-}CH{]}_y{-}OCH_3 \quad (36)$$
$$\qquad\quad OR \qquad\qquad\quad OR' \qquad\qquad\qquad\qquad\qquad\qquad\qquad\quad OR \qquad\quad OR'$$

copolymers are based on biphenyl-mesogenic monomers with a variety of substituents which are copolymerized with either

least 20 mol% of the mesogenic monomer (Fig. 23); the crystalline phase is suppressed at much lower concentrations of the non-

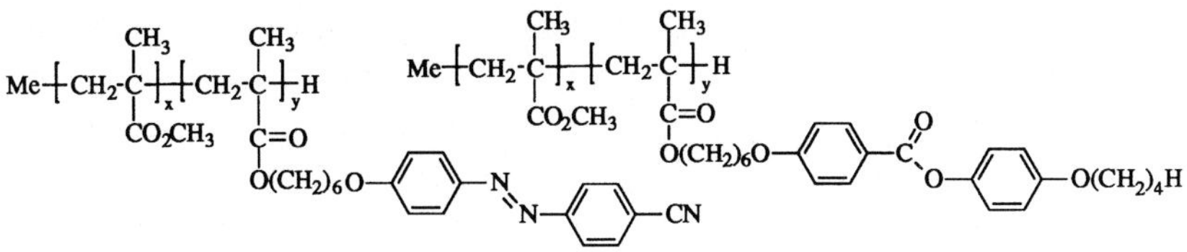

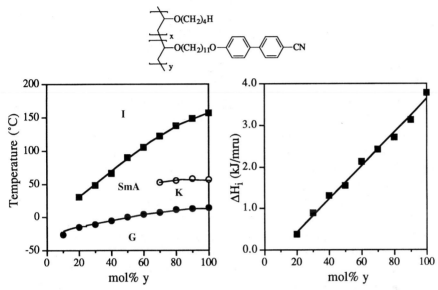

Scheme 36

Figure 23. Phase transition temperatures from the glass (●), crystalline (○) and SmA (■) phases of poly{(*n*-butyl vinyl ether)-*co*-[11-[(4'-(4"-cyanophenyl)phenoxy)undecyl]vinyl ether]} ($DP_n = 17 - 22$, pdi $= 1.07 - 1.15$) [357], and the corresponding changes in enthalpy of isotropization; first heat.

mesogenic units [357]. Figure 23 also demonstrates that all of the transition temperatures decrease as the amount of *n*-butyl vinyl ether in the copolymer increases, with the change in enthalpy of isotropization decreasing accordingly. In this case [357, 358], the glass transition decreases as the copolymer becomes more like poly(*n*-butyl vinyl ether) due to its lower T_g (−55 °C at high molecular weight) [344].

Therefore, copolymerization of a mesogenic monomer with a nonmesogenic monomer whose homopolymer has a lower glass transition temperature can be used to both eliminate side-chain crystallization and to depress the glass transition temperature of the corresponding SCLCP. If the glass transition and/or more ordered phases are depressed, additional mesophases may be revealed [357]. However, the biphasic region or temperature range over which the phase transitions occur increases as the concentration of the nonmesogenic unit increases [357]. The thickness of the SmA layers in similar copolymers with *i*-butyl vinyl ether increases as the concentration of *i*-butyl vinyl ether increases, and apparently changes from interdigitated bilayer to (noninterdigitated) bilayer packing [359].

In contrast, poly{[*n*-butyl vinyl ether]-*co*-[2-[(4'-cyanophenyl-4"-phenoxy)-ethyl]vinyl ether]} must contain at least

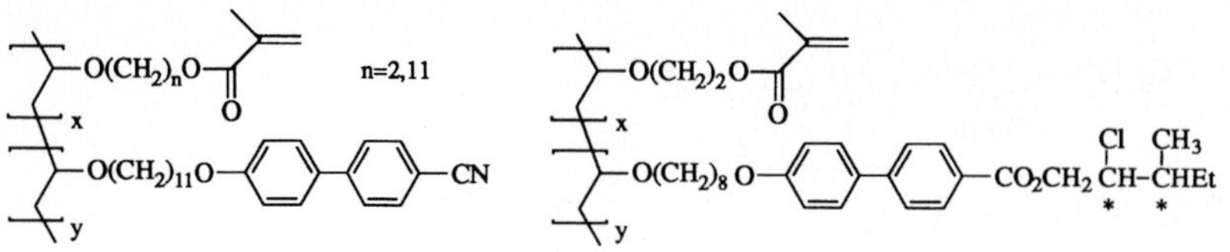

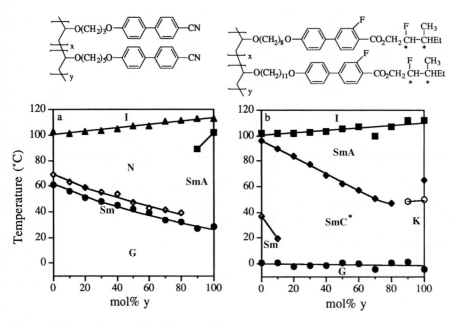

Scheme 37

Figure 24. Dependence of the phase transition temperatures from the glass (●), crystalline (○), unidentified smectic (◇), SmC* (◆), SmA (■) and N (▲) phases of: (a) poly{{3-[(4'-(4''-cyanophenyl)phenoxy)propyl]vinyl ether}-*co*-{5-[(4'-(4''-cyanophenyl)phenoxy)pentyl]vinyl ether}} ($DP_n = 16-21$, pdi = 1.09–1.25) [364]; and (b) poly{{8-[(4'-(2R,3S)-2-fluoro-3-methylpentyloxycarbonyl)-3'-fluorophenyl-4''-phenoxy)octyl]vinyl ether}-*co*-{11-[(4'-(2R,3S)-2-fluoro-3-methylpentyloxycarbonyl)-3'-fluorophenyl-4''-phenoxy)undecyl]vinyl ether}} ($DP_n = 9-15$, pdi = 1.09–1.22) [139]; first heat.

60 mol% of the mesogenic monomer to exhibit a SmC mesophase [357]. Similar copolymers (Scheme 37) containing polymerizable methacrylate pendant groups and at least 40–60% of the mesogenic monomer can be subsequently crosslinked to generate networks which exhibit the SmA or SmC* mesophase of the corresponding homopolymer [360–362]. In addition, the crystalline and/or more ordered smectic mesophases of these homopolymers are eliminated when the networks contain 10–30% of the cross-linkable methacrylate monomer; less 2-vin-

yloxyethyloxy methacrylate is required to eliminate the ordered phases than 11-vinyl-oxyundecanyloxy methacrylate.

Several vinyl ether comonomers based on the same biphenyl mesogen with different spacers and/or different substituents were also copolymerized using the same conditions (Eq. 36). All of the copolymerizations were assumed to be azetropic ($r_1 = r_2 = 1$), and most were based on copolymer pairs containing at least three carbons in the spacer. Although the copolymers prepared by cationic polymerizations were not of

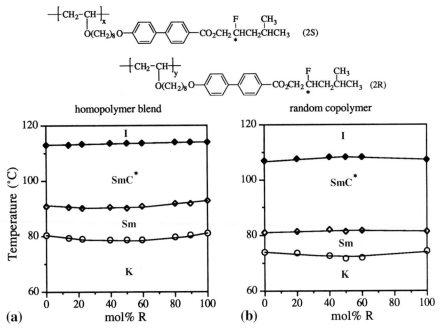

homopolymer blend   random copolymer

Figure 25. Dependence of the phase transition temperatures from the crystalline (○), unidentified smectic (◇) and SmC* (◆) phases of (a) blends of poly{{8-[(4'-(2R)-2-fluoro-4-methylpentyloxycarbonyl)phenyl-4"-phenoxy)octyl]vinyl ether} (DP_n = 16.3, pdi = 1.14) and poly{{8-[(4'-(2S)-2-fluoro-4-methylpentyloxycarbonyl)phenyl-4"-phenoxy)octyl]vinyl ether} (DP_n = 15.2, pdi = 1.14); and (b) poly{{8-[(4'-(2R)-2-fluoro-4-methylpentyloxycarbonyl)phenyl-4"-phenoxy)octyl]vinyl ether}-co-{{8-[(4'-(2S)-2-fluoro-4-methylpentyloxycarbonyl)phenyl-4"-phenoxy)octyl]vinyl ether} (DP_n = 12.5–12.9, pdi = 1.13–1.16) [141]; second heat.

high enough molecular weight for their transitions to be independent of molecular weight [363], they were compared to homopolymers of the same degree of polymerization.

If the copolymer is based on monomers whose homopolymers exhibit identical mesophases, its phase diagram follows ideal solution behavior with a continuous, linear or slightly curved dependence of the transition temperature and enthalpy of at least the highest temperature mesophase on copolymer composition [132, 134, 136–140, 364]. For example, Fig. 24a demonstrates that the two structural units of poly{{3-[(4'-(4"-cyanophenyl)phenoxy)propyl]vinyl ether}-co-{5-[(4'-(4"-cyanophenyl)phenoxy)pentyl]vinyl ether}} are isomorphic within the nematic mesophase, although they are not iso-

morphic within the smectic mesophases [364].

Similarly, the two structural units of poly{{8-[(4'-(2R,3S)-2-fluoro-3-methylpentyloxycarbonyl)-3'-fluorophenyl-4"-phenoxy)octyl]vinyl ether}-co-{11-[(4'-(2R,3S)-2-fluoro-3-methylpentyloxycarbonyl)-3'-fluorophenyl-4"-phenoxy)undecyl]vinyl ether}} are isomorphic within both the SmC* and SmA mesophases, but not the more ordered phases (Fig. 24b) [139]. Therefore, copolymerization can be used to eliminate the crystalline and/or more ordered smectic phases of the corresponding homopolymer(s).

The isotropization temperature shows a slightly positive deviation from the additive values if the hetero-monomer interactions are stronger than the homo-monomer inter-

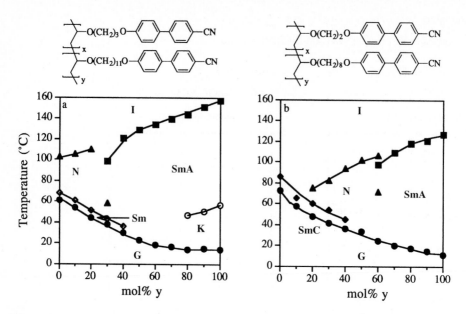

Figure 26. Dependence of the phase transition temperatures from the glass (●), crystalline (○), unidentified smectic (◇), SmC (◆), SmA (■) and nematic (▲) phases of: (a) poly{{3-[(4'-(4"-cyanophenyl)phenoxy)prpyl]vinyl ether}-*co*-{11-[(4'-(4"-cyanophenyl)phenoxy)undecyl]vinyl ether}} ($DP_n = 17-21$, pdi = 1.07 – 1.12) [364]; and (b) poly{{2-[(4'-(4"-cyanophenyl)phenoxy)ethyl]vinyl ether}-*co*-{8-[(4'-(4"-cyanophenyl)phenoxy)octyl]vinyl ether}} ($DP_n = 9 - 11$, pdi = 1.10 – 1.17) [125, 366]; first heat.

actions [141, 365] as shown by the copolymers and blends of homopolymers of 8-[(4'-(2R)-2-fluoro-4-methylpentyloxycarbonyl)phenyl-4"-phenoxy)octyl]vinyl ethers and 8-[(4'-(2S)-2-fluoro-4-methylpentyloxycarbonyl)phenyl-4"-phenoxy)octyl]vinyl ethers based on biphenyl mesogens with diastereomeric substituents (Fig. 25). However, this positive deviation decreases with increasing molecular weight [141], and will apparently vanish as the thermotropic behavior becomes independent of molecular weight (see Sec. 3.1 of this chapter). In contrast, the crystalline melting of such systems shows a negative deviation from ideal solution behavior due to the different physical properties of diastereomers [136, 137, 141].

If the copolymer is based on structural units which are not isomorphic, their respective mesophases are not exhibited over the entire composition (Fig. 26) [125, 133, 135, 364 – 367] and intermediate compositions may exhibit an entirely different [366] phase (Fig. 26 b), or the corresponding chiral mesophase [365] if the comonomer is chiral [368]. The transition temperature and enthalpy of at least the highest temperature mesophase generally also follow a continuous, slightly curved dependence on copolymer composition. Several of the copolymer pairs exhibit a reentrant nematic mesophase at a composition close to that which exhibits both mesophases of the corresponding two homopolymers [125, 364, 366, 367].

Copolymerizations and the resulting phase diagrams can therefore be used to confirm or disprove the tentative assignment of a given mesophase.

Table 20. Thermotropic behavior of 2,5-bis{[4'-n-(perfluoroalkyl)alkoxy-benzoyl]oxy}toluenes [195].

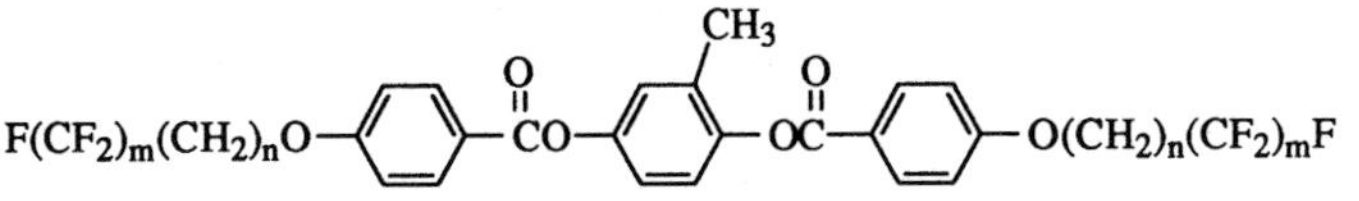

n	m	Phase transitions (°C)
4	6	K 106 SmC 205 SmA 214 I
5	6	K 101 SmC 197 SmA 207 I
6	6	K 99 SmC 197 SmA 203 I
8	6	K 102 SmC 189 SmA 192 I
4	7	K 124 SmC 215 SmA 222 I
5	7	K 119 SmC 208 SmA 216 I
6	7	K 129 SmC 206 SmA 212 I
8	7	K 120 SmC 197 SmA 200 I
4	8	K 130 SmC 218 SmA 226 I
5	8	K 122 SmC 214 SmA 221 I
6	8	K 130 SmC 211 SmA 217 I
8	8	K 124 SmC 201 SmA 205 I

5   Other Factors Controlling the Thermotropic Behavior of SCLCPs as Studied using Living Polymerizations: Induction of Smectic Layering using Immiscible Components

Although liquid crystals were discovered in 1888 [369], chemical concepts are only now being developed for converting the type of mesophase exhibited by a given chemical structure. Compounds which normally form only nematic mesophases can be forced to order into smectic layers by incorporating immiscible fluorocarbon units into their hydrocarbon chemical structure. Comparison of the data in Tables 7, 20 and 21 demonstrate that smectic layering is induced not only in low molar mass liquid crystals, but also in the corresponding side-chain liquid crystalline polynorbornenes with laterally attached mesogens. The latter architecture is the most convincing system possible for demonstrating this concept since lateral attachment of the mesogens to a polymer backbone had previously precluded smectic layering [182, 370–374].

6   The Future

Additional well-defined side-chain liquid crystalline polymers should be synthesized by controlled polymerizations of mesogenic acrylates (anionic or free radical polymerizations), styrenes (anionic, cationic or free radical), vinyl pyridines (anionic), various heterocyclic monomers (anionic, cationic and metalloporphyrin-initiated), cyclobutenes (ROMP), and 7-oxanorbornenes and 7-oxanorbornadienes (ROMP). Ideally, the kinetics of these 'living' polymerizations will be determined by measuring the individual rate constants for termination and

Table 21. Thermotropic behavior of poly{5-[[[2',5'-bis[(4''-n-((perfluoro-alkyl)alkoxy)benzoyl)oxy]benzyl]oxy]carbonyl]bicyclo[2.2.1]-hept-2-ene}s [195].

n	m	Phase transitions (°C)
4	6	G 106 SmC 227 SmA 234 I
5	6	G 96 SmC 228 SmA 231 I
6	6	G 90 SmC 216 SmA 223 I
8	6	G 77 SmC 213 SmA 216 I
4	7	G 90 SmC 24 SmA 251 I
5	7	G 96 SmC 239 SmA 248 I
6	7	G 93 SmC 230 SmA 236 I
8	7	G 97 SmC 228 SmA 232 I
4	8	G 93 SmC 251 SmA 264 I
5	8	G 93 SmC 258 SmA 262 I
6	8	G 98 SmC 250 SmA 261 I
8	8	G 98 SmC 231 SmA 234 I

transfer, or at least the ratios of k_p/k_{tr} and k_p/k_t. In addition, future studies should attempt to detect termination by plotting $\ln[M]_0/[M]$ vs. time in addition to plots of molecular weight vs. conversion or $[M]_0/[I]_0$ to detect transfer.

In addition to conclusively establishing the molecular weight dependence of their thermal transitions, these new SCLCPs will confirm or modify the structure/property correlations that have emerged thus far by synthesizing and studying only SCLCPs with a complete and homologous set of spacer lengths. In particular, the effect of tacticity should be studied using more controlled polymerization systems, rather than only those generated by anionic polymerizations. The effect of polydispersity and/or mixtures of immiscible molecular architectures and chain entanglements should also be determined, perhaps by comparing SCLCPs prepared by controlled radical polymerizations with those first prepared by classic radical polymerizations.

The range of well-defined block and graft copolymers containing at least one mesogenic block should also be expanded in order to elucidate the thermodynamics of phase separation, and perhaps, to discover new morphologies. (Living polymerizations should be used to synthesize both the macromonomers and the graft copolymers.) This will require better characterization of their thermotropic behavior and resulting morphologies. For example, overlapping transitions and glass transitions should be determined by dynamic mechanical analysis and/or dynamic DSC when necessary. The mechanical properties of these copolymers should also be determined in order to establish whether or not they behave as thermoplastic elastomers and if their transverse strength is greater than that of the corresponding homopolymers.

We expect that the range of well-defined statistical copolymers will also increase, primarily to manipulate the phases formed and the transition temperatures of the corresponding homopolymers.

Acknowledgments

Acknowledgment is made to the donors of The Petroleum Research Fund, administered by the ACS, for support of this work. C. P. also acknowledges the National Science Foundation for an NSF Young Investigator Award (1994–1999), and matching funds from Bayer, Dow Chemical, DuPont (DuPont Young Professor Grant), GE Foundation (GE Junior Faculty Fellowship), Pharmacia Biotech and Waters Corporation.

7 References

[1] V. Percec, C. Pugh, in *Side Chain Liquid Crystal Polymers* (Ed.: C. B. McArdle), Chap. 3, Chapman & Hall, New York, **1989**.

[2] V. Percec, D. Tomazos, in *Comprehensive Polymer Science* (Eds.: S. K. Aggarwal, S. Russo), First Supplement, Pergamon Press, Oxford **1992**.

[3] For a general polymer chemistry reference, see: G. Odian, *Principles of Polymerization*, Wiley, New York, **1991**.

[4] K. Matyjaszewski, C. Pugh, in *Cationic Polymerizations*, (Ed.: K. Matyjaszewski), Chap. 1, Marcel Dekker, New York **1996**.

[5] M. Szwarc, *Carbanions, Living Polymers and Electron Transfer Processes*, Interscience Publishers, New York **1968**.

[6] M. Szwarc, *Ionic Polymerization Fundamentals*, Hanser Publishers, New York, **1996**.

[7] R. R. Schrock, *Acc. Chem. Res.*, **1990**, *23*, 158.

[8] K. Matyjaszewski Ed., *Cationic Polymerizations*, Marcel Dekker, New York **1996**.

[9] S. Penczek, P. Kubisa, R. Szymanski, *Makromol. Chem., Rapid Commun.*, **1991**, *12*, 77.

[10] K. Matyjaszewski, *J. Polym. Sci., Polym. Chem. Ed.*, **1995**, *31*, 995.

[11] K. Matyjaszewski, *Macromolecules*, **1993**, *26*, 1787.

[12] R. P. N. Veregin, P. G. Odell, L. M. Michalak, M. K. Georges, *Macromolecules*, **1996**, *29*, 2746; and references therein.

[13] C. J. Hawker, *J. Am. Chem. Soc.*, **1994**, *116*, 11185.

[14] C. J. Hawker, J. L. Hedrick, *Macromolecules*, **1995**, *28*, 2993.

[15] M. Kato, M. Kamigaito, M. Sawamoto, T. Higashimura, *Macromolecules*, **1995**, *28*, 1721.

[16] T. Ando, M. Kato, M. Kamigaito, M. Sawamoto, *Macromolecules*, **1996**, *29*, 1070.

[17] J. S. Wang, K. Matyjaszewski, *J. Am. Chem. Soc.*, **1995**, *117*, 5614.

[18] T. E. Patten, J. Xia, T. Abernathy, K. Matyjaszewski, *Science*, **1996**, *272*, 866; and references therein.

[19] V. Percec, B. Barboiu, *Macromolecules*, **1995**, *28*, 7970.

[20] V. Percec, B. Barboiu, A. Neumann, J. C. Ronda, M. Zhao, *Macromolecules*, **1996**, *29*, 3665.

[21] L. Gold, *J. Chem. Phys.*, **1958**, *28*, 91.

[22] Z. Komiya, C. Pugh, R. R. Schrock, *Macromolecules*, **1992**, *25*, 3609.

[23] A. Heintz, C. Pugh, *ACS Polym. Prepr.* **1996**, *37*(1), 770.

[24] H. L. Hsieh, R. P. Quirk, *Anionic Polymerization*, Marcel Dekker, New York, **1996**.

[25] R. P. Quirk, D. J. Kinning, L. J. Fetters, in *Comprehensive Polymer Science* (Eds.: S. K. Aggarwal and S. Russo), Vol. 7, Chap. 1, Pergamon Press, Oxford, **1992**.

[26] R. V. Figini, *Makromol. Chem.* **1967**, *107*, 170.

[27] G. V. Schultz, *Chem. Techn.* **1973**, 220.

[28] Y. Kawakami, H. Inoue, N. Kishimoto, A. Mori, *Polym. Bull.* **1996**, *36*, 653.

[29] D. M. Wiles, S. Bywater, *Trans. Faraday Soc.*, **1965**, *61*, 150.

[30] B. C. Anderson, G. D. Andrews, P. Arthur, Jr., H. W. Jacobson, L. R. Melby, A. J. Playtis, W. H. Sharkey, *Macromolecules*, **1981**, *14*, 1599.

[31] V. Warzelhan, H. Höcker, G. V. Schulz, *Makromol. Chem.*, **1978**, *179*, 2221.

[32] V. Warzelhan, H. Höcker, G. V. Schulz, *Makromol. Chem.* **1980**, *181*, 149.

[33] H. Jeuck, A. H. E. Müller, *Makromol. Chem., Rapid Commun.* **1982**, *3*, 121.

[34] K. E. Piejko, H. Höcker, *Makromol. Chem., Rapid Commun.* **1982**, *3*, 243.

[35] J.-S. Wang, H. Zhang, R. Jérôme, P. Teyssié, *Macromolecules*, **1995**, *28*, 1758.

[36] S. K. Varshney, J. P. Hautekeer, R. Fayt, R. Jérôme, Ph. Teyssié, *Macromolecules*, **1990**, *23*, 2618.

[37] R. Fayt, R. Forte, C. Jacobs, R. Jérôme, T. Ouhadi, S. K. Varshney, Ph. Teyssié, *Macromolecules*, **1987**, *20*, 1442.

[38] D. Kunkel, A. H. E. Müller, M. Janata, L. Lochmann, *Makromol. Chem., Symp. Ed.*, **1992**, *60*, 315.

[39] K. Hatada, K. Ute, K. Tanaka, Y. Okamoto, T. Kitayama, *Polym. J.* **1986**, *18*, 1037.

[40] K. Hatada, T. Kitayama, K. Ute, E. Masuda, T. Shinozaki, M. Yamamoto, *ACS Polym. Prepr.* **1988**, *29*(2), 54.

[41] Y. Okamoto, T. Asakura, K. Hatada, *Chem. Lett. (Jpn.)* **1991**, 1105.

[42] T. Nakano, T. Hasegawa, Y. Okamoto, *Macromolecules*, **1993**, *26*, 5494.

[43] K. Hatada, T. Kitayama, K. Ute, *Prog. Polym. Sci.*, **1988**, *13*, 189.

[44] M. Yamada, T. Iguchi, A. Hirao, S. Nakahama, J. Watanabe, *Macromolecules*, **1995**, *28*, 50.

[45] R. Bohnert, H. Finkelmann, P. Lutz, *Makromol. Chem., Rapid Commun.* **1993**, *14*, 139.

[46] R. Bohnert, H. Finkelmann, *Macromol. Chem. Phys.* **1994**, *195*, 689.

[47] W. Kreuder, O. W. Webster, H. Ringsdorf, *Makromol. Chem., Rapid Commun.*, **1986**, *7*, 5.

[48] V. Percec, D. Tomazos, C. Pugh, *Macromolecules*, **1989**, *22*, 3259.

[49] R. B. Woodward, R. Hoffmann, *The Conservation of Orbital Symmetry*, Verlag Chemie, Weinheim, **1970**.

[50] K. Fukui, *Acc. Chem. Res.*, **1971**, *4*, 57.

[51] O. W. Webster, W. R. Hertler, D. Y. Sogah, W. B. Farnham, T. V. RajanBabu, *J. Am. Chem. Soc.*, **1983**, *105*, 5706.

[52] P. M. Mai, A. H. E. Müller, *Makromol. Chem., Rapid Commun.*, **1987**, *8*, 99.

[53] P. M. Mai, A. H. E. Müller, *Makromol. Chem., Rapid Commun.*, **1987**, *8*, 247.

[54] I. B. Dicker, G. M. Cohen, W. B. Farnham, W. R. Hertler, E. D. Laganis, D. Y. Sogah, *Macromolecules*, **1990**, *23*, 4034.

[55] A. H. E. Müller, *Makromol. Chem., Macromol. Symp.*, **1990**, *32*, 87.

[56] K. Matyjaszewski, C. Pugh, *Makromol. Chem., Macromol. Symp.* **1993**, *67*, 67.

[57] A. H. E. Müller, G. Litvinenko, D. Yan, *Macromolecules*, **1996**, *29*, 2339.

[58] A. H. E. Müller, G. Litvinenko, D. Yan, *Macromolecules*, **1996**, *29*, 2346.

[59] W. J. Brittain, I. B. Dicker, *Macromolecules*, **1989**, *22*, 1054.

[60] W. R. Hertler, *Macromolecules*, **1987**, *20*, 2976.

[61] A. D. Jenkins, E. Tsartolia, D. R. M. Walton, J. Stejskal, P. Kratochvil, *Polym. Bull.*, **1988**, *20*, 97.

[62] H. Catalgil, A. D. Jenkins, *Eur. Polym. J.*, **1991**, *27*, 651.

[63] M. A. Müller, M. Stickler, *Makromol. Chem., Rapid. Commun.*, **1986**, *7*, 575.

[64] R. P. Quirk, J.-S. Kim, *J. Phys. Org. Chem.* **1995**, *8*, 242.

[65] R. P. Quirk, J. Ren, *Macromolecules*, **1992**, *25*, 6612.

[66] T. Aida, *Prog. Polym. Sci.*, **1994**, *19*, 469.

[67] T. Aida, R. Mizuta, Y. Yoshida, S. Inoue, *Makromol. Chem.* **1981**, *182*, 1073.

[68] T. Aida, S. Inoue, *Macromolecules*, **1981**, *14*, 1162.

[69] T. Aida, S. Inoue, *Macromolecules*, **1981**, *14*, 1166.

[70] T. Yasuda, T. Aida, S. Inoue, *Makromol. Chem., Rapid Commun.*, **1982**, *3*, 585.

[71] T. Yasuda, T. Aida, S. Inoue, *Macromolecules*, **1983**, *16*, 1792.

[72] T. Yasuda, T. Aida, S. Inoue, *Macromolecules*, **1984**, *17*, 2217.

[73] K. Shimasaki, T. Aida, S. Inoue, *Macromolecules*, **1987**, *20*, 3076.

[74] M. Endo, T. Aida, S. Inoue, *Macromolecules*, **1987**, *20*, 2982.

[75] L. Trofimoff, T. Aida, S. Inoue, *Chem. Lett.*, **1987**, 991.

[76] M. Kuroki, T. Aida, S. Inoue, *J. Am. Chem. Soc.*, **1987**, *109*, 4737.

[77] S. Sugimoto, M. Saika, Y. Hosokawa, T. Aida, S. Inoue, *Macromolecules*, **1996**, *29*, 3359.

[78] T. Yasuda, T. Aida, S, Inoue, *Bull. Chem. Soc. Jpn.*, **1986**, *59*, 3931.

[79] S. Asano, T. Aida, S. Inoue, *J. Chem. Soc., Chem. Commun.* **1985**, 1148.

[80] S. Inoue, T. Aida, *Makromol. Chem., Macromol. Symp.* **1986**, *6*, 217.

[81] T. Aida, Y. Maekawa, S. Asano, S. Inoue, *Macromolecules*, **1988**, *21*, 1195.

[82] M. Akatsuka, T. Aida, S. Inoue, *Macromolecules*, **1994**, *27*, 2820.

[83] M. Akatsuka, T. Aida, S. Inoue, *Macromolecules*, **1995**, *28*, 1320.

[84] M. Kuroki, T. Watanabe, T. Aida, S. Inoue, *J. Am. Chem. Soc.*, **1991**, *113*, 5903.

[85] T. Adachi, H. Sugimoto, T. Aida, S. Inoue, *Macromolecules*, **1992**, *25*, 2280.

[86] H. Sugimoto, M. Kuroki, T. Watanabe, C. Kawamura, T. Aida, S. Inoue, *Macromolecules*, **1993**, *26*, 3403.

[87] H. Sugimoto, T. Aida, S. Inoue, *Macromolecules*, **1993**, *26*, 4751.

[88] T. Aida, M. Kuroki, H. Sugimoto, T. Watanabe, T. Adachi, C. Kawamura, S. Inoue, *Makromol. Chem., Macromol. Symp.*, **1993**, *67*, 125.

[89] T. Adachi, H. Sugimoto, T. Aida, S. Inoue, *Macromolecules*, **1993**, *26*, 1238.

[90] T. Kodaira, M. Yamamoto, T. Tanaka, M. Urushisaki, T. Hashimoto, *Polymer*, **1995**, *36*, 3767.

[91] T. Kodaira, K. Mori, *Makromol. Chem., Rapid Commun.* **1990**, *11*, 645.

[92] T. Kodaira, T. Tanaka, K. Mori, *Proc. Jpn. Acad. Ser. B*, **1994**, *70*, 37.

[93] K. Matyjaszewski, C.-H. Lin, C. Pugh, *Macromolecules*, **1993**, *26*, 2649.

[94] See for example: T. Higashimura, M. Miyamoto, M. Sawamoto, *Macromolecules*, **1985**, *18*, 611.

[95] Y. H. Kim, *Macromolecules*, **1991**, *24*, 2122.

[96] M. Kamigaito, M. Sawamoto, T. Higashimura, *Macromolecules*, **1992**, *25*, 2587.

[97] M. Kamigaito, K. Yamaoka, M. Sawamoto, T. Higashimura, *Macromolecules*, **1992**, *25*, 6400.

[98] H. Katayama, M. Kamigaito, M. Sawamoto, T. Higashimura, *Macromolecules*, **1995**, *28*, 3747.

[99] K. Matyjaszewski, M. Teodorescu, C.-H. Lin, *Macromol. Chem. Phys.*, **1995**, *196*, 2149.

[100] L. P. Lorimer, D. C. Pepper, *Proc. Roy. Soc.*, **1976**, *A351*, 551.

[101] T. Kunitake, K. Takarabe, *Macromolecules*, **1979**, *12*, 1067.

[102] H. Mayr, R. Schneider, C. Schade, J. Bartl, R. Bederke, *J. Am. Chem. Soc.*, **1990**, *112*, 4446.

[103] K. Matyjaszewski, R. Szymanski, M. Teodorescu, *Macromolecules*, **1994**, *27*, 7565.

[104] K. Matyjaszewski, C.-H. Lin, A. Bon, J. S. Xiang, *Makromol. Chem., Macromol. Symp.*, **1994**, *85*, 65.

[105] C.-H. Lin, K. Matyjaszewski, *ACS Polym. Prepr.* **1994**, *35*(1), 462.

[106] See for example: T. Higashimura, S. Okamoto, Y. Kishimoto, S. Aoshima, *Polym. J.*, **1989**, *21*, 725.

[107] C. G. Cho, B. A. Feit, O. W. Webster, *Macromolecules*, **1990**, *23*, 1918.

[108] K. Matyjaszewski, C.-H. Lin, *Makromol. Chem., Macromol. Symp.* **1991**, *47*, 221.

[109] See for example: M. Kamigaito, M. Sawamoto, T. Higashimura, *Macromolecules*, **1991**, *24*, 3988.

[110] L. Thomas, A. Polton, M. Tardi, P. Sigwalt, *Macromolecules*, **1995**, *28*, 2105; and references therein.

[111] See for example: S. Aoshima, T. Higashimura, *Macromolecules*, **1989**, *22*, 1009.

[112] See for example: G. Kaszas, J. E. Puskas, J. P. Kennedy, W. G. Hager, *J. Polym. Sci., Polym. Chem. Ed.*, **1991**, *29*, 421.

[113] K. Matyjaszewski, C. Pugh, in *Cationic Polymerizations* (K. Matyjaszewski Ed.), ch. 3, Marcel Dekker, New York, **1996**.

[114] P. Sigwalt, A. Polton, M. Tardi, *J. Macromol. Sci. – Pure App. Chem.*, **1994**, *A31*, 953.

[115] M. Sawamoto, J. Fujimoro, T. Higashimura, *Macromolecules*, **1987**, *20*, 916.

[116] See for example: M. Miyamoto, M. Sawamoto, T. Higashimura, *Macromolecules*, **1984**, *17*, 265.

[117] R. Faust, A. Fehervari, J. P. Kennedy, *Br. Polym. J.*, **1987**, *19*, 379.

[118] K. Matyjaszewski, C.-H. Lin, *J. Polym. Sci., Polym. Chem. Ed.*, **1991**, *29*, 1439.

[119] T. Higashimura, Y. Ishihama, M. Sawamoto, *Macromolecules*, **1993**, *26*, 744.

[120] K. Matyjaszewski, M. Sawamoto, in *Cationic Polymerizations* (Ed.: K. Matyjaszewski), Chap. 4, Marcel Dekker, New York, **1996**.

[121] V. Percec, D. Tomazos, *Adv. Mater.*, **1992**, *4*, 548.

[122] V. Percec, M. Lee, H. Jonsson, *J. Polym. Sci., Polym. Chem. Ed.*, **1991**, *29*, 327.

[123] V. Percec, M. Lee, *J. Macromol. Sci.-Chem.*, **1991**, *A28*, 651.

[124] V. Percec, M. Lee, *Macromolecules*, **1991**, *24*, 1017.

[125] V. Percec, M. Lee, *J. Mater. Chem.*, **1991**, *1*, 1007.

[126] V. Percec, M. Lee, C. Ackerman, *Polymer*, **1992**, *33*, 703.

[127] V. Percec, M. Lee, *Macromolecules*, **1991**, *24*, 2780.

[128] V. Percec, M. Lee, *J. Macromol. Sci., Pure & Appl. Chem.*, **1992**, *A29*, 655.

[129] V. Percec, C.-S. Wang, M. Lee, *Polym. Bull.*, **1991**, *26*, 15.

[130] V. Percec, A. D. S. Gomes, M. Lee, *J. Polym. Sci., Polym. Chem. Ed.*, **1991**, *29*, 1615.

[131] V. Percec, Q. Zheng, M. Lee, *J. Mater. Chem.*, **1991**, *1*, 611.

[132] V. Percec, Q. Zheng, M. Lee, *J. Mater. Chem.*, **1991**, *1*, 1015.

[133] V. Percec, Q. Zheng, *J. Mater. Chem.*, **1992**, *2*, 1041.

[134] V. Percec, Q. Zheng, *J. Mater. Chem.*, **1992**, *2*, 475.

[135] V. Percec, H. Oda, *J. Mater. Chem.*, **1995**, *5*, 1115.

[136] V. Percec, H. Oda, P. L. Rinaldi, D. R. Hensley, *Macromolecules*, **1994**, *27*, 12.

[137] V. Percec, H. Oda, *J. Macromol. Sci., Pure Appl. Chem.*, **1995**, *A32*, 1531.

[138] V. Percec, H. Oda, *J. Polym. Sci., Polym. Chem. Ed.*, **1995**, *33*, 2359.

[139] V. Percec, H. Oda, *J. Mater. Chem.* **1995**, *5*, 1125.

[140] V. Percec, H. Oda, *Macromolecules*, **1994**, *27*, 4454.

[141] V. Percec, H. Oda, *Macromolecules*, **1994**, *27*, 5821.

[142] V. Percec, G. Johansson, *J. Mater. Chem.*, **1993**, *3*, 83.

[143] R. Rodenhouse, V. Percec, *Adv. Mater.*, **1991**, *3*, 101.

[144] G. Scherowsky, U. Fichna, D. Wolff, *Liq. Cryst.*, **1995**, *19*, 621.

[145] V. Percec, M. Lee, P. Rinaldi, V. E. Litman, *J. Polym. Sci., Polym. Chem. Ed.*, **1992**, *30*, 1213.

[146] V. Percec, M. Lee, D. Tomazos, *Polym. Bull.*, **1992**, *28*, 9.

[147] T. Sagane, R. W. Lenz, *Polym. J.*, **1988**, *20*, 923.

[148] T. Sagane, R. W. Lenz, *Polymer*, **1989**, *30*, 2269.

[149] T. Sagane, R. W. Lenz, *Macromolecules*, **1989**, *22*, 3763.

[150] M. Laus, M. C. Bignozzi, M. Fagnani, A. S. Angeloni, G. Galli, E. Chiellini, O. Francescangeli, *Macromolecules*, **1996**, *29*, 5111.

[151] H. Jonsson, V. Percec, A. Hult, *Polym. Bull.* **1991**, *25*, 115.

[152] R. Rodenhouse, V. Percec, A. E. Feiring, *J. Polym. Sci., Polym. Lett. Ed.* **1990**, *28*, 345.

[153] A. J. Amass, in *Comprehensive Polymer Science* (Eds.: G. Allen, J. C. Bevington), Vol. 4, Chap. 6, Pergamon Press, Oxford, **1989**; and references therein.

[154] W. Feast, in *Comprehensive Polymer Science* (Eds. G. Allen, J. C. Bevington), Vol. 4, Chap. 7, Pergamon Press, Oxford, **1989**; and references therein.

[155] L. R. Gilliom, R. H. Grubbs, *J. Am. Chem. Soc.*, **1986**, *108*, 733.

[156] K. C. Wallace, R. R. Schrock, *Macromolecules*, **1987**, *20*, 448.

[157] J. Kress, J. A. Osborn, R. M. E. Greene, K. J. Ivin, J. J. Rooney, *J. Am. Chem. Soc.*, **1987**, *109*, 899.

[158] G. C. Bazan, J. H. Oskam, H.-N. Cho, L. Y. Park, R. R. Schrock, *J. Am. Chem. Soc.*, **1991**, *113*, 6899.

[159] L. F. Cannizzo, R. H. Grubbs, *Macromolecules*, **1987**, *20*, 1488.

[160] R. R. Schrock, J. Feldman, L. F. Canizzo, R. H. Grubbs, *Macromolecules*, **1987**, *20*, 1169.

[161] S. T. Nguyen, L. K. Johnson, R. H. Grubbs, J. W. Ziller, *J. Am. Chem. Soc.*, **1992**, *114*, 3974.

[162] S. T. Nguyen, R. H. Grubbs, J. W. Ziller, *J. Am. Chem. Soc.*, **1993**, *115*, 9858.

[163] D. M. Lynn, S. Kanaoka, R. H. Grubbs, *J. Am. Chem. Soc.*, **1996**, *118*, 784.

[164] R. R. Schrock, R. DePue, J. Feldman, C. J. Schaverien, J. C. Dewan, A. Liu, *J. Am. Chem. Soc.*, **1988**, *110*, 1423.

[165] R. T. DePue, R. R. Schrock, J. Feldman, K. Yap, D. C. Yang, W. M. Davis, L. Park, M. DiMare, M. Schofield, J. Anhaus, E. Walborsky, E. Evitt, C. Kruger, P. Betz, *Organometallics*, **1990**, *9*, 2262.

[166] R. R. Shrock, J. S. Murdzek, G. C. Bazan, J. Robbins, M. DiMare, M. O'Regan, *J. Am. Chem. Soc.*, **1990**, *112*, 3875.

[167] H. H. Fox, K. B. Yap, J. Robbins, S. Cai, R. R. Schrock, *Inorg. Chem.*, **1992**, *31*, 2287.

[168] P. Schwab, M. B. France, J. W. Ziller, R. H. Grubbs, *Angew. Chem. Int. Ed. Engl.*, **1995**, *34*, 2039.

[169] Z. Wu, D. R. Wheeler, R. H. Grubbs, *J. Am. Chem. Soc.*, **1992**, *114*, 146.

[170] Z. Wu, R. H. Grubbs, *Macromolecules*, **1994**, *27*, 6700.

[171] Z. Wu, R. H. Grubbs, *J. Mol. Catal.*, **1994**, *90*, 39.

[172] Z. Wu, R. H. Grubbs, *Macromolecules*, **1995**, *28*, 3502.

[173] R. W. Alder, P. R. Allen, E. J. Khosravi, *J. Chem. Soc., Chem. Commun.*, **1994**, 1235.

[174] M. G. Perrott, B. M. Novak, *Macromolecules*, **1995**, *28*, 3492.

[175] M. G. Perrott, B. M. Novak, *Macromolecules*, **1996**, *29*, 1817.

[176] W. J. Feast, V. C. Gibson, E. Khosravi, E. L. Marshall, J. P. Mitchell, *Polymer*, **1992**, *33*, 872.

[177] W. J. Feast, V. C. Gibson, E. L. Marshall, *J. Chem. Soc., Chem. Commun.*, **1992**, 1157.

[178] J. S. Murdzek, R. R. Schrock, *Macromolecules*, **1987**, *20*, 2640.

[179] G. Bazan, R. R. Schrock, E. Khosravi, W. J. Feast, V. C. Gibson, *Polymer Commun.*, **1989**, *30*, 258.

[180] G. C. Bazan, E. Khosravi, R. R. Schrock, W. J. Feast, V. C. Gibson, M. B. O'Regan, J. K. Thomas, W. M. Davis, *J. Am. Chem. Soc.*, **1990**, *112*, 8378.

[181] G. C. Bazan, R. R. Schrock, *Macromolecules*, **1990**, *24*, 817.

[182] C. Pugh, R. R. Schrock, *Macromolecules*, **1992**, *25*, 6593.

[183] J. P. Mitchell, V. C. Gibson, R. R. Schrock, *Macromolecules*, **1991**, *24*, 1220.

[184] M. A. Hillmyer, W. R. Laredo, R. H. Grubbs, *Macromolecules*, **1995**, *28*, 6311.

[185] P. Schwab, R. H. Grubbs, J. W. Ziller, *J. Am. Chem. Soc.*, **1996**, *118*, 100.

[186] G. C. Bazan, R. R. Schrock, H.-N. Cho, V. C. Gibson, *Macromolecules*, **1991**, *24*, 4495.

[187] B. R. Maughon, M. Weck, B. Mohr, R. H. Grubbs, *Macromolecules*, **1997**, *30*, 257.

[188] Z. Komiya, C. Pugh, R. R. Schrock, *Macromolecules*, **1992**, *25*, 6586.

[189] Z. Komiya, R. R. Schrock, *Macromolecules*, **1993**, *26*, 1393.

[190] B. Winkler, M. Ungerank, F. Stelzer, *Macromol. Chem. Phys.*, **1996**, *197*, 2343.

[191] M. Ungerank, B. Winkler, E. Eder, F. Stelzer, *Macromol. Chem. Phys.*, **1995**, *196*, 3623.

[192] C. Pugh, S. Arehart, H. Liu, R. Narayanan, *J. Macromol. Sci., Pure & Appl. Chem.*, **1994**, *A31*, 1591.

[193] C. Pugh, H. Liu, R. Narayanan, S. V. Arehart, *Macromol. Symp.* **1995**, *98*, 293.

[194] C. Pugh, S. V. Arehart, *ACS Polym. Prepr.* **1996**, *37*(1), 72.

[195] S. V. Arehart, C. Pugh, *J. Am. Chem. Soc.*, **1997**, *119*, 3027.

[196] C. Pugh, J. Dharia, *ACS Polym. Prepr.* **1996**, *37*(1), 772.

[197] C. Pugh, J. Dharia, S. V. Arehart, *Macromolecules*, **1997**, *30*, 4520.

[198] G. Widawski, W. J. Feast, P. Dounis, *J. Mater. Chem.*, **1995**, *5*, 1847.

[199] G. W. Gray, in *Side Chain Liquid Crystal Polymers* (Ed.: C. B. McArdle), Chap. 4, Chapman & Hall, New York, **1989**.

[200] M. A. Hempenius, R. G. H. Lammertink, G. J. Vancso, *Marcomol. Rapid. Commun.*, **1996**, *17*, 299.

[201] J. Adams, W. Gronski, *Makromol. Chem., Rapid. Commun.*, **1989**, *10*, 553.

[202] J. Adams, J. Sänger, C. Tefehne, W. Gronski, *Macromol. Rapid Commun.*, **1994**, *15*, 879.

[203] A. Martin, C. Tefehne, W. Gronski, *Macromol. Rapid Commun.*, **1996**, *17*, 305.

[204] B. Zaschke, W. Frank, H. Fischer, K. Schmutzler, M. Arnold, *Polym. Bull.*, **1991**, *27*, 1.

[205] M. Arnold, S. Poser, H. Fischer, W. Frank, H. Utschick, *Macromol. Rapid Commun.*, **1994**, *15*, 487.

[206] H. Fischer, S. Poser, M. Arnold, W. Frank, *Macromolecules*, **1994**, *27*, 7133.

[207] H. Fischer, *Macromol. Rapid Commun.*, **1994**, *15*, 949.

[208] H. Fischer, S. Poser, M. Arnold, *Liq. Cryst.* **1995**, *18*, 503.

[209] S. Poser, H. Fischer, M. Arnold, *J. Polym. Sci. Polym. Chem. Ed.*, **1996**, *34*, 1733.

[210] H. Fischer, S. Poser, M. Arnold, *Macromolecules*, **1995**, *28*, 6957.

[211] Y. Nagasaki, H. Ito, M. Kato, K. Kataoka, T. Tsuruta, *Polym. Bull.*, **1995**, *35*, 137.

[212] V. Héroguez, M. Schappacher, E. Papon, A. Deffieux, *Polym. Bull.*, **1991**, *25*, 307.

[213] E. Papon, A. Deffieux, F. Hardouin, M. F. Achard, *Liq. Cryst.*, **1992**, *11*, 803.

[214] E. Chiellini, G. Galli, M. C. Bignozzi, S. A. Angeloni, M. Fagnani, M. Laus, *Macromol. Chem. Phys.*, **1995**, *196*, 3187.

[215] P. Masson, B. Heinrich, Y. Frère, P. Gramain, *Macromol. Chem. Phys.*, **1994**, *195*, 1199.

[216] P. Masson, P. Gramain, D. Guillon, *Macromol. Chem. Phys.*, **1995**, *196*, 3677.

[217] D. Navarro-Rodriguez, Y. Frère, D. Guillon, *Makromol. Chem.*, **1991**, *192*, 2975.

[218] D. Navarro-Rodriguez, D. Guillon, A. Skoulios, *Makromol. Chem.*, **1992**, *193*, 3117.

[219] C. G. Bazuin, F. A. Brandys, *Chem. Mater.*, **1992**, *4*, 970.

[220] C. G. Bazuin, F. A. Brandys, T. M. Eve, M. Plante, *Macromol. Symp.* **1994**, *84*, 183.

[221] F. A. Brandys, C. G. Bazuin, *Chem. Mater.* **1996**, *8*, 83.

[222] T. Kato, N. Hirota, A. Fujishima, J. M. J. Fréchet, *J. Polym. Sci., Polym. Chem. Ed.*, **1996**, *34*, 57.

[223] D. Steward, C. T. Imrie, *Liq. Cryst.*, **1996**, *20*, 619.

[224] S. Kobayashi, *Prog. Polym. Sci.*, **1990**, *15*, 751; and references therein.

[225] A. R. Luxton, A. Quig, M. J. Delvaux, L. J. Fetters, *Polymer*, **1978**, *19*, 1320.

[226] K. Ishizu, Y. Kashi, F. Fukutomi, T. Kakurai, *Makromol. Chem.*, **1982**, *183*, 3099.

[227] V. Percec, A. Keller, *Macromolecules*, **1990**, *23*, 4347.

[228] D. Demus, H. Zaschke, *Flüssige Kristalle in Tabellen II*, VEB Deutscher Verlag für Grundstoffindustrie, Leipzig, **1984**.

[229] H. Finkelmann, H. Ringsdorf, J. H. Wendorff, *Makromol. Chem.*, **1978**, *179*, 273.

[230] L. Noirez, P. Keller, J. P. Cotton, *Liq. Cryst.* **1995**, *18*, 129.

[231] U. W. Gedde, H. Jonsson, A. Kult, V. Percec, *Polymer*, **1992**, *33*, 4352.

[232] We have take the mesogen density into account by multiplying ΔH_i and ΔS_i by 0.5 for the poly(vinyl ethers) (1 mesogenic side chain per 2 atoms in the polymer backbone), 0.4 for the disubsituted polynorbornenes (2 side chains/ 5 atoms), and 0.2 for the monosubstituted polynorbornenes (1 side chain/5 atoms).

[233] R. Zentel, H. Ringsdorf, *Makromol. Chem., Rapid Commun.*, **1984**, *5*, 393.

[234] J.-C. Dubois, G. Decobert, P. Le Barny, S. Esselin, C. Friedrich, C. Noël, *Mol. Cryst. Liq. Cryst.*, **1986**, *137*, 349.

[235] V. Percec, D. Tomazos, R. A. Willingham, *Polym. Bull.*, **1989**, *22*, 199.

[236] V. Percec, D. Tomazos, *J. Polym. Sci., Polym. Chem. Ed.*, **1989**, *27*, 999.

[237] V. Percec, D. Tomazos, *Macromolecules*, **1989**, *22*, 2062.

[238] V. Percec, C.-S. Wang, *J. Macromol. Sci., Pure & Appl. Chem.*, **1992**, *A29*, 99.

[239] V. Percec, B. Hahn, *Macromolecules*, **1990**, *23*, 2092.

[240] V. Percec, D. Tomazos, *Polymer*, **1990**, *31*, 1658.

[241] A. A. Craig, C. T. Imrie, *Macromolecules*, **1995**, *28*, 3617.

[242] T. Sunaga, K. J. Ivin, G. E. Hofmeister, J. H. Oskam, R. R. Schrock, *Macromolecules*, **1994**, *27*, 4043.

[243] B. Hahn, J. H. Wendorff, M. Portugall, H. Ringsdorf, *Colloid Polym. Sci.*, **1981**, *259*, 875.

[244] P. L. Magagnini, *Makromol. Chem., Suppl.*, **1981**, *4*, 223.

[245] V. Frosini, G. Levita, D. Lupinacci, P. L. Magagnini, *Mol. Cryst. Liq. Cryst.* **1981**, *66*, 341.

[246] K. Hatada, T. Kitayama, K. Ute, *Prog. Polym. Sci.*, **1988**, *13*, 189.

[247] K. Ono, T. Sasaki, M. Yamamoto, Y. Yamasaki, K. Ute, K. Hatada, *Macromolecules*, **1995**, *28*, 5012.

[248] L. Fritz, H. Springer, *Makromol. Chem.*, **1993**, *194*, 2047.

[249] G. Galli, E. Chiellini, M. Laus, D. Caretti, A. S. Angeloni, *Makromol. Chem., Rapid. Commun.*, **1991**, *12*, 43.

[250] R. J. Sarna, G. P. Simon, G. Day, H.-J. Kim, W. R. Jackson, *Macromolecules*, **1994**, *27*, 1603.

[251] C. Boeffel, B. Hisgen, U. Pschorn, H. Ringsdorf, H. W. Spiess, *Isr. J. Chem.*, **1983**, *23*, 388.

[252] S. G. Kostromin, R. V. Talroze, V. P. Shibaev, N. A. Platé, *Makromol. Chem., Rapid. Commun.*, **1982**, *3*, 803.

[253] M. J. Hill, P. J. Barham, *Polymer*, **1995**, *36*, 1523.

[254] J. Minick, A. Moet, E. Baer, *Polymer*, **1995**, *36*, 1923.

[255] M. J. Hill, P. J. Barham, A. Keller, *Polymer*, **1992**, *33*, 2530.

[256] R. G. Alamo, J. D. Londono, L. Mandelkern, F. C. Stehling, G. D. Wignall, *Macromolecules*, **1994**, *27*, 411.

[257] G. D. Wignall, J. D. Londono, J. S. Lin, R. G. Alamo, M. J. Galante, L. Mandelkern, *Macromolecules*, **19**95, *28*, 3156.

[258] G. D. Wignall, R. G. Alamo, J. D. Londono, L. Mandelkern, F. C. Stehling, *Macromolecules*, **1996**, *29*, 5332.

[259] P. L. Joskowicz, A. Muñoz, J. Barrera, A. J. Müller, *Macromol. Chem. Phys.*, **1995**, *196*, 385.

[260] K. F. Freed, J. Dudowicz, *Macromolecules*, **1996**, *29*, 625.

[261] G. H. Fredrickson, A. J. Liu, F. S. Bates, *Macromolecules*, **1994**, *27*, 2503.

[262] F. S. Bates, G. H. Fredrickson, *Annu. Rev. Phys. Chem.*, **1990**, *41*, 525.

[263] F. S. Bates, M. F. Schulz, A. K. Khandpur, S. Förster, J. H. Rosedale, K. Almdal, K. Mortensen, *Faraday Discuss.*, **1994**, *98*, 7.

[264] L. Leibler, *Macromolecules*, **1980**, *13*, 1602.

[265] G. H. Fredrickson, E. Helfand, *J. Chem. Phys.*, **1987**, *87*, 697.

[266] K. Binder, *Adv. Polym. Sci.*, **1993**, *112*, 181.

[267] E. Helfand, Z. R. Wasserman, *Macromolecules*, **1976**, *9*, 879.

[268] E. Helfand, Z. R. Wasserman, *Macromolecules*, **1978**, *11*, 960.

[269] E. Helfand, Z. R. Wasserman, *Macromolecules*, **1978**, *11*, 960.

[270] G. Hadziioannou, A. Skoulios, *Macromolecules*, **1982**, *15*, 258.

[271] T. Hashimoto, H. Tanaka, H. Hasegawa, *Macromolecules*, **1985**, *18*, 1864.

[272] K. I. Winey, E. L. Thomas, L. J. Fetters, *J. Chem. Phys.*, **1991**, *95*, 9367.

[273] K. I. Winey, E. L. Thomas, L. J. Fetters, *Macromolecules*, **1992**, *25*, 422.

[274] K. I. Winey, E. L. Thomas, L. J. Fetters, *Macromolecules*, **1992**, *25*, 2645.

[275] R. J. Spontak, S. D. Smith, A. Ashraf, *Macromolecules*, **1993**, *26*, 956.

[276] M. M. Disko, K. S. Liang, S. K. Behal, R. J. Roe, K. J. Jeon, *Macromolecules*, **1993**, *26*, 2983.

[277] R. J. Spontak, S. D. Smith, A. Ashraf, *Polymer.* **1993**, *34*, 2233.

[278] D. A. Hajduk, P. E. Harper, S. M. Gruner, C. C. Honeker, G. Kim, E. L. Thomas, L. J. Fetter, *Macromolecules*, **1994**, *27*, 4063.

[279] J. Zhao, B. Majumdar, M. F. Schulz, F. S. Bates, K. Almdal, K. Mortensen, D. A. Hajduk, S. M. Gruner, *Macromolecules*, **1996**, *29*, 1204.

[280] R. J. Spontak, J. C. Fung, M. B. Braunfeld, J. W. Sedat, D. A. Agard, L. Kane, S. D. Smith, M. M. Satkowski, A. Ashraf, D. A. Hajduk, S. M. Gruner, *Macromolecules*, **1996**, *29*, 4494.

[281] M. F. Schulz, F. S. Bates, K. Almdal, K. Mortensen, *Phys. Rev. Lett.*, **1994**, *73*, 86.

[282] I. W. Hamley, K. A. Koppi, J. H. Rosedale, F. S. Bates, K. Almdal, K. Mortensen, *Macromolecules*, **1993**, *26*, 5959.

[283] D. A. Hajduk, S. M. Gruner, S. Erramilli, R. A. Register, L. J. Fetters, *Macromolecules*, **1996**, *29*, 1473.

[284] S. L. Aggarwal, *Polymer*, **1972**, *17*, 938.

[285] D. B. Alward, D. J. Kinning, E. L. Thomas, L. J. Fetters, *Macromolecules*, **1986**, *19*, 215.

[286] D. J. Kinning, E. L. Thomas, D. B. Alward, L. J. Fetters, D. L. Handlin, Jr., *Macromolecules*, **1986**, *19*, 1288.

[287] E. L. Thomas, D. B. Alward, D. J. Kinning, D. C. Martin, D. L. Handlin Jr., L. J. Fetters, *Macromolecules*, **1986**, *19*, 2197.

[288] D. S. Herman, D. J. Kinning, E. L. Thomas, L. J. Fetters, *Macromolecules*, **1987**, *20*, 2940.

[289] H. Hasegawa, H. Tanaka, K. Yamasaki, T. Hashimoto, *Macromolecules*, **1987**, *20*, 1651.

[290] D. A. Hajduk, P. E. Harper, S. M. Gruner, C. C. Honeker, E. L. Thomas, L. J. Fetters, *Macromolecules*, **1995**, *28*, 2570.

[291] M. W. Matsen, F. S. Bates, *Macromolecules*, **1996**, *29*, 1091.

[292] Y. Matsushita, T. Takasu, K. Yagi, K. Tomioka, I. Noda, *Polymer*, **1994**, *35*, 2862.

[293] R. L. Lescanec, D. A. Hajduk, G. Y. Kim, Y. Gan, R. Yin, S. M. Gruner, T. E. Hogen-Esch, E. L. Thomas, *Macromolecules*, **1995**, *28*, 3485.

[294] Y. Tselikas, N. Hadjichristidis, R. L. Lescanec, C. C. Honeker, M. Wohlgemuth, E. L. Thomas, *Macromolecules*, **1996**, *29*, 3390.

[295] Y. Mogi, H. Kotsujï, Y. Kaneko, K. Mori, Y. Matsushita, I. Noda, *Macromolecules*, **1992**, *25*, 5408.

[296] Y. Mogi, K. Mori, Y. Matsushita, I. Noda, *Macromolecules*, **1992** , *25*, 5412.

[297] Y. Mogi, M. Noumra, H. Kotsuji, K. Ohnishi, Y. Matsushita, I. Noda, *Macromolecules*, **1994**, *27*, 6755.

[298] C. Auschra, R. Stadler, *Macromolecules*, **1993**, *26*, 2171.

[299] J. Beckmann, C. Auschra, R. Stadler, *Macromol. Rapid. Commun.* **1994**, *15*, 67.

[300] R. Stadler, C. Auschra, J. Beckmann, U. Krappe, I. Voigt-Martin, L. Leibler, *Macromolecules*, **1995**, *28*, 3080.

[301] U. Krappe, R. Stadler, I. Voigt-Martin, *Macromolecules*, **1995**, *28*, 4558.

[302] J. T. Chen, E. L. Thomas, C. K. Ober, S. S. Hwang, *Macromolecules*, **1995**, *28*, 1688.

[303] J. T. Chen, E. L. Thomas, C. K. Ober, B.-P. Mao, *Science*, **1996**, *273*, 343.

[304] G. Seconi, C. Eaborn, A. Fischer, *J. Organomet. Chem.*, **1979**, *177*, 129.

[305] M. Hefft, J. Springer, *Makromol. Chem., Rapid. Commun.*, **1990**, *11*, 397.

[306] K. Van de Velde, M. Van Beylen, R. Ottenburgs, C. Samyn, *Macromol. Chem. Phys.*, **1995**, *196*, 679.

[307] T. Ohmura, M. Sawamoto, T. Higashimura, *Macromolecules*, **1994**, *27*, 3714.

[308] K. Kojima, M. Sawamoto, T. Higashimura, *Polym. Bull.* **1990**, *23*, 149.

[309] K. Kojima, M. Sawamoto, T. Higashimura, *Macromolecules*, **1991**, *24*, 2658.

[310] M. Sawamoto, T. Hasebe, M. Kamigaito, T. Higashimura, *J. Macromol. Sci., Pure & Appl. Chem.*, **1994**, *A31*, 937.

[311] M. Sawamoto, in *Cationic Polymerizations* (Ed.: K. Matyjaszewski), Chap. 5, Marcel Dekker, New York, **1996**.

[312] J.-M. Oh, S.-J. Kang, O.-S. Kwon, S.-K. Choi, *Macromolecules*, **1995**, *28*, 3015.

[313] Z. Fodor, R. Faust, *J. Macromol. Sci., Pure & Appl. Chem.*, **1994**, *A31*, 1985.

[314] Z. Fodor, R. Faust, *J. Macromol. Sci., Pure & Appl. Chem.*, **1995**, *A32*, 575.

[315] D. Li, R. Faust, *Macromolecules*, **1995**, *28*, 1383.

[316] S. Hadjikyriacou, R. Faust, *Macromolecules*, **1996**, *29*, 5261.

[317] V. Percec, M. Lee, *J. Macromol. Sci., Pure & Appl. Chem.*, **1992**, *A29*, 723.

[318] T. Aida, S. Inoue, *Makromol. Chem., Rapid Commun.*, **1980**, *1*, 677.

[319] T. Yasuda, T. Aida, S. Inoue, *Macromolecules*, **1984**, *17*, 2217.

[320] L. F. Cannizzo, R. H. Grubbs, *Macromolecules*, **1988**, *21*, 1961.

[321] W. Risse, R. H. Grubbs, *J. Mol. Catal.*, **1991**, *65*, 211.

[322] R. R. Schrock, S. A. Krouse, K. Knoll, J. Feldman, J. S. Murdzek, D. C. Yang, *J. Mol. Catal.*, **1988**, *46*, 243.

[323] V. Sankaran, C. C. Cummins, R. R. Schrock, R. E. Cohen, R. J. Silbey, *J. Am. Chem. Soc.*, **1990**, *112*, 6858.

[324] C. C. Cummins, M. D. Beachy, R. R. Schrock, M. G. Vale, V. Sankaran, R. E. Cohen, *Chem. Mater.*, **1991**, *3*, 1153.

[325] V. Sankaran, R. E. Cohen, C. C. Cummins, R. R. Schrock, *Macromolecules*, **1991**, *24*, 6664.

[326] G. C. Bazan, R. R. Schrock, *Macromolecules*, **1991**, *24*, 817.

[327] Y. N. C. Chan, R. R. Schrock, R. E. Cohen, *Chem. Mater.*, **1992**, *4*, 24.

[328] C. C. Cummins, R. R. Schrock, R. E. Cohen, *Chem. Mater.*, **1992**, *4*, 27.

[329] Y. N. C. Chan, G. S. W. Craig, R. R. Schrock, R. E. Cohen, *Chem. Mater.*, **1992**, *4*, 885.

[330] R. S. Saunders, R. E. Cohen, S. J. Wong, R. R. Schrock, *Macromolecules*, **1992**, *25*, 2055.

[331] Y. N. C. Chan, R. R. Schrock, R. E. Cohen, *J. Am. Chem. Soc.*, **1992**, *114*, 7295.

[332] Y. N. C. Chan, R. R. Schrock, R. E. Cohen, *Chem. Mater.* **1993**, *5*, 566.

[333] W. J. Feast, V. C. Gibson, E. Khosravi, E. L. Marshall, *J. Chem. Soc., Chem. Commun.*, **1994**, 9.

[334] K. Nomura, R. R. Schrock, *Macromolecules*, **1996**, *29*, 540.

[335] S. A. Krouse, R. R. Schrock, *Macromolecules*, **1988**, *21*, 1885.

[336] R. S. Saunders, R. E. Cohen, R. R. Schrock, *Macromolecules*, **1991**, *24*, 5599.

[337] G. S. W. Craig, R. E. Cohen, R. R. Schrock, R. J. Silbey, G. Puccetti, I. Ledoux, J. J. Zyss, *J. Am. Chem. Soc.*, **1993**, *115*, 860.

[338] Z. Wu, R. H. Grubbs, *Macromolecules*, **1994**, *27*, 6700.

[339] G. S. W. Craig, R. E. Cohen, R. R. Schrock, A. Esser, W. Schrof, *Macromolecules*, **1995**, *28*, 2512.

[340] S. Kanaoka, R. H. Grubbs, *Macromolecules*, **1995**, *28*, 4707.

[341] D. M. Lynn, S. Kanaoka, R. H. Grubbs, *J. Am. Chem. Soc.*, **1996**, *118*, 784.

[342] M. Weck, P. Schwab, R. H. Grubbs, *Macromolecules*, **1996**, *29*, 1789.

[343] Z. Komiya, R. R. Schrock, *Macromolecules*, **1993**, *26*, 1387.

[344] J. Brandrup, E. H. Immergut (Eds.), *Polymer Handbook*, 3rd edn., Wiley-Interscience, New York, **1989**.

[345] G. Natta, G. Dall'asta, G. Mazzanti, U. Giannini, S. Cesca, *Angew. Chem.*, **1959**, *71*, 205.

[346] E. J. Vandenberg, R. F. Heck, D. S. Breslow, *J. Polym. Sci.*, **1959**, *41*, 519.

[347] J. Höpken, M. Möller, M. Lee, V. Percec, *Makromol. Chem.*, **1992**, *193*, 275.

[348] S. T. Milner, *Macromolecules*, **1994**, *27*, 2333.

[349] M. D. Gehlsen, K. Almdal, F. S. Bates, *Macromolecules*, **1992**, *25*, 939.

[350] B. Benoit, G. Hadziioannou, *Macromolecules*, **1988**, *21*, 1449.

[351] A. Shinozaki, D. Jasnow, A. C. Balazs, *Macromolecules*, **1994**, *27*, 2496.

[352] A. M. Mayes, M. Olvera de la Cruz, *J. Chem. Phys.*, **1989**, *91*, 7228.

[353] See for example: J. P. Kennedy, J. M. Delvaux, *Adv. Polym. Sci.*, **1981**, *38*, 141.

[354] M. Hefft, J. Springer, *Makromol. Chem.*, **1992**, *193*, 329.

[355] M. Blankenhagel, J. Springer, *Makromol. Chem.*, **1992**, *193*, 3031.

[356] P. J. Flory, *Principles of Polymer Chemistry*, Cornell University Press, Ithaca, NY **1953**, pp. 70–71.

[357] V. Percec, M. Lee, *J. Mater. Chem.*, **1992**, *2*, 617.

[358] H. A. Schneider, V. Percec, Q. Zheng, *Polymer*, **1993**, *34*, 2180.

[359] F. Sahlén, M.-C. Peterson, V. Percec, A. Hult, U. W. Gedde, *Polym. Bull.*, **1995**, *35*, 629.

[360] V. Percec, Q. Zheng, *Polym. Bull.*, **1992**, *29*, 485.

[361] V. Percec, Q. Zheng, *Polym. Bull.*, **1992**, *29*, 493.

[362] V. Percec, Q. Zheng, *Polym. Bull.*, **1992**, *29*, 501.

[363] V. Percec, M. Lee, *Polym. Bull.*, **1991**, *25*, 123.

[364] V. Percec, M. Lee, *Macromolecules*, **1991**, *24*, 4963.

[365] V. Percec, M. Lee, Q. Zheng, *Liq. Cryst.*, **1992**, *12*, 715.

[366] V. Percec, M. Lee, *Polymer*, **1991**, *32*, 2862.

[367] V. Percec, M. Lee, *Polym. Bull.*, **1991**, *25*, 131.

[368] The authors refer to these phase diagrams as having a triple point, although the intersection of the two equilibrium lines between the liquid crystalline phases and the isotropic melt should not be unique to the (temperature and) pressure at which the experiments were conducted.

[369] F. Reinitzer, *Monatsch. Chem.*, **1888**, *9*, 421.

[370] F. Hessel, H. Finkelmann, *Polym. Bull.*, **1985**, *14*, 375.

[371] F. Hessel, R.-P. Herr, H. Finkelmann, *Makromol. Chem.*, **1987**, *188*, 1597.

[372] F. Hessel, H. Finkelmann, *Makromol. Chem.*, **1988**, *189*, 2275.

[373] P. Keller, F. Hardouin, M. Mauzac, M. F. Achard, *Mol. Cryst. Liq. Cryst.*, **1988**, *171*, 155.

[374] F. Hardouin, S. Mery, M. F. Achard, M. Mauzac, P. Davidson, P. Keller, *Liq. Cryst.*, **1990**, *8*, 565.

Chapter IV
Behavior and Properties of Side Group Thermotropic Liquid Crystal Polymers

Jean-Claude Dubois, Pierre Le Barny, Monique Mauzac, and Claudine Noel

1 Introduction

Side chain liquid crystal polymers (SCLCP) are formed by attaching rigid units to a flexible chain, as illustrated schematically in Fig. 1. Decoupling of the side group by inserting a flexible spacer allows the main chain to accommodate the anisotropic arrangement of the mesogenic side groups and the polymer may exhibit liquid crystal properties. Thus, side group (or side chain) LCPs are reminiscent of low molecular weight LCs, and their properties are characterized by a combination of LC specific and polymer specific properties. The former include the formation of any of the different types of mesophases. A polymer specific property of importance in SCLCPs is the forma-

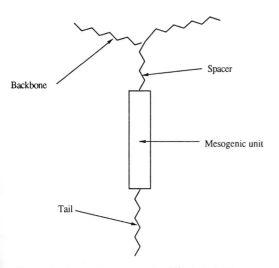

Figure 1. General structure of a side-chain LC homopolymer.

tion of a glass on cooling because of partial decoupling of the flexible backbone from the side chain. The glass transition temperature T_g is partly determined by the chemical constitution of the backbone and in most cases the LC order may be frozen in to yield an anisotropic glass with unique optical properties, e.g. nonlinear optically active side groups can be attached, oriented, and frozen under an electric field. These important electrooptical properties are described in the Sec. 4 of this Chapter.

Many reviews [1–3] of SCLCPs describe the structure of the backbone (main chain), the spacer (flexible linkage), and the side group (mesogenic unit) of the SCLCP. For example, the most widely used backbones include polyacrylates or polymethacrylates, polysiloxanes, and polyphosphazenes; poly-α-chloroacrylates, itaconates, and ethylene oxides have also been reported.

Typical spacer groups consist of 3–12 methylene units. However, oxyalkyl and ester spacer groups are sometimes used to enhance the degree of decoupling through a more flexible spacer. The pendant rigid group is chosen from those that constitute the LMWLCs, including biphenyl, phenyl benzoates, and benzolaniline, also including disc-like moieties such as substituted triphenylene. Combinations of a rigid main chain polymer and a lateral side group have also been described.

Crosslinking through the side group leads to elastomeric LCPs with interesting properties due to the coupling of the elastic properties and mesomorphic behavior. This is studied in Sec. 3 of this Chapter, essentially in terms of mechanical and electromechanical properties.

Chemical modification of the polymer structure allows the obtention of nematic and smectic phases [4, 5]. If the side group and/or the chain are chiral, then cholesteric or chiral smectic C (SmC) phases can be obtained. These can also be obtained by mixing a chiral compound with the SCLCP. SmC* SCLCPs are of particular interest and their behavior is described in Sec. 2 of this Chapter.

2   Ferroelectric Liquid Crystal Polymers' Behavior

2.1   Chemical Structures of SmC* Liquid Crystal Polymers

Since the synthesis of the first chiral smectic C side chain LCP by Shibaev et al. [6], chemists over the last ten years have considerably extended that field. Now, the SmC* mesophase can be exhibited by a variety of polymeric materials including: homopolymers, copolymers and terpolymers, oligomers, combined polymers, and crosslinked polymers.

This synthesis work was not only aimed at obtaining a better knowledge of the structure–properties relationships, but also at designing either polymers with reduced viscosity and improved response time, suitable for display applications, or polymers bearing in their side chains an electron donor-Π system-electron acceptor moiety for second order nonlinear optics.

2.1.1   Homopolymers

The general structure of a homopolymer is shown in Fig. 1. Most of the SmC* LCPs synthesized so far are derived from polyacrylate, polymethacrylate, or polymethylsiloxane backbones. Some polyoxyethylenes [7] as well as some polyvinylethers [8] have also been prepared. The mesogenic core structures, the tails, and the spacers generally used to obtain SmC* LCPs are respectively summarized in Figs. 2–4.

The optically active center essential for obtaining the SmC* phase is generally placed at the tail of the side chain, but it can also be part of the flexible spacer. An SmC* LCP, having the chiral centers located both in the spacer and in the tail, has been reported [21, 52, 53]. Another way of introducing chirality into LCPs is their mixing with chiral dopants [54]. This approach has been used by Ido et al. for getting more suitable materials (mainly in terms of response times) for display applications compared with pure SmC* LCPs [24]. Finally, an example of an SmC* LCP where the chirality is located in the polymer backbone instead of the side chain has been found [55, 56]. It is a polytartrate derivative, as shown in Fig. 5.

2.1.2   Copolymers and Terpolymers

SmC* LCPs can retain their mesomorphic properties when their side chains are diluted by functional groups like dyes [57], crosslinkable moieties [20], or NLO chro-

---

Figure 2. Mesogenic cores used to obtain SmC* LCPs. The molecular formulas have been displayed so that the spacers are linked to the left side of the cores, the tails being linked to their right side.

Two rings systems

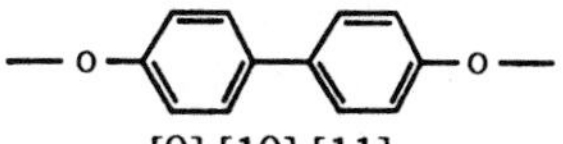

[9] [10] [11]

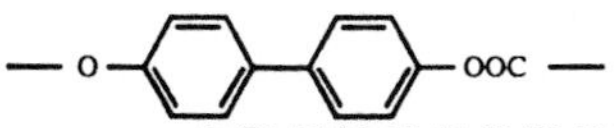

[12] [13] [14] [15] [16]

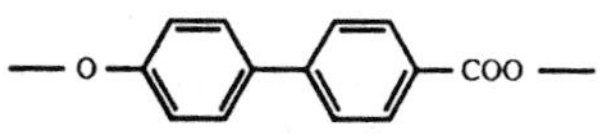

[7] [17] [18] [19]

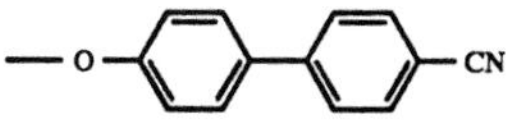

[20]

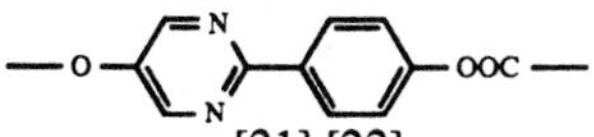

[21] [22]

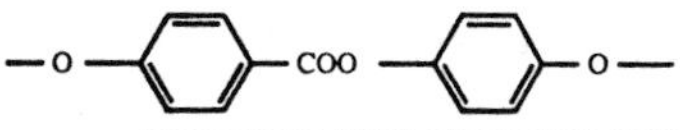

[23] [24] [25] [26] [27] [28]

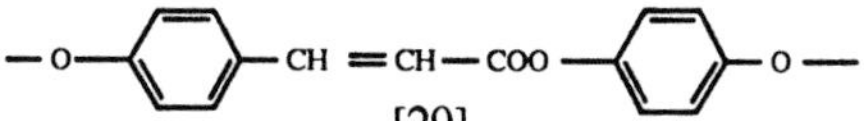

[29]

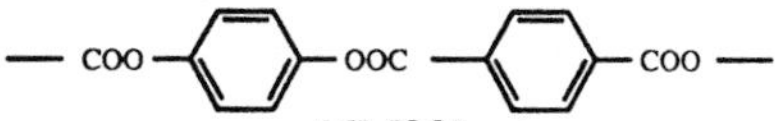

[6] [30]

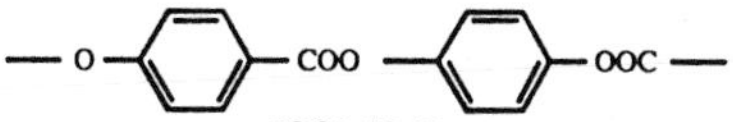

[23] [31]

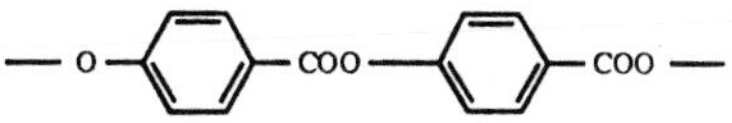

[32] [24] [19]

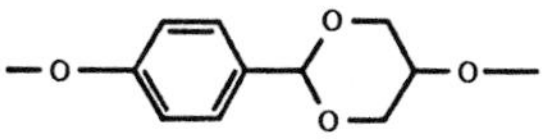

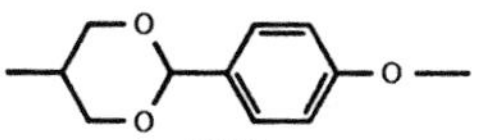

[33]

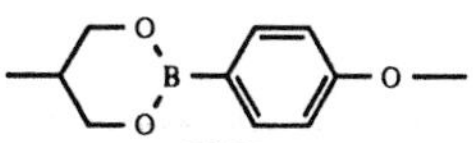

[33]

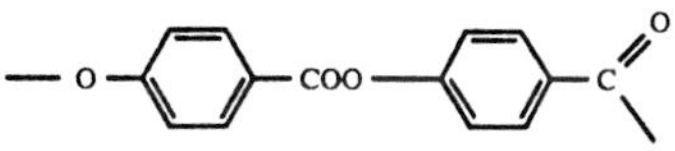

[34]

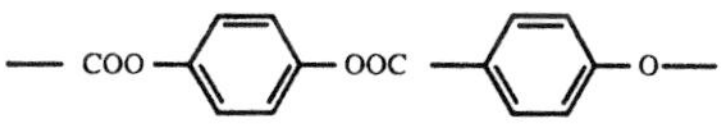

[35]

Three rings systems

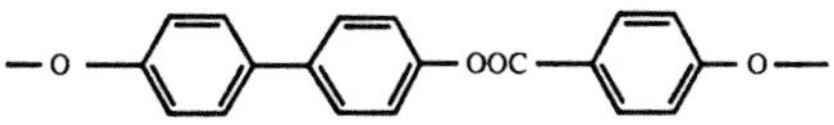

[23]

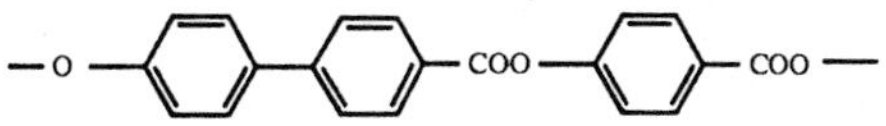

[36] [37]

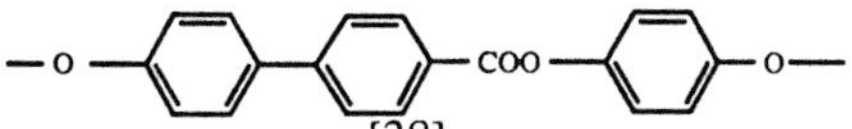

[38]

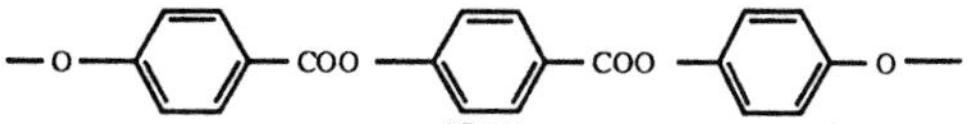

[31]

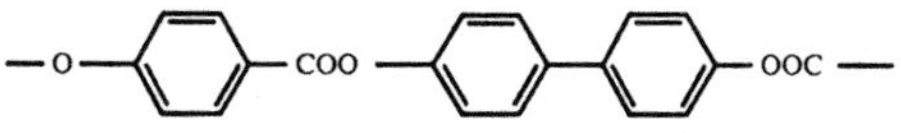

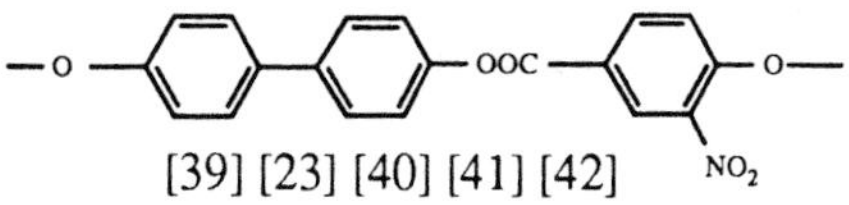

[39] [23] [40] [41] [42] NO_2

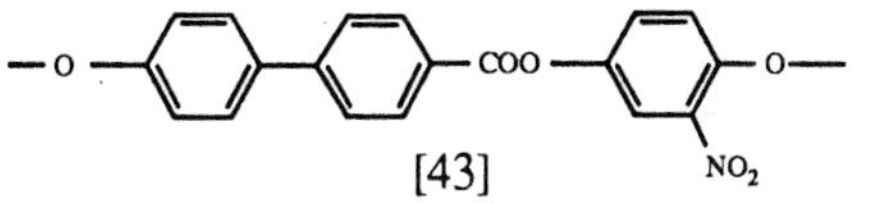

[43] NO_2

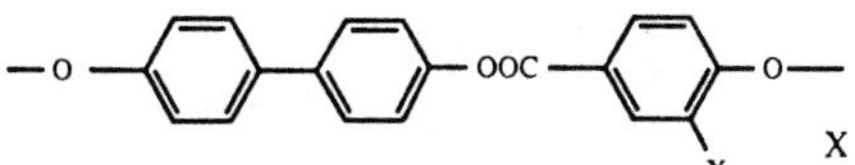

X = Cl, Br
[42]

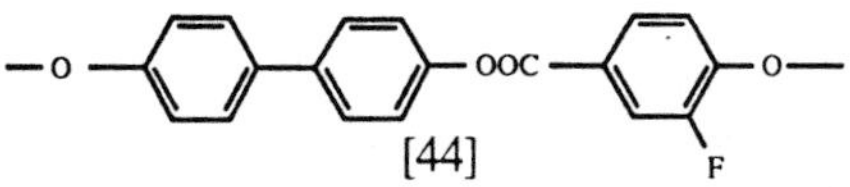

[44]   F

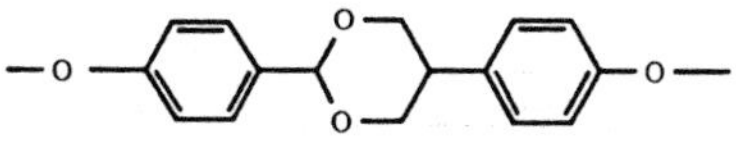

[33]

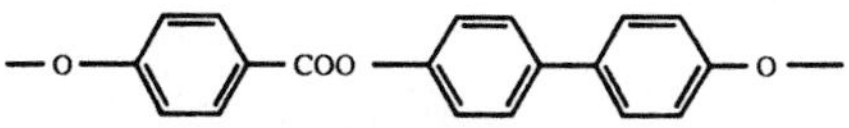

[21] [45] [46]

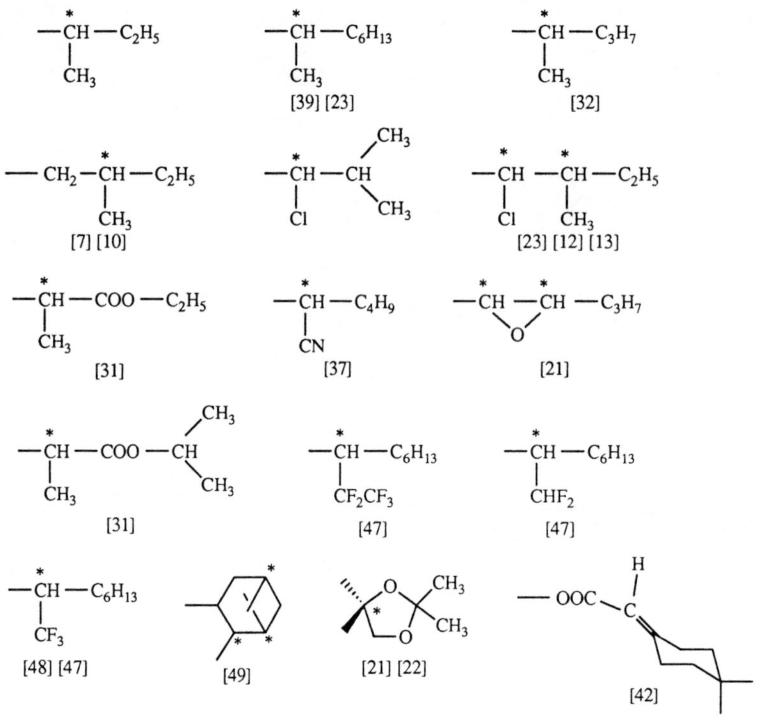

$-CH-C_2H_5$ CH_3 [39] [23]

$-CH-C_6H_{13}$ CH_3

$-CH-C_3H_7$ CH_3 [32]

$-CH_2-CH-C_2H_5$ CH_3 [7] [10]

$-CH-CH$ Cl CH_3

$-CH-CH-C_2H_5$ Cl CH_3 [23] [12] [13]

$-CH-COO-C_2H_5$ CH_3 [31]

$-CH-C_4H_9$ CN [37]

$-CH-CH-C_3H_7$ O [21]

$-CH-COO-CH$ CH_3 CH_3 CH_3 [31]

$-CH-C_6H_{13}$ CF_2CF_3 [47]

$-CH-C_6H_{13}$ CHF_2 [47]

$-CH-C_6H_{13}$ CF_3 [48] [47]

[49]

[21] [22]

$-OOC-C$ H [42]

Figure 3. Examples of chiral units making up the SmC* LCPs' tails.

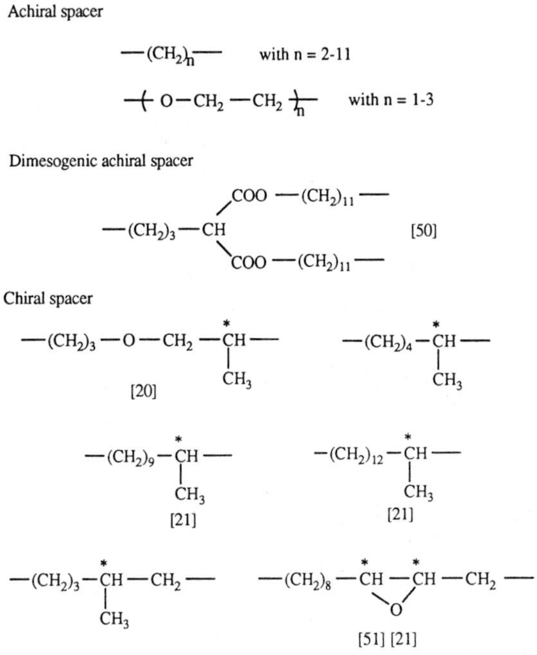

Achiral spacer

$-(CH_2)_n-$ with n = 2-11

$(O-CH_2-CH_2)_n$ with n = 1-3

Dimesogenic achiral spacer

$-(CH_2)_3-CH$ $COO-(CH_2)_{11}-$ $COO-(CH_2)_{11}-$ [50]

Chiral spacer

$-(CH_2)_3-O-CH_2-CH-$ CH_3 [20]

$-(CH_2)_4-CH-$ CH_3

$-(CH_2)_9-CH-$ CH_3 [21]

$-(CH_2)_{12}-CH-$ CH_3 [21]

$-(CH_2)_3-CH-CH_2-$ CH_3

$-(CH_2)_8-CH-CH-CH_2-$ O [51] [21]

Figure 4. Examples of achiral and chiral spacers.

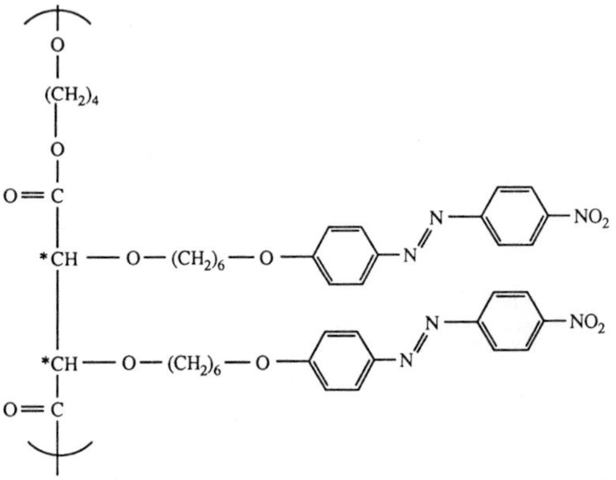

O $(CH_2)_4$ O O=C *CH$-O-(CH_2)_6-O-$ ... $N=N$ ... NO_2 *CH$-O-(CH_2)_6-O-$ ... $N=N$ ... NO_2 O=C

Figure 5. Chemical structure of SmC* polytartrate.

mophores [23], thus giving rise to copolymers or terpolymers with additional properties (Fig. 6).

Another interest in copolymers or terpolymers lies in the possibility of obtaining SmC* LCPs with a reduced viscosity. This

S_C^* LCPs functionalized with a dye

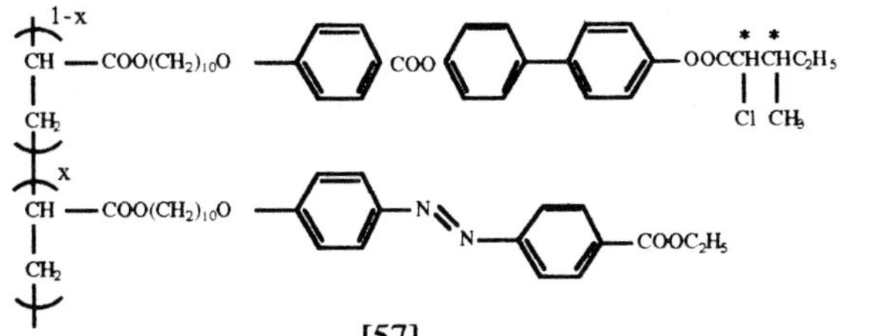

[57]

$$x = 0.05 \quad G\ 75\ S_X\ 100\ S_C^*\ 140\ S_A\ 205\ L$$
$$x = 0.15 \quad G\ 60\ S_X\ 94\ S_C^*\ 110\ S_A\ 190\ L$$

Crosslinkable S_C^* LCP

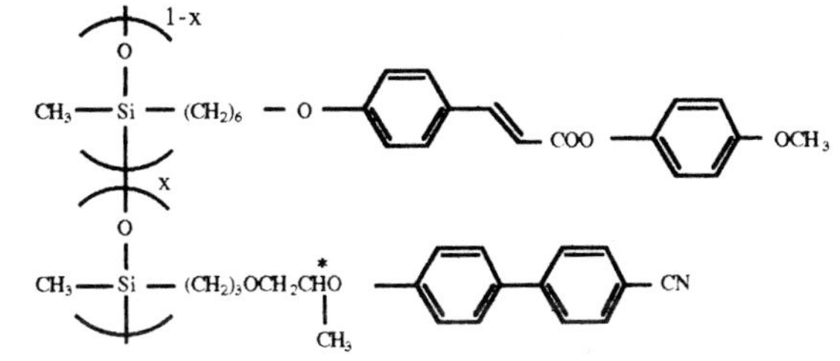

x	G		S_C^*		L
0.1	*	26.6	*	147.3	*
0.2	*	22.6	*	137.6	*
0.4	*	15.9	*	138.8	*
0.6	*	11.3	*	113.1	*
0.95	*	22	*		*

NLO S_C^* LCP

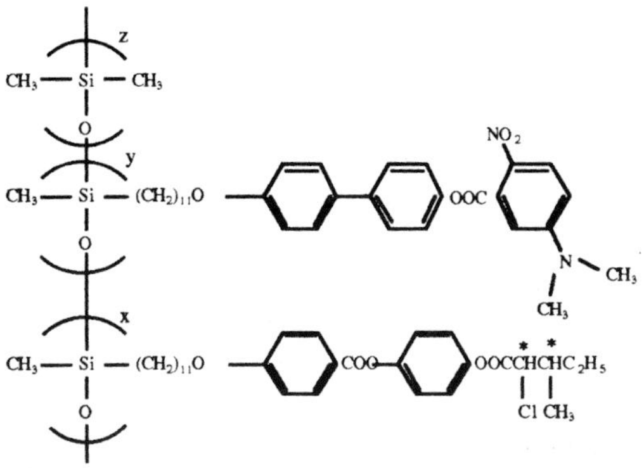

when $x/y = \dfrac{70}{30}$ and $\dfrac{x+y}{z} = \dfrac{1}{2.7}$:

$$G\ -11\ S_C^*\ 67\ S_A\ 80\ L$$

Figure 6. Examples of functionalized SmC* LCPs; S_C^*: SmC; S_X: SmX; S_A: SmA.

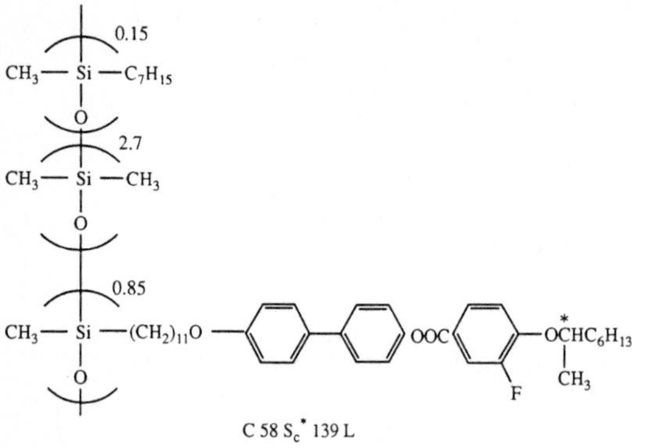

C 58 S$_c^*$ 139 L

Figure 7. Example of 'diluted' SmC* LCP.

is achieved by putting nonmesogenic side chains in place of mesogenic ones [37, 40, 44, 50] (Fig. 7).

Recently, microphase separation has been obtained with new SmC* ABA triblock co-polymers containing crystalline polytetra-hydrofuran (A) and chiral side chain LC (B) blocks [10].

2.1.3   Oligomers

Oligomers can be regarded as a class of ma-terials combining the reduced viscosity of low molecular weight SmC* liquid crystals with the existence of the glassy state at room

temperature of LCP. This newly studied class of SmC* material allows a better align-ment of the mesophases and hence makes their characterization and their use as elec-trooptic or NLO materials easier. Up to now, only cyclic [39, 43] and linear [9] oligosi-loxanes have been published.

2.1.4   Combined Polymers

Combined SmC* LCPs were particularly studied in the late 1980s, mainly by the Mainz group [58–61]. They all derive from a substituted polymalonate obtained by step-growth polymerization (Fig. 8). The occurrence of SmC* with broad temperature ranges was quite easily obtained with this class of compound. It seems that combined polymers are still prepared to obtain SmC* elastomers [62].

2.2   Phase Behavior

2.2.1   Means of Investigation

Basically, as for LMWLCs, structure deter-mination of the LCPs is based on thermal

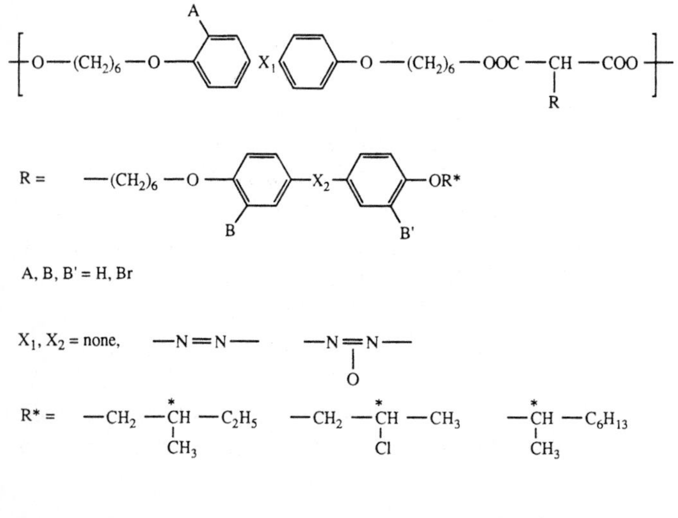

A, B, B' = H, Br

X$_1$, X$_2$ = none,   —N=N—   —N=N—
                                          |
                                          O

R* =   —CH$_2$—$\overset{*}{\text{CH}}$—C$_2$H$_5$ —CH$_2$—$\overset{*}{\text{CH}}$—CH$_3$ —$\overset{*}{\text{CH}}$—C$_6$H$_{13}$
              | | |
              CH$_3$ Cl   CH$_3$

Figure 8. The general formula of combined SmC* LCPs.

analysis, optical microscopy, and X-ray measurements. However, LCPs having a higher viscosity and higher transition temperatures are more difficult to study than LMWLCs. Due to the molar mass distribution, the phase transitions are broadened and, because of the high viscosity, the textures are very often noncharacteristic or poorly defined.

On the other hand, it has been shown on LMWLCs that the well-known SmC, where the molecules are tilted with respect to the layer normal, is no longer the only possibility to obtain a fluid biaxial phase [63]. As a consequence, a strict determination of the chiral smectic phase structure requires not only a careful analysis of the X-ray diagrams obtained on powder as well as on aligned samples, but also a study of the electrooptic response, which allows discrimination between the ferroelectric, the antiferroelectric, and the ferrielectric behavior.

2.2.2   Chemical Structure–Phase Behavior Relationships

When we want to establish detailed chemical structure–phase behavior relationships, we have to take into account many parameters, including:

– the nature of the polymer backbone,
– the nature and the length of the spacer,
– the way the spacer is linked to the mesogenic core (ether, ester),
– the structure of the mesogenic group (2 or 3 ring system, possible existence of linking groups and/or bulky lateral substituents),
– the way the tail is linked to the mesogenic core,
– the nature and the length of the tail,
– the number and position of the chiral centers, and

– the strength of the dipole near the chiral centers.

Due to the limited experimental data reported in the literature, only some general tendencies can be drawn:

• Below the SmC* phase, a more ordered one (generally nonidentified) is often encountered (it is then called SmX).
• Occurrence of the N* phase above the SmC* phase is rare. The phase sequence SmC*–SmA is the most probable.
• The spacer must be long enough to allow the SmC* to appear [13].
• Increasing the flexibility of the polymer backbone enhances the decoupling of the motions of the side and main chains and therefore tends to give rise to a higher thermal stability of the mesophases including the SmC* phase.

The last two trends are well exemplified by a comparison of the mesomorphic properties of two series of polysiloxanes and polymethacrylates having the same side chains. Polysiloxane III-2 exhibits a stable SmC* over a temperature range as wide as 239 °C, including room temperature [36, 38] (Tables 1 and 2).

2.2.3   Influence of the Molecular Weight

The molecular weight and the molecular weight dispersity of the LCPs were not regarded as relevant parameters in the mid 1980s. Later, only the molecular weights of the studied polymers were reported in the literature, sometimes without any specification concerning the type of average ($\bar{M}_w$ or $\bar{M}_n$). Now, in almost all the papers dealing with SmC* LCPs, the average molecular weight and the dispersity are specified. However, all the molecular weights are

Table 1. Phase transitions for polymers I and II.

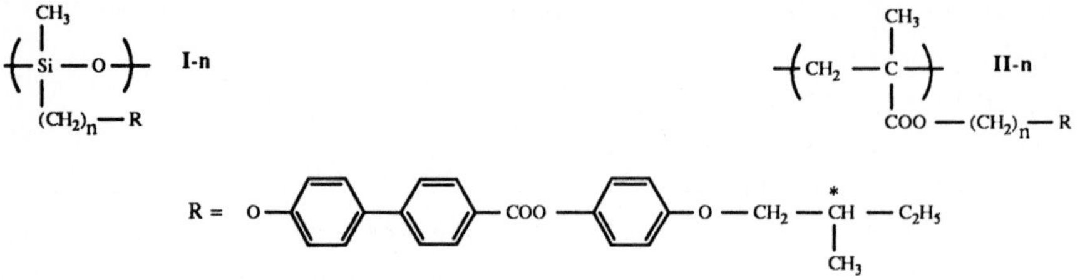

Polymer	Phase transitions	Polymer	Phase transitions
I-3	G 25 SmA 117 L	II-3	G 33 SmB 84 SmA 196 L
I-4	G 25 SmA 117 L		
I-5	G 22 SmB 95 SmA 212 L		
I-6	G 16 SmB 120 SmC* 116 SmA 244 L	II-6	G 31 SmB 131 SmA 181 L
I-11	G 20 SmB 109 SmC* 143 SmA 218 L	II-11	G 30.5 K 101 SmC* 132 SmA 197 L

Table 2. Phase transitions for polymers III and IV.

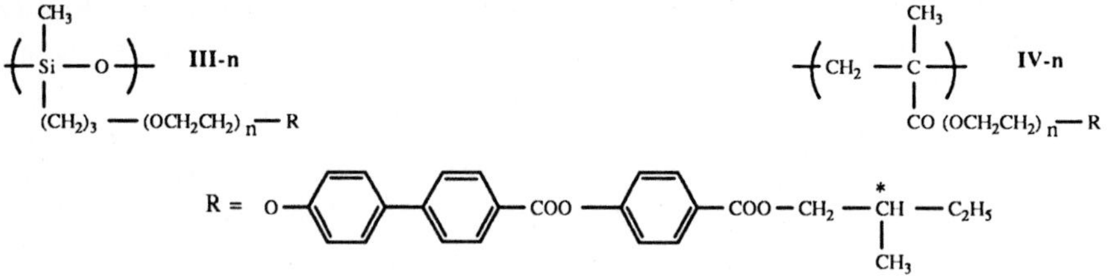

Polymer	Phase transitions	Polymer	Phase transitions
III-1	G 9.8 SmC* 215.2 SmA 234.6 L	IV-1	G 35.3 SmC* 159.4 L
III-2	G −11.2 SmC* 208.2 SmA 211.8 L	IV-2	G 17.8 SmA 130.6 L
III-3	G −25.1 SmC* 168 SmA 190.3 L	IV-3	G −7.7 SmC* 120 SmA 154.8 L

measured by GPC, using polystyrene as a standard, and hence, systematic errors due to the size of the side chains (which are usually >3 nm) can occur. This problem is strongly marked when the molecular weight of the polymers is very low, since its backbone becomes smaller than the side chains.

The influence of the molecular weight on the mesomorphic properties of SmC* polymers has been studied for a few examples [18, 34, 64, 65]. It turns out that the transition temperatures increase with the molecular weight, and in one case, a different phase succession occurs (Table 3).

2.2.4   Influence of the Dilution of the Mesogenic Groups

Most applications of the ferroelectric LCPs require an SmC* polymer having a low viscosity at room temperature. In this regard, copolysiloxanes derived from poly-(methylhydrogen-co-dimethyl)-siloxane, in which only part of the monomer units is functionalized with mesogenic groups, seem very promising. Although the stability of the SmC* phase decreases, the expected lowering of the glass transition temperature and of the viscosity are clearly observed

Table 3. Phase transitions for V-11 polymers as a function of their molecular weight.

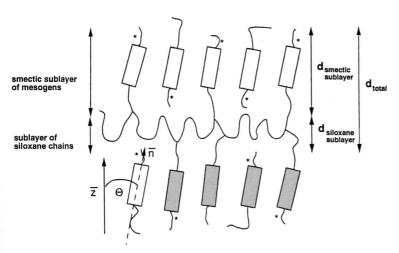

Polymers	$\bar{M}_w$ (g/mol)	$\bar{M}_w/\bar{M}_n$	Phase transitions
V-11-1	15 000	1.5	G 43 SmB$_1$ 89 SmC$^*_{1,x}$ 133 SmC$^*_{1,y}$ 144 L
V-11-2	38 000	2.44	G 40 SmF$^*_2$ 84 U$_2$ 123 SmC$^*_{1,x}$ 141 SmC$^*_{1,y}$ 158 L
V-11-3	240 000	3.07	G 40 SmF$_2$ 88 SmC$^*_2$ 142 SmC$^*_{1,x}$ 160 SmC$^*_{1,y}$ 166 L

[31, 40]. Mesomorphic properties are still recorded at mesogenic contents as low as 17%. This behavior is attributed to the microphase separation between the siloxane backbone and the mesogenic side chains due to immiscibility (see Fig. 9).

The addition of dimethylsiloxane units (dilution) leads to a further swelling of the polysiloxane sublayer without disturbing the preferential interactions between the mesogenic side chains too much, thus retaining the mesomorphic properties.

A variant of this approach is to prepare 'dimesogenic' polysiloxanes, in which two mesogens are linked to the same monomeric unit via a spacer (Fig. 10).

Compared to the corresponding mono-mesogenic copolysiloxane, the dimesogenic copolysiloxane depicted in Fig. 10 has nearly the same SmC*–SmA and SmA–isotropic transition temperature. However, due to an increase of the local concentration of mesogens, the more ordered smectic phase SmX occurs at a higher temperature.

2.2.5 Occurrence of Unusual Mesophases in Chiral Side Chain Polymers

A complex SmC* polymorphism has recently been found in a polyacrylate fam-

Figure 9. Microphase separated structure of a diluted polysiloxane in the SmC* phase according to [40].

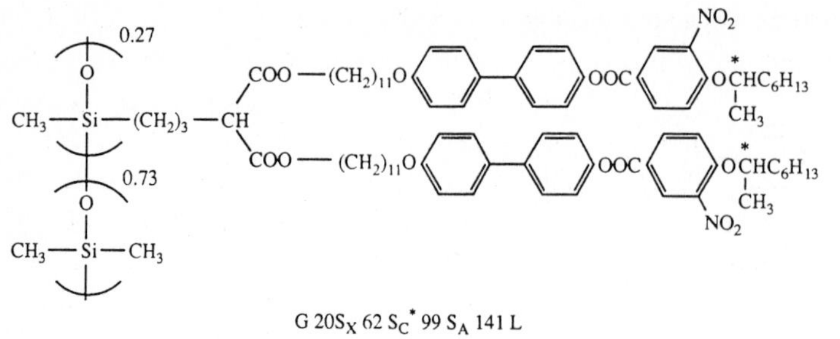

G 20S_X 62 S_C^* 99 S_A 141 L

Figure 10. Structure of the first synthesized SmC* dimesogenic polysiloxane.

ily containing an (R)-4″-(1-methylheptyl-oxy)-4′-biphenylyloxycarbonyl-4-phenoxy moiety in its side chain [45, 46, 64]. The mesomorphic properties of these polymers are reported in Table 4.

Some of the polymers studied show two SmC* phases of similar structure (identical X-ray pattern), but differ in their switching properties ($SmC_{1,x}^*$, $SmC_{1,y}^*$, $SmC_{2,x}^*$, $SmC_{2,y}^*$). In the high temperature phase, ferroelectric switching can be observed and spontaneous polarization can be measured by polarization reversal. In the SmC_x^* type phase, the situation is more complicated, since the behavior depends on the spacer length. When $n=6$ or 11, no ferroelectric switching can be detected, whereas ferroelectric switching is observed in electric fields higher than 25 V μm^{-1} for $n=8$. In the $SmC_{1,x}^*$ phase of polymer V-11-2, electroclinic switching is observed.

A proposed explanation of this unusual behavior, which has not yet been observed with the low molecular weight SmC* liquid crystals, involves:

– the broken tilt cone symmetry (due to the polymer backbone),
– the existence of conformational interaction between the side chains and the polymer backbone, leading to more or less favorable states, and
– a certain long range orientational correlation between different polymer chains.

In addition to these SmC_x^* and SmC_y^* phases, two new phases, called U_1 and U_2 by the authors, are also shown in the polymer V series (Tables 3 and 4). A careful X-ray analysis [46] has shown that U_1 and U_2 phases have a chevron-like ordering where the bilayers made of side-chain pairs are broken by periodic wall defects. More

Table 4. Molecular weights and phase transitions for polyacrylates V-n.

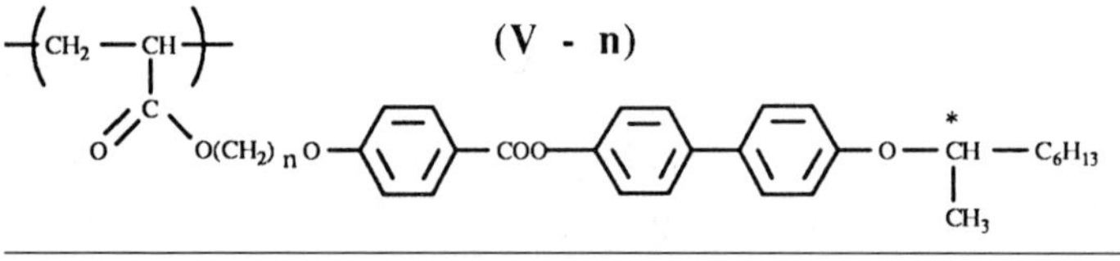

(V - n)

Polymer	n	$\bar{M}_w$ (g/mol)	$\bar{M}_w/\bar{M}_n$	Phase transitions
V-2	2	117 000	2.19	G 74 U_1 192 SmC_2^* 214 SmA_2 277 L
V-6	6	83 000	2.12	G 56 SmF_2^* 80 $SmC_{2,x}^*$ 148 $SmC_{2,y}^*$ 197 SmA_2 216 L
V-8	8	68 000	1.12	G 45 SmF_2^* 59 $SmC_{2,x}^*$ 131 $SmC_{2,y}^*$ 167 L
V-11-2	11	38 000	2.44	G 40 SmF_2^* 84 U_2 $SmC_{1,x}^*$ 141 $SmC_{2,y}^*$ 158 L

precisely, the U_2 phase is a 2-dimensional fluid phase in which the mesogenic side chains are tilted with respect to the normal layer. Nevertheless, the U_2 phase is not of exactly the same nature as the $\tilde{S}mC^*$ encountered in polar low molecular weight SmC* liquid crystals [66]. As a matter of fact, when chirality is left aside, the U_2 phase belongs to the 3-dimensional space group *Cmm2*, whereas the traditional SmC* has the 3-dimensional space group *P2/m*. The structure of the U_1 phase is more complicated; U_1 seems to be a 3-dimensional phase of rather low symmetry (space group *P2*), where a helical superstructure is very unlikely to be present.

2.3 Ferroelectric Properties of SmC* Polymers

The molecular origin of ferroelectricity in FLCs is attributed to a pronounced anisotropy of the angular orientations of the lateral dipole moments, induced by the tilt of the molecular long axes with respect to the normal of the smectic layers. This is supported by the results of broadband dielectric spectroscopy performed on a low molecular weight SmC* liquid crystal and a FLCP with a similar structure [67]. As a matter of fact, in the frequency range between 10^6 Hz and 10^9 Hz, one dielectric relaxation, called the β relaxation, occurs. This process, which is a local relaxation (characterized by a weak dielectric strength) with an Arrhenius-like temperature dependence, does not show any discontinuity in its temperature dependence or its relaxation time distribution at the SmA–SmC* phase transition. This behavior is consistent with a librational motion (hindered rotation) instead of a slowing down of the rotatory movement of the mesogen around its long axis at the SmA–SmC* transition, as was previously believed.

The only symmetry element existing in a monolayer of the SmC* phase is a two-fold axis of rotation C_2 which is parallel to the layer planes [68]. Thus a nonvanishing component of the polarization vector P_s exists in the direction of the C_2 axis (Fig. 11a).

Two directions are possible for the spontaneous polarization to act along. According to the convention of Clark and Lagerwall [69], the polarization is positive $[P_s(+)]$ when P_s, z and n form a right-handed system, and negative $[P_s(-)]$ when P_s, z, and n form a left-handed system (Fig. 11b).

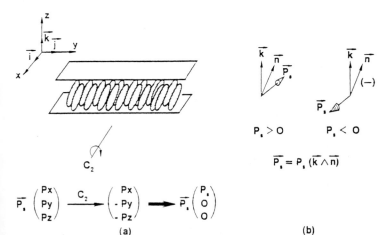

$$\overline{P_s}\begin{pmatrix} Px \\ Py \\ Pz \end{pmatrix} \xrightarrow{C_2} \begin{pmatrix} Px \\ -Py \\ -Pz \end{pmatrix} \Longrightarrow \overline{P_s}\begin{pmatrix} P_s \\ O \\ O \end{pmatrix}$$

(a)

$P_s > O$ $P_s < O$

$$\overline{P_s} = P_s (\overline{k} \wedge \overline{n})$$

(b)

Figure 11. (a) C2 symmetry operation applied to a single SmC* layer, and (b) sign convention for P_s.

However, due to the helical structure of the SmC* mesophase, no net spontaneous polarization results. Macroscopic nonvanishing polarization is only obtained when the helix is unwound. This requirement has been a severe limitation for the determination of P_s, and we had to wait until 1988 for the first observation of ferroelectric switching in a side chain SmC* LCP [65]. Even now, the measurement of P_s remains delicate and its reliability is directly connected to the quality of the uniform alignment achieved.

2.3.1   Uniform Alignment

The general procedure for generating uniform alignment is as follows: the sample under study is sandwiched between two ITO glass plates which are coated with polyimide or nylon and rubbed to favor a planar alignment. Very often, the test cells are filled with the polymer in an isotropic state, by capillary forces. Good planar alignment is achieved by repeated heating and cooling of the sample from the SmA (or isotropic) to the SmC* phase while applying an AC electric field of low frequency (several hertz). Sometimes the polymeric film is sheared in the SmC* phase to promote the expected alignment. The cell gaps are generally in the range of a few micrometers [2].

2.3.2   Experimental Methods for Measuring the Spontaneous Polarization

Several methods exist for measuring the magnitude of the spontaneous polarization [70], but the most widely used is certainly the electric reversal of the polarization obtained when a triangular voltage is applied to the cell. The polarization current is measured as a voltage drop across a series resistance of several kiloohms. This technique requires the applied voltage be sufficiently high and its frequency sufficiently low (lower than the inverse of the response time of the FLCP) in order to obtain saturation of the measured value of P_s. On the other hand, the presence of ionic impurities in the FLCP may disturb and even prevent the P_s measurement. Finally, because of the anchoring effect of the alignment layers, the P_s value is underestimated when the cell thickness is insufficient [14].

2.3.3   Spontaneous Polarization Behavior of FLCPs (Ferroelectric Liquid Crystal Polymers)

Ferroelectric LCPs obey the general rules known for low molecular weight SmC* liquid crystals, namely, that the larger the transverse dipole and the smaller its distance from the chiral center, the higher the spontaneous polarization P_s. Due to a slight increase in the rotational hindrance of the mesogens around their long axes as a result of their link to the polymer backbone, the values of the spontaneous polarization for the FLCPs tend to be higher when compared to the corresponding low molecular weight SmC* liquid crystals [45]. The reported value of $P_s = 420$ nC/cm^2 (at $T = 151$ °C) for polymer VI (Fig. 12) is very likely the highest value so far achieved with FLCPs [42].

The influence of the polymer molecular weight on the spontaneous polarization has been studied for a polyacrylate [21] as well as for a polysiloxane series [18]. The conclusions are somewhat different. Polyacrylates V-11 (Table 3) show a decrease of their spontaneous polarization as their molecular weight increases (Fig. 13a), whereas polysiloxanes VII (Table 5) exhibit a little de-

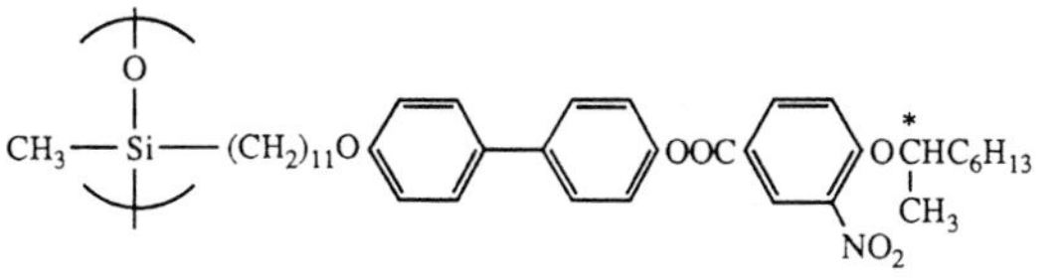

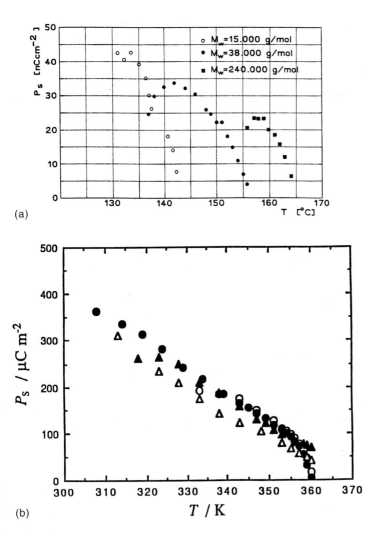

G 21 K 57 S_C^* 161 S_A 183 L

Figure 12. Chemical structure of polysiloxane VI having a very high P_s [42]; K: Cr.

(a)

(b)

Figure 13. (a) Temperature dependence of the spontaneous polarization for samples of polyacrylates V-11 of different molecular weight [21]. (b) Temperature dependence of P_s of polysiloxanes VII [18]. $\bigcirc$ $\bar{M}_w = 49\,300$; $\bullet$ $\bar{M}_w = 12\,700$; $\triangle$ $\bar{M}_w = 8000$; $\blacktriangle$ $\bar{M}_w = 5600$.

pendence of their P_s on molecular weight (Fig. 13b). A possible explanation of these results could lie in the difference of the viscosity of the samples, leading to some differences in the quality of the obtained uniform alignment required to measure the P_s.

Further investigations seem necessary to clarify this point.

Study of the variation of the spacer length in polyacrylates V-n has shown that the spontaneous polarization increases when the spacer is shortened [21]. This could be

Table 5. Molecular weights and phase transitions for polysiloxanes VII.

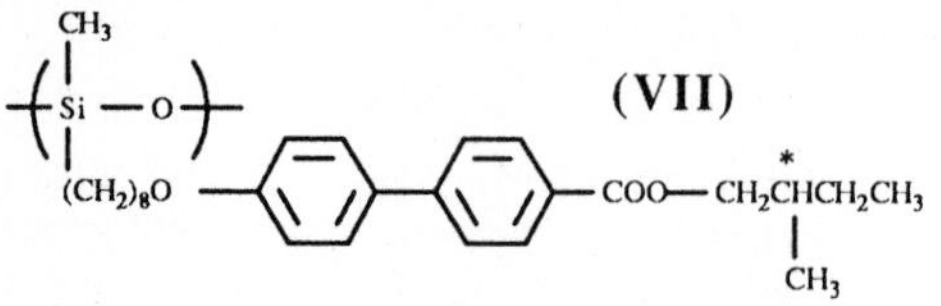

Polymer	$\bar{M}_w$ (g/mol)	$\bar{M}_w/\bar{M}_n$	Transition temperatures (°C)
VII-1	5 600	1.08	G 42 [a] SmC* 90 SmA 110 L
VII-2	8 000	1.16	G 30 SmC* 89 SmA 120 L
VII-3	12 700	1.19	G 30 SmC* 86 SmA 122 L
VII-4	49 300	1.82	G 50 SmC* 88 SmA 132 L

[a] SmX–SmC* transition temperature, SmX: high order smectic phase.

consistent with a decrease of the side chains mobility.

The introduction of a second chiral center of the same absolute configuration R into the spacer of polymer V-11 causes an important shift of the SmC* phase of about 75 °C to lower temperatures and a decrease by half of the spontaneous polarization (Fig. 14). To prove whether there is an effect of partial compensation between the two asymmetric centers, with respect to the polarization, the configuration of the chiral center in the spacer was changed from R to S. It turned out that the ferroelectric properties were completely lost [21].

Cooling down from the SmC* to a more ordered phase should lead to an increase of the spontaneous polarization. This behavior has not been observed in several homopolymers [64, 65] (Fig. 13) nor in mixtures of a racemic SmC polyacrylate with a chiral low molecular weight dopant [54]. The recorded decrease in P_s could be due to an increase of the stiffness of the polymer.

The phenomenon of sign reversal of P_s, which has already been demonstrated by Goodby et al. in a low molecular weight SmC* liquid crystal [71], can also occur in an SmC* polymer [21]. Polymer IX is the first example of an FLCP having such a behavior (Fig. 15). This phenomenon is attributed to a temperature-dependent equilibrium between different conformers of the chiral side chains, which have opposite sign of P_s.

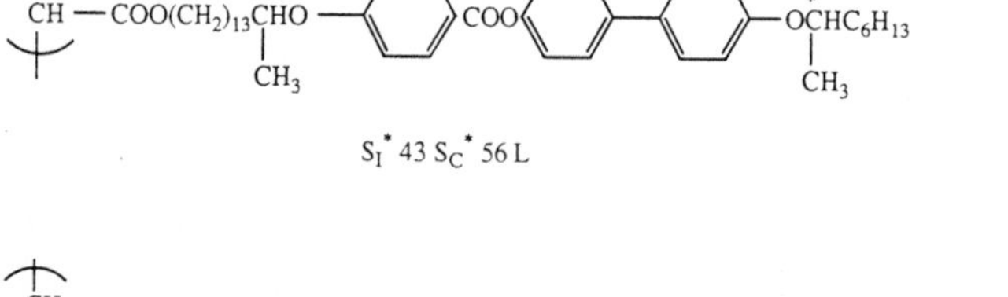

Figure 14. FLC polymer VIII with two remote asymmetric carbons in its side chain; S_I^*: SmI*.

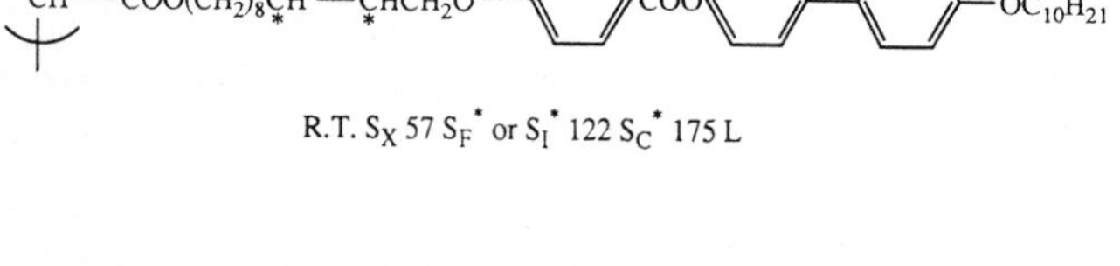

Figure 15. FLC polyacrylate exhibiting a sign reversal of P_s phenomenon at around 132°C; S_F^*: SmF*.

Concerning the 'diluted' polysiloxanes, it appears that the spontaneous polarization is not directly proportional to the chiral mesogen content in one case [31] or to the volume percentage of mesogen in another case [40]. The same tendency is also observed in mixtures of the racemic polyacrylate X with a chiral dopant (Fig. 16) [54].

Finally, close to the SmC*–SmA transition, a voltage dependence of the spontaneous polarization has been observed for the diluted polysiloxane XI [40]. For small electric fields ($3-5$ V$_{pp}$ µm^{-1}), P_s goes to zero at the ferro-to-para electric transition temperature, as expected. But, at higher applied fields ($15-20$ V$_{pp}$ µm^{-1}), a spontaneous polarization can also be measured well above the phase transition in the SmA phase (Fig. 17).

Two interpretations of this effect are suggested by the authors: either a shift in the phase transition temperatures to higher

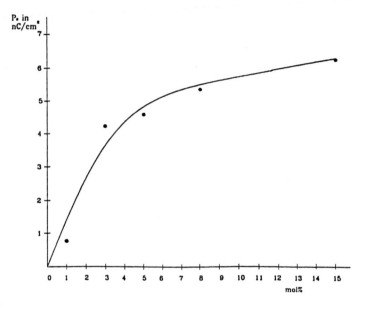

Figure 16. Dependence of the maximum of the P_s of binary mixtures of X and D-1.

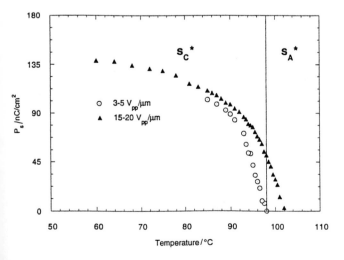

Figure 17. Temperature dependence of the spontaneous polarization for copolymer XI for E field of $3-5$ V$_{pp}$ µm^{-1} and $15-20$ V$_{pp}$ µm^{-1}.

temperatures (phase induction), or a combination of the electroclinic and the cone switching mechanisms allowing the observation of P_s.

2.3.4 Tilt Angle and Spontaneous Polarization

The absolute values of the spontaneous polarization P_s and the tilt angle θ as a function of temperature are well fitted to the following power law equations

$$\theta = \theta_0(T_c - T)^{\alpha'} \qquad (1)$$

$$P_s = P_0(T_c - T)^{\alpha''} \qquad (2)$$

Up to now, α' and α'' have only been determined for the two polysiloxanes XII and XIII, whose structures are depicted in Fig. 18 [12, 14].

As shown in Table 6, α' and α'' are respectively of the same order of magnitude for polymers XII and XIII. It has to be pointed out that in the case of low molecular weight SmC* liquid crystals, both α' and α''

are found to be significantly lower than 0.5 [72, 73].

With α' being different from α'' for both polymers XII and XIII, the relationship between P_s and θ is nonlinear. Such behavior is typical of ferroelectric liquid crystal materials with high P_s, and can be explained on the basis of the generalized Landau model for the free energy density. A complete treatment is available for polymers XII and XIII and the different calculated coefficients.

Depending on the $T_c - T$ value, two different approximate linear relationships between P_s and θ can be derived. When T is close to T_c

$$\frac{P_s}{\theta} \sim \varepsilon C \qquad (3)$$

when $T \ll T_c$

$$\frac{P_s}{\theta} \sim \left(\frac{\Omega}{\eta}\right)^{1/2} \qquad (4)$$

where Ω and C are respectively the coefficients of biquadratic and piezoelectric coupling between P_s and θ, ε is the high frequency permittivity (here, $\varepsilon = 2.7 \times 10^{11}\ C^2\ N\ m^{-2}$), and η is a coefficient introduced to stabilize the system. Ω and C have been found to be significantly lower for copolymer XIII than for homopolymer XII (Table 6).

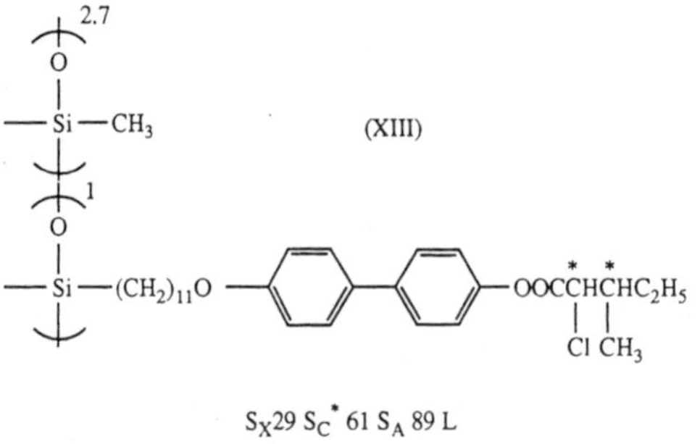

S_X 68 S_C^* 88 S_A 156 L

S_X 29 S_C^* 61 S_A 89 L

Figure 18. Structures of polymers XII and XIII.

Table 6. Exponents of the power law equations fitting the tilt angle and the spontaneous polarization of polysiloxanes XII and XIII, as well as their piezoelectric (C) and biquadratic (Ω) coefficients.

Polymer	α'	α''	C (V/m)	Ω (m/F)
XII	0.35	0.5	6.5×10^7	5.3×10^{11}
XIII	0.35	0.48	3.6×10^7	2.0×10^{11}

2.4 Electrooptic Behavior

Knowledge of the electrooptic behavior of the FLCPs is of the utmost importance for display device applications. One relevant parameter in this respect is the response time. As for the spontaneous polarization, the determination of the response time requires a uniformly aligned sample. The test cell is placed between crossed polarizers so that one tilt direction is parallel to the direction of one polarizer. The electrooptic effect is achieved by applying an external electric field across the cell, which switches the side chains from one tilt direction to the other as the field is reversed. A photodiode measures the attenuation of a laser beam when the cell is switched between the two states. Generally, the electrooptical response time is defined as the time corresponding to a change in the light intensity from 10 to 90% when the polarity of the applied field is reversed (t_{10-90}).

Several switching processes have been identified for FLCPs:

- ferroelectric switching, corresponding to the rotation of the side chains around the cone angle (the Goldstone mode),
- electroclinic switching, where a change of tilt angle θ is observed (the soft mode), and
- antiferroelectric switching.

2.4.1 Ferroelectric Switching

Ferroelectric switching is the most likely process encountered in FLCPs. A solution of the equation of motion of the SmC* director, when the dielectric anisotropy is supposed to be negligible, leads to the introduction of a switching time τ which is related to the rotational viscosity γ and the spontaneous polarization P_s by the following relation [74]

$$\tau = \frac{\gamma}{P_s E} \qquad (5)$$

where E is the external electric field.

As there is no appropriate method to measure directly the rotational viscosity of ferroelectric liquid crystals, γ is generally deduced from the electrooptic response time measurements [12, 18, 44]. The relationship between t_{10-90} and τ is not straightforward and requires the use of a theoretical model for the optical transmission based on the 'bookshelf geometry' briefly summarized in the following.

The transmission T of the SmC* cell placed between crossed polarizers is given as

$$T = \frac{I}{I_0} = \sin^2[2(\Omega - \theta)]\sin^2\left(\frac{\Pi d}{\lambda}\Delta n\right) \qquad (6)$$

where Ω is the angle between the polarizer and the layer normal and θ is the apparent tilt angle expressed by (see Fig. 19)

$$\theta = \tan^{-1}(\tan\theta_0 \cos\varphi) \qquad (7)$$

Δn is the effective optical anisotropy, given by

$$\Delta n = \frac{n_e n_o}{(n_o \sin^2\phi + n_e \cos^2\phi)^{0.5}} - n_o \qquad (8)$$

where ϕ is the angle between the light ray direction and the director

$$\cos\phi = \sin\theta_0 \sin\varphi \qquad (9)$$

T is related to τ through the azimuthal angle φ

$$\varphi = 2\tan^{-1}\left[\tan\left(\frac{\varphi_0}{2}\right)\exp\left(\frac{t}{\tau}\right)\right] \qquad (10)$$

where $\varphi_0 = \varphi(t=0)$ and φ, θ and ϕ are time-dependent.

Fitting the experimental optical response and Eq. (6) allows the determination of φ_0 and the rotational viscosity γ provided P_s is known.

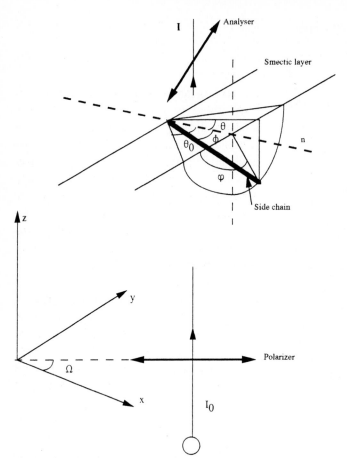

Figure 19. Definition of the different angular parameters involved in ferroelectric switching.

Using this approach, Endo et al. [18] established that the rotational viscosity of the SmC* polysiloxanes VII (Table 5) was approximately proportional to the second power of $\bar{M}_w$ at 60 °C

$$\gamma = \gamma_0 \bar{M}_w^2 \qquad (11)$$

Typical response times of SmC* FLCPs are of the order of a few milliseconds to a few hundred milliseconds, depending (through the viscosity) on the temperature, the chemical nature of the polymer backbone, and the molecular weight of the polymer. The switching time τ increases with the molecular weight and decreases as the temperature increases. On the other hand, the response time measured at the same reduced

temperature decreases with the flexibility of the polymer backbone in the order

$$\gamma_{polysiloxane} < \gamma_{polyoxyethylene} < \gamma_{polyacrylate} \ [7]$$

Fast switching times are also observed when the spacer length increases, thus decreasing the mesogen-backbone coupling [31, 75].

Practical applications of FLCPs require the retention of fast response times at room temperature. Attempts to reduce the response time by increasing the spontaneous polarization failed, since introduction of strong lateral dipole moments in the side chains increases not only P_s but also the viscosity, the overall result being an increase in τ. Hence the following materials have been developed to reduce the viscosity of the FLCPs: 'diluted' polysiloxanes [40, 75], binary mixtures of FLCPs and low molecular weight liquid crystals [31], binary mixtures of SmC LCPs and chiral dopants [24], and SmC* oligosiloxanes [39].

Response times as short as 400 µs at 130 °C (7.5 V_{pp}/µm) have been obtained with an oligosiloxane [39] and, on the other hand, a mixture of diluted polysiloxane with a low molecular weight liquid crystal has led to a response time of 0.5 ms at 40 °C [31].

2.4.2   Electroclinic Switching

The electroclinic effect is an induced molecular tilt observed in the chiral orthogonal smectic phases, such as the smectic A* phase, when an electric field is applied along the smectic layers [76]. The induced molecular tilt θ is a linear function of the applied field E and gives rise to an induced polarization P_i

$$\theta = \frac{\mu E}{\alpha(T - T_c)} \qquad (12)$$

where μ is a structure coefficient, α is the first constant in the Landau free energy expansion, and T_c is the SmX–SmA* transition temperature.

In the linear regime, the electroclinic effect is characterized by a fast field-independent response time τ given by

$$\tau = \frac{\gamma_\theta}{\alpha(T - T_c)} \qquad (13)$$

with γ_θ being the soft mode viscosity.

Electroclinic behavior has been recognized in few SmA* LCPs [77] and response times in the sub-millisecond range have been observed. But, surprisingly, the $SmC^*_{1,x}$ phase of polyacrylate V-11 exhibits an electroclinic switching process with a short response time (200 µs) that is changed into a ferroelectric one as the temperature is raised. Moreover, the electroclinic–ferroelectric transition is shifted towards lower temperatures when the voltage is increased (Fig. 20).

SmC* polyacrylate XIV, having a chiral center in its spacer (Fig. 21), also shows, at

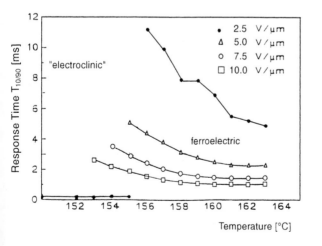

Figure 20. Switching behavior of polyacrylate V-11.

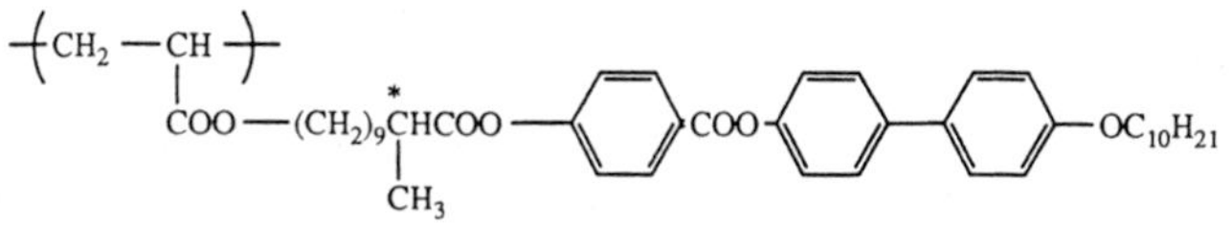

Figure 21. Structure of polyacrylate XIV.

first, electroclinic switching between 70 and 95 °C, but as the temperature increases above 95 °C, the switching depends on the applied voltage. At low voltage ($E = 7$ V/µm), an electroclinic process takes place, whilst at higher voltages ($E = 18$ V/µm), mixed switching is observed. Finally, at high voltages ($E = 25$ V/µm), ferroelectric switching results. Up to now, this behavior has not been explained.

2.4.3   Antiferroelectric Switching

Antiferroelectricity [49, 77–79] was observed in FLCPs only two years after its dis-

covery in low molecular weight SmC* liquid crystals [80]. Figure 22 shows the chemical structure of the tristable FLCPs that have been studied up to now.

The apparent tilt angle of polymer XVI as a function of the applied voltage is depicted in Fig. 23.

The main features of the antiferroelectric switching in FLCPs are: a third state, which shows an apparent tilt angle of zero, a less marked threshold between the three states when compared to the low molecular weight antiferroelectric liquid crystals, a hardly observed hysteresis, and an anomalous behavior of the spontaneous polarization with temperature (Fig. 24), which is not encoun-

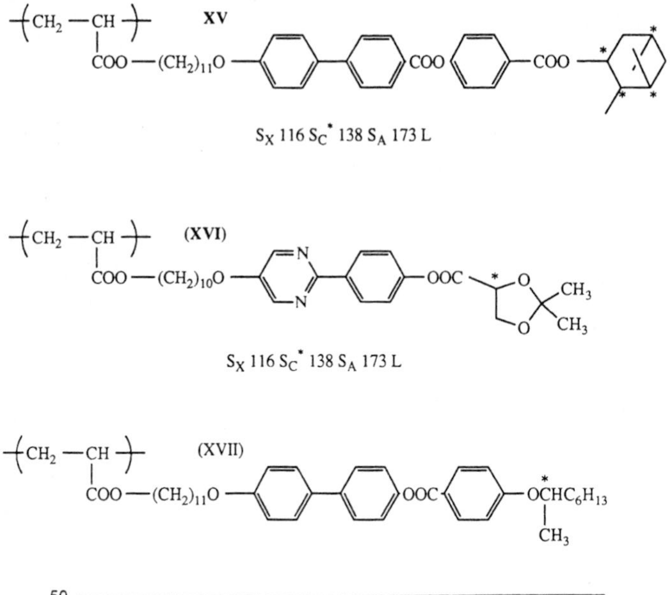

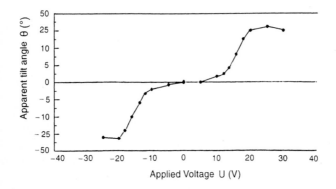

Figure 22. Structures of LCPs exhibiting three states switching (XV–XVII).

Figure 23. Apparent tilt angle of polymer XVI as a function of the applied voltage.

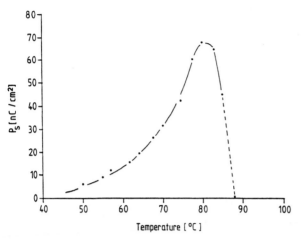

Figure 24. Spontaneous polarization versus temperature for polymer XVI.

tered in low molecular weight antiferroelectric liquid crystals. The existence of a threshold for switching between the antiferroelectric and the ferroelectric states could be of interest in display device applications.

2.4.4 Broadband Dielectric Spectroscopy

Additional information on ferroelectric and electroclinic switching can be obtained with broadband dielectric spectroscopy. It appears that the molecular dynamics of FLCPs are comparable to those of low molecular weight compounds [67]. However, the experimental observations are made more difficult for FLCPs than for low molecular weight SmC* liquid crystals due to the conductivity contribution which takes place at frequencies below 10^4 Hz and to the difficulty to get a macroscopically well-aligned sample.

Nevertheless, several authors, in studying SmC* polyacrylates [22] or SmC* polysiloxanes [14, 41, 67], have observed the two expected collective relaxations in ferroelectric liquid crystals, namely the Goldstone mode and the soft mode. These two relaxations occur at frequencies lower than 10^6 Hz.

2.5 Potential Applications of FLCPs

Chapter V of this Handbook is entirely devoted to the potential applications of the side chain liquid crystal polymers, but it is interesting to mention here the main areas where FLCPs could play a role. As far as we know, the following applications can be considered for SmC* LCPs: nonlinear optics, pyroelectric detectors, and display devices.

2.5.1 Nonlinear Optics (NLO)

Several FLCPs have been designed for NLO [16, 23, 43]. Very low efficiency has been obtained. This point is discussed in Sec. 4.

2.5.2 Pyroelectric Detectors

There is a real need for efficient pyroelectric materials for uncooled infrared detectors. At the moment, low-cost infrared cameras are made with poly(vinylidene fluoride-*co*-trifluoroethylene) [P(VF$_2$-TrFE)] as the pyroelectric material. To take the place of P(VF$_2$-TrFE), FLCPs have to exhibit a better figure of merit than P(VF$_2$-TrFE). The relevant figure of merit for this application being

$$\frac{p}{\sqrt{\varepsilon\,\mathrm{tg}\,\delta}} \qquad (14)$$

with p the pyroelectric coefficient, ε the permittivity, and $\mathrm{tg}\,\delta$ the loss angle.

Comparison of P(VF$_2$-TrFE) with polymer XII [12] (Fig. 25), which is an FLCP having one of the best pyroelectric coefficients, is made in Table 7. It turns out that the figure of merit of polymer XII is an order of magnitude lower than that of P(VF$_2$-TrFE). Three improvements are necessary to FLCPs to compete with P(VF$_2$-TrFe), namely:

– to have an SmC* phase at room temperature,
– to have a lower loss angle,

– and to exhibit a higher pyroelectric coefficient.

At the moment, the two first requirements seem possible to meet, but the last one is still questionable.

2.5.3   Display devices

This is certainly the most advanced potential application of FLCPs. Taking advantage of the processability of FLC polymers, Idemitsu Kosan Co. has fabricated a large-area, static-driven display (12×100 cm) and a dynamically driven simple matrix display (13×37 cm) [7, 19]. The performance of these two demonstrators is summarized in Table 8. The electrooptic material used was plasticized polyoxyethylene XVIII (Fig. 26).

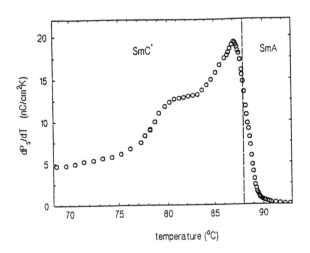

Figure 25. Evolution of the pyroelectric coefficient of polymer XII with temperature.

Table 7. Comparison of the pyroelectric coefficient and the figure of merit for copo(VF2-TrF) and polymer XII.

Polymer	p (nC/cm^2 K)	ε	$\mathrm{tg}\,\delta$	$p/(\varepsilon\,\mathrm{tg}\,\delta)^{0.5}$
Copo(VF$_2$-TrFe) [a]	5	7	0.02	13.36
XII [b]	5	60 [c]	0.2	1.44

[a] At room temperature;   [b] at $T=70$°C;   [c] value deduced from [12].

Table 8. Characteristics of Idemitsu's demonstrators.

Driving type	Static	Dynamic
Size (cm)	12×100	12×37
Pixels		96×288
Pulse width (ms)[a]	5	3
Applied voltage (V)	±20	±30
Viewing angle (deg.)		
(LR)	70	70
(UD)	60	60

[a] Room temperature.

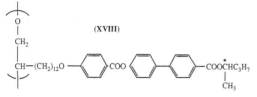

Figure 26. Structure of polyoxyethylene XVIII used by Idemitsu to make their demonstrators.

The displays were fabricated in three steps: first the FLCP was deposited onto an ITO coated polymeric substrate, then the counter electrode (another ITO-coated substrate) was put onto the FLCP film, and the formed sandwich was laminated. Finally, the laminated three-layer film was treated by a bending process to obtain the required alignment of the side chains (Fig. 27). These

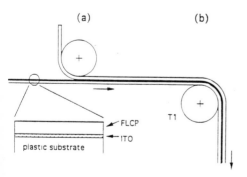

Figure 27. Fabrication of the FLCP display device. (a) An FLCP-coated substrate is laminated with another ITO polymeric substrate. (b) The laminated sandwich is bent on a temperature controlled roll.

results are very promising and we can expect in the very near future the advent of large flexible displays with subsecond rate images.

3 Side Chain Liquid Crystalline Networks and Mechanical Properties

Crosslinking liquid crystalline polymers yield materials with exceptional properties due to the coupling between elastic properties and mesomorphous behavior. These compounds, especially the mesogenic elastomers, have attracted considerable attention in recent years. Like conventional elastomers, they can sustain very large deformations causing molecular extension and orientation, but they can also exhibit spontaneous distortions, some memory effects, an unusual mechanical response, and coupling between mechanical, optical and electric fields.

3.1 Theoretical Approaches

3.1.1 Landau–de Gennes Description of Nematic Elastomers

De Gennes [81, 82] first suggested that remarkable effects, such as mechanical critical points, shifts in the phase transition temperature, jumps in the stress–strain relation, and a spontaneous shape change, would take place in nematic elastomers. His proposal was based on symmetry arguments independent of any microscopic consideration.

According to this model, the free energy expansion, in the case of uniaxial nematic order and incompressible strain of the sample, takes the form [81–83]

$$F(S, e) = F_0(S) - USe + 1/2Ee^2 - \sigma e \quad (15)$$

where S is the order parameter, e the deformation, E the elastic modulus, and σ the stress (force per unit area). $F_0(S)$ represents the nematic energy in the absence of any network effect and $-USe$ the coupling of strain to nematic order.

Without mechanical stress ($\sigma=0$), spontaneous deformation must occur in the nematic phase because of the linear coupling between S and e. Minimizing F with respect to e gives the equilibrium value of $e_m = USl/E$. Inserting this value and the form for F_0, the above equation takes the following form

$$\quad (16)$$

$$F(S, e_m) = \frac{AS^2}{2} + \frac{BS^3}{3} + \frac{CS^2}{4} + \ldots - \frac{U^2 S^2}{2E}$$

With an applied stress, an additional deformation arises and the minimum condition yields the strain $e_m = (\sigma + US)/E$. Neglecting terms in σ^2, the shift in F away from F_0 becomes

$$\Delta F = -\frac{U^2 S^2}{2E} - \frac{\sigma US}{E} \quad (17)$$

The formalism of the influence of the mechanical field on S is similar to the one found with an electric or a magnetic field [84]. Principal results include a shift in the transition temperature [83] and the occurrence of a mechanical critical point. Experimental verification [85] is found by plotting the order parameter as a function of temperature for different nominal stresses.

3.1.2   Models for Nematic Rubber Elasticity

Modifications of the conventional Gaussian elastomer theory were proposed that include short range nematic interactions. These an-

isotropic interactions cause a change in the chain conformation, and hence increase the chain radius of gyration [86–88].

The elastic free energy is written as a function of the probability, $P(R)$, of finding two crosslinks separated by a vector R. If the chains are crosslinked in the isotropic state, the probability P_0, corresponding to the undistorted R_0, is an isotropic Gaussian. In the nematic phase, the system is not able to adopt its natural dimensions and undergoes a bulk deformation λ assumed to be affine ($R = \lambda R_0$). The additional elastic free energy of a strand with length R is given by

$$F_{el} = -k_B T \langle \ln P(R) \rangle P_0 \quad (18)$$

If $l_\parallel$ and $l_\perp$ are the effective step lengths (monomer length) for the random walks parallel and perpendicular to the ordering direction (in the isotropic phase $l_\parallel = l_\perp = l_0$), respectively, and under the assumption that the deformation occurs without a change of volume, the following relation is obtained for the elastic component of the free energy in the nematic phase

$$\frac{F_{el}(\lambda)}{k_B T} = \frac{1}{2} \left(\frac{\lambda^2 l_0}{l_\parallel} + \frac{2l_0}{\lambda l_\perp} \right) - \frac{1}{2} \ln \left(\frac{l_0^3}{l_\parallel l_\perp^2} \right) \quad (19)$$

Warner et al. [88, 89] give a full description of the free energy and recover, by minimization, the spontaneous strain and the mechanical critical point. They also show that, if the network is crosslinked in the nematic phase, a memory of the nematic state is chemically locked. This causes a rise in the nematic–isotropic phase transition temperature compared with the uncrosslinked equivalent. After crosslinking in the isotropic state, the transition temperature (on the contrary) is lowered.

The same idea of nematic interactions favoring the distortion of the chains is also the basis of phenomenological descriptions. Jarry and Monnerie [86] and Deloche and

Samulski [90, 91] described these interactions in the mean field approximation by an additional intermolecular potential from the classical theory of rubber elasticity. A similar expression is proposed for the elastic free energy.

3.2   Characterization of Networks

3.2.1   General Features

Liquid-crystalline networks can be obtained in several ways, either from an isotropic middle or from materials in which the mesogenic groups were first macroscopically aligned prior to the final crosslinking (see Chap. V of this Volume). The latter networks should keep a complete memory of the original orientation and recover it even after been held well above the clearing temperature for extended periods [92, 93].

As long as the crosslinking density is low, the liquid-crystalline phases of the corresponding uncrosslinked polymer are retained for the networks [94–97], with the same structure and without large shifts of the transition temperatures (see Chap. V of this Volume). As the crosslinking density increases, the smectic phases should disappear (Table 9) for the benefit of a less-ordered nematic state [98]; then all the liquid-crystalline phases should be destroyed for the most crosslinked networks, at least for materials crosslinked in an isotropic state [99–101]. Degert et al. [98] pointed out that this evolution in the mesophase stability also depends on the nature (mesogenic or aliphatic) of the crosslinks.

Table 9. Phase transition as a function of the cross-linking density x^a.

$x = b/(a+b)$

	Phase transition temperature (°C)		
x [%]	SmA	N	L
0	• 149		•
5	• 95	• 128	•
10		• 127	•
15		• 116	•

[a]   Molar %, determined from the conditions of synthesis.

3.2.2   Effective Crosslinking Density

In order to compare theories with experiments, the number of effective crosslinks should be determined. Actually, this parameter cannot easily be analyzed, because in any one elastomer, the ratio of linkages that do not contribute to an infinite network is difficult to evaluate. In addition, mesogenic groups induce sterical hindrance, which should modify the chemical reactivity of the netpoints as well as the theories referring to ideal rubbers that are used for the determination of the effective crosslinking density.

Spectroscopic techniques (for example, infrared or solid-state ^{13}C NMR) [102, 103] allow control, with more or less accuracy, of the chemical crosslinking reaction. Greater efficiency of the crosslinking process for materials crosslinked in the isotropic phase over those crosslinked in a mesophase was reported [101, 103], particularly in the case of a smectic phase. This fact can be related to the lack of spacially available sites for crosslinking in the mesophase. Furthermore, the efficiency increases with the length of crosslinking agent [103], which confirms that spacial considerations are important in the crosslinking process.

The effective crosslinking density, i.e., the ratio of linkages taking part in the realization of a network of 'infinite' molecular weight, is strongly dependent on the conditions of the crosslinking reaction. When the reaction is performed in bulk (isotropic or anisotropic state), sterical hindrances play an important role. When the crosslinking reaction takes place in solution, Degert et al. showed that the polymer concentration has a major effect [98]: due to the short length of the backbone (degree of polymerization generally around 100 or 200), complete gel formation needs high concentrations of the polymer in the solvent (see Table 10). At low concentrations, intrachain linkages are favored to the detriment of intermolecular reactions, and the proportion of the former compared with the latter can become very large. Under the concentration corresponding to the gel point, the formation of an 'infinite' network is not achieved, even if the potential branching points are much more than two per chain. Consequently, in this case, the samples either remain soluble or contain extractibles or exhibit abnormal high degree of swelling [98, 100]. These networks should display nonreversible behavior when they are stretched [122].

Table 10. Characteristics of a crosslinked polymer synthesized at various concentrations in the solvent [98].

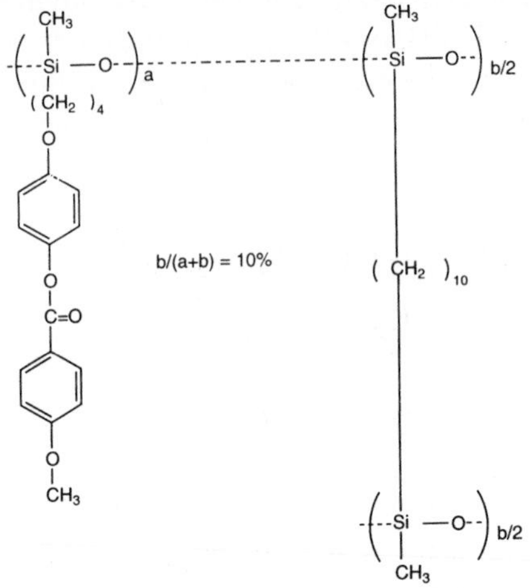

Polymer concentration in the reaction bath		Consistency of the final product	Weight swelling ratio in toluene[a] ±0.3
mmoles of units/cm^3 of toluene	Weight %		
0.24	8	soluble	
0.87	23.5	gel	4.4
1.2	29	gel	4.0
1.6	35	gel	3.6
1.8	37.5	gel	3.6

[a] Weight ratio of the swollen network to the dry network.

Swelling experiments and stress–strain measurements are widely used to estimate the crosslinking density in liquid-crystalline networks [101, 106–108]. For ideal rubbers with tetrafunctional crosslinks, these techniques allow the evaluation of the number average molecular weight between crosslinks M_c [104, 105].

From swelling experiments, Flory theory gives, for high degrees of swelling and polymer chains that are long compared to M_c, the following relation (neglecting the weight of the linkages)

$$M_c = d\,V\,(0.5 - X)^{-1}\,q^{5/3} \tag{20}$$

where d is the density of the network in the unswollen state, V the molar volume of the swelling solvent, and q the ratio of the volumes of the equilibrium swollen network and of the unswollen network. The Flory–Huggins X parameter can be obtained from the second virial coefficient (from light-scattering experiments, for example).

This theory refers to ideal rubbers. The additional entropic component, introduced by the presence of the mesogenic cores [91], is not taken into account, and so the absolute values of M_c are not relevant.

From mechanical measurements, classical rubber elasticity theory gives, for low uniaxial deformations and polymer chains that are long compared to M_c, the following relation [104]

$$\sigma = \frac{d R T \lambda}{M_c} \tag{21}$$

where R is the universal gas constant, σ the stress, T the absolute temperature, and λ the deformation ($\lambda = l/l_0$, l and l_0 being the lengths of the sample after and before deformation, respectively).

As above, this relation, predicted for perfectly formed networks, is not absolutely convenient for mesomorphous elastomers,

Table 11. Average molecular weight $M_c{}^a$ of the network subchains in two networks [b].

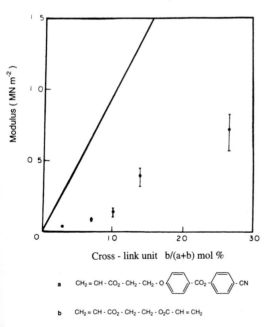

a : x : y : z = 0.96 : 0 : 0.04

b : x : y : z = 0.94 : 0.02 : 0.04

$R_1 = H_{13}C_6O$—⬡—CO_2—⬡—OCO—⬡—OC_6H_{13} with $(CH_2)_{11}$

$R_2 = CN$—⬡—CO_2—⬡—OCO—⬡—CN with $(CH_2)_{11}$

$R_3 =$—— OC—NH—⬡—CH_2—⬡—NH—CO ——

Network	M_c theoretical value	M_c value from swelling experiment	M_c value from mechanical experiment
a	18 300	118 000	70 000
b	18 300	86 000	48 000

[a] In g/mol; [b] [107].

Figure 28. Elastic modulus of liquid crystalline networks obtained from the polymerization of monomers a and b as the function of the composition. The solid line corresponds to the value predicted by Eq. (21) [99].

a $CH_2 = CH \cdot CO_2 \cdot CH_2 \cdot CH_2 \cdot O$—⬡—$CO_2$—⬡—$CN$

b $CH_2 = CH \cdot CO_2 \cdot CH_2 \cdot CH_2 \cdot O_2C \cdot CH = CH_2$

Modulus (MN m^{-2})

Cross - link unit b/(a+b) mol %

even in the isotropic state. Moreover, networks should not be too highly crosslinked if a connection with a Gaussian description is to be sought [83]: If the subchains between links are too short, then they are not flexible enough to be treated by a classical theory. Actually, significant deviations from linearity for σ versus the crosslink density (estimated from the conditions of synthesis) (Fig. 28) were found [99].

Due to the imperfections of the theoretical approaches, the value of M_c obtained from swelling and mechanical experiments widely differ from one another (Tables 11 and 12) and from the values determined either by the conditions of synthesis or by spectroscopy [106–108].

On the other hand, the proportion of permanent entanglements cannot be estimated. The degree of polymerization above which entanglements should appear is unknown for side chain liquid crystal polymers and, because of the coupling of mesomorphous order, the entanglement slippage has to be regarded as dependent on the direction [95].

Lastly, the question of homogeneity in the liquid-crystalline networks remains open. The distribution of the linkages in these materials, which are obtained in bulk or in concentrated solution, would be worth further investigation, especially when the nature of the crosslinking agent is essentially incompatible with the liquid-crystalline organization. Some neutron investigations of networks swollen with deuterated solvent are in progress on different types of elastomers, differing from one another in the nature of crosslinks and in the conditions of synthesis [109]. These studies would help, as was done for another network [110], to analyze the microstructure of liquid-crystalline networks.

3.3   Mechanical Field Effects on Liquid-Crystalline Networks

In the isotropic state, as in classical networks, the elastic behavior is governed by

Table 12. Crosslinking density (cd) of various elastomers (in % repeat units) [a,b].

$CH_2 = CH-CO_2-(CH_2)_2-O-\langle \bigcirc \rangle-CO_2-\langle \bigcirc \rangle-CN$ I

$CH_2 = CH-CO_2-(CH_2)_3-O-\langle \bigcirc \rangle-CO_2-\langle \bigcirc \rangle-CN$ II

$CH_2 = CH-CO_2-(CH_2)_2-OH$ III

cd theoretical value III (%)	cd from IR analysis	cd from modulus [Eq. (21)]	cd from swelling [Eq. (20)]	T_g (°C) [c]	T_{NI} (°C) [d]
6	4.1	1.27±0.13	3.4±0.8	67	100
6	3.9	1.40±0.13	3.8±1.0	67	99
6	3.8	1.47±0.30	3.3±0.5	71	100
6	3.9	1.46±0.13	4.8±1.2	72	99
6	4.2	1.30±0.13	3.3±0.8	77	102
6	5.0	1.40±0.13	3.9±0.8	82	102

[a] [108]; [b] networks synthesized with variable proportions of I and II; III=6%; [c] T_g: glass transition temperature; [d] T_{NI}: nematic–isotropic transition.

the conformational properties of the network chains. The presence of the mesogenic groups should only modify the entropy of the system, and hence the extensibility of the polymer subchains.

In liquid-crystalline phases, the introduction of liquid-crystalline order is the decisive factor in the elastic response. The coupling between mechanical properties and mesomorphous order is the main theme of this chapter (for more details, see Chap. V of this Volume).

3.3.1   Response of the Network as a Function of the Stress

The potential of the mesogenic units for alignment has a marked effect on the stress–strain behavior [85, 111, 112]. Consider a uniaxial stress applied to a liquid-crystalline network synthesized in an isotropic state; this means without any macroscopic orientation.

At low strains, the induced strain increases linearly with the stress (Fig. 29). The polydomain structure remains stable and the X-ray pattern resembles that for powder samples.

Above a strain threshold, macroscopic orientation of the mesogens parallel or perpendicular to the direction of the stress results, as evidenced by X-ray analysis or IR dichroism measurements (see Sec. 3.3.3). Depending on the topology of the network and on the mesogenic units, the polydomain–monodomain transition is more or less sharp, and therefore the threshold deformation is more or less observable [108]. During this orientation, a remarkable change in the dimensions of the sample spontaneously occurs, which depends on the orientation of the mesogens with respect to the stress axis (for the example shown in Fig. 29, the mesogens are in line with the elongation stress axis).

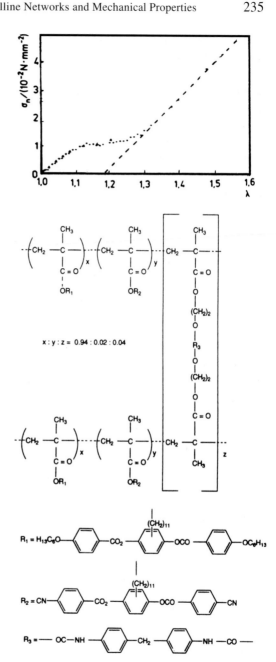

$x : y : z = 0.94 : 0.02 : 0.04$

Figure 29. Nominal stress σ_n versus strain ($\lambda = l/l_0$) for the nematic elastomer shown in [85].

At higher stresses, a linear relationship between the stress and the strain is regained. The sample is uniformly aligned and the orientation process no longer takes place.

3.3.2   Thermal Evolution of the Elastic Modulus

The rheological properties of liquid-crystalline elastomers strongly depend on the mesomorphous order [95, 98, 102, 113–118], even for strains that are too small to induce macroscopic orientation of the mesogens (strain amplitude <10%). This is mirrored by the thermal evolution of the storage modulus (Fig. 30), defined for very small strains. (Fig. 30 shows the change in the shear modulus [118]; the compression modulus evolution is similar [98, 101, 115].) At the glass transition, the elastic modulus falls, as is usual for amorphous rubbery materials. Then the modulus decreases slightly until the isotropic state is reacted. The smectic A–nematic phase transition induces a drop in the modulus at low frequencies [98, 101, 116, 118]. This phenomenon is less and less visible on increasing the fre-

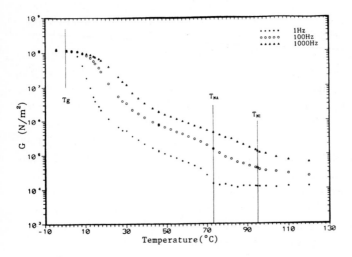

T_g: glass transition temperature; T_{NA}: smecticA - nematic transition; T_{NI}: nematic - isotropic transition

Figure 30. Thermal evolution at three frequencies of the dynamic shear elastic modulus G in an elastomer exhibiting a smectic phase and a nematic phase [118].

quency, which means that the effects associated with this transition have long characteristic times.

Samples that only possess a nematic phase exhibit, at low frequency, a sharper decrease of the modulus than the smectic networks [98, 113]. So the resistance to the deformation appears higher in the layered phase than in a nematic one. Apart from the type of mesophase, the modulus values and their thermal evolutions depend on the chemical nature of the backbone polymer and on the coupling with the mesogenic groups. Different structural parameters play a role in this last point, such as the cross-linking density [101], the proportion of mesogenic moieties [118], and the spacer length [115]. A wider range of materials should induce a more accurate description of these different effects and thus should help the theoretical analysis.

Anisotropic mechanical properties are exhibited by ordered, highly crosslinked polymers obtained by bulk polymerization of oriented monomers [117, 119]. These networks show a high degree of ordering and behave anisotropically in a number of physical properties, such as the refractive index, the thermal expansion coefficient, and the modulus of elasticity. Figure 31 shows the tensile moduli in the two major directions for uniaxially oriented sample, and also for an isotropic sample. As can be observed, the modulus in the direction of molecular orientation is much higher than in the perpendicular direction.

3.3.3   Mechanical Orientability of the Mesogens

The mechanical orientability is the most prominent property of liquid-crystalline elastomers. Above a threshold stress (Sec. 3.3.1), small strains (about 20%–40%) are

enough for the reversible formation of a macroscopically oriented sample [99, 107, 120–125]. This orientation is achieved under equilibrium conditions; it is not necessary to freeze-in an orientation induced by flow, as in uncrosslinked polymer. For most applications, liquid-crystalline phases must be aligned. Mesomorphous networks thus offer a simple way to do this under equilibrium conditions.

The strain-induced orientation is analyzed with the help of birefringence measurements [107, 115, 120, 123] (realized in the isotropic state, close to the transition), measurements of the IR dichroism [124, 126], and X-ray experiments [93, 108, 120, 122, 125]. It was observed that the orientational order parameter, labeled P_2 on Fig. 32, obtained from X-ray experiments [122], quickly increases with strain at the beginning, then saturates at a P2 value of about 0.4–0.6 for a sample synthesized in an isotropic state [99, 107, 121, 122]. On condition that they were realized above the gel point [122], the networks display reproducible and reversible behavior independent of the sample history.

The mechanical orientability of the mesogenic groups in networks is connected with the anisotropic conformation that the polymer adopts in the mesophase. It is clear from small-angle neutron scattering or X-ray studies of side chain liquid crystal polymers [127–131] that the backbone indeed exhibits some ordering as a consequence of the orientation of the mesogenic groups. This ordering is much larger for polymers having smectic phases than for those with nematic ones. In elastomers, the previous effect leads, as theoretically described, to a coupling between the stress-induced directions of the polymer chains and the mesogenic groups. The direction of orientation of the mesogenic groups versus the axis of the stress comes from the preferred position of

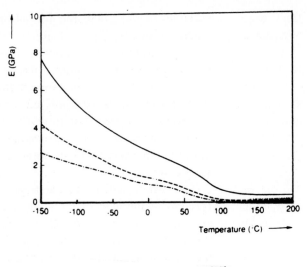

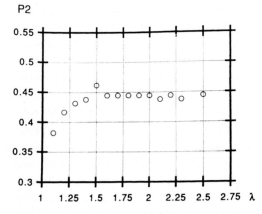

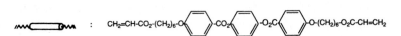

Figure 31. Tensile modulus E at 1 Hz for a network oriented during the synthesis; ─── molecules oriented in the direction of the strain; ─·─·─ molecules perpendicular to the strain; ─ ─ ─ isotropic [109].

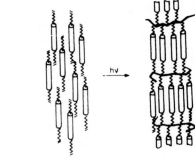

Figure 32. Orientational order parameter, P_2, versus strain ($\lambda = l/l_0$) in the smectic A phase at room temperature for the sample represented in Fig. 3 [122].

the backbone in an oriented liquid crystalline polymer.

For networks that exhibit a smectic phase, mechanical elongation always causes orientation of the director perpendicular to the axis of the stress [122, 125], as shown on Fig. 33. The mesogenic groups therefore become perpendicular to the polymer main chain if we make the reasonable assumption that the polymer chain is extended preferentially in the extension direction. Such an arrangement allows the polymer backbone to occupy the space between the layers.

For samples that only show a nematic phase, both parallel and perpendicular

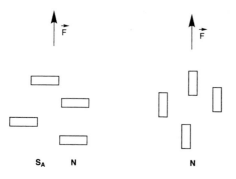

Figure 33. Schematic representation of the orientation of the mesogens in a deformed network as a function of the elongation axis and of the nature of the mesophase.

alignments of the mesogenic units with respect to the stress direction (Fig. 33) have been observed [107, 108, 111, 120–122, 125]. These results confirm the approach described by Warner [83] and Wang and Warner [132], in that a variety of types of nematic phases can exist dependent on the coupling between the side chain and the backbone. Hence the orientation (parallel or perpendicular versus the direction of the deformation) of the mesogens in a deformed nematic network depends on several factors, such as the nature of the polymer backbone [100, 112, 125] or the spacer length [108, 120].

If a smectic phase exists at lower temperatures, a smectic-like short-range order should also influence the orientation in the nematic phase and should induce a perpendicular alignment of the mesogens [122]. On the other hand, when the linkages themselves are mesogenic, the behavior becomes more complicated, because the stress could align these mesogenic linkages in its direction and perhaps the whole nematic field too in a uniform way, whatever the relative natural orientation of the side chains and backbones is. Degert et al. [122] observed the two opposite orientations in two nematic elastomers that only differ from one another

in the nature (aliphatic or mesogenic) of the crosslinking agent.

For an experimental check-up of the theoretical considerations about liquid-crystalline elastomers in a mechanical field, Finkelmann and coworkers [107, 123] studied, in nematic networks, the evolution of the order parameter and of the transition temperature as a function of the stress. The observed results are in full agreement with the predictions of the Landau–de Gennes theory, since an increasing clearing temperature as well as an increasing order parameter are observed with increasing stress. From their results, it was possible to estimate the cross-coupling coefficient U (see Sec. 3.1.1) between the order parameter and the strain of a nematic elastomer [123].

On the other hand, from random copolymers formed by units leading to opposite arrangements in the nematic phase, Guo et al. [108] recently proposed a molecular analysis of the coupling between the side groups and backbone in nematic polymers or in nematic elastomers. Using a main-field model derived from the description of Warner [83], they gave a quantitative estimation of the relative intensity of the two components of the coupling defined as 'nematic-like interactions' between the side chain and the polymer chain (inducing a parallel orientation), and as 'chemical hinge effects' (forcing, for example, the mesogens to lie perpendicular to the backbone).

Networks realized in a mesophase exhibit specific behavior: It was found that crosslinking in a liquid-crystalline phase results in an increase in the stability of this phase [92, 103, 133], which is in agreement with the theoretical predictions (Sec. 3.1.2). This stabilization was noted as considerably greater in the case of a smectic phase than for a nematic one [103]. Coupling between mesogenic order and the main chain results in a preferred conformation arrangement of the polymer in the mesophase (parallel or perpendicular with locally aligned mesogens). In a layered phase, these constraints are more drastic than in a nematic phase. So the enhanced stability of the mesophase, with regard to networks prepared in an isotropic state, arises from the additional elastic energy required, at the phase transition, to distort the network from its initial crosslinked state to an isotropic coil, which is more costly energetically in the presence of a smectic phase than in the presence of a nematic state. In addition to these considerations, Kupfer and Finkelmann [126] introduced in this analysis the influence of the orientational distribution of the crosslinks.

When the liquid-crystalline phase is ordered (for example, by applying a magnetic field on the monomer or the intermediate polymer [93, 119], or by applying a mechanical stress on a 'soft' intermediate elastomer [92]) prior to the final crosslinking, the polymer chain exhibits macroscopic alignment which is locked-in by the crosslinking process. Any deviation away from this configuration involves a distortion of the network for which the required energy depends on the temperature and on the crosslinking density. The so-synthesized elastomers show very good orientation of the mesogenic groups expressed by a high transparency, and a remarkable memory of the original director orientation [92, 93, 119]. They recover this orientation in the liquid-crystalline state, with the same order parameter, even after being heated for a long time in the isotropic state [93]. On the other hand, Kupfer and Finkelmann [92] showed, in the case of mechanically oriented networks obtained by a two-step crosslinking process, that these anisotropic networks are susceptible to a reorientation process by mechanical stress. The efficiency of this reorientation of the mesogenic groups depends on the relative directions and intensities of the orientation and reorientation mechanical fields. In the same way, Hikmet and Higgins [134] proved that, in lightly crosslinked materials, the mesogenic groups could be reoriented in the presence of electric fields, reverting back to the initial orientation on removal of the field at a rate comparable with those observed in the monomeric state.

The coupling of network elasticity to the liquid-crystalline phase thus induces unusual behavior, such as mechanically created optical properties or temperature-induced memory.

3.3.4   Electromechanical Properties

The properties described in the previous section apply to networks exhibiting helical structures. This type of mesophase is obtained by introducing side mesogenic substituents with chiral end groups on the backbone [94, 138–141, 145–147], or by doping an achiral network with a chiral molecule [144, 145]. As a result of the interaction between orientational (liquid-crystalline) and translational (network) degrees of freedom, these noncentrosymmetric systems exhibit rich electromechanical effects, such as piezoelectricity.

As was previously described for achiral elastomers, in chiral networks, the mesogenic groups can be mechanically oriented up to a 'threshold deformation', which defines the deformation needed by the sample in order to obtain a macroscopically aligned system [140]. As a consequence, above this threshold the helical axis becomes aligned in a direction that depends on the orientation of the mesogenic units versus the strain. For fairly large elongations (several hundred percent), it has been shown that it is possible to untwist reversibly the helical superstructure and thus to transform the chiral structure into the corresponding achiral structure [138–139].

Theoretical investigations by Brand [135] and Brand and Pleiner [136] predicted that a monodomain liquid-crystalline elastomer exhibiting a cholesteric or a chiral smectic C phase should display piezoelectric properties due to a modification of the pitch of the helix under strain. So, a piezoelectric voltage should be observed across the sample when a mechanical field is applied parallel to the helicoidal axis. In this description, the crosslinking density is supposed to be weak enough to allow the motion of the director, and deformations of the sample (compression, elongation, etc.) are assumed to be much smaller than those that should lead to a suppression of the helix. The possibility of a piezoelectric effect do not only concern cholesteric and chiral smectic C phases, but was also theoretically outlined for more exotic chiral layered systems such as chiral smectic A mesophases [137].

Preceding the reports on elastomers, piezoelectricity in chiral smectic C phases of low-molar weight molecules or of polymers has usually been observed. The special property is that the system possesses macroscopic electrical polarization without an external field, so it is classified as ferroelectric.

Experimentally, piezoelectricity in cholesteric and chiral smectic C phases was reported for liquid-crystalline networks [140–147]. Multidomain lightly crosslinked systems were synthesized, then the orientation is obtained by mechanical strain [140] or by poling [147]. In other samples this orientation is performed prior to the crosslinking process [144, 146]. Macroscopically oriented samples were subjected to either a static or a periodically varying strain. Open circuit voltages across the samples were measured that are linear functions of the applied strain [140–142, 144, 145].

The sign and magnitude of the piezoelectric constant were largely dependent on the direction of the applied strain [144]. In the direction parallel to the helical axis, where the effect is maximum, the piezoelectric constant was found to be about $2.5–5$ pC N^{-1} [144, 146, 147], depending on the nature of the sample. The piezosignal does not vanish after annealing at temperatures that are even much higher than the Curie point, and invariably reappears on cooling by a memory effect for one polar state [146, 147].

In their theoretical study, Brand [135] and Brand and Pleiner [136] discussed the other electromechanical effects that should occur in chiral elastomers, such as electrostriction, flexoelectricity, and pyroelectricity. Pyroelectricity and its analogs, which are closely related to piezoelectricity, should be present in these systems and have been reported in chiral smectic C elastomers [147]. On the other hand, the above-mentioned linear relation between the piezoelectric voltage and the intensity of the compression in both cholesteric and chiral smectic C elastomers implies that the electrostrictive contribution is fairly weak. In addition, Meier and Finkelmann [145] showed that changing the handedness of the cholesteric helix changes the sign of the voltage, which proves that flexoelectricity is negligible [135, 136]. In fact, it seems that in perfect alignment (i.e., a cholesteric or chiral smectic monodomain with its helical axis exactly parallel to the compression axis), a quasi-pure piezoelectrical effect is observed. In the case of nonperfect alignment, however, any applied stress can also bend the layers and gives rise to flexoelectric contributions. Lastly, for smallest compressions in cholesteric phases, the question of possible electroclinic-like effects remains open [136]; initially, cholesteric domains with different orientations of the helix exist in the sample, and the compression could first induce a reorientation of these domains without any compression of the layers. After a quasi-monodomain is reached, the piezoelectric effect should start.

From these theoretical and experimental works, it emerges that both chiral smectic C and cholesteric elastomers are piezoelectric and can lead to a piezoelectric voltage comparable to that of classical piezoelectric crystals, such as quartz. Thus they can be used potentially as piezoelectric elements, which can be produced in any shape needed.

4   Nonlinear Optical Properties

4.1   Introduction

Asymmetric charge-transfer organic molecules containing charge correlated π electrons have shown nonlinear coefficients larger than those found in organic crystals [148, 149]. Especially in the case of frequency doubling, organic materials appeared to be more effective than the traditional inorganic crystals, owing to the purely electronic character of the nonlinear effects in organic materials, permitting sizeable effects in the optical frequency range. However, organic crystals are difficult to grow and not easy to process. Polymers are increasingly being recognized as the materials of the future because of their ability to provide thin films of high optical quality and excellent mechanical properties, which can be produced rapidly at a reduced cost and can be applied over very large areas and over various substrates. By attaching sufficient optically nonlinear chromophores to a polymer backbone, large nonlinear coefficients will be achieved if the material can be properly poled by a DC electric field [150–154].

This may be accomplished by heating the film above the glass transition temperature (T_g) of the polymer, then cooling the film below T_g in the presence of the external fied. The versatility of synthetic chemistry can be used to modify and optimize the molecular structure in order to maximize the nonlinear optical (NLO) response and other properties (processing, mechanical, optical, etc.).

Recently, it has been recognized that the axial ordering present in nematic and smectic A polymers can be used to enhance field-induced polar ordering by increasing the orientational distribution function along the electric field direction [155–159]. Depending on the values of the microscopic order parameters $\langle P_2 \rangle$ and $\langle P_4 \rangle$, the performance may be improved by a factor of $1-4$ by using liquid-crystalline polymers (LCPs) instead of ordinary amorphous polymers for second harmonic generation (SHG).

This section is devoted to liquid-crystalline polymers that may find applications in frequency doubling or in electrooptic modulation and in waveguide configuration. The structural principles underlying the design of LC systems with large second-order nonlinearity will be reviewed. As new materials are most easily evaluated by SHG measurements, the theoretical basis of the nonlinearity will be introduced in this context. Nevertheless, the material parameters for SHG and the electrooptic effect are closely related. Electric field induced ordering will be treated using theoretical models.

4.2   Theoretical Approach

4.2.1   Second Harmonic Generation (SHG)

When a medium is subjected to an external field, which may be an externally applied potential, the electric field of a light beam, or a combination of these, the polarization can be expressed in a power series of the field strength E in terms of the dipole approximation

$$P = P_0 + \chi^{(1)} \cdot E + \chi^{(2)} \cdot EE$$
$$+ \chi^{(3)} \cdot EEE + \ldots \qquad (22)$$

where P_0 is the permanent polarization of the medium, $\chi^{(1)}$ is the linear susceptibility, and the higher order χ_s are the NLO susceptibilities. $\chi^{(1)}$ gives rise to linear optical processes including the propagation of light in dielectric media, absorption, and surface reflection. Nonlinear processes arise from the higher order terms. The coefficient $\chi^{(2)}$, relating the polarization to the square of the field strength E is called the second-order nonlinear susceptibility of the medium and is a third-rank tensor. Its magnitude describes the strength of second-order processes. It is worth noting that, because of their symmetry properties, odd-ranked tensors such as $\chi^{(2)}$ are nonzero only in non-centrosymmetric systems.

The manifestation of NLO behavior can easily be seen by substituting a sinusoidal field equation into the polarization expansion equation [Eq. (22)]. This gives

$$P = P_0 + \chi^{(1)} E_0 \cos(\omega t - kz)$$
$$+ \frac{1}{2} \chi^{(2)} E_0^2 [1 + \cos(2\omega t - 2kz)]$$
$$+ \chi^{(3)} E_0^3 \left[\frac{3}{4} \cos(\omega t - kz) \right.$$
$$\left. + \frac{1}{4} \cos(3\omega t - 3kz) \right] \qquad (23)$$

where ω is the angular frequency and t is the time.

Equation (23) clearly shows the presence of new frequency components due to the nonlinear polarization. The second-order term gives a frequency independent contribution as well as one at 2ω. The former sug-

gests that a DC polarization should appear in a second-order nonlinear material when it is appropriately irradiated. This phenomenon is referred to as optical rectification. The latter term corresponds to the best known and highly utilized effects in NLO, i.e., SHG [150, 159, 160]. In this section we will illustrate this process in side chain liquid crystalline polymers with the emphasis on identifying and understanding the relevant physical phenomena that are occurring.

In analogy to Eq. (22), the polarization per molecule can be expanded in terms of the molecular nonlinear response in a power series of the field strength E

$$\qquad (24)$$
$$\mu = \mu_0 + \alpha \cdot E + \beta \cdot EE + \gamma \cdot EEE + \ldots$$

where μ_0 is the molecular ground state dipole moment, α the linear polarizability, and β and γ the two lowest-order molecular NLO susceptibilities or hyperpolarizabilities. Also, here the coefficients α, β, γ, etc. are represented by tensors. Similarly as in the macroscopic case, a symmetry constraint exists on the molecular level with respect to the occurrence of even-order NLO effects. For the first hyperpolarizability β not to be equal to zero, no centrosymmetry is allowed in the molecule.

Neglecting dispersion and assuming a one-dimensional molecule, where β_{zzz} is the only nonvanishing component, the relationship(s) between macroscopic and molecular second-order NLO properties in polymer films of $C_\infty v$ or ∞mm symmetry reduces to the following two relations

$$\chi_{ZZZ}^{(2)} = N F \beta_{zzz} \langle \cos^3 \theta \rangle \qquad (25)$$
$$\chi_{XXZ}^{(2)} = \chi_{YYZ}^{(2)} = \chi_{XZX}^{(2)} = \chi_{YZY}^{(2)} = \chi_{ZXX}^{(2)}$$
$$= \chi_{ZYY}^{(2)} = N F \beta_{zzz} \langle 0.5 \cos\theta \sin^2 \theta \rangle \, (26)$$

where the indices X, Y, Z denote the laboratory frame (Z or 3-unit vector along the electric poling field) and xyz the molecular frame (z along the permanent dipole moment). N is the number density of active molecules, F the local field correction factor, and θ the angle between the permanent dipole moment and the poling field. The expressions in angle brackets describe the degree of polar order obtained during the poling process.

4.2.2 Theoretical Models for the Electric Field Poling

In addition to the polar order produced by the electric field, E_p, axial order may be present due to either intrinsic ordering (e.g., in liquid-crystalline polymers) or mechanically induced ordering (e.g., uniaxial elongation). For rod-like molecules with dominant dipole and hyperpolarizability directed along the molecular axis, i.e. μ_z and β_{zzz} only, in uniaxial systems, the orientational distribution function $G(\theta)$ can be expanded in terms of Legendre polynomials

$$G(\theta) = \sum_{n=0}^{\infty} \frac{(2n+1)}{2} \langle P_n(\cos\theta) \rangle P_n(\cos\theta) \tag{27}$$

where the expansion coefficients $\langle P_n(\cos\theta) \rangle = \langle P_n \rangle$ are given by

$$\langle P_n \rangle = 2\pi \int_0^\pi G(\theta) P_n(\cos\theta) \sin\theta \, d\theta \tag{28}$$

These coefficients, known as microscopic order parameters in the liquid crystal literature, are convenient quantities to define the degree of order. The even-order terms give the degree of axial ordering. Of these,

$$\langle P_2 \rangle = \frac{\langle 3\cos^2\theta - 1 \rangle}{2} \tag{29}$$

and

$$\langle P_4 \rangle = \frac{\langle 35\cos^4\theta - 30\cos^2\theta + 3 \rangle}{8} \tag{30}$$

are the most used axial order parameters. The odd-order terms characterize the field-induced polar order. The terms $\langle \cos^3\theta \rangle$ and $\langle 0.5\cos\theta \sin^2\theta \rangle$ in Eqs. (25) and (26) are linear combinations of $\langle P_1 \rangle$ and $\langle P_3 \rangle$.

In Boltzmann statistics, the orientational distribution function is given by

$$G(\theta) = \frac{e^{-\frac{U}{k_B T}}}{2\pi \int_0^\pi e^{-\frac{U}{k_B T}} \sin\theta \, d\theta} \tag{31}$$

where k_B is the Boltzmann constant, T the absolute poling temperature, and U the potential energy of the molecule.

For the calculation of order parameters, four statistical models have been developed:

1) The isotropic model [155, 161] (no initial axial order, $\langle P_2 \rangle = 0$ at $E_p = 0$).
2) The Ising model [155, 161] (perfect axial order, $\langle P_2 \rangle = 1$ without a field).
3) The Singer–Kuzyk–Sohn (SKS) model [156].
4) The self-consistent Maier–Saupe model, as extended by Van der Vorst and Picken [157, 158] (MSVP model).

The main difference between the four models is in the choice of the expression for the molecular energy. The dominant energy term, taken into account by all four models, is the energy of the permanent dipole moment μ_0 in the electric field E_p

$$U_{E_p}(\theta) = -\mu_0 E_p \cos\theta \tag{32}$$

Singer et al. [156] consider the intrinsic axial order of a liquid crystalline host, which is transferred to NLO guest molecules, without introducing an extra energy term. The MSVP model [157, 158] takes the intrinsic mesogenic properties of the NLO molecules themselves into account. The tendency of such strongly anisotropic, rodlike molecules to have their long axes in mutually parallel alignment is described by the energy term

$$U_0(\theta) = -\xi \langle P_2 \rangle P_2(\cos\theta) \qquad (33)$$

If this tendency is large enough to overcome thermal motion, a liquid-crystalline phase may result with spontaneous (at $E_p=0$) axial order. The expression in the MSVP model for $U_0(\theta)$ was used originally by Maier and Saupe [162] in their theory for the axial order in the nematic phase and for the nematic–isotropic phase transition. The parameter ξ scales the magnitude of the mutual interactions. The Maier–Saupe theory predicts that the temperature at which the nematic–isotropic transition takes place, T_C ('clearing' temperature in the absence of a field), is related to the parameter ξ through $k_B T_C \cong 0.22\,\xi$ [162]. The energy term connected with the linearly induced dipole moment in the poling field is taken into account only in the MSVP model [157, 158].

The expressions for $\langle\cos^3\theta\rangle$, $\langle 0.5\cos\theta\sin^2\theta\rangle$, and $\langle P_2\rangle$ for the various models can be found in Table 13.

4.2.3 The Electrooptic Effect

The changes in refractive index produced by an electric field E are given by

$$\delta n_o = -\frac{n_o^3\, r_{13}\, E_3}{2} \qquad (34)$$

$$\delta n_e = -\frac{n_e^3\, r_{33}\, E_3}{2} \qquad (35)$$

where n_o and n_e are the ordinary and extraordinary indices for light polarization perpendicular and parallel to the poling field direction (Z direction), and r_{13} and r_{33} are the electrooptic or Pockels coefficients. The susceptibilities $\chi^{(2)}$ are related to the electrooptic coefficients by

$$\chi^{(2)}_{113} = -\frac{n_o^4\, r_{13}}{2} \qquad (36)$$

$$\chi^{(2)}_{333} = -\frac{n_e^4\, r_{33}}{2} \qquad (37)$$

Combination of Eqs. (25), (26), (36), and (37) yields the relations between the electrooptic coefficients and the molecular parameters

$$r_{13} = -\frac{2\,N\,F\,\beta_{zzz}\,\langle 0.5\cos\theta\sin^2\theta\rangle}{n_o^4} \qquad (38)$$

$$r_{33} = -\frac{2\,N\,F\,\beta_{zzz}\,\langle\cos^3\theta\rangle}{n_e^4} \qquad (39)$$

From Eqs. (25), (26), (36), and (37) it is also seen that the isotropic model predicts

Table 13. Expressions for $\langle\cos^3\theta\rangle$, $\langle 0.5\cos\theta\sin^2\theta\rangle$, and $\langle P_2\rangle$ in the four models.

Model	$\langle\cos^3\theta\rangle$	$\langle 0.5\cos\theta\sin^2\theta\rangle$	$\langle P_2\rangle$
Isotropic	$u/5$ [a]	$u/15$	$\approx u^2/15$
Ising	u	0	1
SKS	$u(1/5+4/7\langle P_2\rangle+8/35\langle P_4\rangle)$	$u(1/15+1/21\langle P_2\rangle-8/70\langle P_4\rangle)$	[b]
MSVP	No analytical formulae [c]		[c]

[a] $u=\mu_0 E_p/(k_B T)$ where μ_0 is the ground state permanent dipole moment, E_p the poling field strength, and T the poling temperature. [b] In the SKS model, $\langle P_2\rangle$ and $\langle P_4\rangle$ are input parameters which must be determined experimentally at zero field strength. [c] The MSVP model does not provide analytical expressions. First, a self-consistent value of $\langle P_2\rangle$ satisfying the Eqs. (28) and (31) must be calculated by numerical methods. After substituting the $\langle P_2\rangle$ value obtained in the expressions for $U_0(\theta)$, $\langle\cos^3\theta\rangle$ and $\langle 0.5\cos\theta\sin^2\theta\rangle$ can be calculated.

$r_{33}/r_{13}=3$, if the two refractive indices n_o and n_e are equal.

4.3   Liquid Crystalline Systems

In the early 1980s, Meredith et al. [155] first explored the possibility of using SCLCPs as NLO media. Since then there has been considerable interest in polymers of this type. The relevance of LCPs is threefold. Firstly, the molecular design requirements, which will optimize the NLO parameters, are those that will lead to straight, rigid molecules and thus liquid crystallinity. Secondly, a mesophase is an attractive medium in which molecules can be aligned and poled. Thirdly, it is relatively easy to quench a polymer to a clear glass that is tough and can be formed into a film. Academic interest has focused largely on:

- solid solutions of NLO active molecules in an LCP host,
- LC copolymers containing both nematogenic (or smectogenic) and NLO active side groups, and
- SCLCPs wherein the NLO active moieties possess mesogenic properties themselves.

4.3.1   Molecular Design for High NLO Activity

The optimization of NLO behavior in organics depends on two main factors: the design of the active molecular elements to maximize the hyperpolarizability, β, and the organization of these elements within the structure so that their activity is transmitted in the most efficient manner to the material itself. Attention will be focused first on the design of the active molecular elements.

The basic principles for designing materials for second-order nonlinear processes are well understood [152, 163–167]. The molecular structure must have a dipole with delocalized electron density. The dipole is created by substituting electron-donating and electron-accepting groups at either end of a conjugated length of molecule, which serves as an electron conduit between the donor and acceptor to allow perturbation of the local electron density. The nonlinear activity of the molecular structure, measured in terms of the second-order hyperpolarizability (β) is determined by the choice of donor and acceptor groups and the length of the conduit. Higher activity is achieved with stronger donors and acceptors and longer conduits.

Examples of the effects of molecular structure on the magnitude of β are given in Table 14. Molecules 1 and 2 show the result of improving the electron donation and withdrawal efficiency. As a result of the extended conjugation, the biphenyl derivative 4, and to a greater extent the terphenyl derivative 5, are found to give higher β values than the benzene analog 1. However, since the phenyl rings are able to rotate, breaking up the conjugation pathway for electron transfer, the biphenyl structure cannot be very effective for optical nonlinearity. For a given set of donor and acceptor groups, the azomethine linkage (molecule 6) twists the two phenyl rings with respect to each other, while the stilbene linkage (molecule 7) ensures planarity, effectively doubling β. It is of interest to note that the factors which enhance the second order NLO activity of the molecules (high axial ratio, planarity, charged end groups) are essentially those factors that encourage liquid crystallinity in low molar mass compounds.

Table 14. Molecular structure control of β (from [152] and [163]).

Molecule number	Structure	β at 1.9 µm ($\times 10^{-30}$ esu [a])
1	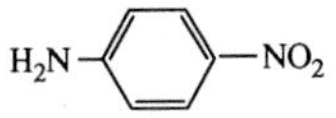	5.7
2	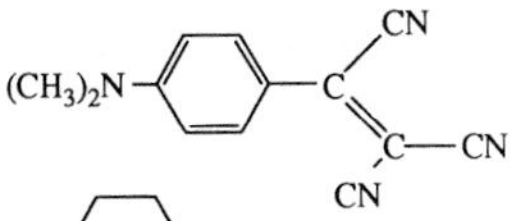	21.4
3	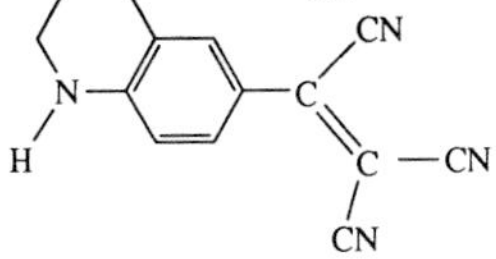	41.8
4	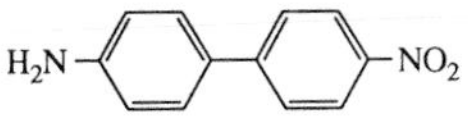	20.1
5	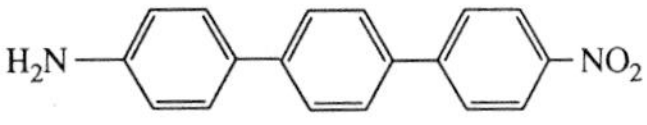	50.1
6	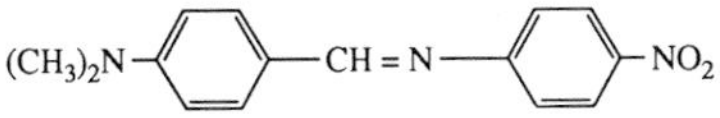	23.4
7	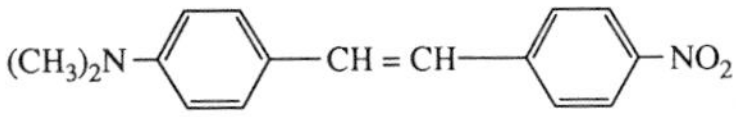	60.0
8	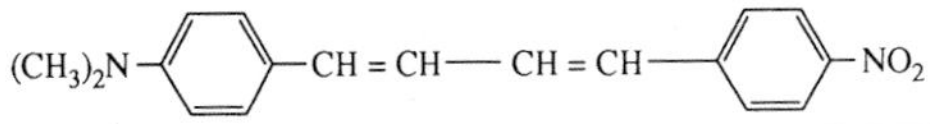	111.2

[a] 10^{-7} esu $= 4 \times 10^{-11}$ m/V.

4.3.2 NLO Liquid Crystalline Systems

4.3.2.1 Guest–Host Materials

A possible route for preparing polymeric films with large second-order NLO coefficients is to remove the orientational averaging of a dopant molecule with large hyperpolarizability (β) by the application of an external DC electric field to a softened film. This may be accomplished by heating the film above the glass transition temperature, then cooling the film to ambient temperature before the electric field is removed, thus leaving a noncentrosymmetric medium which should have a nonvanishing $\chi^{(2)}$, provided the polymer can permanently prevent relaxation of the alignment. The magnitude of $\chi^{(2)}$ is determined by the degree of polar order, the concentration of active molecules, and the magnitude of β.

The fundamental properties of the host phase are high transparency, a reasonable T_g to minimize diffusion and relaxation from poling, high solubility of the guest, and processability without degrading the guest. It has been shown theoretically [155–159] that substantial improvements in alignment along the Z direction might be anticipated in a liquid-crystalline matrix. The first example of this approach was described by Meredith et al. [155]. In that study, films consisting of p-(dimethylamino)-p'-nitro-stilbene **7** dissolved up to 2% by weight in a nematic copolymer **9** were solvent cast

where $R_1 = -(CH_2)_6-O-\bigcirc-COO-\bigcirc-CN$

and $R_2 = -(CH_2)_6-O-\bigcirc-COO-\bigcirc-OCH_3$

onto interdigital electrode patterns photo-lithographically delineated on glass substrates. A number of important findings were reported in this initial study. First, it was shown that polar $C_{\infty v}$ symmetry was induced along the poling direction, which has been verified in many subsequent studies, and that homogeneous transverse alignment of the nematic host could be achieved, free from domain-induced light scattering ($\langle P_2 \rangle = 0.3$). SHG was used to measure the $\chi^{(2)}$ tensor elements. A comparison with **7** doped at 2% in poly(methylmethacrylate) showed a large enhancement (~100-fold) in the liquid-crystalline system. This result was interpreted as being due to the correlation between the dopant molecules and the principal axis of the nematic LC. Excellent agreement was obtained between the experimentally measured value of $\chi^{(2)}_{333}$ (3×10^{-9} esu [1]) for $E_p = 1.3$ V/μm, $t_p = 6$ h, and $T_p \cong T_g - 15\,°C$, and that predicted by a simple model based on statistical dipolar alignment of the guest in the oriented host.

The polar alignment that was induced in this system was shown to be stable for long periods of time. However, several complexities in the poling process were uncovered in this study. It was found that the induced SHG dropped precipitously as T_g was approached from below. This was speculated as being due to the establishment of a mono-

mer–dimer or higher order aggregate equilibrium.

In addition to this SCLCP host, MCLCPs have also been used. Li et al. [168] dissolved molecules of an oligomeric fragment of poly(p-phenylenebenzobisthiazole) **10**, which was terminated by opposing nitro and dimethylamino groups, in the polymer itself. The resultant material has a $\chi^{(2)}$ value of 6×10^{-9} esu. Stupp et al. [169, 170] carried out a careful analysis of the system of Disperse Red 1 (**11**) dissolved in a chemically aperiodic nematic polymer containing the following three structural units (**12**).

Electrical poling ($E_p \sim 67$ kV/cm) of the dye–polymer alloys below the melting point of the nematic host resulted in strong SH intensities from an infrared laser pulse passing through the medium. Interestingly, a significant increase in the SHG signal was observed in alloys that had been aligned by an external magnetic field prior to poling by the electric field. $\chi^{(2)}_{333}$ was tripled in 5% dye samples when the high molar mass solvent was macroscopically ordered ($\langle P_2 \rangle \sim 0.6-0.7$) in the magnetic field ($\chi^{(2)}_{333} = 4.53 \times 10^{-9}$ esu instead of 1.51×10^{-9} esu). This susceptibility enhancement points to coupling between polar ordering of dye molecules in the electric field and the uniaxial organization of the nematic solvent in the magnetic field. In the extreme of a macroscopic order pa-

---

[1] Note that 10^{-7} esu $= 4 \times 10^{-11}$ m/V.

rameter approaching 1.0, the magnetic field would produce an 'Ising' solution of nematic polymer and dye in which external poling would simply bias each chromophore's dipole moment parallel rather than antiparallel to the electric field (Fig. 34). Additional evidence of this coupling is revealed in magnetically aligned samples by a more rapid rise of the SH intensity with increasing temperature under the electric field. Indeed, this difference supports the premise that dye molecules are themselves preordered in the uniaxial environment of the solvent, thus facilitating the removal of an inversion center by the electric field. It is of interest to note that the thermal stability of the polar dye network was also greater in the aligned environment.

The trade-off between concentration and β limits the applicability of guest–host materials for useful devices. The higher β molecules tend to be larger and less soluble in the hosts. Exceptionally high concentrations of up to 20 mol% of 4-hydroxy-4'-nitrostilbene (**13**) have been achieved in the polymer **14** without destroying the LC for-

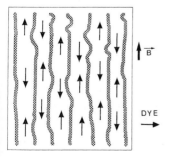

HO—◯—CH = CH—◯—NO₂   **13**

14

$-(CH_2-\underset{\underset{C=O}{\overset{CH_3}{\underset{|}{C}}})_n$

O—(CH₂)₆—O—◯—CH = CH—◯—NO₂

Figure 34. Schematic representation of the Ising solution of nematic polymer and dye formed in a magnetic field [170].

mation [171]. However, limiting values of the order of 5–10 wt% are more common. In addition, in guest–host materials a plasticizing effect was often observed, which lowered the glass transition temperature. More recently, the NLO molecules have been incorporated as part of the polymer structure. Attaching the NLO groups to the polymer reduces their mobility and tendency to crystallize or phase-separate at high concentrations, and leaves the glass transition temperature relatively unaffected.

4.3.2.2 LC Copolymers Containing Both Nematogenic (or Smectogenic) and Active Side Groups

The design requirements for synthesizing copolymers containing both nematogenic (or smectogenic) and active side groups are a stable nematic or smectic phase for alignment and poling, the ability to quench to a glass, and long term stability of the glass. Suitable structures may be obtained by appropriately adjusting the flexibility of the polymer backbone and the length of the flexible spacer. If the mesogenic groups are attached directly and rather rigidly to the backbone, the dynamics of the backbone dominate and no LC phases are formed. If the mesogenic groups are appended to the backbone by flexible spacers such as $-(CH_2)_n-$, the mesogenic groups may now adopt the anisotropic ordering of an LC phase. However, long spacers increase the degree of anisotropic LC ordering from nematic to smectic, favor the tendency of the polymer to crystallize, and lower the T_g. This does not mean that the properties of the polymer are uninfluenced by the backbone. The broadest range of mesophase thermal stability is exhibited by polymers containing the most flexible backbone, namely, polysiloxanes. However, compared to more rigid polymers, polymers containing extremely flexible backbones usually exhibit additional mesophases or even side chain crystalline phases and have lower T_g values. The review in [172] by Percec and Pugh may be of value to readers who are interested in this subject. Other difficulties which can be overcome, but may have been less than fully appreciated in the past are:

(1) Retention of SC precursor in the polymer.
(2) The danger of cross-linking involving SC substituents such as CN.
(3) The need to demonstrate that the ratio of the mesogenic and active side groups is in fact the same as that in the reaction mixture. With bulky SCs, this may not be so, and side chain ratios can be affected by reaction conditions.

Factors (1), (2), and (3) clearly affect product reproducibility, which is vital if the properties of these polymers are to be correlated with structure and if they are to play a role in applications.

A number of copolymers having nematogenic (or smectogenic) and NLO active groups have been synthesized [155, 173–181]. Table 15 lists a number of chemical groups that have been used for their mesogenic properties. As can be seen from their formulas, highly polar terminal substituents

Table 15. Typical mesogenic groups.

Molecule no.	Mesogenic group	Reference
15	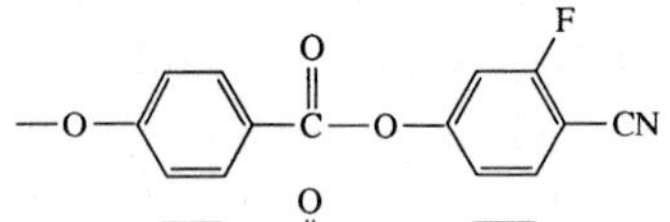 (structure: —O—phenylene—C(=O)—O—phenylene(F)—CN)	[174]
16	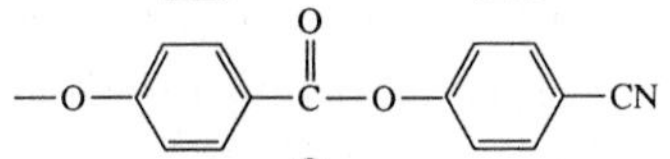 (structure: —O—phenylene—C(=O)—O—phenylene—CN)	[155, 173, 177, 178, 180, 181]
17	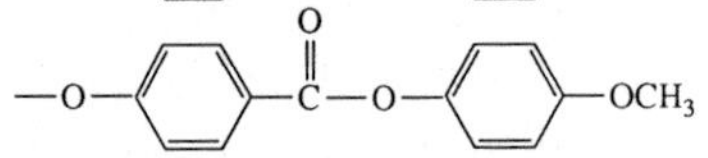 (structure: —O—phenylene—C(=O)—O—phenylene—OCH$_3$)	[155, 178]
18	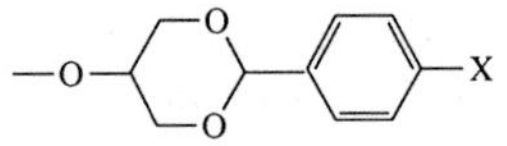 (structure: —O—dioxane—phenylene—X, X = CF$_3$, CN)	[175]
19	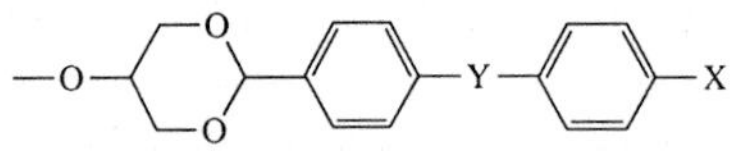 (structure: —O—dioxane—phenylene—Y—phenylene—X, X = CF$_3$, CN, OCH$_3$; 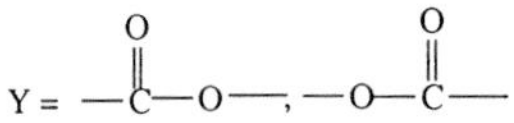 Y = —C(=O)—O—, —O—C(=O)—)	[175]
20	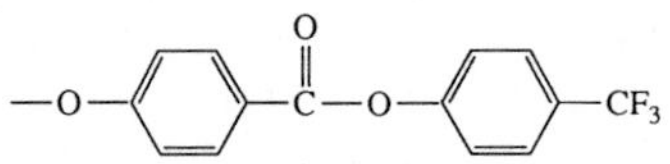 (structure: —O—phenylene—C(=O)—O—phenylene—CF$_3$)	[173, 177]
21	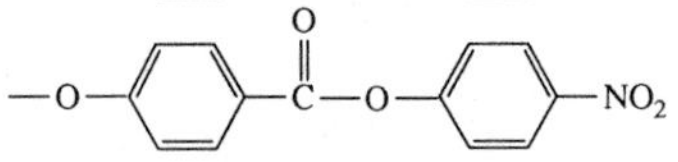 (structure: —O—phenylene—C(=O)—O—phenylene—NO$_2$)	
22	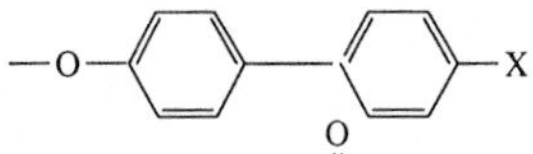 (structure: —O—phenylene—phenylene—X, X = CN, OCH$_3$)	[176, 177]
23	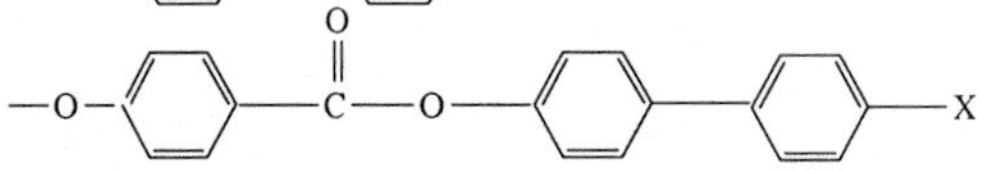 (structure: —O—phenylene—C(=O)—O—phenylene—phenylene—X, X = CN, OCH$_3$)	[181]
24	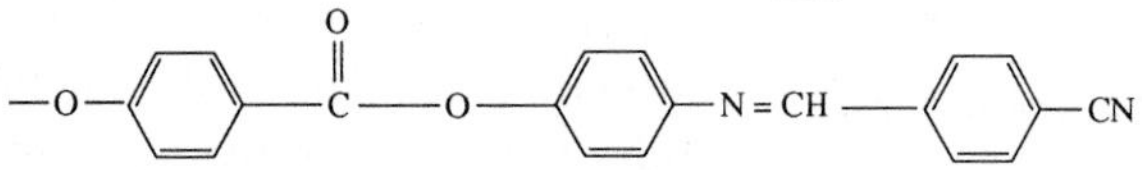 (structure: —O—phenylene—C(=O)—O—phenylene—N=CH—phenylene—CN)	[181]

have been incorporated to allow the pendant groups to be aligned by applying an electric field. The most common systems so far considered are polyacrylates and polymethacrylates with medium-length spacers of two to six methylenic units.

The ability to gain molecular control of the LC character through synthesis is clearly outlined in Table 16 for copolyacrylates having the structure **25** [173].

$R_1 = -(CH_2)_6-O-\langle\bigcirc\rangle-COO-\langle\bigcirc\rangle-Z$

$R_2 = -(CH_2)_y-X-\langle\bigcirc\rangle-Y-\langle\bigcirc\rangle-NO_2$

25

These copolymers were selected according to the above outlined principles. Phenyl ring–CN terminal substituents are known to enhance nematic thermal stability, while terminal CF_3 groups favor smectic mesophase formation. The R_2 side groups possess the basic structure associated with large hyperpolarizability (β). They contain a strong electron accepting group ($-NO_2$) and an electron-donating group ($X=-O-, >N-$) linked in a *trans*-stilbene ($Y=-CH=CH-$) or an azobenzene ($Y=-N=N-$) fashion. As shown in Table 16, up to 50 mol% of active R_2 groups can be incorporated in these copolymers without destroying LC formation. It should be noted, however, that $X=-N(CH_3)-$ lowers the LC stability. This

Table 16. Molecular structural control of the LC character[a].

	Polymer			x (mol%)	Transition temperatures (°C)
Z	Y	X	Y		
CN	—	—	—	0	N 106 I
CN	6	—O—	—N = N—	7.1	N 119.8 I
CN	6	—O—	—N = N—	21.3	N 111.4 I
CN	6	—O—	—N = N—	50.3	SmAd 90 N 107.5 I
CN	6	—O—	—CH = CH—	12	N 110 I
CN	6	—O—	—CH = CH—	34	N 91 I
CN	5	—COO—⟨N⟩—	—CH = CH—	5.5	N 113 I
CN	5	—COO—⟨N⟩—	—CH = CH—	19.7	N 103.5 I
CN	5	—COO—⟨N⟩—	—CH = CH—	32.1	N 109.5 I
CN	6	—N(CH₃)—	—CH = CH—	2.6	I
CF₃	—	—	—	0	SmA₁ 99 I
CF₃	6	—N(CH₃)—	—CH = CH—	3.8	SmA₁ 87 I
CF₃	6	—O—	—N = N—	8.1	SmA₁ 110 I

[a] [173].

reflects the fact that the molecular axes of the side chains are forced apart by the methyl group, thus reducing the intermolecular forces of attraction.

Terminal cyano groups are commonly used to obtain LCPs with positive dielectric anisotropy. However, such highly polar groups induce strong antiparallel near-neighbor correlations and favor an antiparallel arrangement of the permanent dipoles. This results in the formation of antiparallel interdigitated dimers [173]. Since these dimers have no net dipole moment, they will not contribute any more to the poling process and will thus lead to a reduction in the effective number of polable dipoles [182]. In addition, Dalmolen et al. [183] have given some evidence that high dimer fractions may affect the order parameters. The introduction of a substituent next to the cyano group is expected to markedly reduce the tendency of compound to form dimers. This is supported by data obtained by Le Barny et al. [174] for mesogenic group **15** in Table 15.

$$R = \ \ —(CH_2)_2 \ \ —N \underset{\substack{|\\ C_2H_5}}{\overset{}{}}— \!\!\!\!\! \langle\text{ring}\rangle \!\!\!\!\! —N=N— \!\!\!\!\! \langle\text{ring}\rangle \!\!\!\!\! —NO_2$$

26

Worboys et al. [181] reported that the preparation and characterization of a number of copolyacrylates containing the NLO side group, Disperse Red 1 **26**, which, depending upon the concentration of active groups, exhibited either isotropic or LC character. Using a sandwiched electroded structure with poling fields of up to 180 V μm^{-1}, they were able to obtain an electrooptic coefficient $r_{eff}(=r_{33}-r_{13})$ value of 45 pm/V at 633 nm. Comparison with isotropic polymers showed that an enhancement of 2.8 at low poling fields and 1.4 at high (>70 V μm^{-1}) poling fields could be achieved in r_{eff} in the liquid-crystalline systems. The polar align-

ment induced in liquid-crystalline materials was shown to be stable for long periods of time. An initial analysis of the data indicated that this class of NLO LCPs could exhibit lifetimes of >5 years with only a 20% decay in the electrooptic coefficient if used at temperatures ~ 90 °C below T_g.

In a subsequent study, Koide et al. [176] reported on experiments on copolymers containing nonmesogenic azo side groups having Disperse Red 1 as the NLO component and the mesogenic side groups **22**. In this example, the films were poled above the clearing temperature with fields of up to 0.125 MV/cm. The ratio $\chi_{33}^{(2)}/\chi_{31}^{(2)}$ was found to be significantly greater than the theoretical value of three predicted for isotropic systems, thus confirming that it is possible to gain in the net polar ordering by starting from a liquid-crystalline system. Using azo aromatic groups to provide the long conjugation between the donor and acceptor groups, other side chain LC copolymers have been synthesized and their NLO properties have been tested. Again it has been shown that liquid crystallinity enhanced the NLO properties upon poling under similar conditions. The reader is referred to a recent review by Xie et al. [184].

Smith and Coles [177] looked at a number of liquid-crystalline polymers. The contact electric field poling method was used to achieve noncentrosymmetric ordering of the NLO side groups. The materials were orientationally aligned in an alternating field at a temperature just below the transition to isotropic liquid. This alignment was stored by cooling into the glass phase before applying the poling field of 10 kV/cm, at a temperature a few degrees above T_g for a period of 24 h. The second order susceptibility was typically improved by a factor of three by the orientational alignment process when compared with the application of the poling field to a material with no alignment history. Again the results indicated that the liquid crystallinity has some effect on the polar ordering, since $\chi^{(2)}_{31} < \chi^{(2)}_{33}/3$ for many of the liquid-crystalline materials.

Recently, workers at the Weizmann Institute of Science reported photochromic side chain liquid crystal copolymers containing spiropyran units **27** [178, 180, 185, 186].

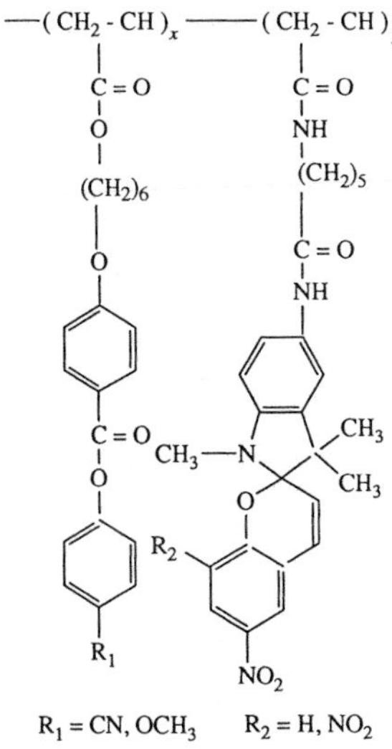

$R_1 = CN, OCH_3$ $\quad R_2 = H, NO_2$

In this study, after alignment in an electric field of $\cong 10$ kV/cm at a temperature above T_g but below the clearing point, thin films were cooled to room temperature in the pres-

ence of the field with simultaneous UV irradiation [180]. The investigators demonstrated that this procedure provides a means of inducing SHG by photochromic spiropyran–merocyanine conversion, and that this effect can be enhanced by blending the copolymer with thermochromic quasi-liquid crystals (QLCs) containing the merocyanine chromophore as the high β moiety (**28**) [178] or by using the copolymer as host for DANS **7** and another related substituted stilbene guest molecule (**29**) [178].

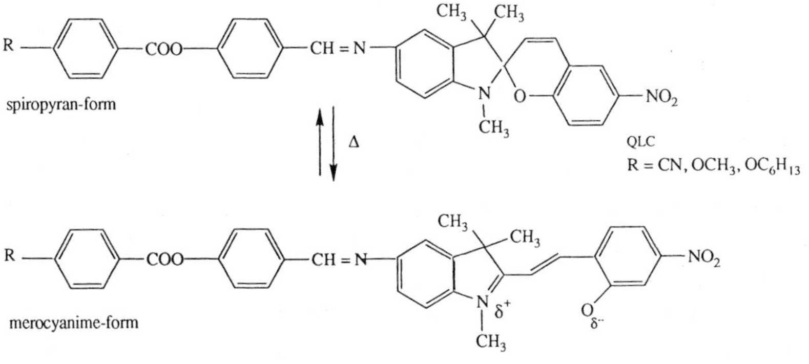

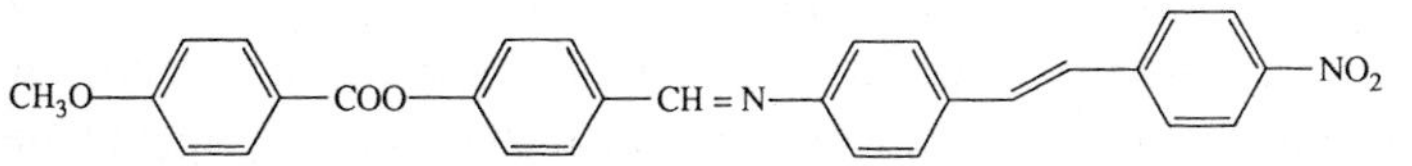

They also showed that in these samples the SHG via the dominant $\chi^{(2)}$ component parallel to the direction of electric field alignment decreases over a time scale of several weeks which can be attributed to both orientational relaxation and the spontaneous merocyanine fading to the spiropyran form, which has a much smaller β. Interestingly, it was found that the polar ordering could be enhanced by reapplying a DC electric field to the glassy films parallel to the aligning direction at room temperature. However, these special poled films exhibited asymmetry not only in the poling direction, but also perpendicularly to it. This resulted in more nonzero components of the second-order susceptibility tensor than are obtained through the usual poling techniques. The $\chi^{(2)}$ component perpendicular to the poling direction was the strongest. This nonlinearity was attributed to merocyanine aggregates stacked normal to the film substrate. The effect could be large, for example, a $\chi^{(2)}$ of 6×10^{-8} esu/cm^3 (25 pm/V) was reported for a QLC/LCP blend containing 2% by weight of DANS, poled at 5×10^6 V/m, but it was short-lived with decay over about one hour. Longer term stability was, however, reported in the case of the larger substituted stilbene when used in a larger concentration of 10% by weight, but here the coefficient was 1.1 pm/V for a poling field of 10^6 V/m [178].

Ou et al. [179] reported on the successful fabrication of X-deposited multilayer films using the Langmuir–Schaefer deposition technique [187]. They used a polysiloxane copolymer containing both a mesogenic and an NLO side chain (**30**).

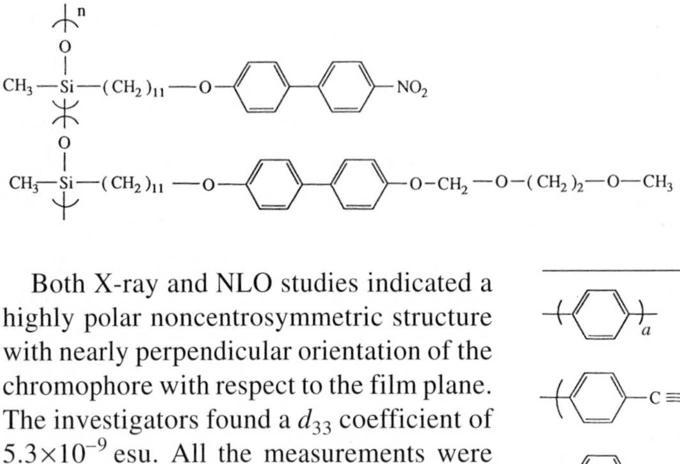

sentially those factors that encourage liquid crystallinity in low molar mass compounds. Therefore considerable effort has been devoted to the synthesis of NLO polymers in which the NLO moieties themselves possess mesogenic properties. Examples of NLO active mesogenic groups are given in Table 17. They are depicted below in a schematic form along with examples of chemical structural subunits where X and W

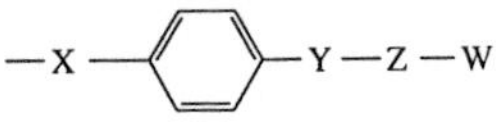

represent a range of electron-donating and electron-withdrawing substituents and Y–Z is a conjugated π electron system with Y$=-$CH$=$CH$-$, $-$N$=$N$-$, $-$CH$=$N$-$, and Z a substituent selected from

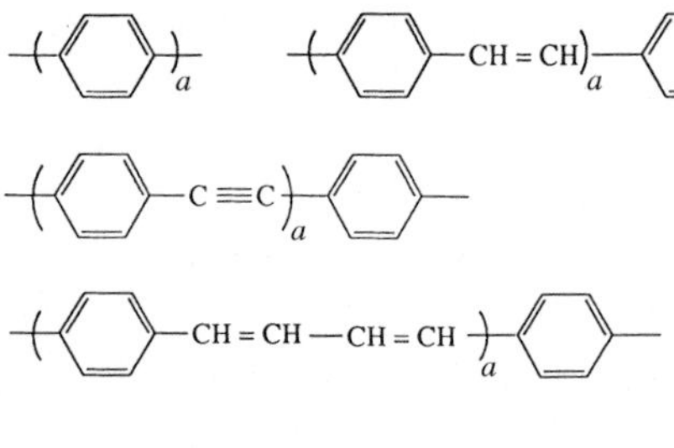

30

Both X-ray and NLO studies indicated a highly polar noncentrosymmetric structure with nearly perpendicular orientation of the chromophore with respect to the film plane. The investigators found a d_{33} coefficient of 5.3×10^{-9} esu. All the measurements were found to be reproducible over a span of several weeks, indicating a stable structure.

4.3.2.3   SCLCPs Wherein the NLO Active Moieties Possess Mesogenic Properties Themselves

The factors that enhance the second order NLO activity of the molecules (high axial ratio, planarity, charged end groups) are es-

where a has small integral value.

Examples of electron-donating substituents are divalent radicals such as $-$O$-$, $-$S$-$, and $-$NR$'-$, where R$'$ is hydrogen or methyl. Typical acceptor substituents are $-$NO$_2$, $-$CN, and $-$CF$_3$. The majority of SC polymers with NLO mesogenic groups have the donor substituent in the spacer chain and the

Table 17. Examples of NLO active mesogenic groups.

Structure no.	NLO active group	References
31	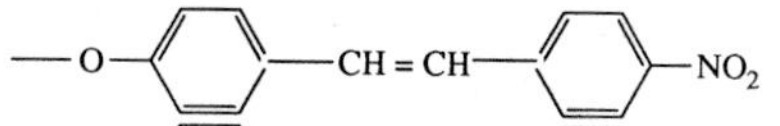 —O—⬡—CH = CH—⬡—NO₂	[171, 177, 182, 189, 191–193, 195, 196, 198–200, 202]
32	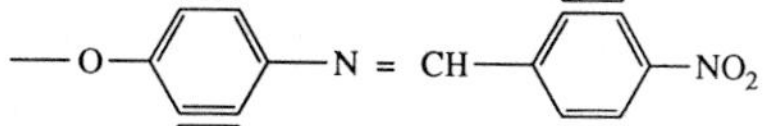 —O—⬡—N = CH—⬡—NO₂	[189, 191–193, 195, 199]
33	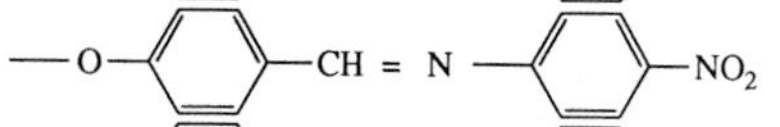 —O—⬡—CH = N—⬡—NO₂	[177]
34	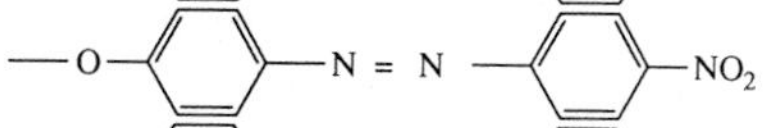 —O—⬡—N = N—⬡—NO₂	[177, 189, 195]
35	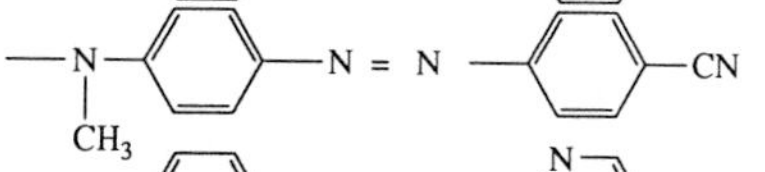 —N(CH₃)—⬡—N = N—⬡—CN	[177]
36	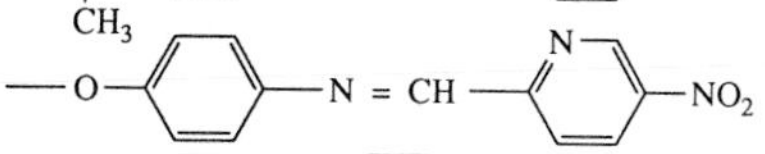 —O—⬡—N = CH—(pyridine)—NO₂	[191, 192]
37	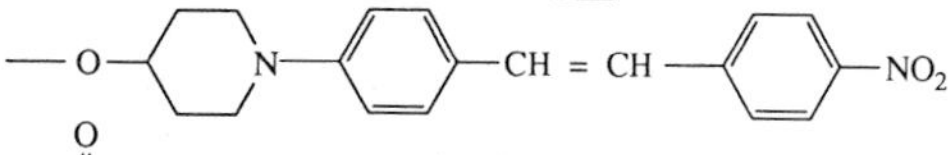 —O—(piperazine N)—⬡—CH = CH—⬡—NO₂	[177]
38	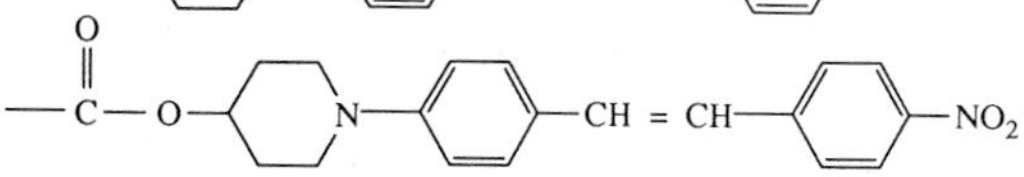 —C(=O)—O—(piperazine N)—⬡—CH = CH—⬡—NO₂	[193, 196, 199]
39	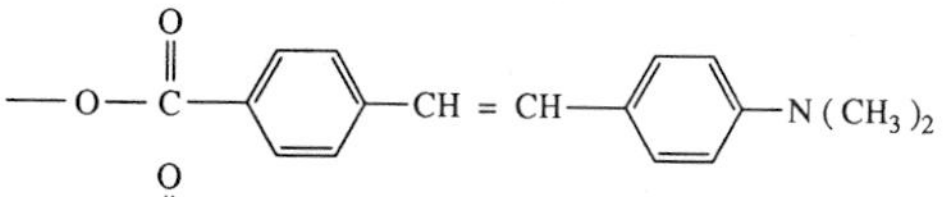 —O—C(=O)—⬡—CH = CH—⬡—N(CH₃)₂	[190, 194, 197]
40	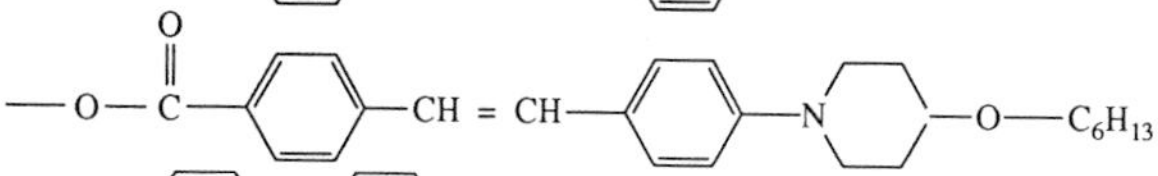 —O—C(=O)—⬡—CH = CH—⬡—(piperazine N)—O—C₆H₁₃	[197]
41	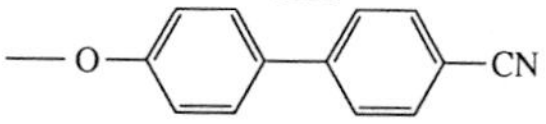 —O—⬡—⬡—CN	[177, 188]
42	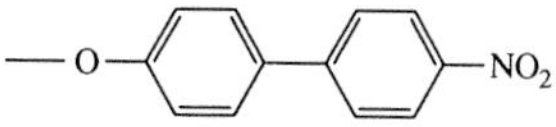 —O—⬡—⬡—NO₂	[152, 200, 201, 203]
43	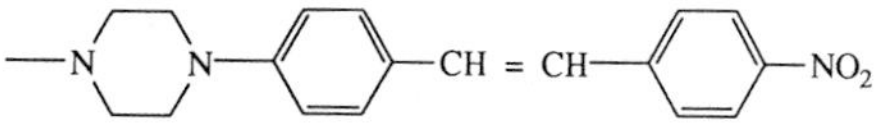 —(piperazine N)—⬡—CH = CH—⬡—NO₂	[204]
44	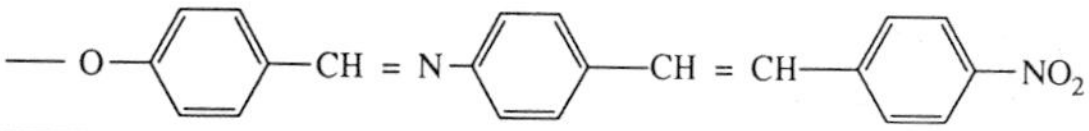 —O—⬡—CH = N—⬡—CH = CH—⬡—NO₂	[205]

acceptor at the end of the side chain, as in compounds **31–37** (Table 17). However, Zhao et al. [194] and Bautista et al. [197] have prepared a number of SC polyacrylates and polysiloxanes with 4-(dimethylamino)-4′-stilbene carboxylic ester mesogens (**39** and **40** in Table 17), which have the donor substituent at the end of the side chain and the acceptor in the spacer chain.

There have been two main approaches toward the synthesis of SCLCPs containing NLO mesogenic groups:

1) Side groups with suitable polymerizable functional ends can be synthesized first, with subsequent polymerization.

Free radical polymerization of a vinyl monomer, e.g., acrylate, methacrylate, or styrene derivative, has been widely used to produce NLO single-component SCLCPs. This type of monomer is suitable as it is relatively easy to prepare and the conditions for free radical polymerization of acrylates, methacrylates, and styrene derivatives have been well documented. There are, however, special difficulties associated with the free radical polymerization of an interesting class of NLO materials, namely, nitroaromatics: the nitroaromatic group is a retarder of free radical polymerization, intercepting the propagating radicals by reaction on the aromatic ring or on the nitro group itself.

Nitrostilbenes are desirable for NLO, but the stilbene unit itself is potentially reactive with a free radical center. The presence of the Ar–CH=CH–Ar double bond may lead to an uncontrolled crosslinking reaction leading to a gel [193, 194]. This could prevent a good polarization of the material. As reported by Griffin and Bhatti [196], polymerization of nitroaromatic stilbene methacrylates in solution under nitrogen using a single charge of initiator (AIBN) can provide an acceptable route to NLO SCLCPs. The analogous polymerization of acrylates produces lower yields with crosslinking. This difference in the extent of crosslinking is attributed to the greater stability of the propagating radical from methacrylate monomer (tertiary) and its concomitant greater selectivity.

Alternatively, polycondensation appears to be a facile synthetic entry into NLO polymers having nitroaromatic-containing SCLC groups. Griffin et al. [191, 192] were successful in preparing three series of malonate SCLCPs by using a polycondensation route first described by Reck and Ringsdorf [206]. This synthetic approach obviates the difficulties described above for free radical polymerization, as the nitro group is not an inhibitor of polycondensations.

2) The preformed polymer can be reacted with the side groups. This has the advantage of starting with well-defined polymer parameters of the molecular weight and distribution. However, incomplete reaction of all possible sites on the polymer can lead to unwanted crosslinking at a later stage of processing, unless they are mopped up by a more reactive component after the NLO group has been attached.

Bautista et al. [197] used the hydrosilylation reaction to synthesize NLO SCLC polysiloxanes. They prepared a homopolymer from poly(hydrogen methylsiloxane) and a copolymer from poly(hydrogen-methyldimethylsiloxane). However, the higher T_g values that are sought for use in NLO need the use of a backbone that is inherently less flexible than the Si–O linkage. Saturated carbon–carbon backbones fulfill this requirement, and McCulloch and Bailey [189, 195] grafted alkylbromochromophore units onto a preformed poly-4-hydroxystyrene polymer.

It should be noted that SC copolymers based on nonmesogenic SCs and NLO mesogenic SCs offer the opportunity to fine-tune the polymer properties by varying the ratio of the NLO mesogenic SCs, although care must be taken to characterize the product, since the final ratios of the SCs may not correspond to the feed concentrations. An interesting example is the series of copolyethers prepared by chemical modification of atactic polyepichlorohydrin with the sodium salt of 4-cyano-4′ hydroxybiphenyl [188, 207] (45).

The data in Table 18 show that the thermal behavior of these copolyethers strong-

45

Table 18. Thermal properties of copolyethers derived from polyepichlorohydrin [207].

$x\%$	0	25.5	40	65	74	83	91
T_g (°C)	−26.5	10	26	66.6	79.5	83	86
T_{NI} (°C)				105	133.5	143	153.5

ly depends on the degree of substitution. There is a minimum substitution limit of approximately 65% necessary for LC formation. Below this limit, the copolymers all resemble typical amorphous copolymers. There is a sharp linear increase in the glass transition temperature with increasing amount of substitution, indicating that the bulky side groups cause severe hindrance to main chain motions. Once 65% substitution is reached, the increase in T_g drastically decelerates and the copolymers exhibit a nematic phase. The copolyethers with a degree of substitution of about 50–60% display no liquid crystallinity at rest. However, miscibility studies demonstrated that they exhibit a virtual nematic phase, the virtual nematic–isotropic liquid transition point being a few degrees below T_g [208].

The backbones that have been most commonly employed are those of the acrylate [177, 190, 194, 196], methacrylate [152, 171, 196, 198–200], and siloxane [152, 177, 197] types. Polyethers [207–209], polyesters [182, 191, 192], and polystyrenes [177, 189, 195] have also been reported. Typical spacer groups consist of between 3 and 12 methylene units. The phase transitions of a number of SCLCPs containing NLO mesogenic groups are collected in Table 19. Unfortunately, the molecular masses of many of these polymers have not been determined, and the influence of the polymer structure on the phase transitions can not therefore be quantitatively discussed. However, the general points to emerge from these data are as follows:

- As the spacer length increases so T_g tends downwards.
- Any N phases normally arise at shorter spacer lengths; longer spacers tend to encourage smectic phases (see **41** PA 2–6; **42** PMA 3–12).
- T_g decreases as the backbone flexibility increases (see **41** PA 2–**41** PSi 3, **42** PMA 5–**42** PSi 5), while $\Delta T = T_c - T_g$ increases with increasing flexibility.
- The highest clearing temperatures are obtained with the most flexible backbones (**42** PSi 5 > **42** PMA 5, **41** PSi 3 > **41** PA 2, **41** PA 6 > **41** PMA 6, **39** PSi 11 > **39** PA 10).

Similar trends have been noted in many other SCLCP series involving different polymer components [172].

The first question to be decided is whether it is possible to polymerize a molecular sequence, effectively adding the NLO mesogenic groups onto a backbone, without degrading the optical and NLO properties of the active moiety.

The UV/visible properties of SCLCPs containing NLO mesogenic groups are similar to those of their low molar mass analogs. In all cases, the chromophore appears unchanged by the polymerization. The absorption maximum wavelength (λ_{max}) of these polymers lies in the range 285–420 nm. Polystyrenes [189, 195], polyacrylates [193], and polyesters [191, 193] based on the NLO mesogenic groups **31**, **32**, and **34** (see Table 17) have similar absorption maxima (375–381 nm in THF or CHCl$_3$).

Table 19. Thermal behavior of SCLCPs containing NLO mesogenic groups.

Chromo-phore	Back-bone [a]	$+CH_2{+}_n$	$\overline{DP}$	$\bar{M}_n$ $(\times 10^3)$	$\bar{M}_w$ $(\times 10^3)$	T_g (°C)	Mesophase	T_c (°C)	Ref.
31	PS	6	250			85	SmA	165	[189, 195]
31	PS	8	250			65	SmA	159	[189, 195]
31	PS	10	250			50	SmA	155	[188, 195]
31	PMA	6				61	Sm?	156	[200]
31	PMA	6				52	SmA	106	[199]
31	PMA	6				71	unspecified	153	[198]
31	PMA	6				unspecified	N	unspecified	[171]
31	PEsa	1				60	N	85–90	[182]
31	PEsb	6				–	N	106.5	[191, 192]
31	PEsc	6				30	SmA	88.5	[191, 192]
31	PEsd	6				–	Ch	71.5	[191, 192]
32	PS	6	250			75	SmA	118	[189, 195]
32	PS	8	250			65	SmA	119	[189, 195]
32	PS	10	250			40	SmA	101	[189, 195]
32	PMA	6				55	SmA	116	[199]
32	PEsb	6				–	N	62	[191, 192]
32	PEsc	6				5	Ch	52.9	[191, 192]
32	PEsd	6				10	Ch	33.5	[191, 192]
34	PS	6	250			85	SmA	151	[189, 195]
34	PS	8	250			70	SmA	151	[189, 195]
36	PEsb	6					SmA	131.2	[192]
36	PEsc	6					SmA	71.4	[192]
36	PEsd	6					SmA	68.0	[192]
38	PMA	6				80	?		[199]
39	PA	10		13.3	186	79	unspecified	128	[190]
39	PA	2		5.7	14.2	129	unspecified	129	[194]
39	PA	4		2.8	7	119	unspecified	130	[194]
39	PA	6		3.9	9.7	109	unspecified	125	[194]
39	PA	8		3.4	9.7	91	unspecified	121	[194]
39	PA	10		5.2	10.11	81	unspecified	117	[194]
39	PSi	11	60			$T_m = 167$	N	215	[197]
40	PSi	11	60			$T_m = 110$	SmB 192 N	258	[197]
41	PMA	6				64.4	N	115	[176]
41	PA	2				84	N	113.5	[177]
41	PA	4				42	N	110	[210]
41	PA	5				35	SmA$_d$ 120 N	124.4	[210]
41	PA	6				32	N$_{re}$ 80 SmA$_d$ 124.5 N	132	[210]
41	PSi	3				30	SmA	145	[177]
42	PMA	3				85	N	100	[152, 201]
42	PMA	5				45	N	72	[152, 201]

Table 19. (continued).

Chromo-phore	Back-bone [a]	$+(CH_2)_n$	$\overline{DP}$	$\overline{M}_n (\times 10^3)$	$\overline{M}_w (\times 10^3)$	T_g (°C)	Mesophase	T_c (°C)	Ref.
42	PMA	6				40	N	64	[152, 201]
42	PMA	8				35	N	75	[152, 201]
42	PMA	11				20	SmA	95	[152, 201]
42	PMA	12				10	SmA	80	[152, 201]
42	PMA	6			147	51	Sm	75	[200]
42	PMA	6			777.5	47	Sm	77	[200]
42	PMA	6			849.8	47	Sm	77	[200]
42	PMA	6			1241	54	Sm	79	[200]
42	PSi	5	55			18	SmA	165	[152, 201]

[a] PS, PA, PMA, PEsa, PEsb, PEsc, PEsd, PSi structures shown below.

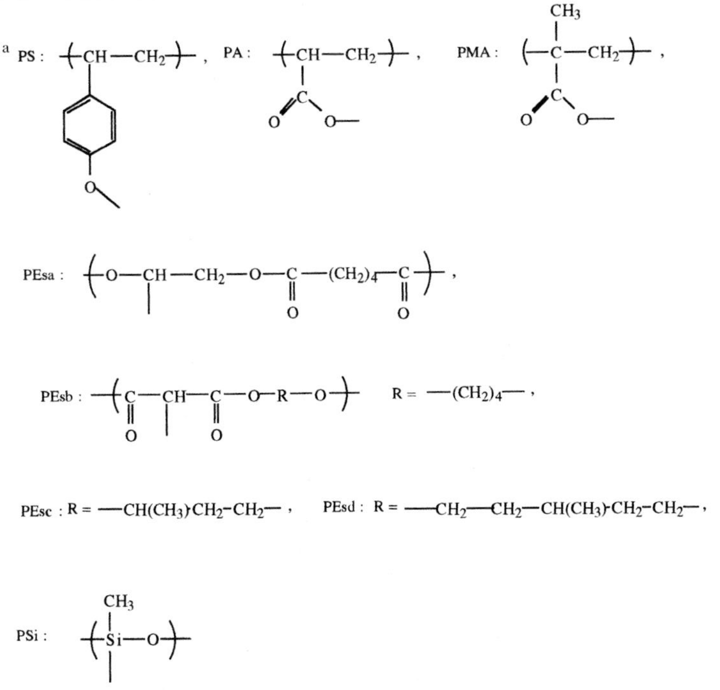

Replacement of  —⬡— by —⬡—(N=) (**36**) shifts λ_{max} to 393–394 nm. Similarly, changing X=—O— by X=—COO—⬡—N— (**38**) results in a shift of λ_{max} to longer wavelengths ($\cong 412$ nm) [193].

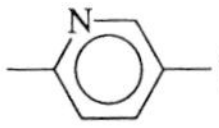

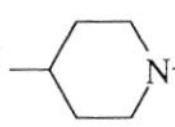

The effect of increasing the length of the conjugated linkage between the donor and the acceptor is clearly seen from compounds **42**, **31**, **32**, and **34**. A shift of ca. 40 nm towards longer wavelengths can be realized on going from nitrobiphenyl ($\lambda_{max} \cong 338$ nm) [201] to nitrostilbene or nitroazobenzene [189, 191].

One particular comparison of the monomer and polymer properties has been done by De Martino et al. [201], who measured β solvatochromically for 4-(12-dodecyloxy)-4'-nitrobiphenyl, its methacrylate, and the polymer. Their results are shown in Table 20. All the compounds showed an absorption band at about 338–340 nm. However, a noticeable decrease of β was observed after polymerization.

McCulloch and Bailey [189, 195] reported β values for a series of SCLC polymers based on a polystyrene backbone (Table 21). They showed that both the azo and stilbene link units exhibit a greater solvatochromic shift than their imine counterpart, which manifests itself as a larger hyperpolarizabil-

Table 20. Comparison of β values measured for a monomer and its polymer[a].

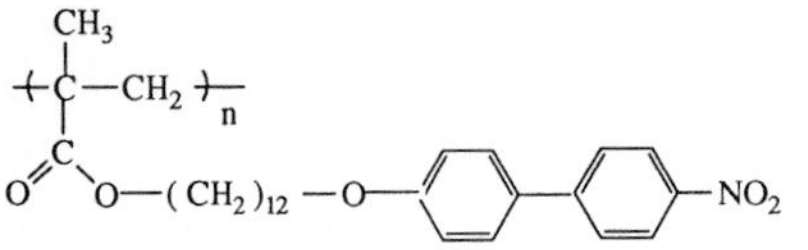

Compound	λ_{max} (nm)	β (10^{-30} esu)
Monomer	340	11
Polymer	338	8

[a] [201].

Table 21. Values of λ_{max} and β for polymers of the following structure[a].

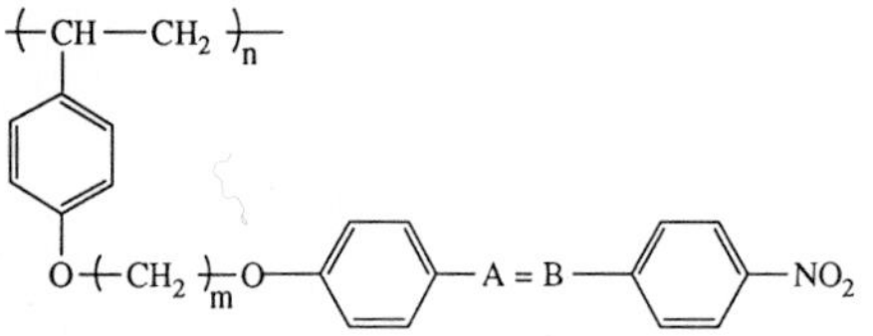

A	B	m	λ_{max} (nm)	β (10^{-30} esu)[b]
N	N	3	378	31.2
CH	CH	6	378	29.1
N	CH	6	377	21.6
N	N	6	378	28.9
CH	CH	8	378	29.3
N	CH	8	377	16.0
N	N	8	378	28.5
CH	CH	10	378	24.2
N	CH	10	377	17.2
N	N	10	378	23.5

[a] [189, 195]. [b] Estimated by a solvatochromic technique using the solvent shift between THF and hexane.

ity. This difference is mainly attributable to the two aromatic rings being twisted out of plane, thus reducing the π-orbital overlap. Also, an increase in the length of the flexible spacer is accompanied by a slight decrease of β.

In a subsequent study, Wijekoon et al. [199] carried out a careful analysis of a series of LC polymethacrylates and corresponding monomers based on the NLO mesogenic groups **31**, **32**, and **38**. A noticeable trend in their results is that, even though the hyperpolarizabilities of the monomers and the corresponding polymers have similar magnitudes, the β values obtained for the monomers are somewhat higher than those for the polymers. This effect may be attributed partially to the different local molecular environment resulting from the solvent–solute interaction. Several additional interesting effects were noted in this study. First, it was observed that for the same donor–acceptor pair, replacement of an ethylene linkage $-CH=CH-$ between two aromatic rings by an imine $-CH=N-$ linkage is damaging, so reducing the β value. This is in agreement with the data reported by McCulloch and Bailey [189, 195] and confirms that the extended and planar structure of the stilbene moiety is very effective for optical nonlinearity. Second, it was found that replacement of an ether linkage $-O-$ by

an amino linkage 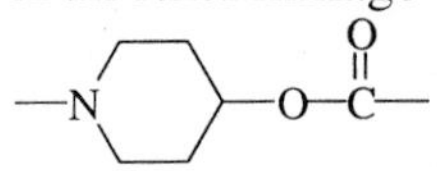 results

in an increase in β, the donor strength of the former being inferior to that of the latter.

It is difficult to compare the second order nonlinear response of different materials because of its dependence on processing and the electrical poling process. Nevertheless, we have collected the SHG values of $d_{33}\,(=\chi^{(2)}_{33}/2)$ tensor components and electrooptic r_{33} coefficients for a number of SCPs containing NLO mesogenic groups in Table 22.

Table ... Linear, poling, and nonlinear optical characteristics of some SCLCPs listed in Table ...

Polymer	Ref.	Chromophore β (10^{-30} esu at 1.06 μm)	Characteristics λ_{max} (nm)	E_p (kV/cm)	T_p (°C)	t_p	$\langle P_2 \rangle$	$\chi_{33}^{(2)}$, d_{33}, or r_{33}^a (pm/V at 1.06 μm)	Evaluation of NLO properties enhancement by LC order
31 PMA 6	[198]	74	370	150	$T_g = 71$	2 min		pyroelectric coefficient γ $d_v/dE_p = 20$ μCV^{-1} m^{-1} K^{-1}	
31 PMA 6	[171]	74	370	Corona poling	$T_p > T_g$		$\Delta n = 0.38$	$\chi_{33}^{(2)} = 3.4$	Enhancement of $\chi_{33}^{(2)}$ by a factor of 1.5
31 PS 6	[189, 195]	74	370 (378 in THF)	Corona poling	$T_g - 35 = 50$			d_{33} (1.319 μm) = 8.4	
31 PEs a	[182]	74	370	100	$T_c + 5 = 95$ $T_c - 10 = 80$	a few min	$0.6 < \Delta\varepsilon < 0.95$ with $\Delta\varepsilon = n_e^2 - n_o^2$	$r_{33} - r_{31} = 3.6$ at 95°C (iso) $r_{33} - r_{31} = 0.2$ at 80°C (N)	Decrease of the electrooptic effect on cooling through the isotropic/nematic transition
32 PS 6	[189, 195]	18	383 (377 in THF)	Corona poling	$T_g - 25 = 50$			d_{33} (1.319 μm) = 1.1	
38 PMA 6	[199]			Corona poling	$T_g + 25 = 105$	20 min	0.26	$\chi_{33}^{(2)} = 30.6$	No difference between $\chi_{33}^{(2)}$ measured after poling in the isotropic or the LC state
39 PA 10	[190]			800	$T_c - 5 = 123$	24 h		$d_{33} = 0.15$ (0.36×10^{-9} esu)	
41 PA 2	[177]	17	292	100	$T_p > T_g$	24 h		$\chi_{33}^{(2)} = 0.6$	$\chi_{33}^{(2)}/\chi_{31}^{(2)} = 5.65$
41 PA 6	[177]	17	292	100	$T_p > T_g$	24 h		$\chi_{33}^{(2)} = 1.4$	$\chi_{33}^{(2)}/\chi_{31}^{(2)} = 5.4$
41 PSi 3	[177]	17	292	100	$T_p > T_g$	24 h		$\chi_{33}^{(2)} = 0.23$ ($0.5 - 5 \times 10^{-9}$ esu)	$\chi_{33}^{(2)}/\chi_{31}^{(2)} = 4.2$
33 PS 11	[177]	18 (at 1.910 μm)	383	100	$T_p > T_g$	24 h		$\chi_{33}^{(2)} = 2.6$	$\chi_{33}^{(2)}/\chi_{31}^{(2)} = 3.3$
35 PSi 6	[177]			100	$T_p > T_g$	24 h		$\chi_{33}^{(2)} = 0.8$	$\chi_{33}^{(2)}/\chi_{31}^{(2)} = 4$

^{a} $d_{33} = \chi_{33}^{(2)}/2$, $r_{33} = 2\chi_{33}^{(2)}/n^4$, where n is the refractive index.

Griffin and co-workers developed SCLCPs [196] that demonstrate some interesting SHG phenomena [211]. They are easily poled at room temperature, having a significant $\chi^{(2)}$ in the mesophase. With the poling field continuously applied, the $\chi^{(2)}$ diminishes as the temperature increases, going to zero at the mesophase–isotropic transition (125–145 °C). Upon cooling back into the mesophase, the $\chi^{(2)}$ reappears and grows larger as the temperature decreases. That is, the $\chi^{(2)}$ tracks with the orientational order parameter of the liquid-crystalline phase. A maximum value of d_{33} of 35×10^{-9} esu ($\cong 14.7$ pm/V) is reached at room temperature in polymer **38** PA 5, which also shows excellent temporal stability of SHG (Fig. 35). The preliminary interpretation of this observation is that dipoles in the nitroaromatic pendant groups that are aligned by the electric field can relax back easily in the isotropic phase where the pendant rods are 'loosely' organized. However, they cannot relax back in the mesophase with its 'closely packed' parallel orientation of neighbor-ing rods. That is, the motions of the pendant NLO dipoles are retarded and $\chi^{(2)}$ is enhanced significantly.

Conformation of the better temporal stability of SCLCPs compared to that of isotropic SCPs that have similar chromophores has been made by Carr et al. [198]. These workers found that the polarization of the SCLCP **31** PMA 6 is much more stable than that of the isotropic copolymer (**31** PMA 3)$_{0.9}$–(**31** PMA 6)$_{0.1}$ in spite of the fact that the T_g of the latter polymer is about 117 °C compared with 71 °C for the LC homopolymer. However, in this work, the better temporal stability of the LC polymer was related to the absence of β relaxation, as shown by dielectric measurements.

Quite recently, Gonin et al. [188, 208, 212] and Guichard [209] conducted a systematic study of factors influencing SH efficiency. In this work, SHG was measured for five polymers containing the NLO mesogenic groups **41**: a polyacrylate **41** PA 3, a polymethacrylate **41** PMA 3, two copolyethers derived from polyepichlorohydrin **45** ($x=0.8$) and **45** ($x=0.55$) and a copolyether **46** derived from poly(3,3-bis(chloromethyl)oxetane).

Optical texture observations and X-ray investigations showed that the polyacrylate **41** PA 3 and the copolymer **45** ($x=0.8$) form nematic phases. The polymethacrylate **41** PMA 3 and the copolyether **45** ($x=0.55$) display no liquid crystallinity at rest, but exhibit a virtual isotropic liquid–nematic transition a few degrees below T_g, as evidenced by miscibility studies. Copolyether **46** does not show any threaded or schlieren texture, which would be characteristic of a nematic.

d$_{33}$(t)/d$_{33}$(t=0) vs TIME (hours)

Figure 35. Temporal stability of polymer **38** PA 5.

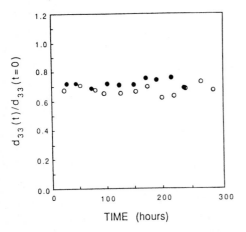

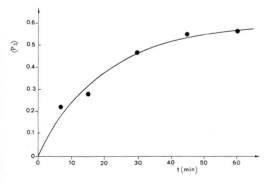

(mole ratio $x : y : z$ 0.14 : 0.43 : 0.43)

46

The glass transition is associated with weak birefringence only. A number of important findings were reported in this study. First, it was shown that homeotropic alignment of the NLO mesogenic groups, free from domain-induced light scattering, could be achieved by using the corona poling technique. As shown in Fig. 36, the order parameter $\langle P_2 \rangle$ increases with poling time before leveling off. A saturation limit for $\langle P_2 \rangle$ of about 0.65 was observed for the nematic polymers **41** PA 3 and **45** ($x=0.8$). It is of interest to note that the response times of the polymers increased as the flexibility of the polymer backbone was increased. Clearly, the openness of the polymer structure is an important factor governing the axial order. The microscopic order parameters $\langle P_2 \rangle$ derived for **41** PMA 3 and **45** ($x=0.55$) were smaller (0.4–0.5), but larger than those usually found for covalently functionalized isotropic polymers (0.1–0.3) [213], indicating that the poling field induced an isotropic/nematic transformation in these polymers, which are isotropic at rest but exhibit a vir-

tual isotropic liquid/nematic transition. This was predicted by the MSVP model [157–159].

SHG was used to monitor the polar alignment that was induced in these systems. Values of the second-order NLO coefficients d_{31} and d_{33} are given in Table 23. The salient conclusions of the results are as follows:

1) The experimental values of d_{33} are much larger than those reported for other materials suitable for blue conversion applications, like potassium titanyl phosphate, and are approximately equal to those determined for lithium niobate [153].

2) The thermodynamic model predicts that $d_{31}=d_{33}/3$ for a poled isotropic material and $d_{31}=0$, if the liquid crystallinity of the material is accounted for using the Ising model (Table 13). In agreement with these theoretical predictions, the results listed in Table 23 show that liquid crystallinity results in higher anisotropy for the $\chi^{(2)}$ tensor, since d_{31} is much lower than $d_{33}/3$ for all the polymers investigated.

3) The observed difference of behavior between the polyacrylate **41** PA 3 and the polymethacrylate **41** PMA 3 and the two copolyethers **45** ($x=0.8$) and **45** ($x=0.55$) is probably due to the smaller axial ordering induced by the poling field in the films prepared from the isotropic polymethacrylate **41** PMA 3 and copolyether **45** ($x=0.55$), even though these polymers undergo an isotropic/nematic phase transformation during poling.

Figure 36. Microscopic order parameter $\langle P_2 \rangle$ as a function of poling time for **41** PA 3.

Table 23. NLO coefficients d_{31} and d_{33} for **41** PA 3, **41** PMA 3, **45** ($x=0.8$), **45** ($x=0.55$), and **46**[a].

Polymer	T_g (°C)	t_p (min)	T_p (°C)	$\langle P_2 \rangle$	d_{33} (pm/V)	d_{31} (pm/V)	$r=d_{33}/d_{31}$
41 PA 3	51.5	15	57	0.17	9.5	1.2	8 ± 1.5
		60	57	0.57	35.6	2.1	17 ± 3
41 PMA 3	97	15	97	0.45	33.5	3.3	10 ± 2
45 ($x=0.8$)	79.5	15	81	0.57	17.8	0.97	15 ± 3
		60	81	0.62	23.4	1.65	14 ± 3
45 ($x=0.55$)	56	15	57	0.45	7.9	0.7	11 ± 2
46	106	30	107	0.19	20	2.4	8 ± 1.5

[a] Corona poling, SHG measurements at 1.06 μm.

The values of d_{33} and $r=d_{33}/d_{31}$ predicted by the different theoretical models and the resulting evaluation of SHG enhancement with liquid crystalline axial order are given in Table 24 for polyacrylate **41** PA 3 and copolyether **45** ($x=0.8$). The data in Table 24 confirm that it is possible to gain in net polar ordering by starting from a liquid-crystalline system. The enhancement of the NLO performance with liquid-crystalline order is by a factor of 2.3–3.3 in qualitative agreement with the SKS [156] and MSVP [157–159] model results. It should be noted, however, that the measured order parameters $\langle P_2 \rangle$ are lower than the calculated ones, which may be due to the antiparallel dipole–dipole correlation evidenced in the nematic state by X-ray diffraction.

Several additional interesting effects were noted in this study. First, it was shown that the NLO SCLCPs investigated exhibited enhanced stability over isotropic polymers. Second, it was found that the temperature at which the polymeric films were stored below T_g was important in determining the rate of degradation of d_{33}. Excellent thermal stability was observed for **46** at room temperature, while a decrease in d_{33} was evidenced for **45** ($x=0.8$), and, to a greater extent, **45** ($x=0.55$). Sub-T_g mobil-

Table 24. Calculated and experimental values of $\langle P_2 \rangle$, $\langle P_4 \rangle$, d_{33}, $r=d_{33}/d_{31}$, and A = calculated (or experimental) d_{33}/calculated isotropic d_{33}.

	Model	$\langle P_2 \rangle$	$\langle P_4 \rangle$	d_{33} (pm/V)	r	A
45 ($x=0.8$)	Isotropic	0	0	10.0	3	1
	SKS	0.68[b]	0.33[b]	33.3	10.8	3.3
	MSVP	0.77[c]	0.44[c]	37.0	13.0	3.7
	Ising	1	1	50.1	∞	5
	Experimental[a]	0.62	–	23.4	14 ± 3	2.34
41 PA 3	Isotropic	0	0	10.7	3	1
	SKS	0.61[b]	0.26[b]	32.3	9.2	3.0
	MSVP	0.75[c]	0.40[c]	38.7	11.9	3.6
	Ising	1	1	53.2	∞	5
	Experimental[a]	0.57	–	35.6	17 ± 3	3.3

[a] At 1.064 μm; [b] calculated with the MSVP model at zero field; [c] calculated with the MSVP model at $E_p = 2.9$ MV/cm ($f(o)=2$).

ity of the polymer chain segments might account for these relaxation phenomena. However, it was found that the absorption spectra of the poled films did not change with time, implying that the axial order is frozen in the glassy state. It was suggested that the δ relaxation process (reorientation of the side chains around the polymer backbone) could play a role in the stability of induced polar ordering. It can therefore be concluded that the temporal and thermal characteristics of the poling process must be well understood and controlled to maximize poling-induced nonlinearity and stability.

4.3.2.4 Ferroelectric LC Polymeric Systems

Up to now we have considered the uniaxial nematic and smectic A phases. As already mentioned, these mesophases are centrosymmetric and in order to achieve second order NLO effects it is necessary to remove this centrosymmetry by inducing a net polar ordering through electric field poling. A challenge therefore is to modify an otherwise centrosymmetric system having a large first hyperpolarizability β to render it capable of exhibiting second-order NLO response. Molecular asymmetry imparts form chirality to LC phases, which results in the formation of helical ordering of the constituent molecules of the phase and imposes a reduction in the space symmetry [214], which leads to some phases having unusual NLO properties. Essentially, this section deals with chiral smectic C phases that have polar symmetry $C2$ and possess an easily measured macroscopic polarization once the helical structure is unwound. For low molar mass liquid crystals and SCLCPs, the common practice is to apply an external electric field or to use surface forces.

Since the early days of ferroelectric low molar mass LC research, the exploration of the NLO properties of these substances has been a topic of interest. For a recent review on SHG in ferroelectric LCs, see [215]. Unfortunately, the SHG efficiency of most ferroelectric LCs investigated so far [216–219] is orders of magnitude below that of state-of-the-art inorganic crystals, such as lithium niobate. The main reason for the low electronic NLO activity is that the ferroelectric LC compounds and mixtures used in these investigations were optimized for display applications rather than for NLO performance. Quite recently, some research groups started the development of ferroelectric LC compounds specifically devoted to NLO applications, and indeed improved the SHG efficiency by orders of magnitude [220–224].

In order to create ferroelectric LC materials with improved second-order NLO activity, implanting well-known NLO chromophores into the core of standard ferroelectric LC molecules might be considered. However, since the direction of dipolar orientation of the ferroelectric LC molecules is parallel to the C2 axis (perpendicular to the director), these NLO active molecular groups have to be incorporated in such a fashion that the direction of maximum second-order polarizability is polarly oriented along this axis. This is expected to occur if the chromophore is coupled to the chiral center of the molecule, because the specific nature of the chiral group defines the tilt direction and hence the direction of the polar axis. This requirement is automatically fulfilled if the chirality is part of the NLO group. If this is not the case, it seems to be advantageous to locate the two groups as close to each other as possible.

As already mentioned, NLO chromophores with large β values normally consist of a donor and an acceptor group which are linked by a conjugated spacer, often in the form of aromatic rings. The direction of large second-order polarizability is then roughly along the charge-transfer axis. In view of the symmetry requirements discussed above, the NLO chromophores have to be integrated in the ferroelectric LC structure with their long axis perpendicular to the long axis of the LC molecule. It follows that liquid crystallinity can only then be compatible with such a functionalization if the NLO group is sufficiently compact and the rigid core of the mesogen sufficiently long.

On the basis of this design approach, Walba et al. [222] succeeded in synthesizing the (*o*-nitroalkoxy)phenyl biphenylcarboxylate **47** exhibiting a $d_{22}=d_{23}$ value of 0.6 pm/V ($d_{eff}=0.23$ pm/V).

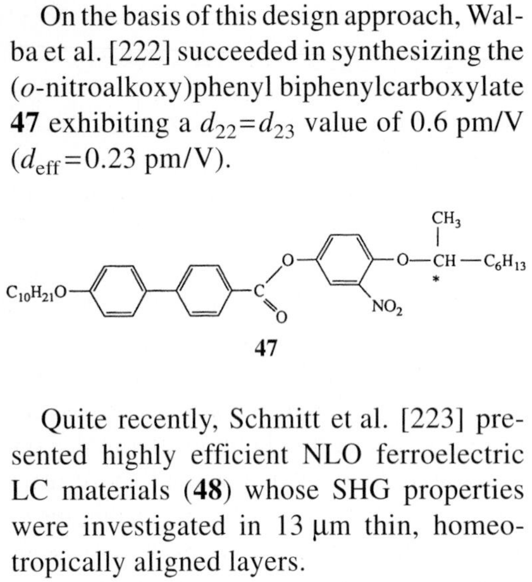

47

Quite recently, Schmitt et al. [223] presented highly efficient NLO ferroelectric LC materials (**48**) whose SHG properties were investigated in 13 μm thin, homeotropically aligned layers.

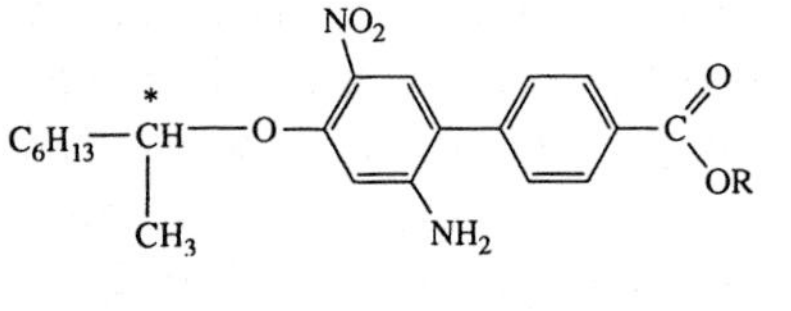

48

Positioning the chiral center in the alkoxy terminal chain adjacent to the nitroaniline dipole moment strongly enhances both P_s and the SHG efficiency, leading to the largest SHG coefficients (e.g., $d_{22}=5$ pm V^{-1}

when R = re-

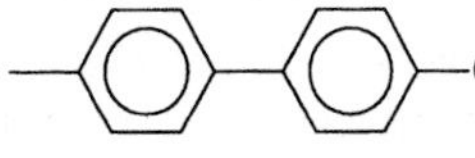

ported so far for ferroelectric low molar mass LCs.

In order to achieve amorphous polar solids, some research groups have explored ferroelectric SCLCPs, and have shown that the basic rules governing the relationships between molecular structure and macroscopic ferroelectric LC properties are the same [225–228]. The main difference between low molar mass and polymer ferroelectric LCs, however, is the existence of stable glassy phases in most of the latter.

Recently, Wischerhoff et al. [229] reported the synthesis, the thermal behavior, and the ferroelectric properties of liquid-crystalline copolysiloxanes **49** containing chiral

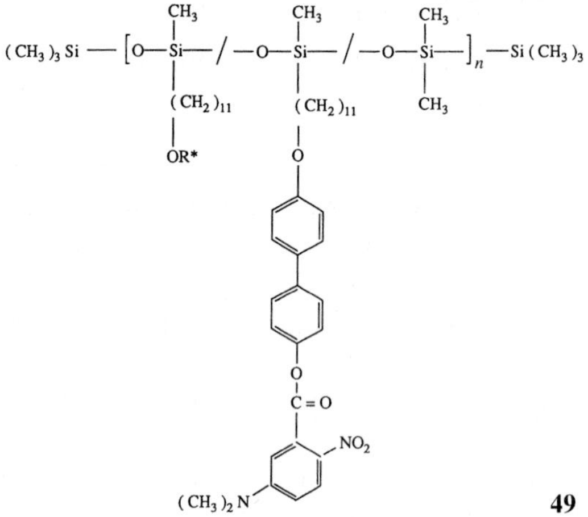

49

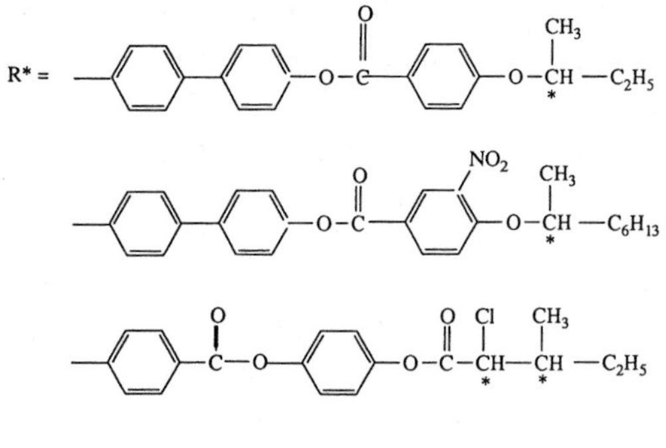

mesogens and NLO chromophores. The investigators showed that in all the polymer series a noticeable amount of chromophore (up to about 30 mol%) could be incorporated without losing the chiral smectic C phase. However, approaching high chromophore contents, a tendency for the copolymers to form the nonferroelectric smectic A phase could be detected. SHG measurements were performed on thin homeotropically aligned polymer films, with the polar axis perpendicular to the incoming beam. However, in the geometry used, the applied electric field was necessarily limited and did not allow the helix to untwist completely, leading to low SHG efficiency.

The techniques used to obtain the untwisted SmC* phase structure in low molar mass LCs and SCLCPs are limited to thin layers. In contrast to this, LC elastomers can be macroscopically uniformly oriented by mechanical deformations [230], and this orientation process is not limited to thin samples or suitable dielectric anisotropy of the material. Furthermore, for LC elastomers the oriented structure can be chemically locked in by crosslinking, resulting in the so-called liquid single crystal elastomers [231].

Quite recently, Finkelmann and co-workers [232, 233] showed that an appropriate mechanical deformation of an SmC* elastomer **50** yields a permanent macroscopically uniform orientation. This process also unwinds the helicoidal superstructure, and accordingly, frequency doubling is observed where the intensity of the SHG is directly related to the perfection of the uniform smectic layer orientation. The d_{22} and $d_{23}=d_{34}$ coefficients for a highly oriented sample were reported to be 0.1 pm/V and 0.15 pm/V, respectively. Taking into account that only 50% of the mesogenic units in the LC elastomer are active groups, these values are of the same order as those reported for low molar mass LCs containing similar chromophores [222].

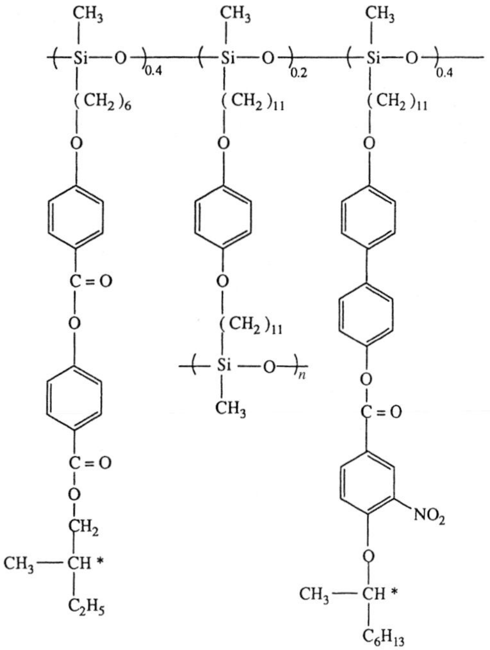

50

4.3.2.5 Rigid Rod-Like Polymers

Rigid rod-like polymers represent a special class whose phase behavior differs distinctly from that observed in flexible polymers. They are able to form LC phases, and highly ordered structures are generated due to their anisotropic molecular shape. In addition, they exhibit superior intrinsic mechanical and electrical properties and have specific advantages over polymers with a flexible backbone, such as high temperature and dimensional stability, toughness, high modulus, and tensile strength. This unique combination of profitable properties makes rigid-rod polymers particularly interesting for NLO applications. However, investigations have been mainly limited to third-order NLO properties and applications [234, 235].

Another concept, in its early stages of exploration, for obtaining large values of $\chi^{(2)}$ through the poling process as well as long-term stability, is rigid rod-like aromatic polyesters and poly(esteramide)s having long alkyl or alkoxy side chains and NLO chromophores that are either directly incorporated into the main chain [236, 237], or covalently linked to the backbone [236, 238–241] by flexible spacers. An overview including discussions on synthesis and properties is given in [242].

Three rigid-rod polymers with chromophores in the main chain (**51**, **52**, and **53**)

to an Arrhenius-like rather than a WLF-like behavior. This indicated that the relaxation process is based on a local mechanism, which is not coupled to a collective motion within the main-chain layers.

A number of rigid rod copolymers, **54** and **55**, with chromophores in the side chains were synthesized. According to the X-ray diffraction results, these polymers exhibit-

51

52

53

were used to cast transparent films with good mechanical and optical properties. Well-ordered structures were observed with polymer main-chain layers parallel to the substrate and side chains, extending normal to the layer plane (Fig. 37). One such polymer (**52**) exhibited a $\chi^{(2)}_{33}$ value of 15 pm/V at 0.632 μm for a poling field of $\cong 40$ V/μm. These new polymers showed promise for further development in that the materials once poled retained the NLO activity over long periods of time (up to 920 h at 100 °C). It was also found that the temperature dependence of the relaxation times conformed

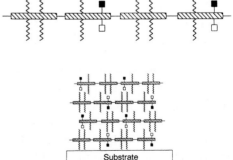

〜〜 Alkyl side chain   ⧄⧄⧄⧄ Rigid-rod main chain

□⊣, ■⊢ Donor and acceptor chromophores, respectively

Figure 37. Suggested structure of polymers **51–53**.

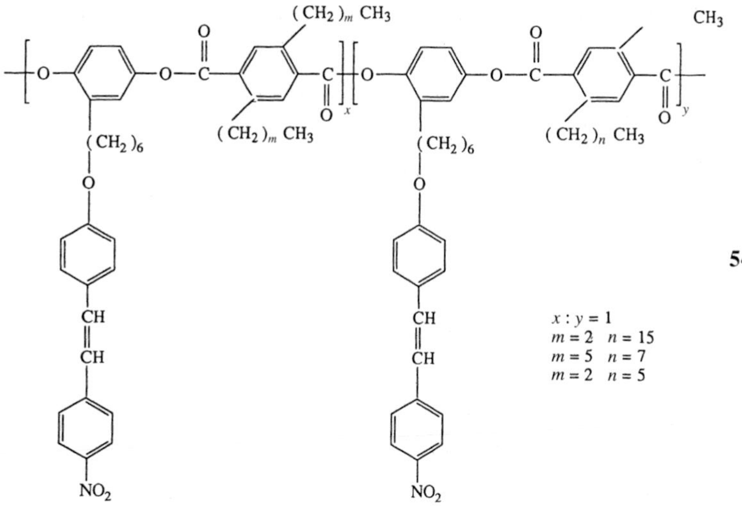

Structure **54**:

(CH$_2$)$_m$ CH$_3$ ... CH$_3$

(CH$_2$)$_6$ (CH$_2$)$_m$ CH$_3$ (CH$_2$)$_6$ (CH$_2$)$_n$ CH$_3$

CH=CH — NO$_2$ (left and right substituents)

$x : y = 1$
$m = 2 \quad n = 15$
$m = 5 \quad n = 7$
$m = 2 \quad n = 5$

Structure **55**:

OC$_n$H$_{2n+1}$ ... OC$_n$H$_{2n+1}$

C=O ... (CH$_2$)$_2$ ... N—CH$_2$CH$_3$ ... N=N ... NO$_2$

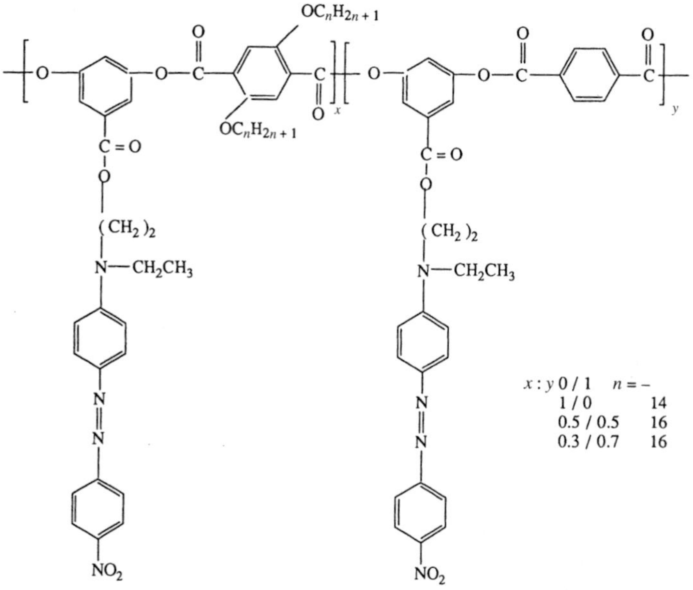

$x : y \; 0 / 1 \quad n = -$
$\quad\quad 1 / 0 \quad\quad 14$
$\quad 0.5 / 0.5 \quad 16$
$\quad 0.3 / 0.7 \quad 16$

ed a layered structure, the space between the layers formed by the main chains being filled by the interdigitated side chains (Fig. 38). Polymer solutions were spin cast into optical quality thin films and the NLO properties were measured by the attenuation total reflection (ATR) method. For series **54**, it was found that $\chi^{(2)}_{33}$ strongly depends on the length of the alkyl chains. A maximum value of $\chi^{(2)}_{33} = 11.8$ pm/V (at 0.632 μm) was obtained for polymer with $m = 5$ and $n = 7$ at the poling field of 40 V/μm. In series **55**, the second-order susceptibilities $\chi^{(2)}_{33}$ and $\chi^{(2)}_{31}$ as well as the piezoelectric coefficient, were linearly proportional to the poling voltage at a given temperature. The maximum values of $\chi^{(2)}_{33}$ were 140 (at 0.632 μm) and 91 pm/V (at 0.685 μm) at a poling field of 79 V/μm. The thermal relaxation behavior investigated at various temperatures demonstrated the enhanced stability of polymers

(a)

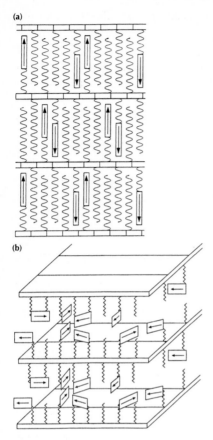

(b)

Figure 38. Suggested structures for (a) polymer **54** ($m = 2$, $n = 15$) with long side chains and (b) polymers **54** ($m = 5$, $n = 7$ and $m = 2$, $n = 5$) with short side chains.

55 compared to side chain polymer systems with flexible main chains. This enhanced stability was ascribed to hindering of the local mobility of the chromophores by the stiff main chains and the matrix of long alkoxy side groups in which the chromophores are tightly embedded.

5   References

[1] C. B. McArdle, *Side Chain Liquid Crystal Polymers*, Blackies, **1989**.

[2] J. C. Dubois, *Phys. Sci.* **1988**, *23*, 299–305.

[3] C. Noel, P. Navard, *Prog. Polym. Sci.* **1991**, *16*, 55–110.

[4] M. V. Kozlovsky, L. A. Beresnev, *Phase Trans.* **1992**, *40*, 129–169.

[5] E. Chiellini, G. Galli, *Mol. Cryst. Liq. Cryst.* **1994**, *254*, 17–36.

[6] V. P. Shibaev, L. A. Beresnev, L. M. Blinov, N. A. Platé, *Polym. Bull.* **1984**, *12*, 299–301.

[7] S. Hachiya, K. Tomoike, K. Yuasa, S. Togawa, T. Sekiya, K. Takahashi, K. Kawasaki, *J. SID* **1993**, *1/3*, 295–299.

[8] V. Percec, Q. Zheng, *Polym. Bull.* **1992**, *29*, 501–508.

[9] B. Gallot, G. Galli, E. Dossi, E. Chiellini, *Liq. Cryst.* **1995**, *18*, 463.

[10] E. Chiellini, G. Galli, E. I. Serhatli, Y. Yagci, M. Laus, A. Sante Angeloni, *Ferroelectrics* **1993**, *148*, 311–322.

[11] E. Chiellini, G. Galli, F. Cioni, E. Dossi, B. Gallot, *J. Mater. Chem.* **1993**, *3 (10)*, 1065–1073.

[12] A. Kocot, R. Wrzalik, J. K. Vij, R. Zentel, *J. Appl. Phys.* **1994**, *75 (2)*, 728–733.

[13] G. H. Hsiue, P. J. Hsieh, S. L. Wu, C. S. Hsu, *Polym. Bull.* **1994**, *33*, 159–166.

[14] A. Kocot, R. Wrzalik, J. K. Vij, M. Brehmer, R. Zentel, *Phys. Rev. B* **1994**, *50 (22)*, 16346–16356.

[15] R. Zentel, M. Brehmer, *Adv. Mater.* **1994**, *6 (7/8)*, 598–599.

[16] M. Ozaki, M. Utsumi, K. Yoshino, K. Skarp, *Jpn. J. Appl. Phys.* **1993**, *32*, L852–L855.

[17] T. Ikeda, O. Zushi, T. Sasaki, K. Ichimura, H. Takezoe, A. Fukuda, K. A. W. Skarp, *Mol. Cryst. Liq. Cryst.* **1993**, *225*, 67–79.

[18] H. Endo, S. Hachiya, T. Sekiya, K. Kawasaki, *Liq. Cryst.* **1992**, *12 (1)*, 147–155.

[19] K. Yuasa, S. Uchida, T. Sekiya, K. Hashimoto, K. Kawasaki, *Proc. SPIE* **1992**, *1665*, 154–165.

[20] L. C. Chien, L. G. Gada, *Macromolecules* **1994**, *27*, 3721–3726.

[21] G. Scherowsky, *Polym. Adv. Technol.* **1992**, *3*, 219–229.

[22] M. Pfeiffer, L. A. Beresnev, W. Haase, G. Scherowsky, K. Kühnpast, D. Jungbauer, *Mol. Cryst. Liq. Cryst.* **1992**, *214*, 125–141.

[23] E. Wischerhoff, R. Zentel, M. Redmond, O. Mondain-Monval, H. Coles, *Macromol. Chem. Phys.* **1994**, *195*, 1593–1602.

[24] M. Ido, K. Tanaka, S. Hachiya, K. Kawazaki, *Ferroelectrics* **1993**, *148*, 223–232.

[25] G. Decobert, F. Soyer, J. C. Dubois, *Polym. Bull.* **1985**, *14*, 179.

[26] G. Decobert, J. C. Dubois, S. Esselin, C. Noël, *Liq. Cryst.* **1986**, *1*, 307.

[27] S. Esselin, L. Bosio, C. Noël, G. Decobert, J. C. Dubois, *Liq. Cryst.* 1987, *2*, 505.

[28] S. Esselin, C. Noël, G. Decobert, J. C. Dubois, *Mol. Cryst. Liq. Cryst.* **1988**, *155*, 371.

[29] J. M. Guglielminetti, G. Decobert, J. C. Dubois, *Polym. Bull.* **1986**, *16*, 411.

[30] V. Shibaev, M. Kozlovski, N. Platé, L. Beresnev, L. Blinov, *Vissokolol Soedin* **1987**, *29 (7)*,
1470.

[31] M. Dumon, H. T. Nguyen, *Polym. Adv. Technol.*
1992, *3*, 197–203.

[32] H. Endo, S. Hachiya, K. Kawasaki, *Liq. Cryst.*
1993, *13 (5)*, 721–728.

[33] B. Hahn, V. Percec, *Macromolecules* **1987**, *20
(12)*, 2961.

[34] S. Ishibashi, K. Takahashi, S. Ishizawa, F. Yamamoto, *Ferroelectrics* **1993**, *148*, 277–283.

[35] P. Keller, *Ferroelectrics* **1988**, *85*, 425.

[36] C.-S. Hsu, L.-J. Shih, G.-H. Hsiue, *Macromolecules* **1993**, *26*, 3161–3167.

[37] Z. Z. Zhong, D. E. Schuele, W. L. Gordon,
L.-C. Chien, A. J. Walz, *Mol. Cryst. Liq. Cryst.*
1994.

[38] C.-S. Hsu, J.-H. Lin, L.-R. Chou, G.-H. Hsiue,
Macromolecules **1992**, *25*, 7126–7134.

[39] H. Poths, E. Wischerhoff, R. Zentel, A. Schönfeld, G. Henn, F. Kremer, *Liq. Cryst.*, in press.

[40] H. Poths, R. Zentel, *Liq. Cryst.* **1994**, *16 (5)*,
749–767.

[41] A. Schönfeld, F. Kremer, H. Poths, R. Zentel,
Proc. SPIE **1994**, *2175*, 60–70.

[42] H. Poths, A. Schönfeld, R. Zentel, F. Kremer, K.
Siemensmeyer, *Adv. Mater.* **1992**, *4 (5)*,
351–354.

[43] D. M. Walba, D. A. Zummach, M. D. Wand,
W. N. Thurmes, K. M. Moray, K. E. Arnett,
Proc. SPIE **1993**, *1911*, 21–28.

[44] H. Poths, R. Zentel, D. Hermann, G. Andersson, K. Skarp, *Ferroelectrics* **1993**, *148*, 285–
295.

[45] H. Kühnpast, J. Springer, G. Scherowsky, F.
Gießelmann, P. Zugenmaier, *Liq. Cryst.* **1993**,
14 (3), 861–869.

[46] P. Davidson, K. Kühnpast, J. Springer, G. Scherowsky, *Liq. Cryst.* **1993**, *14 (3)*, 901–910.

[47] T. Kitazume, T. Ohnogi, K. Ito, *J. Am. Chem.
Soc.* **1990**, *112*, 6608–6615.

[48] T. Kitazume, T. Ohnogi, *J. Flur. Chem.* **1990**, *47*,
459–466.

[49] J. Bömelburg, G. Heppke, J. Hollidt, *Makromol.
Chem., Rapid Commun.* **1991**, *12*, 483–488.

[50] H. Poths, R. Zentel, *Macromol. Rapid. Commun.* **1994**, *15*, 433–440.

[51] G. Scherowsky, B. Brauer, K. Grünegerg, U.
Müller, L. Komitov, S. T. Lagerwall, K. Skarp,
B. Stebler, *Mol. Cryst. Liq. Cryst.* **1992**, *215*,
257–270.

[52] H. Coles, R. Simon, H. Gleeson, J. Bone,
G. Scherowsky, A. Schliwa, U. Müller, *Polym.
Prep. Jpn. (Engl. Ed.)* **1990**, *39*, 1.

[53] S. U. Vallerien, F. Kremer, G. Scherowsky,
A. Schliwa, K. Kühnpast, E. W. Fischer, *Liq.
Cryst.* **1990**, *8 (5)*, 719–725.

[54] G. Scherowsky, K. Grüneberg, K. Kühnpast,
Ferroelectrics **1991**, *122*, 159–166.

[55] S. Uiie, K. Iimura, *Polym. J.* **1991**, *23*, 1483–
1488.

[56] S. Uiie, K. Iimura, *Ferroelectrics* **1993**, *148*,
263–269.

[57] H. J. Coles, H. F. Gleeson, G. Scherowsky,
A. Schliwa, *Mol. Cryst. Liq. Cryst.* **1990**, *7*, 117.

[58] R. Zentel, G. Reckert, B. Reck, *Liq. Cryst.* **1987**,
2 (1), 83.

[59] S. Bualek, R. Zentel, *Makromol. Chem.* **1988**,
189, 797.

[60] H. Kapitza, R. Zentel, *Makromol. Chem.* **1988**,
189, 1793.

[61] R. Zentel, *Liq. Cryst.* **1988**, *3*, 531.

[62] R. Zentel, H. Poths, F. Kremer, A. Schönfeld, D.
Jungbauer, R. Twieg, C. G. Willson, D. Yoon,
Polym. Adv. Technol. **1992**, *3*, 211–217.

[63] P. E. Cladis, H. R. Brand, *Liq. Cryst.* **1993**, *14
(5)*, 1327–1349.

[64] K. Kühnpast, J. Springer, P. Davidson, G. Scherowsky, *Makromol. Chem.* **1992**, *193*, 3097–3115.

[65] S. Uchida, K. Morita, K. Miyoshi, K. Hashimoto, K. Kawasaki, *Mol. Cryst. Liq. Cryst.* **1988**,
155, 93–102.

[66] F. Hardouin, A. M. Levelut, M. F. Achard,
G. Sigaud, *J. Chim. Phys.* **1983**, *80*, 53.

[67] A. Schönfeld, F. Kremer, R. Zentel, *Liq. Cryst.*
1993, *13 (3)*, 403–412.

[68] R. B. Meyer, L. Liebert, L. Strzelecki, P. Keller,
J. Phys. **1975**, *36*, L-69.

[69] N. Clark, S. Lagerwall, *Ferroelectrics* **1984**, *59*,
25.

[70] P. Martinot-Lagarde, *Ferroelectrics* **1988**, *84*, 53.

[71] J. W. Goodby, E. Chin, J. M. Geary, J. S. Patel,
P. L. Finn, *J. Chem. Soc., Faraday Trans.* **1987**,
83, 3429.

[72] J. S. Patel, J. W. Goodby, *Chem. Phys. Lett.*
1987, *137*, 91.

[73] S. K. Prasad, G. G. Nair, *Mol. Cryst. Liq. Cryst.*
1991, *202*, 91.

[74] S. Kimura, S. Nishiyamo, Y. Ouchi, H. Takezoe, A. Fukuda, *Jpn. J. Appl. Phys.* **1987**, *26*,
L255.

[75] S. Pfeiffer, R. Shashidhar, J. Naciri, S. Mery,
SPIE **1992**, *1665*, 166–174.

[76] S. Garoff, R. Meyer, *Phys. Rev. Lett.* **1977**, *38*,
848.

[77] E. Chiellini, G. Galli, *Mol. Cryst. Liq. Cryst.*
1994, *254*, 17–36.

[78] G. Scherowsky, K. Kühnpast, J. Springer, *Makromol. Chem., Rapid Commun.* **1991**, *12*, 381–
385.

[79] K. Skarp, G. Andersson, F. Gouda, S. T. Lagerwall, H. Poths, R. Zentel, *Polym. Adv. Technol.*
1992, *3*, 241.

[80] A. D. C. Chandani, T. Hagiwara, Y. Suzuki,
Y. Ouchi, H. Takezoe, A. Fukuda, *Jpn. J. Appl.
Phys.* **1988**, *27*, L729.

[81] P. G. de Gennes, *C. R. Acad. Sci. Paris* **1975**,
281, 101.

[82] P. G. de Gennes in *Polymer Liquid Crystals* (Eds.: A. Ciferri, W. R. Krigbaum, R. B. Meyer), Academic Press, New York, U.S.A. **1982**, Chap. 5.

[83] M. Warner in *Side Chain Liquid Crystal Polymers* (Ed.: C. B. McArdle), Blackie, Glasgow, U. K. **1989**, Chap. 2.

[84] E. Gramsgergen, L. Longa, W. de Jeu, *Phys. Rep.* **1986**, *135*, 195.

[85] J. Schätzle, W. Kaufhold, H. Finkelmann, *Makromol. Chem.* **1989**, *190*, 3269.

[86] J. P. Jarry, L. Monnerie, *Macromolecules* **1979**, *12 (2)*, 316.

[87] B. Erman, L. Monnerie, *Macromolecules* **1989**, *22 (8)*, 3348.

[88] M. Warner, K. P. Gelling, T. A. Vilgis, *J. Chem. Phys.* **1988**, *88*, 4008.

[89] M. Warner, *Phys. Scr.* **1991**, *T35*, 53.

[90] B. Deloche, E. T. Samulski, *Macromolecules* **1981**, *14*, 575.

[91] B. Deloche, E. T. Samulski, *Macromolecules* **1988**, *21*, 3107.

[92] J. Küpfer, H. Finkelmann, *Makromol. Chem., Rapid Commun.* **1991**, *12*, 717.

[93] C. H. Legge, F. J. Davis, G. R. Mitchel, *J. Phys. II France* **1991**, *1*, 1253.

[94] H. Finkelmann, H. J. Koch, G. Rehage, *Makromol. Chem., Rapid Commun.* **1981**, *2*, 317.

[95] W. Gleim, H. Finkelmann in *Side Chain Liquid Crystal Polymers* (Ed.: C. B. McArdle), Blackie, Glasgow, U. K. **1989**, Chap. 10.

[96] R. Zentel, *Angew. Chem., Adv. Mater.* **1989**, *10*, 1437.

[97] R. Zentel, M. Benalia, *Makromol. Chem.* **1987**, *188*, 655.

[98] C. Degert, H. Richard, M. Mauzac, *Mol. Cryst. Liq. Cryst.* **1992**, *214*, 179.

[99] G. R. Mitchell, F. J. Davis, A. Ashman, *Polymer* **1987**, *28*, 639.

[100] F. J. Davis, A. Gilbert, J. Mann, G. R. Mitchell, *J. Polym. Sci., Polym. Chem.* **1990**, *28*, 1455.

[101] M. Brehmer, R. Zentel, *Mol. Cryst. Liq. Cryst.* **1994**, *243*, 353.

[102] M. H. Litt, W. T. Whang, K. T. Yen, X. J. Qian, *J. Polym. Sci., Polym. Chem.* **1993**, *31*, 183.

[103] A. J. Symons, F. J. Davis, G. R. Mitchell, *Liq. Cryst.* **1993**, *14 (3)*, 853.

[104] P. J. Flory in *Principles of Polymer Chemistry,* Cornell University Press, Ithaca, New York **1953**.

[105] L. Treloar in *The Physics of Rubber Elasticity,* Clarendon, Oxford **1975**.

[106] R. Löffler, H. Finkelmann, *Makromol. Chem., Rapid Commun.* **1990**, *11*, 321.

[107] J. Schätzle, W. Kaufhold, H. Finkelmann, *Makromol. Chem.* **1989**, *190*, 3269.

[108] W. Guo, F. J. Davis, G. R. Mitchell, *Polymer* **1994**, *35 (14)*, 2952.

[109] C. Degert, Ph. D. Thesis, Université de Bordeaux I, France **1991**.

[110] J. Bastide, F. Boue, M. Buzier in *Proc. in Physics-N° 42: Molecular bases of polymer networks,* Springer, Berlin **1989**, p. 48.

[111] F. J. Davis, *J. Mater. Chem.* **1993**, *3 (6)*, 551.

[112] R. V. Talroze, T. I. Gubina, V. P. Shibaev, N. A. Platé, V. I. Dakin, N. A. Shmakova, F. F. Sukhov, *Makromol. Chem., Rapid Commun.* **1990**, *11, 67*.

[113] W. Oppermann, K. Braatz, H. Finkelmann, W. Gleim, H. J. Kock, G. Rehage, *Rheol. Acta* **1982**, *21*, 423.

[114] H. J. Kock, H. Finkelmann, W. Gleim, G. Rehage, *Polym. Sci. Technol.* **1985**, *28*, 275.

[115] W. Gleim, H. Finkelmann, *Makromol. Chem.* **1987**, *188*, 1489.

[116] T. Pakula, R. Zentel, *Makromol. Chem.* **1991**, *192*, 2410.

[117] R. A. M. Hikmet, D. J. Broer, *Polymer* **1991**, *32 (9)*, 1627.

[118] J. L. Gallani, L. Hillion, P. Martinoty, F. Doublet, M. Mauzac, *J. Phys. II France* **1996**, *6*, 443.

[119] D. J. Broer, R. G. Gossink, R. A. M. Hikmet, *Angew. Makromol. Chem.* **1990**, *45*, 3235.

[120] H. Finkelmann, H. J. Kock, W. Gleim, G. Rehage, *Makromol. Chem., Rapid Commun.* **1984**, *5*, 287.

[121] G. R. Mitchell, F. J. Davis, W. Guo, R. Cywinski, *Polymer* **1991**, *32 (8)*, 1347.

[122] C. Degert, P. Davidson, S. Megtert, D. Petermann, M. Mauzac, *Liq. Cryst.* **1992**, *12 (5)*, 779.

[123] W. Kaufhold, H. Finkelmann, H. R. Brand, *Makromol. Chem.* **1991**, *192*, 2555.

[124] J. Schatzle, H. Finkelmann, *Mol. Cryst. Liq. Cryst.* **1987**, *142*, 85.

[125] R. Zentel, M. Benalia, *Makromol. Chem.* **1987**, *188*, 665.

[126] J. Kupfer, H. Finkelmannn, *Macromol. Chem. Phys.* **1994**, *195 (4)*, 1353.

[127] R. G. Kirste, H. G. Ohm, *Makromol. Chem., Rapid Commun.* **1985**, *6*, 179.

[128] L. Noirez, G. Pepy, P. Keller, L. Benguigui, *J. Phys. France* **1991**, *1*, 821.

[129] F. Moussa, J. P. Cotton, F. Hardouin, P. Keller, M. Lambert, G. Pepy, M. Mauzac, H. Richard, *J. Phys. France* **1987**, *48*, 1079.

[130] S. Lecommandoux, L. Noirez, M. Mauzac, F. Hardouin, *J. Phys. II France* **1994**, *4*, 2249.

[131] P. Davidson, A. M. Levelut, *Liq. Cryst.* **1992**, *11 (4)*, 469.

[132] X. J. Wang, M. Warner, *J. Phys. A: Math. Gen. (U. K.)* **1987**, 713.

[133] G. G. Barclay, C. K. Ober, *Prog. Polym. Sci.* **1993,** *18*, 899.

[134] R. A. M. Hikmet, J. A. Higgins, *Liq. Cryst.* **1992**, *12 (5)*, 831.

[135] H. R. Brand, *Makromol. Chem., Rapid Commun.* **1989**, *10*, 441.

[136] H. R. Brand, H. Pleiner, *Makromol. Chem., Rapid Commun.* **1990**, *11*, 607.

[137] E. M. Terentjev, M. Warner, *J. Phys. II France* **1994**, *4*, 111.

[138] R. Zentel, *Liq. Cryst.* **1988**, *3 (4)*, 531.

[139] R. Zentel, G. Reckert, S. Bualek, H. Kapitza, *Makromol. Chem.* **1989**, *190*, 2869.

[140] W. Meier, H. Finkelmann, *Makromol. Chem., Rapid Commun.* **1990**, *11*, 599.

[141] S. U. Vallerien, F. Kremer, E. W. Fischer, H. Kapitza, R. Zentel, H. Poths, *Makromol. Chem., Rapid Commun.* **1990**, *11*, 593.

[142] A. Schonfeld, F. Kremer, S. U. Vallerien, H. Poths, R. Zentel, *Ferroelectrics* **1991**, *121*, 69.

[143] H. Hirschmann, W. Meier, H. Finkelmann, *Makromol. Chem., Rapid Commun.* **1992**, *13*, 385.

[144] R. A. M. Hikmet, *Macromolecules* **1992**, *25*, 5759.

[145] W. Meier, H. Finkelmann, *Macromolecules* **1993**, *26*, 1811.

[146] M. Brehmer, R. Zentel, *Macromol. Chem. Phys.* **1994**, *195*, 1891.

[147] M. Mauzac, H. T. Nguyen, F. G. Tournilhac, S. V. Yablonsky, *Chem. Phys. Lett.* **1995**, *240*, 461.

[148] A. F. Garito, K. D. Singer, C. C. Teng in *Nonlinear Optical Properties of Organic and Polymeric Materials* (Ed.: D. J. Williams), ACS Symp. Ser. 233, Washington, DC **1983**, pp. 1–26.

[149] D. J. Williams, *Angew. Chem., Int. Ed. Engl.* **1984**, *23*, 690.

[150] K. D. Singer in *Polymers for Lightwave and Integrated Optics. Technology and Applications* (Ed.: L. A. Hornak), Marcel Dekker, New York **1992**, pp. 321–342.

[151] M. Goodvin, D. Bloor, S. Mann in *Special Polymers for Electronics and Optoelectronics* (Eds.: J. A. Chilton, M. T. Goosey), Chapman and Hall, London **1995**, pp. 131–185.

[152] T. M. Leslie, R. N. Demartino, E. Won Choe, G. Khanarian, D. Haas, G. Nelson, J. B. Stamatoff, D. E. Stuetz, C. C. Teng, H. N. Yoon, *Mol. Cryst. Liq. Cryst.* **1987**, *153*, 451.

[153] P. Pantelis, J. R. Hill, S. N. Oliver, G. J. Davies, *Br. Telecom Technol. J.* **1988**, *6*, no. 3.

[154] J. C. Dubois, P. Le Barny, P. Robin, V. Lemoine, H. Rajbenbach, *Liq. Cryst.* **1993**, *14*, 197.

[155] G. R. Meredith, J. G. Van Dusen, D. J. Williams, *Macromolecules* **1982**, *15*, 1385.

[156] K. D. Singer, M. G. Kuzyk, J. E. Sohn, *J. Opt. Soc. Am. B: Opt. Phys.* **1987**, *4*, 968.

[157] C. P. J. M. Van der Vorst, S. J. Picken, *Proc. SPIE* **1987**, *866*, 99.

[158] C. P. J. M. Van der Vorst, S. J. Picken, *J. Opt. Soc. Am.* **1990**, *B7*, 320.

[159] G. R. Möhlmann, C. P. J. M. Van der Vorst in *Side Chain Liquid Crystal Polymers* (Ed.: C. B. McArdle), Blackie, Glasgow **1989**, pp. 330–356.

[160] P. N. Prasad, D. J. Williams, *Introduction to Nonlinear Optical Effects in Molecules and Polymers*, Wiley, New York **1991**.

[161] G. R. Meredith, J. G. Van Dusen, D. J. Williams in *Nonlinear Optical Properties of Organic and Polymeric Materials* (Ed.: D. J. Williams), ACS Symp. Ser. 233, Washington, DC **1983**, pp. 109–133.

[162] W. Maier, A. Saupe, *Z. Naturforsch.* **1958**, *13A*, 564; **1959**, *14A*, 882; **1960**, *15A*, 287.

[163] J. B. Stamatoff, A. Buckley, G. Calundann, E. W. Choe, R. De Martino, G. Khanarian, T. Leslie, G. Nelson, D. Stuetz, C. C. Teng, H. N. Yoon, *Proc. SPIE Int. Soc. Opt. Eng.* **1987**, *682*, 85.

[164] J. C. Dubois in *Conjugated Polymeric Materials: Opportunities in Electronics, Optoelectronics and Molecular Electronics* (Eds.: J. L. Brédas, R. R. Chance), Kluwer Academic, Dordrecht **1990**, pp. 321–340.

[165] L. T. Cheng, W. Tam, G. R. Meredith, G. L. J. A. Rikken, E. W. Meijer, *SPIE Nonlinear Optical Properties of Organic Materials II* **1989**, *1147*, 61.

[166] R. A. Huijts, G. L. J. Hesselink in *Nonlinear Optical Effects in Organic Polymers* (Eds.: J. Messier et al.), Kluwer Academic, Dordrecht **1989**, pp. 101–104.

[167] I. Ledoux, J. Zyss, A. Jutand, C. Amatore, *Chem. Phys.* **1991**, *150*, 117.

[168] D. Li, T. J. Marks, M. A. Ratner, *Chem. Phys. Lett.* **1986**, *131*, 370.

[169] J. S. Moore, S. I. Stupp, *Macromolecules* **1987**, *20*, 273.

[170] S. I. Stupp, H. C. Lin, D. R. Wake, *Chemistry of Materials* **1992**, *4*, 947; *Bull. Am. Phys. Soc.* **1990**, *35*, 560.

[171] M. Amano, T. Kaino, F. Yamamoto, Y. Takeuchi, *Mol. Cryst. Liq. Cryst.* **1990**, *182A*, 81.

[172] V. Percec, C. Pugh in *Side Chain Liquid Crystal Polymers* (Ed.: C. B. McArdle), Blackie, Glasgow **1989**, pp. 30–105.

[173] C. Noël, C. Friedrich, V. Léonard, P. Le Barny, G. Ravaux, J. C. Dubois, *Makromol. Chem., Macromol. Symp.* **1989**, *24*, 283.

[174] P. Le Barny, G. Ravaux, J. C. Dubois, J. P. Parneix, R. Njeumo, C. Legrand, A. M. Levelut, *Proc. SPIE Int. Soc. Opt. Eng.* **1986**, *682*, 56.

[175] C. Legrand, C. Bunel, A. Le Borgne, N. Lacoudre, N. Spassky, J. P. Vairon, *Makromol. Chem.* **1990**, *191*, 2971.

[176] N. Koide, S. Ogura, Y. Aoyama, M. Amano, T. Kaino, *Mol. Cryst. Liq. Cryst.* **1991**, *198*, 323.

[177] D. A. McL. Smith, H. J. Coles, *Liq. Cryst.* **1993**, *14*, 937.

[178] S. Yitzchaik, G. Berkovic, V. Krongauz, *Opt. Lett.* **1990**, *15*, 1120.

[179] S. H. Ou, J. A. Mann, J. B. Lando, L. Zhou, K. D. Singer, *Appl. Phys. Lett.* **1992**, *61*, 2284.

[180] S. Yitzchaik, G. Berkovic, V. Krongauz, *Chem. Mater.* **1990**, *2*, 162.

[181] M. R. Worboys, N. A. Davies, M. S. Griffith, D. G. McDonnell, *Proc. of 2nd Int. Conf. on Electrical, Optical and Acoustic Properties of Polymers*, Canterbury, UK **1990**, 18/1–18/6.

[182] C. P. J. M. Van der Vorst, S. J. Picken, *Proc. SPIE – Int. Soc. Opt. Eng.* **1993**, *2025*, 243.

[183] L. G. P. Dalmolen, S. J. Picken, A. F. de Jong, W. H. de Jeu, *J. Physique* **1985**, *46*, 1443.

[184] S. Xie, A. Natansohn, P. Rochon, *Chem. Mater.* **1993**, *5*, 403.

[185] I. Cabrera, V. Krongauz, *Nature* **1987**, *326*, 582; *Macromolecules* **1987**, *20*, 2713.

[186] S. Yitzchaik, G. Berkovic, V. Krongauz, *Adv. Mater.* **1990**, *2*, 33.

[187] I. Langmuir, V. J. Schaefer, *J. Am. Chem. Soc.* **1938**, *57*, 1007.

[188] D. Gonin, C. Noël, A. Le Borgne, G. Gadret, F. Kajzar, *Makromol. Chem., Rapid Commun.* **1992**, *13*, 537.

[189] I. A. McCulloch, R. T. Bailey, *Mol. Cryst. Liq. Cryst.* **1991**, *200*, 157.

[190] W. T. Ford, M. Bautista, M. Zhao, R. J. Reeves, R. C. Powell, *Mol. Cryst. Liq. Cryst.* **1991**, *198*, 351.

[191] A. C. Griffin, A. M. Bhatti, R. S. L. Hung, *Proc. SPIE* **1986**, *682*, 65.

[192] A. C. Griffin, A. M. Bhatti, R. S. L. Hung, *Mol. Cryst. Liq. Cryst.* **1988**, *155*, 129.

[193] A. C. Griffin, A. M. Bhatti, R. S. L. Hung in *Nonlinear Optical and Electroactive Polymers* (Eds.: P. N. Prasad, D. R. Ulrich), Plenum, New York **1988**, p. 375.

[194] M. Zhao, M. Bautista, W. T. Ford, *Macromolecules* **1991**, *24*, 844.

[195] I. A. McCulloch, R. T. Bailey, *Proc. SPIE* **1989**, *1147*, 134.

[196] A. C. Griffin, A. M. Bhatti, *Org. Mater. Nonlinear Opt.* **1989**, *69*, 295.

[197] M. O. Bautista, W. T. Ford, R. S. Duran, M. Naumann, *Div. Polym. Chem., Am. Chem. Soc., Polym. Prep.* **1992**, *33*, 1172.

[198] P. L. Carr, G. R. Davies, I. M. Ward, *Polymer* **1993**, *34*, 5.

[199] W. M. K. P. Wijekoon, Y. Zhang, S. P. Karna, P. N. Prasad, A. C. Griffin, A. M. Bhatti, *J. Opt. Soc. Am.* **1992**, *B9*, 1832.

[200] R. B. Findlay, T. J. Lemmon, A. H. Windle, *Mater. Res. Soc. Symp. Proc.* **1990**, *175*, 305; *J. Mater. Res.* **1991**, *6*, 604.

[201] R. N. De Martino, E. W. Choe, G. Khanarian, D. Haas, T. Leslie, G. Nelson, J. Stamatoff, D. Stuetz, C. C. Teng, H. Yoon in *Nonlinear Optical and Electroactive Polymers* (Eds.:

P. N. Prasad, D. R. Ulrich), Plenum, New York **1988**, pp. 169–187.

[202] M. Schulze, *Trends Polym. Sci.* **1994**, *2*, 120.

[203] T. M. Leslie, US Patent 4 807 968, **1989**.

[204] R. N. De Martino, European Patent Application 0 294 706, **1988** (Priority 1987 US 58 414).

[205] R. N. De Martino, H. N. Yoon, J. B. Stamatoff, European Patent Application 0 271 730, **1987** (Priority: 1986 US 933 425).

[206] B. Reck, H. Ringsdorf, *Makromol. Chem., Rapid Commun.* **1985**, *6*, 291.

[207] S. Piercourt, N. Lacoudre, A. Le Borgne, N. Spassky, C. Friedrich, C. Noël, *Makromol. Chem.* **1992**, *193*, 705.

[208] D. Gonin, Thèse de Doctorat de l'Université Paris, Paris **1994**.

[209] B. Guichard, Thèse de Doctorat de l'Université Paris, Paris **1995**.

[210] J. C. Dubois, G. Decobert, P. Le Barny, S. Esselin, C. Friedrich, C. Noël, *Mol. Cryst. Liq. Cryst.* **1986**, *137*, 349.

[211] D. R. Ulrich, *Mol. Cryst. Liq. Cryst.* **1990**, *189*, 3.

[212] D. Gonin, G. Gadret, C. Noël, F. Kajzar in *Nonlinear Optical Properties of Organic Materials VI* (Ed.: G. Möhlmann), *Proc. SPIE* **1993**, *2025*, 129.

[213] R. H. Page, M. C. Jurich, B. Reck, A. Sen, R. J. Twieg, S. D. Swalen, G. C. Bjorklund, C. G. Willson, *J. Opt. Soc. Am.* **1990**, *B7*, 1239.

[214] J. W. Goodby, *J. Mater. Chem.* **1991**, *1*, 307.

[215] I. Drevensek, R. Blinc, *Condensed Matter News* **1992**, *1*, 14.

[216] J. Y. Liu, M. G. Robinson, K. M. Johnson, D. Doroski, *Opt. Lett.* **1990**, *15*, 267.

[217] A. Taguchi, Y. Oucji, H. Takezoe, A. Fukuda, *Jpn. J. Appl. Phys.* **1989**, *28*, L-997.

[218] M. Ozaki, M. Utsumi, T. Gotou, Y. Morita, K. Daido, Y. Sadohara, K. Yoshino, *Ferroelectrics* **1991**, *121*, 259.

[219] K. Yoshino, M. Utsumi, Y. Morita, Y. Sadohara, M. Osaki, *Liq. Cryst.* **1993**, *14*, 1021.

[220] D. M. Walba, M. B. Ros, N. A. Clark, R. Shao, K. M. Johnson, M. G. Robinson, J. Y. Liu, D. Doroski, *Mol. Cryst. Liq. Cryst.* **1991**, *198*, 51.

[221] J. Y. Liu, M. G. Robinson, K. M. Johnson, D. M. Walba, M. B. Ros, N. A. Clark, R. Shao, D. Doroski, *J. Appl. Phys.* **1991**, *70*, 3426.

[222] D. M. Walba, M. B. Ros, N. A. Clark, R. Shao, M. G. Robinson, J. Y. Liu, K. M. Johnson, D. Doroski, *J. Am. Chem. Soc.* **1991**, *113*, 5471.

[223] K. Schmitt, R. P. Herr, M. Schadt, J. Fünfschilling, R. Buchecker, X. H. Chen, C. Benecke, *Liq. Cryst.* **1993**, *14*, 1735.

[224] M. Schadt, *Liq. Cryst.* **1993**, *14*, 73.

[225] H. Kapitza, R. Zentel, R. J. Twieg, C. Nguyen, S. U. Vallerien, F. Kremer, C. G. Wilson, *Adv. Mater.* **1990**, *2*, 539.

[226] H. Kapitza, H. Poths, R. Zentel, *Makromol. Chem., Macromol. Symp.* **1991**, *44*, 117.

[227] R. Zentel, H. Poths, F. Kremer, A. Schönfeld, D. Jungbauer, R. Twieg, C. G. Wilson, D. Yoon, *Polym. Adv. Technol.* **1992**, *3*, 211.

[228] D. M. Walba, P. Keller, D. S. Parmar, N. A. Clark, M. D. Wand, *J. Am. Chem. Soc.* **1989**, *111*, 8273.

[229] E. Wischerhoff, R. Zentel, M. Redmond, O. Mondain-Monval, H. Coles, *Makromol. Chem. Phys.* **1994**, *195*, 1593.

[230] W. Gleim, H. Finkelmann in *Side Chain Liquid Crystalline Polymers* (Ed.: C. B. McArdle), Blackie, Glasgow **1989**, pp. 287–308.

[231] J. Küpfer, H. Finkelmann, *Makromol. Chem., Rapid Commun.* **1991**, *12*, 717.

[232] K. Semmler, H. Finkelmann, *Polym. Adv. Technol.* **1994**, *5*, 231.

[233] I. Benné, K. Semmler, H. Finkelmann, *Macromol. Rapid. Commun.* **1994**, *15*, 295.

[234] B. A. Reinhardt, *Trends Polym. Sci.* **1993**, *1*, 4.

[235] H. S. Nalwa, *Adv. Mater.* **1993**, *5*, 341.

[236] G. Wegner, D. Neher, C. Heldmann, H. J. Winkelhahn, T. Servey, M. Schulze, C. S. Kang, *Mater. Res. Soc. Symp. Proc.* **1994**, *328*, 15.

[237] C. Heldmann, Ph. D. Thesis, University of Mainz, Germany **1994**.

[238] T. K. Servay, H. J. Winkelhahn, L. Kalvoda, M. Schulze, C. Boeffel, D. Neher, G. Wegner, *Ber. Bunsenges. Phys. Chem.* **1993**, *97*, 1272.

[239] H. J. Winkelhahn, T. K. Servay, L. Kalvoda, M. Schulze, D. Neher, G. Wegner, *Ber. Bunsenges. Phys. Chem.* **1993**, *97*, 1287.

[240] C.-S. Kang, H. J. Winkelhahn, M. Schulze, D. Neher, G. Wegner, *Chem. Mater.* **1994**, *6*, 2159.

[241] Z. Sekkat, C.-S. Kang, E. F. Aust, G. Wegner, W. Knoll, *Chem. Mater.* **1995**, *7*, 142.

[242] M. Schulze, *Trends Polym. Sci.* **1994**, *2*, 120.

Chapter V
Physical Properties of Liquid Crystalline Elastomers

Helmut R. Brand and Heino Finkelmann

1 Introduction

In the preceding chapters the synthesis properties of linear liquid crystalline polymers are described, where different approaches exist to obtain the liquid crystalline state: rod-like or disc-like mesogenic units are either incorporated in the polymer backbone or are attached as side groups to the monomer units of the main chain. Following conventional synthetic techniques these linear polymers can be converted to polymer networks. Compared to low molar mass liquid crystals and linear liquid crystalline polymers, these liquid crystalline elastomers exhibit exceptional new physical and material properties due to the combination and interaction of polymer network elasticity with the anisotropic liquid crystalline state.

In this chapter we review the physical properties of liquid crystalline elastomers using publications until 1996. For a review of the physical properties of strongly cross-linked liquid crystalline polymers, which give rise to anisotropic nonliquid crystalline materials (duromers or anisotropic solids) we refer to the recent review by Hikmet and Lub [1].

To organize this chapter efficiently we have split it into two sections: the behavior near phase transitions and the physical properties of the bulk phases. Accordingly, subsections contain material on one type of phase transition or on the bulk behavior of (typically) one specific liquid crystalline phase.

2 Phase Transitions

2.1 The Nematic–Isotropic Phase Transition

2.1.1 Mechanical Properties of Polydomains

Much of the early physical work on liquid crystalline elastomers (LCE) focused on the mechanical properties of polydomain samples. Gleim and Finkelmann [2] studied the length of the sample as a function of temperature for a number of values of the applied load over a large temperature range including the vicinity of the nematic–isotropic transition. In addition the temperature dependence of the nominal stress was evaluated for constant sample deformation. As a result it was found that the length of the sample increased when cooling through the phase transition indicating an increasing degree of order in the orientation of the mesogens.

In the pretransitional region above the transition it has been shown that a stress–strain relation similar to that of common rubbers occurs [3]. From the behavior of stress–strain curves just below the phase transition, but above the transition to a monodomain state it has been demonstrated that one can linearly extrapolate to stress zero and extract the length l_0 of a corresponding monodomain sample [3]. When this length is plotted as a function of reduced temperature (Fig. 1), one finds a behavior that is reminiscent of the temperature behavior of the order parameter, a feature that will be discussed in more detail below.

The change in length of an elastomer under a constant external load has been studied in detail in the immediate vicinity of the phase transition [4]. In Fig. 2 the strain

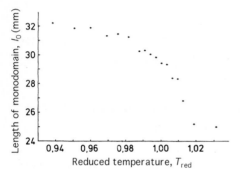

Figure 1. Length l_0 of a monodomain elastomer vs. the reduced temperature T_{red} (reproduced with permission from [3]).

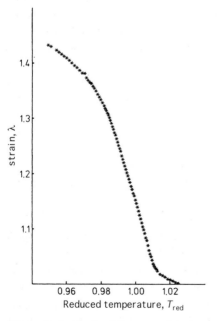

Figure 2. Strain $\lambda = L/L_0$ vs. reduced temperature T_{red} at constant nominal stress (reproduced with permission from [4]).

$\lambda = L/L_0'$ is plotted as a function of the reduced temperature T_{red} at constant nominal stress $\sigma_n = 2.11 \times 10^{-2}$ N mm^{-2}. Here L_0' is the loaded sample length at $T_{red} = 1.03$. These results will also be used below to establish a close connection between the strain tensor and the nematic order parameter. It has also been shown that a quadratic stress–strain relation yields in the isotropic phase above the nematic–isotropic phase transition a good description of the data for elongations up to at least 60% [4].

It has been demonstrated that a uniaxial compression of cylindrical samples along their axis leads to a reorientation of the mesogenic groups orthogonal to their orientation under strain [5], thus complementing experiments done earlier [2]. Furthermore, it has been shown that this reorientation of the mesogenic side groups takes place when compression changes into strain. Measurements of the length and the diameter of cylindrical samples as a function of temperature were used to determine the linear expansion coefficients parallel $(\beta_{||})$ and perpendicular $(\beta_\perp)$ to the cylinder axis. It is found that both, $(\beta_{||})$ and $(\beta_\perp)$ peak sharply in the vicinity of the nematic–isotropic transition (Fig. 3).

Measurements of the temperature dependent elongation λ over a large temperature range including the nematic–isotropic transition for fixed load have been reported [6]. Depending on the compound studied it was found that the elongation either peaks or dips at this phase transition [6].

Only comparatively few dynamic mechanical measurements in the vicinity of the nematic–isotropic transition have been reported; these include some for side-chain liquid crystalline elastomers [7, 8], and some for liquid crystalline elastomers with mesogenic groups in the main chains and in the side-chains (combined LCE) [9]. The latter reference describes the evaluation of

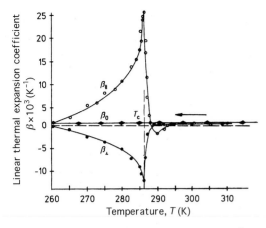

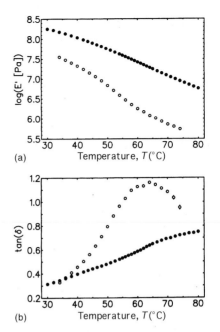

Figure 3. Linear thermal expansion coefficient β as a function of the temperature T of a cylindrical network determined parallel ($\beta_{\parallel}$) and perpendicular ($\beta_{\perp}$) to the longitudinal axis of the unloaded sample (reproduced with permission from [5]).

Figure 4. Temperature dependence of $\log(E')$ and $\tan(\delta)$ for nematic elastomers measured at 30 Hz (reproduced with permission from [8]).

the shear storage modulus G' and the shear loss modulus G'' as a function of temperature across various liquid crystalline phase transitions including the nematic–isotropic transition. It is found that the storage modulus always stays higher than the loss modulus in the vicinity of the nematic–isotropic transition. A torsional oscillator was used to measure G' and G'' in a side-chain LCE and it was found that both, G' and G'' show a jump at the phase transition and that G'' is smaller than G' on both sides of the nematic–isotropic phase transition [7].

Dynamic mechanical and dynamic stress–optical measurements have been performed on side-chain LCEs with different crosslinking densities [8]. It was found that the relaxation strengths depend strongly on the crosslinking density demonstrating that the vicinity of the crosslinking points perturbs the liquid crystalline order. In this experiment the samples were exposed to a time-dependent stress $\sigma(t)$, and the time-dependent responses, that is the time-dependent strain $\varepsilon(t)$ and the time-dependent birefringence $\Delta n(t)$, were measured. The real part

of the dynamic elastic modulus, E', and the loss factor, δ, obtained at a frequency $f = 30$ Hz are shown in Fig. 4. From this figure it emerges that the dynamic mechanical properties reveal no special feature around the nematic–isotropic transition ($T_{N-I} \approx 60\,^{\circ}\text{C}$) for the frequency chosen.

2.1.2 Mechanical Properties of Monodomains: Liquid Single Crystal Elastomers

Static mechanical properties in the vicinity of the nematic–isotropic transition in liquid single crystal elastomers (LSCEs) have been investigated [10, 11]. In Fig. 5 the deformation $L/L_0(\text{mon})$ is plotted as a function of the reduced temperature T_{red}. Here $L_0(\text{mon})$ denotes the length of the LSCE at T_{N-I}, the phase transition temperature of the nematic–isotropic phase transition and

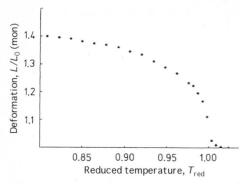

Figure 5. Deformation (L/L_0(mon)) of a LSCE as a function of the reduced temperature T_{red}. L_0(mon), length of the elastomer at T_{N-I} (reproduced with permission from [10]).

$T_{red} = T/T_{N-I}$. The continuous behavior of the data in this graph in the vicinity of T_{N-I}, as well as the fact that there is a finite deformation above T_{N-I}, reflects the influence of the internal field, which is due to the synthetic process of LSCE described in section 3 of this chapter. In this process the mechanical stress applied during the second crosslinking step is frozen in giving rise to the internal strain field observed.

The influence of the crosslinking density and annealing temperatures on the physical properties of LSCEs has been studied [11]. In Fig. 6 we have plotted the relative strength L_T/L_{298}, where L_{298} is the length of the sample at the reference temperature $T_{ref} = 298$ K, as a function of temperature for various annealing temperatures. It can be seen from the curve taken at the highest annealing temperature ($T = 150\,°C$) that there is a much smaller length change in the vicinity of T_{N-I} for the higher crosslinking density reflecting the fact that actually a duromer is investigated. The nematic properties of the latter are confined to yielding an anisotropic material, which does not show, however, the director variability and dynamics of the more weakly crosslinked LSCE samples. We also note that one can infer from the other curves in the same figure that the crosslinking process is far from completion at lower annealing temperatures.

Using X-ray investigations and optical measurements the reorientation behavior of nematic monodomain samples has been studied in detail by Kundler and Finkelmann [12]. This reorientation behavior has been modeled using a bifurcation analysis of the macroscopic dynamic description by Wei-

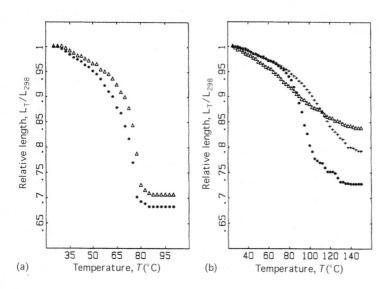

Figure 6. The relative length (L_T/L_{298}) of elastomers as a function of temperature:
(a) Annealing temperature $\Delta = 150\,°C$, $* = 60\,°C$,
(b) annealing temperature $\Delta = 150\,°C$, $+ = 75\,°C$, $* = 40\,°C$ (reproduced with permission from [11]).

lepp and Brand [13] and, subsequently, by Verwey et al. [14] using rubber elasticity for an anisotropic rubber.

2.1.3   Optical Properties of Polydomains

The first stress–optical measurements above the nematic–isotropic transition have been described shortly after the synthesis of side-chain LCE [15]. In this publication it was demonstrated that the stress–optical coefficient shows strong pretransitional effects, and that it changes sign when the spacer length between the mesogen and the polymeric backbone is changed from 3 to 4 indicating that the mesogenic groups are oriented predominantly parallel to the axis of deformation in one case and perpendicular to this axis in the other.

Similar results for this stress–optical coefficient were obtained for a different family of LCEs [2]. In addition it was shown that the birefringence, Δn, is linear in the applied stress for the range of temperatures investigated, a result one would expect on the basis of linear stress optics; these results are summarized in Fig. 7.

Early measurements of the IR-dichroism in stretched samples of LCEs have been described [16]. The dependence of the IR dichroism of the CN-stretching mode was used to obtain the absorption coefficients and to extract the order parameter as a function of temperature. However, while these measurements yield reliable results inside the nematic phase, the sensitivity of this method turns out to be not high enough to analyze quantitatively the influence of the applied stress on the order parameter and on T_{N-I}.

Stress–optical experiments on cylindrical samples under uniaxial compression have also been performed [5]. It was found

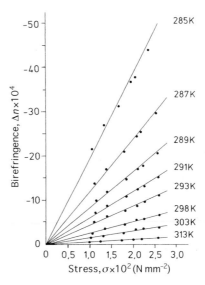

Figure 7. Birefringence Δn versus true stress σ of a nematic elastomer for different temperatures above clearing temperature T_{N-I} as indicated (reproduced with permission from [2]).

that the stress–optical coefficient behaves qualitatively similarly to that in strained samples when approaching the nematic–isotropic transition. There is, however, one important qualitative difference for the cylindrical samples under uniaxial compression: the reorientation of the mesogenic side groups is parallel (perpendicular) with respect to the direction of the mechanical force when it is perpendicular (parallel) to this direction for strain. These observations complement earlier results on the same family of compounds [2]. In addition it has been shown, using measurements of the birefringence Δn as a function of temperature, that unloaded cylindrical samples behave optically like biaxial crystals provided the director **n** of the nematic phase is oriented in the plane perpendicular to the cylinder axis [5].

The results and the analysis of detailed stress–optical measurements in the immediate vicinity of the phase transition have been reported [3, 4]. In Fig. 8 the reciprocal

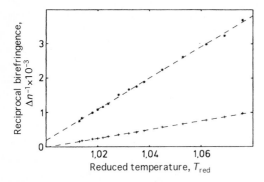

Figure 8. Reciprocal birefringence Δn^{-1} vs. reduced temperature T_{red} of a nematic elastomer for different loads ($* = 1.0 \times 10^{-2}$, $+ = 4.5 \times 10^{-2}$ N mm^{-2}) (reproduced with permission from [3]).

birefringence is plotted as a function of the reduced temperature T_{red}. As one can see from this figure the extrapolated intersection of the curve through the experimental data with the abscissa depends on the applied stress. This implies in turn that the hypothetical second order phase transition temperature T^* also depends on the applied stress as can be seen in Fig. 9. Using IR dichroism measurements in addition below T_{N-I} it was shown that an increase in external stress leads to a shift of T_{N-I} to higher temperatures, and to an increase in the nematic order parameter [3]. Both results are

in agreement with the Landau model for this phase transition by de Gennes [17].

Detailed stress–optical measurements have been analyzed to yield further information [4]. In Fig. 10 the birefringence (order parameter) was plotted as a function of reduced temperature for several nominal stresses σ_n. These results were combined with the predictions of the Landau model and static stress–strain curves and led to a number of interesting consequences. In Fig. 11 the shift in the phase transition temperature is plotted as a function of nominal stress and shifts of up to 7.5 K were found; compared to maximum displacements by electric and magnetic fields of about 5 mK in low molecular weight materials. In Fig. 12 the birefringence Δn is shown as a function of strain $\lambda = L/L_0'$ at constant nominal stress $\sigma_n = 2.11 \times 10^{-2}$ N mm^{-2}. A strictly

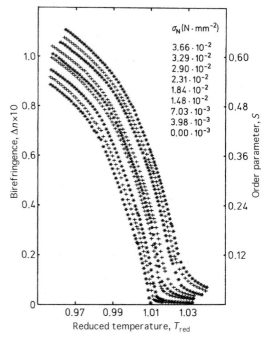

Figure 10. On the left side birefringence Δn and on the right side order parameter S vs. $T_{red} = T/T_{N-I(\sigma=0)}$ for different nominal stresses σ_n (reproduced with permission from [4]).

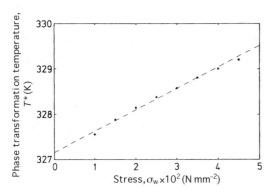

Figure 9. Phase transformation temperature T^* vs. true stress σ_w for a nematic elastomer (reproduced with permission from [3]).

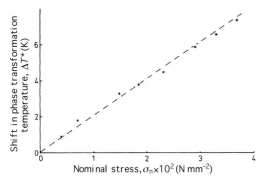

Figure 11. Shift of the phase transformation temperature $\Delta T = T_{\mathrm{N-I}(\sigma)} - T_{\mathrm{N-I}(\sigma=0)}$ vs. nominal stress σ_n (reproduced with permission from [4]).

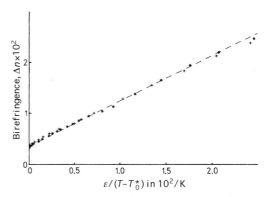

Figure 13. Birefringence Δn vs. $\varepsilon/(T - T_0^*)$ at two different constant stresses (reproduced with permission from [4]).

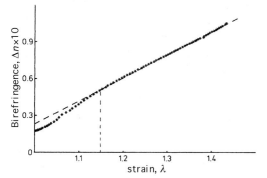

Figure 12. Birefringence Δn vs. the strain $\lambda = L/L_0$ at constant nominal stress $\sigma_\mathrm{n} = 2.11 \times 10^{-2}$ N mm^{-2} (reproduced with permission from [4]).

linear relation is obtained for temperatures in the nematic phase as expected for quantities that can be used equivalently to characterize the state of order in a system [4]. The observed behavior in the isotropic phase can be analyzed using a Landau energy expression in the isotropic phase. A linear relation holds if the order parameter or the birefringence are plotted as a function of $[(T - T_0^*)]^{-1}$ (Fig. 13). This plot has also been used to determine the bilinear cross-coupling coefficient U between the strain and the order parameter in the Landau energy.

By making use of the form of the Landau energy proposed by de Gennes [17], all coefficients in this expression have been evaluated from experimental data [4]. Using these data to calculate U, one obtains a value that is considerably smaller than the one determined from the linearized analysis outlined above. As has been noted, however, a consistent analysis should not only include terms quadratic in the strains and bilinear coupling terms of the order parameter and the strain, but rather also nonlinear effects as well as nonlinear coupling terms between strain and the nematic order parameter [4].

The only existing publication describing dynamic optical properties in LCEs in the vicinity of the nematic–isotropic transition appear to be the measurements of the stress–optical coefficient [8]. It is found that there are two relaxation frequencies and strengths characterizing the dynamic birefringence of the LCE samples analyzed in the pretransitional regime. The authors ascribe the slow and fast processes observed to the lateral coupling of the spacers to the mesogens resulting in a difference in the rotational mobility for the two short axes. In addition they find that a higher crosslinking density leads to lower relaxation strengths reflecting the

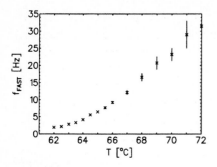

Figure 14. Temperature dependence of the relaxation frequency, f_{fast} (reproduced with permission from [8]).

perturbations of the nematic order in the vicinity of the crosslinking points. Both relaxation frequencies show critical behavior and it has been pointed out that this is in qualitative agreement with de Gennes' mean field analysis [8]. In Fig. 14 we have plotted the frequency of the fast relaxation process as a function of temperature in the isotropic phase. As will be discussed in the subsection on modeling the curvature in this figure can be interpreted in the framework of a mode-coupling theory.

2.1.4   Optical Properties of Monodomains: Liquid Single Crystal Elastomers

Optical techniques to investigate the isotropic–nematic transition in liquid single crystal elastomers have been used [10, 11, 18]. The nematic order parameter S_N (the modulus of the order parameter Q_{ij}) has been determined by measuring the absorbances $A_{\parallel}$ and $A_{\perp}$ using the integrated intensities of the CN absorption band with the incident light polarized parallel and perpendicularly to the director of the monodomain [10, 18]. The result is plotted in Fig. 15 as a function of reduced temperature. It is similar to these obtained for polydomains or-

dered macroscopically by an external mechanical field (compare the last subsection). The behavior in the immediate vicinity of the nematic–isotropic transition and, in particular, the presence of paranematic order above T_{N-I} reflects the influence of the internal field of LSCE acquired in their synthetic process.

In Fig. 16 we show the nematic order parameter S_N obtained from the type of IR dichroism measurements just described as a function of the deformation ($L/L_0(mon)$) of the same elastomer, where $L_0(mon)$ is the length of the LSCE at T_{N-I}. As one can see

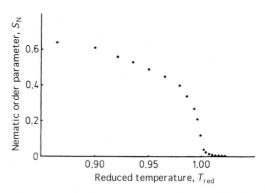

Figure 15. Nematic order parameter S_N vs. reduced temperature T_{red} for a LSCE (reproduced with permission from [18]).

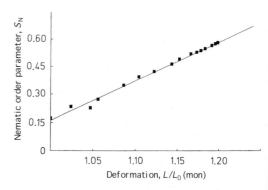

Figure 16. Nematic order parameter S_N as function of the deformation ($L/L_0(mon)$). $L_0(mon)$, length of the LSCE at T_{N-I} (reproduced with permission from [10]).

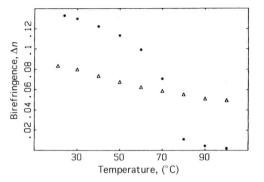

Figure 17. The birefringence, Δn, as function of temperature for differently crosslinked networks, * = weak, Δ = strong crosslinking density (reproduced with permission from [11]).

from this graph very clearly there is also for LSCE a strictly linear relationship between the deformation and the degree of nematic order underscoring once more the fact that the deformation can be used as an order parameter for liquid crystalline elastomers.

The influence of the crosslinking density on the birefringence has been studied [11]. For weakly crosslinked materials the results obtained coincide with the previous ones while for strongly crosslinked liquid crystalline elastomers the birefringence is only weakly affected in the vicinity of the phase transition as can be seen in Fig. 17. We stress, however, that the strongly crosslinked LSCE are duromers for which there are no independent macroscopic director degrees of freedom anymore, which are characteristic for the nematic liquid crystalline phase in low molecular weight materials, in sidechain polymers and in weakly crosslinked liquid crystalline elastomers.

2.1.5 Electric and Dielectric Properties

While the effect of static electric and magnetic fields on the bulk of the nematic phase

has been investigated previously, such investigations, corresponding to the measurement of electric and magnetic birefringence, do not seem to exist in the pretransitional region of the nematic–isotropic transition. Their effect on equilibrium properties such as the transition temperature T_{N-I} can be expected to be even smaller than those for low molecular weight materials, for which T_c can be shifted at best by about 10^{-3} K [19], since they also involve elastic deformations of the network via coupling effects between the fluctuations of the nematic order parameter and the network. Concerning the detection of fluctuations we note that the measurement of strain birefringence discussed earlier in this section yields much bigger effects. We close this short subsection by noting that there also appear to be neither dielectric nor electric conductivity measurements in the vicinity of the nematic–isotropic transition in liquid crystalline elastomers.

2.1.6 NMR Investigations of Monodomains

While there are not many NMR investigations of the isotropic–nematic transition in sidechain liquid crystal elastomers [20, 21], they turned out to be crucial in answering the question whether strained LCEs show a first order phase transition with a classical two phase region or whether there is a continuous change of the local degree of nematic order, since one is in a state beyond the critical point. As already mentioned earlier in this section this question could not be settled conclusively with the mechanical and optical investigations described above.

In Fig. 18 we show the ^{2}H-NMR spectra of both an uncrosslinked linear liquid crystalline polymer and a monodomain elasto-

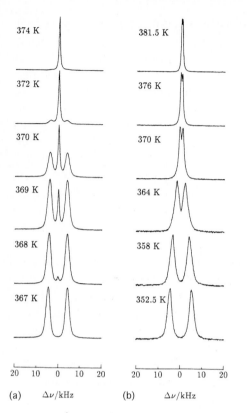

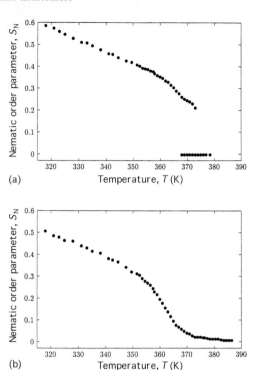

Figure 19. Order parameter as a function of temperature: (a) linear polymer; (b) LSCE (reproduced with permission from [20]).

Figure 18. ^{2}H-NMR spectra at different temperatures: (a) linear polymer; (b) LSCE (reproduced with permission from [20]).

mer as a function of temperature. Comparing the two sets of spectra as the temperature is varied from a value above to one below the transition in several steps, one sees immediately, that the spectra of the polymer show a coexistence of the isotropic peak and of the two symmetric nematic peaks over a finite temperature range. This behavior is qualitatively different from that observed in the LSCE, where the peak in the isotropic phase is found to split symmetrically when passing into the nematic phase. In Fig. 19 we see that the degree of local nematic order (which is proportional to the quadrupolar splitting) shows a jump as a function of temperature as well as a two phase coexistence for the liquid crystalline polymer,

while the LSCE with its built-in mechanical field shows a continuous decrease of the local degree of order as the temperature is increased and no two phase coexistence region. It is thus clear, that one is in a situation beyond a critical point for the LSCE due to the mechanical stress that was applied to the LSCE during its synthetic process.

It should be pointed out that, in addition to the built-in mechanical field discussed so far, there is also a contribution from finite size effects on the nematic–isotropic transition in liquid crystalline elastomers due to the finite average distance between crosslinking points. This question has been discussed, and it has been pointed out that varying the crosslinking density aids in disentangling these two different types of effects [20–22].

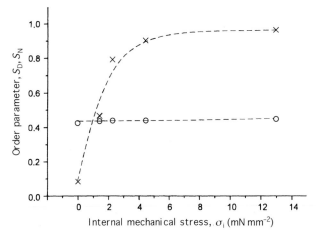

Figure 20. Nematic order parameter S_N (o) and director order parameter S_D (×) as function of internal mechanical stress σ_i, obtained from ^{2}H NMR lineshape analysis (after [21]).

In addition it was also stressed recently that one can deduce the degree of alignment of the director orientation from NMR measurements using a quantitative line shape analysis [21]. This is brought out in Fig. 20 where both, the degree of local nematic order (the order parameter modulus) and the degree of alignment of the director orientation are plotted as a function of an externally applied mechanical stress. While the degree of local nematic order is essentially constant as a function of the applied mechanical stress, the degree of alignment of the director orientation increases for low values of the external stress rather rapidly and saturates for higher values.

2.1.7   Models for the Nematic–Isotropic Transition for Poly- and Monodomains

As the experimental investigations have focused predominantly on static properties of the nematic–isotropic transition, most of the theoretical papers have used a Ginzburg–Landau description involving an expansion in the nematic order parameter to describe static properties [4, 17, 22–25].

The most important coupling to deformations of the network is the one that is linear in both the strain of the network and the nematic order parameter. As has been discussed earlier in this section this leads to the consequence that the strain tensor can be used as an order parameter for the nematic–isotropic transition in nematic sidechain elastomers, just as the dielectric or the diamagnetic tensor are used as macroscopic order parameters to characterize this phase transition in low molecular weight materials. But it has also been stressed that nonlinear elastic effects as well as nonlinear coupling terms between the nematic order parameter and the strain tensor must be taken into account as soon as effects that are nonlinear in the nematic order parameter are studied [4, 25]. So far, no deviation from classical mean field behavior concerning the critical exponents has been detected in the static properties of this transition and correspondingly there are no reports as yet discussing static critical fluctuations.

To account for the mechanical reorientation properties of monodomains (LSCE), which differ depending on whether the crosslinking has been done in the nematic or the isotropic phase, the concept of frozen order has been suggested [22]. The essential idea is to introduce a splitting of the macroscopic nematic order parameter into two contributions, one associated with the vicinity of the crosslinking points and one with the conventional nematic order of the mesogenic units in the sidechains in the regions of the LSCE sufficiently far away from the crosslinking points. This decomposition turns out to be useful to account for a number of static and dynamic experimental results of these spatially inhomogeneous systems. As the vicinity of the crosslinking cannot reorient as easily, the elastic and the optic responses to external mechanical stress depend on the direction of the applied mechanical force as well as on the crosslinking density used.

The polydomain–monodomain transition in nematic LCE was investigated [26] by incorporating a local anchoring interaction into the Ginzburg–Landau description of the nematic–isotropic transition in LCE [4, 17, 25].

Since there are very few dynamic experimental investigations of pretransitional effects [8], not much modeling has been reported to date either. Based on work for the macroscopic dynamics of the nematic–isotropic transition in sidechain polymers [27–29], it has been suggested [28] that the non-meanfield exponent observed in dynamic stress–optical experiments [8] can be accounted for at least qualitatively by the mode-coupling model [28, 29]. Intuitively this qualitatively new dynamic behavior can be traced back to static nonlinear coupling terms between the nematic order parameter and the strain tensor.

2.1.8   X-ray Investigations

While there have been numerous X-ray investigations of the bulk nematic phase, a systematic investigation of the nematic–iso-

tropic transition (or of any phase transition in liquid crystalline elastomers) using X-rays does not exist. We note, however, that already the second publication describing the properties of sidechain liquid crystalline elastomers compared X-ray data obtained in the isotropic phase above the phase transition with those in the nematic phase of a stretched elastomer [15]. It was pointed out that, in addition to the usual wide angle halo associated with short range positional order in both phases, there is a second halo which is associated with the elastomer, since it vanished upon swelling in a low molecular weight material. A similar cooling procedure was used to show that the mean value of the angle between the molecular longitudinal axis of the mesogenic sidechains and the director of the nematic phase is strongly reduced as one reduces the temperature indicating an increasing degree of order with decreasing temperature in the nematic phase [2].

2.2   The Cholesteric–Isotropic Phase Transition

While the first paper on liquid crystalline elastomers [30] already reports the detection of a cholesteric–isotropic transition using differential calorimetry and polarizing microscopy, comparatively little work has been done to characterize thy physical properties in the vicinity of this phase transition (compare, however, also the discussion of electromechanical effects in the next section) [9, 30, 31]. Combined liquid crystalline elastomers have been synthesized and various of these materials show a cholesteric–isotropic transition using X-ray scattering, polarizing microscopy and differential scanning calorimetry [31]. Dynamic mechanical investigations have been carried out [9] for a number of main-chain liquid crystalline elastomers. The measurements of the shear storage and the shear modulus both reveal an increase in the vicinity of the isotropic–cholesteric phase transition.

2.3 The Smectic A–Isotropic Phase Transition

A SmA–I phase transition has been identified in a large number of liquid crystalline elastomers using differential scanning calorimetry, polarizing microscopy, and X-ray diffraction, both for side-chain liquid crystalline elastomers [30, 32] and for main-chain/combined liquid crystalline elastomers [31, 33]. However, until very recently no detailed investigations of physical properties in the vicinity of the SmA–I transition have been reported [34]. Birefringence measurements under a static external mechanical stress were carried out above the SmA–I phase transition for a side-chain liquid crystalline elastomer for a range of values of the external stress [34]. The plot of the inverse birefringence versus temperature (Fig. 21) shows two different regimes

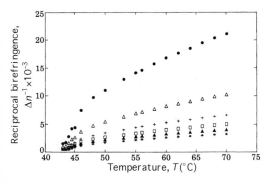

Figure 21. The inverse of the birefringence plotted as function of the temperature for different values of the true stress (reproduced with permission from [34]).

connected by a smooth cross-over. Using a simple model, it is suggested that these two regimes correspond to nematic fluctuations (at higher temperatures further above the transition) and SmA fluctuations (close to the SmA–I transition), respectively. It appears that such a cross-over behavior has never been seen before for low molecular weight and polymeric materials.

2.4 The Discotic–Isotropic Phase Transition

While there is already a substantial body of work on phase transitions in liquid crystalline elastomers composed of rod-like molecules, much less has been done on columnar phases typically composed of disk-like molecules. In all cases published to date [35, 36] discotic–isotropic phase transitions were found at higher temperatures followed by a glass transition at lower temperatures. Applying the technique developed by Küpfer and Finkelmann [18], Disch et al. [36] were able to produce discotic monodomains as elucidated by X-ray diffraction and, in particular, by measuring the thermal expansion across the transition for a monodomain and by comparing these results with those obtained for a polydomain. As expected from the change in the degree of macroscopic order (compare the detailed discussion for the isotropic–nematic transition for poly- versus monodomains), one finds a drastic change in length in the vicinity of the isotropic–discotic transition for the monodomain while there is a small and very smooth change for the polydomain (Fig. 22).

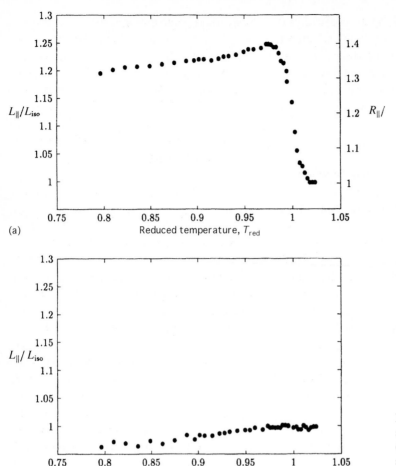

(a)

(b)

Figure 22. Thermal expansion of discotic networks: (a) LSCE; (b) polydomain sample (reproduced with permission from [36]).

2.5   Other Phase Transitions in Liquid Crystalline Elastomers

In addition to the phase transitions described so far, a number of other phase transitions involving liquid crystalline phases has been reported. These include for side-chain liquid crystalline elastomers: SmC–I [32], N–SmA [6, 30], N–SmF [32], SmA–SmC [32], SmA–SmB [32] and SmC–SmF [32]. Furthermore, a phase transition between a SmA phase and a re-entrant nematic phase has been described [6]. The identification of these phase transitions was performed using differential scanning calorimetry, observations in the polarizing microscope, and X-ray scattering. The elongation was measured as a function of temperature for different loads and it was found that there appears to be a change in slope near the SmA–reentrant nematic phase transition [6]. In the field of main-chain and combined liquid crystalline elastomers the phase transitions observed include: cholesteric–SmC* [9, 37], N–SmA [9, 38], SmA–SmC [9, 38], N–SmC [39], SmA–SmB [38], cholesteric–SmA [31] and cholesteric–SmC* [31]. The detection of all these phase transitions was based on polarizing microscopy, X-ray diffraction, and differential scanning calorimetry. In addition dynamic mechanical measurements were used [9] to show that the loss factor peaks below the cholesteric–SmC* and the SmA–SmC phase transition, respectively.

Quite recently electromechanical and electro-optic effects have been studied in some detail for the SmA–SmC* transition in sidechain LCE [40]. The authors account for their observation using a Landau model, which contains an additional elastic energy associated with the tilt, when compared to the description of low molecular weight materials.

Generalizing the concept of side-chain liquid crystalline elastomers from thermotropic to lyotropic materials, various phase transitions have been observed in lyotropic liquid crystalline elastomers. A hexagonal to isotropic phase transition was described using polarizing microscopy and X-ray diffraction; in addition stress–strain measurements in the hexagonal phase demonstrated the rubber-like behavior of these lyotropic materials [41]. A material showing an isotropic–lamellar (L_α) phase transition in an elastomer has been examined [42]. The phase transition was identified with polarizing microscopy and NMR was used to analyze the change in the lineshape when one is passing from the isotropic into the lamellar phase in this elastomer. In addition it was shown by X-ray diffraction that the liquid crystalline phase could be oriented by a uniaxial mechanical compression.

3   Macroscopic Properties

3.1   Nematic Phase

3.1.1   Mechanical Properties of Polydomains

It has been shown, for a number of fixed values of the compression, that the length of a cylindrial sample under compression shrinks as one cools from the nematic–isotropic transition temperature towards the glass transition [5]. Conversely the diameter of these samples grows throughout the nematic temperature range.

Stress–strain measurements in the nematic phase showed that above a certain threshold stress associated with the polydomain–monodomain transition, a linear stress–strain relation results [3]. Extrapolating these linear curves to zero stress one was able to extract the fictitious length l_0 of a monodomain sample. When this length l_0 was plotted as a function of reduced temperature, it was found that l_0 increased in the nematic phase similarly to the order parameter.

The elongation, λ, was measured as a function of temperature for fixed load for various nematic phases as well as for a reentrant nematic phase in sidechain LCEs [6]. In the reentrant nematic phase the observed elongation λ increased monotonically (by up to 50%) as the temperature was decreased from the smectic A–reentrant nematic phase transition [6].

X-ray diffraction measurements have been used to extract the orientation parameter $\langle P_2 \rangle$ as a function of the extension ratio [43]. $\langle P_2 \rangle$ was found to increase rapidly first and then to saturate for higher deformations. These studies were generalized by varying the crosslinking density and the length of the flexible spacer [44]. It was found that the orientation parameter $\langle P_2 \rangle$ is almost the same for different spacerlengths. In addition it ws shown that $\langle P_2 \rangle$ decreases when the nematic–isotropic transition is approached from well inside the nematic phase.

Small angle neutron scattering measurements of the anisotropy of the main chain conformation of a side-on sidechain LCE under a mechanical stress are described in the isotropic and in the nematic phase [45]. It is found that a weak anisotropy of the main chain conformation exists only in the nematic phase and for sufficiently high values of the elongation.

Very recently the first synthesis of a main chain LCE has been described [46]. It turns out, as was to be expected due to the completely different location of the mesogenic units, that the stress strain properties of these novel materials are rather different from what is known from side chain LCEs. For example, the main chain LCEs require a stretching by a factor of about 5 to get macroscopic alignment of the director. Such an extension is not even an option for side chain LCEs, which typically rupture when elongated by a factor between about 2 and 3. Thus in many ways the main chain LCEs might resemble more closely a classical rubber than the sidechain LCEs.

Laser induced phonon spectroscopy has been used to study the ultrasonic properties of nematic sidechain LCEs [47, 48]. It has been shown that both the ultrasonic velocity and the attenuation show a strong anisotropy and it was concluded that the coupling between density changes and reorientational motion of the sidechains takes place on a time scale of $10^{-7} - 10^{-9}$ s [47].

Dynamic measurements of the storage (G') and the loss modulus (G'') in the nematic phase of a sidechain LCE have been described [7]. It was found that throughout the nematic phase $G' > G''$ and that both were increasing monotonically as the temperature was lowered towards the glass transition.

3.1.2   Mechanical Properties of Monodomains: Liquid Single Crystal Elastomers

Memory effects in LCE have been investigated [49]. Small monodomain LCE films were prepared by aligning the nematic polymer phase in a magnetic field prior to crosslinking. These LCE films preserved their original director alignment as well as their degree of preferred orientation, even when they were held for over two weeks in the isotropic phase and then cooled back down into the nematic phase. The degree of alignment was measured by wide angle X-ray scattering and was found to show a similar temperature dependence as in earlier investigations by the same group [44].

The influence of the spacer length on the nature of the coupling between sidechains and the elastomeric network has been investigated [50]. It turns out that a lengthening of the spacer is accompanied by a sign change of the coupling between the polymeric backbone and the mesogenic units. In addition it emerges from these investigations that the magnitude of this coupling decreases with increasing spacerlength, a result which is in agreement with one's intuition.

Strain-induced transitions of the director orientation in nematic LCEs were studied [51]. These abrupt transitions with a director reorientation of 90° [51] were found to have a characteristic threshold strain and show qualitative similarities to predictions from a model based on rubber elasticity [52]. However, it should be noticed that the extension ratio was varied by 0.05 at a time and that the actually observed variation of the director angle for this step size was 65° and not 90°, thus rather suggesting a steep, but continuous change. Clearly further more detailed experiments are necessary to clarify this point.

Nematic monodomains have been prepared using a variant of the method introduced by Küpfer and Finkelmann [10, 18]: instead of keeping the stress constant in the second crosslinking step the strain was held constant in the second crosslinking step [53]. As for the case studied by Küpfer and Finkelmann, it is found that the monodomain character is retained after heating the sample into the isotropic phase upon return into the nematic phase. Stress strain relation has been measured for two classes of polysiloxane elastomers.

Ultrasonic experiments using laser induced phonon spectroscopy have been performed in a nematic liquid single crystal elastomer [48]. The experiments reveal a dispersion step for the speed of sound and a strong anisotropy for the acoustic attenuation constant in the investigated frequency range (100 MHz–1 GHz). These results are consistent with a description of LCEs using macroscopic dynamics [54–56] and reflect a coupling between elastic effects and the nematic order parameter as analyzed in detail previously [48].

3.1.3 Linear and Nonlinear Optical Properties of Polydomains

It has been shown that changing the length of the flexible spacer, m, by one unit can change the optical properties of a deformed nematic LCE qualitatively [15]. While for $m=3$ the deformed LCE is optically biaxial and the mesogenic units are oriented on average perpendicularly to the axis of defor-

mation, for $m=4$ an optically uniaxial configuration arises in which the director is oriented parallel to the axis of deformation. These optical investigations have been complemented by X-ray investigations leading to the same conclusions. Further X-ray investigations on a similar family of compounds also showed the change from uniaxial to biaxial behavior as the spacer length was changed from $m=4$ to $m=3$ [2].

IR-dichroism measurements of the CN-stretching mode revealed that the order parameter S extracted from the absorption coefficients along and perpendicular to the long axis of the molecules has a similar temperature dependence as in low molecular weight materials [16].

3.1.4 Linear and Nonlinear Optical Properties of Monodomains

The reorientation behavior has been studied, which results when a nematic monodomain LCE is exposed to a mechanical field perpendicular to the preferred direction of the monodomain [18]. For this configuration the dichroic ratio $R=A_{\parallel}/A_{\perp}$ of the CN-absorption band as well as the external mechanical stress, σ_e, are plotted versus the elongation, L/L_0, perpendicular to the director axis of the unstrained sample in Fig. 23. While the dichroic ratio saturates for small and large values of the external stress with a nearly linear change for intermediate values of the mechanical force, the nematic order parameter, S, changes from a value of $S=0.53$ in the unstrained configuration to a value of $S=0.36$ for high values of the external stress thus indicating that the reorientation process does not involve all parts of the monodomain LCE.

This last point has been investigated further and monodomain LCEs crosslinked in

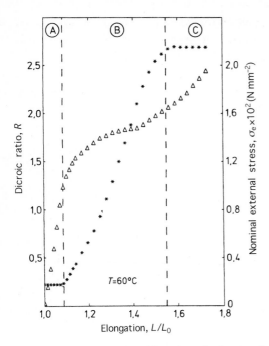

Figure 23. Dicroic ratio $R(*)$ and nominal external stress $\sigma_e(\triangle)$ as function of the elongation L/L_0 of a nematic LSCE (reproduced with permission from [18]).

the nematic and in the isotropic phase, respectively, were examined [10]. X-ray investigations performed for both types of samples show that the monodomains go for intermediate values of the elongation through a state with domains for which the director is either ordered by $\Theta=45°$ or $\Theta=-45°$ to the stress axis. While for the nematic LCE crosslinked in the isotropic phase a complete reorientation is achieved, this is not the case for the sample crosslinked in the nematic phase for which the X-ray pattern shows, even for an elongation $\lambda=1.6$, clearly four distinct maxima. This qualitative difference in behavior between samples crosslinked in the isotropic and in the nematic phase is also clearly brought out by measurements of the dichroic ratio. While the order parameter S in the reoriented sample crosslinked in the nematic phase

is drastically reduced from $S=0.5$ in the unstrained sample to $S=0.3$, this reduction is not observed for the sample crosslinked in the isotropic phase. From these observations one concludes that the state of order in the vicinity of the crosslinking points also affects the macroscopic behavior of the nematic monodomain LCE [10], a feature that has been modeled using a Landau approach and the concept of a frozen-in field [22].

3.1.5   Effects of Electric and Magnetic Fields

Deformations of nematic elastomers in electric fields have been described [44, 57, 58]. Reversible shape variations of a nematic sidechain LCE swollen with low molecular weight (LMW) materials in an electric field were reported, provided the elastomer is freely suspended [57]. If it is compressed between glass slides, no field effect is observed. Shape changes in mainchain LCEs swollen with a LMW material have been investigated in an electric field [44]. Similarly, reversible shape changes were observed for freely suspended LCEs on a time scale of one second and no shape changes were obtained for the unswollen elastomers [57]. More recently further experiments of the same type were done on swollen sidechain LCEs [58]. In addition it was checked that there was no response to an electric field if both, LMW material and LCE, were in their isotropic phase [58].

No reports on deformations and shape changes of nematic elastomers in external magnetic fields appear to exist yet.

3.1.6   Diffusion Effects

It has been shown that the diffusion coefficient for salicyclic acid in a nematic side-chain LCE can be controlled by varying the mass fraction of the mesogens thus demonstrating the potential of LCE for pharmaceutical applications [59].

3.1.7 Macroscopic Equations

Macroscopic properties of nematic elastomers have been discussed [56, 60]. De Gennes focused on the static properties, emphasizing especially the importance of coupling terms associated with relative rotations between the network and the director field [60]. The electrohydrodynamics of nematic elastomers has been considered generalizing earlier work by the same authors [54, 55] on the macroscopic properties of nematic sidechain polymers [56]. The static considerations of earlier work [60] were extended to incorporate electric effects; in addition a systematic overview of all terms necessary for linear irreversible thermodynamics was given [56].

3.1.8 Rubber Elasticity

An extension of rubber elasticity (i.e. of the description of large, static and incompressible deformations) to nematic elastomers has been given in a large number of papers [52, 61–66]. Abrupt transitions between different orientations of the director under external mechanical stress have been predicted in a model without spatial nonuniformities in the strain field [52, 63]. The effect of electric fields on rubber elasticity of nematics has been incorporated [65]. Finally the approach of rubber elasticity was also applied recently to smectic A [67] and to smectic C* [68] elastomers. Comparisons with experiments on smectic elastomers do not appear to exist at this time. Recently a rather detailed review of the model of an-

isotropic rubber elasticity has been given by Warner and Terentjev [69].

3.1.9 X-ray Investigations

It has been shown using X-ray fiber patterns, that a weakly crosslinked sidechain LCE can be oriented by an external mechanical stress with the polymer chains oriented parallel to the stress axis and the director associated with the mesogenic units perpendicular to the stress axis [38]. X-ray investigations also have been done for nematic phases of combined elastomers [33].

X-ray investigations on strongly stretched and rather well-oriented samples using a two-dimensional detector have been reported for a polysiloxane series of LCEs [70].

3.2 Cholesteric Phase

3.2.1 Mechanical Properties

Static mechanical measurements to evaluate the stress–strain relationship in cholesteric sidechain LCEs have been described [71, 72]. In [72] it has been found, for example, that for $0.94 < \lambda < 1$ (where $\lambda = 1$ corresponds to the unloaded sample), the nominal stress σ_n is nearly zero as the polydomain structure must be converted first into a monodomain structure. For deformations $\lambda < 0.94$, the nominal stress increases steeply. Similar results have also been reported elsewhere [71]. The nominal mechanical stress as a function of temperature for fixed compression has also been studied for cholesteric sidechain elastomers [71]. It turns out that the thermoelastic behavior is rather similar as that of the corresponding nematic LCE [2, 5].

The temperature dependence of the reciprocal length $1/L_{mon}$ of a macroscopically uniform, oriented cholesteric LCE has been measured. Using independent optical measurements to determine the orientational order parameter S, it has been shown [73], that S is directly proportional to $1/L_{mon}$, similar to what has been found for nematic LCE [4], thus demonstrating that the deformation can also be used as an order parameter in ordered samples of cholesteric LCEs.

The mechanical properties of polydomains of a main-chain cholesteric LCE, namely of crosslinked hydroxypropyl cellulose, have been studied [74]. The orientation parameter $\langle P_2 \rangle$, as extracted from X-ray measurements at fixed temperature, first increases as a function of extension ratio and then saturates at higher strains. As a function of temperature for fixed extension ratio, $\langle P_2 \rangle$ decreases monotonically.

It has been shown for cholesteric combined LCE that polydomain samples can be untwisted by a sufficiently large external mechanical force leading to a strain of 150% [75]. This has been checked by using X-ray fiber patterns. The same type of investigations has also been applied to other combined cholesteric LCE [31].

Dynamic mechanical measurements of the storage modulus G' and the loss modulus G'' in polydomain samples of cholesteric LCE have been described [9]. It turns out that $G' > G''$ throughout the whole cholesteric range and that both G' and G'' increase when approaching the cholesteric–isotropic and cholesteric–chiral SmC* transition, respectively, thus indicating the presence of substantial pretransitional effects at both phase transitions.

3.2.2 Electromechanical Properties

The electromechanical response in a cholesteric sidechain LCE under an external mechanical force has been studied [71, 73]. For sufficiently high values of the mechanical compression, needed to achieve a cholesteric monodomain with the helix axis parallel to the axis of the cylindrical sample, a linear electromechanical effect is found. When the temperature-dependent electric response is plotted for sufficiently high, but fixed compression, the resulting electric signal closely resembles the plot of the orientational order parameter as a function of temperature (Fig. 24).

In addition to the electromechanical observations described so far, which have been confirmed [73], the dependence on the cholesteric pitch, p, has been studied [73]. It is found that the electromechanical response coefficient is directly proportional to $1/p$ and vanishes for a racemic sample.

Static measurements of the change in sample thickness, $\Delta L/L_0$ (with L_0 the thickness in the mechanically loaded state, but without external electric field), as a function of an applied electric field (parallel to the helical axis of a cholesteric monodomain sample) have been carried out [72]. To eliminate effects that are quadratic in the electric

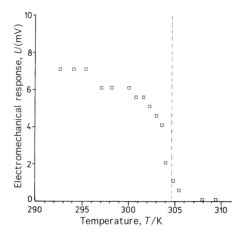

Figure 24. Electromechanical response, U, vs. temperature for a compression of 25% of a cholesteric elastomer.

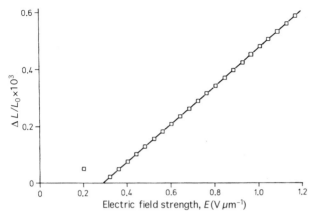

Figure 25. Relative fitted change of sample thickness of a cholesteric elastomer minus the relative fitted change of sample thickness of the racemic elastomer $\Delta L/L_0$ as function of the electric field strength E (reproduced with permission from [72]).

field such as the attractive force between the plates bounding the sample and electrostrictive effects, the same experiments have also been done for the corresponding racemic compound. When the change in thickness of the cholesteric elastomer minus the change in thickness of the racemic elastomer $\Delta L/L_0$ is plotted as a function of the electric field strength, a linear relationship results above a threshold field, E_{th} (Fig. 25).

Alternating current measurements were used to investigate the electromechanical properties in combined photo-crosslinked cholesteric LCEs. In a compound showing both, a cholesteric and a chiral SmC* phase, the electromechanical response in the cholesteric phase was considerably higher.

Various electromechanical effects in cholesteric and chiral SmC* phases were presented including a discussion of how to distinguish piezoelectricity from flexoelectricity and electrostriction [77, 78]. It was examined for cholesteric structures in general for which structures one can get longitudinal piezoelectricity [79].

3.3   Smectic A and Smectic C Phases

In contrast to the extensive experimental investigations of nematic, cholesteric and chiral SmC* phases, comparatively little work has been done on the characterization of the physical properties of SmA and SmC elastomers. The elongation, λ, has been measured as a function of temperature for constant external load for a number of different loads in SmA sidechain elastomers [6]. It was found that λ increases monotonically as a function of temperature for a material, which has a SmA–I transition. In addition it was shown that the elasticity modulus, E, decreases monotonically with temperature. X-ray investigations on SmA phases in sidechain LCE have been performed [70]. It was found that for the family of compounds studied, the orientation of the mesogenic groups was always perpendicular to the direction of stretching.

Second harmonic generation of a poled SmA elastomer has been demonstrated [80]. A uniform director orientation was achieved by a simple mechanical deformation of the elastomer, poling was done in an electric field and the NLO active chromophores (nitrogroups) were bounded covalently to the network.

The layer spacing in the SmA and SmC phases of combined LCE has been investigated using X-ray diffraction [53]. The storage modulus G' and the loss modulus G'' have been studied in the SmA and SmC phases of combined elastomers and it was found that $G'>G''$ throughout the whole temperature range of these two smectic phases [9].

The diffusion of gases through films of SmA elastomers has been analyzed [81, 82]. It was shown experimentally that the logarithm of the diffusion coefficient depends linearly on inverse temperature throughout all SmA phases and that the transport behavior is stress-dependent [81]. The diffusion of hydrocarbon gases through SmA elastomers was studied in detail [82]. It turns out that these LCE materials are suitable for the separation of gases due to their high selectivity and wide range of permeabilities.

Physical properties of the higher ordered smectic phases SmF, SmI, SmB, SmH etc. have not been investigated as yet.

Various physical properties of the layered L_α-phase in a lyotropic LCE have been studied [42]. Using NMR and X-ray investigations it was demonstrated that a lamellar phase can be oriented macroscopically by applying a uniaxial compression to the disordered swollen network.

3.4   Smectic C* Phase

3.4.1   Mechanical Properties of Polydomains

X-ray investigations were used to study the orientation behavior of a sidechain SmC* elastomer under mechanical stress [83]. If a uniaxial mechanical stress is applied to a swollen C* sample after the first crosslinking step, an X-ray pattern, which shows four distinct maxima in the small angle regime results (Fig. 26), indicating a layer orientation of the layer normal under an angle $\pm\alpha$ to the stress axis. In addition, the azimuthal wide angle intensities indicate an average alignment of the long molecular axis of the mesogens parallel to the stress axis. Thus, while the helical structure is untwisted, a monodomain is not obtained when applying only a uniaxial external stress. To come closer to achieving this goal a second deformation was carried out under an angle compared to the previous one in order to lift the remaining azimuthal degeneracy. As it turns out (Fig. 27) from analyzing the resulting X-ray pattern, a sample that is oriented to more than 95% (and thus close to a monodomain) results. This technique has been discussed and studied in more detail [84], where it is also emphasized that the samples resulting after the second crosslinking have macroscopically C_2-symmetry and thus possess dipolar order giving rise to pyroelectricity and piezoelectricity.

The mechanical behavior of combined SmC* polydomain samples was investigat-

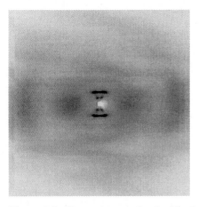

Figure 26. X-ray pattern of a SmC*-elastomer after uniaxial deformation in the swollen state (reproduced with permission from [83]).

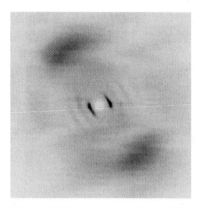

Figure 27. X-ray pattern of a SmC*-elastomer after second uniaxial deformation (reproduced with permission from [83]).

axis. This line of investigations was extended shortly thereafter to other compounds with similar results [37].

Dynamic mechanical measurements of the storage, G', and the loss modulus, G'', in combined LCE [9] show, that both, G' and G'', fall monotonically as a function of temperature throughout the entire range of existence of the C*-phase and that $G' > G''$ is always valid.

3.4.2 Nonlinear Optical Properties

Using the technique described earlier [83] to get samples that are nearly monodomain, second harmonic generation has been investigated using a Q-switched Nd-YAG laser [85]. Since the helix is unwound in these samples and since they are close to a monodomain, they are macroscopically of C_2-symmetry and thus suitable for second harmonic generation. In Fig. 28 the intensity of the second harmonic signal is plotted as a function of the angle around the direction in the plane of the film, which is perpendicular to the first deformation direction. The results of the polarization dependent measure-

ed for large mechanical deformations [75]. To detect the effect of the external stress on the helix, X-ray fiber patterns were used. It was found that for moderate values of the strain which was necessary to orient the fiber, $\geq 50\%$, the smectic layers were perpendicular to the fiber axis and the helix was still present. At very large values of the strain, $> 400\%$, however, X-ray data showed that the mesogenic groups were oriented on average parallel to the fiber axis and that thus the helix was unwound and that the layers were tilted with respect to the fiber

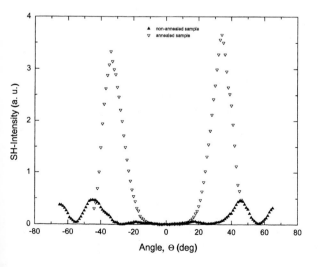

Figure 28. Marker fringe diagram for an annealed (Δ) and a non-annealed SmC*-elastomer (reproduced with permission from [85]).

ments are consistent with overall C_2-symmetry. The quality of the orientation was checked independently by X-ray measurements revealing the high degree of orientation. In contrast to other polymer networks, including centrosymmetric LCE such as nematic LCE, which require external poling fields, the unwound, uniformly oriented SmC* LCE has built-in noncentrosymmetry and thus possesses an intrinsic macroscopic electric polarization.

3.4.3   Electromechanical Properties of Polydomains

Alternating current measurements were used to investigate the piezoelectric effect in combined photocrosslinked SmC* LCEs [76]. A linear relationship was obtained for the measured piezo voltage as a function of the mechanical deformation applied. In addition it was found that the measured piezo voltage is a monotonically decreasing function of temperature for a compound showing an I–SmC* phase transition.

Piezoelectric and pyroelectric properties of polydomain sidechain LCEs in the chiral SmC* phase have been investigated [86]. The pyroelectric response was generated by heat pulses from a YAG laser. While the extracted saturation polarization falls monotonically as a function of temperature throughout the SmC* phase, the pyroelectric coefficient, $\gamma = \partial P / \partial T$, goes through a maximum several degrees below the SmC*–I phase transition temperature. An AC piezoelectric effect has been measured throughout the entire temperature interval of the SmC* phase as well for the unoriented sample. An additional DC poling field led to an increase in the observed piezoresponse, probably due to a better degree of orientation of the macroscopic polarization throughout the sample.

Making use of the technique by Semmler and Finkelmann [83, 84] to obtain a highly oriented SmC*–LCE film with untwisted helix and macroscopic C_2-symmetry, Eckert et al. [87] investigated piezoelectricity and inverse piezoelectricity. They find that in these highly oriented and untwisted SmC* films a linear behavior results for the charge as a function of the applied force as well as for the displacement as a function of applied AC voltage.

3.5   Discotic Phases and Hexagonal Lyotropic Phases

There are only two papers describing the physical properties of discotic LCEs [35, 36]. X-ray diffraction showed that a narrower azimuthal intensity distribution can be obtained when a second crosslinking step is done under an external load reflecting a macroscopic alignment parallel to the direction of the external stress [35]. X-ray diffraction experiments on monodomains (LSCE) show the occurrence of sharp maxima in the azimuthal distribution. Thermal expansion has also been studied for polydomain LCEs and for LSCEs in [36]. It is found that well inside the discotic phase, that is sufficiently far below the discotic–isotropic transiton, the length parallel to the stress axis of both, poly- and monodomain samples, increases approximately linearly with increasing temperature as expected from the isobaric expansion of the network.

In addition to the stress–strain measurements already mentioned, the hexagonal lattice constant was measured in a lyotropic LCE as a function of the water concentration [41]. Furthermore it was shown, using X-ray scattering, that a hexagonal phase can be oriented parallel to the preferred direction by applying an external mechanical field.

4 References

[1] R. A. M. Hikmet, J. Lub, *Progr. Polym. Sci.* **1996**, *21*, 1165–1209.

[2] W. Gleim, H. Finkelmann, *Makromol. Chem.* **1987**, *188*, 1489–1500.

[3] J. Schätzle, W. Kaufhold, H. Finkelmann, *Makromol. Chem.* **1989**, *190*, 3269–3284; **1991**, *192*, 1235–1236.

[4] W. Kaufhold, H. Finkelmann, H. R. Brand, *Makromol. Chem.* **1991**, *192*, 2555–2579.

[5] K. Hammerschmidt, H. Finkelmann, *Makromol. Chem.* **1980**, *190*, 1089–1101.

[6] R. V. Talroze, T. I. Gubina, V. P. Shibaev, N. A. Plate, V. I. Dakin, N. A. Shmakova, F. F. Sukhov, *Makromol. Chem., Rapid Commun.* **1990**, *11*, 67–71.

[7] W. Oppermann, K. Braatz, H. Finkelmann, W. Gleim, H. J. Kock, G. Rehage, *Rheol. Acta* **1982**, *21*, 423–426.

[8] R. Sigel, W. Stille, G. Strobl, R. Lehnert, *Macromolecules* **1993**, *26*, 4226–4233.

[9] T. Pakula, R. Zentel, *Makromol. Chem.* **1991**, *192*, 2401–2410.

[10] J. Küpfer, H. Finkelmann, *Makromol. Chem.* **1994**, *195*, 1353–1367.

[11] J. Küpfer, E. Nishikawa, H. Finkelmann, *Polym. Adv. Technol.* **1994**, *5*, 110–115.

[12] I. Kundler, H. Finkelmann, *Macromol. Chem., Rapid Commun.* **1995**, *16*, 679–686.

[13] J. Weilepp, H. R. Brand, *Europhys. Lett.* **1996**, *34*, 495–500.

[14] G. C. Verwey, M. Warner, E. M. Terentjev, *J. Phys. II (France)* **1996**, *6*, 1273–1290.

[15] H. Finkelmann, H. J. Kock, W. Gleim, G. Rehage, *Macromol. Chem., Rapid Commun.* **1984**, *5*, 287–293.

[16] J. Schätzle, H. Finkelmann, *Mol. Cryst. Liq. Cryst.* **1987**, *142*, 85–100.

[17] P. G. de Gennes, *C. R. Acad. Sci. (Paris)* **1975**, *B281*, 101–103.

[18] J. Küpfer, H. Finkelmann, *Macromol. Chem., Rapid Commun.* **1991**, *12*, 717–726.

[19] E. Gramsbergen, L. Longa, W. de Jeu, *Phys. Rep.* **1986**, *135*, 195–257.

[20] S. Disch, C. Schmidt, H. Finkelmann, *Macromol. Chem., Rapid Commun.* **1994**, *15*, 303–310.

[21] S. Disch, C. Schmidt, H. Finkelmann, *Polymer Enzyclopedia*, CRC Press, Boca Raton, FL, **1995**.

[22] H. R. Brand, K. Kawasaki, *Macromol. Chem., Rapid Commun.* **1994**, *15*, 251–257.

[23] P. G. de Gennes in *Polymer Liquid Crystals* (Eds.: A. Ciferri, W. R. Krigbaum, R. B. Meyer), Academic Press, NY **1982**, pp. 115–131.

[24] M. Warner, P. Geiling, T. A. Vilgis, *J. Chem. Phys.* **1988**, *88*, 4008–4013.

[25] H. R. Brand, *Macromol. Chem., Rapid Commun.* **1989**, *10*, 57–61, 317.

[26] A. ten Bosch, L. Varichon, *Makromol. Theory Simul.* **1994**, *3*, 533–542.

[27] H. R. Brand, K. Kawasaki, *J. Phys. II (France)* **1992**, *2*, 1789–1795.

[28] H. R. Brand, K. Kawasaki, *J. Phys. II (Paris)* **1994**, *4*, 543–548.

[29] K. Kawasaki, H. R. Brand, *Physica A* **1994**, *208*, 407–422.

[30] H. Finkelmann, H.-J. Kock, G. Rehage, *Macromol. Chem., Rapid Commun.* **1981**, *2*, 317–322.

[31] R. Zentel, G. Reckert, S. Bualek, H. Kapitza, *Makromol. Chem.* **1989**, *190*, 2869–2884.

[32] Y. S. Freidzon, R. V. Talroze, N. I. Boiko, S. G. Kostromin, V. P. Shibaev, N. A. Plate, *Liq. Cryst.* **1988**, *3*, 127–132.

[33] S. Bualek, J. Meyer, G. F. Schmidt, R. Zentel, *Mol. Cryst. Liq. Cryst.* **1988**, *155*, 47–56.

[34] M. Olbrich, H. R. Brand, H. Finkelmann, K. Kawasaki, *Europhys. Lett.* **1995**, *31*, 281–286.

[35] H. Bengs, H. Finkelmann, J. Küpfer, H. Ringsdorf, P. Schuhmacher, *Macromol. Chem., Rapid Commun.* **1993**, *14*, 445–450.

[36] S. Disch, H. Finkelmann, H. Ringsdorf, P. Schuhmacher, *Macromolecules* **1995**, *28*, 2424–2428.

[37] R. Zentel, *Angew. Chem. Adv. Mater.* **1989**, *101*, 1437–1445.

[38] R. Zentel, G. Reckert, *Makromol. Chem.* **1986**, *187*, 1915–1926.

[39] S. Bualek, R. Zentel, *Makromol. Chem.* **1988**, *189*, 791–796.

[40] M. Brehmer, R. Zentel, F. Gießelmann, R. Germer, P. Zugenmaier, *Liq. Cryst.* **1996**, *21*, 589–596.

[41] R. Löffler, H. Finkelmann, *Macromol. Chem., Rapid Commun.* **1990**, *11*, 321–328.

[42] P. Fischer, C. Schmidt, H. Finkelmann, *Macromol. Chem., Rapid Commun.* **1995**, *16*, 435–447.

[43] F. J. Davis, G. R. Mitchell, *Polymer Commun.* **1987**, *28*, 8–11.

[44] N. R. Barnes, F. J. Davis, G. R. Mitchell, *Mol. Cryst. Liq. Cryst.* **1989**, *168*, 13–25.

[45] H. Finkelmann, H. R. Brand, *Trends Polym. Sci.* **1994**, *2*, 222–226.

[46] G. H. F. Bergmann, H. Finkelmann, V. Percec, M. Zhao, *Macromol. Chem., Rapid Commun.* **1997**, *18*, 353–360.

[47] F. W. Deeg, K. Diercksen, G. Schwalb, C. Bräuchle, H. Reinecke, *Phys. Rev. B* **1991**, *44*, 2830–2833.

[48] F. W. Deeg, K. Diercksen, C. Bräuchle, *Ber. Bunsenges. Phys. Chem.* **1993**, *97*, 1312–1315.

[49] C. H. Legge, F. J. Davis, G. R. Mitchell, *J. Phys. II (France)* **1991**, *1*, 1253–1261.

[50] G. R. Mitchell, M. Coulter, F. J. Davis, W. Guo, *J. Phys. II (France)* **1992**, *2*, 1121–1132.

[51] G. R. Mitchell, F. J. Davis, W. Guo, *Phys. Rev. Lett.* **1993**, *71*, 2947–2950.

[52] R. Bladon, M. Warner, E. M. Terentjev, *Phys. Rev. E* **1993**, *47*, R3838–R3840.

[53] C. J. Twomey, T. N. Blanton, K. L. Marshall, S. H. Chen. S. D. Jacobs, *Liq. Cryst.* **1995**, *19*, 339–344.

[54] H. Pleiner, H. R. Brand, *Mol. Cryst. Liq. Cryst.* **1991**, *199*, 407–418.

[55] H. Pleiner, H. R. Brand, *Macromolecules* **1992**, *25*, 895–901.

[56] H. R. Brand, H. Pleiner, *Physica A* **1994**, *208*, 359–372.

[57] R. Zentel, *Liq. Cryst.* **1986**, *1*, 589–592.

[58] R. Kishi, Y. Suzuki, H. Ichijo, O. Hirasa, *Chem. Lett.* **1994**, *12*, 2257–2260.

[59] H. Loth, A. Euschen, *Macromol. Chem., Rapid Commun.* **1988**, *9*, 35–38.

[60] P. G. de Gennes in *Liquid crystals of one- and two-dimensional order* (Eds.: W. Helfrich, G. Heppke), Springer, NY **1980**, pp. 231–237.

[61] P. Bladon, M. Warner, *Macromolecules* **1993**, *26*, 1078–1085.

[62] P. Bladon, M. Warner, E. M. Terentjev, *Macromolecules* **1994**, *27*, 7067–7075.

[63] R. Bladon, E. M. Terentjev, M. Warner, *J. Phys. II (France)* **1994**, *4*, 75–91.

[64] E. M. Terentjev, *Europhys. Lett.* **1993**, *23*, 27–32.

[65] E. M. Terentjev, M. Warner, P. Bladon, *J. Phys. II (France)* **1994**, *4*, 667–676.

[66] M. Warner, R. Bladon, E. M. Terentjev, *J. Phys. II (France)* **1994**, *4*, 93–102.

[67] E. M. Terentjev, M. Warner, *J. Phys. II (France)* **1994**, *4*, 849–864.

[68] E. M. Terentjev, M. Warner, *J. Phys. II (France)* **1994**, *4*, 111–126.

[69] M. Warner, E. M. Terentjev, *Progr. Polym. Sci.* **1996**, *21*, 853–891.

[70] C. Degert, P. Davidson, S. Megtert, D. Petermann, M. Mauzac, *Liq. Cryst.* **1992,** *12*, 779–798.

[71] W. Meier, H. Finkelmann, *Macromol. Chem., Rapid Commun.* **1990**, *11*, 599–605.

[72] H. Hirschmann, W. Meier, H. Finkelmann, *Macromol. Chem., Rapid Commun.* **1992**, *13*, 385–394.

[73] W. Meier, H. Finkelmann, *Macromolecules* **1993**, *26*, 1811–1817.

[74] G. R. Mitchell, W. Guo, F. J. Davis, *Polymer* **1992**, *33*, 68–74.

[75] R. Zentel, *Liq. Cryst.* **1988**, *3*, 531–536.

[76] S. U. Vallerien, F. Kremer, E. W. Fischer, H. Kapitza, R. Zentel, H. Poths, *Macromol. Chem., Rapid Commun.* **1990**, *11*, 593–598.

[77] H. R. Brand, *Macromol. Chem., Rapid Commun.* **1989**, *10*, 441–445.

[78] H. R. Brand, H. Pleiner, *Macromol. Chem., Rapid Commun.* **1990**, *11*, 607–612.

[79] H. Pleiner, H. R. Brand, *J. Phys. II (Paris)* **1993**, *3*, 1397–1409.

[80] H. Hirschmann, D. Velasco, H. Reinecke, H. Finkelmann, *J. Phys. II (France)* **1991**, *1*, 559–570.

[81] H. Modler, H. Finkelmann, *Ber. Bunsenges. Phys. Chem.* **1990**, *94*, 836–856.

[82] H. Reinecke, H. Finkelmann, *Makromol. Chem.* **1992**, *193*, 2945–2960.

[83] K. Semmler, H. Finkelmann, *Polym. Adv. Tech.* **1994**, *5*, 232–235.

[84] K. Semmler, H. Finkelmann, *Makromol. Chem.* **1995**, *196*, 3197–3208.

[85] I. Benné, K. Semmler, H. Finkelmann, *Macromol. Chem., Rapid Commun.* **1994**, *15*, 295–302.

[86] M. Mauzac, H.-T. Nguyen, F. G. Tournilhac, S. V. Yablonsky, *Chem. Phys. Lett.* **1995**, *240*, 461–466.

[87] T. Eckert, H. Finkelmann, M. Keck, W. Lehmann, F. Kremer, *Macromol. Chem., Rapid Commun.* **1996**, *17*, 767–773.

Part 3:
Amphiphilic Liquid Crystals

Chapter VI
Amphotropic Liquid Crystals

Dirk Blunk, Klaus Praefcke and Volkmar Vill

1 Introduction, Remarks on History

Numerous organic compounds exhibit liquid crystalline properties in the pure state. In a defined temperature range between the melting and the clearing temperature the order parameter of such a system is between that of true crystals and that of isotropic liquids. This is what one usually describes with the term *thermotropic* liquid crystal. But temperature is not the only term which may be variable. The intrinsic property of liquid crystals having an order parameter between those of real crystals and liquids includes the possibility of other ways of creating that interesting 'fourth state of matter' than just by changing the temperature.

Another possible way of generating this intermediate state is by the addition of solvents in appropriate amounts to convenient compounds leading to *lyotropic* liquid crystals. Under suitable conditions, especially in well defined concentration ranges, the molecules tend to organize themselves forming aggregates of various well defined geometries. These supramolecular assemblies behave as mesogenic units forming lyotropic liquid crystalline phases. In this type of liquid crystals the *concentration range* in which the system is liquid crystalline can be compared with the *temperature range* between the melting- and the clearing point of thermotropic liquid crystals; however, it must be mentioned here that

in case of lyotropic liquid crystals the temperature is an additional degree of freedom.

In contrast to the technologically extremely important thermotropic mesogens known for about 110 years and quickly growing in number, lyotropic analogues are as yet poorly studied and are much more complicated as their mesophase units do not consist of single molecules, but are made up of more or less flexible and variable aggregates of molecules in (so far mostly aqueous) solution.

Hitherto, most solvents used to form lyotropic liquid crystals have been polar in character. By far the most conventional (and also from the biological point of view certainly the most important) solvent is water. However, other solvents can be used ranging from methanol at the polar end of the scale up to long chained alkanes on the other, apolar side of this classification of solvents.

Chemical compounds are said to act *amphotropically* if they give rise to *both* kinds of liquid crystal formation. Such particular compounds show thermotropic liquid crystalline behavior in their pure state on heating or the formation of lyotropic mesophases on the addition of a further component, mostly of an inorganic or organic solvent in certain amounts.

Nearly all the compounds of interest for this chapter, beside their conventional form anisotropy possess general amphiphilic behavior. The hydrophilic (polar) part of a molecule, for example a carboxyl group or a multiol substructure, is well segregated from the lipophilic (apolar) moiety.

As will be discussed, three main factors of the molecular architecture have a significant influence on the thermomesomorphic properties of such compounds:

- the ratio of the space demands of the hydrophilic and the hydrophobic part of a molecule,
- the sterical arrangement of the substituents at the hydrophilic headgroup, and
- geometrical factors such as molecular shape or symmetry.

Beside this, in the recent past some investigations on amphotropic materials of nonamphiphilic structure were made as well (see 'Further Molecular Architectures' in this Chapter). Additionally, it is somewhat difficult to draw a sharp separation line between amphiphilic and nonamphiphilic compounds: the ability to act as an amphiphile depends mainly on the size and the polarity of the hydrophilic and hydrophobic parts of a molecule. This 'internal equilibrium' can be varied in a broad range. Thus, in this chapter, although mainly directed toward mesogens of the amphiphilic type, some apolar mesogens will be discussed as well.

Due to their importance for life sciences, for example the formation of biological membranes, and their technical importance as surfactants and soaps, potential lyotropic properties of amphiphilic compounds are much better investigated than other ones of different molecular types; the broad field of *surfactant liquid crystals* is discussed in Chapter VII of this volume.

Some reviews focusing on amphotropy have been published; they take the development of liquid crystal research in this field into account [1–12]. With regards to the history of amphotropic liquid crystals it seems that in 1854 (about 35 years before O. Lehmann and F. Reinitzer established liquid

crystallinity as the fourth state of matter) the first observation of a liquid crystalline state had been described [13, 14] though not understood. The magnesium salt of myristic acid (tetradecanoic acid) is described here as '... ein lockeres, aus mikroscopischen Nadeln bestehendes Pulver, welches über 100 °C zu einer durchsichtigen, aber nicht flüssigen Masse wird; noch vor dem eigentlichen Zusammenfließen wird es zersetzt' [13]. Although this magnesium salt is described here as 'a loose powder' this could be taken as a strong hint towards possible liquid crystalline properties. In spite of this strong hint, its technological significance (about 20 patents related to this salt have been issued since) and also the knowledge about the liquid crystallinity of other magnesium soaps, e.g., magnesium stearate, that is magnesium octadecanoate, nobody has yet investigated the thermotropic properties of magnesium myristate more thoroughly. Although Müller [15, 16] looked somewhat closer on this historically very interesting amphiphile by carrying out temperature dependent X-ray measurements, differential thermal analysis, and thermogravimetry, clear statements about its thermotropic properties are still missing. Looking back to 1854, no planned and detailed research on self organizing molecules was done; at that time, this phenomenon was not yet known. Nevertheless, a few scientists started working on this topic in those early times. For example, R. Virchow studied biological systems by means of polarizing microscopy; he observed and described myelin figures [17], a type of texture of lyotropic liquid crystals. The phenomena of birefringence of such myelin structures was investigated by Mettenheimer in 1857 [18]. The book by Valentin [19] published about 140 years ago can be considered as an early milestone about investigations of materials of various kind by this optical method.

The connection between observed myelin figures and mesomorphism was investigated much later. For instance, Quinke described the formation of myelin figures of salts of fatty acids [20].

Further milestones in the field of lyotropic liquid crystals are investigations of binary systems of soaps and detergents with water [21] and work on the lyomesomorphism of polypeptides and nucleic acids [22]. In 1933 the first review on lyotropic liquid crystals appeared [23]. The author of this review investigated the *amphotropic* properties of various compounds. He suggested that in amphotropic materials raising the temperature or adding a solvent (water) lead to the same partial break up of the crystal lattice, however, a certain order is maintained. Furthermore, colloidal solutions showing partial supramolecular structures are also described here. But, due to different parameters of order these latter systems differ from lyotropic liquid crystals.

How the lateral alkyl chains are organized in the liquid crystal state of matter was already a relevant subject of studies starting in the early thirties initiating discussions in which way alkyl chains are conformed in it (e.g., static- or liquid-like all-trans) [24, 25].

Somewhat later, even biological systems were investigated from a liquid crystal research point of view (e.g., the tobacco-mosaic-virus) [26], see also Chap. VIII of this volume. The main progress of lyotropic liquid crystal research in these times is connected to works on soap/water mixtures [27, 28] and the investigation of thermotropic mesophases of soaps [29]. The amphotropic character of such compounds was also studied. In such systems, no continuous transitions were observed between thermotropic and lyotropic mesophases, but always biphasic regions could be seen [30, 31]. Thereafter, the interest in understanding biological cell membranes inspired researchers to study watery lipid systems as models for such membranes [32]. Here, for the first time, the 45 nm reflex from X-ray investigations was identified to stem from disordered paraffinic chains.

Up to this time, research work was characterized by compiling facts originating from investigations of different amphotropic liquid crystals, but a systematization of research in this field began only in the mid-forties [33–35].

Helpful tools for this structurization of liquid crystal research were temperature dependent X-ray investigations [36] of natural and synthetic lipids, and the discovery that mesophases may be identified by their different textures appearing in the microscope using crossed polarizers [37]. In the decade starting in about 1957 systematic screening of the concentration and temperature dependency of the major lyotropic mesophases was done and models of the molecular arrangement in the different phases were developed [38–45] (e.g., the so-called 'middle' or 'neat' phases [38], the cholesteric phase of polypeptides and nucleopeptides [44]).

During this work Luzzati found hitherto unnoticed phases (e.g., the rectangular columnar mesophase in the concentration range between the so-called 'neat' and 'middle' phases [39]). In the year 1958 he observed an optically isotropic appearing mesophase. Although, he could not clear up the structure of this interesting mesophase by X-ray experiments in detail, he predicted a cubic geometry of the lattice of this liquid crystal phase. The first investigation of the inverted hexagonal phasetype in the quaternary system lecithin, cephalin, phosphatidylinositol and water in 1962 [41] was the next milestone in the field of 'new phases'.

During the same period Ekwall studied the mesophases previously discovered by McBain, in order to clear up their molecular architecture. He investigated mainly ternary systems of soaps, alcohols and water [46], describing for instance the coexistence of two phases in one area of a mixture of sodium caprylate, decanol and water [47].

The early sixties were golden times for the progress of investigation techniques in this field of science: the first NMR studies in the anisotropic media of lyotropic liquid crystals [48] were done, the introduction of the freeze fraction technique for the preparation of electron microscopy probes (e.g., of cell membranes [49]) and the separation of different lyomesophases by centrifugation [47] were introduced all between 1960 and 1962.

Since the beginning of the 20th century the research on liquid crystals has been closely connected with molecules of amphiphilic character.

As early as 1910 Vorländer described the sodium salt of diphenyl acetic acid as a liquid crystal [50]. Much later, in 1970, Demus characterized its mesophase as being of the *columnar type*, which was the first description of a columnar arrangement in a thermotropic liquid crystal [51].

By 1919 agaric acid was explicitly described to be liquid crystalline [52] and later has been characterized as SmA [53].

An important example of a thermotropic liquid crystal possessing a columnar mesophase is diisobutylsilandiol, which originally was reported to possess an unknown mesophase [54]; its columnar type was determined later [55].

The interest in mesogens based on *carbohydrates* started only recently, although since 1911 several reports on 'double melting points' of sugar based surfactants are documented in the literature. For the first review on this particular field see [4].

Fischer reported in 1911 the double melting behavior of hexadecylglucopyranoside [56], two years later Salway verified

Fischers observations and extended the studies to steroidal glycosides [57]. In 1938 hexadecylglucopyranoside was described explicitly as being *liquid crystalline* [58] by Noller and Hori extended this term to various other families of glycolipids [59, 60]. At this time *Chemical Abstracts* refused to use the expression 'liquid crystal' in their abstract about this reference, which may be the reason why the work of Hori has not been recognized.

For amphotropic liquid crystals it is obvious that the so-called surfactants (including for example soaps) are the best investigated, as documented in two reviews [61, 62]. Therefore, an own chapter of this book is dedicated to that field of supramolecular chemistry (see Chapter VII of this volume). Since the lyotropic properties of this type of amphotropic liquid crystals are discussed there in detail, we focus now mainly on the thermotropic behavior of such compounds.

Amphotropic behavior can be found for a large number of different chemical structures. Additional information is given in other chapters of this Handbook. Typical classes of amphotropic materials are for instance classical soaps (see lyotropics), transition metal soaps (see metallomesogens), viologens, quarternary amines and other ionic surfactants (see lyotropics), block copolymers (see polymer liquid crystals), cellulose derivatives (see cellulose liquid crystals) and partially fluorinated paraffines, diols, peptide surfactants, lecithins, lipids, alkylated sugars and inositols, naturally occurring glycosides and silanols, which are discussed in this chapter.

Some other compounds may also have the potential to possess amphotropic behavior, but because of too high melting points these properties can not be observed, for example polyamides such as Kevlar show lyotropic behavior only, and are not discussed here. Furthermore, one has to take into account

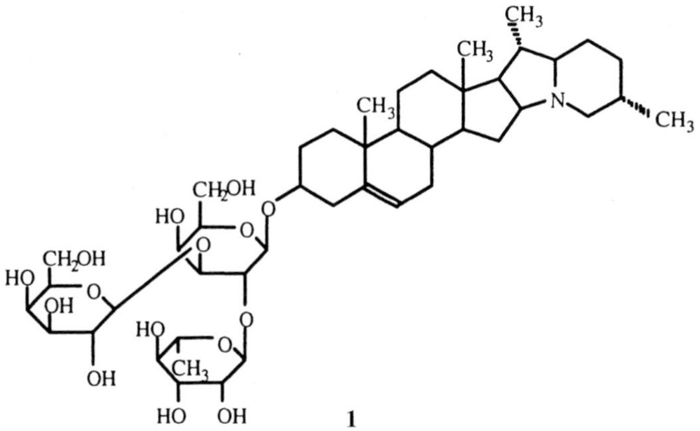

Figure 1. Molecular structure of solanin, **1**, the poisonous compound in green potatoes recently proven to exhibit a SmA phase [63, 64].

that many existing chemicals have never been tested for amphotropic properties.

An example for such a compound is solanin (**1**, the structure in Fig. 1) which for many years has been reported to possess two different melting points [63]. Only recently was it shown that these two temperatures are the melting and clearing points of a smectic liquid crystal [64].

Other compounds which might possess amphotropic behavior, but which have not been studied hitherto, can be found within tannines, saponines, gangliosides, naturally occurring membrane components or biopolymers. Also many synthetic surfactants have never been studied for their lyotropic and thermotropic liquid crystal properties.

As a result, not all the compounds discussed in the following are already tested for both, lyo- *and* thermomesomorphism. Thus, in some cases their *amphotropic* behavior is not yet proved but nevertheless likely. On the other hand, it becomes more and more clear that the strong separation of both kinds of liquid crystallinity is fading [65]. The amphiphilic/amphotropic compounds seem to open a view on an uniform field of liquid crystals, regarding both lyo- and thermotropic behavior of compounds together as one intrinsic principle of nature.

2 Principles of Molecular Structures

Most liquid crystals can be classified as being monophilic or amphiphilic in character, but the borderline between these extremes is not sharp, see Table 1.

A comparison of the examples selected in Fig. 2, see also Table 2, may emphasize this difference. While **2** is a classical, monophilic liquid crystal, **3** with the same rigid core has an intramolecular contrast due to the perfluoration of one of its chains. This weak amphiphilic character leads to a higher stability, i.e. to a higher clearing temperature of the mesophase (84.5 °C → 134 °C) exhibited by this mesogen and to a change in phase (nematic to smectic). The same effect could be observed for the cholesterol derivative **4** which exhibits a SmA phase with a clearing temperature above 110 °C caused by its amphiphilic character. All the other cholesterol derivatives, monophilic in character, have only a cholesteric (and thus less ordered) mesophase with clearing points below 100 °C [69].

The thermotropic properties of amphotropic materials are determined by the equilibrium of the hydro- and lipophilic sections

Table 1. Comparison of the general properties of monophilic and amphiphilic mesogens.

Properties	Monophilic mesogens	Amphiphilic mesogens
Origin of mesomorphic behavior	Molecular shape, space filling	Intramolecular contrast, phase separation
Molecular requirements	Rod-like or disc-like shape, rigid central part	Geometrically separated, hydrophilic and lipophilic sections
Nematic phases	Very common	Very seldom, because phase separation is not possible here
Polymorphism	Very common, stepwise degradation of the order from the crystalline to the isotropic phase	Seldom, normally only the least ordered type of a packing principle occurs
Thermotropic behavior	Normal	Normal
Lyotropic behavior	Rarely observed	Normal
Chiral macroscopic order	Common (BP, N*, TGB$_A$)	Seldom
Disordered tilted lamellar phases	Common (SmC)	Seldom
Influence of configuration	Determinating, normally only one isomer is mesogenic	Small, normally all isomers show comparable mesogenic properties
Influence of conformation	High	Very small
Relevance	Technical applications (e.g., displays)	Biological systems, membranes, Langmuir-Blodgett-films

Table 2. Phase transition temperatures (°C) of three mesogens (**2**–**4**) different in character (see Fig. 2).

Mesogen	Cr			M		Iso	Ref.
2	•	57.5	N	84.5		•	[66]
3	•	64	Sm	134		•	[67]
4	•	96	SmA	114.2		•	[68]

Cr = crystalline, M = mesophase, Iso = isotropic.

of their constitute molecules. A simple sketch depicting the influence of this equilibrium on the types of mesophases developed by such substances is shown in Fig. 3; this behavior is similar to the concentration dependent lyomesomorphic properties of surfactants.

In order to exhibit *amphotropic* properties, liquid crystals require low melting points and a suitable polar/apolar ratio of their molecular parts. Thus, this group of mesogens is a subset of *amphiphiles* showing also lyotropic properties.

In the case of a balanced relation between the different molecular parts a SmA phase is observed. Excessive space demand of one of these parts results in the formation of a columnar mesophase or, in extreme situations, even in discontinuous cubic mesophases. Bicontinuous cubic mesophases may occur in the area between the SmA and the columnar regions.

For pure unbranched paraffins which, of course, are monophilic in character, a dense packing of the SmB type of phase (rotator phase) is observed.

A pure amphiphilic compound resembles a vertical line in the sketch of Fig. 3, thus, normally it only exhibits *one* mesophase. Unusual geometries of the hydrophilic headgroup may lead to a deviation from that vertical orientation, thus, in exceptions even

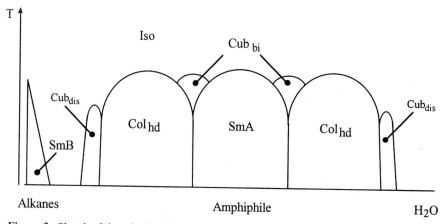

Figure 2. Examples for the influence of differences in polarity of the molecular mesogen structures **2–4**: whereas the monophilic ester **2** forms a nematic mesophase [66], the related amphiphilic compound **3** is smectic [67]; the amphiphilic cholesteryl derivative **4** exhibits a smectic mesophase on heating [68] in contrast to simple, nonamphiphilic cholesteryl compounds which are cholesteric [69].

polymorphism with two or three different mesophases can be observed.

In case of comparable sizes of the headgroup and the lipophilic part, more dense packings of the molecules in the mesophases may occur leading to higher ordered liquid crystals (e.g., to a SmB phase).

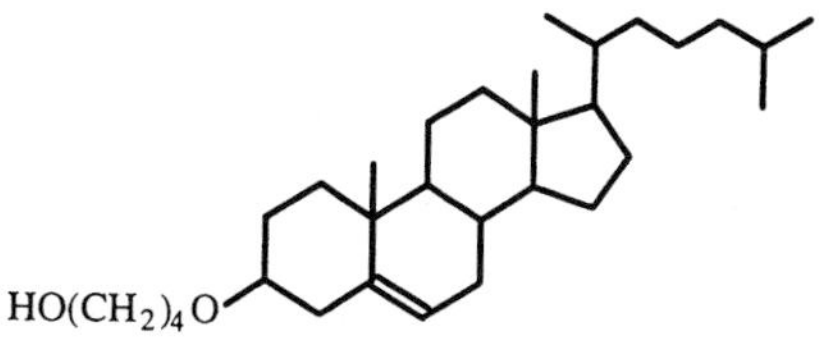

Addition of paraffinic or protic solvents shifts the system to the left or right side, respectively, in the scheme of Fig. 3. Thus, regarding the concentration gradient (a horizontal line in the sketch) one compound in different compositions, in principle, can show all the mesophases mentioned there.

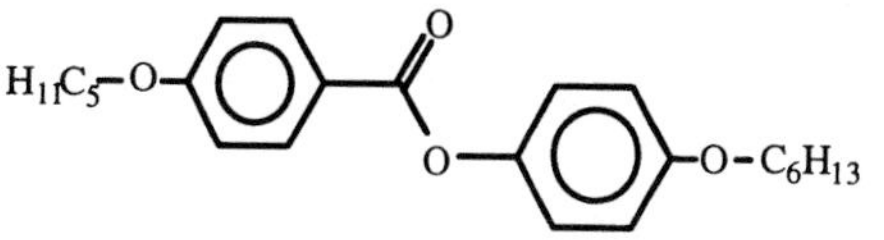

2.1   Simple Amphiphiles

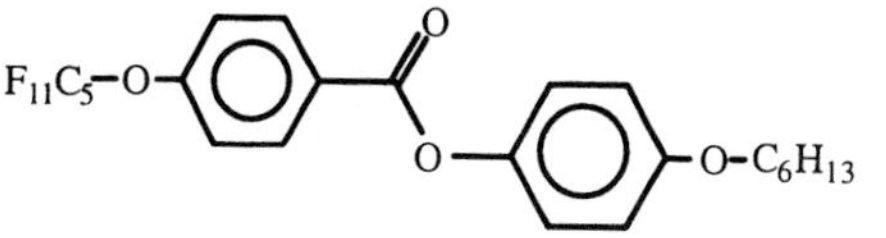

The term 'simple amphiphile' describes the principle of molecular constitution of compounds which will be discussed in this section. The simplest molecular architecture of an amphiphile consists of a hydrophilic group connected to a lipophilic substructure as shown in Fig. 4.

Figure 3. Sketch of the principal phase behavior of amphiphilic compounds. Usual amphiphiles are represented by a vertical line in this scheme; they exhibit only one type of mesophase. Extreme geometries of one of their molecular parts or the addition of solvents (linear alkanes or water) may lead to a deviation from the vertical orientation of that line, thus, amphiphilic compounds in such situations may form various types of liquid crystal phases: T = temperature, SmB phase (rotator phase), Cub_{dis} and Cub_{bi} = cubic discontinuous or cubic bicontinuous phases, respectively, Col_{hd} = columnar hexagonal phase, Iso = isotropic phase.

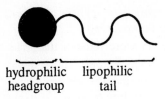

hydrophilic   lipophilic
headgroup   tail

Figure 4. Schematic drawing of the structure of a monotailed amphiphile. Very often, its polar head contains several hydroxyl groups whereas the lipophilic part, for example, is an alkyl chain.

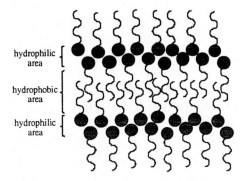

hydrophilic
area

hydrophobic
area

hydrophilic
area

Figure 5. Schematic representation of the architecture of a bilayer SmA phase of intercalated amphiphilic molecules.

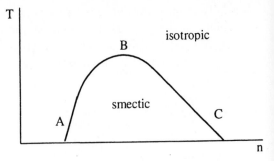

Figure 6. The principal phase behavior of monotailed amphiphiles given by the clearing temperature T vs. the chain length n of the lipophilic molecular part. With increasing chain length liquid crystallinity appears (region A), reaches a maximum stability (plateau B), and vanishes (region C) in the case of very long substituents.

Amphiphiles of this type, thermotropically, form most often layered mesophase structures with different arrangements of the molecules within their layers. In these layers the molecules are arranged in such a way that the hydrophilic heads point towards each other as, similarly, the hydrophobic tails do. This special type of smectic arrangement is called *bilayered* because, on average, each layer in the direction of the layer-normal consists of two molecules. The notation of this phase type is often done by adding a subscript 2 to the appropriate index, that is SmA_2 for the smectic A bilayered phase type. A schematic representation of such an arrangement is shown in Fig. 5.

Often, the layer distance in such mesophases is not twice the molecular length but less. This might be due to a tilt of the molecules with respect to the layer normal resulting in a SmC type of phase, but the same effect may occur because of an interdigitation of molecular parts. Such an interdigitation is indicated by the subscript d in the index of the phase notation, SmA_d. With regard to amphotropic liquid crystals, this SmA_d phase is by far the most observed type of phase, whereas SmC, SmA, and SmA_2 phases are found only rarely.

For the arrangement of molecules within the layers the usual notation is used: in the SmA phase the molecules have no two-dimensional order within their layers, in the SmB phase the molecules within their layers are set in a hexagonal arrangement, characteristically in the SmC the molecules are tilted with respect to the layer normal but possess no other degree of order. An interesting fact is that the smectic phases of such amphiphiles are not miscible with the mesophases of the same type of nonamphiphilic smectic mesogens.

The general phase behavior of mono-tailed amphiphiles is shown in Fig. 6. Starting at a certain chain length of the lipophilic part of the molecule, a thermotropic smectic phase

T / [°C]

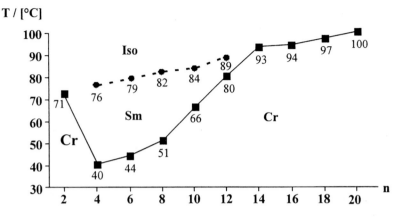

Figure 7. Phase transition temperatures of the fluor organyl series $F(CF_2)_{12}(CH_2)_nH$ (type **5**) over the length of their alkyl part [74, 76, 77].

occurs. With increasing chain length the stability of this mesophase enhances rapidly (see Fig. 6, region *A*). The maximum clearing temperature is observed for a chain length appropriate to the size of the hydrophilic headgroup (plateau in region *B*); for simple monotailed monosaccharides the optimum is reached with $n = 16$ (i.e., hexadecyl). Further increase of the chain length of the lipophilic part leads to a decrease of the clearing temperature. Thus, the equilibrium between the hydrophilic and the lipophilic parts of the molecules determines the liquid crystalline behavior. Principally, the occurrence of liquid crystallinity in such amphiphiles can be compared to the miscibility gap of, for example, glycerol and alkanes. In the region where separation occurs in such two component systems, mesomorphic properties are observed in case of *chemically linked* multiolalkyl amphiphiles; since they can not separate, macroscopically, 'microphase separation' takes place leading to the formation of liquid crystalline mesophases.

Lyotropically, several arrangements of their supramolecular aggregates (formations of layered, columnar and cubic phases) are observed for the simple amphiphilic type of molecules.

One way of synthesizing a simple aliphatic carbon compound amphiphilic in character results from the combination of a polyfluorinated alkyl group with a normal alkyl chain yielding $F(CF_2)_m(CH_2)_nH$. Several studies on such amphiphiles with perfluorocarbon segments have been published [70–76].

An overview of the transition temperatures of series **5** consisting of one $F(CF_2)_{12}$ end and a hydrocarbon residue of variable length is given in Fig. 7 and in Table 3.

In the low temperature phase the fluorinated parts of these molecules are helically arranged whereas the hydrocarbon parts display a trans-planar conformation [74]. Raman spectra show that the hydrocarbon segments are loosely packed in a hexagonal structure as observed for alkanes in the rotator phase. This implies that the hydrocarbon segments of these compounds are not packed closely as in the polyethylene lattice, they are separated enough from each other allowing motion/rotation around their axis [70]. Above the transition temperature of compounds with $n = 4-12$ the fluorocarbon segments remain ordered whereas the hydrocarbon chains in this high temperature phase become disordered [74] leading to a

Table 3. Phase transition temperatures (°C) of multifluoroalkane amphiphiles (**5a–j** and **6**).

Amphiphile	Perfluorodecyl-	Cr		Sm[a]		Iso	Ref.
5a	ethane	•	71.0	–		•	[74]
5b	butane	•	40	•	76.0	•	[74]
5c	hexane	•	44	•	79.0	•	[74]
5d	octane	•	51	•	82.0	•	[74]
5e	decane	•	66	•	84.0	•	[74]
5f	dodecane	•	80	•	89.0	•	[74]
5g	tetradecane	•	90	•	93	•	[76]
5h	hexadecane	•	94	–		•	[76]
5i	octadecane	•	97	–		•	[76]
5j	eicosane	•	100[b]	–		•	[74, 76]
6	Perfluorodecyldecane	•	38	SmB	61	•	[71]

[a] The mesophase of **5b–g** was characterized in reference [77] as *smectic*.
[b] This melting temperature depends on the thermal history of the sample.

smectic phasetype. It has to be quoted from literature that this phase transition is often denoted a solid–solid transition [70, 72–76] although, for example, ^{13}C NMR spectra show that the 'hydrocarbon segments have already gained a nearly liquid-like gauche/trans ratio' [76].

Some years after *liquid crystalline* properties were described for the (perfluorodecyl)decane **6** [71], the *mesomorphic* nature of the phase exhibited by **5e** and **5f** at temperatures above their 'solid–solid transitions' were also recognized [77]. In case of **6**, polarizing microscopy and X-ray experiments prove the existence of a SmB mesophase in the temperature range 38–61 °C, see Table 3.

A different thermal behavior is observed for those examples in which the hydrocarbon chain length exceeds the length of the fluorocarbon part. Here, no phase transitions other than that to the isotropic liquid occur at elevated temperatures. Indeed, in case of **6**, $F(CF_2)_{12}(CH_2)_{16}H$, the thermally induced molecular motion of the long hydrocarbon chain is comparable to that of short chained examples at room temperature, shown by Raman spectra, for

example, $F(CF_2)_{12}(CH_2)_8H$ compared to $F(CF_2)_{12}(CH_2)_{16}H$ [74].

Generally, the miscibility of fluorocarbon and hydrocarbon molecules is very poor [76]. This fact is seen as one reason for the 'microphase separation' discussed above. However, mixtures of semifluorinated compounds with hydrocarbon solvents [72] or fluorocarbon solvents [76] form opaque, birefringent gels [72] which are highly viscous and have been suggested to consist of interdigitated crystallites enclosing large amounts of solvent [76] between the layers of the perfluoralkyl segments. It has also been shown [76] that micelle formation occurs in nonwatery systems offering the way to many interesting supramolecular assemblies. Thus, semifluorinated compounds act like tensides in water: they are surface active compounds with various types of micellar associates.

More complex molecules are those carrying functional (e.g., hydroxyl) groups or other headgroups; such compounds have been studied in a great number, see some reviews [4, 7, 9–11].

Examples of the simplest class of them are the 1,2-dihydroxyalkanes **7a–h** [78,

Table 4. Phase transition temperatures (°C) of selected alkane-1,2-diols, **7a–h** [78, 79]. T_{Cl}=clearing temperatures of these compounds saturated with water (50 wt%), { }=monotropic mesophases.

Compound	Alkane-1,2-diol	Cr		SmB		SmA		Iso	T_{Cl}	Ref.[a]
7a	octane-	•	27	{•	−4	•	16}	•	39	[78]
7b	nonane-	•	26	{•	12	•	25}	•	63	[78]
7c	decane-	•	46	{•	31	•	35}	•	72	[78]
7d	undecane-	•	49	{•	42}	–		•	77	[78]
7e	dodecane-	•	58	{•	50}	–		•	80	[78]
7f	tridecane-	•	61	{•	57}	–		•	80	[78]
7g	tetradecane-	•	63	{•	60}	–		•	76	[78]
7h	pentadecane-	•	68	{•[b]	65}	–		•	86[c]	[79]

[a] In reference [78] the temperatures are given in K.
[b] In reference [79] it is classified only as smectic. Since the longer chained members **7d–g** of this series exhibit only the SmB phase [78], therefore, we feel it reasonable also to expect this type for **7h**.
[c] This temperature value is the maximum clearing temperature in a contact preparation of **7h** with water.

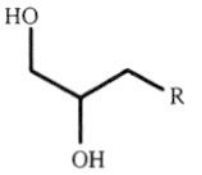

Figure 8. Structure of 1,2-dihydroxyalkanes (**7**, R=alkyl). Thermotropically, these diols exhibit layered phases of the SmA and SmB type [78].

79], shown in Fig. 8. These carbinols are *amphotropic*, that is they exhibit both types of mesomorphism (thermotropic as well as lyotropic) [78–81].

The ethylene glycol derivatives exhibit their thermotropic mesophases only monotropically, see Table 4; interestingly, their mesomorphic phase range can be stabilized remarkably by addition of water. In the latter case, the clearing temperature increases whereas the transition temperature of the SmB–SmA transition decreases. The authors conclude, here, that the formation of hydrogen bonds between adjacent molecules is responsible for their mesophase formation [78].

By 1972 the thermal properties of monoalkylated glycerols (e.g., 1-alkoxy-2,3-propandiol) had been investigated [82] of which the mesophase was later identified as smectic [79].

The extension of the α-diol type, shown in Fig. 8, by insertion of a rigid, rod-like

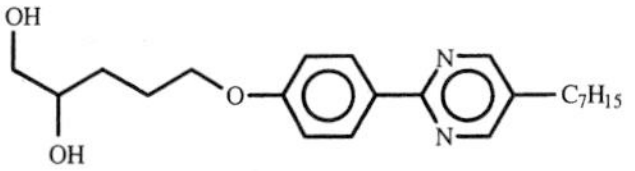

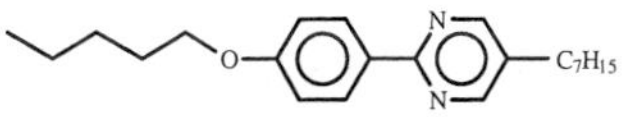

Figure 9. Molecular structure of the hybrid amphiphile **8** and the nonamphiphilic, ordinary calamitic phenylpyrimidine **9** [83].

Table 5. Phase transition temperatures (°C) of the two phenylpyrimidine derivatives **8** and **9** (see Fig. 9).

Compound	Cr		SmA		N		Iso	Ref.[a]
8	•	94	–		-		•	[83]
9	•	49	•	52	•	66	•	[83, 84]

[a] In reference [83] the temperatures are given in K.

phenylpyrimidine unit is illustrated in Fig. 9 [83], yielding the new 1,2-diol **8** of which the hydrophilic head is linked to it by a propyloxy spacer. Opposite this phenyl substituent an alkyl chain is attached as the flexible lipophilic part of the molecule. The nonamphiphilic analogue **9** without the hydroxyl groups, a typical calamitic liquid crystal, is also known. Thus, a direct comparison of these two compounds is possible showing that the incorporation of, for instance, hydroxyl functions into a liquid crystalline molecule leads to remarkable changes of its mesomorphic behavior.

In contrast to **9** which exhibits a SmA and a nematic mesophase on heating, the lyotropic **8** is thermotropically not liquid crystalline. After two solid–solid transitions the diol **8** melts at 94 °C directly into the isotropic liquid. The thermotropic data of both compounds are given in Table 5.

Addition of water to the amphiphile **8** induces several lyomesophases. With growing water content N, SmA, and SmC mesophases are observed [83]. Fig. 10 shows the phase diagram of **8** with water [83]. The uptake of water into the liquid crystalline phase is limited to 10 to 20 water molecules per diol molecule. This is slightly more than for other diol amphiphiles where the thermotropic mesophases could be stabilized by addition of water and where the amount of water in the mesophases is also limied [85, 86].

The binary system of **8** and water has several eutectica and its melting temperature as well as that of the upper solid–solid transition $(Cr_2 \rightarrow Cr_3)$ decreases with growing amounts of water, however, the lower solid–solid transition $(Cr_1 \rightarrow Cr_2)$ remains constant. With growing water concentration the Cr_3 phase is replaced by the N mesophase which occurs in a very narrow concentration range from 1 to 2 wt% of water in the mixture (see Fig. 10). This was the first observation of an inverse nematic N_{2L} phase in a binary system of an amphiphile with water [83]. At higher water concentrations lamellar phases occur: at first, starting at 2 wt% of water, the SmA and then, above 3 wt% of water, the SmC type of lyomesophase. The clearing temperature of the SmA phase increases with growing water concentration until about 16 wt% of water above which it remains constant. For the first time, the inverse phase sequence $Cr \rightarrow N \rightarrow$ SmA $\rightarrow$ SmC was observed here in lyotropic systems of this type of binary mixture [83]. The authors discuss three different models for the possible architecture of its nematic phase [83]:

– The nematic phase could represent the first example of an inverse nematic N_{2L} phase, because it appears in the concentration range between the crystalline phase and the lamellar mesophases. In this case, the mesogenic structures would consist of small and flat, disc-like supramolecular units in which the molecules cling together by cooperative hydrogen bonds between their headgroups whereas their lipophilic tails point outwards of these associates. The geometry of this phase is similar to the N_D phase in thermotropic liquid crystals.

– The nematic phase could consist of linear associations which are formed by hydrogen bonds between the diol groups. The formation of large hydrogen bond networks in the phase is hindered because of the small amount of only about 1.5 wt% of water.

– The nematic phase could occur because the hydrogen bond interactions between the diol groups are disturbed by the nitrogen atoms of the heteroaromatic ring, also hydrogen bond acceptors; a molecular situation which disfavors the formation of bilayers. In the water-free state the crystalline phase is so stable that the (monotropic) nematic phase is, so to speak, hidden in the crystalline phase and could not develop due to rapid recrystallization during cooling. The addition of small amounts of water decreases the melting temperature and the nematic phase get more stable than the crystalline state.

Owing to the very tiny concentration range, it is difficult to investigate this nematic phase in detail [83].

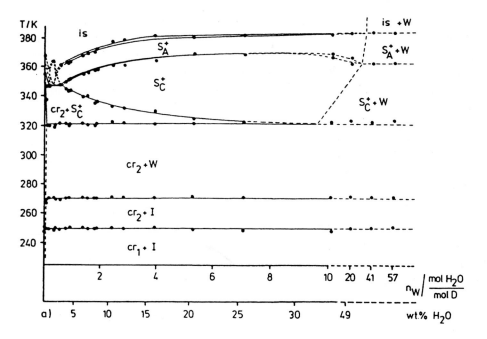

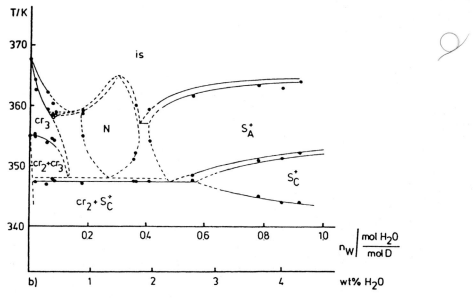

Figure 10. Phase diagrams of the binary mixture of **8** and water [83]. The lower drawing shows a detailed, enlarged section of the upper diagram; cr_1, cr_2, cr_3 = different crystalline phases, S_A^+, S_C^+ = smectic phases, N = nematic phase, W = water, is = isotropic phase, for further details see [83].

A more complex and larger headgroup is found in the family of monotailed glucopyranosides **10a–l** of which the phase transition data are compiled in Table 6. In this series, the hexadecylglucopyranoside **10k**, shown in Fig. 11, is of historical interest. From our viewpoint today, it is one of the longest known amphiphilic mesogens: in 1911, E. Fischer already described the behavior of this amphiphile [56]. Although the phenomenon of liquid crystallinity has been known since about 1890 [87, 88] it was not until 1938 that the thermotropic mesomorphic character of this multiol (**10k**) was discovered [58]. This compound is, so to speak, the prototype of amphiphilic (and amphotropic) liquid crystals of the great family of monotailed carbohydrate liquid crystals. Their hydrophilic headgroup consist of the carbohydrate moiety and their alkyl chain represents the lipophilic part of this amphiphile. In the meantime, a long homologous series of this family of carbohydrate derivatives has been studied. Some of them have gained great interest and are commercially available due to their ability to act as biodetergents, for example, for the solubilization of membrane proteins [89].

The general interest in this particular field of amphotropic liquid crystals is represented in several review articles [4, 7, 9–11].

Whereas the stability of the crystalline phase of members of series **10** remains roughly constant (their melting points deviate only in a temperature range of about 30 K), the behavior of the liquid crystal into isotropic phase transition is more complex. Their clearing temperatures rise with increasing length of the alkyl chain up to a maximum clearing temperature at the tetradecyl derivative. Afterwards they drop again and the longest (tridecyl) member of this family is no longer thermotropically liquid crystalline.

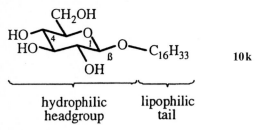

Figure 11. Hexadecyl-β-D-glucopyranose [56], see Table 6.

Table 6. Phase transition temperatures of various β-glucopyranosides (**10a–l**) and of one selected α-glucopyranoside (**10m**).

Structure	Compound	R	Cr		SmA		Iso	Ref.
	10a	C_2H_5	•	73	–		•	[90]
	10b	C_4H_9	•	66–67	–		•	[91]
	10c	C_6H_{13}	•	88–91	–		•	[58]
	10d	C_7H_{15}	•	×	•	85[a]	•	[92]
10a–l	**10e**	C_8H_{17}	•	67.07	•	106.38	•	[93]
	10f	C_9H_{19}	•	68	•	113	•	[93]
	10g	$C_{10}H_{21}$	•	70.26	•	133.53	•	[93]
	10h	$C_{11}H_{23}$	•	×	•	139[a]	•	[92]
	10i	$C_{12}H_{25}$	•	80.4	•	144.9	•	[94]
	10j	$C_{14}H_{29}$	•	×	•	155	•	[95]
	10k	$C_{16}H_{33}$	•	81.5	•	147.5	•	[56]
	10l	$C_{18}H_{37}$	•	81.5	•	120	•	[60]
the α-anomer	**10m**	$C_{12}H_{25}$	•	81.4	•	149.5	•	[94]

[a] In reference [92] this temperature is given in K.
× = this temperature is not given in the literature.

Figure 12. Schematic representation of three different lyomesophases, lamellar, cubic, or columnar hexagonal in type [97].

Table 7. Phase transition temperatures of the α- and β-anomers of 1-O-dodecyl glucose with a pyranosidic and furanosidic headgroup, respectively (see Figs. 11 and 13).

Sugar	Compound	Cr		SmA		Iso	Ref.
α-D-Glucopyranoside	**10m**	•	81.4	•	149.5	•	[94]
β-D-Glucopyranoside	**10i**	•	80.4	•	144.9	•	[94]
α-D-Glucofuranoside	**11a**	•	69.5	•	114.2	•	[94]
β-D-Glucofuranoside	**11b**	•	74.4	•	147.8	•	[94]

The lyotropic behavior of these carbohydrate derivatives has been studied very carefully [96]. It follows the classical sequence of mesophases established for the ionic detergents such as soaps. In contact preparations of 1-O-octyl-β-D-glucopyranoside (**10e**) with water at room temperature three types of lyomesophases are observed (from high to low amphiphile concentration): lamellar, cubic and columnar hexagonal, their principal structures are depicted in Fig. 12.

To explore the relationship between molecular structure and properties of this type of mesogens in detail several modified structures, different in their hydrophilic headgroups, have been synthesized and their mesogenic properties investigated [98].

For example, the glucopyranoside has formally been replaced by the furanoside as

Figure 13. Molecular structure of the dodecyl glucofuranoside **11**. Both, the α- and β-anomer have been studied in their pure states [94].

depicted for the dodecyl glucofuranosides **11a** and **11b** in Fig. 13 [94]. The properties of both families **10** and **11** (see Table 7) are quite similar causing the authors to conclude that the increased flexibility of the five-membered furanose ring compared to the more rigid one of the pyranose is not lowering the stability of the mesophase [94]. This phenomenon is generally valid for am-

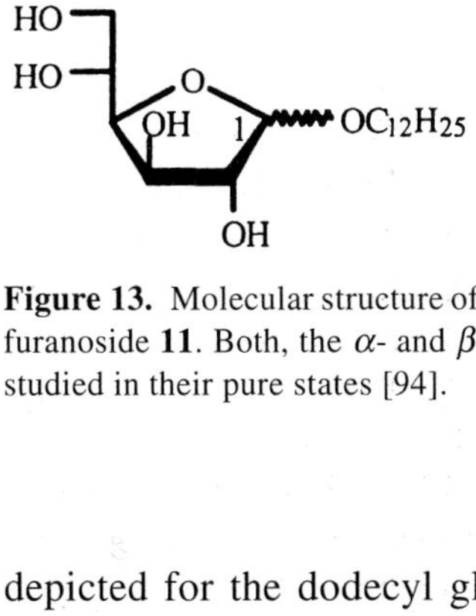

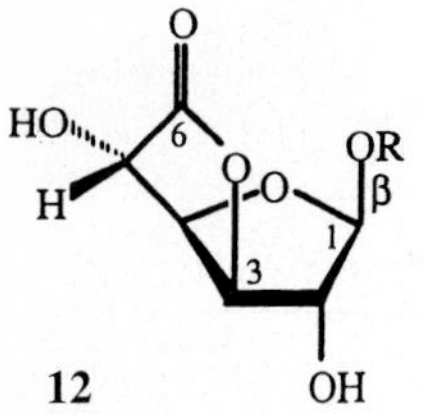

a) R = C_8H_{17}
b) R = C_9H_{19}
c) R = $C_{10}H_{21}$
d) R = $C_{11}H_{23}$
e) R = $C_{12}H_{25}$

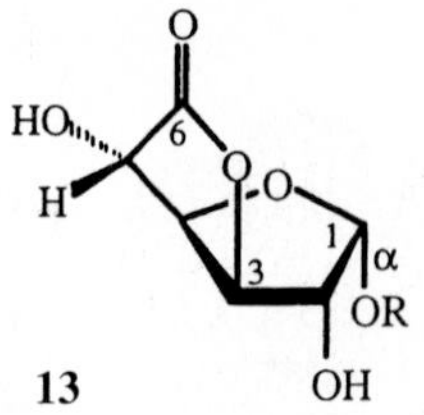

Figure 14. Molecular structures of the two anomeric amphiphiles **12** and **13** with each the bicyclic glucofuranosiduro-no-3,6-lactone headgroup [99].

Table 8. Phase transition temperatures (°C) of the *O*-alkyl-D-glucofuranosidurono-3,6-lactone series **12** and **13** [99].

Compound	Alkyl, R	Cr		SmA		Iso
12a	octane-	•	71.8	{•	41.6}	•
12b	nonane-	•	77.4	{•	70.9}	•
12c	decane-	•	82.4	•	90.9	•
12d	undecane-	•	86.8	•	104.9	•
12e	dodecane-	•	89.5	•	116.0	•
13a	octane-	•	91–92 [a]	–		•
13b	nonane- [b]	•	91.1	–		•
13c	decane- [b]	•	94.8	–		•
13e	dodecane- [b]	•	95.8	–		•

{ } = monotropic mesophases.
[a] This temperature range was measured by polarizing microscopy.
[b] Solid–solid transition temperature given in the literature is skipped here.

phiphilic liquid crystals: the rigidity of molecules is not as important as it is for classical calamitic liquid crystals. The α-anomer of the furanosidic glucose possess a lower clearing temperature then the β-anomer. Whereas this effect is rather small in case of the glucopyranosides, the clearing temperatures of furanosides display a difference of up to 30 K. However, an explanation for this sterical influence could not yet be given by simple models.

An interesting variation of the furanosidic headgroup is realized in the amphiphiles **12a–e** [99] shown in Fig. 14. These 1-*O*-alkyl-D-glucofuranosidurono-3,6-lactones consist of a lipophilic tail joined to a bicyclic headgroup which, in comparison to those already discussed, is more rigid and space demanding.

The phase transition temperatures of the two *O*-alkyl series **12** and **13** are given in Table 8 [99]. Whereas the 2,5-diols of the β-anomers **12a** to **12e** are thermotropically liquid crystalline each exhibiting a SmA phase either monotropically (**12a** and **12b**) or enantiotropically (**12c** to **12e**), the whole α-series **13** is thermotropically not liquid crystalline. The clearing temperatures of series **12** are related to the length of the alkyl chains in the usual way: their clearing temperatures rise with increasing chain length. A possible explanation for the absence of thermomesogenic properties of series **13** (compared to series **12**) could be their more stable crystalline phases covering the possible mesophase [99].

The isotropic melts of members of series **13** can not be supercooled very much: crystallization occurs just below their melting points. Moreover, the anomeric configuration probably prevents the formation of strong intermolecular hydrogen bond networks and an efficient intercalation of the alkyl chains [99]. These examples clearly demonstrate the importance of phase stability and stereochemistry for the formation of mesophases among these amphiphiles and prove, for instance, the effect that a simple change at one chiral center can have on the thermotropic behavior of members of both anomeric series.

The solubility of the humpbacked lactones **12a–e** in water is very low (less than 1 mM at 60 °C in the case of **12e**) [99]; in this respect, no comparison has been made with regard to the structurally stretched ster-

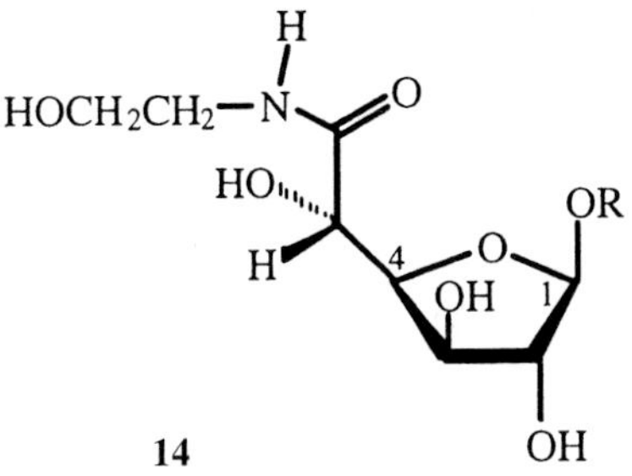

a) R = C$_{10}$H$_{21}$
b) R = C$_{11}$H$_{23}$
c) R = C$_{12}$H$_{25}$

Figure 15. Basic molecular structure of some *O*-alkyl-*N*-(2-hydroxyethyl)-*β*-D-glucofuranosiduronamides [99].

eoisomers **13**. The essential stereochemical differences between **12** and **13** have also not been considered with regard to their thermotropic behavior. It is surprising that the more bowl-shaped examples of **12** are thermomesomorphic whereas those of the more linear ones of **13** are not. The cleavage of the lactone ring in **12** with 2-aminoethanol leads to members of the corresponding series **14** which possess a more hydrophilic headgroup (see Fig. 15). Their liquid crystalline properties are summarized in Table 9.

Further modifications of the hydrophilic headgroup of such cyclic multiols have been realized. For example, the introduction of a pyranose ring in which the ring-oxygen atom is replaced by nitrogen offering for the first time the possibility of substitution at this particular position of the cyclic headgroup of an amphiphile (see Fig. 16) [100].

The thermal data of these interesting *N*-alkylated 1-deoxynojirimycin derivatives **15** are summarized in Table 10. It is noteworthy to stress that these compounds are biologically active [100]: they possess antibiotic properties and inhibit various glucosidases. Several patents on antiviral effects of 1-deoxynojirimycin derivatives on HIV and other retroviruses have been registered.

As expected for single-tailed rod-shaped amphiphiles, four of the investigated eight 1-deoxynojirimycin derivatives **15 a–h** exhibit an interdigitated SmA phase on heat-

Table 9. Phase transition temperatures (°C) of the *O*-alkyl-*N*-(2-hydroxyethyl)-*β*-D-glucofuranosiduronamides **14 a–c** [99]. The type of mesophase (M) exhibited by them was not determined explicitly.

Compound	Alkyl, R	Cr		M		Iso
14 a	decyl-	•	39–41	•	83–85	•
14 b	undecyl-	•	38–41	•	108	•
14 c	dodecyl-	•	58–60	•	133	•

ing. Furthermore in contact preparations with water, the long chained tetrols, **15 e–h**, develop lyotropic mesophases of layered, cubic, and columnar hexagonal architectures.

In order to investigate the influence of the position of the alkoxy chain at the hydrophilic headgroup on the mesomorphic properties of such monotailed carbohydrate liquid crystals, two research groups systematically varied the structure of two series of model compounds (**16** and **17**) [9, 101].

Comparing the transition temperatures of the ten pentols **16 a–e** and **17 a–e**, shown in Fig. 17, there is a strong dependence of their melting temperatures on

– the position of this (lipophilic) alkyl ether function, and on
– the fixed stereochemistry of the hydrophilic headgroups.

On the other hand, the clearing temperatures of at least three mono-ethers of each series **16** and **17** (**16 b**, **16 c**, and **16 e** as well as **17 a–c**) are very similar, at around 141 or

Table 10. Phase transition temperatures (°C) obtained by DSC of eight *N*-alkyl derivatives of 1-deoxynojiri-mycin [100].

Structure	Compound	R	Cr		SmA		Iso
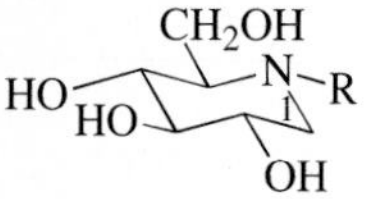	**15a**	methyl	•	151.0	–		•
	15b	ethyl	•	153.4	–		•
	15c	phenylethyl	•	179.7	–		•
	15d	cinnamyl	•	161.4	–		•
	15e	nonyl	•	105.6	•	149.5	•
	15f	decyl	•	88.3	•	155.3	•
	15g	dodecyl	•	98.1	•	164.6	•
	15h	tetradecyl	•	101.9	•	166.7	•

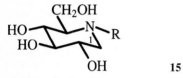

15

Figure 16. Molecular structure of some antibiotic derivatives of 1-deoxynojirimycin. Here, the pyranosidic oxygen atom of glucose is substituted by nitrogen allowing derivatizations of the hydrophilic head group at this particular position of the heterocycle [100].

220 °C, respectively. As, in particular, can be seen in the bottom part of Fig. 17, relatively small structural/sterical changes at the inositol ring cause dramatic changes in the melting temperatures of these amphiphiles.

For instance, the formal migration of the dodecyloxy group around the cyclohexane ring, away from the axial hydroxy function changes drastically the liquid crystalline properties of these natural product derivatives. Their melting points rise by about 90 to 123 K. Interestingly, the arrangements of the substituents in **17b** and **17c** of this carbocyclic series possessing the weakest molecular symmetry give rise to the formation of very broad, stable SmA phases with almost the same clearing temperatures. The three mono-ethers of highest molecular symmetry (**17a**, **17d**, and **17e**), however, do not or nearly not exhibit a mesophase due to their very high melting points.

This relationship can also be detected in the sugar series **16**, e.g., considering **16c**. Of much lower molecular symmetry but very similar constitution are in this heterocyclic series both **16a** and **16d** showing almost the same melting and clearing temperatures and also the broadest and most stable SmA phases.

Thus, symmetry of these molecules is an important factor for their thermotropic properties: the higher their symmetry the more stable is the crystalline phase or the less probable is the appearance of a thermotropic mesophase.

Another way of varying the molecular structure of such amphiphiles is the modification of their lipophilic part by formal substitution of an alkyl chain by an 4-alkylphenyl group [102a], see **18** and **19** in Fig. 18, or by an 4-alkoxyphenyl group [102b]. The phenyl ring makes the sidechain more rigid. It was found here, that this structural feature increases the transition temperatures, in particular, the melting points of the α-series **18**.

Since a phenyl group is equal to about four to five methylene groups in length a correspondence of the temperature increments with regard to simple alkyl glucosides (e.g., series **10**) is discussed [102a]. For all of the six 4-alkylphenyl glucopyranosides of the series **18** and **19** investigated, lyotropic mesomorphism has been observed at room temperature. In contrast, this is not

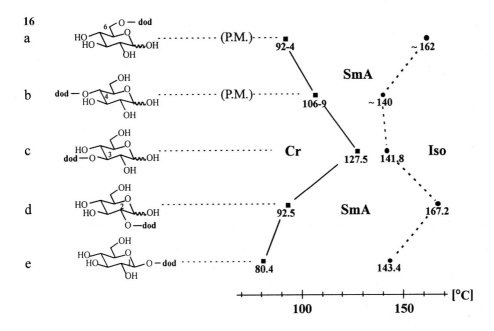

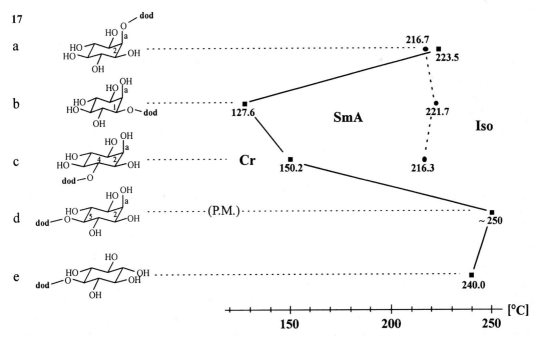

Figure 17. Plot of the phase transition temperatures, obtained by DSC or polarizing microscopy, against the structure of each five mono-dodecyl ethers of the two series of D-glucopyranose [101] (top) or *myo-/scyllo*-inositol [9] (bottom). Obviously, the appearance of the mesophase on heating is determined by their relative molecular symmetry originating from the different localization of the ether group in both series, hetero- or carbocyclic, respectively, in their molecular structures; Cr = crystalline, SmA = smectic A phase, Iso = isotropic liquid.

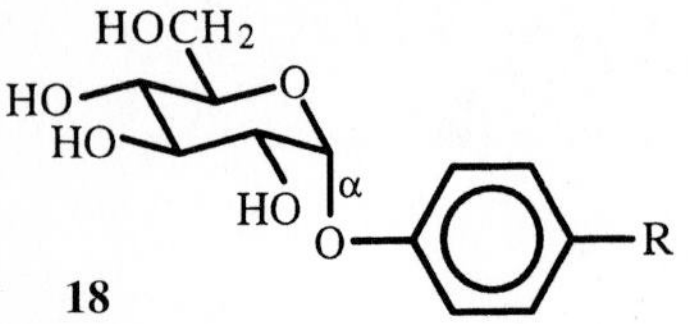

Figure 18. Molecular structures of the 4-alkylphenyl α- and β-glucopyranosides **18** and **19** [102a]; R = C$_3$H$_7$, C$_4$H$_9$, or C$_7$H$_{15}$.

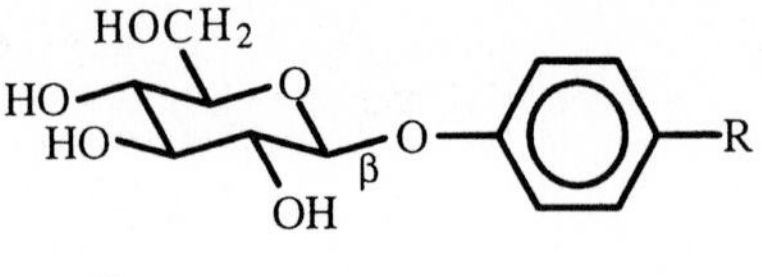

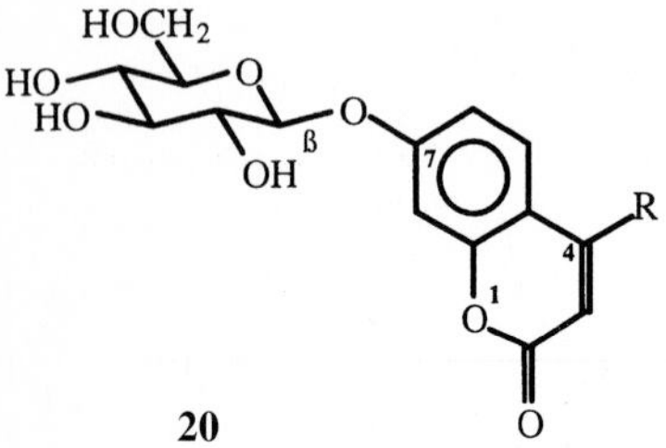

Figure 19. Molecular structure of the 4-alkyl umbelliferyl β-D-glucoside derivatives **20** [103]; R = C$_7$H$_{15}$, C$_9$H$_{19}$, or C$_{13}$H$_{27}$.

Table 11. Phase transition temperatures (°C) obtained by DSC of the three derivatives **20a–c** of umbelliferyl β-D-glucoside (see Fig. 19).

Tetrol	Alkyl, R	Cr		SmA		Iso
20a	C$_7$H$_{15}$	•	107.2	–		•
20b	C$_9$H$_{19}$	•	163.2	{•	158.8}	•
20c	C$_{13}$H$_{27}$	•	163.4	•	198.0	•

{ } = monotropic mesophase.

concentration) was found [102b]. Furthermore the nonyloxy derivative forms long ribbons in dilute aqueous solutions [102b].

Another series of carbohydrate liquid crystals incorporating a rigid group in their lipophilic part of the mesogen is shown in Fig. 19; the phase transition data of these umbelliferyl derivatives **20** [103] are given in Table 11. The middle part of these glucosides consists of a flat, rigid coumarin unit which is linked to a hydrophilic tetrol (D-glucopyranose) and a lipophilic, aliphatic tail. Due to the location of the alkyl chain (4-position of the lactone ring) the overall molecular shape is somewhat bent, humpbacked in the region of the carboxylic group. In spite of this steric situation these coumarin derivatives can obviously exist in a conformation which is stretched enough and allow mesogenity [103].

the case for short chained alkyl glucosides (chains shorter than nonyl) [102a].

In conclusion it is emphasized that the introduction of a rigid phenyl group does not significantly affect the liquid crystalline properties, only the transition temperatures rise [102a].

Studies of the mesomorphic properties of 4-alkoxyphenyl-β-D-glucopyranosides [102b] reveal that these amphiphiles form a SmA phase on heating. Lyotropically, the usual phase sequence for single tailed carbohydrates (isotropic, columnar hexagonal, cubic, and lamellar with decreasing water

2.2   Bolaamphiphiles

The term bolaamphiphile [104, 105] describing the shape of this class of mesogens very pictorial: it is related to the South American hurl weapon consisting of two stones or leather balls linked by a rope. Bolaamphiphiles consist of two hydrophilic headgroups at both ends of the lipophilic section of the molecules.

Formally, one could compare the structure with two simple amphiphiles (see Fig. 4) which are linked at the ends of their lipophilic parts. The principal structure of

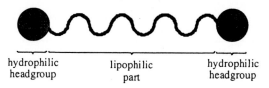

hydrophilic   lipophilic   hydrophilic
headgroup   part   headgroup

Figure 20. Schematic drawing of a bolaamphiphilic molecular structure. The polar heads consist, for example, of cyclic or acyclic multiols or carbohydrate substructures which are separated by an alkyl chain.

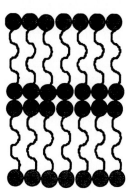

Figure 21. Schematic drawing of a SmA phase built up by monolayers of bolaamphiphiles.

HO, HO—...(CH$_2$)$_n$...—OH, OH

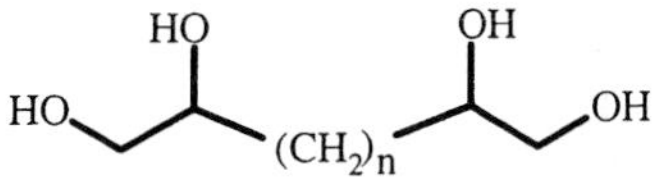

21

Figure 22. Molecular structure of members of the bolaamphiphilic family **21** [108]; $n \geq 4$, see Table 12.

such bolaamphiphiles is sketched in Fig. 20. For instance, the polar headgroups may each be formed by a diol substructure [80, 81] or by carbohydrate moieties [106]. As expected, these compounds form layered SmA or SmC phases on heating: see the remark upon the nomenclature used here [107] and the note with regard to the $L_{\beta'}$-phase in section 3, General Phase Behavior, at the end of this chapter. In contrast to single tailed simple amphiphiles forming *bilayers*, for bolaamphiphiles *monolayers* are characteristic.

Since bolaamphiphilic molecules form hydrogen bond networks at both of their molecular ends the stability of their mesophase is remarkably increased compared to simple amphiphiles of half the length.

The α,ω-bisdiols **21** depicted in Fig. 22 are characteristic examples of this kind of amphiphile. Both ends of the molecules carry a hydrophilic, vicinal bisdiol unit. Except for the one with the shortest alkyl middle section (**21a**, $n=4$), all these multiols are thermotropic liquid crystalline showing at least a SmC phase [80, 81, 108]. The thermotropic phase transition temperatures of members of this amphiphilic family are compiled in Table 12. Multimesomorphism is observed here for examples of which $n \geq 11$; their SmC phase is followed by a narrow (1–5 K broad) SmA phase on heating. The mesophase stability increases with growing length of the alkyl spacer. X-ray investigations show that in the low tempera-

ture smectic phase the alkyl chains are tilted by 28° with respect to the layers of the hydrogen bond network, hence, this mesophase is of the SmC type. The authors suggest a model of the molecular interactions on the basis of their experimental data and assume that the primary OH-groups interact with the secondary OH-groups of adjacent molecules by hydrogen bond formation leading to a tilt angle of about 30° [108]. In the high temperature phase, SmA, the alkyl part between the hydrophilic ends is liquid-like, unordered, and untilted. Owing to the big difference in the molecular order the transition between both smectic phases is indicated by relatively large transition enthalpies between ≈10 and ≈40 kJ mol^{-1}. Since the melting of the polymethylene chains is accompanied by a reorganization of the hydrogen bond network, the energy for this process is part of the enthalpy measured here.

Table 12. Phase transition temperatures (°C) of the bolaamphiphilic tetrol series **21** [108]. Solid–solid transition temperatures on some of these tetrols, given in the literature [108], are skipped here.

Tetrol	n	Cr		SmC [a]		SmA [a]		Iso
21a	4	•	79	–		–		•
21b	5	•	77	{•	36}	–		•
21c	6	•	101	{•	75}	–		•
21d	7	•	118	{•	90}	–		•
21e	8	•	114	{•	110}	–		•
21f	9	•	110	•	115	–		•
21g	10	•	85	•	124	–		•
21h	11	•	101	•	127	•	128	•
21i	12	•	87	•	132	•	134	•
21j	13	•	90	•	131	•	136	•
21k	14	•	79	•	134	•	137	•
21l	15	•	89	•	135	•	139	•
21m	16	•	84	•	138	•	142	•
21n	17	•	61	•	139	•	143	•
21o	18	•	87	•	140	•	143	•
21p	20	•	111	•	140	•	141	•

{ } = monotropic mesophase.
[a] See the note on the nomenclature used here [107].

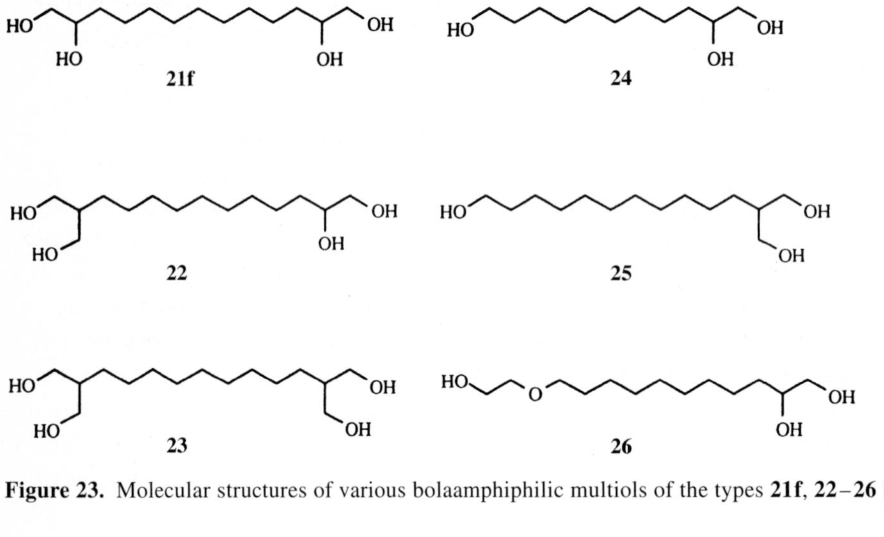

Figure 23. Molecular structures of various bolaamphiphilic multiols of the types **21f**, **22–26** [80].

The hydrophilic parts of such tetrols have also been modified leading to structures shown in Fig. 23 [81]. The bolaamphiphilic tetrol **22** with its two different headgroups has the most extended polymorphism of these multiols. One of its mesophases is hitherto unidentified due to rapid recrystallization.

All the multiols shown in Fig. 23 are liquid crystalline except the tetrol **23** with its bis-1,3-diol headgroups [81]. Apparently, this situation causes a high melting point (103 °C [81]) suppressing the occurrence of a mesophase. The same effect was discussed in subsection 2.1 'Simple Amphiphiles' of this chapter, for example, for members of a monotailed pentol series derived from inositols [9]. The other multiols depicted in Fig. 23 form layered mesophases of various types in the water-free state which are partly monotropic, however, in some cases even multimesomorphism is observed [81]. Addition of water to bolaamphiphiles of this type with long hydrophobic middle parts leads to a stabilization of mesophases, whereas mesophases of such multiols with short middle sections get destabilized. In all these cases, the addition of water leads to the formation of the *less ordered* SmA type of phase instead of lamellar mesophases of *higher order* described above [81, 108].

2.3   'Y'-Shaped Materials

The terms *'Y'-shaped* [109], *'tripodal'*, *'1,1-double-tailed'* or *'peg-shaped'* [110] describe the molecular geometry of this type of amphiphile. At the hydrophilic head two hydrophobic tails are joined close together or even by a common link to the headgroup. Very famous examples for this molecular architecture are the biologically active phospholipids and sphingolipids [111]. Lipids occur in all biological cells and have common solubility properties: generally they are water-insoluble, amphiphilic molecules which may form colloids, micelles or liquid crystalline phases in water. They fulfill important functions in cell membranes and signal transduction and even seem to play a tremendous role in the origin of certain kinds of cancer; see an overview on the biological aspects of glycosphingolipids [112]. One of the best investigated phospholipids carries a cholin residue in its hydrophilic part and is, therefore, named phosphatidylcholin. Related phospholipids are phosphatidylethanolamine, phosphatidylinositol and phosphatidylserine. All these phospholipids have the same molecular architecture except for their hydrophilic substituent at the phosphate link, see Fig. 24.

The degree of unsaturation of their fatty acyl chains and the chemically nature of their headgroups, the latter being responsible for the ability and strength of the hydrogen bond formation both between the lipid as well as the lipid and water molecules, do play a major role for the amphotropic behavior of such lipids.

In water, these compounds spontaneously aggregate to bilayers in which the hydrophilic headgroups point outwards at each side whereas the lipophilic tails form the hydrophobic middle part of the layer. These bilayers may also form more complex aggregates, such as micelles, vesicles or lyome-

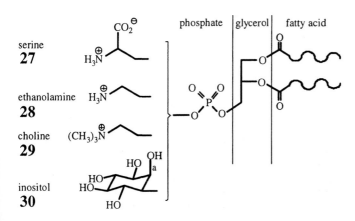

Figure 24. Sketch of the principal molecular structures of phospholipids. These molecules are the main components of cell membranes and may form complex supramolecular aggregates in water.

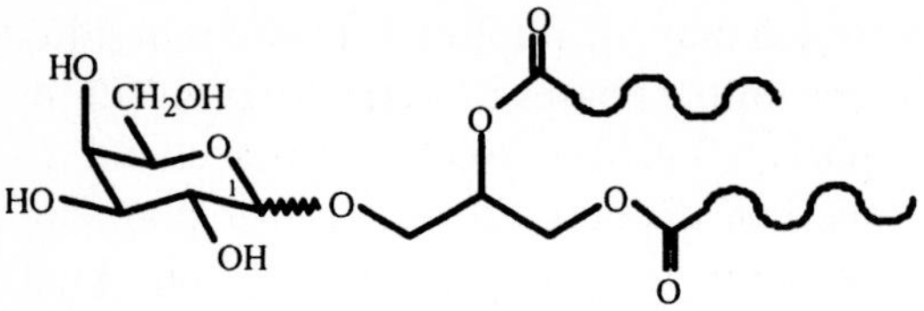

31

Figure 25. Molecular structure of the diacyl type **31** of galactosylglycerols [113–118].

sophases. Without such supramolecular structures life would not be possible, a fact best describing the basic importance of such supramolecular structures.

Up to 75% of the total lipid content of plant chloroplast membranes consists of galactoglycerolipids **31**, see Fig. 25. They have been isolated and studied in their natural form and as hydrogenated derivatives. Synthetic analogues and homologues have also been prepared and studied in detail [113–118]. These materials exhibit a very complex lyomesogenic behavior including metastable and time dependent phase equilibria. Their chemical structures, types of sugar, and chain length have significant influence on their mesogenic properties. In summary, one could make the following remarks about this very important class of

multiols: lamellar (L_β, L_α), cubic ($Q_{||}$) and inverse columnar hexagonal ($H_{||}$) lyomesophases were observed [113], in annealed samples even an L_C phase. At temperatures above their L_β/L_α phase transitions, energetically weak bilayer–nonbilayer transitions may occur but, compared to the corresponding diacyl phosphatidylethanolamines, these phase transitions (L_β–L_α and bilayer–nonbilayer) occur at lower temperatures [113].

X-ray investigations of especially the L_C phase of galactopyranosyl- and glucopyranosyl-glycerols reveal that not simply the size or orientation of the hydrophilic headgroups regulate the supramolecular packing and, thus, the structure of the mesophases, but rather a combination of these supported interactions of their alkyl chains are responsible in this matter [113].

A discussion of the thermotropic properties of this type of compounds began in 1983 [119] with naturally occurring lipids. Again, the heterogeneity of the materials made syntheses of uniform derivatives desirable [114a]. Pure 1,2-di-O-acyl-3-O-(α-D-glucopyranosyl)-*sn*-glycerols (**32**) were investigated in view of their thermotropic properties. The thermotropic mesophase of the members of series **32** has been denoted as

Table 13. Phase transition temperatures (°C) of some selected α-D-glucopyranosyl-diacylglycerols of type **32** [114a]. The temperature ranges in the first column (data for the transitions Cr→M) are indicated in reference [114a] as so-called 'softening point'.

Structure	Lipid	n	Cr		M[a]		Iso
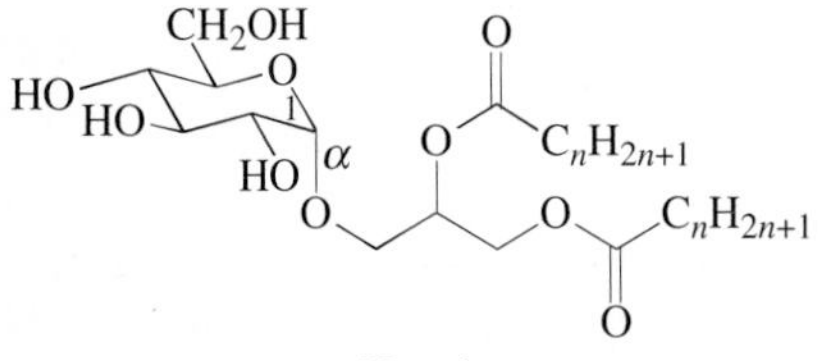	**32a**	12	•	40–44	•	120–122	•
	32b	13	•	58–60	•	128–130	•
	32c	14	•	67–69	•	129–131	•
	32d	15	•	67–71	•	139–140	•
	32e	16	•	76–77	•	142–143	•
	32f	17	•	79–81	•	143–144	•
	32g	18	•	80–82	•	144–145	•
	32h	19	•	81–83	•	145–146	•
32a–i	**32i**	20	•	89–91	•	141–142	•

[a] The type of this mesophase has originally been denoted as probably smectic [114a] which later (without evidence) has been emphasized as columnar [10, 12].

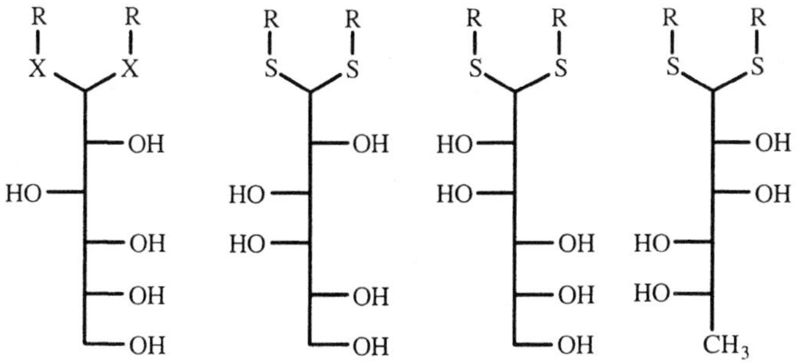

33   **34**   **35**   **36**

X = O, S

Figure 26. Acetals and thioacetals possessing a tripodal, Y-shaped structure; R = alkyl [123].

'probably smectic' [114a]; later it has been referred to be of a columnar type [10, 12] which would be in better accordance with the molecular structure of these amphiphiles. It should be pointed out here, that this di-O-acyl series **32** was discussed in ref. [10 and 12] as di-O-alkyl compounds synthesized by other workers [114b]. Selected phase transition data of this family of multiols are compiled in Table 13.

An interesting example of a synthetic bioactive representative of this family is 1,2-di-O-9′-octadecynyl-3-O-β-D-galactopyranosyl-sn-glycerol [120] described by the authors of having a 'sintering point' at 55 °C and a melting point at 76 °C; here, a liquid crystalline phase is much likely. Its lyotropic phase behavior and also that one of mono-galactosyl- or -glucopyranosyl-glycerol derivatives [121, 122] is very complex: a gel-phase and lyomesophases of the L_α, $Q_{\parallel}^a$, $Q_{\parallel}^b$ and $H_{\parallel}$ phases have been observed. Here again, differences in the stereochemistry of the headgroups, especially in the interfacial region between the polar and the apolar part of the molecules, are discussed to cause differences in the phase behavior, see the discussion for the monotailed amphiphiles [121].

Completely different in structure, but also belonging to this group of Y-shaped am-

photropic materials, are the thio- and oxyacetals **33–36** of carbohydrates shown in Fig. 26 [94, 123–127]. Interestingly, these amphiphiles exhibit a *columnar* thermotropic mesophase [123, 126]. The two space demanding lipophilic chains enforce a curvature of the interfacial region between the hydrophilic and the hydrophobic parts of the mesophase resulting in a columnar type of their mesophase. The architecture of this phase is of an inverse micellar kind: the hydrophilic headgroups point inwards of the columns whereas the lipophilic chains do the opposite.

A substitution of the CH_2OH part by the nonpolar methyl group, illustrated in **36** of Fig. 26, destroys the mesomorphic properties totally which is obvious since the hydrogen bond network (inside the columns) is severely perturbed, even disturbed in this case [123, 126].

All the Y-shaped mesogens described hitherto have two alkylchalcogeno chains as the upper part and a sugar moiety as the lower part of that 'Y'. A different architecture is found, for example, for N-dodecyl-D-galacturonamide-didodecyl-dithioacetal [128] which also has a 'Y'-shaped molecular structure, however, carrying now the sugar part in the center of that 'Y' surround-

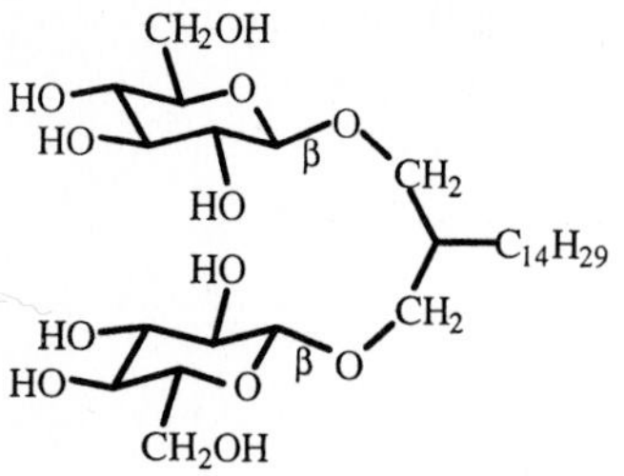

Figure 27. Formula of the unusual, *Y*-shaped amphiphile **37** with two hydrophilic headgroups [129].

2.4 Disc-shaped Amphiphiles

Research on liquid crystals with a disc-like shape started in 1977 [130] when hexa(heptanoyl)benzene (**38**, first synthesized forty years earlier [131]) was proved to form a columnar mesophase (m.p.=I_m 81.2°C, cl.p.=I_{cl} 87.0°C) [132]. Disc-shaped amphiphilic molecules with *saturated* cores, such as the hexaesters **39** or even the hexaethers **40** of the naturally occurring *scyllo*-inositol [9, 109, 133–137] form columnar liquid crystals much more easily with mesophases more stable and very much wider in range than known for the mentioned benzene derivatives.

Table 14 compiles the phase transition data of some of these cyclic model compounds, **38–42** [133–139] discussed here and shown in Fig. 28.

The stability of the mesophase decreases drastically when the carbonyl functions of **39** are replaced by CH$_2$-groups; this formal

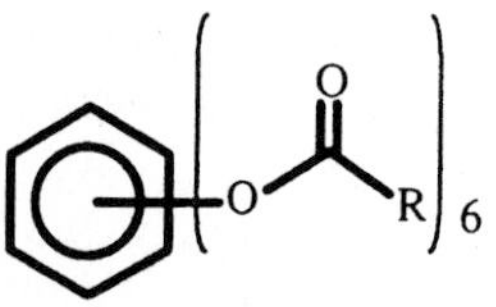

ed by three alkylhetero chains. This compound behaves similarly to disc-shaped amphiphiles, see below.

An 'exotic' '*Y*'-shaped mesogen is **37** [129], depicted in Fig. 27, possessing two hydrophilic (*β*-glucopyranosidic) units as the upper parts of the '*Y*', both terminally linked to a branched (lipophilic) alkyl chain. The mesophase of this interesting glycoside is of a cubic type (Cr 111°C Cub 205°C I) [129].

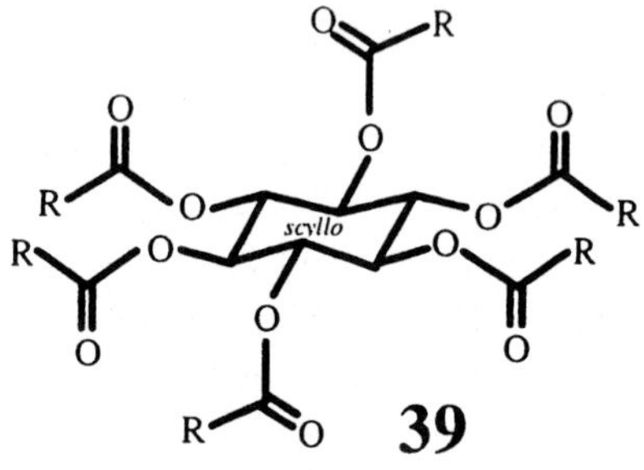

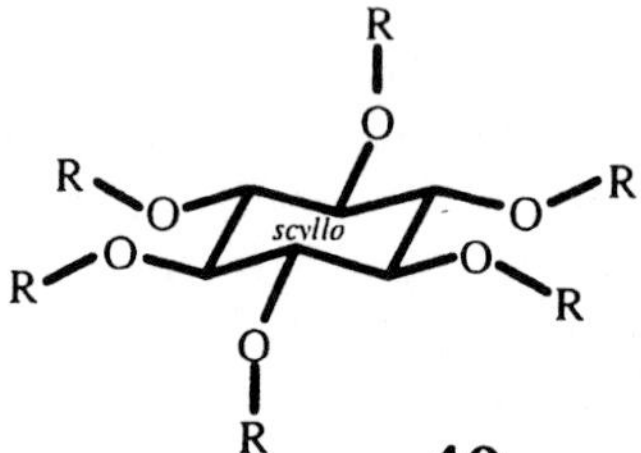

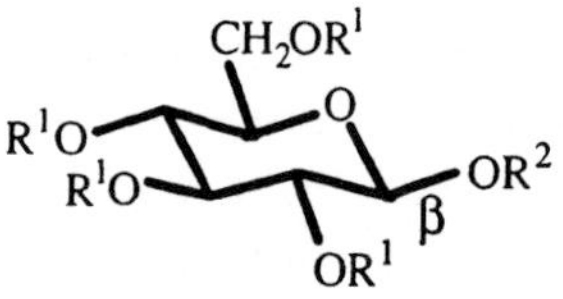

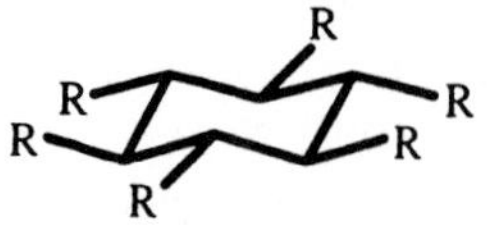

Figure 28. Some disc-shaped model compounds (**38–42**) amphiphilic and mesomorphic in character (**38–41**) or monophilic and not liquid crystalline (**42**) [9, 109, 131–137]; R=alkyl, R^1=–CO-alkyl, R^2=–O-alkyl or only alkyl.

Table 14. Phase transition temperatures (°C) of selected disc-shaped molecules (**38–42**), different in their amphiphilic character (see Fig. 28).

Compound	R	Cr		Col$_h$		Iso	Ref.
38	C$_6$H$_{13}$	•	81.2	•	87.0	•	[132]
39a	C$_5$H$_{11}$	•	68.5	•	199.5	•	[133]
39b	C$_6$H$_{13}$	•	68.0	•	200.0	•	[134]
39c	C$_9$H$_{19}$	•	84.0	•	188.7	•	[133]
40	C$_6$H$_{13}$	•	18.4	•	90.8	•	[134]
41	C$_9$H$_{19}$	•	37.5–39.5	–		•	[134]
		•	37.5–40.0	{•	31.5–32.5}	•	[138]
42	C$_6$H$_{13}$	•	66.2	–		•	[137]

transformation from esters into ethers weakens the amphiphilic character desisively. Moreover the star-shaped alkane **42** without any heteroatom is not liquid crystalline at all, apparently, due to a missing possibility of 'microphase separation'.

Again, an intramolecular contrast of hydrophilic and lipophilic parts is necessary for the occurrence of liquid crystallinity of such compounds, for which also a nearly perfect space filling of the substituents plays an important role [134, 137].

Peracylated sugar derivatives of type **41**, studied independently by two groups [134, 138], possess the required amphiphilicity, but owing to the gap around the pyranosidic oxygen the space-filling of the molecular periphery by only five substituents is not optimal and complicates the molecular situation. On the one hand [138, 139], these compounds have been described as monotropically mesomorphic whereas on the other thermomesomorphism was not observed [134].

The monotropic mesomorphism is described to be very durable and allow detailed studies, for example by polarizing microscopy, DSC, X-ray diffraction, and circular dichroism spectra [139].

Both the α- and the β-anomer of type **41** form columnar mesophases of which the type depends on the molecular structure and on the thermal history as well as the thickness of the sample [139].

With regard to the architecture of these phases the authors suggest a helical arrangement of the molecules in the columns. This chiral arrangement is more pronounced in case of the α-anomers, because the axially oriented chain at the anomeric position mesh into the gap at the pyranosidic oxygen of the neighboring molecule in the column [139]. In case of the β-anomer this 'ratcheting' is not expected, and indeed not observed, since all five substituents are in an equatorial position [139].

A comparative interpretation of the results of both research groups and a discussion of them with regard to tetraacylated-1-O-alkylated glucosides is given in [140]. The latter materials form a highly viscous anisotropic paste which shows all the properties of a mesophase. Owing to this investigation [140] it was suggested that the strong anomeric effect of the ester groups in **41** should be considered. It reduces the population of the all-equatorial conformation and weakens the stabilizing *gauche* interactions between all the ester groups compared to the hexaesters of *scyllo* inositol. It has to be mentioned, here, that short chained *scyllo*-inositol hexaesters **39** exhibit highly ordered columnar mesophases and a cubic phase as well [135].

In view of the chirality of compounds like **41** and other mostly carbohydrate based materials their ability to *induce* chirality in a

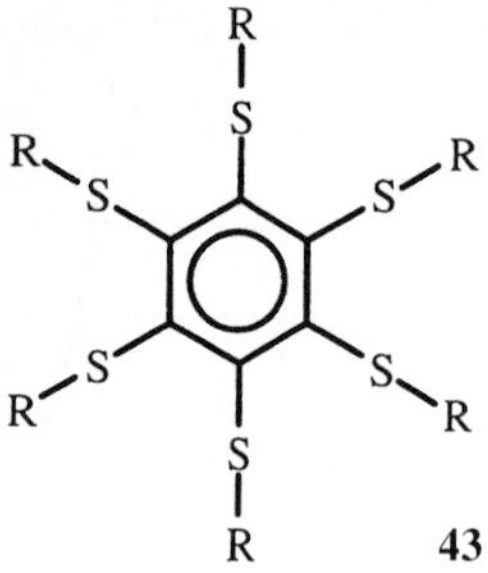

Figure 29. Molecular structures of the nonmesomorphic hexakis(thioether) series **43** and their hexasulfones **44** exhibiting a columnar type of mesophase [142, 143]; R = for example, nonyl, undecyl, or tridecyl.

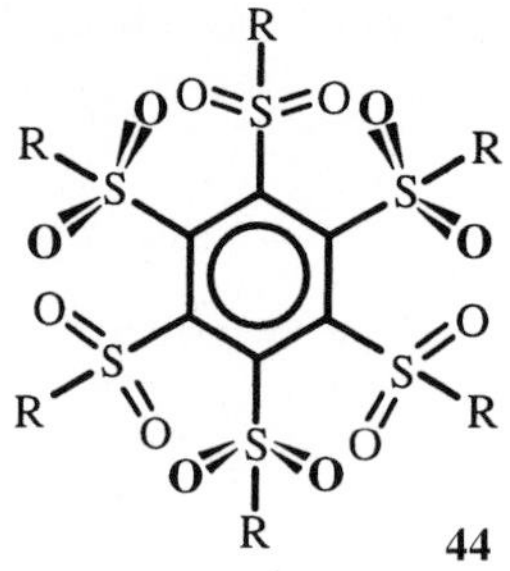

nematic mesophases was discussed elsewhere [141].

The phenomenon of increasing ability to form a mesophase as a function of amphiphilicity was also observed when the benzene hexakis(thioethers) **43** were oxidized into their hexakis(sulfones) **44** [142–144] (Fig. 29). In contrast to the nonliquid crystalline thioether series **43**, the radial-polar hexasulfones **44** with chains containing between 7 and 15 carbon atoms exhibit a columnar hexagonal type of mesophase established by various methods, including X-ray diffractometry [142–144]. Their lyotropic mesomorphic behavior is also of interest and currently under investigation [145].

Single crystal X-ray studies of derivatives of series **44** [144] proved that the very dense packing of the sulfono groups form a broad, wavy quasi-macroheterocycle surrounding the benzene ring in a tight, space-filling manner. The diameter of this superimposed covalently nonbonded 'belt-structure' of this new 'core' is about 107 nm [144].

2.5   Metallomesogens

A completely different type of amphotropic materials is made up by various metal organyls, such as the linear ones illustrated by the di-palladium organyl **45**, the twinned rod-like organyl **47**, or the disc-like organyls **48** and **49** (Figs. 30 to 32). These and related metallomesogens exhibit thermo-

tropic properties and are lyomesogenic in *apolar organic solvents*, such as pentadecane [146–157]. A survey about the mesomorphic behavior of these compounds including their amphotropic nature and their ability to form charge transfer complexes is given in [151, 152, 158].

The linear palladium organyl **45** is nonmesomorphic in its pure state, but a highly viscous thermotropic mesophase is induced by formation of charge transfer complexes with electron acceptors, such as trinitrofluorenone (**46**, TNF) [148]. In ternary systems of **45** with TNF and apolar solvents two types of mesophases, a *nematic* and a *columnar hexagonal* one, are observed; the latter one exhibits a 'herring-bone' texture, typical for an *M*-chromonic type of phase. Owing to strong attracting forces in donor acceptor complexes a columnar nature of the nematic lyomesophase is reasonable in these cases [148].

Twin-like, bis-linear di-palladium mesogens of type **47** which thermotropically exhibit smectic phases [159] form lyotropically different variants of supramolecular packings depending on the length and type of their four substituents as well as on the chain length of the organic solvent [154, 155]. In contrast to **45** such *H*-shaped metallomesogens are lyotropic liquid crystalline only without TNF.

The formal addition of lateral substituents to the chemical structure of **47** (ringed in the formula of **48** in Fig. 31) leads to disc-

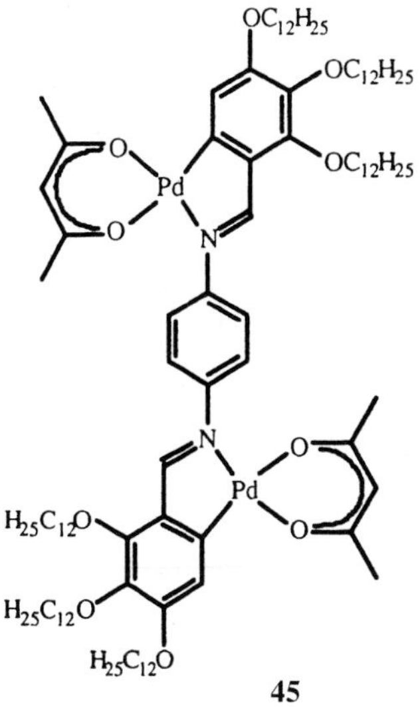

Figure 30. Molecular structures of the dinuclear palladium complexes **45** [148] and the electron acceptor trinitrofluorenone (**46**).

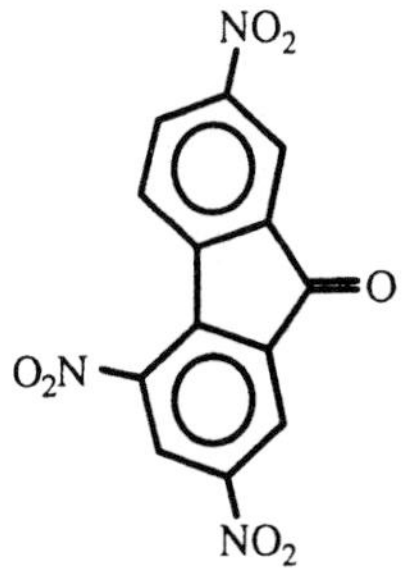

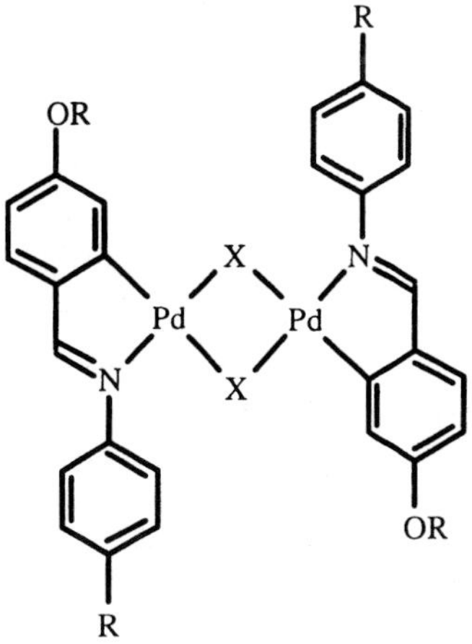

Figure 31. Molecular structures of the twinned rod-like palladium organyl **47** [159] and the disc-like analogue **48** [149, 156]. The ringed lateral substituents of **48** are responsible for the change of the meso-morphic behavior; M = Pd or Pt, X = Cl, Br, I (**47**) or X = Cl, Br, I, SCN, N$_3$ (**48**).

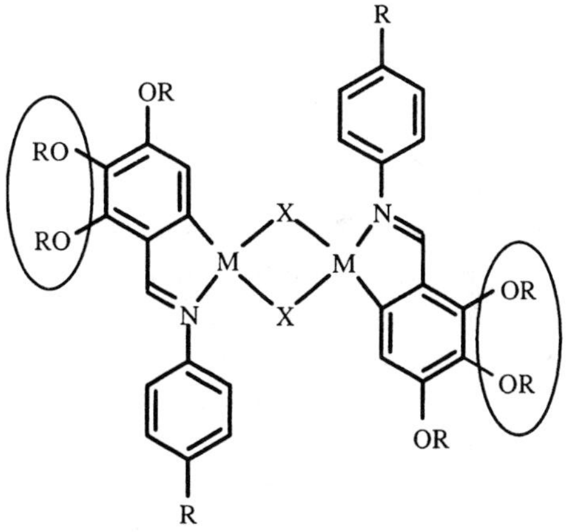

shaped bis-metal complexes of type **48** with either palladium [146] or platinum [149, 156] in their molecular centers. In contrast to the bis-linear metalorganyl family 47 which behaves as *calamitic* mesogens (exhibiting mostly smectic phases), those ones of type **48** form a monotropic nematic-discotic (N$_D$) phase in their pure states, the palladium organyls are the first metallomesogens with this property. However, mixtures of **48** with 2,4,7-trinitrofluorenone (**46**, TNF) shows also charge transfer complexes and the induction of columnar mesophases (Col$_{ho}$ or N$_{col}$) [157].

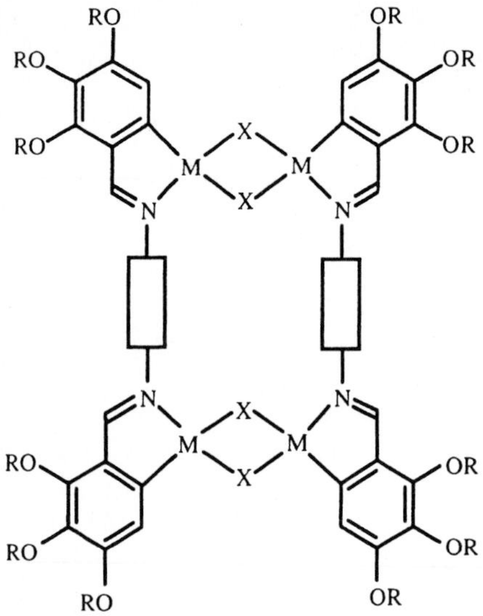

Figure 32. Molecular structure of the tetrametallo-mesogen **49** [146, 149, 156]; M = Pd or Pt, X = Cl, Br, I. The spacer unit (–□–) may be phenylene, stilbenylene or terphenylene.

The lath-like tetrapalladium or -platinum liquid crystals of structure **49** form columnar (Col$_{ob}$) mesophases on heating. Usually, the range of the mesophase this class of compounds exhibits is quite broad, typically about 200–250 K wide [155].

At high concentrations of **49** in solvents, such as in long-chain alkanes, a viscous two dimensionally ordered columnar mesophase of a similar type as under only thermal conditions is observed. At lower concentrations the formation of either one or *two* different nematic columnar mesophases appears [147, 148, 150].

In studies on such metallomesogens it became clear that the metal located deeply inside, quasi 'hidden', in the molecules has only a relatively weak influence on the mesogenic behavior of these organyls [149, 160].

Nevertheless, the amphiphilic character of these metallomesogens arises from the in-

compatibility of their organometallic aromatic core with the surrounding alkyl chains. Thus, the latter has been denoted as 'internal solvent' [158].

2.6 Polyphilic Liquid Crystals

Most of the amphiphiles discussed hitherto have only *two* distinct parts in their molecules which are different in their polarity. Polyphilic compounds possess more than two of them (see Fig. 33) depicting the calamitic biphenyl derivative **50** with *three* different parts: a polyfluorinated one, an alkoxy chain, and the aromatic (biphenyl) section. The mesomorphic properties of this triphilic examples are compiled in Table 15.

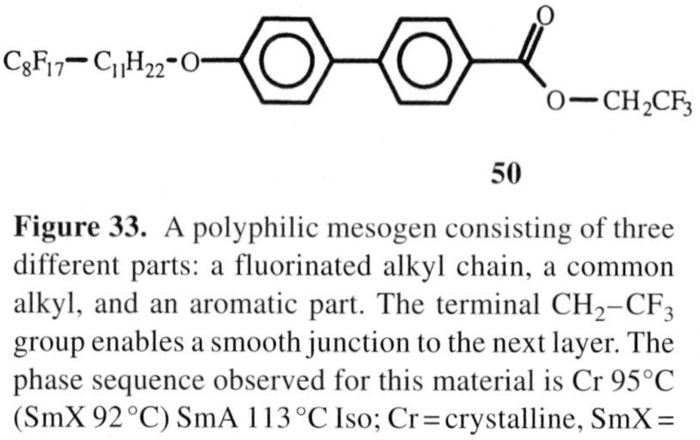

$$C_8F_{17}-C_{11}H_{22}-O-\!\!\!\bigcirc\!\!-\!\!\bigcirc\!\!-\!\!\overset{O}{\underset{O-CH_2CF_3}{||}}$$

50

Figure 33. A polyphilic mesogen consisting of three different parts: a fluorinated alkyl chain, a common alkyl, and an aromatic part. The terminal CH$_2$–CF$_3$ group enables a smooth junction to the next layer. The phase sequence observed for this material is Cr 95°C (SmX 92°C) SmA 113°C Iso; Cr = crystalline, SmX = a ferroelectric smectic phase [162].

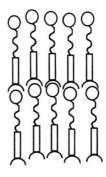

Figure 34. Sketch of the structure of a directed SmA phase of polyphilic mesogens.

Table 15. Phase transition temperatures (°C) of the triphilic biphenyl derivative **50** [162] (see Fig. 33).

Compound	Cr		SmX		SmA		Iso
50	•	95	{•	92}	•	113	•

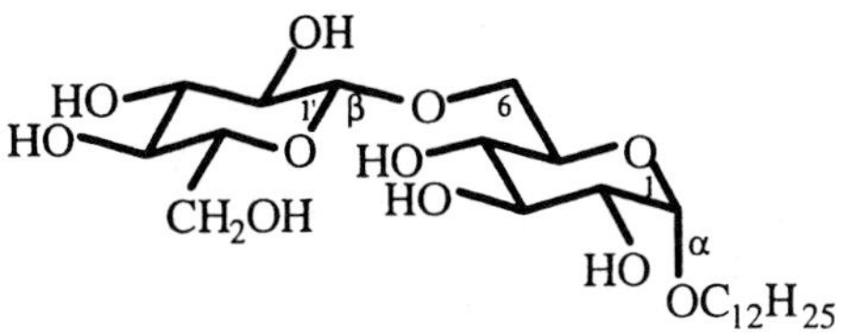

51

Figure 35. Molecular structure of a gentiobioside (**51**) possessing a bent geometry [95, 129].

The polyphilic quality leads to a further ordering of the molecules in their mesophase: Fig. 34 gives a simple schematic drawing of a directed lamellar phase. The molecules have a polar orientation within the layers and long range correlations between them. This special mesophase structure causes macroscopically polar properties. With regard to this kind of molecular arrangement some polyphilic compounds have been successfully studied in respect of *ferroelectric properties* [161–165].

2.7 Further Molecular Architectures

Other types of interesting, sometimes even curious, amphotropic materials with unusual molecular architectures have also been found. Whereas *Y*-shaped [109] molecular structures have flexible wings, *T*-shaped compounds are made of rigid lipophilic units [166]. In the *T*-shaped case, smectic instead of columnar phases are observed. While mono-alkylated amphiphiles usually show only SmA phases, an interesting exception are so-called 'banana shaped' amphiphiles of which **51** in Fig. 35 may be an example. This gentiobioside exhibits a bicontinuous cubic mesophase [129].

Hitherto, only monomeric, unimolecular compounds have been discussed in this overview article of which the *amphotropic behavior* is based on a *balanced amphiphilic character*. This is also true for polymeric materials [5, 11, 167, 168] which, however, are not included here.

3 General Phase Behavior

According to Fig. 3, classical thermotropic smectic phases of amphotropic liquid crystals are (SmA$_d$), columnar hexagonal (Col$_h$), bicontinuous cubic (Cub$_{bi}$), or discontinuous cubic (Cub$_{dis}$) [169]. All these mesophases include a disclination surface between the hydrophilic and the lipophilic parts of the unordered molecules. This surface can be uncurved (SmA), curved in one direction (columnar), curved in two directions with the same sign (discontinuous cubic), or curved in two directions with opposite sign (bicontinuous cubic).

Whereas monophilic liquid crystals can show a high diversity of smectic phases (SmA–SmQ), the amphotropic liquid crystals normally exhibit only the SmA$_d$ phase. Tilted smectic phases are only observed in a few cases. The first indication of possibly tilted phases was given in 1933 for thallium stearate [170]. A disordered SmC phase was also clearly described for mesogens containing a classical calamitic core aside to their amphiphilic structure [171]. Monophilic liquid crystals can show various ordered tilted smectic phases, for example, smectic I, F, G, J, H, and K. In the case of lipids only one mesophase, the *β'* phase, with an ordered tilted lamellar structure was reported. A detailed analysis of this β' phase in relation to the tilted smectic phases was given by Clark et al. [172]. Ordered orthogonal lamellar phases were reported for lecithins (β phase) and for diols [78, 82]. In these mesomorphic amphiphiles the diameter of each of the polar headgroups is comparable with the diameter of the paraffinic part. This allows a very dense packing of the molecules in their mesophase, thus, these phases are related to the rotator phase of simple paraffins.

The columnar phase of amphiphilic liquid crystals is most often of a hexagonal disordered type. Only a very few examples are known for rectangular [173–175] or monoclinic [173] columnar phases or for ribbon phases [176].

4 Summary

In this chapter we described the functional principles of mainly amphiphilic compounds possessing the ability to form both lyo- and thermotropic liquid crystals. Although amphiphilicity and amphotropy are not synonymic, these properties are often coupled. Owing to the tremendous number of potentially amphotropic materials it is impossible to give a complete listing of all of them in such a chapter; but we have tried to show the functional principles which allow compounds to behave the way they do. Both, the lyo- and thermomesogenity are of high importance. While the technological breakthrough for thermotropic liquid crystals is obvious (for example, liquid crystal displays or thermography), some promising recent findings in the field of lyotropic liquid crystals concern the probable observation of a *biaxial* nematic phase (for the existence or doubts related to biaxial nemat-

ics see [177–182]), and the spontaneous formation of a chiral order in lyomesomorphic compositions of nonchiral materials [183].

5 References

[1] I. G. Chistyakov, *Kristallografiya (russ.)* **1960**, *5*, 962.
[2] V. A. Usol'tseva, I. G. Chistyakov, *Uspekhi Khim. (russ.)* **1963**, *32*, 1124.
[3] H. Kelker, *Mol. Cryst. Liq. Cryst.* **1973**, *21*, 1.
[4] G. A. Jeffrey, *Acc. Chem. Res.* **1986**, *19*, 168.
[5] H. Ringsdorf, B. Schlarb, J. Venzmer, *Angew. Chem.* **1988**, *100*, 117; *Angew. Chem., Int. Ed. Engl.* **1988**, *27*, 113.
[6] A. S. Sonin, *100 Years – The History of Discovery and Research on Liquid Crystals (russ.)* (Ed.: B. K. Vainshtein), Nauka Publishers, Moscow, Russia, **1988**.
[7] G. A. Jeffrey, L. M. Wingert, *Liq. Cryst.* **1992**, *12*, 179.
[8] N. V. Usol'tseva, *Lyotropic Liquid Crystals – Chemical and Supramolecular Structure (russ.)*, Ivanovo University, Ivanovo, Russia, **1994**.
[9] K. Praefcke, D. Blunk, J. Hempel, *Mol. Cryst. Liq. Cryst.* **1994**, *243*, 323.
[10] H. Prade, R. Miethchen, V. Vill, *J. Prakt. Chem.* **1995**, *337*, 427.
[11] C. M. Paleos, D. Tsiourvas, *Angew. Chem.* **1995**, *107*, 1839; *Angew. Chem., Int. Ed. Engl.* **1995**, *34*, 1696.
[12] C. Tschierske, *Proc. Polym. Sci.* **1996**, *21*, 775.
[13] W. Heintz, *Ann. Chem.* **1854**, *92*, 291.
[14] W. Heintz, *J. Prakt. Chem.* **1855**, *66*, 1.
[15] B. W. Müller, *Seifen-Öle-Fette-Wachse* **1976**, *102*, 493.
[16] B. W. Müller, *Arch. Pharm. (Weinheim, Ger.)* **1977**, *310*, 693.
[17] R. Virchow, *Virchow's Arch. Pathol. Anatom. Physiol. Klin. Med.* **1854**, *6*, 571.
[18] C. Mettenheimer, *Corr. Blatt des Vereins für gem. Arbeitz. Ford. der Wissensch.* **1857**, *24*, 331.
[19] G. Valentin, *Die Untersuchung der Pflanzen und Thiergewebe im polarisierten Lichte*, Engelmann, Leipzig, Germany, **1861**.
[20] G. Quinke, *Ann. Phys. Chem.* **1894**, *53*, 593.
[21] J. W. McBain, *Kolloid Z.* **1926**, *40*, 1.
[22] F. C. Bouden, N. W. Pirie, *Proc. Roy. Soc. Lond.* **1931**, *123*, 274.
[23] A. S. C. Lawrence, *Trans. Faraday Soc.* **1933**, *29*, 1008.
[24] G. S. Hartley, *Aqueous Solutions of Parafin Chain Salts*, Hermann et Cie, Paris, France, **1936**.

[25] R. S. Bear, K. J. Palmer, F. O. Schmitt, *J. Cell. Comp. Physiol.* **1941**, *18*, 355.

[26] J. D. Bernal, I. Fankuchen, *J. Gen. Physiol.* **1941**, *25*, 111.

[27] J. Stauff, *Kolloid Z.* **1939**, *89*, 224.

[28] H. Kiessig, W. Philippoff, *Natur Wiss.* **1939**, *27*, 593.

[29] R. D. Vold, M. J. Vold, *J. Am. Chem. Soc.* **1939**, *61*, 808.

[30] R. D. Vold, *J. Phys. Chem.* **1939**, *43*, 1213.

[31] J. W. McBain, R. D. Vold, M. Frick, *J. Phys. Chem.* **1940**, *44*, 1013.

[32] F. O. Schmitt, R. S. Bear, K. J. Palmer, *J. Cell. Comp. Physiol.* **1941**, *18*, 31.

[33] S. Ross, J. W. McBain, *J. Am. Chem. Soc.* **1946**, *68*, 296.

[34] J. W. McBain, S. S. Marsden, *Acta Cryst.* **1948**, *1*, 270.

[35] S. S. Marsden, J. W. McBain, *J. Am. Chem. Soc.* **1948**, *70*, 1973.

[36] J. B. Finean, *Experimentia* **1953**, *9*, 17.

[37] F. B. Rosevear, *J. Am. Oil Chem. Soc.* **1954**, *31*, 628.

[38] V. Luzzati, H. Mustacchi, A. Skoulios, *Nature* **1957**, *180*, 600.

[39] V. Luzzati, H. Mustacchi, A. Skoulios, F. Husson, *Acta Cryst.* **1960**, *13*, 660.

[40] F. Husson, H. Mustacchi, V. Luzzati, *Acta Cryst.* **1960**, *13*, 668.

[41] A. E. Skoulios, V. Luzzati, *Acta Cryst.* **1961**, *14*, 278.

[42] P. A. Spegt, A. E. Skoulios, V. Luzzati, *Acta Cryst.* **1961**, *14*, 866.

[43] V. Luzzati, F. Husson, *J. Cell Biol.* **1962**, *12*, 207.

[44] V. Luzzati, A. Nicolaieff, *J. Mol. Biol.* **1963**, *7*, 142.

[45] P. A. Spegt, A. E. Skoulios, *Acta Cryst.* **1964**, *17*, 198.

[46] P. Ekwall, I. Danielsson, *Acta Chem. Scand.* **1951**, *5*, 973.

[47] K. Fontell, P. Ekwall, L. Mandell, I. Danielsson, *Acta Chem. Scand.* **1962**, *16*, 2294.

[48] R. F. Grant, B. A. Dunell, *Can. J. Chem.* **1960**, *38*, 2395.

[49] H. Moor, K. Mühlenthaler, H. Waldner, A. Frey-Wyssling, *J. Biophys. Biochem. Cytol.* **1961**, *10*, 1.

[50] D. Vorländer, *Ber. Dtsch. Chem. Ges.* **1910**, *43*, 3120.

[51] D. Demus, H. Sackmann, K. Seibert, *Wiss. Z. Univ. Halle, Math.-Nat. R.* **1970**, *19 (5)*, 47.

[52] P. Gaubert, *Compt. Rend.* **1919**, *168*, 277.

[53] V. Vill, *Mol. Cryst. Liq. Cryst.* **1992**, *213*, 67.

[54] C. Eaborn, *J. Chem. Soc.* **1952**, 2840; C. Eaborn, N. H. Hartshorne, *J. Chem. Soc.* **1955**, 549.

[55] J. D. Bunning, J. W. Goodby, G. W. Gray, J. E. Lydon, *Liquid Crystals of One- and Two-Dimensional Order, Springer Ser. Chem. Phys.*, Vol. 11 (Eds.: W. Helfrich, G. Heppke), Springer, Berlin, **1980**.

[56] E. Fischer, B. Helferich, *Liebigs Ann. Chem.* **1911**, *383*, 68.

[57] A. H. Salway, *J. Chem. Soc.* **1913**, *103*, 1022.

[58] C. R. Noller, W. C. Rockwell, *J. Am. Chem. Soc.* **1938**, *60*, 2076.

[59] R. Hori, *Yakugaku Zasshi* **1958**, *78*, 523; *ibid.*, **1958**, *78*, 999; *ibid.*, **1958**, *78*, 1171; R. Hori, T. Koizumi, *ibid.*, **1958**, *78*, 1003.

[60] R. Hori, Y. Ikegami, *Yakugaku Zasshi* **1959**, *79*, 80.

[61] G. J. T. Tiddy, *Phys. Rep.* **1980**, *57*, 1.

[62] H. Hoffmann, *Ber. Bunsenges. Phys. Chem.* **1994**, *98*, 1433.

[63] W. L. Porter, *Am. Potato J.* **1972**, *49*, 403.

[64] V. Grassert, V. Vill, *Liq. Cryst. Today* **1994**, *4*, 4.

[65] M. A. Marcus, P. L. Finn, *Mol. Cryst. Liq. Cryst. (Lett.)* **1985**, *2*, 159.

[66] D. Demus, H.-J. Deutscher, D. Marzotko, H. Kresse, A. Wiegeleben, *Liq. Cryst.: Proc. Int. Conf.*, Bangalore (Ed.: S. Chandrasekhar), **1980**, p. 97.

[67] T. I. Zverkova, E. I. Kovshev, L. I. Moklyachuk, Y. A. Fialkov, L. M. Yagupolski, *Adv. Liq. Cryst. Res. Appl.* (Ed.: L. Bata) **1980**, 991.

[68] J. A. Szabo, A. I. Zoltai, *4th Liq. Cryst. Conf. Soc. Countries*, Tbilisi, **1981**, abstract C40.

[69] V. Vill, *LiqCryst – Database of Liquid Crystalline Compounds*, LCI Publisher, Hamburg, Germany, Fujitsu Kyushu Systems (FQS) Ltd., Fukuoka, Japan, **1995**.

[70] J. F. Rabolt, T. P. Russell, R. J. Twieg, *Macromolecules* **1984**, *17*, 2786.

[71] W. Mahler, D. Guillon, A. Skoulios, *Mol. Cryst. Liq. Cryst. (Lett.)* **1985**, *2*, 111.

[72] R. J. Twieg, T. P. Russell, R. Siemens, J. F. Rabolt, *Macromolecules* **1985**, *18*, 1361.

[73] R. J. Twieg, J. F. Rabolt, T. P. Russell, *Polym. Prepr. (Am. Chem. Soc., Div. Polym. Chem.)* **1985**, *26*, 234.

[74] T. P. Russell, J. F. Rabolt, R. J. Twieg, R. L. Siemens, B. L. Farmer, *Macromolecules* **1986**, *19*, 1135.

[75] J. F. Rabolt, T. P. Russell, R. Siemens, R. J. Twieg, B. Farmer, *Polym. Prepr. (Am. Chem. Soc., Div. Polym. Chem.)* **1986**, *27*, 223.

[76] J. Höpken, C. Pugh, W. Richtering, M. Möller, *Makromol. Chem.* **1988**, *189*, 911.

[77] J. Höpken, M. Möller, *Macromolecules* **1992**, *25*, 2482.

[78] C. Tschierske, G. Brezesinski, F. Kuschel, H. Zaschke, *Mol. Cryst. Liq. Cryst. (Lett.)* **1989**, *6*, 139.

[79] C. Tschierske, G. Brezesinski, S. Wolgast, F. Kuschel, H. Zaschke, *Mol. Cryst. Liq. Cryst. (Lett.)* **1990**, *7*, 131.

[80] C. Tschierske, H. Zaschke, *J. Chem. Soc., Chem. Commun.* **1990**, 1013.

[81] F. Hentrich, C. Tschierske, H. Zaschke, *Angew. Chem.* **1991**, *103*, 429; *Angew. Chem., Int. Ed. Engl.* **1991**, *30*, 440.

[82] A. Seher, *Chem. Phys. Lipids* **1972**, *8*, 134. For a discussion of 1-alkoxy-2,3-propandiol and related thioethers and amines see also H. A. van Doren, R. van der Geest, R. M. Kellogg, H. Wynberg, *Recl. Trav. Chim. Pays-Bas* **1990**, *109*, 197.

[83] N. Pietschmann, A. Lunow, G. Brezesinski, C. Tschierske, F. Kuschel, H. Zaschke, *Colloid Polym. Sci.* **1991**, *269*, 636.

[84] D. Demus, H. Demus, H. Zaschke, *Flüssige Kristalle in Tabellen*, Deutscher Verlag für Grundstoffindustrie, Leipzig, **1974**, compound 4142, p. 260.

[85] G. Brezesinski, A. Mädicke, C. Tschierske, H. Zaschke, F. Kuschel, *Mol. Cryst. Liq. Cryst. (Lett.)* **1988**, *5*, 155.

[86] C. Tschierske, A. Lunow, D. Joachimi, F. Hentrich, D. Girdziunaite, H. Zaschke, A. Mädicke, G. Brezesinski, F. Kuschel, *Liq. Cryst.* **1991**, *9*, 821.

[87] F. Reinitzer, *Monatsh. Chem.* **1888**, *18*, 421.

[88] O. Lehmann, *Z. Phys. Chem.* **1889**, *4*, 462.

[89] W. J. de Grip, P. H. M. Bovee-Geurts, *Chem. Phys. Lipids* **1979**, *23*, 321.

[90] W. Schneider, J. Sepp, O. Stiehler, *Ber. Dtsch. Chem. Ges.* **1918**, *51*, 220.

[91] L. C. Kreider, E. Friesen, *J. Am. Chem. Soc.* **1942**, *64*, 1482.

[92] A. Loewenstein, D. Igner, U. Zehavi, H. Zimmermann, A. Emerson, G. R. Luckhurst, *Liq. Cryst.* **1990**, *7*, 457.

[93] J. W. Goodby, *Mol. Cryst. Liq. Cryst.* **1984**, *110*, 205.

[94] L. F. Tietze, K. Böge, V. Vill, *Chem. Ber.* **1994**, *127*, 1065.

[95] V. Vill, Ph. D. Thesis, Universität Münster, **1990**.

[96] Y. J. Chung, G. A. Jeffrey, *Biochim. Biophys. Acta* **1989**, *985*, 300.

[97] G. H. Brown, P. P. Crooker, *Chem. Eng. News* **1983**, Jan. 31, 24.

[98] V. Vill, T. Böcker, J. Thiem, F. Fischer, *Liq. Cryst.* **1989**, *6*, 349.

[99] H. W. C. Raaijmakers, B. Zwanenburg, G. J. F. Chittenden, *Recl. Trav. Chim. Pays-Bas* **1994**, *113*, 79.

[100] D. Blunk, K. Praefcke, G. Legler, *Liq. Cryst.* **1994**, *17*, 841.

[101] R. Miethchen, J. Holz, H. Prade, A. Liptak, *Tetrahedron* **1992**, *48*, 3061.

[102] a) L. M. Wingert, G. A. Jeffrey, Jahangir, D. C. Baker, *Liq. Cryst.* **1993**, *13*, 467; b) E. Smits, J. B. F. N. Engberts, R. M. Kellogg, H. A. van Doren, *Liq. Cryst.* **1997**, *23*, 481.

[103] D. Blunk, K. Praefcke, G. Legler, *Liq. Cryst.* **1995**, *18*, 149.

[104] J.-H. Fuhrhop, J. Mathieu, *Angew. Chem.* **1984**, *96*, 124; *Angew. Chem., Int. Ed. Engl.* **1984**, *23*, 100.

[105] J.-H. Fuhrhop, H.-H. David, J. Mathieu, U. Liman, H.-J. Winter, E. Boekema, *J. Am. Chem. Soc.* **1986**, *108*, 1785.

[106] W. V. Dahlhoff, *Z. Naturforsch.* **1988**, *43b*, 1367.

[107] In their paper the authors follow the nomenclature by Luzzati and indicate the observed mesophases L$_\alpha$ or L$_{\beta'}$, respectively. Here, however, to separate between lyotropic and thermotropic mesophases the indication was changed into SmA and SmC, respectively.

[108] F. Hentrich, S. Diele, C. Tschierske, *Liq. Cryst.* **1994**, *17*, 827.

[109] K. Praefcke, P. Marquardt, B. Kohne, W. Stephan, *J. Carbohydr. Chem.* **1991**, *10*, 539.

[110] N. H. Hartshorne, *The Microscopy of Liquid Crystals, The Microscope Series*, Vol. 48, Microscope Publications Ltd., London **1974**, pp. 105–109; N. H. Hartshorne in *Liquid Crystals and Plastic Crystals*, Vol. 2, Chap. 2 (Eds.: G. W. Gray, P. A. Winsor), Ellis Horwood Ltd., Chichester, Halsted Press a div. of John Wiley & Sons, Inc., New York, London, Sydney, Toronto, **1974**, pp. 51–54; K. Praefcke, B. Kohne, A. Eckert, J. Hempel, *Z. Naturforsch.* **1990**, *45b*, 1084.

[111] In sphingolipids the glycerin component of the lipid is replaced by sphingosin. Main families of sphingolipids are glycosphingolipids (cerebrosides, gangliosides and sulfatides) and sphingophospholipids (sphingomyelins and inositolsphingophospholipids).

[112] S. Hakomori, *Sci. Am.* **1986**, *254 (5)*, 32.

[113] D. A. Mannock, R. N. McElhaney, *Biochem. Cell Biol.* **1991**, *69*, 863.

[114] a) D. A. Mannock, R. N. A. H. Lewis, R. N. McElhaney, *Chem. Phys. Lipids* **1990**, *55*, 309; b) L. Six, K.-P. Rueß, M. Liefländer, *Tetrahedron (Lett.)* **1983**, *24*, 1229 and references cited therein.

[115] H. Kuttenreich, H.-J. Hinz, M. Inczedy-Marcsek, R. Koynova, B. Tenchov, P. Laggner, *Chem. Phys. Lipids* **1988**, *47*, 245.

[116] D. A. Mannock, R. N. A. H. Lewis, R. N. McElhaney, *Chem. Phys. Lipids* **1987**, *43*, 113.

[117] E. Heinz, *Biochim. Biophys. Acta* **1971**, *231*, 537.

[118] H. P. Wehrli, Y. Pomeranz, *Chem. Phys. Lipids* **1969**, *3*, 357.

[119] P. J. Quinn, W. P. Williams, *Biochim. Biophys. Acta* **1983**, *737*, 223.

[120] E. Heinz, H. P. Siebertz, M. Linscheid, *Chem. Phys. Lipids* **1979**, *24*, 265.

[121] D. A. Mannock, A. P. R. Brain, W. P. Williams, *Biochim. Biophys. Acta* **1985**, *817*, 289.

[122] D. A. Mannock, R. N. A. H. Lewis, R. N. McElhaney, M. Akiyama, H. Yamada, D. C. Turner, S. M. Gruner, *Biophys. J.* **1992**, *63*, 1355.

[123] A. Eckert, B. Kohne, K. Praefcke, *Z. Naturforsch.* **1988**, *43b*, 878.

[124] W. V. Dahlhoff, *Z. Naturforsch.* **1987**, *42b*, 661.

[125] H. van Doren, T. J. Buma, R. M. Kellogg, H. Wynberg, *J. Chem. Soc., Chem. Commun.* **1988**, 460.

[126] K. Praefcke, A.-M. Levelut, B. Kohne, A. Eckert, *Liq. Cryst.* **1989**, *6*, 263.

[127] H. A. van Doren, R. van der Geest, C. A. Keuning, R. M. Kellogg, H. Wynberg, *Liq. Cryst.* **1989**, *5*, 265.

[128] C. Vogel, U. Jeschke, V. Vill, H. Fischer, *Liebigs Ann. Chem.* **1992**, 1171.

[129] S. Fischer, H. Fischer, S. Diele, G. Pelzl, K. Jankowski, R. R. Schmidt, V. Vill, *Liq. Cryst.* **1994**, *17*, 855.

[130] S. Chandrasekhar, B. K. Sadashiva, K. A. Suresh, *Pramana* **1977**, *9*, 471.

[131] H. J. Backer, S. van der Baan, *Recl. Trav. Chim. Pays-Bas* **1937**, *56*, 1161.

[132] S. Chandrasekhar, B. K. Sadashiva, K. A. Suresh, N. V. Madhusudana, S. Kumar, R. Shashidhar, G. Venkatesh, *J. Phys. (Paris)* **1979**, *40*, C3–120.

[133] B. Kohne, K. Praefcke, *Angew. Chem.* **1984**, *96*, 70; *Angew. Chem., Int. Ed. Engl.* **1984**, *23*, 82.

[134] B. Kohne, K. Praefcke, *Chem.-Ztg.* **1985**, *109*, 121.

[135] B. Kohne, K. Praefcke, J. Billard, *Z. Naturforsch.* **1986**, *41b*, 1036.

[136] K. Praefcke, B. Kohne, W. Stephan, P. Marquardt, *Chimia* **1989**, *43*, 380.

[137] K. Praefcke, B. Kohne, P. Psaras, J. Hempel, *J. Carbohydr. Chem.* **1991**, *10*, 523.

[138] R. G. Zimmermann, G. B. Jameson, R. G. Weiss, G. Demailly, *Mol. Cryst. Liq. Cryst. (Lett.)* **1985**, *1*, 183.

[139] N. L. Morris, R. G. Zimmermann, G. B. Jameson, A. W. Dalziel, P. M. Reuss, R. G. Weiss, *J. Am. Chem. Soc.* **1988**, *110*, 2177.

[140] V. Vill, J. Thiem, *Liq. Cryst.* **1991**, *9*, 451.

[141] V. Vill, F. Fischer, J. Thiem, *Z. Naturforsch.* **1988**, *43a*, 1119.

[142] W. Poules, K. Praefcke, *Chem.-Ztg.* **1983**, *107*, 310.

[143] K. Praefcke, W. Poules, B. Scheuble, R. Poupko, Z. Luz, *Z. Naturforsch.* **1984**, *39b*, 950.

[144] N. Spielberg, Z. Luz, R. Poupko, K. Praefcke, B. Kohne, J. Pickardt, K. Horn, *Z. Naturforsch.* **1986**, *41a*, 855; M. Sarkar, N. Spielberg, K. Praefcke, H. Zimmermann, *Mol. Cryst. Liq. Cryst.* **1991**, *203*, 159.

[145] K. Praefcke, D. Blunk, N. Usol'tseva, unpublished studies.

[146] K. Praefcke, D. Singer, B. Gündogan, *Mol. Cryst. Liq. Cryst.* **1992**, *223*, 181.

[147] N. Usol'tseva, K. Praefcke, D. Singer, B. Gündogan, *Liq. Cryst.* **1994**, *16*, 601.

[148] N. Usol'tseva, K. Praefcke, D. Singer, B. Gündogan, *Mol. Mater.* **1994**, *4*, 253.

[149] K. Praefcke, B. Bilgin, N. Usol'tseva, B. Heinrich, D. Guillon, *J. Mater. Chem.* **1995**, *5*, 2257.

[150] N. Usol'tseva, G. Hauck, H. D. Koswig, K. Praefcke, B. Heinrich, *Liq. Cryst.* **1996**, *20*, 731.

[151] K. Praefcke, J. Holbrey, N. Usol'tseva, D. Blunk, *Mol. Cryst. Liq. Cryst.* **1997**, *292*, 123; see also the *Festschrift* in honor of Prof. A. Saupe, editors P. E. Cladis and P. Palffy-Muhoray, in press.

[152] K. Praefcke, J. D. Holbrey, *J. Incl. Phen. Mol. Recogn. Chem.* **1996**, *24*, 19.

[153] N. Usol'tseva, K. Praefcke, D. Singer, B. Gündogan, *Liq. Cryst.* **1994**, *16*, 617.

[154] N. Usol'tseva, K. Praefcke, P. Espinet, D. Blunk, J. Baena, *poster B2P.21 at the 16th International Liquid Crystal Conference*, Kent, USA, June 24–28, **1996**, see the abstract volume, p. P-69.

[155] K. Praefcke, J. D. Holbrey, N. Usol'tseva, D. Blunk, *Mol. Cryst. Liq. Cryst.* **1997**, *292*, 123.

[156] K. Praefcke, B. Bilgin, J. Pickardt, M. Borowski, *Chem. Ber.* **1994**, *127*, 1543.

[157] D. Singer, A. Liebmann, K. Praefcke, J. H. Wendorff, *Liq. Cryst.* **1993**, *14*, 785.

[158] K. Praefcke, J. D. Holbrey, N. Usol'tseva, *Mol. Cryst. Liq. Cryst.* **1996**, *288*, 189.

[159] J. Barbera, P. Espinet, E. Lalinde, M. Marcos, J. L. Serrano, *Liq. Cryst.* **1987**, *2*, 833.

[160] S. A. Hudson, P. M. Maitlis, *Chem. Rev.* **1993**, *93*, 861 (p. 882).

[161] F. Tournilhac, L. M. Blinov, J. Simon, S. V. Yablonskii, *Nature* **1992**, *359*, 621.

[162] F. G. Tournilhac, L. Bosio, J. Simon, L. M. Blinov, S. V. Yablonsky, *Liq. Cryst.* **1993**, *14*, 405.

[163] F. Tournilhac, L. M. Blinov, J. Simon, D. B. Subachius, S. V. Yablonsky, *Synth. Metals* **1993**, *54*, 253.

[164] P. Bassoul, J. Simon, *New J. Chem.* **1996**, *20*, 1131.

[165] P. Kromm, M. Cotrait, J. C. Rouillon, P. Barois, H. T. Nguyen, *Liq. Cryst.* **1996**, *21*, 121.

[166] F. Hildebrandt, J. A. Schröter, C. Tschierske, R. Festag, R. Kleppinger, J. H. Wendorff, *Angew. Chem.* **1995**, *107*, 1780; *Angew. Chem., Int. Ed. Engl.* **1995**, *34*, 1631.

[167] H. Finkelmann, M. A. Schafheutle, *Colloid Polym. Sci.* **1986**, *264*, 786.

[168] Z. Ali-Adib, A. Bomben, F. Davis, P. Hodge, P. Tundo, L. Valli, *J. Mater. Chem.* **1996**, *6*, 15.

[169] V. Vill, H. Kelkenberg, J. Thiem, *Liq. Cryst.* **1992**, *11*, 459.

[170] K. Herrmann, *Trans. Faraday Soc.* **1933**, *29*, 972.

[171] D. Joachimi, C. Tschierske, H. Müller, J. H. Wendorff, L. Schneider, R. Kleppinger, *Angew. Chem.* **1993**, *105*, 1205; *Angew. Chem., Int. Ed. Engl.* **1993**, *32*, 1165.

[172] E. B. Sirota, G. S. Smith, C. R. Safinya, R. J. Plano, N. A. Clark, *Science* **1988**, *242*, 1406.

[173] P. Marquardt, K. Praefcke, B. Kohne, W. Stephan, *Chem. Ber.* **1991**, *124*, 2265.

[174] K. Praefcke, P. Marquardt, B. Kohne, W. Stephan, A.-M. Levelut, E. Wachtel, *Mol. Cryst. Liq. Cryst.* **1991**, *203*, 149.

[175] H. Fischer, V. Vill, C. Vogel, U. Jeschke, *Liq. Cryst.* **1993**, *15*, 733.

[176] C. Tschierske, J. A. Schröter, N. Lindner, C. Sauer, S. Diele, R. Festag, M. Wittenberg, J.-H. Wendorff, *Lecture O-36 at the European Conference on Liquid Crystals, Science and Technology*, Zakopane, Poland, March 3–8, **1997**; see the abstract volume, p. 61.

[177] K. Praefcke, B. Kohne, B. Gündogan, D. Singer, D. Demus, S. Diele, G. Pelzl, U. Bakowsky, *Mol. Cryst. Liq. Cryst.* **1991**, *198*, 393.

[178] S. M. Fan, I. D. Fletcher, B. Gündogan, N. J. Heaton, G. Kothe, G. R. Luckhurst, K. Praefcke, *Chem. Phys. (Lett.)* **1993**, *204*, 517.

[179] S. Chandrasekhar, *Mol. Cryst. Liq. Cryst.* **1994**, *243*, 1.

[180] J. M. Goetz, G. L. Hoatson, *Liq. Cryst.* **1994**, *17*, 31.

[181] J. S. Patel, K. Praefcke, D. Singer, M. Langner, *Appl. Phys.* **1995**, *B60*, 469.

[182] With regard to discussions of nematic Schlieren pattern and possible two- or four-brush disclinations considered a diagnostic sign for the axiality of nematic phases see S. Chandrasekhar, Geetha G. Nair, K. Praefcke, D. Singer, *Mol. Cryst. Liq. Cryst.* **1996**, *288*, 7; S. Chandrasekhar, Geetha G. Nair, D. S. Shankar Rao, S. Krishna Prasad, K. Praefcke, D. Blunk, *Liq. Cryst.* **1997**, and *Mol. Cryst. Liq. Cryst.* **1997**, both in press.

[183] N. Usol'tseva, K. Praefcke, D. Blunk, *Lecture O-60 at the European Conference on Liquid Crystals, Science and Technology*, Zakopane, Poland, March 3–8, **1997**; see the abstract volume, p. 85.

Chapter VII
Lyotropic Surfactant Liquid Crystals

Claire E. Fairhurst, Stuart Fuller, Jason Gray, Michael C. Holmes,
Gordon J. T. Tiddy

1 Introduction

Surfactants, or **surface active agents**, occur widely in nature where they are usually classified as lipids. They have been used for over 1000 years in everyday applications as emulsifiers in cleaning and in foods. Their functionality derives from a unique feature in the molecular structure: they are molecules that consist of two distinctly different regions. One is a polar (hydrophilic) moiety termed the head group, and the other a nonpolar (hydrophobic) part, referred to as the chain. The head group conveys water-solubility, while the hydrophobic chain drives the formation of self-assembled aggregates that are the subject of this article. A wide variety of structures can act as head groups. They can be ionic (anionic or cationic; single or multiple charges), zwitterionic, or nonionic or any mixture of these. The chain can be one or more alkyl groups (by far the most common), a perfluorocarbon group, or a polydimethyl siloxane. Examples of types of surfactant together with their commercial use are shown in Table 1.

Table 1. Common surfactants and their applications.

Name	Structure	Application
Sodium dodecyl sulfate	$CH_3(CH_2)_{11}OSO_3Na$	Detergents, emulsifier
Sodium dodecanoate (laurate)	$CH_3(CH_2)_{10}CO_2Na$	Soap Bars
Hexa-ethylene glycol monododecyl ether	$CH_3(CH_2)_{11}(OCH_2CH_2)_6OH$	Detergents, emulsifier
1-octadecanoyl-sn-glycerol (monostearin)	$CH_3(CH_2)_{16}CO_2CH_2CH(OH)CH_2OH$	Food emulsifier
Hexadecyltrimethyl ammonium chloride	$CH_3(CH_2)_{15}N^+(CH_3)_3Cl^-$	Hair conditioner
Dioctadecyldimethyl ammonium chloride	$CH_3(CH_2)_{17}\diagdown \overset{+}{N} \diagup CH_3$ Cl^- $CH_3(CH_2)_{17}\diagup \diagdown CH_3$	Fabric conditioner
Dodecyl dimethyl amine oxide	$CH_3(CH_2)_{11}\underset{O^-}{\overset{}{N}}{}^+(CH_3)_2$	Speciality surfactant
1,2-dioctadecyl-sn-glycero-3 phosphatidylcholine (distearoyl lecithin)	$CH_3(CH_2)_{16}CO_2CH_2$ $CH_3(CH_2)_{16}CO_2CH$ $CH_2-\overset{O}{\overset{\|}{P}}-O-CH_2CH_2N(CH_3)_3$ O	Food emulsifier, membrane lipid
Perfluorooctanoic acid	$CF_3(CF_2)_6CO_2H$	Speciality surfactant
Octaethylene glycol mono tri siloxyl propyl ether	$((CH_3)_3SiO)_2Si(Me)(CH_2)_3(OCH_2CH_2)_8OH$	Cosmetics, wetting agent

When water-miscible surfactants are dissolved in water they form aggregates termed micelles above a well defined concentration. These aggregates are the building blocks of the liquid crystalline phases that occur at still higher concentrations. An understanding of micelle formation and micelle properties provides a good basis for a description of the liquid crystals, hence we begin with a discussion of this topic. This is followed by sections describing the various liquid crystal phases, the way that the mesophase type varies with surfactant chemical structure, and what happens with mixed surfactants. We also include the effects of additives such as electrolytes and other solutes. Because of the enormous number of publications in this area, space considerations, and the limited time of the authors, it simply is not possible to document all the available information here. Instead, it is our intention to give an introduction to the concepts and phenomena that underlie the formation of surfactant mesophases and to summarize the present knowledge of the area. As well as providing an up-to-date description our objective is to provide sufficient information so that anyone with a practical problem concerning mesophase formation or structure can see the general context. This should allow them to find some indication of experiments that may provide a solution.

As a further comment we draw attention to several books [1–8] which provide a good general introduction to the area of surfactants and colloid science. Most of the books have been published in the last few years and summarize recent research.

2   Surfactant Solutions: Micelles

When surfactants dissolve in water at low concentrations they exist as monomers (ionic surfactants are dissociated). As the concentration is increased aggregates termed micelles are formed. These appear at a well-defined concentration known as the 'critical micelle concentration' (CMC). This is *not* a critical point in the sense of modern physics since micelle formation occurs over a *very narrow range* of concentrations. This range is so small that for almost all practical purposes it can be represented by a specific value, the CMC. For pure single surfactants below the CMC, all of the dissolved surfactant exists as monomers, while above the CMC all added surfactant forms micelles. With surfactant mixtures the phenomenon is more complex because the components usually have different individual CMCs, but the same considerations apply [2].

Micelles are aggregates of at least $15-20$ monomers. Typical surfactants such as SDS (sodium dodecyl sulfate) or $C_{12}EO_8$ (Table 1) form micelles with aggregation numbers (q) in the range $50-100$. The aggregation numbers can become very large up to 10^4 or more, according to surfactant type, aggregate shape, temperature, and surfactant concentration (see below).

Micelle formation arises from the *hydrophobic effect* [4, 9–13]. This is the term used to describe the interaction between nonpolar solutes and water. It is well known that nonpolar solutes are almost insoluble in water, with the limited degree of solubility decreasing rapidly with increasing solute size. A thermodynamic analysis of the process shows that the introduction of a hydrocarbon into water at ambient temperature is always associated with a decrease of entropy, and an enthalpy of about zero, resulting in a large and positive free energy [10–14]. 'Structuring' of water molecules in the neighborhood of the alkane, akin to the formation of clathrate hydrates [15], is frequently cited as the origin of the entropy change. Unfortunately, attempts to locate the 'clathrate structure', for example by self-diffusion measurements of water and organic solutes such as tetraalkyl ammonium ions [16], have consistently failed to show that more than a single layer of water around the solute differs from bulk water: hardly evidence to support the 'water structure' concept.

In a series of recent papers Kronberg et al. [10–12] have discussed this problem in detail. They view the hydrophobic effect as the resultant of two contributions, one arising from the 'ordering' of water molecules around the solute, the second from the energy required to make a cavity in the water large enough to accommodate the nonpolar solute. The first contribution is associated with a negative entropy because water molecules next to a nonpolar solute have fewer conformations available than 'free' water: they can not H-bond to the solute. It also gives a negative enthalpy, presumably because vicinal water molecules make stronger H-bonds. The second and opposite contribution arises from the large energy required to form a cavity to accommodate the solute. This is large, owing to both the high cohesion in water arising from H-bonding connectivity, and the small size of water molecules compared to, for example, alkanes. An important consequence of this mechanism is that the magnitude of the hydrophobic effect is proportional to the *area* of hydrophobic contact between water and the solute. As will be seen below, using this concept it is possible to estimate an approximate CMC for almost any novel surfactant, given that a few simple guidelines are followed.

Micelles can only form when the surfactant solubility is equal to or greater than the CMC. In general this occurs only above a particular temperature known as the Krafft point (temperature). Below this temperature surfactant solubility increases slowly with increasing temperature because the surfactant dissolves as monomers. The limit to monomer solubility occurs when the chemical potential of the monomers is equal to that of the pure (usually crystalline) surfactant. Above this temperature the solubility increases *very* rapidly because the surfactant dissolves as micelles: the contribution of each micelle to the surfactant chemical potential being the same as that of a monomer.

Micelle formation involves a dynamic equilibrium between monomers and aggregates represented by Eq. (1) [17–21]:

$$S_1 \Leftrightarrow S_q \tag{1}$$

(where S_q represents an aggregate of n monomers). In fact micelles form by a series of step-wise reactions from monomers, Eq. (2).

$$S_1 + S_1 \Leftrightarrow S_2$$
$$S_2 + S_1 \Leftrightarrow S_3$$
$$S_{q-1} + S_1 \Leftrightarrow S_q$$
$$S_q + S_1 \Leftrightarrow S_{q+1} \tag{2}$$

Clearly, there is a range of aggregate sizes present in solution, with an average value of q. However, the size distribution is usually narrow (say $\pm 10\%$) particularly for globular micelles. Although the micelle formation process involves dimers, trimers, etc., the actual concentration of these species is very, very small. Hence, once formed, micelles have a reasonably large, but finite, lifetime. For dilute solutions, using fast relaxation techniques it is possible to measure both the exchange rate between monomers and micelles, and the rate of micelle formation/breakdown. Both of these processes are related to the CMC, becoming faster as the CMC increases. Typically, the monomer–micelle exchange rate is in the range 10^3-10^6 s^{-1}, while micelle breakdown/formation rates are about $10^{-1}-10^2$ s^{-1} (i.e. micelle lifetimes of $10^{-2}-10$ s). Note that from the kinetic analysis of fast relaxation studies an upper limit of $<10^{-10}$ mol dm^{-3} can be placed on the concentration of dimers, trimers etc. (i.e. aggregates with association numbers much smaller than q) [21, 22].

Figure 1 shows a schematic diagram of a small globular micelle. The rapid, continuous exchange of monomers with bulk solution means that the micelle is a very mobile, disordered, aggregate. There is a continuous movement of molecules jumping part way in and out of the interface [21] (this has recently been termed 'protrusion' [23, 24]). Also, diffusion around the micelle surface is rapid [self-diffusion coefficient $(D) \simeq 10^{-10}$ m^2 s^{-1}], as is the exchange between the various possible conformations of the alkyl chain (correlation time $(\tau_c) \sim 10^{-9}-10^{-10}$ s). With an ionic surfactant, typically about 70–80% of counter-ions reside close (within ≈ 1 nm) to the micelle surface in a loosely 'bound' state due to the very high surface charge density. The exchange

between 'bound' and free counter-ions occurs on a time scale of $\approx 10^{-9}$ s. There also exists a single layer of 'bound' water molecules associated with polar groups and ions. Again, the exchange between bound and free water occurs on a time scale of about 10^{-9} s [25].

We have discussed the formation of micelles at a 'critical micelle concentration', and the absence of significant concentrations of small aggregates. In fact, micelle formation occurs over a very narrow *range* of concentrations, the *range* being too narrow to detect for all but the shortest chain (highest CMC) surfactants. The almost complete absence of small aggregates at concentrations above the CMC makes their occurrence *below* this highly unlikely. There are numerous claims in the literature of 'pre-CMC' surfactant aggregates formed due to a contribution from the hydrophobic effect. To date, none of these have withstood a thorough examination of the evidence. Usually, the deviations from expected behavior are due to the presence of impurities. A good illustration of how minor constituents can influence surfactant properties is the recent study of Thomas and Penfold who demonstrate [26] that doubts concerning the application of the Gibbs adsorption equation apply to the *systems* (i.e. the presence of impurities, which should be included in the equation) rather than the *thermodynamics*.

Arguably the most important parameters for any surfactant is the CMC value. This is because below this concentration the monomer level increases as more is dissolved hence the surfactant chemical potential (activity) also increases. Above the CMC the monomer concentration and surfactant chemical potential are approximately constant, so surfactant absorption at interfaces and interfacial tensions show only small changes with composition under most

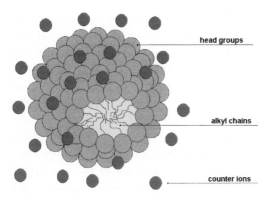

Figure 1. Schematic representation of a spherical micelle.

head groups

alkyl chains

counter ions

conditions. For liquid crystal researchers the CMC is the concentration at which the building blocks (micelles) of soluble surfactant mesophases appear. Moreover, with partially soluble surfactants it is the concentration above which a liquid crystal dispersion in water appears. Fortunately there are well established simple rules which describe how CMC values vary with chain length for linear, monoalkyl surfactants. From these, and a library of measured CMC values [27–29], it is possible to estimate the approximate CMC for branched alkyl chain and di- (or multi-) alkyl surfactants. Thus most materials are covered. This includes the 'gemini' surfactants, a 'new' *fashionable* group where two conventional surfactant molecules are linked by hydrophobic spacer of variable length [30].

The major determinant of the CMC is the hydrophobic effect, which is proportional to the area of the nonpolar chain exposed to water (see above) [10–12]. This leads to a logarithmic relationship between the CMC and alkyl chain length for linear surfactants:

$$\log(CMC) = A n + B \tag{3}$$

(where n=alkyl chain length; A, B are constants).

The second factor is the valency (s) of the head group charge ($s=0, 1, 2$ ... for non-ionic/zwitterionic, monoionic, di-ionic ... etc.). According to the value of s, the constant A takes the values: $A=0.5$ ($s=0$), $A=0.3$ ($s=1$), $A=1.5–1.8$ ($s=2$). Finally, the valency of the counterions is also important as this influences the value of B. Essentially, the CMC is reduced for multivalent counterions because fewer ions are required close to the micelle surface to (partially) balance the high surface charge density.

Table 2 lists examples of CMC values for various surfactants. While there is some temperature dependence of the CMC [9–12] with many materials showing a shallow

Table 2. Typical CMC values from [2, 27–29].

Surfactant	Temp. °C	CMC mol dm^{-3}
C_8EO_6	25	9.9×10^{-3}
$C_{10}EO_6$	20	9.5×10^{-4}
$C_{12}EO_6$	25	6.8×10^{-5}
$C_{12}EO_8$	25	7.1×10^{-5}
$C_{14}EO_8$	25	9.0×10^{-6}
C_8SO_4Na	25	1.3×10^{-2}
$C_{12}SO_4Na$	25	8.3×10^{-3}
$C_{16}SO_4Na$	25	2.1×10^{-4}
C_8NMe_3Br	25	1.4×10^{-1}
$C_{12}NMe_3Br$	25	1.4×10^{-2}
$C_{16}NMe_3Br$	25	3.3×10^{-4}
$C_{12}NMe_3Cl$	25	1.7×10^{-2}
$C_{12}NMe_2O$	27	2.1×10^{-3}
$C_{12}NMe_2CH_2SO_3$	25	3.6×10^{-3}
di C_6 lecithin	unknown	1.6×10^{-2}
di C_7 lecithin	unknown	1.6×10^{-3}
$(C_{12})_2NMe_2Cl$	unknown	1.0×10^{-4}
$C_{12}SO_3Ca$	70	3.3×10^{-3}

minimum, the effect is small below 100 °C. It is clear that for nonionic and zwitterionic surfactants the CMC values reduce by about a factor of 10 for the addition of two CH_2 groups to the alkyl chain, while for monovalent ionic surfactants the factor is 4. Branched and multichain surfactants can be treated by estimating the equivalent linear chain having the same area of hydrophobic contact with water. There is a free energy penalty for constraining the conformations of multichain surfactants at the micelle surface [31, 32] which leads to a higher than expected CMC. This effect is small: equivalent to the loss of about one CH_2 group [32].

With mixed surfactants the CMC of the mixed micelle varies according to the CMCs of the individual surfactants, and their concentrations. Clearly, micelle composition varies with concentration since the micelles that form at the lowest concentration are rich in the lowest CMC surfactant, while the higher CMC materials become more abundant in micelles as the overall concentration is increased. The detailed dependence of CMC values on mixed surfactant composition varies according to whether there are specific interactions between head groups which lead to nonideal mixing in the micelle. This applies particularly with mixtures of nonionic and ionic surfactants, and ionic surfactants of opposite charge. Various treatments are available to describe the behavior (which are outside the scope of this article), for example as outlined in Clint [2], Ch 5 and 6.

Given that micelles are present in solution above the CMC, the most important consideration for those concerned with liquid crystal phases is the micelle shape. There are three major types: spheres, rods and 'discs'. They can be described using the *packing constraint* concepts [4, 33]. These give a simple description of the relationship between micelle shape and molecular shape. The micelles are assumed to be smooth, with only the hydrophobic volume in the micelle interior. The main molecular parameters are the hydrophobic group volume, usually taken as being equal to the alkyl chain volume (v), the area that the molecule occupies at the micelle surface (a) and the maximum length of the alkyl chain (taken as the all-*trans* length, l_t). For a spherical micelle having a hydrophobic volume (V) with radius r, a total surface area A and aggregation number q:

$$A = qa = 4\pi r^2$$
$$V = qv = \frac{4}{3}\pi r^3$$

Hence

$$a = 3\frac{v}{r} \qquad (4)$$

Ignoring end or edge effects for circular cylinders (radius$=r$) and bilayer/disc (thickness$=2r$) shapes, the equivalent relationship are:

$$a = 2\frac{v}{r} \quad (\text{rod}) \qquad (5)$$

$$a = \frac{v}{r} \quad (\text{disc}) \qquad (6)$$

The value of r can not be larger than l_t, hence there are limitations on the lowest value of a for a given shape:

$$a \geq 3\frac{v}{r} \; (\text{sphere});$$

$$a \geq 2\frac{v}{r} \; (\text{rod});$$

$$a \geq \frac{v}{r} \; (\text{disc}) \qquad (7)$$

Clearly, a surfactant with a given chain length can pack into spheres, rods or discs according to the size of the head group, with all three shapes being possible for the largest a values and only disc micelles for small a values. Entropy favors the formation of the smallest possible aggregate at the CMC (i.e. spheres over rods and rods over discs). The present authors are not aware of any exceptions to this. Thus large headgroup surfactants form spherical micelles, smaller headgroups give rods, and smaller headgroups still give discs. (Because of the flexibility of alkyl chains there does not appear to be a lower limit on the values of r for conventional surfactants, hence maximum values for a are not known.)

It is a simple matter to estimate the volumes of hydrophobic groups from published density data for alkanes (normal and fluorinated) and polydimethyl siloxanes [4, 5, 34]. One simply sums the group volumes (see Table 3). Similarly, the maximum length of the hydrophobic group can be calculated from known bond lengths. Thus, the maximum micelle radius and hence the limiting a values for the various aggregates can be calculated. Note that there are small dif-

Table 3. Molecular sizes of hydrophobic groups (25 °C).

Hydrophobic group	Fragment volumes (Å)			Bond length (Å)	
Alkane	$V(CH_3)$ 54.2	$V(CH_2)$ 27.0	$V(CH)$ ~0	$L(C-C-C)$ 2.54	$L(CH_3)$ 1.53
Fluorocarbon	$V(CF_3)$ 92.5	$V(CF_2)$ 36.0		$L(C-C)-C$ 2.54	$L(CF_3)$ 1.33
Polydimethyl siloxane	$V\left(\begin{array}{c} Me \\ | \\ O-Si \\ | \\ Me \end{array}\right)$ 123			$V\left(\begin{array}{c} Me \\ | \\ O-Si \\ | \\ Me \end{array}\right)$ 2.4	

	Limiting a values [nm$^2 \times 10^{-2}$]		
	Sphere	Rod	Disc
Alkane	68	46	23
Fluorocarbon	102	68	34
Polydimethylsiloxane	~160	~106	~53

ferences between the parameters of Table 3 and those of other authors [4–6]. These are unimportant since they result only in small differences between the limits to a values. Moreover, we recall the known roughness of the micelle surface (typically 0.2–0.3 nm), hence we emphasize that this model gives only an approximate description.

An important consequence of the above model is that a simple increase in alkyl chain length should *not* alter micelle shape because both chain length and volume increase by a constant increment. However, in practice it is often observed that short chain (e.g., C_{12}) surfactants form globular ('spherical') micelles while higher chain length materials with the same head group form long rod micelles. This probably arises from the influence of surface roughness on micelle shape and aggregation numbers.

Micelle size (aggregation number) varies according to micelle shape and alkyl chain length. Spherical micelles always have low aggregation numbers, due to the micelle ra-

dius being limited by the all-*trans* alkyl chain length, hence the surface area is limited. For rod and disc micelles, where the fraction of 'end' or 'edge' molecules plays a significant part in micelle size, it is useful to consider a thermodynamic description employed by Israelachvili and collaborators [6, 33, 35]. In a solution of aggregates with a range of aggregation numbers at equilibrium, the chemical potential (μ) of all identical molecules is the same, whatever the aggregation number

$$\mu_n = \mu_n^0 + \frac{kT}{n} \log \frac{X_n}{n} = \text{constant},$$
$$n = 1, 2, 3, \ldots \tag{8}$$

where μ_n is the mean chemical potential of a molecule in aggregates of aggregation number n, μ_n^0 is the standard part of the chemical potential (i.e. the mean interaction free energy per molecule) and X_n the concentration of molecules for the n-aggregates. Taking simple models for the interaction free energy within aggregates of various shapes, it can be shown that:

$$\mu_n^0 = \mu_{\text{infinity}}^0 + \frac{\alpha kT}{n^p} \qquad (9)$$

where p takes the value 1/3, 1/2 or 1 for spheres, discs or rods respectively. The monomer–monomer bond energy within an aggregate is described by αkT. Above the CMC, with $\alpha > 1$ (a reasonable assumption) and $p < 1$ (i.e. for spheres and discs), this approach leads to expressions which predict vanishingly low concentrations of aggregates having n values which are not small (say $n > 10–20$). Thus spherical and disc micelles remain small (or increase to 'infinite' aggregation numbers) i.e. they phase-separate. Only rod micelles can have large aggregation numbers ($n > \approx 100$). In practice this does appear to be generally true. In any case we have already seen that spherical micelles are prevented from becoming large by the alkyl chain packing constraints. Hence the major conclusion of this approach is that disc micelles either remain small, or grow to infinite size to form a lamellar liquid crystalline phase (see below).

In real surfactant systems the interactions are much more complex that the simple picture used above. However, the general formalism still holds, but observed aggregation numbers for small micelles are often larger than expected for spheres, while the micelles do *not* grow very large as expected for rods. This is almost certainly due to micelle surface roughness. A surface roughness of 1–2 C–C bonds allows the micelle radius to be slightly larger than the all-*trans* chain length: with a significant increase in n. The roughness also allows repulsive interactions between adjacent head groups to be relaxed by the formation of a thick interfacial layer rather than a smooth surface. In addition, shape fluctuations can occur, which can also lead to larger aggregation numbers. However, if the alkyl chain lengths is long

enough, the micelles that have a rod shape do become very long, up to 100+ nm. This is not usually seen for common ionic surfactants because the longer chain homologues have high Krafft temperatures, hence they are insoluble at normal temperatures.

3   Liquid Crystal Structures

There are six classes of liquid crystal phases, most of which now have well-established structures. These are the lamellar, hexagonal, cubic, nematic, gel, and intermediate phases. All except the intermediate phases have been recognized for many years, hence a *comprehensive* literature review is outside the scope of this article. In all of the states except the gel phases, both surfactant and water have a liquid-like molecular mobility, that is short-range rotational and translational diffusion on a time scale of 10^{-12} s. They differ in the long-range symmetry of the surfactant aggregates and in the curvature of the micelle surface. Except for the phases with flat aggregate surfaces, each class can occur with either the polar regions or the nonpolar regions as the continuous medium, the former being referred to as *normal*, while the latter are *reversed*. As will be seen later, the phases occur in a particular composition sequence, occupying a specific region of the phase diagram. Each class of mesophases is usually labelled by a particular letter (see below) with the symbols having subscripts 1 or 2 to distinguish between the *normal* or *reversed* forms. Unfortunately there is no universally accepted nomenclature as is the case with thermotropic mesophases. In this review we employ the system that we have used in previous papers [36–38]. As with thermotropic mesophases, the most important technique to identify the mesophase type is polarizing microscopy. The birefringent phases usually have typical textures, while cubic phases have none, but they are very viscous.

3.1 Lamellar Phase (L$_\alpha$)

By far the most common surfactant mesophase is the lamellar phase (L$_\alpha$), also known as the neat phase from its occurrence during soap manufacture. In this phase, the surfactant molecules are arranged in bilayers frequently extending over large distances (a micron or more), which are separated by water layers (Fig. 2). This phase is similar to the thermotropic SmA phases. Its major repeating unit, the bilayer, forms the basic structural matrix of biological membranes [5, 7, 8]. While the lamellar phase does not usually flow under gravity, it has a fairly low viscosity, with the material being easily shaken into a container. It is readily identified from its characteristic optical textures (Fig. 3).

The surfactant bilayer thickness can vary from about 1.0–1.9 times the all-*trans* alkyl chain length (l_t) of the surfactant. Within this layer the 'fluid-like' characteristic of

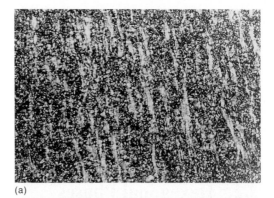

(a)

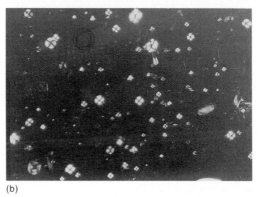

(b)

Figure 3. Typical optical textures of a lamellar phase: (a) mosaic oily streaks, (b) Maltese crosses.

the alkyl chains is shown by a diffuse wide angle X-ray diffraction peak corresponding to a Bragg reflection of 0.45 nm [39]. The difference in layer thickness arises from differences in head group areas and gives rise to differing degrees of disorder within the alkyl chain region. For the bilayers of thickness $l_t = 1.0$ the disorder is large. Alternative suggestions that these lamellar phases consist of 'interdigitated monolayers' are incorrect. By contrast, the water layer thickness varies over a much larger range, usually within the limits 0.8 and 20+ nm. The water thickness is the same throughout the sample, except at very high water contents [36] where low energy fluctuations can occur. The minimum water content is often that required to hydrate the polar groups, but

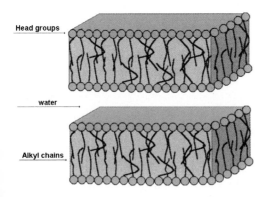

Head groups

water

Alkyl chains

Figure 2. Schematic representation of a lamellar phase.

very low or zero water content can occur with surfactants that form thermotropic lamellar phases. Sharp reflections in the ratio $d : d/2 : d/3 \ldots$ are observed in the low angle X-ray region due to the regular alternating layer structure with repeat spacings being the sum of the water and the alkyl chain layer dimensions.

3.2   Hexagonal Phases (H_1, H_2)

The next most common mesophase type is the hexagonal phase. There are two distinct classes of hexagonal phase, these being a 'normal hexagonal' (H_1) also known as the middle phase (again from the soap industry) and a 'reversed hexagonal' (H_2). The *normal* phase (H_1) is water-continuous, while the *reversed* (H_2) is alkyl chain-continuous. They consist of indefinitely long circular aggregates packed on a hexagonal lattice (Fig. 4).

The normal micelles have a diameter of 1.3–2.0 times the all-*trans* alkyl chain length, with a typical intermicellar separation being in the region of 0.8–5 nm. The reversed micelles have a polar region diameter in the same range, but values above 3 nm are rare.

(a)

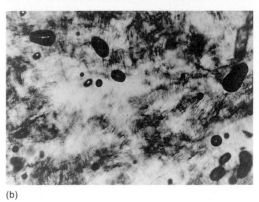

(b)

Figure 5. Typical optical textures of a hexagonal phase: (a) fan-like, (b) nongeometric.

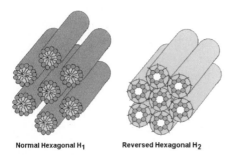

Normal Hexagonal H_1 Reversed Hexagonal H_2

Figure 4. Schematic representation of a normal (H_1) and reversed hexagonal (H_2) phase.

X-ray diffraction of both phases shows Bragg reflections in the ratio $1 : 1/\sqrt{3} : 1/\sqrt{4} : 1/\sqrt{7} : 1/\sqrt{12} \ldots$, again with a diffuse reflection at 0.45 nm. Both phases are rather viscous, much more so than the L_α phase. It is usually not possible to shake a sample into a container by hand. The optical textures are similar for both types, again being distinctly different from those of L_α phase (Fig. 5).

3.3   Cubic Phases

A third category of mesophases formed by surfactants comprises the cubic phases. They are also known as 'viscous isotropic'. As the name implies, these phases are based

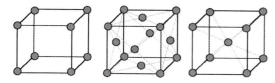

Figure 6. Schematic representations of a primitive, face-centered and body-centered cubic lattices.

around one of three cubic lattices, namely the primitive, face centred and body centred (see Fig. 6). Therefore, they have no optical texture under polarized light. It is still not certain which symmetries form for the different cubic phase, but the overall picture has become much clearer during the past few years [37, 39–47] with more and more cubic structures being identified. There are at least four classes of cubic phase, these being the normal and reversed forms of two very different structures. One set of structures (I) is comprised of small globular micelles while the second (V) consists of a 3-D micellar network. Within each class several different structures occur.

The simplest cubic phase is that of the I type where the surfactant aggregates are small globular micelles. For the water-continuous I_1 phases, primitive, body centred and face centred lattices [41] have all been proposed. Pm3n, Im3m and Fm3m lattice structures have been proposed [46, 47] (see Fig. 8). In the nonionic $C_{12}EO_{12}$/water system [46] all three symmetries are reported. Here the micelles have diameters similar to normal micelles with a separation similar to that found in the H_1 phase. The structure of the Pm3n symmetry has been the attention of much debate [45–49]. However, it is now thought that there are two different sizes of micelle present, one being slightly larger than the other [45]. Whether these micelles are short rods or flattened spheres is still a matter for debate. The short rod model fits better with their position in the phase diagram (between L_1 and H_1). However, a model composed of two spherical micelles and six disc shaped micelles has been proposed [49]. It is this structure which is now thought to be the more accurate. As for the Im3m and Fm3m cubic lattices only one micelle type, a quasispherical structure, is proposed for the lattice [49] (see Fig. 7).

A reversed I_2 structure of globular aggregates packed in a cubic array also has been reported recently for some surfactants [46–49], despite previous doubts about their existence. Here it appears that the micelles are spherical, but of two different sizes (Fig. 8). For reversed micelles the alkyl chain packing constraints no longer limit the micelle diameter, hence the coexistence of two spherical micelles of different sizes is more plausible than with the I_1 phases. The Fd3m phase is well established now [48, 49], but again, there appear to be several other distinct symmetries possible in the I_2 region [48]. It is likely that the confusion in this area will be resolved in the next few years as the surfactant structures necessary for both I_1 and I_2 phases are now clear, hence we can obtain the phases with many more surfactant types.

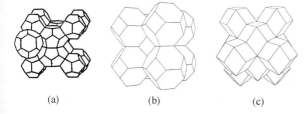

(a)   (b)   (c)

Figure 7. Polyhedral representations showing the micellar arrangements for (a) Pm3n, (b) Im3m and (c) Fm3m of I_1 cubic phases (reproduced from [46]). Micelles are located at the centres of the polyhedra

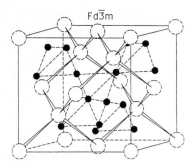

Figure 8. Schematic representation of the Fd3m I$_2$ phase (reproduced from [42]).

The second set of cubic phases has a 'bicontinuous' aggregate structure. The basic structures for these bicontinuous cubics have received more attention in previous years, with the structures being well established. With the three main spacegroups Pn3m, Im3m and Ia3d being well reported. The aggregates form a 3-D network extending throughout the sample, having curvature towards water (V$_1$) or towards oil (V$_2$), with many different structures being reported [41–44]. These phases are structurally similar to the intermediate phases (see below) and although organized around the same cubic lattices as the I phases, have a completely different aggregate structure. The first one to be detailed was based on a body centred lattice, namely Ia3d [50] (Fig. 9). This structure, originally proposed by Luzzati [50] was thought to be composed of rod-like aggregates joined three by three to form two independent networks. However, it is now believed that these structures are composed of infinite periodic minimal surfaces [51] with the surfactant and water cylinders being interwoven and connected three by three. Another body centred cubic structure proposed for lipid/water systems [52], that of the Im3m (Fig. 9), is composed of a network of water channels connected six by six.

There are two reported cubic phases composed around a primitive lattice. The first is the Pn3m, which first proposed to be composed of rod-like aggregates connected four by four at tetrahedral angles forming two independent diamond lattices. However, it is now believed to be composed around a minimal surface of two interwoven tetrahedral networks arranged on a double diamond structure. The second structure composed around a primitive lattice is that of the Pm3n. Its structure is still under debate, Luzzati and Tardieu [52] proposed a rod-like network of micelles forming a cage in which a spherical micelle is enclosed. However, this is now thought to be unlikely [43, 53].

The cubic phases are all optically isotropic (as are micellar solutions), but they have

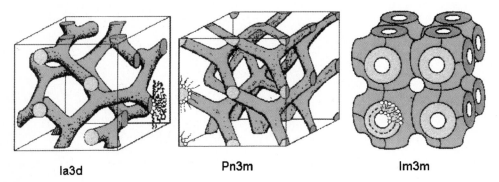

Ia3d   Pn3m   Im3m

Figure 9. Schematic representations of cubic phases (reproduced from [42]).

very high viscosities hence are easily distinguished from micellar solutions. The two classes of cubic phase, I and V, are distinguished from each other by their location in the phase diagram. I phases occur at compositions between micellar solutions and hexagonal phases, whilst V phases occur between hexagonal and lamellar phases. The factors that determine the particular structure within any set of I or V phases are not understood.

3.4   Nematic Phases

Lyotropic nematic phases were first reported by Lawson and Flautt [54] for mixtures of C_8 and C_{10} alkyl sulfates together with their corresponding alcohol in water. They are some what less common than the mesophases discussed so far. When they do form they occur at the boundary between an isotropic micellar phase (L_1) and the hexagonal phase (H_1) or between L_1 and the lamellar phase (L_α). As their name implies they have a similar long range micellar order to that of the molecules in a thermotropic nematic phase. They are of low viscosity, possessing long range micellar orientational order but reduced translational order compared to the other lyotropic phases described above, and like the thermotropic phase can be aligned in a magnetic field. It is possible to identify nematic phases optically because of their characteristic schlieren texture.

Lyotropic nematic phases are generally found for short chain surfactants, for both hydrocarbon or fluorocarbon derivatives [55, 56]. Two different micelle shapes can occur (Fig. 10) [57]. One type (N_c) is thought to be composed of small cylindrical micelles and is related to the hexagonal phase, while the other type of nematic (N_d) is composed of planar disc micelles and is related to the lamellar phase. Note that the 'disc' micelles are likely to be 'matchbox' or 'ruler shaped', rather than the circular discs. Hence the 'disc' nematic phase can have the director along the long axis of 'ruler' micelles or along the shortest micelle dimension, as with 'match box' micelles, while with N_c phases the director always lies along the rod axis.

As with thermotropic nematics, the addition of optically active species to lyotropic nematic phases gives lyotropic cholesteric phases. Whilst details of their structures are not fully established they appear to follow the general pattern outlined above. The cholesteric 'twist' would appear to derive from the packing of optically active mole-

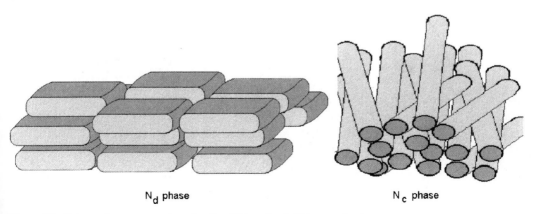

N_d phase N_c phase

Figure 10.   Schematic representation of a disc (N_d) and rod (N_c) nematic phase.

cules within micelles, leading to a twisted micellar structure, rather than to the transmission of the anisotropic forces via solvent mediated forces. A recent report has shown evidence for the occurrence of cholesteric 'blue' phases, a remarkable observation [58].

3.5   Gel Phases (L_β)

The gel phase (L_β) closely resembles the lamellar phase in that it is comprised of surfactant layers, but it differs in its very high viscosity. The term 'gel' again originates from industry where these systems were observed to have a gel-like rheology. However, these states should not be confused with polymer gels or gels formed by hydrocolloid systems, since they are single phases in terms of the phase rule rather than being multiphase systems like polymer and colloid gels.

Within the gel phase, the bilayers have rigid, mostly all-*trans* alkyl chains, as shown by a sharp, wide angle X-ray spacing of about 0.42 nm and a large transition heat on melting, typically 25–75% of the crystalline surfactant melting transition. This indicates restricted chain motions, mostly limited to rotation about the long axis only. By contrast the water (polar medium) is in a 'liquid-like' state, with fast rotational and translational mobility. There are commonly reported to be three different structures of the gel phase as shown in Fig. 11. The first structure, with the bilayer normal to the liquid crystal axis, is the structure most commonly found in dialkyl lipid systems [59]. Here, the alkyl layer thickness is found to be approximately twice the all-*trans* alkyl chain length of the surfactant. The second structure shown, the tilted bilayer, is found in systems where the polar head group is larger than the width of the alkyl chain. This structure has been reported for monoglyceride systems [60]. The third structure, the interdigitated form, is found with long chain mono-alkyl systems such as potassium stearate [61].

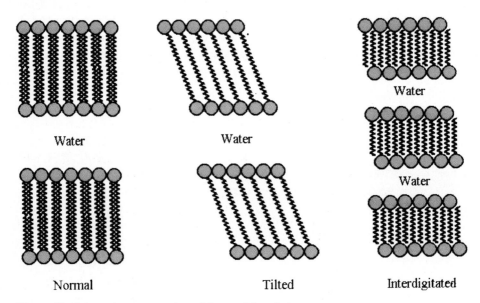

Figure 11. Schematic representation of the possible gel phases.

Whilst the occurrence of gel phases is commonly recognized for long chain dialkyl surfactants it is also a common occurrence for monoalkyl surfactants. Because of the lower packing order within the alkyl chain region in than the normal crystals, different chain length derivatives (usually up to four carbons) can mix within the L_β phase. Additionally, different head groups can also mix within the L_β phases. Indeed, L_β phases can occur for mixed systems where none is observed for the individual constituents. For example sodium dodecyl sulfate and dodecanol form a mixed gel phase where as none is observed for SDS alone [62]. (Note that dodecanol does form a stable L_β phase termed the 'α-crystalline phase' with about 0.2 mol fraction of water [5, 63, 64].)

Whilst L_β phases have been accepted for years recently there has been debate about whether the state really does exist as a true thermodynamic equilibrium phase, based on very reasonable criticism of deficiencies in their location on properly determined phase diagrams [65]. However, in at least one case (the nonionic surfactant trioxyethylene hexadecyl ether [66]) the L_β phase in water melts at a higher temperature than the crystalline surfactant. It forms on mixing water and the liquid surfactant just above the crystalline surfactant melting point, clear proof that it is the equilibrium state.

In fact, as Small has discussed in detail [5], the stability of the L_β state is determined by the packing of the alkyl chains. Polyethylene does not melt until ca. 140°C [67]. Long chain alkanes do form stable 'α-crystalline' (rotator) phases where the alkyl chain packing and mobility is similar to that of the L_β phase. The melting of alkanes can be regarded as being driven by the mismatch between CH_2 and CH_3 sizes within the crystal (2.3–2.6 nm and ca. 120 Å respectively). Thus the transition tempera-

tures for the sequence:

$$\text{crystal} \xrightarrow{T_A} \text{'rotator' phase} \xrightarrow{T_B} \text{melt}$$

increase with hydrocarbon chain length, the minimum chain size required to form a rotator phase being about C_{22}. The value of T_B for alkanes of chain length $2n$ represents an upper temperature limit for the L_β phase of surfactants with chain length n. Usually, the L_β phases melt at a lower temperature than the limit T_B because the head groups are more hydrated in the molten phases than the L_β phase, and this free energy contribution is larger than the chain packing energy.

To date the hydration of headgroups in the L_β phase has always been found to be lower than that of the higher temperature molten phases (usually lamellar). Hence L_β phases do not swell in water to the same extent as the L_α phases. Moreover, the size (area) of the hydrated head group determines which of the three structures occur (see Fig. 11). The perpendicular bilayer requires a head group area (a) of about 22 Å^2, whilst the tilted bilayer occurs with $a = 22-40$ Å^2, and the monolayer interdigitated structure with $a \geq 44$ Å^2. Hence increasing m in the series of polyoxyethylene surfactants C_nEO_m ($n > 16$, $m = 0-3$), either singly or in mixtures, one expects to encounter all three phases. Whilst the monolayer and perpendicular bilayer structures are known, the tilted phase has not been reported.

In fact, a careful examination of properties such as heat of melting, high angle X-ray data and phase behavior for a series of closely related surfactants shows that considerable anomalies exist in both assumed alkyl chain packing structure and phase stability with the conventional picture. The area deserves a systematic broad study to give a proper molecular based understanding.

3.6   Intermediate Phases

We have already described the occurrence of bicontinuous cubic phases having an aggregate curvature between those of hexagonal and lamellar phase. Over the past 40 years there have been sporadic reports of other structures with similar '*intermediate*' curvature. A number of these so-called 'intermediate' phases have now been identified. They replace V_1 bicontinuous cubic phases for surfactants with longer or more rigid hydrophobic chains. It is likely that they replace V_2 phases under some conditions, but this area has yet to receive the same systematic attention as the V_1/intermediate phases. Unlike the V_1/V_2 phases, intermediate phases are anisotropic in structure, and consequently birefringent; also they are often much more fluid than cubic phases. There is still discussion in the literature as to which structures are possible. The observed or proposed structures divide topologically into three broad types according to symmetry; *rectangular ribbon* structures, layered *mesh* structures, and *bicontinuous* structures which do not have cubic symmetry. Ribbon phases may be regarded as a distorted hexagonal phase (Fig. 12 a). Mesh phases are distorted lamellar phases in which the continuous bilayers are broken by water filled defects which may or may not be correlated from one layer to the next (Fig. 12 b). The bicontinuous phases are distorted cubic structures.

Phases with noncubic structures were first identified by X-ray scattering from aqueous soap mixtures [68, 69] and anhydrous soap melts in a series of papers by Luzzati and Skoulios [69–73]. In the first of these papers [68] the term intermediate was applied to a rectangular structure found in aqueous mixtures of potassium and sodium oleates and potassium laurate and palmitate. The structures found in the anhydrous soaps were reinterpreted as intermediate, tetragonal, rhombohedral, and ribbon structures in the paper by Luzzati and coworkers [74] in 1968. It was not until the early eighties that interest in these unusual phase structures was rekindled.

Ribbon phases have been the most comprehensively studied of the intermediate phases. They occur when the surfactant molecules aggregate to form long flat ribbons with an aspect ratio of about 0.5 located on two dimensional lattices of oblique, rectangular (primitive or centred), or hexagonal symmetry. Ribbon phases were first proposed by Luzzati [68, 69] in aqueous sur-

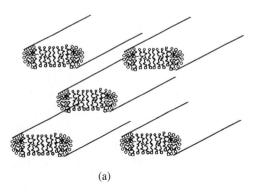

(a)

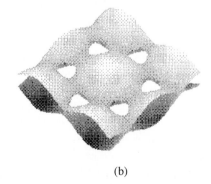

(b)

Figure 12. Schematic representations of (a) centered rectangular phase, (b) representation of six connected rhombohedral mesh phase.

factant systems and by Skoulios [71] in sodium stearate and tetradecane or cyclohexane. Hagslätt et al. [75] have investigated ribbon phases in a number of ternary systems. In [75] they review the results from a wide variety of studies on ribbon phases and show that these studies are consistent with the conclusion that all ribbon phases index to a centred rectangular cell, *cmm*. A 'hexagonal-rod' model of the ribbon cross section was suggested in which the ribbon structure is controlled by the competition between the requirement for a constant water layer thickness around each ribbon, the surface area per molecule and the minimization of total surface area. Note that given the anisotropic nature of the interaction forces between the ribbons, the assumption of a constant water layer must be an approximation.

The first identification of intermediate *mesh* phase structures was by Luzzati [74] from the measurements by Spegt and Skoulios [70–73] in anhydrous soap melts. It was not until the work of Kékicheff and others [62, 76–81] on sodium dodecyl sulfate (SDS)/water and on lithium perfluoroocta-

noate (LiPFO)/water [82] that intermediate mesh phases were recognized in aqueous surfactant water systems. SDS/water has been extensively studied by optical microscopy [78], X-ray scattering [68, 77, 78], SANS [79] and NMR [77, 78] techniques. More recently, mesh intermediate phase structures have been identified in a number of nonionic surfactants with long alkyl chains [83–85]. These reveal a very rich intermediate behavior [77–80]. There are a number of possible mesh phase structures with both tetragonal and rhombohedral symmetry. These generate X-ray scattering patterns which, although they may show up to ten lines (Fig. 13) must be indexed with care because a variety of structures may be possible.

Many intermediate phase regions are bounded by lamellar phases which contain water filled defects; the nonuniform curvature is retained although there is no longer any ordering of the defects within the bilayer and there are no correlations between the bilayers. These defected lamellar phases may also be regarded as random mesh phases. They have been seen in the SDS/water

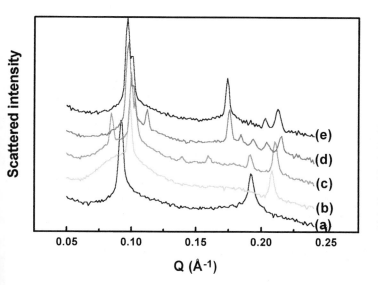

Figure 13. The X-ray scattering from a 52 wt% $C_{16}EO_6$/ water sample. (a) Lamellar phase; (b) random mesh phase; (c) rhombohedral mesh intermediate phase; (d) Ia3d cubic phase; and (e) hexagonal + gel two phase region.

and LiPFO/water systems and as independently occurring phases in decylammonium chloride/NH$_4$Cl/water [86, 87] and in caesium perfluorooctanoate/water [88–91]. These phases are characterized by lamellar like Bragg reflections in the ratio $1:1/2:1/3$ ..., but with a broad liquid-like reflection from the intralamellar defects (Fig. 13).

Whilst ribbon and mesh phases are now fairly well established, the bicontinuous noncubic structures are still elusive. The identification of tetragonal or rhombohedral phases of a mesh or bicontinuous type is ambiguous because there usually is insufficient information to make a definitive identification. There are only a few examples where authors have identified a bicontinuous phase usually because of its association with adjacent bicontinuous cubic phases [79, 92]. Theoretically, Hyde has cast doubt on noncubic bicontinuous phases because periodic minimal surfaces with tetragonal or rhombohedral symmetries are expected to have a higher associated bending energy cost than their cubic phase counterparts. He suggests that the phase transition from ribbons to V$_1$ can be achieved by extra tunnels connecting the mesh layers in a rhombohedral mesh phase. Anderson and coworkers [93–95] have proposed that both rhombohedral and tetragonal bicontinuous structures of the type first proposed by Schoen [96] are not only possible but are the most likely intermediate phase structures with these symmetries. The existence of bicontinuous noncubic phases still remains an open question.

Hyde et al. [97, 98] have considered the origin of intermediate phases in detail. All intermediate phase structures are characterized by nonuniform interfacial curvature. The forces between the polar head groups and those between the lipidic alkyl chains tend to impose a certain value for the interfacial curvature. This curvature may not be compatible with the molecular length. The problem is resolved by either the system phase separating or by the formation of structures with nonuniform surface curvature. However, it is still not clear why increasing chain length or rigidity should favor the formation of these intriguing phases over bicontinuous cubic structures.

Whilst the early reports of intermediate phases concerned systems with *reversed* curvature [73–76] these were for surfactants where some residual short range order in the polar groups was probably present. There are few definitive reports of fully molten intermediate phases with reversed curvatures. In fact the pattern of how intermediate phases replace the normal bicontinuous cubic phase as alkyl chain size increase only became recognized as systematic studies on homologous series were carried out [37, 66]. Here it has required a combination of microscopy, multinuclear NMR and X-ray diffraction to elucidate the structures. Such studies on reversed phases have yet to be carried out, particularly where *small* variations in alkyl chain structure are made.

The key factor in studies of the normal systems was a gradual increase of chain length. Conformational restrictions on the chains in normal aggregates appear to be responsible for the reduced stability of the V$_1$ phase for long chain derivatives. With reversed structures the presence of water in the aggregate cores allows conformational freedom. Hence it is likely that reversed intermediate phases will be found for systems with low water contents and bulky head groups: inevitably with multiple alkyl chain compounds. Where the chains have identical lengths such surfactants have high melting temperatures. Hence multiple unequal chains are probably required to observe reversed intermediate phases.

4 Phase Behavior of Nonionic Surfactants

In principle the aggregation properties and mesophases of nonionic surfactants should be easier to understand than the behavior of ionic surfactants because there are no long range electrostatic forces. As the intra- and intermicellar head group interactions operate over a much shorter range than for ionic compounds, the head groups can pack close together as is required for the transformation from highly curved aggregates such as those in I_1 and H_1 phases to geometries with a smaller or negative curvature as in the reversed phases [37, 99–101]. The effective head group volume is given by its actual size, together with about one hydration layer.

The most widely studied group of nonionic surfactants is that of the poly(ethylene oxide) alkyl ethers [n-$C_nH_{2n+1}(OCH_2CH_2)_xOH$; C_nEO_x] [37, 101]. One reason for this is that it is possible to study the phase behavior whilst systematically varying the length of either the hydrophobic alkyl chain or the hydrophilic ethylene oxide groups. Surfactants with large head groups form cubic (I_1) and hexagonal (H_1) phases. With increasing temperature the head group dehydrates and reduces in effective size, often until the interfacial curvature becomes zero or even negative. This results in the formation of the lamellar (L_α) phase and a subsequent temperature increase (or further decrease in hydration and head group size) can lead to the formation of reversed phases.

The variation of the phase behavior with head group size is illustrated with reference to the $C_{12}EO_x$ homologous series [37, 46, 102, 103], where $x=2–6, 8, 12$. The simplest phase behavior is shown by $C_{12}EO_6$ (Fig. 14) which has H_1, V_1 and L_α phases and a region of partial miscibility or clouding ($W+L_1$). The cloud temperature (critical point) is defined as the lowest temperature at which the region of partial miscibility is seen. The maximum temperatures of the H_1 and V_1 phases are 37° and 38 °C respectively while the L_α phase exists up to

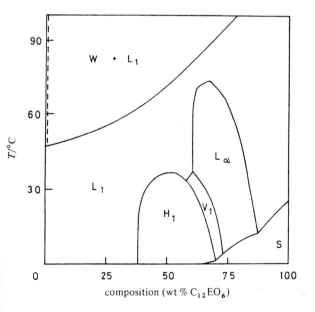

Figure 14. Phase diagram of $C_{12}EO_6$/water. Note that most two phase regions are small and not shown (L_1, aqueous surfactant solution; W, very dilute surfactant solution; H_1, normal hexagonal phase; V_1, normal bicontinuous cubic phase; L_α, lamellar phase; S indicates the presence of solid surfactant) (reproduced from [37]).

73 °C. Note that the pure surfactant forms a liquid, which is miscible with water over a certain composition/temperature range. The clouding region arises from the partial miscibility of the $C_{12}EO_6$ micellar solution with water above about 49 °C. This is caused by a net intermicellar attraction arising from EO–EO interactions between adjacent micelles. It is discussed further in [103] and the references therein. Whilst the clouding phenomenon is important for many micellar properties of these surfactants, its only influence on the mesophases is to determine the limit to which they can swell in water, that is it fixes the boundary between mesophases and the dilute aqueous solution (Fig. 14).

On increasing the EO size, an I_1 cubic phase is observed for $C_{12}EO_8$ between ≈ 30 and 43 wt% surfactant on the dilute side of the large H_1 region (Fig. 15). A V_1 cubic is seen over a narrow concentration range and the L_α phase is reduced to an even smaller concentration and temperature range. With a further increase in EO size to $C_{12}EO_{12}$,

only I_1 and H_1 phases are observed with the cubic phase existence range increasing to $\approx 35-53$ wt% surfactant (Fig. 16). Recently it has been shown that there are in fact three distinct I_1 phases present [46]. With increasing surfactant concentration these have spacegroups Fm3m, Im3m and Pm3m. As one might expect, the partial miscibility region shifts to higher temperatures as EO size increases. With very long EO groups the I_1 phase becomes even more dominant. However, eventually increasing the EO size is expected to lead to 'micelles' with small aggregation numbers and a tiny micelle core, which will not form ordered phases.

There is a distinct change in the behavior as the number of ethylene oxide groups decreases below 6. For $C_{12}EO_5$ [102], H_1, V_1 and L_α phases are all clearly observed (Fig. 17), but with the H_1 and V_1 phases existing over much narrower concentration ranges and to lower temperatures than for $C_{12}EO_6$ (to 18.5 and 20 °C respectively). The L_α phase is formed over a much wider concentration range. Note that above the

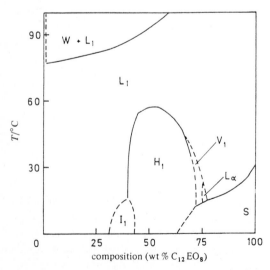

Figure 15. Phase diagram of $C_{12}EO_8$/water (reproduced from [37]) (I_1, close-packed spherical micelle cubic phase; others as for Fig. 14).

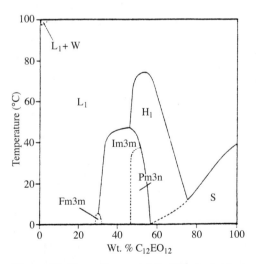

Figure 16. Phase diagram of $C_{12}EO_{12}$/water (reproduced from [46]) (Fm3m, Im3m, Pm3n represent different I_1 cubic phases; others see Fig. 14).

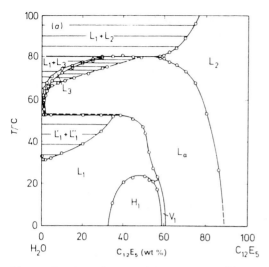

Figure 17. Phase diagram of $C_{12}EO_5$/water (L_1' and L_2', surfactant solutions; L_2, liquid surfactant containing dissolved water, not fully miscible with water; L_3, sponge phase, isotropic solution not fully miscible with water or surfactant, otherwise as for Fig. 14) (reproduced from [102]).

cloud point the lamellar phase coexists with a dilute aqueous phase and just above this the L_3 so called 'sponge phase' occurs. This phase has received much attention from academic researchers, both because of its unique position in the phase diagram and because it is closely related to the occurrence of bicontinuous microemulsions [104–107]. It is comprised of large aggregates having a net negative curvature (the opposite of normal micelles). The aggregates extend throughout the phase, hence it is continuous in both the aqueous and surfactant regions, so it is bicontinuous. Also the 'aggregates' are extremely labile, hence L_3 phases have low viscosities but occasionally show shear birefringence at higher concentrations. Their formation can be slow, so the establishment of exact phase boundaries is sometimes tedious and time consuming. In some of the earlier mesophase studies it was not recognized just how slowly the L_3 forms under some conditions, hence

L_3 regions may be underestimated (e.g. Fig. 18, below). In fact, the surfactant aggregate structure is closely related to that of the reversed bicontinuous cubic phases which are frequently found on the phase diagram adjacent to L_3 regions (e.g. $C_{12}EO_2$, see below). Whilst there is a dramatic difference between the mesophase regions of $C_{12}EO_6$ and $C_{12}EO_5$, the difference arises just from the small difference in head group size, and the existence of partial miscibility above the cloud temperature. Note the marked decrease in the miscibility of $C_{12}EO_5$ and water above 80 °C. This is a common feature with these surfactants.

The phase behavior of $C_{12}EO_4$ is very similar to that of $C_{12}EO_5$ with the L_α phase existing over a similar concentration range (25–80 wt%), but to a slightly lower temperature (68 °C). Clouding occurs at 4 °C and the H_1 phase exists only at temperatures below -2 °C. The occurrence of a V_1 phase, however, can not be definitely proved. An L_3 phase is observed at similar concentrations but about 10 °C below that of $C_{12}EO_5$.

For $C_{12}EO_3$ no micellar phase occurs (Fig. 18 a), at least above 0 °C, while the only mesophase shown on the phase diagram is L_α which exists over a concentration range of $\approx 47–85\%$ surfactant and to a maximum temperature of 52 °C. The L_3 phase is shown on the phase diagram as being continuous with L_2. In many of these systems it is difficult to observe coexistence of L_2 and L_3 phases at high concentrations because of the similar refractive index of the two phases. Laughlin has observed that L_3 and L_2 are not continuous in this system [108]. In the above three systems there are also a number of two phase regions in which various phases coexist with very dilute surfactant solution ($W+L_1$, $W+L_2$, $W+L_3$, $W+L_\alpha$).

For $C_{12}EO_2$ [103], in addition to L_α and L_3 phases there are two V_2 reverse bicon-

(a)

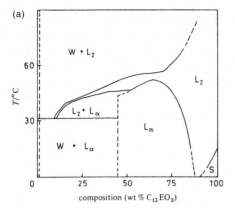

composition (wt % $C_{12}EO_3$)

(b)

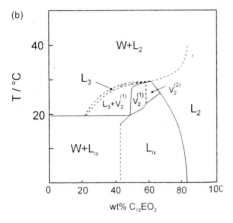

wt% $C_{12}EO_2$

Figure 18. (a) Phase diagram of $C_{12}EO_3$/water (reproduced from [37]). (b) Phase diagram of $C_{12}EO_2$/water (reproduced from [112]). Symbols as for Figs. 14 and 17.

tinuous cubic phases which exist between the other two phases over a temperature range of $24-36\,°C$ (Fig. 18b). X-ray diffraction [109] has shown the cubic phases to have space groups Ia3d and Pn3m. Because of the reduced hydrophilicity of the head group, no mesophases exist above $36\,°C$. Note that these occur at higher water concentrations than the L_α phase over a small temperature region: so that addition of water promotes negative aggregate curvature. This is consistent with the view of L_3 as having aggregates with negative curvature, and is a common occurrence with many surfac-

tants that form reversed phases. It underpins the important general observation *that there is not a similar regular phase sequence (L_α, V_2, H_2 ...) with concentration for the reversed structures as occurs for normal mesophases.*

For the remaining members of the series $C_{12}EO_1$ and $C_{12}EO_0$ (dodecanol) no mesophases occur, only separate water and amphiphile phases which show slight miscibility. With all the C_{12} surfactants no mesophase exists above $80\,°C$, while the maximum mesophase 'melting' temperature varies much less than the cloud temperatures.

Like behavior is observed for homologous series of surfactants of different alkyl chain lengths, but the mesophase regions extend to higher temperatures with increasing n. The effect of curvature increases with EO number as expected. So for all alkyl chain lengths ($n > 10$), I_1 cubic phases only exist for EO_x when $x > 8$ and extensive L_α regions occur for small EO groups ($x \leq 5$).

The influence of alkyl chain length can be illustrated using C_nEO_6 surfactants [37]. C_8EO_6 gives just an H_1 phase which melts at $\approx 12\,°C$ [110, 111]. $C_{10}EO_6$ shows a substantial area of H_1 phase but there is contention as to whether it additionally shows a low temperature L_α phase [111, 112]. No V_1 phase occurs with these two materials. The phase behavior of $C_{12}EO_6$ is reported above. $C_{14}EO_6$ is very similar to $C_{12}EO_6$ only with the L_α phase having a significantly higher melting temperature ($95\,°C$ as opposed to $73\,°C$). On increasing the chain length to C_{16} there is a dramatic alteration in the phase behavior similar to that occurring between $C_{12}EO_6$ and $C_{12}EO_5$. The behavior is made even more complex because a monolayer interdigitated gel (L_β) phase occurs below about $25\,°C$. The H_1 and V_1 melting temperatures are slightly decreased (both $34\,°C$ compared to 37 and $40\,°C$ for $C_{14}EO_6$) and the L_α melting temperature slightly increased (102 °C). This is the first member of the C$_n$EO$_6$ series to show an L$_\alpha$+W region and an L$_3$ phase (90–>102 °C). Moreover, with an increase in chain size to C$_{16}$, the bicontinuous cubic phase begins to be replaced by intermediate phases. In fact C$_{16}$EO$_6$ exhibits several phases not in the original report, including a rod micelle nematic phase, a random-mesh-lamellar phase and another intermediate phase, the latter being metastable [113].

Whilst the existence range of the nematic phase is small for C$_{16}$EO$_6$, with C$_{16}$EO$_8$ a low viscosity long rod micelle nematic phase has been observed (Fig. 19) over a narrow concentration range at ≈ 34 wt% surfactant over the temperature range 28–54 °C between the H$_1$ and L$_1$ phases [114]. Here the increase in EO size removes the intermediate phases and the L$_\alpha$+W and L$_3$ regions, but a gel (L$_\beta$) phase is present at temperatures below the molten phases. For much longer chain surfactants such as C$_{22=}$EO$_6$ (hexaethylene glycol *cis*-13-docosenyl ether) [115] or 'C$_{30}$EO$_9$' (nonaethylene glycol mono(11-oxa-14,18,22,26-tetramethylheptacosyl)ether [85] the corre-

lated defect mesh phases replace the V$_1$ region. No gel phases are expected here because of the disrupted alkyl chain packing. Note that several general conclusions are illustrated by the above behavior. First, increasing surfactant chain length has little effect on the composition ranges of phases with positive curvature, the major influence is to raise the upper temperature limit of the mesophases. Hence very short chain (C$_n$, $n < 8$) surfactants do not form mesophases at all. Moreover, long rod nematic phases occur at the L$_1$/H$_1$ boundary for long chain surfactants. This pattern of behavior is true for all surfactant types. The reason why few N$_c$ phases are reported in the literature is because such long chain surfactants are usually insoluble. The C$_{22=}$ and 'C$_{30}$' derivatives referred to above have their molecular packing in the crystal disrupted by the presence of methyl side groups or unsaturation, both of which increase miscibility with water.

Conroy et al. [103] investigated the effect of replacing the terminal OH in three polyoxyethylene surfactants by OMe (C$_{12}$EO$_m$OMe, $m = 4, 6, 8$). For $m = 4$ the L$_\alpha$ phase melts at 27 °C for OMe (68°C for OH) and the L$_3$ phase is seen at lower temperatures (24–27° as opposed to 51.5–70 °C). For $m = 6$ the H$_1$, V$_1$ and L$_\alpha$ phases all melt at lower temperatures for OMe (24, 28, 43 °C as opposed to 37, 38, 73 °C for OH). For $m = 8$ the V$_1$ and L$_\alpha$ phases are not seen for OMe and the I$_1$ and H$_1$ melting temperatures reduced to 15 and 41 °C respectively from 16 and 59 °C for OH. This shows the importance of the terminal OH of the EO groups, a factor often neglected.

Commercial polyoxyethylene surfactants contain a wide range of EO sizes, often with a reasonably defined alkyl chain. Bouwstra et al. [116] studied a commercial sample of C$_9$=C$_9$EO$_{14}$ (i.e. a C$_{18}$ chain with 9–10 cis double bond and a polydisperse EO group). It exhibits phase behavior similar to C$_{12}$EO$_8$

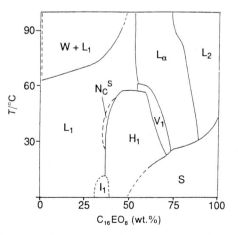

Figure 19. Phase diagram of C$_{16}$EO$_8$/water (reproduced from [114]). (N$_c^s$, rod nematic phase; otherwise as for Figs. 14, 17 and 18).

with I_1, H_1, V_1 and L_α phases. The H_1 phase exists between 35.6 and 74.8 wt% surfactant and up to 84 °C. The other phases exist over narrower concentration ranges and to lower temperatures.

The commercial Synperonic A surfactants are a mixture of C_{13} (66%) and C_{15} (34%) alkyl chains while the hydrophilic chains are composed of polydisperse ethylene oxide groups. Investigations into the phase behavior of A7, A11 and A20 (containing an average of 7, 11 and 20 EO groups respectively) have been achieved using a number of experimental techniques [117, 118]. A7 shows H_1 and L_α phases, with the H_1 existing between 30 and 62 wt% surfactant up to a maximum of $\approx 30°C$, while the L_α exists between 51 and 86 wt% surfactant to a temperature greater than 60 °C. With an increase of EO size to the A11 compound, a small area of V_1 cubic phase appears between the H_1 and L_α phases, while the latter is greatly reduced in stability (73–83 wt% with a maximum temperature of ≈ 35 °C). The H_1 phase exists at slightly higher concentrations than for A7 and up to ≈ 64 °C.

The phase behavior of the A20 compound is radically different to the shorter chain homologues, with only I_1 and H_1 phases present. The I_1 phase exists between 24 and 62 wt% surfactant up to ≈ 70 °C and the H_1 between 60 and 81 wt% to over 80 °C. Unlike the other two surfactants the A20 compound shows no cloud point (below 100 °C) at low surfactant concentrations.

Triton X-100® and X-114® are industrial *p-tert*-octylphenol polyoxyethylene surfactants with about 9.2 and 7.5 ethylene oxide groups per molecule respectively. They are used in biochemical studies because of their ability to disrupt biological membranes without denaturing integral membrane proteins [119]. TX-100 [120] forms an H_1 phase between 37 and 63% surfactant from 0 to

28 °C and a L_α phase between 65 and 78% up to 6 °C. TX-114 [121] exhibits a larger area of L_α phase from 35 to 83% surfactant and from ≈ -20 to 65 °C, but does not form H_1.

Essentially these studies allow two general conclusions to be made when comparing pure and commercial nonionic surfactants. First, the stability of the V_1 phase is much reduced. Secondly, a commercial surfactant of average formula C_nEO_m resembles a pure surfactant with a slightly smaller EO groups such as C_nEO_{m-1}. (An example is given in the section on mixed surfactants below.) Thus the surfactant molecules with different EO sizes mix together in the same aggregates, rather than the long and short EO components separating into phases with large and small aggregate curvatures respectively. The small difference in hydrophilicity caused by changing the head group size by one EO unit is responsible for this. With other commercial surfactants where the head group is a small polymer of strongly hydrophilic residues, such as with alkyl polyglycerols, the difference in hydration and size between head groups differing by 1–2 polymer units is very often sufficient to cause the coexistence of phases with different curvatures (e.g. $L_\alpha + H_1$) over a wide range of water contents. Obviously, this is only a rough guide for materials where the EO distribution follows some regular pattern. Mixed commercial nonionic surfactants can show marked deviations from that expected for the 'average' EO size if the distribution of EO sizes is bimodal (i.e. for mixed commercial surfactants).

Whilst studies on linear alkyl surfactants are common, in recent years branched-chain nonionic surfactants have been studied as well as surfactants with novel head groups and surfactant mixtures.

The phase behavior of a series of mid-chain substituted surfactants with the general structure $CH_3(CH_2)_4CHEO_x(CH_2)_5CH_3$ where $x=3, 4, 6, 8, 10$ (denoted s-$C_{12}EO_x$) has been studied by optical microscopy using the penetration technique [122]. These have much more bulky hydrophobic groups and show clear differences from the behavior of the linear materials described above, as expected from the packing constraints. No liquid crystal phases are seen for s-$C_{12}EO_3$, just the coexistence of W and L_2. For s-$C_{12}EO_4$, a V_2 reversed bicontinuous cubic and a L_α phase are seen, both of which melt at low temperature (8.0 and 0.5 °C respectively) in addition to L_2 and W. On increasing the EO size to that of s-$C_{12}EO_6$, L_α and L_3 phases are observed. The L_3 phase exists between 35.3 and 53 °C on the lower surfactant concentration side of the L_α phase which exists up to 48.8 °C. L_1 micellar solution is seen only below 5.9 °C with a cloud temperature below 0 °C.

The next compound s-$C_{12}EO_8$ (Fig. 20), shows similar phase behavior to s-$C_{12}EO_6$

with the melting temperatures of the L_α and L_3 phases increased (to 63.5 and 66.7 °C respectively). A micellar solution with a cloud point at 19.2 °C is present. Finally, s-$C_{12}EO_{10}$ (Fig. 21) gives H_1, V_1 and L_α phases on penetration at 0 °C. The H_1 and V_1 phases melt to micellar solution at <10 °C, while the L_α phase exists up to 74.2 °C. An L_3 phase also exists between 69.0 and 79.0 °C with the cloud point being 46.4 °C. Clearly the pattern of behavior resembles that of the linear surfactants, but an increase EO size is required to observe a particular phase sequence. It is initially surprising that mesophases do form with the small (maximum) hydrophobic chain of C_7, where none are seen with simple linear surfactants of this chain length. This may indicate that the micelles are more monodisperse in size for dialkyl surfactants than for monoalkyl surfactants because local molecular 'protrusion' of monomers out of the micelle is damped by the hydrophobic effect on two

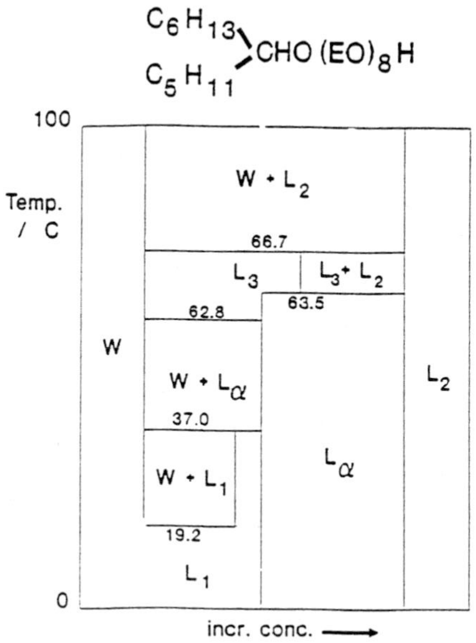

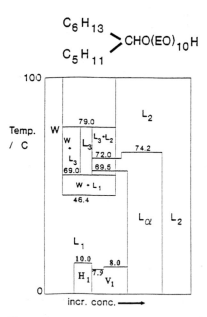

Figure 20. Schematic phase diagram of s-$C_{12}EO_8$/water (reproduced from [122]). Symbols as for Figs. 14 and 17.

Figure 21. Schematic phase diagram of s-$C_{12}EO_{10}$/water (reproduced from [122]). Symbols as for Figs. 14 and 17.

alkyl chains, rather than one. Thus the micelle surface is smoother for dialkyl surfactants then for monoalkyl derivatives.

Another series of 'branched' (dialkyl) nonionic surfactants has been studied [123] having the general formula $C_kC_nGE_8M$, where C_k and C_n denote different alkyl chains, with $k=4$ for n-butyl (C_4) and tert-butyl (C_{4-t}) and $n=10$ or 12. G denotes a glyceryl unit and E_8M denotes octaoxyethylene monomethyl ether. This work follows on from earlier work on similar compounds [124–126]. The structural differences between n- and tert-butyl chains within the asymmetrical V-isomers lead to a different phase behavior as expected by the packing constraints. Both C_{10} isomers show H_1, V_1 and L_α phases, with the existence range and maximum temperature of the L_α much greater for the n-butyl compound (8–80 wt% surfactant compared to 68–77 wt% and 50.1 °C compared to 11.6 °C). The H_1 and V_1 phase ranges are similar for the two compounds, but with the V_1 phase existing to higher temperature for the n-butyl compound (Fig. 22). This compound also shows a L_3 phase above the dilute L_α phase up to 50.1 °C. The n-butyl isomer of C_{12} has very similar phase behavior to the C_{10} compound with all the phases existing to slightly higher temperatures.

Previous to the above work, the authors had investigated a homologous series of nonionic surfactants containing two hydrophilic chains [127], and having the general formula $C_nG(E_mM)_2$ where C_n denotes the alkyl chain ($n=10–16$), G=glycerol and E_mM=oligo-oxyethylene mono-methyl ether ($m=3–5$) (Fig. 23). The compound $C_{10}G(E_4M)_2$ exhibits no mesophases but has a cloud temperature of 54.2 °C. The other compounds studied (for which $n=12$, 14, 16) all exhibit two I_1 cubic phases and a hexagonal phase (Fig. 23). The temperature at which the phases melt increases with

increasing length of alkyl chain (the melting temperatures of the H_1 phase in $C_nG(E_4M)_2$ for $n=12$, 14, 16 are 2.1, 31.6 and 38.6 °C respectively). The cloud point is virtually unaffected. However, on increasing the length of the head group the cloud point increases (38.7, 53.0 and 61.8 °C for $C_{14}G(E_mM)_2$ where $m=3$, 4, 5 respectively.

All of these studies confirm that the pattern of phase behavior shown for the conventional surfactants generally holds at least qualitatively whatever the head group or chain structure provided that a proper consideration of the alkyl chain packing conditions is made.

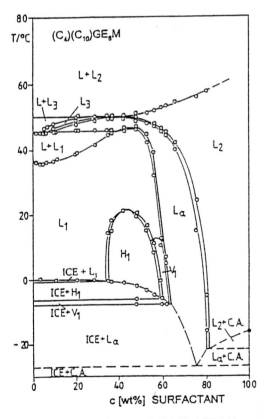

Figure 22. Phase diagram of $(C_4)(C_{10})GE_8M$/water (reproduced from [123]). (**L**, very dilute surfactant solution; **C. A.** indicates the presence of solid surfactant; otherwise as Figs. 14 and 17.)

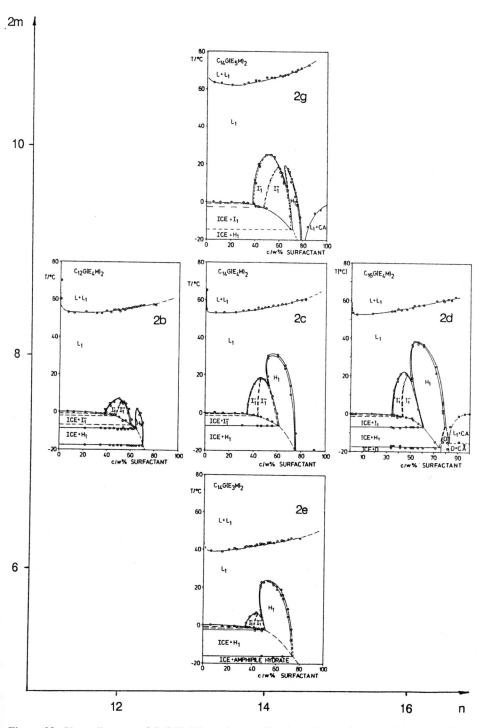

Figure 23. Phase diagrams of $C_nG(E_mM)_2$ and water. The dependence of phase behavior on alkyl chain length n and the number of oxyethylene units $2m$ per surfactant (reproduced from [127]). (**D**, lamellar phase; otherwise as Figs. 14, 15 and 22).

A severe test of this is where the flexible alkyl chain is replaced by the rigid hydrophobic steroid skeleton. When sufficiently long ethoxy chains are attached to the cholesterol OH, phase behavior similar to more common alkyl derivatives is observed [128]. For a polydisperse EO_{13} chain L_1, H_1, L_α and L_2 phases are seen along with a region of clouding above 85 °C. The H_1 phase exists between 18 and 60% surfactant from at least room temperature to 65 °C. The L_α phase exists from 35 to over 85% and to at least 100 °C. When the EO length is increased to 35 no lamellar phase occurs but an I_1 cubic phase is seen between 23 and 67% surfactant and to over 100 °C. The H_1 phase is seen between 67 and 83% and up to ≈ 80 °C. For a chain length of 50 ethoxy groups only the I_1 phase is seen between 27 and 93% up to 100 °C. Estimating the cholesterol length as being equivalent to about C_{15}, and a volume of about C_{20}, this is well in agreement with the behavior of the normal surfactants. It might be thought that the bulky nature of the cholesterol moiety would prevent packing into spherical micelles, but this is clearly not the case.

An exciting development in recent years has been the systematic study of polymeric surfactants rather than the occasional studies reported previously. Polysurfactants can be made via polymerization of the chain termini (head groups or tails) or from hydrophilic and hydrophobic blocks (called side-chain or block copolymer surfactants respectively). Over the years, Finkelmann and Lühmann have reported the phase behavior of a variety of nonionic surfactants and polymers formed from them. In the latter category, they report [129] on the liquid crystalline properties of side chain polymers comprising a poly[oxy(methylsilyl)] backbone ('P' = $O-SiCH_3-$, average degree of polymerization 95) with side chains of $(CH_2)_{10}C(O)O(CH_2CH_2O)_8CH_3$. These are produced by esterifying the polyglycol with 10-undecanoic acid which forms the hydrophobic part of the amphiphilic molecule. In water (Fig. 24), the polymer exhibits large regions of H_1 (36–75%, $-15 \rightarrow 49$ °C) and L_α (64–90%, $-5 \rightarrow 64$ °C) phases separated by a narrow band of V_1 phase ($-10 \rightarrow 45$ °C). A region of clouding is seen up to 90% polymer. The clouding temperature increases with concentration from a minimum of 53 °C. The changes induced by polymerization are similar to those expected for increasing the alkyl chain size to about C_{14}, a rather small alteration. The high flexibility of the polyoxysilyl backbone is probably responsible for this. A more rigid backbone would lead to larger changes.

When the side chain is modified to include a rigid rod-like biphenyl moiety [130] $[(CH_2)_3-\Phi-\Phi-O-(CH_2CH_2O)_9CH_3]$ and the degree of polymerization reduced to 55 ($P_{55}C_3BiE_9$), subtle differences are seen in the phase behavior. H_1 (34–75%, $-2 \rightarrow 68$ °C) and L_α (60–95%, $20 \rightarrow 100$ °C) phases are still seen but with no V_1 phase between them. Clouding is observed up to 58% polymer between 66 and 82 °C. Above this a biphasic $L_1 + L_\alpha$ region is seen to $\approx 80\%$.

The phase behavior of related polymers and monomers containing the rigid biphenyl moiety are studied in a later paper [131]. The phase behavior of the monomeric surfactants is generally compatible with that of common nonionic surfactants (especially ethylene oxide alkyl ethers). They exhibit I_1 (sometimes two), H_1 and L_α phases as well as clouding. The polymers, which have an average degree of polymerization of 55, nearly all exhibit H_1 and L_α phases, whereas the I_1 phase is only seen in one ($PC_3BiE_{11})_{55}$. One major difference between the polymer and monomer phase behavior is the appearance of a nematic phase (N_c) built up of rod-like micelles in a num-

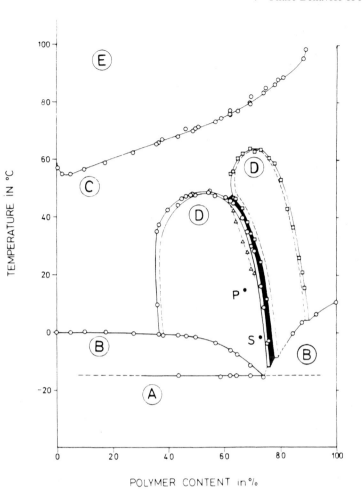

Figure 24. Phase diagram of $P_{95}C_{10}EO_8$/water. (A) Heterogeneous mixed crystals, (B) heterogenous melt, (C) homogenous isotropic solution, (D) homogenous mesomorphous phases, (E) heterogenous isotropic liquids, miscibility gap with lower consolute point (reproduced from [129]).

ber of the polymers. This is seen over a narrow concentration range between the L_1 and H_1 phases, melting at temperatures below the H_1 phase. Also reported is the phase behavior of oligomers of $P_rC_3BiE_9$ for $r=3–6$ and 13.4. H_1 and L_α phases are seen for all, but the I_1 is not seen for $r=13.4$ (or $r=55$, i.e. the polymer). The nematic phase is seen for $r=6$ ($<50.0–52.2\,°C$) and for $r=13.4$ ($38.6–58.2\,°C$) and $r=55$ ($34.8–58.2\,°C$).

Lühmann and Finkelmann also reported what was the first nematic phase observed in a binary nonionic/water system [132]. It was formed by the surfactant $H_2C{=}CH{-}CH_2{-}O{-}\Phi{-}\Phi{-}O{-}CH_2{-}COO(CH_2CH_2O)_7CH_3$.

It is observed between the L_1 and L_α in a narrow band of 34–38% surfactant between 7.5 and 23.4 °C and is made up of disc shaped micelles (Fig. 25). The L_α phase exists up to $\approx72\%$ surfactant (and to $\approx80\%$ in a biphasic region with water). Clouding is seen over a wide concentration range (up to $>90\%$ surfactant) with a lower critical temperature of 33.2 °C.

Polyhydroxy surfactants have recently received some attention as alternatives to EO materials. Polyhydroxy compounds are more strongly hydrated than EO groups hence the mesophases exist to high temperatures. Also, the intermolecular H-bonding

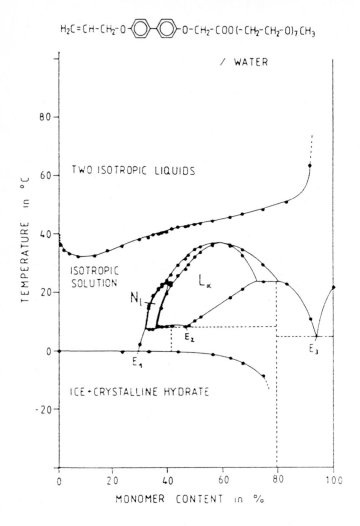

Figure 25. Phase diagram of H₂C=CHCH₂OBiEO₇/water (reproduced from [132]). (**N₁**, disc nematic phase; **E₁**, **E₂** and **E₃**, eutectic points; otherwise as Fig. 14).

results in the occurrence of thermotropic mesophases to above 100 °C. Generally one requires at least a hydrophilicity of about 2x OH groups for surfactant properties, but di- or trihydroxy compounds usually form reversed phases. For water continuous phases a larger polyhydroxy group is required. An easily synthesized class of materials are the alkyl glucamides. Unfortunately single chain materials have high Krafft temperatures.

The properties of new surfactants possessing two *n*-alkyl chains and two glucamide head groups have been reported [133]. These have the general formula $(C_nH_{2n+1})_2C[CH_2NHCO(CHOH)_4CH_2OH]_2$ with $n=5-9$ (abbreviated to di-(C_n-Glu). Di-(C_5-Glu) shows H_1, V_1 and L_α phases on penetration at $\approx 5°$. The H_1 and V_1 phases melt above 80° with the L_α remaining to >100 °C. Di-(C_6-Glu) shows the same sequence of phases as above with the H_1 phase first seen at 56 wt% surfactant (coexisting with L_1 micellar solution). A single H_1 region exists between 62 and 71 wt% surfactant and the V_1 region between 71 and 75 wt%. Both phases melt between 70 and 85 °C. The L_α phase exists at higher concentrations and to >90 °C.

Di-(C$_7$-Glu) also has a biphasic H$_1$/L$_1$ phase forming at 62 wt% surfactant existing to ≈60 °C. The H$_1$ phase melts over the range 62–80 °C. Between the H$_1$ and L$_\alpha$ phases is another biphasic region between 71 and 75 wt%. This, and the L$_\alpha$ phase exist to >90 °C. A two phase loop occurs at low surfactant concentrations and temperatures (<8 wt% and <50 °C). The authors believe this to be the first observation of an upper critical solution temperature for a nonionic surfactant. The shape of the upper consolute loop is extremely unusual suggesting remarkable changes in micellar interactions … or the presence of impurities. One would be pleased to see these observations verified by other groups. Di-(C$_8$-Glu) shows only a L$_\alpha$ phase but with a wide miscibility gap to a dilute aqueous solution. This is not remarkable.

Seddon et al. have studied the lyotropic (and thermotropic) phase behavior of n-octyl-1-O-β-D-glucopyranoside and its thio derivative n-octyl-1-S-β-D-glucopyranoside [134]. Both form L$_1$ micellar solutions and V$_1$ (Ia3d) and L$_\alpha$ phases in water with the β-OG also forming an H$_1$ phase. For β-OG the L$_\alpha$ phase exists from 80 to 100% surfactant and to over 120 °C, the V$_1$ between 73 and 80% and up to ≈60 °C and the H$_1$ between 64 and 73% up to 35 °C. All melt to L$_1$ micellar solution. The L$_\alpha$ phase of β-thio-OG also exists up to 100% surfactant (from 75%) and to ≈135 °C. The V$_1$ phase exists between 72 and 90% surfactant (i.e. it melts to give the lamellar phase above 75%). The temperature at which it melts varies from 0° at 90% to 65 °C at 75%. N-octyl-β-D-glucoside exhibits similar phase behavior to β-OG in water [135] with H$_1$, V$_1$ and L$_\alpha$ phases existing over very similar concentration and temperature ranges.

Hall et al. studied a series of 1-(alkanoyl-methyl amino) 1-deoxy-D-glucitols (C$_8$–C$_{12}$, C$_{18=}$) [136]. These are also called N-meth-

yl glucamides, the presence of the N-methyl group increasing solubility compared to the glucamides referred to above. They form both thermotropic and lyotropic mesophases. Schematic phase diagrams were produced for all the compounds using the phase penetration technique with a complete phase diagram being produced for the C$_{10}$ compound (Fig. 26). The C$_8$ derivative (C$_8$G) forms an H$_1$ phase when contacted with water. It exists between 0 and 39 °C, melting to a micellar L$_1$ phase. C$_9$G forms an extensive range of mesophases, H$_1$ (15–69 °C), V$_1$ (36–56 °C) and L$_\alpha$ (44–76 °C). All melt to an L$_1$ phase. C$_{10}$G observes the same phase sequence as C$_9$G. The H$_1$ phase forms between 42 and 75% surfactant (30–81 °C), the V$_1$ between 67 and 79% (42–80 °C) and the L$_\alpha$ between 75 and 100% (46–>100 °C). C$_{11}$G exhibits an H$_1$ phase between 33 and 86 °C and V$_1$ and L$_\alpha$ phases from 39.5 °C and 49 °C respectively to >100°C. Similarly for C$_{12}$G, H$_1$ (45.5–75 °C), V$_1$ and L$_\alpha$ (50 °C and 56 °C respectively to >100°C). The phase behavior for C$_{18=}$G is much simpler exhibiting only an L$_\alpha$ phase from 22 °C to >100 °C.

Raaijmakers et al. studied the mesogenic properties of some 3-O-alkyl derivatives of D-glucitol and D-mannitol [137]. Phase penetration scans have shown that the C$_{10–16}$ glucitol derivatives and the C$_{12}$ mannitol derivative all exhibit lyomesophases. 3-O-decyl-D-glucitol (C$_{10}$G) gives an H$_1$ phase on penetration at room temperature, and on heating a V$_1$ phase at 34 °C and a L$_\alpha$ phase at 46 °C. The solid bulk melts to L$_\alpha$ at 55 °C and all phases exist to at least 98 °C. C$_{12}$G gives an I$_1$ phase at 36 °C, H$_1$ (37 °C), V$_1$ (40 °C) and L$_\alpha$ (46 °C), with the solid melting to L$_\alpha$ at 65 °C. All phases exist to >100 °C with the exception of the H$_1$ which melts to an I$_1$ phase at 62 °C. C$_{14}$G gives only V$_1$ and L$_\alpha$ phases from 48° to >100 °C. Below T_{pen} the L$_\alpha$ phase cools to give a

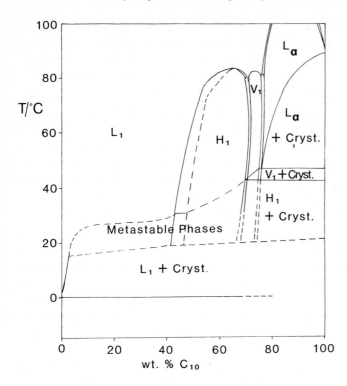

Figure 26. Phase diagram of C_{10} glucitol/water (reproduced from [136]). Symbols as for Fig. 14.

L_β gel phase, a transition that is reversible on reheating. $C_{16}G$ gives a L_α phase from 52 to >100 °C. A reversible L_α–L_β transition is observed on cooling at 44 °C. The pattern of phase formation in 3-O-dodecyl-D-mannitol ($C_{12}M$) is similar to $C_{12}G$ with all transition temperatures approximately 20 °C higher, apart from the melting temperature of the solid (97 °C).

Finkelmann and Schafheutle studied monomeric and polymeric amphiphiles containing a monosaccharide headgroup [138]. The monomer, N-$D(-)$-gluco-N-methyl-(12-acryloyloxy)-dodecane-1-amide, exhibits L_1 (>52 °C) and L_α phases (65–88%, 57–87 °C), with a biphasic region separating the two. The polymer, poly-(N-$D(-)$-gluco-N-methyl-(12-acryloyloxy)-dodecane-1-amide) again exhibits L_1 and L_α phases but these are formed at −0.1 °C. The lamellar phase exists between 37 and 100% polymer, and up to 184.8 °C. It is separated from

the L_1 phase by a narrow biphasic region. There is a region of clouding (L_1+L_2) reaching from ≈ 0 to 31% polymer with a lower critical temperature of 17.9 °C at 4.6% and an upper critical temperature of 105.5 °C at 9.6% polymer.

Most of the above studies involve changes in alkyl chain size, with no alteration of the head group. In a very important early study, Sagitani et al. reported the behavior of pure alkyl polyglycerol $[C_{12}O(CH_2CHOHCH_2O)_nOH]$ surfactants with an ether-linked C_{12} alkyl chain, with $n=1$–4 [138a]. The compounds with $n=1$, 2 give just a lamellar phase to over 120 °C, while with $n=4$, a micellar solution and hexagonal phase occur. For $n=3$ there is a cloud point at ca. 53 °C, with micellar solution and lamellar phase being present. The shape of the L_1/L_α phase boundary indicates the presence of H_1 below the cloud point, remarkably similar behavior to $C_{12}EO_6$ as the authors point out. In fact, poly-OH compounds do resemble EO nonionics, but with a very much reduced influence of temperature.

Much attention has been given recently to commercial polyglucoside surfactants (alkyl polyglucosides, APG's) [138b, c, d], where the average number of 'poly'-glucoside units (termed d.p.) falls within 1–3. From the above studies, one expects short chain derivatives with d.p. = ca. 1 to form L_1, H_1 and L_α phases, while the low stability of the H_1 phase suggests that long chain compounds will form L_α dispersions. Compounds with d.p. = 2 will give H_1 and L_α phases, while with d.p. = 3 or more the I_1 cubics will appear. Multi-phase coexistence is likely, along with the occurrence of metastable, sticky, viscous phases at high concentrations.

5 Block Copolymer Nonionic Surfactants

Block copolymer surfactants show qualitatively similar phase behavior with temperature to conventional materials. However, solubility requires the presence of branched chains and/or ether links to reduce the occurrence of crystalline polymer with a high melting point. It must be emphasized that commercial polymeric surfactants are polydisperse in both alkyl chain and EO blocks, hence a range of molecular weights and head group sizes will be present. This will certainly modify the behavior compared to pure low molar mass surfactants, but should not qualitatively change it. It could lead to the occurrence of multiphase regions such as H_1 and L_α or cubic regions, rather than single phases and 'intermediate' structures that might occur with very monodisperse polymers.

Alexandridis et al. have studied a number of block copolymers of ethylene oxide (EO), propylene oxide ($-OCH_2CHCH_3-$, PO) and butylene oxide ($-OCH_2CH(CH_2CH_3)-$, BO) in water and in ternary systems with p-xylene. The polypropylene oxide and butylene oxide blocks are the hydrophilic portions. Obviously, the short methyl and ethyl branches will need to be included in any packing constraints considerations.

The phase behavior of three ethylene oxide/propylene oxide ABA copolymers, $(EO)_6(PO)_{34}(EO)_6$ (L62), $(EO)_{13}(PO)_{30}(EO)_{13}$ (L64) and $(EO)_{37}(PO)_{58}(EO)_{37}$ (P105) has been studied [139]. The number of mesophases formed increases with the poly(ethylene oxide) content (Fig. 27). L62 exhibits only a L_α phase, between 51 and 76% polymer and from < 20° up to 65 °C. Two phase regions of lamellar phase with both L_1 and L_2 solution phases exist, with the former first seen at < 30% polymer. L64 has a small region of H_1 phase between 46 and 54% polymer. It forms at 22 °C and melts at 46 °C. The L_α phase exists between 48 and 82% polymer, from below 10° to 85 °C. Again, the two phase regions are fairly wide. P105 exhibits an I_1 cubic phase between 26 and 44% polymer, melting at 60 °C. The H_1 (47–67%) and L_α (67–88%) are separated by a two phase region and exist to > 85 °C. Note that simple AB block copolymers are expected to resemble even more closely the conventional surfactants. (The ABA blocks are likely to assume a U type conformation for the hydrophobic block.)

The ternary system of L64/P105/water was also studied [140] (Fig. 28). At 25 °C a lamellar phase is formed along the L64–water axis between 61 and 79% polymer, and between 73 and 87% polymer on the P105–water axis, and extends all the way from one axis to the other. Upon increasing the water concentration the hexagonal phase is formed. This extends between

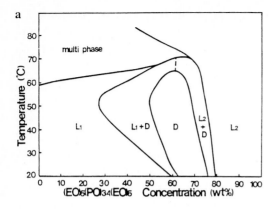

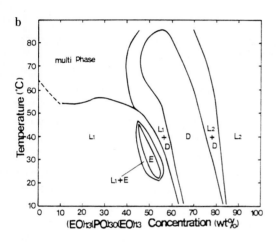

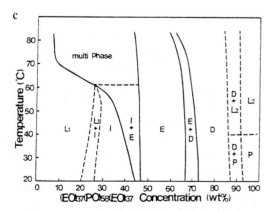

Figure 27. Phase diagrams of (a) $(EO)_6(PO)_{34}(EO)_6$ (L62), (b) $(EO)_{13}(PO)_{30}(EO)_{13}$ (L64), and (c) $(EO)_{37}(PO)_{58}(EO)_{37}$ (P105) block copolymers and water (reproduced from [139]). (**E**, normal hexagonal phase; **I**, cubic phase; **P** indicates presence of solid polymer; otherwise as for Figs. 14, 17 and 23.)

52–55% on the L64–water axis and 47–67% on the P105–water axis. Additionally there is an I_1 cubic phase between 28 and 43% P105, and a narrow melted V_1 phase between 58 and 59% L64 between the H_1 and L_α regions.

The L64/water (D_2O)/p-xylene ternary system exhibits L_1, L_2 (containing a high p-xylene to water ratio), L_α, H_1, H_2 and V_2 phases at 25 °C [141]. Most of the phase diagram consists of two and three phase regions (10 and 6 respectively). The L_1 solution phase forms up to $\approx 46\%$ polymer, and can solubilize up to 4% xylene. The polymer is miscible with xylene in all proportions along the whole water-lean side of the phase diagram (the L_2 region): 12% water can be incorporated into the L_2 phase. The L_α phase forms on the polymer–water axis between 61 and 81% polymer and extends well into the phase diagram, accommodating up to 20% xylene. The H_1 phase is formed on the polymer–water axis between the L_1 and L_α phases (47–57% polymer). Like the L_1 phase, it can solubilize only a small amount of xylene. There is a fairly extensive H_2 phase between the L_α and L_2 phases at polymer concentrations of 43–78%, and with water concentrations of 13–22%. The V_2 phase forms in a small region between the L_α and H_2 phases at $\approx 70\%$ polymer. It has the Ia3d space group. One other phase seen is an isotropic liquid phase (at 58–59% polymer) denoted here as L'. It has low viscosity and is an L_3 sponge phase.

Pluronic 25R4 has the structure $(PO)_{19}(EO)_{33}(PO)_{19}$, and has similar ternary phase behavior in water/p-xylene as Pluronic L64 which has the opposite block sequence and a comparable molecular weight (see above). At 25 °C the 25R4/water/p-xylene ternary phase diagram exhibits the same phases as L64 [142]. Again, the phase diagram is dominated by two and

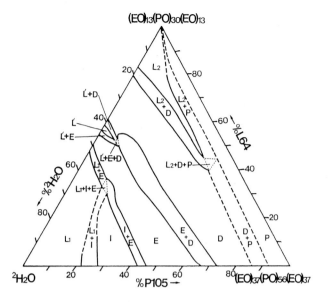

Figure 28. Phase diagram of L64/P105/ water (reproduced from [140]). (**L′**, **L₃**, sponge phase; otherwise as for Figs. 14, 17, 23 and 27.)

three phase regions. The L_α phase is formed along the polymer–water axis at polymer concentrations between 61 and 79%, and can accommodate up to 25% p-xylene. The H_1 phase is formed between the L_1 and L_α phases at polymer concentrations of between 43 and 56%: 3–4% xylene is actually required for the phase to form. The H_2 phase is formed between the L_α and L_2 phases, at polymer concentrations of between 48 and 73% polymer, with the water concentration varying in the range 13–18%. The V_2 phase forms in a narrow concentration range of 65–68% polymer. It is thought to have a space group of Ia3d.

In the $(EO)_{17}(BO)_{10}/D_2O/p$-xylene Alexandridis et al. claimed the first, to their knowledge, reverse micellar cubic phase in a 'typical' ternary amphiphile/water/oil system [143] before studying and reporting the system in more detail [144]. They report six mesophases and two solution phases at 25 °C. On increasing the polymer concentration along the polymer–water axis the phase sequence L_1 (<22% polymer), I_1 (23–37%), H_1 (42–54%), L_α (62–84%) is observed. At even higher concentration a polymer-rich paste-like phase is seen. The L_α phase is the most extensive of the mesophases and can accommodate up to 16% xylene. The I_1 phase most likely has the Im3m space group. Along the polymer–xylene axis the same sequence of structures (sphere → cylinder → plane) is seen for the reversed phases. The L_2 phase exists below 75% polymer, the I_2 between 47 and 62%, the H_2 between 45 and 84%, the V_2 at ~80% polymer between the H_2 and L_α phases. The I_2 phase has the Fd3m space group, and the V_2 Ia3d. Note that all the reverse mesophases require the presence of water to form (whereas the normal mesophases can form without xylene).

A comparison of the phase behavior of PEO/PPO and PEO/PBO copolymers in butanol/water was recently reported [145]. Note that in this paper poly(butylene oxide) is referred to as poly(tetrahydrofuran). Pluronic F127, $E_{100}P_{70}E_{100}$, exhibits L_1, L_2, I_1, H_1 and L_α phases. The L_1 phase exists along the polymer–water axis up to 20% polymer, accommodating up to 10% butanol. The I_1

cubic phase supercedes the L_1 phase and exists up to 65% polymer. Its ability to solubilize butanol decreases with increasing polymer concentration. The H_1 phase is formed at above $\approx 20\%$ polymer but at least 15% butanol is required for it to form. A maximum of $\approx 23\%$ butanol can be accommodated in the H_1 phase. As the polymer concentration is increased the butanol concentration required to form the H_1 phase decreases and between 70 and 80% polymer no butanol is required. The L_α phase is stable in the 20–30% polymer and 25–30% butanol ranges. A large L_2 region extends from the butanol rich corner down to 20% butanol. Note that when butanol is replaced by hexanol to L_1, L_2 regions are reduced and the I_1 phase is still observed along with a birefringent phase that was not investigated.

$E_{100}B_{27}E_{100}$ is of a similar molecular weight to $E_{100}P_{70}E_{100}$ but with a more hydrophobic middle block. Only a single one-phase region is observed extending from the water-rich corner over a middle region to the butanol-rich corner. No liquid crystalline phases are observed. The $E_{17}B_{27}E_{17}$/butanol/water system also shows only a single one-phase region and no mesophases.

6 Zwitterionic Surfactants

Zwitterionic surfactants broadly resemble nonionic materials but because two bulky charged groups are involved the head groups are large. An exception is dimethyldodecyl amine oxide [146] which has a compact head group. This forms H_1, V_1 and L_α phases. The H_1 (termed middle phase in this paper) exists between 35 and 65% surfactant and from <20 °C to >105 °C. The V_1 (viscous isotropic) phase has a narrow range

of existence (65–70% surfactant) and melts at a slightly higher temperature than the H_1. The L_α (neat) phase exists between 70 and 80% surfactant at 20° and up to 95% surfactant at 90 °C. It melts at >140 °C. The C_{14} compound exhibits similar phase behavior [147] but also has a nematic phase (N_d) between 31 and 35% surfactant up to 60 °C.

A more typical surfactant is (dodecyldimethylammonio)propanesulfonate [148]. This forms an I_1 cubic phase between 49.5 and 63% surfactant from ≈ 20 °C up to ≈ 85 °C. A hexagonal phase (65–87% surfactant) forms at a slightly higher temperature but exists up to >160 °C. A lamellar phase is seen at very high concentrations and above 170 °C. Below this is a hydrated solid surfactant phase which when viewed under a polarizing microscope has the optical appearance of a gel phase.

7 Ionic Surfactants

Ionic surfactants have been studied over a much longer period than nonionic materials, but the long-range electrostatic interactions and the insolubility of long chain materials have resulted in the general picture of mesophase behavior emerging only rather slowly. Typical surfactants with a single charge ionic group usually give small globular micelles and H_1 as the low concentration mesophase. As chain length increases (and if solubility allows), then rod micelles form and a rod nematic phase can occur at the L_1/H_1 boundary. Occasionally, if the counter-ion is highly dissociated, an I_1 phase can occur as the first mesophase. Otherwise two charged groups are required for I_1 formation. At higher concentrations the sequence $H_1/V_1/L_\alpha$ changes to H_1/Intermediate/L_α (or a combination of the two) as the alkyl chain size increases. In this region there can be long lived metastable phases that are difficult to work with and prevent easy determination of equilibrium boundaries.

Sodium dodecyl sulfate is a very commonly used surfactant, yet for many years its mesophase behavior was not studied in detail. One reason for this is the number of phases to be dealt with. Besides the lamellar and hexagonal phases there are a number of intermediate phases existing over narrow concentration ranges [78]. The complete phase diagram was published by Kékicheff et al. [62] (Fig. 29). (Note that in this paper the hexagonal phase is represented by the symbol H_α.) Between the H_α and L_α phases they identify four intermediate phases, with biphasic regions separating most of the single phase regions. The H_α phase is first seen at 37% (with L_1) and at 40% in a single phase region at 25.7 °C. A two-dimensional monoclinic phase (M_α) exists with the H_α phase from 57–59% up to 58 °C. Note that this is the only mesophase region that does not exist up to ≈ 100 °C. On increasing surfactant concentration between 59 and 69% the phase sequence is M_α (40.1 °C), $M_\alpha + R_\alpha$ (rhombohedral), R_α (43.0 °C), Q_α (cubic, 47.8 °C), $Q_\alpha + T_\alpha$ (tetragonal), T_α (48.5 °C), $T_\alpha + L_\alpha$, L_α (49.8 °C). The L_α phase exists up to 87% surfactant above which it coexists with a crystal phase. Despite the extensive work involved in this diagram, there remain questions. For example, the disappearance of the $H_\alpha + M_\alpha$ two phase region is difficult to understand. Note that this system typifies those where equilibration between different states is expected to be slow, as mentioned above.

Another system showing an intermediate phase between the H_1 and L_α phase is the caesium tetradecanoate–water [149]. The phase behavior between 24 and 80 °C was studied (Fig. 30). The H_1 phase is first seen in a two phase region with L_1 micellar phase at 33% surfactant and in a single phase region between 37 and 67%. Above 65 °C the H_α phase is replaced by a two phase $H_\alpha + V_1$ region. However, below this temperature, a ribbon phase (R) is formed. The most likely structure of this phase is rodlike aggregates with an elliptic cross section. At 24 °C the R phase exists from 62–72% surfactant, with the concentration range decreasing with increasing temperature. Below 39 °C a biphasic $R + L_\alpha$ region exists between 72 and 75% surfactant, and between 39 and 65 °C the two phase region is $R + V_1$. The V_1 region is very narrow ($\approx 2\%$ surfactant) and is bordered by a two phase $V_1 + L_\alpha$ region up to 75% surfactant. The lamellar phase exists to over 90% surfactant. That the sequence of phases can be very complex and varies with chain size is illustrated by the 'penetration scan' table reported by Rendall et al. [150]. Here the 'intermediate' phase types have not been identified, but the general pattern and complexity of behavior is clear. One should also caution that minor levels of surface active impurities are likely to have a marked influence on the cubic/intermediate phases, particularly if the impurities are uncharged cosurfactants (often likely to be present from the chemical synthesis).

One of the first major surveys into the phase behavior of cationic surfactants was done by Blackmore and Tiddy who did penetration scans on a variety of surfactants including the $-NH_3^+$, NMe_3^+, NEt_3^+, NPr_3^+ and NBu_3^+ series [151]. Previous studies had mostly been limited to alkylammonium and alkylmethylammonium salts [152, 153]. These show a qualitatively similar behavior to that of the anionic systems, with the sequence $H_1/V_1/L_\alpha$ for short chain derivatives being replaced by $H_1/Int/L_\alpha$ for long chain compounds. For the intermediate chain lengths, V_1 phases replace the intermediate structures as temperature increases.

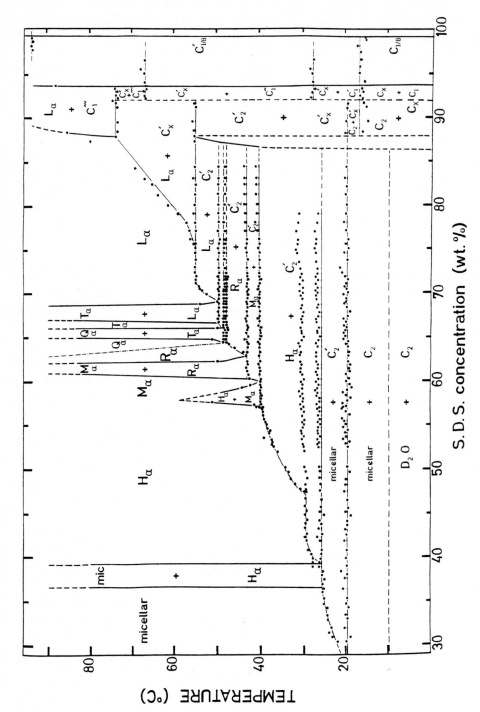

Figure 29. Phase diagram of SDS/water (reproduced from [62]). (H_α, hexagonal phase; M_α, two-dimensional monoclinic phase; R_α, rhombohedral phase; Q_α, cubic phase; T_α, tetragonal phase; C, C' and C'' refer to different polymorphic varieties for the same SDS hydrate; otherwise as for Fig. 14.)

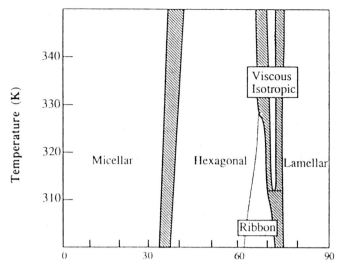

Figure 30. Phase diagram of CsTD/water (reproduced from [149]).

Weight % Cesium n-Tetradecanoate

Surfactants containing divalent headgroups have received little attention. Hagslätt et al. have investigated a number of these. Dodecyl-1,3-propylene-bis(ammonium chloride) (DoPDAC) has a headgroup consisting of two protonated amine groups separated by a propylene group [154]. The surfactant shows conventional phase behavior in water (D_2O) (Fig. 31) forming L_1, I_1

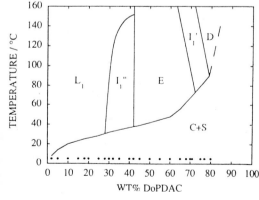

Figure 31. Phase diagram of DoPDAC/water (reproduced from [154]). (C+S indicates the presence of solid surfactant; otherwise as for Figs. 14, 15, 23 and 27.)

cubic (denoted I_1'), H_1 hexagonal (denoted E), V_1 cubic (denoted I_1') and L_α lamellar (denoted D) phases. The I_1 cubic phase forms at 30% surfactant at $\approx 30\,°C$ and exists up to $\approx 150\,°C$. The hexagonal phase is seen at 42% surfactant from 38 to $>160\,°C$. Between 65 and 72% surfactant the H_1 phase melts to the V_1 cubic phase (which forms above $\approx 75\,°C$). Similarly above 72% surfactant the V_1 phase melts to a L_α lamellar phase. At 79% the L_α forms at $90\,°C$. Note that the Krafft boundary has a rather large slope. Usually for monovalent surfactants the Krafft boundary is flat or has only a small slope [40].

They also studied the ternary phase diagram of the divalent surfactant dipotassium dodecylphosphate (K_2DoP), monovalent surfactant potassium tetradecanoate (KTD) and D_2O [155]. At $25\,°C$ K_2DoP exhibits L_1 (0–37% surfactant), I_1 (37–67%), H_1 (67–75%) and two phase liquid crystal/hydrated surfactant crystal (75–100%). In the ternary system I_1 cubic is seen between 20 and 60% K_2DoP. Up to 30% KTD can be incorporated into the phase at lower concentrations of K_2DoP. On decreasing the percentage of D_2O along the D_2O/KTD axis, hexagonal (E), ribbon (R), intermediate (possibly orthorhombic) and lamellar (D) phases are seen. The hexagonal phase exists right down to the K_2DoP/D_2O axis (70–75% K_2DoP), whereas the maximum K_2DoP concentrations for the ribbon, intermediate and lamellar phases are 45, 24 and 22% respectively.

Amphitropic (also referred to as amphotropic) surfactants are compounds containing both lyotropic (e.g. quaternary ammonium headgroup) and thermotropic (e.g. oxycyanobiphenyl group) moieties [156]. Fuller et al. report on two such compounds. 10-(4′-cyano-4-biphenyloxy)decyltriethylammonium bromide exhibits only a L_α phase [157, 158] from 22 to >90% surfactant with a temperature range of 33–100 °C. The second compound N,N'-bis(5-(4′-cyano-4-biphenyloxy)pentyl)-(N,N,N',N')-tetramethyl hexane diammonium dibromide (5-6-5 OCB) exhibits two L_α phases with a re-entrant L_1 micellar phase between them (Fig. 32). The low concentration lamellar phase (L_α) exists between 18 and 64% surfactant, up to 65 °C. Between 35 and 47% surfactant the melting temperature of the L_α phase decreases from 65 to 50 °C with increasing surfactant concentration. At this concentration the high concentration phase (L'_α) appears and exists to >80% surfactant and >90 °C. Note that the absence of H_1 and V_1 phases is due to the difficulty of packing the bulky oxycyanobiphenyl group into spherical or rod-like micelle interiors.

The second of the above compounds is an example of a so-called 'gemini' surfactant. Recently there has been considerable interest in these surfactants which are double-headed cationic compounds in which two alkyl dimethyl quaternary ammonium groups per molecule are linked by a hydrocarbon spacer chain. (These are denoted as n-m-n

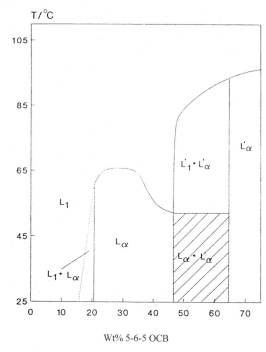

Figure 32. Phase diagram of 5-6-5 OCB/water (reproduced from [152]). Symbols as for Fig. 14.

surfactants where n is the length of the terminal alkyl chains and m the length of the spacer chain.)

In a second paper [158] Fuller et al. report the phase behavior of further m-n-m OCB surfactants and some straight chain 15-n-15 surfactants (n = 1, 2, 3, 6). Note that these compounds have terminal hydrophobic chains of the same length as the oxycyanobiphenyl compounds. The m-n-m OCB surfactants all give just a lamellar phase from <18°–>100 °C. Penetration scans on the 15-n-15 surfactants show that they all exhibit H_1, V_1 and L_α phases to >100 °C. Additionally, a nematic phase is seen for m = 1, 2 and intermediate phases for m = 1, 2, 3.

Zana et al. have investigated various properties of a series of gemini surfactants including their thermotropic and lyotropic phase behavior. They report that the 12-4-12 and 12-8-12 compounds form hexagonal and lamellar phases in water [159, 160]. For the latter compound the H_1 phase exists between ≈ 55 and 78% surfactant and the L_α phase between 82 and 97% surfactant. The phase behavior of a wide range of gemini surfactants ($n = 12-18$, $m = 2-20$) has now been studied and a wide range of phases observed [161]. Generally the phases are qualitatively in accord with packing constraint expectations, where the spacer group is included in the hydrophobic volume. However, unusual partial miscibility regions involving hexagonal, lamellar or concentrated surfactant solutions in coexistence with dilute solutions are also observed.

Pérez et al. [162] have studied a series of novel gemini surfactants with guanidyl headgroups (from the amino acid arginine) and are referred to as bis(Args). They are made up of two symmetrical N^α-acyl-L-arginine residues of 10 [$C_n(CA)_2$] or 12 [$C_n(LA)_2$] carbon atoms linked by covalent bonds to an α,ω-alkenediamine spacer chain of varying length (n, $n = 3, 6, 9$). In the $C_3(LA)_2$ and $C_6(LA)_2$/water systems an H_1 phase is seen at very low surfactant concentrations (lower than 5%) between 4 and 20 °C. Penetration scans have shown that in $C_9(LA)_2$ the H_1 phase is replaced by an L_α phase.

8   The Influence of Third Components

There is a vast body of data concerning the influence of third components on surfactant liquid crystals. Because of the possible great complexity of the inherent mesophase behavior this array of data can appear to be enormously difficult to rationalize. However, if we can consider the simple concepts described above (micelle formation, micelle shape/packing constraints and the nature of intermicellar interactions) then a reasonably simplified picture emerges, at least for the water continuous phases. This section does not attempt to be comprehensive, it simply reports selected examples of behavior to illustrate general concepts. The simplest way to show the changes in mesophase behavior is to employ ternary phase diagrams. The reader should recall that the important factors are the behavior as a function of surfactant/additive ratio, and the volume fraction of surfactant + amphiphile (where present) for mesophase formation.

8.1   Cosurfactants

Cosurfactants are 'surfactants' that are insufficiently hydrophilic to form micelles or mesophases with water alone, but can have dramatic effects when mixed with normal surfactants. Examples of such materials are alcohols, fatty acids, and long chain aldehydes. Depending on the strength of the polar group hydration there is greater or smaller incorporation of the cosurfactant into mesophases. The extent to which the polar group resides at the micelle surface also has a profound effect on the mesophases. Weakly polar groups such as methyl esters can both occupy the micelle interior and reside with the ester group at the surface. They could be classified as 'polar oils'. Thus cosurfactants span the range of properties from oils to surfactants. To illustrate the effects we show the behavior of a simple surfactant, sodium octanoate, with various additives taken from the extensive body of research produced by Ekwall, Fontell and coworkers [40, 163]. Sodium octanoate forms only hexagonal phase at room temperature, but with V_1 cubic and lamellar phases occurring at higher temperatures.

Figures 33 to 35 show the ternary phase behaviors of sodium octanoate with decanol, octyl aldehyde (octanal), and methyl octanoate.

All three additives have roughly the same alkyl chain length and volume as sodium octanoate (decanol is just a bit bigger). The phase behavior of octanol/sodium octanoate/water is available but the decanol system has received the most attention by far. We see that decanol mixes in the system to form a large lamellar region, and even an inverse hexagonal phase. This is because the alcohol group always resides at the water/

alkyl chain surface. Its contribution to the micelle area is ca. 12 Å^2, and since sodium octanoate has a head group size of ca. 58 Å^2 in the mesophases, packing constraint calculations suggest a hexagonal/lamellar transition at a sodium octanoate/decanol weight ratio of 7:3 in excellent agreement with the phase diagram. (A theoretical calculation of this type of phase behavior has been described by Wennerström et al. [164]). The other two additives are dissolved within the hexagonal region, presumably because they occupy the micelle interior to some extent. Octanol can reside at the micelle surface to

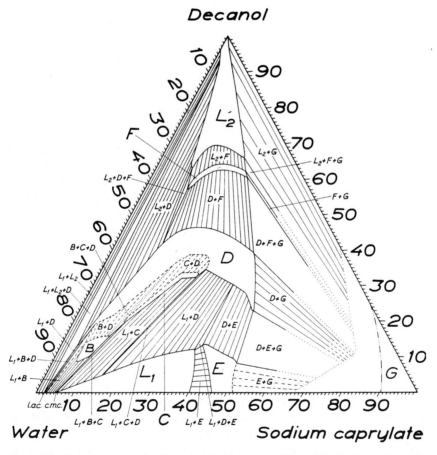

Figure 33. Phase diagram of sodium caprylate (octanoate)/decanol/water (reproduced from [40]). (**B**, 'mucous woven' lamellar phase; **C**, tetragonal phase; **F**, reversed hexagonal phase; **G**, isotropic phase; otherwise as for Figs. 14, 17, 23 and 27.)

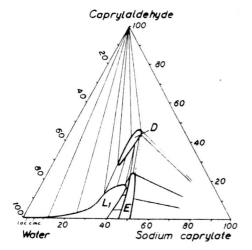

Figure 34. Phase diagram of sodium caprylate (octanoate)/caprylaldehyde (octanal)/water (reproduced from [40]). Symbols as for Figs. 14, 23, and 27.

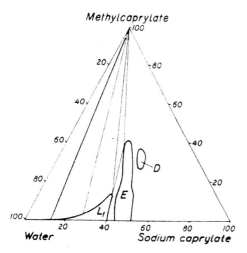

Figure 35. Phase diagram of sodium caprylate (octanoate)/methylcaprylate (octanoate)/water (reproduced from [40]). Symbols as for Figs. 14, 23 and 27.

some extent because it forms a lamellar phase which is in equilibrium with L_1, hence disc micelles can occur. Methyl octanoate mainly resides in the micelle interior, thus the lamellar phase does not have a boundary with L_1. The fact that the lamellar phase

appears rather than a V_1 cubic phase is a general observation. Both intermediate and bicontinuous cubic phases are much less frequently encountered in multicomponent systems than in binary surfactant/water mixtures.

8.2 Mixed Surfactants

The behavior of mixed surfactants (at least in water-rich regions) can be understood from considering the nature of interactions between head groups and packing constraints. A simple example is that of the commercial nonionic surfactants reported by Bouwstra et al. [165]. They compared the phase behavior of technical grade $C_{12}EO_{\langle 7 \rangle}$ and pure $C_{12}EO_6$. The ternary phase behavior with water and decane were studied and compared. Both exhibited I_1, H_1 and L_1 phases but only the $C_{12}EO_6$ gives a very small region of V_1 cubic phase. Note that the I_1 phase is not seen in the binary $C_{12}EO_6$/water system [37]. Also the H_1 phase in $C_{12}EO_6$ is composed of infinite long rods, whereas short interrupted rods were found in the hexagonal phase of $C_{12}EO_{\langle 7 \rangle}$.

Where ionic surfactants are involved, like charges show only small changes in phase structures but overall solubility can be greatly increased because generally surfactants do not form mixed crystals, hence each increases the solubility of the other. For surfactants with opposite charges, usually the mixed salt is insoluble. Where solubility does occur, the average head group size is much reduced because of the neutralized electrostatic repulsions. Typically cationic and anionic surfactants, which alone form hexagonal or I_1 cubic phases, form lamellar phases in 1:1 mixtures [166a]. Obviously, with dialkyl surfactants more complex behavior can occur.

8.3 Oils

Oils are solubilized into the interior of micelles where they allow the micelle to swell to a larger radius, hence giving rise to cubic (I_1) and hexagonal (H_1) phases at smaller a values than for the surfactant alone. Polar oils can also reside at the micelle surface to some extent, reducing micelle curvature and inducing the occurrence of lamellar and inverse phases. This behavior is typified by the behavior of the commercial nonionic surfactant nonylphenol-(probably branched)-decaethylene oxide with hexadecane and p-xylene [40].

Figures 36 and 37 show that the hexadecane induces the formation of an I_1 cubic phase, whereas the much more polarizable p-xylene, which can reside at the chain/water interface, induces a lamellar phase. As expected with the increase in micelle diameter, all the phases have their boundaries shifted to higher volume fractions with hexadecane addition.

For surfactants having small polar groups and bulky chains there can be extensive effects with the addition of oils. Alone with water these surfactants form reversed micelles and/or reversed mesophases. Large volumes of oil can be incorporated into these systems because of the possibility of swelling the alkyl chain regions in these oil-continuous phases (L_2, H_2, V_2). While extensive research has been carried out on this area, it appears to be much more complex than for the water-continuous phases. Each different surfactant type can show individual behavior according to the curvature properties of the surfactant layer.

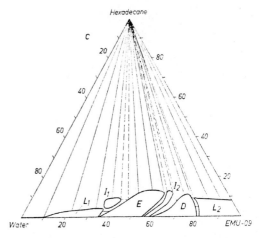

Figure 37. Phase diagram of EMU-09/hexadecane/water (reproduced from [40]). Symbols as for Figs. 14, 15, 23 and 27.

8.4 Hydrotropes

Hydrotropes are small, highly water soluble additives that increase markedly the solubility of other components, including surfactants, in water. They are employed very widely in industry. In fact, they only work when the 'insoluble' phase is a mesophase with high molecular mobility (e.g. polymer coacervate or surfactant mesophase). They include weakly self-associating elec-

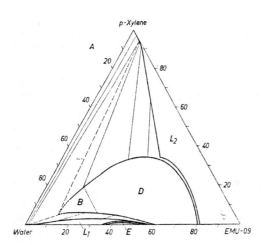

Figure 36. Phase diagram of EMU-09/p-xylene/water (reproduced from [40]). Symbols as for Figs. 14, 17, 23 and 28.

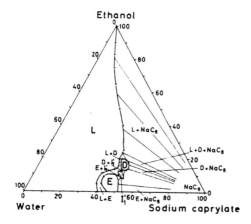

Figure 38. Phase diagram of sodium caprylate (octanoate)/ethanol/water (reproduced from [40]). Symbols as for Figs. 15, 22, 23 and 27.

trolytes such as sodium xylene sulfonate and short chain alchols. Their influence can be illustrated by the effect of ethanol on the hexagonal mesophase of sodium octanoate (Fig. 38).

Ethanol 'solubilizes' the hexagonal phase, shifting the boundary from ≈40% to ≈45% although it initially stabilizes it. There is almost no increase in maximum solubility of crystalline surfactant. Where *mesophases* occur over a wider concentration range (say 10–40%) then hydrotropes can sharply increase the isotropic solution range by removing the mesophase.

8.5 Electrolytes

The influence of electrolytes divides into two areas, the influence on nonionic (uncharged) surfactants and that on ionic materials. For nonionic surfactants the effects are either 'salting out' or 'salting in', in line with observations from the Hofmeister series of electrolyte effects on protein precipitation. Whilst there has been much discus-

sion over the past 50 years on the molecular mechanism involved, including many words on the 'structure' of water. Ninham and Yaminsky have recently shown in a landmark paper that the phenomena can be explained by dispersion interactions [166b]. Essentially electrolytes that are adsorbed to surfactants increase their solubility, hence mesophases dissolve. Those that are desorbed from the aggregates raise the chemical potential and decrease solubility. Like hydrotropes, large effects are seen only with mesophase or 'coacervate' precipitates. For ionic surfactants there are two additional effects. Where a counter-ion that produces an insoluble surfactant salt is present then any mesophases are removed by an increase of the Krafft temperature. Also, the well known common-ion effect raises Krafft temperatures. Additionally, electrolytes have a marked influence on CMC values, reducing the relatively high CMC of ionic surfactants to the low values of zwitterionic derivatives. This happens typically when the added electrolyte level is of the same order as the CMC range (≈1–300% of the CMC without electrolyte). Otherwise, where solubility is maintained, at high electrolyte levels, one observes salting in or salting out effects. These can be illustrated by the phase behavior of sodium soaps with water (Fig. 39) [167].

Added electrolyte induces coexistence between a dilute aqueous phase and micellar solutions (similar to a nonionic surfactant cloud temperature) and then changes hexagonal phase to the lamellar phase. Clearly, the ionic surfactant mesophase swells in water because of electrostatic repulsions. High salt levels negate the electrostatic repulsions, changing the behavior to that of a moderately polar surfactant (e.g. $C_{12}EO_3$). Salting-in electrolytes, such as sodium thiocyanate are likely to cause the lamellar phase to swell possibly forming mi-

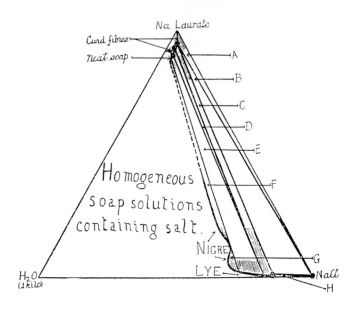

Figure 39. Phase diagram of sodium laurate (dodecanoate)/sodium chloride/water (reproduced from [167]). (A) Salt + various curds, (B) Salt + curd + saturated lye electrolyte solution, (C) Grained soap (curd) on lye, (D) Neat soap (lamellar, L_α) + lye + curd, (E) Neat soap on lye, (F) Neat soap on nigre (L_1), (G) Nigre on laye, (H) Salt + saturated solution containing traces of soap.

celles again at very high levels. This merits further study.

8.6   Alternative Solvents

Over recent years the question of whether or not surfactants form aggregates similar to micelles and mesophases in other polar solvents has received considerable attention [168–191]. The solvents concerned are polar liquids such as glycerol, ethylene glycol, formamide or ethyl ammonium nitrate (a molten electrolyte at ambient temperature). The answer is *yes*, but the thermodynamics of the aggregation process is somewhat different. Figures 40 and 41 show measurements of EMF in hexadecylpyridinium bromide solutions where water and ethylene glycol are solvents.

These results clearly show that aggregates form over a very narrow concentration range in ethylene glycol, similar to micelle formation. However, the aggregation concentration is higher by two orders of magnitude: simply because the 'solvophobic' effect is much smaller than the hydrophobic effect. In fact a useful 'rule of thumb' is that a surfactant with a C_{16} chain in the alternative solvents resembles a C_{12} surfactant in water. It seems that at least a C_{10} or C_{12} chain is required for micelle-like aggregation. The mesophase behavior is remarkably similar

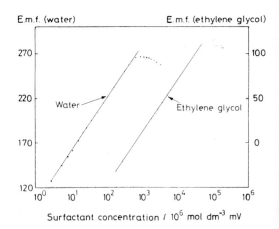

Figure 40. EMF as a function of concentration for the cetylpyridinium bromide electrode in water and ethylene glycol (reproduced from [168]).

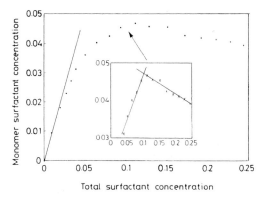

Figure 41. The monomer concentration of cetylpyridinium bromide plotted against total concentration (reproduced from [168]).

to that in water, again with a C_{16} chain derivative in the solvent resembling that of a C_{12} surfactant in water. There are, of course, detailed differences. Thus cubic phases are stabilized over the intermediate phases (as expected for shorter chain surfactants). Polyoxyethylenes are not miscible with glycerol, so form no mesophases. In the main though, given the large differences between the nature of the polar solvents and water (molecular size, symmetry, polarity H-bonding ability, conformational freedom, influence on electrostatics) this similarity is remarkable. Table 4 shows the mesophase

Table 4. Phase penetration data of anhydrous $C_{18}P_9Br$, showing the temperature range of the mesophases.

Solvent	I_1	L_1'	H_1	L_1''	V_1	L_α
Water	–	–	43–100+	–	54–100+	59–100+
EG	–	–	44–108	99	54–106	56–150+
GLY	–	–	31–150+	–	48–150+	56–150+
FA	48–61	57	50–100+	–	61–100+	61–100+
EAN	–	–	49–120+	–	74–120+	73–120+

The sequene of the columns indicates the order of the phases with increasing surfactant concentration. The first temperature given is T_{pen}; the second indicates the upper temperature limit of the mesophase. A '+' indicates that the mesophase is stable above the temperature given (the limit of measurement). L_1' indicates L_1 intrusion between the I_1 and H_1 phases. Similarly, L_1'' indicates an L_1 intrusion between the H_1 and V_1 phases. The anhydrous surfactant forms L_α at 74 to 75 °C.

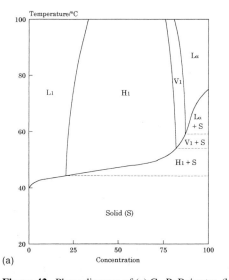

(a)

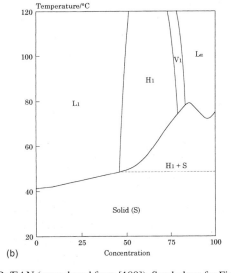

(b)

Figure 42. Phase diagram of (a) $C_{18}PyBr$/water, (b) $C_{18}PyBr$/EAN (reproduced from [189]). Symbols as for Fig. 14.

behavior of octadecylpyridinium bromide in a range of solvents.

The phase sequence and types are all remarkably similar. Only formamide shows a real difference with I_1 cubic phases being observed. Even so, these are formed by shorter chain derivatives (the chloride salts) in water. The phase diagrams (Fig. 42) show clearly that in ethylammonium nitrate as solvent the boundaries are shifted to much higher concentrations, again resembling a shorter chain surfactant.

9   Conclusions

It is clear that a framework based on packing constraints, surfactant chemical type, and water (solvent) – surfactant interactions now exists, so that the pattern of solvent-continuous surfactant mesophases can be rationalized for a wide range of materials, both for water and other polar solvents. Moreover, many data exist on the influence of additives, which can be seen to fit into the general picture. Whilst no mention of theoretical computer modelling has been made here, this is an area that is evolving rapidly at present. Both molecular and mesoscopic models are under development that should enable semi-quantitative computer predictions of phase behavior to be made in the next few years. These should be able to shed light on remaining problems such as the structures of intermediate and gel phases. However, further experimental work is required on the structures of intermediate phases and gel phases. In addition, the knowledge of mesophase glasses and particularly other semi-solid mesophases is at a rather primitive level. Moreover, the general framework can now be employed to rationalize surfactant mesophase behavior with more complex additives such as well-defined biomolecules (e.g. proteins) as well as synthetic polymers of all types. Membrane protein crystals (where the protein hydrophobic region is covered by surfactant) certainly fall into this class.

It is now timely to consider problems where mesophase kinetics are important.

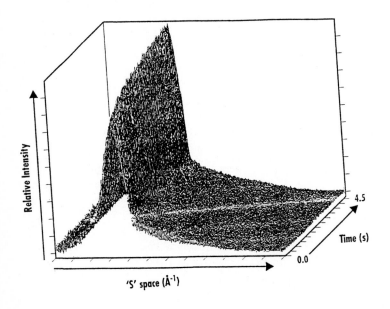

Figure 43. Series of X-ray patterns of a commercial nonionic lamellar phase, formed by fast mixing (reproduced from [192]).

Mesophase formation is involved in the applications of many specialist formulated products, including detergents. Recent advances in instrumental techniques allow the initial events that occur on mixing surfactant and water to be monitored, for example using synchrotron X-ray diffraction and a fast mixing cell (Fig. 43) [192]. To develop a proper understanding of the behavior will require a consideration of diffusion processes in labile anisotropic media, an exciting opportunity for computational modelling.

Rheology is perhaps the most important user property of surfactant mesophases, since almost every product is pumped, poured and stirred. Recent developments mean that it is now possible to make X-ray, neutron, optical microscope and other measurements on flowing systems to relate macroscopic orientation/morphology to rheological properties at the same time as molecular/aggregate reorientation is also studied. This offers the exciting prospect that in future not only will it be possible to formulate products for control of the mesophase structure through an understanding of ideas such as packing constraints etc., but rheological properties and dissolution rates will be controlled to deliver optimum benefit to customers. The future developments offer exciting intellectual challenges which will have an important scientific impact as well as being of widespread practical use.

10   References

[1] D. F. Evans, H. Wennerström, *The Colloid Domain Where Physics, Chemistry, Biology and Technology Meet*, VCH, New York, **1994**.

[2] J. H. Clint, *Surfactant Aggregation*, Blackie, Glasgow, **1990**.

[3] R. G. Laughlin, *The Aqueous Phase Behavior of Surfactants*, Academic Press, London, **1994**.

[4] C. Tanford, *The Hydrophobic Effect: Formation of Micelles and Biological Membranes*, Wiley Interscience, New York, **1980**.

[5] D. M. Small, *The Handbook of Lipid Research – The Physical Chemistry of Lipids*, Plenum Press, New York, **1988**.

[6] J. Israelachvili, *Intermolecular and Surface Forces*, Academic Press, London, **1992**.

[7] R. Lipowsky, E. Sackman, Eds., *Handbook of Biological Physics, Vol. 1a, Structure and Dynamics of Membrane, From Cells to Vesicles*, Elsevier Science, Netherlands, **1995**.

[8] R. Lipowsky, E. Sackman, Eds., *Handbook of Biological Physics, Vol. 1b, Structure and Dynamics of Membrane, Generic and Specific Interactions*, Elsevier Science, Netherlands, **1995**.

[9] D. F. Evans, *Langmuir* **1988**, *4*, 3.

[10] M. Costas, B. Kronberg, R. Silveston, *J. Chem. Soc. Faraday Trans. 1* **1994**, *90 (11)*, 1513.

[11] B. Kronberg, M. Costas, R. Silveston, *Pure Appl. Chem.* **1995**, *67 (6)*, 897.

[12] B. Kronberg, M. Costas, R. Silveston, *J. Disp. Sci. Technol.* **1994**, *15 (3)*, 333.

[13] K. Shinoda, M. Fujihara, *Bull. Chem. Soc. Jpn.* **1968**, *41*, 2162.

[14] K. Shinoda, *J. Phys. Chem.* **1977**, *81*, 1300.

[15] D. N. Glew, *J. Phys. Chem.* **1962**, *66*, 605.

[16] P. Eriksson, G. Lindbom, E. Burnell, G. J. T. Tiddy, *J. Chem. Soc. Faraday Trans. 1* **1988**, *84 (9)*, 3129.

[17] E. A. G. Aniansson, S. N. Wall, M. Almgren, H. Hoffmann, I. Kielmann, W. Ulbicht, R. Zana, J. Lang, C. Tondre, *J. Phys. Chem.* **1976**, *80*, 905.

[18] C. Elvingson, S. Wall, *J. Colloid. Int. Sci.* **1981**, *121*, 414.

[19] R. Palepu, D. G. H. Hall, E. Wyn-Jones, *J. Chem. Soc. Faraday Trans. 1* **1990**, *86*, 1535.

[20] D. M. Bloor, E. Wyn-Jones, unpublished data.

[21] H. Gharabi, N. Takisawa, P. Brown, M. A. Thomason, D. M. Painter, D. M. Bloor, D. G. Hall, E. Wyn-Jones, *J. Chem. Soc. Faraday Trans. 1* **1991**, *87*, 707.

[22] E. A. G. Aniansson, *J. Phys. Chem.* **1978**, *82*, 2805.

[23] J. Israelachvili, H. Wennerström, *Langmuir* **1990**, *6*, 873.

[24] J. Israelachvili, H. Wennerström, *J. Phys. Chem.* **1992**, *96*, 520.

[25] G. J. T. Tiddy, M. F. Walsh, E. Wyn-Jones, *J. Chem. Soc. Faraday Trans. 1* **1982**, *78*, 389.

[26] Z. X. Li, E. M. Lee, R. K. Thomas, J. Penfold, *J. Colloid Int. Sci.* **1997**, *187*, 492.

[27] P. Mukerjee, K. J. Mysels, *Critical Micelle Concentrations of Aqueous Surfactant Systems*, NSRDS-NBS 36, **1971**.

[28] V. Degiorgio, *Physics of Amphiphiles: Micelles, Vesicles and Microemulsions* (Eds.: V. Degiorgio, M. Corti), NPC, **1985**.

[29] R. J. M. Tausk, Ph. D. Thesis, *Physical Chemical Studies of Short-Chain Lecithin Homologues*, State University, Utrecht, Netherlands, **1974**.

[30] M. J. Rosen, *Chemtech.* **1993**, *23 (3)*, 16–27.

[31] V. V. Kumer et al., *J. Phys. Chem.*, in press.

[32] T. J. Madden, Unilever Research Port Sunlight Laboratory, Personal Communication, **1990**.

[33] D. J. Mitchell, B. W. Ninham, J. Israelachvili, *J. Chem. Soc. Faraday Trans. 1* **1976**, *72*, 1525.

[34] *CRC Handbook of Chemistry and Physics*, CRC Press, Boca Raton, FL.

[35] D. J. Mitchell, B. W. Ninham, J. Israelachvili, *Biochem. Biophys. Acta* **1977**, *470*, 605.

[36] G. J. T. Tiddy, *Physics Report* **1980**, *57*, 1.

[37] D. J. Mitchell, G. J. T. Tiddy, L. Waring, T. Bostock, M. P. McDonald, *J. Chem. Soc. Faraday Trans. 1* **1983**, *79*, 975.

[38] P. A. Winsor, *Chem. Rev.* **1968**, *68*, 1.

[39] V. Luzzati, *Biological Membranes. Physical Fact and Function*, Academic Press, New York, **1968**.

[40] P. Ekwall in *Advances in Liquid Crystals*, Vol. 1 (Ed.: G. H. Brown), Academic Press, New York, **1975**.

[41] J. M. Seddon, *Biochim. Biophys. Acta* **1990**, *1031*, 1–69.

[42] J. M. Seddon, R. H. Templer, Ch. 3 in *Structure and Dynamics of Membranes: From Cells to Vesicles* (Ed.: A. J. Hoff), Elsevier Science, Netherlands, **1995**.

[43] G. Lindblom, L. Rilfors, *Biochim. Biophys. Acta* **1989**, *998*, 221.

[44] K. Fontell, K. K. Fox, E. Hanser, *Mol. Cryst. Liq. Cryst.* **1989**, *1*, 9.

[45] J. Charvolin, J. F. Sadoc, *J. Phys. (Paris)* **1988**, *49*, 521.

[46] P. Sakya, J. M. Seddon, R. H. Templer, R. J. Mirkin, G. J. T. Tiddy, submitted to *Langmuir*, **1997**.

[47] A. Gulik, H. Delacroix, G. Kischner, V. Luzzati, *J. Phys. II France* **1995**, *5*, 445.

[48] G. J. T. Tiddy, J. M. Seddon, J. Walsh, unpublished data.

[49] J. M. Seddon, E. A. Bartle, J. Mingins, *J. Phys. Condens. Matter 2*, suppl. A, **1990**.

[50] V. Luzzati, P. A. Spegt, *Nature* **1967**, *215*, 707.

[51] L. E. Scriven, *Nature*, **1976**, *263*, 123.

[52] A. Tardieu, V. Luzzati, *Biochim. Biophys. Acta* **1970**, *219*, 11.

[53] P. Eriksson, G. Linblom, G. Arvidson, *J. Phys. Chem.* **1985**, *89*, 1050.

[54] K. D. Lawson, T. J. Flautt, *J. Am. Chem. Soc.* **1967**, *89*, 5489.

[55] B. J. Forrest, L. W. Reeves, *Chem. Rev.* **1981**, *81*, 1.

[56] N. Boden, P. H. Jackson, K. McMullen, M. C. Holmes, *Chem. Phys. Lett.* **1979**, *65*, 476.

[57] B. Luhmann, H. Finkelmann, G. Rehage, *Makromol. Chem.* **1985**, *186*, 1059.

[58] K. Radley, *Liq. Cryst.* **1995**, *18*, 1, 151.

[59] D. Chapman, R. M. Williams, B. D. Labroke, *Chem. Phys. Lipids* **1967**, *1*, 445.

[60] K. Larsson, *Z. Phys. Chem.* **1967**, *56*, 173.

[61] J. M. Vincent, A. E. Skoulios, *Acta Cryst.* **1966**, *20*, 432.

[62] P. Kékicheff, C. Grabielle-Madelmont, M. Ollivan, *J. Colloid Interface Sci.* **1989**, *131*, 112.

[63] A. S. C. Lawrence, M. P. McDonald, *Liq. Cryst.* **1966**, *1*, 205.

[64] A. S. C. Lawrence, *Mol. Cryst. Liq. Cryst.* **1969**, *7*, 1.

[65] R. C. Laughlin, Paper presented to the SIS, Jerusalem, 1996.

[66] C. D. Adam, J. A. Durrant, M. R. Lowry, G. J. T. Tiddy, *J. Chem. Soc. Faraday Trans. 1* **1984**, *80*, 789.

[67] L. Mandelkern in *Comprehensive Polymer Science*, Vol. 2 (Ed.: C. Booth, C. Price), Pergamon Press, Oxford **1989**, p. 415.

[68] V. Luzzati, H. Mustacchi, A. Skoulios, F. Husson, *Acta Crystallogr.* **1960**, *13*, 660.

[69] F. Husson, H. Mustacchi, V. Luzzati, *Acta Crystallogr.* **1960**, *13*, 668.

[70] A. Skoulios, V. Luzzati, *Acta Crystallogr.* **1961**, *14*, 278.

[71] A. Skoulios, *Acta Crystallogr.* **1961**, *14*, 419.

[72] P. A. Spegt, A. Skoulios, *Acta Crystallogr.* **1963**, *16*, 301.

[73] P. A. Spegt, A. Skoulios, *Acta Crystallogr.* **1964**, *17*, 198.

[74] V. Luzzati, A. Tardieu, T. Gulik-Krzywicki, *Nature* **1968**, *217*, 1028.

[75] H. Hagslätt, O. Söderman, B. Jönsson, *Liq. Cryst.* **1992**, *12*, 667.

[76] O. Söderman, G. Lindblom, L. B.-A. Johansson, K. Fontell, *Mol. Cryst. Liq. Cryst.* **1980**, *59*, 121.

[77] I. D. Leigh, M. P. McDonald, R. M. Wood, G. J. T. Tiddy, M. A. Trevethan, *J. Chem. Soc. Faraday Trans. 1* **1981**, *77*, 2867.

[78] P. Kékicheff, B. Cabane, *J. de Phys.* **1987**, *48*, 1571.

[79] P. Kékicheff, B. Cabane, *Acta Crystallogr.* **1988**, *B44*, 395.

[80] P. Kékicheff, *J. Colloid Interface Sci.* **1989**, *131*, 133.

[81] P. Quist, K. Fontell, B. Halle, *Liq. Cryst.* **1994**, *16*, 235.

[82] P. Kékicheff, G. J. T. Tiddy, *J. Phys. Chem.* **1989**, *93*, 2520.

[83] C. E. Fairhurst, M. C. Holmes, M. S. Leaver, *Langmuir*, **1997**, *13*, 4964.

[84] C. E. Fairhurst, M. C. Holmes, M. S. Leaver, *Langmuir* **1996**, *12*, 6336.

[85] J. Burgoyne, M. C. Holmes, G. J. T. Tiddy, *J. Phys. Chem.* **1995**, *99*, 6054.

[86] M. C. Holmes, J. Charvolin, *J. Phys. Chem.* **1984**, *88*, 810.

[87] M. C. Holmes, J. Charvolin, D. J. Reynolds, *Liq. Cryst.* **1988**, *3*, 1147.

[88] M. C. Holmes, P. Sotta, Y. Hendrikx, B. Deloche, *J. de Phys. II* **1993**, *3*, 1735.

[89] M. C. Holmes, A. M. Smith, M. S. Leaver, *J. de Phys. II* **1993**, *3*, 1357.

[90] M. S. Leaver, M. C. Holmes, *J. de Phys. II* **1993**, *3*, 105.

[91] M. C. Holmes, M. S. Leaver, A. M. Smith, *Langmuir* **1995**, *11*, 356.

[92] X. Auvray, T. Perche, R. Anthore, C. Petipas, I. Rico, A. Lattes, *Langmuir* **1991**, *7*, 2385.

[93] D. M. Anderson, P. Ström in *Polymer Association Structures: Microemulsions and Liquid Crystals, ACS Symposium Series 384* (Ed.: M. A. El-Nokaly), American Chemical Society, Washington **1989**, p. 204.

[94] D. M. Anderson, *Colloq. de Phys.* **1990**, *51*, C7-1.

[95] D. M. Anderson, H. T. Davis, L. E. Scriven, J. C. C. Nitsche, *Advances in Chemical Physics*, Vol. LXXVII (Eds.: I. Prigogine, S. A. Rice), John Wiley, New York **1990**, p. 337.

[96] A. H. Schoen, *Infinite Periodic Minimal Surfaces Without Self-intersections*, NASA Technical Note D-5541, Washington **1970**.

[97] S. T. Hyde, *Pure Appl. Chem.* **1992**, *64*, 1617.

[98] S. T. Hyde, *Colloq. de Phys.* **1990**, *51*, C7-209.

[99] B. Lindman in *Surfactants* (Ed.: Th. F. Tadros), Academic Press, London, **1984**.

[100] V. Degiorgio, M. Conti in *Proceedings of the International School of Physics, 'Enrico Fermi'*, Italian Physical Society, **1985**.

[101] J. Sjöblom, P. Stenius, I. Danielsson, Ch. 7 in *Nonionic Surfactants* (Ed.: M. J. Schick), Marcel Dekker, New York, **1987**.

[102] R. Strey, R. Schomäker, D. Roux, F. Nallet, U. Olsson, *J. Chem. Soc. Faraday Trans.* **1990**, *86*, 2253.

[103] J. P. Conroy, C. Hall, C. A. Lang, K. Rendall, G. J. T. Tiddy, J. Walsh, G. Lindblom, *Prog. Colloid. Polym. Sci.* **1990**, *82*, 253.

[104] B. P. Binks, *Annual Reports Section C*, The Royal Society of Chemistry **1996**, *92*, 97.

[105] H. Kellay, B. P. Brinks, Y. Hendrikx, L. T. Lee, J. Meunier, *Adv. Colloid Int. Sci.* **1994**, *49*, 85.

[106] M. S. Leaver, U. Olsson, H. Wennerstrom, R. Strey, U. Wurz, *J. Chem. Soc. Faraday Trans.* **1995**, *91*, 4369.

[107] U. Olsson, H. Wennerstrom, *Adv. Coloid Int. Sci.* **1994**, *49*, 85.

[108] R. G. Laughlin, *Personal Communications* (1996).

[109] J. S. Clunie, J. M. Corkill, J. F. Goodman, P. C. Symons, J. R. Tate, *Trans. Faraday Soc.* **1967**, *63*, 2839.

[110] J. S. Marland, B. A. Mulley, *J. Pharm. Pharmacol.* **1971**, *23*, 561.

[111] B. A. Mulley, A. D. Metcalf, *J. Colloid Sci.* **1964**, *19*, 501.

[112] S. S. Funari, G. Rapp, *J. Phys. Chem. B* **1997**, *101*, 732.

[113] S. S. Funari, M. C. Holmes, G. J. T. Tiddy, *J. Phys. Chem.* **1994**, *98*, 3015.

[114] J. Corcoran, S. Fuller, A. Rahman, N. N. Shinde, G. J. T. Tiddy, G. S. Attard, *J. Mater. Chem.* **1992**, *2*, 695.

[115] S. S. Funari, M. C. Holmes, G. J. T. Tiddy, *J. Phys. Chem.* **1992**, *96*, 11029.

[116] H. Jousma, J. A. Bouwstra, F. Spies, H. E. Junginger, *Colloid Polym. Sci.* **1987**, *265*, 830.

[117] G. T. Dimitrova, Th. F. Tadros, P. F. Luckham, *Langmuir* **1995**, *11*, 1101.

[118] G. T. Dimitrova, Th. F. Tadros, P. F. Luckham, M. R. Kipps, *Langmuir* **1996**, *12*, 315.

[119] C. Tanford, J. A. Reynolds, *Biochim. Biophys. Acta* **1976**, *457*, 133.

[120] K. Beyer, *J. Colloid Int. Sci.* **1982**, *86*, 73.

[121] R. Heusch, F. Kopp, *Ber. Bunsenges. Phys. Chem.* **1987**, *91*, 806.

[122] L. Thompson, J. M. Walsh, G. J. T. Tiddy, *Colloids and Surfaces A: Physicochemical and Engineering Aspects* **1996**, *106*, 223.

[123] K. Kratzat, H. Finkelmann, *J. Colloid Int. Sci.* **1996**, *181*, 542.

[124] K. Kratzat, H. Finkelmann, *Colloid Polym. Sci.* **1994**, *272*, 400.

[125] K. Kratzat, C. Schmidt, H. Finkelmann, *J. Colloid Int. Sci.* **1994**, *163*, 190.

[126] H. Kratzat, C. Stubenrauch, H. Finkelmann, *Colloid Polym. Sci.* **1995**, *273*, 257.

[127] H. Kratzat, H. Finkelmann, *Liq. Cryst.* **1993**, *13*, 691.

[128] H. Söderlund, J. Sjöblom. T. Wärnheim, *J. Disp. Sci. Technol.* **1989**, *10*, 131.

[129] H. Finkelmann, B. Lühmann, G. Rehage, *Colloid Polym. Sci.* **1982**, *260*, 56.

[130] H. Finkelmann, B. Lühmann, G. Rehage, *Angew. Makromol. Chem.* **1984**, *123*, 217.

[131] B. Lühmann, H. Finkelmann, *Colloid Polym. Sci.* **1987**, *265*, 506.

[132] B. Lühmann, H. Finkelmann, *Colloid Polym. Sci.* **1986**, *264*, 189.

[133] J. Eastoe, P. Rogueda, A. M. Howe, A. R. Pitt, R. K. Heenan, *Langmuir* **1996**, *12*, 2701.

[134] P. Sakya, J. M. Seddon, R. H. Templer, *J. Phys. II* **1994**, *4*, 1311.

[135] F. Nilsson, O. Södermann, I. Johansson, *Langmuir* **1996**, *12*, 902.

[136] C. Hall, G. J. T. Tiddy, B. Pfannemüller, *Liq. Cryst.* **1991**, *9*, 527.

[137] H. W. C. Raaijmakers, E. G. Arnouts, B. Zwanenburg, G. J. F. Chittenden, H. A. van Doren, *Recueil des Travaux Chimiques des Pays-Bas* **1995**, *114*, 301.

[138] H. Finkelmann, M. A. Schafheutle, *Colloid Polym. Sci.* **1986**, *264*, 786.

[138a] H. Sagihari, Y. Hayashi, M. Ochiai, *J. Am. Oil Chem. Soc.* **1989**, *66*, 146.

[138b] D. Balzer, *Langmuir* **1993**, *9*, 3375.

[138c] G. Platz, C. Thunig, J. Policke, W. Kirchhoff, D. Nickel, *Colloids and Surfaces A: Physicochemical and Engineering Aspects* **1994**, *88*, 113.

[138d] G. Platz, J. Policke, C. Thunig, R. Hofman, D. Nickel, W. Vonrybinski, *Langmuir* **1995**, *11*, 4250.

[139] P. Alexandridis, D. Zhou, A. Khan, *Langmuir* **1996**, *12*, 2690.

[140] D. Zhou, P. Alexandridis, A. Khan, *J. Colloid Int. Sci.* **1996**, *183*, 339.

[141] P. Alexandridis, U. Olsson, B. Lindman, *Macromolecules* **1995**, *28*, 7700.

[142] P. Alexandridis, U. Olsson, B. Lindman, *J. Phys. Chem.* **1996**, *100*, 280.

[143] P. Alexandridis, U. Olsson, B. Lindman, *Langmuir* **1996**, *12*, 1419.

[144] P. Alexandridis, U. Olsson, B. Lindman, *Langmuir* **1997**, *13*, 23.

[145] P. Holmqvist, P. Alexandridis, B. Lindman, *Langmuir* **1997**, *13*, 2471.

[146] E. S. Lutton, *J. Am. Oil Chem. Soc.* **1966**, *43*, 28.

[147] H. Hoffmann, G. Oetter, B. Schwandner, *Progr. Coll. Polym. Sci.* **1987**, *73*, 95.

[148] C. La Mesa, B. Sesta, M. G. Bonicelli, G. F. Ceccaroni, *Langmuir* **1990**, *6*, 728.

[149] J. C. Blackburn, P. K. Kilpatrick, *J. Colloid Int. Sci.* **1992**, *149*, 450.

[150] K. Rendall, G. J. T. Tiddy, M. A. Trevethan, *J. Chem. Soc., Faraday Trans. 1* **1983**, *79*, 637.

[151] E. S. Blackmore, G. J. T. Tiddy, *J. Chem. Soc. Faraday Trans. 2* **1988**, *84*, 1115.

[152] A. Khan, K. Fontell, G. Lindblom, *J. Phys. Chem.* **1982**, *86*, 383.

[153] R. G. Laughlin in *Cationic Surfactants, Surfactant Science Series, Vol. 37* (Eds.: D. N. Rubingh, P. M. Holland), Marcel Dekker, New York, **1990**.

[154] H. Hagslätt, O. Söderman, B. Jönsson, L. B.-Å. Johansson, *J. Phys. Chem.* **1991**, *95*, 1703.

[155] H. Hagslätt, O. Söderman, B. Jönsson, *Langmuir* **1994**, *10*, 2177.

[156] S. Fuller, J. Hopwood, A. Rehman, N. N. Shinde, G. J. T. Tiddy, G. S. Attard, O. Howell, S. Sproston, *Liq. Cryst.* **1992**, *12*, 521.

[157] S. Fuller, Ph. D. Thesis, University of Salford **1995**.

[158] S. Fuller, N. N. Shinde, G. J. T. Tiddy, G. S. Attard, O. Howell, *Langmuir* **1996**, *12*, 1117.

[159] E. Alami, H. Levy, R. Zana, A. Skoulios, *Langmuir* **1993**, *9*, 940.

[160] R. Zana, Y. Talmon, *Nature* **1993**, *362*, 228.

[161] S. Fuller, G. J. T. Tiddy, R. Zana, unpublished results, 1996.

[162] L. Pérez, J. L. Torres, A. Manresa, C. Solans, M. R. Infante, *Langmuir* **1996**, *12*, 5296.

[163] H. Hagslätt, K. Fontell, *J. Colloid Int. Sci.* **1994**, *165*, 2, 431.

[164] H. Wennerström, B. Jönsson, *J. Phys. Chem.* **1987**, *91*, 338.

[165] J. A. Bouwstra, H. Jousma, M. M. van der Meulen, C. C. Vijverberg, G. S. Gooris, F. Spies, H. E. Junginger, *Coll. Polym. Sci.* **1989**, *267*, 531.

[166a] C. A. Barker, D. Saul, G. J. T. Tiddy, B. A. Wheeler, E. Willis, *J. Chem. Soc. Faraday Trans. 1* **1974**, *70*, 154.

[166b] B. W. Ninham, V. Yaminski, *Langmuir* **1997**, *13*, 2097.

[167] J. W. McBain, A. J. Burnett, *J. Chem. Soc.* **1922**, *121*, 1320.

[168] H. Gharabi, R. Palepu, G. J. T. Tiddy, D. G. Hall, E. Wyn-Jones, *J. Chem. Soc. Chem. Commun.* **1990**, *2*, 115.

[169] D. F. Evans, A. Yamauchi, R. Roman, E. Z. Casassa, *J. Coll. Interf. Sci.* **1982**, *88*, 89.

[170] D. F. Evans, A. Yamauchi, G. Wei, A. Bloomfield, *J. Phys. Chem.* **1983**, *87*, 3537.

[171] W. Binani-Limbele, R. Zana, *Coll. Polym. Sci.* **1989**, *267*, 440.

[172] M. Sjöberg, U. Henriksson, T. Wärnheim, *Langmuir* **1990**, *6*, 1205.

[173] M. Jonströmer, M. Sjöberg, T. Wärnheim, *J. Phys. Chem.* **1990**, *94*, 7549.

[174] I. Ricco, A. Lattes, *J. Phys. Chem.* **1986**, *90*, 5870.

[175] A. Couper, G. P. Gladden, B. T. Ingram, *Faraday Discuss. Chem. Soc.* **1975**, *59*, 63.

[176] A. Ray, *J. Am. Chem. Soc.* **1969**, *91*, 23, 6511.

[177] P. D. I. Fletcher, P. J. Gilbert, *J. Chem. Soc. Faraday Trans. 1* **1989**, *85*, 147.

[178] N. Moucharafieh, S. E. Friberg, *Mol. Cryst. Liq. Cryst.* **1979**, *49*, 231.

[179] M. A. El Nokaly, L. D. Ford, S. E. Friberg, D. W. Larsen, *J. Coll. Interf. Sci.* **1981**, *84*, 228.

[180] L. Ganzui, M. A. El Nokaly, S. E. Friberg, *Mol. Cryst. Liq. Cryst.* **1982**, *72*, 183.

[181] S. E. Friberg, P. Laing, K. Lockwood, M. Tadros, *J. Phys. Chem.* **1984**, *88*, 1045.

[182] S. E. Friberg, C. Solans, L. Ganzuo, *Mol. Cryst. Liq. Cryst.* **1984**, *109*, 159.

[183] D. F. Evans, E. W. Kaler, W. J. Benton, *J. Phys. Chem.* **1983**, *87*, 533.

[184] A. Balmajdoub, J. P. Marchal, D. Canet, I. Rico, A. Lattes, *Nouv. J. Chem.* **1987**, *11*, 415.

[185] T. Wärnheim, A. Jönssen, *J. Colloid. Interf. Sci.* **1988**, *125*, 627.

[186] X. Auvray, C. Petipas, R. Anthore, I. Rico, A. Lattes, *J. Phys. Chem.* **1989**, *93*, 7458.

[187] Z. Lin, H. T. Davies, L. E. Scriven, *Langmuir* **1996**, *12*, 5489.

[188] T. A. Bleasdale, G. J. T. Tiddy, E. Wyn-Jones, *J. Phys. Chem.* **1991**, *95*, 5385.

[189] T. A. Bleasdale, G. J. T. Tiddy in *Organised Solutions* (Eds.: S. E. Friberg, B. Lindman), Marcel Dekker, New York **1992**.

[190] X. Auvray, C. Petipas, I. Rico, A. Lattes, *Liq. Cryst.* **1994**, *17*, 109.

[191] H. D. Dörfler, *Z. Phys. Chem.* **1994**, *187*, 135.

[192] CLCR Synchrotron Radiation Department, Daresbury Laboratory Annual Report, 1995–1996.

Chapter VIII
Living Systems

In memoriam Hans Kelker (21. 9. 22 – 25. 6. 92) and Horst Sackmann (3. 2. 21 – 2. 11. 93), who pioneered the essentials of liquid crystal research and human coherences in a bipartite world. This chapter is also dedicated to the memory of my parents, of which, may father was a pupil of Daniel Vorländer for a short fortunate period before World War I ended all.

Siegfried Hoffmann

1 Introduction

Living systems, like the universe from which they are descended, are mainly understandable as a process [1–8]. They are characterized preferentially by their dynamics rather than by their statics. As "structure" and "phase" correspond in some way to "molecular biology" (Fig. 1) and "liquid crystals" (Fig. 2), the two fields will appear in our days as two scientific aspects of much more general duality phenomena (Fig. 3) that govern quite different stages and hierarchies of developmental processes within our universe. They both represent essentials in our somewhat inadequate attempts to cope with the complexity of life patterns, which can be only approximately – if at all – comprehended by the only partially adequate complexity of our different scientific approaches.

The consideration of structure – phase dualities in pre-life transitions and in the development of life will thus appear as a continuation and amplification of basic quantum dualities – with their complementarities of "subject" and "relatedness" – into the transient order – disorder patterns of supra-molecular organizations and their biomesogenic regulations, bringing them forward, at least in our spacetime, to the as-yet final achievements and destinations of the fields of self-awareness of our species within a creative universe.

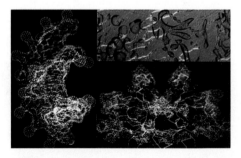

Figure 1. "Structure": Dickerson dodecamer [9] (left) and prealbumin dimer [10] (right bottom) (Langridge's first examples [11] of electrostatic potential visualizations in combination with skeletal and transparent CPK presentation) as molecular biological structural representatives of the nucleoprotein system, but also as biological representations of "individual" and "relatedness" within the frame of structure-phase transitions; familiarities between early universe structure computer simulations [12] and cholesteric texture paintings by Lehmann [13, 14] (right top).

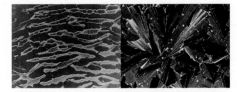

Figure 2. "Phase": from ammonium oleate (left), Lehmann's [14] first "seemingly living crystals" (original photograph as a kind gift of Hans Kelker) [13], and life's informational component's liquid crystalline DNA texture [15] (right), the present realization of Lehmann's early visions.

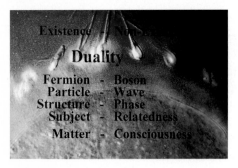

Figure 3. "Duality"-phenomena in life onto- and phylogenesis [1 – 8, 16, 17, 18] within universe developments and continuations from quantum vacuum up to consciousness pattern creativities together with the intriguing background design modified from [16].

2   Biomesogens and the Grand Process

Biomesogens (Fig. 4) evolved as an early amphiphilic pattern from the interface dialectics between order and disorder on the primordial earth [5 d, 7 a, 17, 18]. Their holographic molecular designs provided decisive prerequisites for the projection of individual molecular facilities into the structural and functional amplifications of cooperative and dynamic mesogen domain ensembles and their spatiotemporal coherences. By directional nonlinear dynamizations of the supramolecular organizations and their (bio)mesogenic operation modes, they arrived at a state of recognition and responsiveness. Competing individualizations drove autocatalytic propagation, self-reproduction, information processing, and optimization. Far from thermal equilibrium, the biomesogenic   order – disorder   patterns gained intelligence by reality adaptation and variation. Experiencing optimization strategies, they developed self-awareness and creativity.

Figure 4. "Biomesogen" introduction in the 1970s, mesogenic nucleic acid/protein interplays framed by cholesteric color variations of polydisperse evolutionary DNA [7 a, 17, 18, 19, 63 a, 75].

2.1 Historical Dualities

The dualities of supramolecular biomesogenic organizations [7, 17, 18] reemerge
in the process of their scientific discovery.
The history of "liquid crystals", lovingly retraced by Kelker [13] (Fig. 5), parallels the
development of "molecular biology" [5–8].
"Liquid crystals are beautiful and mysterious, I am fond of them for both reasons",
states de Gennes [19]. And, indeed, how can
we not feel attracted to what in the inanimate world seems most closely related to
us?

Notwithstanding the priority roles of lyotropics and amphotropics both in native
standards and by their (decades) earlier
detection [13], notwithstanding Virchow's
soft, floating substance from the nerve
core that he named myelin in 1854 [20a],
Mettenheimer's discovery of the birefringence of this substance in 1857 [20c],
Heintz's statements in Halle in the early
1850s that "stearin becomes cloudy at
51–52 °C, opalizes at higher temperature,
being opaque at 58 °C, and molten and clear
at 62.5 °C" [20b], it was the "two melting
points" of thermotropic cholesterol derivatives that started the scientific history of liquid crystals in the early "correspondence
cooperation" between Reinitzer and Lehmann in 1888 [20–27].

Reinitzer's letter to Lehmann, dated
March 14, 1888 [21, 22], which introduces
from rather hidden beginnings the history of
liquid crystals, is very impressed by their
unique appearances. Reinitzer sends two
substances "with the request to investigate
more closely their physical isomerism". He
describes previously unnoticed and unknown phenomena: "Cholesteryl acetate
melts to a clear liquid. On cooling the liquid down, a bright color variation moves
through the liquid before crystallization.
Bright violet, blue, green, yellow and

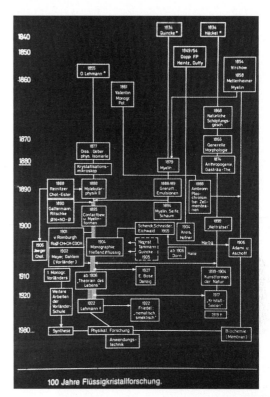

Figure 5. 100 years of liquid crystal research as depicted by Kelker [13].

orange-red colors appear. If one observes
the molten compound with crossed nicols,
on cooling suddenly starlike groups of crystals appear colored and brightly shining on
a dark background. Thereby the rest of
the substance remains liquid, which might
be demonstrated by slightly touching
the cover-slide, which makes the crystals
flow".

Lehmann [14] investigates the preparations with a heated-stage polarization microscope constructed by himself. They re-
mind him of his earlier studies on "iodine-
silver", and, even more closely, of his endeavors concerning "ammonium oleate hydrate". He coins the term "liquid crystals".
The scientific development of "liquid crystals" or "cystalline liquids" (Fig. 6) – terms

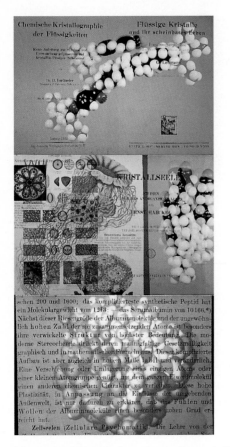

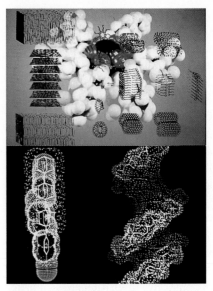

Figure 7. Liquid crystal abstractions (left to right and top to bottom): thermotropics and lyotropics with structural visualizations of formerly alien discoid [28 a, b], the well-known calamitic artificial liquid crystal 1-[*trans*-4-(alk-3-en-1-yl)cyclohex-1-yl]-4-cyanobenzene [28 c], and long disregarded native helical DNA [28 d].

Figure 6. Historical views (top to bottom): Vorländer's *Crystallography of Liquids* [24] beside Lehmann's *Seemingly Living Crystals – Presented by a Film* [14]; Haeckel's *Artful Presentations of Nature* [23], related to Lehmann's *Liquid Crystals*; his rather modern view of protein properties, illustrated by ribonuclease with a water cover at the active site [29].

that will compete in the time following Lehmann and Vorländer [14, 24] in the search for more precise and sophisticated differentiations and that will ultimately be integrated into the more comprehensive "mesogen" concept of Kelker [13] – commences with contradictions similar to those that seem to be typical of their molecular patterns. Remarkably, it is the two biomolecules that, with the thermotropy of cholesterol and the lyotropy of oleic acid derivatives, not only provoke a first dissent concerning priority but also seem to direct further research in quite different directions (Fig. 7) [25].

The thermotropics celebrated their first period of academic dominance in the school of Vorländer [24] in Halle. Within three decades at the beginning of the 20th century, more than 2000 synthesized liquid crystals were used to establish the basis of early systematizations and structure – function correlations. Much later, in the 1950s, Vorländer's substance-inheritance – cigar-like abstracted molecules dedicated to posterity in cigarboxes – will mediate the historical "reentrant" phenomenon in the old Vorländer institute building [26, 27], preceded only by a forerunning reawakening of phase chemistry – a sleeping-beauty role – in the search for a miscibility rule for imidazole derivatives [26 d – f].

Vorländer [24] anticipates many of the synthesis strategies used today on the basis of what might now be called stereoelectronic structure – function elucidations. Even his errors will offer intriguing suggestions. Much more recent polymer research [25] is foreseen in the "nature of the infinitely long molecules"; "star-, cross- and platelet-shaped molecules" are offered for discussion. The "optically anisotropic building of packet-like stacked molecular platelets resembling a Volta-column" describes the stacks of today's discoid arrangements [25, 28]. In love, however, with his own hundredfold-proven rod model, Vorländer treats these ideas about liquid crystals with scepticism. Vorländer "does not see any possibility to use the crystalline-liquid substances for technical applications" and "feels himself much more attracted by pure scientific desire" for the "crystalline-liquid appearances" in which "even more pronounced than in the case of solid crystals, the original, sensitive and yet powerful nature of molecules and their constituents will be expressed". He does not ignore the relations to life processes, but shrinks back from any interpretation. So he would not approve "the data of Lehmann as to the crystalline-liquid characteristics of ammonium oleate hydrate and similar soaps that – molten together with water – should yield crystalline liquids", nor would he follow his general hypotheses. Lehmann's "comparisons of liquid crystals with living beings" appear to him "unscientific" and "contadictory", in their last stages even "mystical" and "degrading the very object". Looking back, however, to the first liquid crystal empire, created by him and his school, Vorländer takes at last somewhat restrained pride "in having yet arrived at a state where crystalline-liquid substances might be synthesized

– as many and as different ones as might be expected or wanted". And – going beyond all his own doubts – his "Nevertheless, the soft crystals of organic compounds are the housing of life; without soft crystals and colloids no living beings" forms a bridge towards de Gennes's "soft matter" in particular and "life sciences" in general [24, 25, 28].

But a generation will pass before liquid crystals will be really wanted. Kelker's "MBBA", the first liquid crystal at ambient temperature, will provide the impetus for applicational research in liquid crystals [13].

Lehmann [14] in Karlsruhe, however, in close mental vicinity to and mutual stimulation by the "crystal souls" of Haeckel [23] sets out for the first grand romantic period of a general view of mesogenic behavior, with special emphasis also on its lyotropic aspects. "The analogies between the shaping and driving forces of both liquid crystals and living beings" had caused him "to characterize the myelin crystals of paraazoxycinnamic acid ester as seemingly living crystals". Proposed by his own intuition, this early impression will intensify and, towards the end of his way, widen and amplify to general considerations on "liquid crystals and life theories". Lehmann believes "that physiology once might be able to prove that those analogies are really caused by intimate relationship or even identity of the governing forces". It "appears to him of outstanding interest that the so-called biocrystals of living substances will orientate themselves regularly to the shapes of organisms, so as if the molecules of living substances would, indeed, exert directional forces". Lehmann who unravels the close similarities of his liquid crystals to the world of biologically relevant mono-, oligo- and polymers at that time, who enriches his horizons with their foreseen technical applications, and who uses – in a way comparable to quite recent approaches in molecular dynamics – the simulative power of films to illustrate his results and derived implications [14 h], this Lehmann leaves to us, like the life processes that he tried to reduce to their fundaments scientifically in his first early biomesogen approaches, a system that is open in all aspects. The ambitious attempt to bridge the gap between the phenomenology of liquid crystals and their biomolecular structural prerequisites ends, owing to the times, in nothingness. What the chemist Vorländer was able to elucidate scientifically in terms of the thermotropic liquid crystal empire erected by him and his pupils, the lone physicist Lehmann had to do without. His "biocrystals" are nebulous entities. Proteins – also philosophically – are associated with life [5, 7a]. Their structure, however, remains uncertain, although Haeckel's intuitive perception [23] appears dramatically modern [5–8, 17, 18, 29, 30]. Miescher announced in a first prophetical view an early overture to molecular biology [29 a]. But nobody will take his nucleic acids as informational components of life processes seriously. Lehmann had already outlined a lecithin model. The structure–function relationships veiled by membranes, nevertheless, will still have to await our insights into the nucleation of informational and functional components, before the compartmental components can be elucidated. Lehmann's investigations, thoughts, and hypotheses will be valued by the future. And also not forgotten – within the creative Halle circle – will be Schenck's [24 d, e] ingenious visions on liquid crystals, in which he attempted to create a modern synthesis of Lehmann and Vorländer's contradictory theses and antitheses, and Staudinger's pioneering path into the field of large molecules, which opened up to chemistry not only the first views of future prosperous fields, but established, morevoer, the scientific basis for an understanding of life patterns.

In the meantime, however, there is the lonely prophecy of Schroedinger [29 c] that the genetic material is an aperiodic crystal, which not only anticipates with its outstanding prediction the later – elucidated reality, but raises, moreover, questions about the meaning and content of terms such as order, disorder, entropy and negentropy [1–8, 17], and the philosophical categorizations, inferences and deductions around at a time that prefers merely static views. When Watson and Crick [29 f], following the laying of foundation stones in the field of protein geometries by Pauling [29 d], hypothesize their double helical matrix structure, it proves, with regard to its linguistics, indeed to be an aperiodic arrangement. With regard to its state, however, it appears to be a liquid rather then a proper solid crystal. And all biooligomers and biopolymers that have since been discovered likewise occupy fertile meso-positions between the sterile statics of solids and the vain dynamics of liquids. They will appear to be mesogens in the broad productive ranges between the extrema (Fig. 7) [5, 7, 17, 18, 28–30]. The time had come to award Lehmann and his grand hypotheses their long-denied rehabilitation. But liquid crystal people on one hand and molecular biologists on the other were mainly concerned with their own problems and appeared, moreover, to be split within their own disciplines. The possible unifying (bio)mesogen view in the dialectical synthesis of the divergent approaches of Lehmann in Kalsruhe and Vorländer in Halle had to continue to await its realization.

2.2   "Genesis" and Phase Transitions

A first grand phase transition from quantum vacuum to existence within an instability; probabilistics rather than deterministics in all the incomprehensible beginnings, irreversibilities and the time arrow of entropy guiding further developments; consecutive freezing of originally unified forces and symmetry breaking; "matter" developments between fermion–boson, particle–wave, and structure–phase dualities; interfacial routes to mesogens leading out of the contradictions – will these suggestions reveal valid pictures [1–8, 17, 18, 29–33]?

"Somewhat forlorn between the glows of genesis and the shadows of decline: endangered, tentative structures at fluid borders, is such the origin of the long journey of life? Do the death rhythms of an extinguishing, cooling-down satellite of a middle-sized, middle-aged sun – somewhere in the intermediary populations of a nameless galaxy – inspire functional and organizational patterns of new existences, do the basic matrices of future life patterns – influenced and imprinted by a chaos of interchanging surroundings – within a nearly unique life-expectant situation at these coordinates of a universe dare a completely new game, does a sequence of unique steps from first molecular coming together, forgetting, recognizing, not-completely forgetting, remembrance and discrimination, understanding and learning, of trial and error, failure and success, over the landmarks of cooperation, self-organization and self-reproduction, individualization, metabolism, cell-predecessors, cell-differentiation, the appearance of organism and species landscapes, up to the collective consciousness patterns of a thinking, speaking, abstracting, simulation power and creative activities developing

being lead to the present final outcome of a grand dynamic process, of ridge-climbing of selecting necessities over the endless plains of statistical coincidences, a principle, a plan, a movement of matter that optimizes itself in the growth and decay of its individual constituents, that gives the coming what it had to experience in the past, that suffers in each individual death the threatening end of its own existence and, nevertheless, sets out in each individual birth to new and unknown horizons?" [7 a]. The picture from the early 1970s is filled-in in our days.

Monod's "vagrant at the borders of a universe that remains silent to his complaints" [5 a] finds himself within the overwhelming changes of overall transitions. He channels his thoughts of loneliness and despair, the question of his destiny and his longing dreams of fellow existences within the universe into scientific approaches. He extends his probes to the depths of space and time, retraces the complexity of his ways, spells the codes that govern his existences and advances his capabilities in approaches to creating in partial simulations some sort of minima vita on his own. In his strange meso-position between elementary particles on one hand and cosmic dimensions on the other he tries to cover with quantum mechanics and general relativity theory the extremes, so far, however, being unable to formulate his desired unifying view of a final theory. Having changed his considerations from a more static, infinite universe to the more dynamic expectations of its beginning, developing, and ending, he questions again the laboriously achieved convictions by the more recent picture of grand fluctuating universes that combine dynamic developmental processes with the new–old quality of a forwarding eternal being [1–8].

He tries to answer the questions about the "before and behind of space and time" by

Figure 8. Universal processings versus processing of universes? [1–8]. Background modified from [3].

Figure 9. Birth of universe – a first phase transition between nothingness and existence? [1–8].

the eternal dynamics of self-reproducing inflationary universes (Fig. 8). He tends to describe the birth of "his" universe in nonclassical versions as a phase transition between quantum vacuum and material existence, realizing instability-mediated possibilities of events, rather than in the former classical terms of an incomprehensible singularity obeying deterministic laws. He delights in the idea that all this might have been brought about by a first grand spontaneous fluctuation that induced by matter–gravitation compensation an almost-zero-energy transition (Fig. 9). Realizing his self-awareness and self-consciousness, he would look upon his maternal universe as a likewise unique individuality, and sometimes he indulges in the hybris of an anthropological

principle that favors his singularity as the final destination of a universe. His dreams, however, would prefer the intriguing suggestion that it is his self-consciousness that awakened a universe [1–8].

Anticipating at least the developmental aspects in a compromise between the tensions produced by these theses and antitheses, independently of whether they arose, together with space and time, in a first dramatic singularity, or in the new qualities of a grand fluctuation creating the universe as a spatiotemporal part of an eternally developing being, the grand process of our evolution – at least within our scope of space and time – seems to have endeavored for a period of 15–20 billion years to gain a certain consciousness and understanding of itself [1–8].

Our roots thus reach back to the depths of the past. Together with the universe (or parts of it), life patterns originated from an alien phase transition in the incomprehensibilities of a first grand expansion. We were part of it at the very outset, and we will share the final destiny. Asymmetries of the developing patterns produced dynamic directionalities. Together with general amphiphilicities, they provided new qualities of dynamic order far from thermal equilibrium, advantages that constituted prerequisites for recognition, self-organization, self-reproduction, information processing and optimization [5–8]. Leaving the mysteries of the first beginning of our universe fully intact, the following scenario might perhaps provide the framework for the evolutionary developments (Fig. 10).

Partial freezing of originally unified forces, accompanied by corresponding symmetry breaking, seems to have transformed the primarily symmetric grand unifications of our universe into the diversifications of its present appearance [1–4]. By the subsequent freezing of gravity, strong and elec-

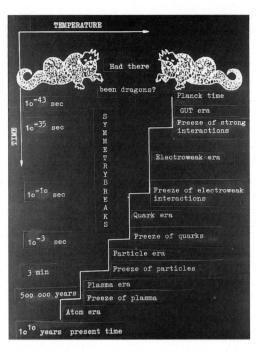

Figure 10. Expansion physics and freezing of originally unified forces [2].

troweak interactions, the grand process evolved through the GUT, the electroweak, the quark, the plasma and elementary particle eras into the current period of atoms. A multitude of heavier atoms, burnt in the hellfires of stars and liberated in their catastrophes, engaged in quite different chemical interactions and in this way created hierarchical patterns of increasing complexity [1 – 8, 17]. The chaos, however, appears to be predetermined by string inhomogeneities and a strange message from the inherencies of the universe. Among the four forces (gravity, strong, electromagnetic, and weak interactions) governing the world of elementary particles, the weak interaction and its unification with the electromagnetic interaction to the electroweak force exhibit a strange characteristic. Contrary to the other forces, which display parity-conserving

symmetries, the electroweak force – mediating by $W^{+/-}$ and Z^0 bosons both weak charged and neutral currents – endows the whole process with a parity-violating asymmetric component [31]. Not only atomic nuclei, but also atoms and molecules, as well as their multifarious aggregations, are sensitized to the special message. Static and dynamic states of enantiomeric species are distinguished from the beginning by a minute but systematic preference for one enantiomer and discrimination against its mirror-image isomer. Amplification mechanisms traveling long evolutionary roads elaborated the first weak signals into dominant guiding patterns [7, 17, 181–n, 31].

The freezing of strong chemical interactions at the interfaces of phase boundaries – and by this also the freezing of the special characteristics of their spatiotemporal coherencies into the individualities of chirally affected mesophases – liberated the richness of their folds into the directionalities of dynamic order patterns (Fig. 11). By the freezing of self-reproduction and self-amplification conditions along water-mediated autocatalytic bundles of nucleation trajectories, the appearing systems – governed by Lehmann's "shaping, driving and directional forces" – gained, far from thermal equilibrium, abilities of information generation, adaptation, storage, processing, transfer and, finally, optimization (Fig. 12). Based on the unique amphiphilic design of their constituents, the biomesogenic patterns evolving from there developed complex structure – motion linguistics and promoted their more and more homochirally based contents in synergetic regulations. Within their adaptational and spatiotemporal universalities, the evolving life patterns retraced the impetus of the early dynamics and reflected within the developing consciousness of their chirally instructed and determined organismic organizations the grand

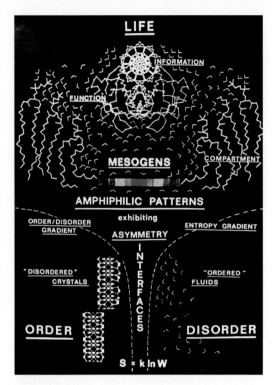

Figure 11. Evolution of amphiphilic patterns – end to the 1970s [7a, 17, 18].

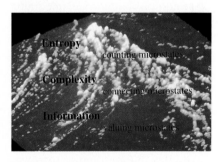

Figure 12. Entropy–complexity–information relationships around the $A = k \ln Z$ postulates, superimposed on evolutionary polydisperse DNA as elucidated by scanning force microscopy (SFM) investigations [1–8, 75].

unifications that dominated their origins. Their creativity, however, somehow aims at beyond the boundaries.

2.3   Evolution of Amphiphilic Patterns

- Views derived from the structure–phase duality centering on solid–liquid-phase interfacial birth-zones and their contradictions.
- The liberation of blueprinting holographic constituents of amphiphilic transient order–disorder patterns and their mesophase coherences, order–disorder (rigidity–flexibility) gradients of molecular individuals and their amplification into phase/domain (transition) characteristics.
- Ways from molecules to macromolecules and from phases to microphases centering on supramolecular biomesogenic organizations.
- Supramolecular structure and systemic biomesogen relationships.
- Polyelectrolyte/water patterns competing for informational, functional, compartmental, and mediating components and combining individual contributions into integrative systems.
- The interchange of transiently acting multi-solvent–solute systems between non-classical chaos foundations and fundamental complexity goals, waiting for energy input to set all their enthalpy- and entropy-determined functional and informational richness "alive".
- Entropy counting, complexity connecting, and information evaluating microstates.
- Guidelines between symmetry and asymmetry, ambivalence and directionality; autopoiesis and system evolution.
- Transition to life and spatiotemporal coherences.

These examples show that it also seems to be possible to describe evolution in mesomorphic terms [5–8, 13, 17, 18, 30–35].

Understandable only as the as-yet last and most highly sophisticated derivatives of our universe, life patterns developed and displayed in their growing complexity all the abilities from which they had originated and that contributed to their further development. The predecessors of the highly advanced life patterns of our days learnt their first lessons while endeavoring to blueprint their maternal inorganic matrix systems [32] in all their informative and functional richness, and they have been further educated in their attempts to follow the changes in their environment. Between the thesis of solid order and the antithesis of liquid disorder, capabilities of optimizable information and function processing seem to have been opened to amphiphilic systems that were able to develop flexible and adjustable structure – function correlation of their own by virtue of the multitude of their transient interchanging phase-domain organization. They gained individuality in the successful integrative handling of complex informational, functional, and compartmental patterns together with sensitive assistant water media. They reached developmental facilities by molecular recognition, intra- and inter-phase/domain relatedness, self-organization and self-reproduction, information processing and optimization far from thermal equilibria.

Among the molecular species screened by evolution in a Darwinian selection for suitable constituents of first dynamic reality-adaptation and, later, reality-variation and creation patterns, amphiphiles with specific hydrophilic – hydrophobic and order – disorder distributions – sensitized to the chiral message of the electroweak force – were the preferred survivors of the grand process (Fig. 13) [17, 18, 33].

At the beginning, a rather omnipotent biopolyelectrolyte pool, dependent on the phase dimensionalities of its outsets, pro-

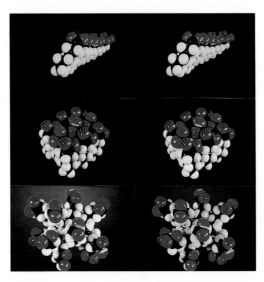

Figure 13. Evolutionary strategies according to the dimensionalities of amphiphilic patterns (top to bottom): one-dimensional: information; two-dimensional: compartmentation; three-dimensional: function [5, 6, 7 a, 17, 18].

vided informational, functional, and compartmental components. By preintelligently developing their individual molecular asymmetries into dynamic system directionalities, the mesogenic constituents of the developing chiral amphiphilic patterns succeeded in a fertile and creative synthesis. Avoiding hyperstatics and hyperdynamics, the disadvantages of the extreme states that contradicted their origins, they developed the creative meso-positions of ongoing dynamic order. Optimizable free-energy stategies on the basis of their molecularly imprinted affinity patterns selected, by preintelligently handling the entropic order – disorder gradients, patterns of chiral mesogenic backbone structures (Fig. 14) [17]. It has been the exceptional usefulness of these side-chain backbone arrangements, the elite successors of the first amphiphilic blueprints of maternal inorganic matrices and their inherent informational and functional capabilities (Fig. 15), that allowed

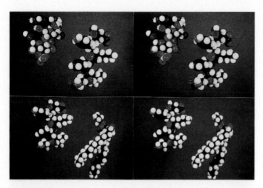

Figure 14. Chiral biomesogenic backbone structures as survivors of a Darwinian selection for optimized homochiral transient order–disorder distributions: (top left) stacked nucleic acid single strand, (top right and bottom left) extended protein single strand, (bottom right) membrane component with a small but nevertheless chiral backbone [5, 6, 7, 17, 18, 33].

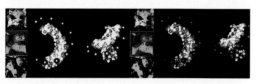

Figure 15. Orientational presentation of liberation of solid-phase information content into self-replicating biomesogenic information patterns, neglecting complex intermediate states of highly condensed information channel designs: (left top to bottom) solid-phase space-partitioners: $Ca_3[Al_2Si_{12}O_{12}]$, $Li_{14}MgSi_4$, Ta_6Cl_{15} [32 c]; (right) replicative DNA, including water cover and counterion clouds [33 a, c, p, q].

interactive division of labor, within mediating water swarms, into the specializations of information, function, and compartmentation, conserving, however, beneath the skin of their special adjustments, the continued primitive universalities of their origin [7 a, 17, 18]. Thus, while a first swift glance might connect the structural features of nucleic acids with information, that of proteins with (information realizing) function, and the remaining characteristics of membrane components with (information and function securing) compartmentation

[5–8, 17, 18], a closer look at the three major components of biopolymeric and biomesogenic organization reveals much broader ranges of different abilities. For instance nucleic acids exhibit functional capabilities in the widespread landscape of catalytic RNA species [34], proteins deal with information, not only in their instruction of certain old protein-production lines of their own [35] but also in the skillful handling of diverse partner molecules and molecular organizations – including themselves [7 a, 33], and compartmental membranes demonstrate additional informational and functional potentials in the complex tools available to them [7 a, 33]. Their self-organizational and self-reproductive facilities, together with their ability to cooperate, created by nonlinear dynamics a rich variety of dissipative structures far from thermal equilibrium and caused – by symmetry breaking – the transition of the whole process from its mainly racemic prebiotic period into the optimizable biotic patterns of homochirality. They regained, as expressed within their final organisms on the basis of these homochiral molecular and supramolecular designs, the symmetry beauty and operation optimization of the whole world of bichirality (Fig. 16). The grand process, however, remained subject to the dialectics from which it originated, the general chiral approach between order and disorder, and the permanent renewal and achievement of selecting a forward-directed path out of these contradictions.

2.4   Transition to Life and (Bio)mesogenic Reflections

Transitions to life and preintelligent operation modes; self-replication and self-reproduction between square-root laws and autocatalysis; transitions from hetero- to homo-chirality; dissipative structures and auto-poieses; amphiphilicity and holographic designs; nonlinear dynamics and self-organizational forces; spatial order – disorder distributions and spatiotemporal coherences; phase/domain transition strategies and cooperativity; reentrant phenomena, molecular hystereses and memory imprints, oscillations and rhythm generation; signal transduction and amplification; information processing and optimization; self/non-self recognition and discrimination; structure – motion linguistics and system intelligence; the possibility of non-classical chaos and the emergence of reality-adaptation, variation, and creation patterns of critical, subcritical, and fundamental complexity; nucleoprotein system developments by metabolism, mutation, and selectivity; synergetics and system optimization; a transition to life almost as mysterious as the universes transition to existence – are such the development of the first blueprints of "$\Gamma\alpha\iota\alpha$ interface phenomena" to life? [5 – 8, 13, 14, 17, 18, 23, 30 – 35].

For a long time, science has tackled constructionist life processes with rather reductionist approaches. Our present investigations of pre-life states, transitions to life, and the developments of life patterns, unraveling pictures of utmost complexity, seem incapable of suitable elucidating the extensive coherences. "The whole being not only more than the sum of its parts, but exhibiting a superior new quality", represents, amidst multiple guiding uncertainty principles, crucial but so far unsolved challenges of scientific endeavors [1, 5 – 7].

Chemistry, and especially organic chemistry, going against the long-held, preferentially reductionistic aims of a purely artificial carbon chemistry, and remembering its long-disregarded original intentions, has moved structures from the level of molecule

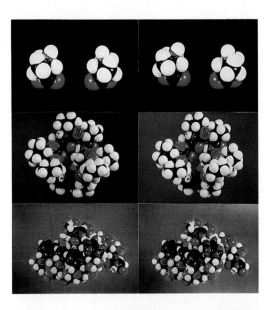

Figure 16. Evolution of molecular and organismic chirality (top to bottom): L- and D-amino acid enantiomers [31]; valinomycin K$^+$-ion carrier [35] as a highly sophisticated outcome of independent amphiphilic bichirality protein developments; transitions from hetero- to homochirality via selection of enantiomers in building up suprachiral structures of RNAs and their self-replication [34]; development of structural and functional asymmetric organismic bichirality on the basis of molecular homochirality [7 a, 17, 18, 31].

to macromolecule to the creative fields of supramolecular organizations, where it has met the attempts of physics to elucidate idealized supreme phase characteristics in the underlying dynamics of microphase and domain systems. Thus, the area of preintelligent self-organisational, self-informational, and self-reproducible pattern as a result of structure–(micro)phase coevolutions [7a, 17], anticipates with first scienfic excursions [36–50] to "minimal replication" [44] and "minima vita" models [49], within the complexity of their interacting molecular individual "parts", presentiments of the "whole".

The biomesogenic approach [7a, 17], proposed nearly two decades ago, tries to shed light on (pre)biologically relevant, highly condensed systems. Within the irreversibilities of non-equilibrium processes, supramolecularly structured and biomesogenically amplified sensitive, transient order–disorder patterns, due to their inherent entropy gradients, appear to be capable of properly utilizing energy dissipation for transformating entropic possibilities from a chaos of creative facilities into sense-giving information patterns. The apparent inconsistencies in the description of the animate and inanimate worlds uncover insufficiencies in our underlying categorizations. The current attempts to redefine terms such as entropy and information, chaos and complexity, matter and consciousness within equilibrium/non-equilibrium contractions might promote not only a better understanding of pre-life states, transitions to life, and life processes themselves, but might also found a more consistent picture of life singularities within the evolving creative universe. "Structure" rather than "phase", however, maintained its leading role for scientific progress for a long time [1–8, 17, 30].

2.4.1   Molecular developments

Life-pattern developments have been promoted by structure–phase dualities. Structural contributions, however, rather than (bio)mesogenic approaches, constituted the "golden age of molecular biology". Basic geometries of biopolymeric species, nevertheless, correspond to classical structural liquid-crystal motifs. Structural interplays, resulting from the earliest outcomes of proteins' own information and production lines up to the nucleation of nucleoprotein systems and their further destinations, display common mesogenic features and mesomorphic operation modes [7a, 17].

Miescher's prophetic view of his "Nuclein" as the genetic material [29a] and Virchow, Mettenheimer, and Heintz's independent findings of behavioral abnormalities [20] could not have been valued at a time when proteins ($\Pi\rho\omega\tau o\gamma$ – the first) were the favored species and structure the ultimate goal of people engaged in "life sciences" [36]. And although some early intuitive perceptions of protein statics and dynamics (not to mention their intimate relatedness) appear convincingly modern [23], the real structural designs of the patterns of life remained obscure for nearly a century.

Somehow indicative of the "optical disposition" of the species *Homo sapiens*, it was not the admirable achievements of the classical period of natural products chemistry [36], beginning, with Fischer [36a] and accompanying with its very subjects also the historical birth of liquid-crystal phase chemistry at the dawn of the century, it was not the highlighting work of Eschenmoser [36d] and Woodward [36c] half a century later and not the lonely prophecy of a physicist that an aperiodic crystal is the genetic material [29c], and it was not even the mystic secret formulae of Chargaff [29e] concerning the base-pairing schemes of nucleic acids and anticipating the whole story to be further elucidated that caused the dramatic scientific "phase transition" in the 1950s. It was the perception of the two "holy" chiral structures: Pauling's protein α-helix [29 d] and Watson and Crick's (as well as Wilkins and Franklin's!) DNA double helix [29 f] that introduced a new age (Fig. 17). The first elucidations of basal chiral geometries of the nucleoprotein system and their immediate projection into the multifold molecular landscapes around was an event that irrevocably altered our views of life patterns in nearly all aspects. In these chiral structures and their biomesogenic relatedness molecular biology set out for far horizons, and the beacon of their structural beauty and informational and functional transparencies has enlightened scientific approaches up to our day.

Since those times the landscape of nucleoprotein geometries has been increasingly enriched with quite different structural motifs, classified into the basal units of "secondary" structure, determined mainly by hydrogen-bond pattern, their "supersecondary" combination schemes, their "tertiary" adaptations to the stereoelectronic prerequisites of real three-dimensional entities and, finally, their aggreagations into supramolecular "quaternary" organizations and beyond [5 – 8, 29, 33].

The realms of both nucleic acids and proteins (Figs. 17, 18) display a certain predilection for the flexible mesogenic filigree of helices – the dynamization of the supreme symmetry of a circle into the asymmetries of space and time, thus connecting both structural and developmental aspects. In the case of proteins, the helical designs vary from preferred right-handed single-stranded versions, via more restricted right- and left-handed single- and double-stranded arrays to special forms of left-handed helix triples. For nucleic acids, the structural landscape is dominated by right-handed antiparallel double-helix motifs, enriched by differently intertwisted helical triples and quadruples and contrasted by alien

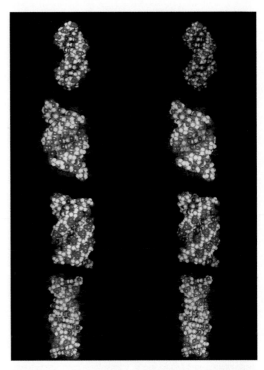

Figure 17. Biomesogenic nucleic acid helix families (top to bottom): ds B-DNA; ds A-RNA; ts DNA/RNA; ds Z-DNA/RNA [7 a, 33 a, c, f, p, q].

versions of left-handed antiparallel double helices. The antitheses of more rigid basic secondary-structure motifs appear in the chirally twisted parallel and antiparallel β-sheets of proteins and will find some correspondence in the cylindrically wound-up double-helix design of the nucleic acid A-families as well as in the so far somewhat dubious versions of suprahelical Olson-type arrangements [5 – 8, 17, 18, 33, 50 g].

In the rivalry between the D- and L-enantiomers of protein and nucleic acid pattern constituents as well as in their chiral amplifications into larger supramolecular motifs, the "less fortunate" enantiomers, energetically discriminated by the chiral instruction of the electroweak force, fought a losing battle from the beginning of the grand evolutionary competition [31]. Ab initio

a)

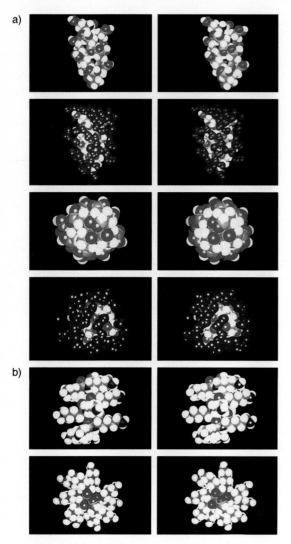

b)

Figure 18. Biomesogenic peptide helices of a) poly(L-glutamic acid) and b) poly(L-lysine) with partial water cover, views along and perpendicular to helix axes [7a, 33g–l, p, q].

for the α-helical and β-sheet patterns of L-α-amino acids in comparison to the structural mirror-images of their D-antipodes. Absolute weighting by the electroweak force, further selections by different amplification mechanisms, as well as the rejection of mirror-image species in building up supramolecular chiral arrangements, together with energy minimizations of the side-chain tools in both proteins (the 20-l proteinogenic side-chain versions) and nucleic acids (the four-letter alphabet of the nucleobases) selected, finally, for the preferential appearance of L-α-amino acids' right-handed protein-α-helices and β-sheets and favored (2'-deoxy)-D-ribose moieties' right-handed antiparallel double helices of nucleic acid families [7a, 17, 18, 31]. And it is also due to these selection and optimization procedures that poly-L-proline backbones can be forced into left-handed helical versions, and special alternating sequences of nucleic acid designs tend to adopt the strange double-helical left-handed Z-motif.

All these chiral structural standards and the corresponding families around them in proteins and nucleic acids represent, however, only the main building blocks upon which evolution has been operating. The colored loop- and knot-stretches and all the more disordered parts of protein and nucleic acid structural designs are of primary importance for the evolutionary aspects of informational and functional processings in our evolving life patterns.

It fits nicely into the picture of dual structure–phase views of biomesogenic organizations that objects of "rod-like appearance", for instance the little "world" of the tobacco mosaic virus (which – its overall design reduced to a simple rod-like entity – became the starting point of Onsager's theory [37]), as well as much simpler protein and nucleic acid helices are typical mesophase formers in the classical liquid-crystal phase

calculations of the preferred aqueous-solution conformations of some α-amino acids and glyceraldehyde, the classification-ancestor of all sugar-moieties, are indicative of a stabiliztion of L-α-amino acids and D-glyceraldehyde relative to their discriminated mirror-image enantiomers by some 10^{-14} J mol^{-1}. The same seems to hold true

sense [7 a, 17, 25]. As the transient final "or-der" in self-recognition, self-organization, and self-reproduction is approachable only by means of mediating "disorder" steps, Fischer's more static lock-and-key model [36 a, d–f] turns by its own inherent "key"-dynamics into Koshland's "induced fit" [36 g]. It is within the frame of its close connections with the preinstruction of biomesogenic species by the chiral message of the electroweak force that the transformation of chiral molecular units into the power of "supra-chiral" macromolecular patterns, and from here to the further extended chiral expression of supramolecular phase organizations and their autocatalytic self-reproductions, provided a useful sequence for the rescue and elaboration of an early message of our universe from the noise of its first hidden appearances up to the homochirality as a matter of course nowadays [7 a, 17, 30, 31, 47].

The "central dogma" [8 c, 29 f, 33 p, q], proposed by the creators of the DNA double-helix model in order to channel their considerations of the information flow in biological systems, was partially inverted in historical sequence. Its alpha and omega determined the evolutionary action scheme: variation in the information content of the genotype (DNA) and subsequent functional selection in the phenotype (protein). "Dynamic proteins" and "static nucleic acids" competed, as in the "chicken and egg" problem, for historical evolutionary priority [5 b]. But while the original alternative found its solution in both the informational and functional "hypercycle" [5 b], developing its space-symmetry beauties into time-helical creativities, and while dual structure–phase views – contrary to classical perceptions – foresaw much closer spatio-temporal coherencies for the evolving mesophase systems [7, 17], while RNA turned out to be in some cases DNA-informative

[8 c, 33 p, q], the old chicken and egg problem presented a futher surprising aspect of the "meso"-positions of the so-far unfortunate and disregarded RNAs, a view of what might be called an early RNA world [34].

The evolutionary highways in the development of the DNA–RNA-protein triad, however, seem to have been preceded by a peculiar intermezzo. The ambitious attempt of proteins of cover self-consistently both informational and functional capacities has survived up to our days in some unusual and, despite all their structural restrictions, admirable molecular systems [35]. A group of depsipeptide and peptide carriers and channels as well as their possible subunit structures – produced on old, both informational and functional protein lines – display enchanting achievements in the skillful handling of biomesogenic operation modes [7 a, 17, 35]. Interestingly, all this was brought about within any preference of a special homochirality. On the contrary, it is just the surprising selection of alternating different chirality patterns from the prebiotic racemic pool and the careful and intelligent order–disorder design of rather small molecular entities with adjusted alternating chiral codes that made such beautiful arrangements as, for example, the "bracelet" of valinomycin work (Fig. 19) [18 l–n]. In the impossibility of bringing the giant information content not only functionally but also informationally alive, the ambitious enterprise of proteins, the "first" to create by the power of their own legislative and executive intelligent "bichiral" patterns, ended up in an evolutionary impasse – notwithstanding all the convincing achievements that match not only intriguingly later-to-be-developed operational modes of the DNA–RNA–protein triad but also our own intelligence.

While proteins thus failed in their omnipotence, the apparently less qualified RNAs seem to have built up a preliminary, medi-

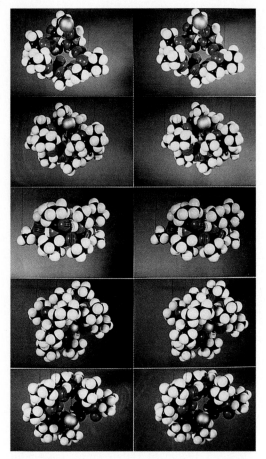

Figure 19. Bacterial representatives of active ion-transport through membranes, modeling phase-transition strategies of protein information function lines: stereo-presentation of a CPK-valinomycin "movie", mediating by highly sophisticated biomesogenic interplays a K^+-ion membrane passage [7a, 33p, q, 35].

did they all develop a more and more interesting structure – phase "eigenleben", RNAs also proved to be both informational and functional. With their special characteristic of molecular hysteresis [48], they were among the first to provide a basic understanding of memory records, oscillations, and rhythm generation in biological systems, and they supplied – starting with their catalytical facilities in self-splicing and ending up with general "ribozyme" characteristics – successful predecessors and competitors for proteinic enzyme functions in both basic and applied research. RNAs – combining in their early forms both geno- and phenotypic aspects – seem to have mediated the grand DNA – RNA – protein triad into its present expressions and survived up to now as an inherent principle of life patterns [17, 34, 39 – 44, 48, 50].

The final ways to utmost complexity in informational and functional processings are illustrated by impressive landmarks. The early prebiotic interactions of the shallow groove of sheet-like right-handed A-type RNAs with the right-handed twisted antiparallel β-sheet of a peptide partner might have established the first fertile intimacies between nucleic acids and proteins [38]. Within this hypothetical first productive organization, the archetypic protein-β-sheet, with its suitability for easily sorting polar – apolar and hydrophilic – hydrophobic distributions, as well as the archetype A-RNA, with its partially complementary informational and functional matrix patterns, could "live" first vice versa polymerase capabilities and thus mutually catalyze early reproduction cycles, connected with chiral amplifications and informational and functional optimizations (Fig. 20) [7a, 17, 18, 38, 43]. It proves possible to further extend this model to protein-β-sheets fitting into the minor groove of B-DNA, connecting by this geometry and functionality the

ating, and progressing RNA world [34]. "A tRNA looks like a nucleic acid doing the job of a protein." Crick's pensive considerations became reality: mRNAs – the differentiating blueprints of the DNA information store, tRNAs – the structurally and phase/domain disrupted mediators of nucleoproteinic interrelationships, and rRNAs – the long-blamed "structural support" of ribosome protein-synthesis machinery, not only

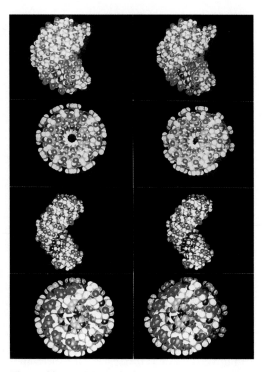

Figure 20. Nucleic acid and protein versus polymerase efficiencies. (top) RNA and (bottom) DNA with chirally twisted β-sheets in the shallow and minor grooves, respectively [38], perhaps as an outcome of Katchalsky's catalytic montmorillonite solid-phase patterns [7a, 32a].

A/B-alternatives of nucleic acid families by mediating β-protein design [38].

The last step to liberating all the inherent potential of these early nucleoproteinic systems of RNA and protein was found in the dual conformational abilities of the small 2′-deoxy-D-ribose cycle to account for a conformationally amplifying switch between A- and B-type nucleic acid versions [7a, 8c, 33]. The discovery of the information store of DNA enabled the systems to separate the replication of a more densely packed DNA message from the informational and functional instruments needed in the hitherto existing nucleic-acid–protein interaction schemes and promoted by this

transcriptional and translational operation modes with far-reaching maintenance of so far elaborated RNA–protein cooperation. The new qualities were brought about by the structural deepening of the shallow RNA groove into the minor groove of DNA, which allowed the continuation of the successful, however, rather unspecific, RNA–protein contacts together with newly established small effector regulations, and, still more important, by the opening of the so-far hidden huge information content of the RNA deep groove into the much more exposed major groove of B-type DNA, which admitted specific DNA–protein recognitions [7a, 17, 18, 33]. All these newly developed and achieved operational capabilities brought about a dramatic transition from the unspecific contacts between proteins and nucleic acids in their shallow and minor grooves, respectively, to specific recognition and functional information processing, mostly exemplified by protein helices in the B-DNA major groove.

The resulting DNA–RNA–protein triad – mediated by intelligently handled clusters of water molecules – promoted the development of universally applicable informational and functional patterns, which not only developed their order–disorder distributions into optimizable operational modes of biomesogenic systems, but responded, moreover, intelligently to the early chiral message from the inherent qualities of the universe and its translation into "molecular creativity" [7a, 17, 30, 33, 43].

2.4.2   Molecular Matrices

The projection of individual capabilities into collective cooperation modes; self-replication and self-reproduction in highly condensed states – suggestions for naturally occurring systems, simulations by artificial

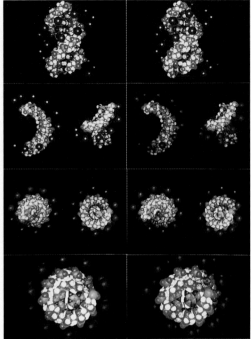

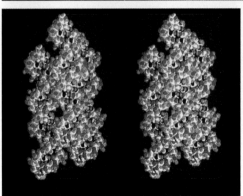

Figure 21. Biomesogen regulations between structure and phase (left to right and top to bottom): visualization of biomesogen polyelectrolyte regulations exemplified by DNA/water-shell/(hydrated) counterion-cloud pattern statics and dynamics; arbitrary DNA/counterion-cloud/water-domain arrangements symbolizing operative biomesogen nucleations "between structure and phase" [7 a, 29, 33 a, c, f, p, q].

phase duality miracles abound in the pre-transitions and transitions to life [5 – 8, 17, 31, 39 – 43].

It was the new quality of dynamic order of coherent matrix systems that established, within the transiently changing patterns between flexibility and rigidity, order and disorder, growth and decay, guidelines for self-organizational behavior and self-reproductive developments. The crucial step that had to be achieved by the candidate patterns of life, however, was the evolutionary development of self-reproduction systems that could not only propagate their dynamic order states within the surrounding disorder but would even make use of the creative possibilities of uncertainty principles within non-classical chaos for their own informational optimizations [7 a, 43], a first decisive step that unravels as well all the other life developmental processes to follow that make use of mesogenic capacities for mediation and continuation of the grand process. "It has not escaped our notice that the specific pairing we have postulated immediately suggests a possible copying mechanism for the genetic material" – rarely has anything fascinated science as much as the "holy" structure of DNA, born in that classical understatement of Watson and Crick [29 f], in which the prophecy of Schrödinger that an "aperiodic crystal" was the genetic material became reality (Fig. 21) [29 c]. The discovery of the structure of DNA irrevocably altered our view of life and represents one of the really major landmarks in the development of science. The prototype of a dynamic matrix became the beacon that illuminated research in the fields of molecular biology and redirected organic chemistry, which had, contrary to its first ambitions, developed mainly as a chemistry of artificial carbon compounds, back to its early aims. The duplication of the holy structure represented the key mechanism for prebiot-

systems; square-root laws and cooperation of partners versus autocatalysis and decisive selections of competitors; inherent coherences and judging bifurcations – structure –

ic systems on passing the boundary between the inanimate and animate worlds. The accompanying scientific views, however, elucidated mosaics of processing of forms rather than integrative visions of forms of processing.

Todd [39a] sought an answer for the somewhat excluded organic chemistry in a landmark appeal: "The use of one molecule as a template to guide and facilitate the synthesis of another … has not hitherto been attempted in laboratory synthesis, although it seems probable that it is common in living systems. It represents a challenge, which must, and surely can, be met by organic chemistry". And this grand concept, indeed, appeared to be a good sign. Quite soon Schramm et al. [39b] aroused enthusiasm with the "noenzymatic synthesis of polysaccharides, nucleotides and nucleic acids, and the origin of self-replicating systems." Schramm's experiments – interestingly, brought about in highly condensed states – had to push the evolutionarily optimized standards back into the high error rates of their prebiotic beginning. Purely artificial template and matrix experiments that evaluated all the interactive possibilities of both covalent and non-covalent nature, however, were devoid of even this possible relationship to evolutionary developments. The first great "chemical" departure into the temptations of matrix reactions [39] ran aground on innumerable difficulties. Only a few special cases of matrix reactions found successful applications. The use of coordination matrices reminiscent of solid-phase informational and functional patterns, and already known from stereopolymerizations, advanced research on complicated natural products, for example, chlorophyll and vitamin B_{12} [50a, c]. Nucleic-acid-derived recognitions enriched the informational and functional landscapes of mesogenic systems [17, 18]. The main goal, however, to follow

evolution on the supposed way to make polymers on suitable templates in zipperlike reactions – ranging from covalent to noncovalent variants – failed due to the insurmountable complexity of the systems involved. In vain, all attempts to model nucleic acid self-replication properties by nucleic acid analogous polycondensation, polyaddition, polymerization, and polymer analog reactions [7a, 17, 18, 43]. In vain, for instance, the once so intriguing idea of blueprinting natural patterns independently of natural polycondensations by the help of artificial matrix radical reactions of N-vinylnucleobases that linked together man-made polymerization groups and evolution's recognition patterns [39f–i].

The dominant element in scientific progress turned out not to be the tremendous variety of chemical matrices and the distant aim of building up systems capable of self-replication [39]. Rather, fascination with the seemingly effortless elegance of the native structure–phase prototypes stimulated the vanguards of chemistry and biology in their campaign of molecular penetration of biological systems. Rapid scientific breakthroughs introduced the "golden age" of molecular biology [40]. Oparin and Haldane's [40b, c] heirs, Eigen and Kuhn [40f, h], gave these events a time perspective and with their "information" described the vector of the grand process of evolution. Along also with their hypercycle-view [5b, 40h], a multitude of rival problems, which had been simmering in the collective subconsciousness, resolved themselves out of the complexity of the original goals of template experiments. Self-replication, mutation, and metabolism, as prerequisites for selection, made up the list of criteria, and through these, information and its origin, its evaluation and processing, and finally its optimization governed the evolutionary history of prebiotic and biotic systems. The two ways of scientifically approaching the problems, the use of natural systems [40, 41] and the "ab initio" chemical outsets [42], were to find their followers [43 – 50].

In connection with Spiegelman's initial experiments [40 e], Eigen and Schuster [5 b, 40 h], Joyce, and others [41] attempted to "bring to life" the theoretical premises behind the laboratory realities of enzyme-catalyzed RNA replications and evolutionary experiments. Orgel's group, by contrast, exerted itself of transform diverse matrix relationships into artificial enzyme-free nucleic acid formation [42, 43, 44 b]. When Altmann [34 a] and Cech [34 b, c] – confirming earlier speculations by Crick, Orgel and Woese [7 a] – raised RNA to the throne of an archaic informational and functional omnipotence, it amounted to a late justification of the toil and trouble of this "nucleic acid first" [34, 42] route. Self-splicing RNA complexes, polymerase activities, ribozymes and hypothetical RNAsomes (views partially extendable also to DNA) afforded unusual insights into the genotypic and (!) phenotypic behavior of a single nucleic acid species [34 a – e].

And with the supposedly greater understanding of this "RNA world" fresh impetus was given even to the purely chemical approaches, once so promising and now almost forgotten. A rapidly growing number of artificial self-replication models, covering nucleic acids as well as their near and distant analogs, but also successors of the nearly forgotten Fox microspheres [40 d] in the field of membrane components, are representative of the wealth of variety of the creative outburst and clarify with their different degrees of abstraction one of the dynamic peripheries of our times [43]. Through elucidation of the static and dynamic principles of replicative information systems, all these models endeavor to gain insights into both the transitional stages

between chemical and biological evolution and the possibilities of artificial developments of their chemical simulations.

After innumerable attempts to explain the matrix relationships of mono- and oligonucleotides to orientating and catalytic oligo- and polynucleotide templates, two self-replicating nucleic acid models (a DNA-analogous hexamer system by von Kiedrowski et al. [44 a, c – d, f, g – i] and an RNA-analogous tetrameric assembly from the Orgel group [44 b]) opened new ways (Fig. 22) [44]. Von Kiedrowski's first successful model – hexadeoxyribonucleotide duplex – became a leitmotif in the detailed treatment of growth kinetics and anticipated later self-replicating "minimal systems" in many ways. By using the ligation of oligonucleotides, he exploited their higher cooperativity to attain increased stability of the matrix duplex and the ternary formation complex. The choice of palindromic systems reduced the complexity of biological replication experiments to a simplified kinetic measurement of its identical (since self-complementary) matrix components. Demonstrating the autocatalytic behavior of his systems, he was able to obtain direct proof of self-replication. The surprising square root law of matrix growth kinetics (curiously even computer viruses seem to be in love with it) with its ideal case of a parabolic reaction course, derived from the hexadeoxynucleotide duplex, which was confirmed slightly later by the Zielinski/Orgel system [44 b] and, finally, also verified in the Rebek dimer assembly [45], was recognized as autocatalytic system behavior of self-replicating oligonucleotide templates under the constraints of isothermal conditions, where stability relationships of matrix reaction partners might exclude the expected exponential growth kinetics. Later functional and reactive refinements overcame certain shortcomings of autocatalytic reaction

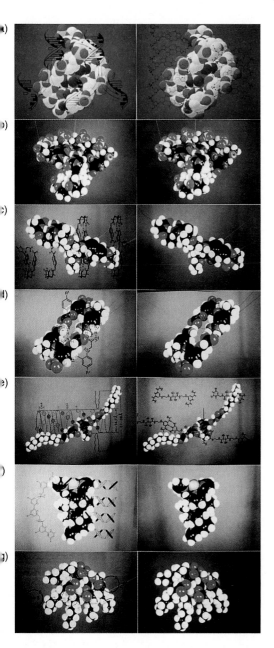

channels that were too heavily weighted by spontaneous noninstructed syntheses. A hexameric system, constructed from two trimeric blocks by phosphoamidate coupling, revealed sigmoidal growth behavior and stood out on account of its extraordinary autocatalytic efficiency. Its present extension to a three-matrix block will not only enrich the informational content of the model system but seems for the first time to invite studies of selection behavior [40h, k, 44].

The patterns of autocatalysis with respect to parabolic and exponential reaction courses, which strongly affect the conclusions of Eigen's evolution experiments concerning the decision criteria for mutant selection and coexistence [5b, 40h, k], can now be derived from the thermodynamic data of the matrix patterns and their reactivities, and offer quite new views, with autocatalytic cooperation between competitive species [40k]. Separate from "enzyme-catalyzed" evolution experiments with RNA and DNA systems, basic questions of prebiotic behavior can for the first time become the subject of detailed experimental research [40k, 43]. While continuing their studies on more complex autocatalysis patterns, von Kiedrowski et al. diagnosed modulation of molecular recognition as an operational deficit of earlier artificial self-replicational nucleic acid systems with regard to exponential reaction courses, and identified it as an ideal aim for future models [44]. On its way to the nucleoprotein system, evolution must have

Figure 22. (Bio)mesogens approaching pre-life states [7a, 17, 18, 43]: a) von Kiedrowski's and Orgel's "minimal models of replication" on the basis of self-complementary oligonucleotide DNA and RNA systems [44a–d, f–h]; b) distant nucleic acid strand-analogs as matrix reaction models [7a, 18, 19, 39f–i]; c) Rebek's self-replicational and evolutionary nucleoside analog model [45]; d) von Kiedrowski's self-replicational amidinium-carboxylate model, being suggestive of exponential growth kinetics [44e]; e) Lehn's

successful mesomorphic system for demonstrating surprisingly efficient homochiral selection in building up supramolecular arrangements [47a]; f) Lehn's helicates, commemorative of early information transfers between inorganic matrices and organic ligands [47b–e]; g) Luisi's first "minima vita" approach of chemical autopoiesis, modeling self-replicational evolutionary states on the basis of compartmental and functional arrangements [49].

had a similar view of the problem, when, in Lucretius's "external game" [40 a], it endowed nuclei acids, which are somewhat insufficient in this respect, with proteins, experienced in phase and domain regulation control, and thus achieved an ideal milieu for directing modulations of recognition [5 d, 7 a, 17, 18 m, n]. And, indeed, at this point something of what might have favored nucleoprotein systems so extremely in comparison to replicative states of nucleic acids themselves will become apparent. While, presumably, the both informational and functional RNAs achieved the first successful self-replications on their own [43], they seem to have been outclassed in future developments by the cooperative efforts of nucleic acids and proteins. The urgent need to establish suitable and reliable regimes of strand-recognition, annealing, separation, and reannealing, which had thus far only been easily brought about by drastic variations of reaction temperature, seem to have been accounted for under stringent isothermal conditions only by the complex patterns of nucleoprotein systems [7 a, 17, 18]. With no directing man-made artificial chemistry and physics at hand, the general evolutionary breakthroughs in self-replication had to await the play-educated abilities of complex biomesogenic organization. It was especially the impressive capabilities of proteins – both functional and informational – which display with more then 20 side chains a great variety of affinity and entropy variations [7 a, 33], that could provide suitable system-inherent isothermal conditions. They did this, in close cooperation with their preferentially informational, but also functional, nucleic-acid matrix mates and mediating water patterns by the highly sophisticated transiently acting mixing of domain-modulated paths along desirable reaction coordinates. Only the integrative efforts of nucleic acids and proteins, which sub-

merged their structural individualities into the biomesogenic unifications of the functionally and informationally completely new characteristics of transiently acting order – dis-order patterns of nucleoprotein systems, reached the desired optima in self-organization and self-replication, in general information-processing and optimizations [17].

Matrix growth kinetics as known for oligonucleotides were followed by a drastically abstracted artificial replication system developed by Rebek's group [45]. This constitutes, in some sort of distant nucleic acid/protein analogy, with the two nucleic-acid systems the pioneering triad in this field, and seems to augur, after all that has gone before, a fresh departure for the far horizons of artificial chemical evolution systems. The dimer assembly developed from host – guest relationships combined the interactive and cognitive possibilities of native nucleic acid matrices with the more general biopolymer relationships of the amide bond formation involved in the matrix reaction. "Kuhn divergences" [40 f] generate a tremendous number of new matrix and replication variations, and from this breadth develops an impressive range of impending convergences. 2-Pyridone derivatives [46 c], already foreseen in mesogenic and polymeric nucleic acid analog approaches [46 a, b], reoccur in new guises [46]. The successors to elite nucleic acid templates are surprisingly the apparently simpler amidinium-carboxylate matrices [44 e]. Astoundingly, however, in the attempt to reduce generalizing principles as far as possible, desirable operational modes become accessible. The dramatically abstracted amidinium-carboxylate systems, which, nevertheless, cover certain essentials of complex nucleic acid – protein interactions, have proved to be susceptible to molecular recognition modulation and even seem to display – to the delight of their examiners – exponential growth kinetics. Beautiful chiral main-chain LC-polymers – built up as impressive supramolecular helical arrangements from bifunctional recognition units – convincingly confirm in the hands of Lehn [47 a] the selection and discrimination capabilities of supramolecular organizations in transitions from hetero- to homochirality [7 a, 17, 18, 31]. Detailed aspects, such as the possibilities of coordination matrices in native nucleic acid assemblies, make themselves felt in helicates [47 b–d], whose structures reflect also relationships between our life process and its basic matrices [7 a, 32]. Matrix studies of triplex systems [48 a–q, t] model possible strategies of nucleic acid organization and by this detect not only protein-like behavior of RNA-Hoogsteen strands in reading informational DNA-duplex patterns [48 l, o] but also bridge the gap of the basic hysteretic mechanism of information processing in more highly condensed systems [48 a–q]. The paths of chemistry – from the chemical bond to the chemical system, from molecularity to supramolecularity, from statics to dynamics – are united with the spatiotemporal coherences of phase and domain regulation strategies from physics in the asymmetries and nonlinearities of life sciences, affording processes that are stabilized by their own dynamics. From the earliest prelude to the unknown intimacies of nucleation of the nucleoprotein system up to today's complex evolutionary character, only the to some extent adequate complexity of the scientific approaches might provide sufficient stimulation potential to approximate the grand process.

It is fully within this context when, in addition to the nucleic acid–nucleic acid analog pioneering triad of self-replication systems, a fourth innovative approach adds to the "minimal systems" of preferentially informational replicators [43, 44] the new view of a "minima vita" challenge [40 l] related mainly to functional and compartmental aspects. It appears somehow as a reincarnation of Fox's microspheres [40 d], when micelles, as representatives of the compartmental partner in life games, are proposed by Luisi et al. as first examples of "minimal life" models [49], where, independently of all historical outcomes, the chemical autopoiesis is taken as a minimum criterion for not only self-reproductive but also in some way life-bearing systems. With the quite recent implantation of informational nucleic acids and the future goal of a nucleoprotein core [49 f], they might really develop as first intriguing artificial nucleations of primitive protocell design. An approach that might perhaps be able to realize Lucretius's "quacumque inter se possent congressa creare" [40 a] of nucleic acids, proteins, and membrane components that governed – together with suitable mediating water networks – the origin of life patterns.

It is, indeed, just this complexity that for today's chemistry provides provocation and stimulation, intimidation and temptation, love and hate and fate together. The present artificial systems still remain utterly outclassed by even the most primitive life forms such as RNA viruses [5 b, 17, 18, 40 h, 43]. The possibilities of describing natural selection behavior according to quasi-species distributions in the extreme multidimensionalities of sequence spaces are, for artificial systems, at best a very distant utopia. However, independent of the respective chemical character, basic patterns of autocatalytic activities once again make the original image of that "notational replicative" DNA double helix the center of hopeful expectations. With all its early primitiveness, but also with its promising inherent potential of "minimal models" of self-replication [43–48, 50], and – just to follow –

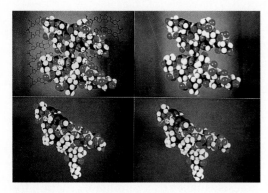

Figure 23. Eschenmoser's homo-DNA [50c–g] playing the game of artificial evolution and simulating from nucleic acid viewpoints aspects of general carbohydrate codes [60, 61]; (top) Olson-DNA and RNA arrangements [50g] in comparison to comparable appearances of homo-DNA; bottom) Olson-RNA –peptide interactions modeling early intimacies in the nucleation of the nucleoprotein system [7a, 17, 18] and by this excluding more rigid hexose-DNA/RNAs from evolutionary trends [50].

"minima vita" models [49] of life, chemistry enters the fields of biological complexity in what should be the most decisive region of evolutionary formative processes, and in doing so is gaining new qualities. Qualities, however, that emphasize in continuation of Schramm's starting point [39b] the importance of still disregarded mesogenic reaction behaviors in particular and inherent "phase-chemistry" in general [7a, 17, 18].

And then there is finally – besides the artificial evolutions of selecting functional nucleic acids for specific interactions with cooperative mates, high-affinic recognitions, and catalytic activities from random pools – the "evolution" of an individual scientific life's work [50], which itself follows decisive stages of the grand process: the exploration of early chemical requirements, the development of prebiotic ligand systems, the fixation into the ordered structures of informational inorganic matrix patterns

and, finally, the liberation of their inherent wealth of design and information into the order–disorder of today's nucleoprotein system. Using basic hexoses – in some respects the successor molecules of evolution – a never attempted, or perhaps only forgotten, "evolutionary step" is taking place [50c–f, h] – once again in the area of replicative (homo)nucleic acid systems (Fig. 23), and intriguingly not without relevance to the so far unsolved problems of carbohydrate recognition codes [7a, 8c, 33p, q].

But this is another whole story, and another great game. A game that is representative in all its loving utilizations and impressive manifestations of today's chemistry standards and facilities for our future ways of modeling of what has created us – without any chance, however, to renew artificially the "whole" on our own.

2.4.3   (Bio)mesogenic Order–Disorder Patterns

Transient order–disorder coherences converting a chaotic variety of quantum-duality-derived structural prerequisites and mesogenic capabilities – far from thermal equilibrium – into complexity patterns fundamental to life; artificial systems in their independent enterprises and in the attempts to retrace evolution's (bio)mesogenic developments; fundamental complexity irreducibilities contradicting reductionistic scientific approaches; quantitative limitations standing in contrast to qualitative comprehension; entropic gradients achieving equilibria self-organization and mediating nonequilibria spatiotemporal coherences into life-pattern developments – transiently acting (bio)mesogenic patterns intriguingly add to a consistent view of life-pattern developments within the grand process [7a, 17, 51–52].

Reflections of life's biomesogenic modes of operation remained outside mainstream molecular biology for rather a long time. The rapidly enlarging range of thermotropic and lyotropic liquid crystals repeated in their second heyday the long journey of native mesogen developments from monomers to oligomers and polymers almost independently of molecular biological achievements (Figs. 24 a–h) [7 a, 17, 18, 51, 52]. The last few decades have brought about the end of simplicity by bizarre stereoelectronic variations and the accompanying difficulties of corresponding phase systematization [51, 52]. The theses of rod-like thermotropics and head–tail designed lyotropies (Fig. 7), which governed liquid crystal developments for a century, have lost their uniqueness. Complex molecular interactions submerge the extreme conceptions of thermotropy and lyotropy. Dynamic order–disorder patterns are creating overlying interconnections. Artifical mesogens are approaching the complexity of their biomesogen standards and encountering all the advanced-material aspects evolution had already operationally integrated. But the same difficulties that Lehmann experienced are being met by the complex biomesogen systems [17, 18, 51–53].

The creative principles of "subject" and "relatedness", having already traveled across the laborious plains of fermion–boson and particle–wave dualities [1–8], reach the complementaries of "structure" and "phase" [17, 18], culminating in the beloved species of "side-chain polymers" and their outstanding mesogenic capabilities [7 a, 17, 18, 51, 72]. Biopolymeric and biomesogenic species seem further integrable into supreme guiding fractal motifs of organismic morphogenetic expressions [78] and even into quantum-field patterns of human consciousness fields [72 x–z, 79, 80]. From the discontinuities of rudimentary, but nevertheless chiral backbone-stumps of membrane components up to the continuations of the beautiful organic/inorganic strand-repetition of nucleic acids, the functional backbone-filigrees of proteins and the carbon-substitution-framed water-cluster individualizations of carbohydrates, evolution has developed nearly infinite action mode capabilities from a limited selection of building and functional blocks within narrow operating temperature ranges [5–8, 17, 18, 29, 30, 43, 51–71]. Impressive side-chain-pattern selections, adjusted to the individualities of quite different backbone variations, complete the energetic and entropic instruments for creating life as an outstanding new quality of universe accelerations. The primitive amide group – a vision of horror to all classical liquid-crystal scientists – embedded with all its hydrogen-bond capabilities into the exclusivities of heteroaromatic horizontal and vertical interaction modes and reappearing also in the intra- and interstrand recognition and stabilization patterns of proteins, has developed as an essential part of information origin, processing, reproduction functionalization, and optimization. Amides – partially modified by their imino substitutes – frame the first successful information channels and mediate the first replication schemes, probably of RNA species that gain additional functionality from the excess patterns of residual hydroxy-groups within the conventional nucleic acid backbone designs. And amide groups also provide for the three-dimensional intra- and interstrand "intelligence" of proteins that functionalize the amide sender–signal–channel–receiver patterns of the informational component. The developmental one-dimensional information patterns of nucleic acids, wound-up for purposes of information storage and transfer within the elegance of preferably double-helices and their suprahelical packages, find some

a)

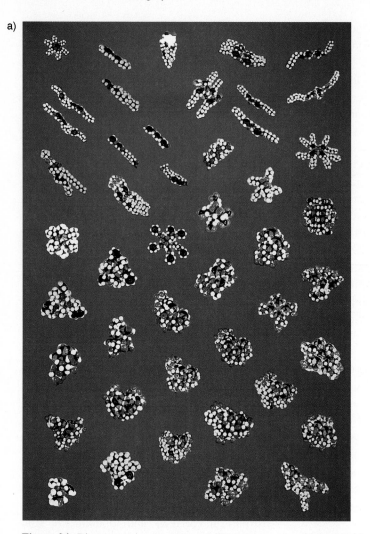

Figure 24. Biomesogenic structures: a) (Bio)mesogens displaying order – disorder distributions in CPK-presentation (left to right and top to bottom); hexa-*n*-alkanoyl-oxybenzene discoid – Chandrasekar's first non-rodlike liquid crystal [28 a, 51 c]; enantiomeric cholesteric estradiol- and estrone-derivatives [17 a, c, d, 26 f, 51 a, s, u]; Reinitzer's cholesterolbenzoate [21, 22] – together with the acetate the foundation stones of liquid crystal history [21, 22]; Kelker's MBBA – first liquid crystal "fluid" at ambient temperature [13 f, g]; Gray's cyanobiphenyl nematics for electrooptic displays [25 a, 51 e]; lyotropic lecithin membrane component [7 a, 14, 27 d, 52 a] and valinomycin-K⁺-membrane carrier [7 a, 35]; thermotropic cholesteryl-side-chain-modified polysiloxanes with the combination of flexible main-chain and side-chain spacers [51 a, h]; thermotropic azoxybenzene polymers with flexible main-chain spacers [51 a]; thermotropic cya-nobiphenyl-side-chain-modified polyacrylates with flexible side-chain spacers [52 b]; lyotropic stiff-chain polyamide [52 c, d]; lyotropic cellulose triacetate [52 e]; smectic diacylhydrazines [17 a, c, d, 26 f, 51 t]; thermotropic multiple-phase long-chain vinamides displaying intra- and intermolecular hydrogen bonds [52 f]; *n*-alkoxybenzoic-acid dimer smectics [52 g]; thermotropically designed 2-pyridone-dimer smectics [17 a, c – g, 27 g, 39 g, 46]; Lehn's supramolecular-helix building-blocks on the basis of nucleic-acid-base-analog recognitions [47 a]; gramicidin-A-K⁺-membrane channel [52 h]; (tentoxin)₃-K⁺-membrane subunit carrier/channel/store [52 l]; lyotropic poly(γ-benzyl-L-glutamate) [51 d, 52 k]; poly(L-glutamic acid) [51 d, 52 l]; poly(L-lysine) [51 d, 52 l]; discoid metallophthalocyanine [51 o, 52 m]; (A)ₙ-stack as operating in messenger processings [52 n]; 2–5(I)ₙ as structural analog of the oligonucleotide regulides

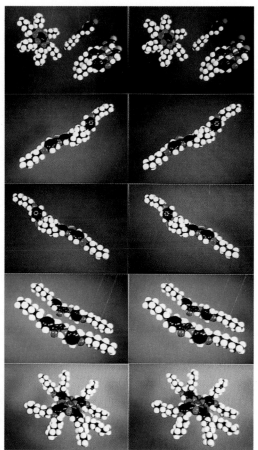

c)

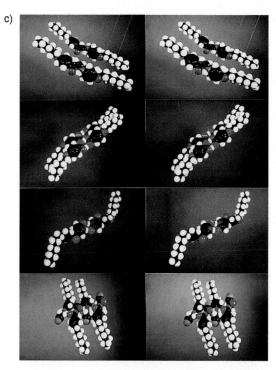

d)

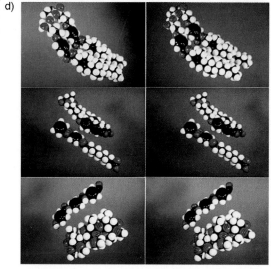

2-5-A [52 o]; unusual pairings of ds $(U)_n \cdot (U)_n$, ds $(C)_n \cdot (CH^+)_n$, ds $(AH^+)_n \cdot (AH^+)_n$ [52 n, u]; bisacridinyl/DNA bisintercalator complex as an advanced model for biomesogenic nucleic acid/small ligand interactions [7 a, 18, 52 p]; ds B-DNA as life's informational component [7 a, 17, 18, 28 d, 29 f–h, 33 a, c, f, p, q, 52 o–v]; ds Z-DNA as regulation facility within genome [33 a, c, f, p, q]; ds A-RNA with simulation potential for viruses and stimulators of cytokine-induction [7 a, 17, 18, 33 a, c, p, q, 52 o–v]; ts DNA/RNAs as logic-blocks of hysteretic biomesogenic regulation systems [7 a, 17, 18, 33 a, c, f, p, q, 48 a–q]; qs $(G)_n$ with teleomeric genome regulation functions [7 a, 17, 18, 33 a, c, p, q, 52 x]; ds RNA/analog semiplastic complex with interferon induction and reverse transcriptase inhibition activities [7 a, 18, 33 a, c, 52 w]; β-casomorphin messenger between opioid and cardiovascular-cardiotonic regulation programs [17, 18, 52 y]; substance P as messenger of central and peripheral neuronal programs [17, 18, 52 z]; $(L-Lys)_n$-adaptation to B-DNA major and minor, and A-RNA

deep and shallow grooves [7 a, 17, 18, 33 a, c, p, q]; mismatched GGU-base-triple presumably mediating self-splicing of tetrahymena rRNA [18 k]; specific recognition of α-helical Cro-repressor in the major B-DNA groove of Cro-operator [33 m, n]; unspecific recognition of protein-β-sheet and B-DNA minor groove as modeling early polymerase efficiencies of nucleo-

e)

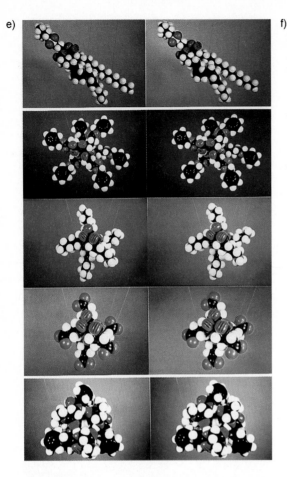

f)

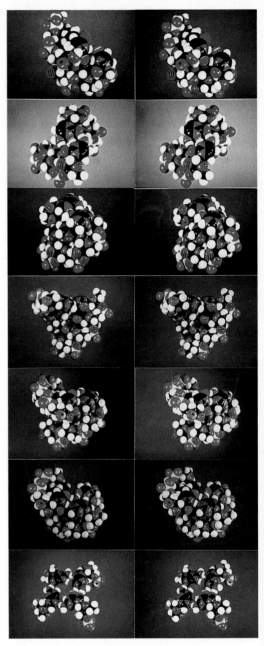

protein intimacies [38]; ds Olson-RNA/(L-Lys)$_n$ interplay dynamics [17, 18, 50 h]. b) Thermotropic artificial systems illustrating geometric and functional variabilities around the common motif of longitudinal and circular order – disorder patterns (top to bottom): hexa-n-alkanoxyloxybenzene discoid [28 a, 51 c], cyanobiphenyl nematic [25 a, 51 e]; diacylhydrazine smectic [17 a, c, d, 26 f, 51 t]; enantiomeric cholesteric estradiol derivatives [17 a, c, d, 26 f, 51 a, s, u]; vinamide smectics [52 f]; discoid metallophthalocyanine [51 o, 52 m]. c) Thermotropic hydrogen-bond species partially modeling drastically abstracted nucleic acid and protein β-sheet design (top to bottom): long-chain alkanolyoxybenzene substituted vinamides [52 f]; n-alkoxybenzoic acid dimer smectic [52 g]; thermotropically designed 2-pyridone-dimer smectic [17 a, c – g, 39 g, 46]; diacylhydrazine smectic [17 a, c, d, 26 f, 51 t]. d) Polymer variations between natural and artificial systems, between side- and main-chain variations (top to bottom); thermotropic cholesteryl-side-chain-modified polysiloxane with the combination of flexible main-chain and side-chain spacers [51 a, h]; thermotropic cyanobiphenyl-side-chain-modified polyacrylate with flexible side-chain spacers [51 b]; thermotropic azoxybenzene polymer with flexible main-chain spacer [51 a]; lyotropic stiff-chain aromatic polyamide [52 c, d]; lyotropic cellulose triacetate [52 e]. e) Native lyotropics around membrane-component-head/tail and peptide-helical and platelet motifs

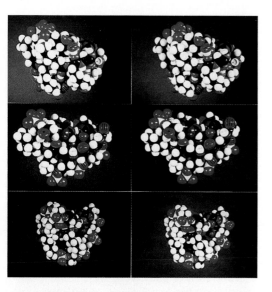

structural correspondence in the fundamentally deformed, but nevertheless comparable individual expressions of two-dimensional ambiguous arrangements in membranes. On the way from nucleic acids to membranes, the precise coding of nucleic acid informational side-chain designs flexibilizes into the loose recognitional and functional interaction modes of long-chain aliphatics, while the geometrical beauties of nucleic acid backbone phosphates dissolve into the scattered mosaics of individualized head-group arrangements of flexible membrane components. Covalently bound counterion patterns of membrane components constitute an additional aid for tightly interwoven molecular carpets, while rigid nucleic acid backbones preserve their ability to regulate by easily shiftable counterion clouds. Two-dimensional membrane patterns rediscover their global geometries in the wound-up cylinders of nucleic acid A-families and in the chirally twisted bipolar adjustments of protein β-sheet designs. And it is within the pictures of long-favored protein/membrane and long-disregarded nucleic acid/membrane intimacies that nucleic acid/membrane-passages, already suggested from the interferon-inducing facilities of ds RNAs and later used for artificial electric-field-pulse mediated nucleic acid

(top to bottom): lecithin [7 a, 14, 27 d, 52 a]; poly(γ-benzyl-L-glutamate) [51 d, 52 k]; poly(L-lysine) [51 d, 52 l]; poly(L-glutamic acid) [51 d, 52 l]; (tentoxin)$_3$K$^+$-membrane-carrier subunit complex [52 i]. f) Lyotropic nucleic acid strand variations between cholesteric and columnar phase design (top to bottom): (U)$_n$ · (U)$_n$ duplex; (C)$_n$ · (CH$^+$)$_n$ duplex; (AH$^+$)$_n$ · (AH$^+$)$_n$ duplex; B-DNA duplex, A-RNA duplex; DNA/RNA triplex; (G)$_n$ quadruplex [7 a, 17, 18, 33 a, c, p, q, 52o–w, 75]. g) Mesogenic nucleoprotein models around helical motifs: (L-Lys)$_n$ fit into RNA (top) and DNA (middle) shallow and minor grooves; (bottom) (L-Lys)$_n$–DNA/RNA-major/deep groove fits [7 a, 17, 18, 33 a, c, p, q, 38, 75]. h) Low- and high-molecular-weight biomesogen inter-

relationships – simple artificial thermotropics modeling operation modes of complex native amphotropics: recognition and modulation in selfreplication modeled by mesophase behaviour of cis- and trans-amide patterns of 2-pyridones and 1,2-diacylhydrazines (left to right and top to bottom): hypothetical recognition and modulation within first vice-versa polymerase facilities between nucleic-acid DNA and RNA duplexes and protein β-sheets; isopotential space partitioners (red: 4.184, blue: –4.184, yellow: 0 kJ/mol) visualizing prerequisites for cis-amide recognition in nucelobases and trans-amide modulation facilities in β-sheets; mesophase simulations by recognition systems of 2-pyridones and modulation systems of 1,2-diacylhydrazines.

membrane passages in genetics, are now opening prospering fields of "gene-immunization" on the basis of general membrane acceptances for nucleic acid patterns. Carbohydrates, their oligo- and polysaccharide variants as well as their extensive derivatizations, have developed – besides rudimentary polar – apolar distinctions – massive hydrogen-bond donor/acceptor capabilities of carbon-substitution-framed ordered water clusters into adjustable mediating patterns of widespread dynamic motifs, ranging from impressive helical designs to circles, holes and stretches, and emulating the mediating facilities of "water order – disorder patterns" in their emphathy roles of recognition and function. Finally, proteins have developed the giant set of instruments of at least 20 side-chain variants within nearly omnipotent threedimensional extensions of one-dimensional backbone stretches into a legislatively ungovernable, nearly infinite information content, but also into the executively incomparable functional universalities of intelligent playmates for everybody else.

The inevitable problems of solubilization [53], which threatened to exile all these molecular achievements in the dead end of solid-phase immobilities – Fischer's ingenuity ended at the level of around 20 residues in his first synthetic approaches to the kingdom of proteins at the beginning of the 20th century [36 a], and Guschlbauer, despite his farsightness [33 a, b], still looked upon DNA as "a hopeless mess" at the beginning of the 1970s – were solved by the ability of these "side-chain-polymer polyelectrolytes" to occupy the fertile fields of mesophase solvent – solute solubilizations between the extrema of solid and liquid phases [7 a, 17, 18]. Water clusters not only liberated suitably designed biopolymeric species to structural mobility, but also sensitized them to the supreme control of liquid-crystalline states.

Helical motifs of informational and functional nucleic acids [52 n – y, 62 – 69, 74, 75] and peptides [52 k, l, 58, 59, 74], which might also be comprehended as wound-up discoids, display with the individual geometries of their constituents cholesteric phase variants of both calamitic and columnar arrangements and even present blue-phase-textures of cubic phase design [62]. While the informational and functional representatives are thus reminiscent of ancient thermotropic guiding motifs, the compartmental membranes, on the other hand, are easily reducible to the basal abstractions of primitive head – tail lyotropics [7 a, 17, 18, 54 – 57]. Carbohydrates [52 e, 60, 61], finally, have developed calamitic prerequisites into more sanidic designs of laterally and longitudinally extended main-chain arrangements.

While water – and to a certain degree also other solvent partners and solvation aids – thus succeeds in the liberation of mesogenic side-chain polymers [7, 53], it forms by this, however, only sterile repetitions of overall phase symmetries. For the transition to life, the mediating abilities, of water patterns prove necessary, but not sufficient. The transition to life was not brought about by the mere existence of supramolecular organizations, nor by the phase beauties of banal basal repetition patterns. Approaches to life had to detect phase/domain regulation facilities within the nonlinear dynamics of supramolecular mesogenic organizations [18 n]. They had to develop the richness of informational and functional capabilities out of creative chaotic possibilities as well as the fleeting permanence of phase/domain transition strategies withing narrowing ranges of operation temperatures far from thermal equilibrium. Ways into the nucleation centers of the transition to life imply ways into fundamental complexity [6 b, 7 a, 17], the dominance of instabilities and probabilities rather than deterministic laws [1], the subjection to irreversibilities along irreducible bundles of nucleation trajectories and the development of spatial order – disorder distributions into spatiotemporal coherences [1–8, 17, 18]. Ways of following them up scientifically imply the retreat from mere reductionistic approaches, with the shortcomings, however, of the dominance of qualitative inference rather than quantitative scientific precision.

One of the real breakthroughs in transition-to-life capabilities seems to have been the establishing of connectivity lines of coherent solute – solvent relationships within the biomesogenic partners themselves [7 a, 17, 18, 53 b, 72 r, s]. The decisive development of replicational nucleic acid systems – with their square root laws of propagation, but also their inviting cooperativity of partners [40 h, k, 43] – had to await the overall assistance of proteins, displaying for exponential autocatalysis – and the decisive selection of competitors – not only suitable solvents but also the intelligent stereoelectronically mediating and promoting milieu [7 a, 17, 18]. Only the nucleation centers of nucleoprotein systems, individualized by early compartmental membrane walls, could provide all the enthalpic and entropic prerequisites to set the new qualities of the universe "alive" far from thermal equilibrium [1 –8, 17, 18]. The extreme coherences of such fundamental complex systems contradict our advanced theories of phase and phase-transition strategies [76]. The findings in the field of biomesogenic species reflect the situation [7 a, 17, 18, 51 –72].

In the classical view only membranes and their components seem tolerable, with their close bilayer similarities to thermotropic smectics and head/tail abstracted lyotropics [7 a, 33 p, q, 54 –57]. While nucleic acids and proteins represent both structure and system individualities, membranes, as huge collectives of highly differentiated membrane components with specifically adjusted lipid, protein, carbohydrate and perhaps even nucleic acid patterns, exhibit far more complex system hierarchies [54 b, c]. The slogan of the 1970s that "life is at membranes, in membranes and through membranes" [7 a, 48 b – e, r, s] characterizes a situation where the fundamental complexity of life has to meet up with the critical complexity of localized compartment organizations, which turn out to be completely irreducible to simple bilayer characteristics in structural design and mesogenic regulation capabilities.

From the involvement of the fully expressed diversities of membrane components [7 a, 8 c, 33 p, q, 54 a], their asymmetric designs and nonlinear modes of operation, to different ligand, especially cholesterol, regulations [54 b, c], passive and especially active transport (to and from the system) by cooperative protein-carriers, pores, channels and complex subunit organizations [7 a, 33 p, q, 35, 52 h, i], the diversity of peripheral and integrative proteins [7 a, 33 p, q, 56, 58, 59], the cell-coat glycocalix made up of sugar side chains that are joined to intrinsic membrane glycoproteins and glycolipids as well as to glycoproteins and proteoglycans [7 a, 33 p, q, 54 – 59], carbohydrate – protein patterns in selectin and integrin adhesion molecules [59, 60], lock-and-key glycoprotein – lectin interactions [6 b, 7 a], signal recognition, acceptance, transduction [56 b] and amplification [7 a] in cellular self/nonself recognitions and discriminations [7 a, 8 c, 17, 18, 33 p, q], cellular uptakes and propagations, cellular communications and differentiations [8 c, 33 p, q], molecular linguistics vertically in genetic intraspecies information transfer and horizontally in the free-floating immunological extensions of individual CNS-peripheries, up to CNS as the operational basis of individual intelligence and the origin of consciousness [5 a, 8, 17, 18, 78–80] – all this constitutes, on the basis of highly differentiated membrane components, proteins, carbohydrates and nucleic acid cooperation partners, hierarchies of increasing complexity, resulting intracellularly in supreme organizational units and intercellularly in morphogenetic programs, organismic expressions, and system and species evolution. A picture that reveals in all its insufficiently sketched coherences not only the fundamental complexity of life patterns but also their irreducibilities [1–8, 17, 18, 51–61, 78].

Proteins present like problems. Completely neglecting the utmost structural and functional complexity involved [7 a, 8 c, 33], endeavors in the field of protein mesogen behavior have mainly concentrated on poly(L-glutamic acid) derivatives with their intimate relationships to rod-like helical designs [58, 59]. More sophisticated than the usual cholesteric systems, poly(γ-benzyl-L-glutamate) [51 d, 52 x] responds to environmental influences. It covers within its very shape patterns of a "pre-Pauling" helix [7 a, 29 d, 51 d] and anticipates with its aromatic stacks the later discoid arrangments [28 a, 51 c]. The replacement of chloroalkane-solubilized benzyl substituents by an alkyl disorder belt, mobilizing the helical-order cores, shifts the system from lyo- to thermotropic characteristics and, intriguingly, makes visible in this way the dominance of order–disorder distributions in the establishment of mesomorphic behavior [58 e]. The protein-mesogen system – although primitive in comparison with native protein organizations – continues to be a source of inspiration, without so far being extendable to the real complexity of native protein patterns [51, 52, 58, 59].

Later, together with carbohydrates [7 a, 60, 61], the liquid crystals of the much more transparent nucleic acids [52 n–w] join

these circles. Nucleic acids, when treated under stringent conditions as idealized helices [62, 64–69] – and even small building-block composites fit into this picture [65] – reveal impressive pictures of classical phase textures with relationships also to overlying, biomesogenic operation modes in chromatin (re)orientations [29 g, h, 62, 77 h] and mitosis [8 c]. Independently of intercalator, groove-binder or further nucleic-acid-ligand characteristics [7 a, 17, 18, 52 t, 67], small ligand-interaction capabilities up to small-peptide partners [67–75] are integrable into the common picture of classical liquid-crystal design, and it was just within the field of biopolymer/ligand interaction that the "Vorländer effect" of directing mesophases by the "chiral infection" with small amounts of doping compounds [70] advanced as a model of hormonal signal direction and amplification (Fig. 25) [7 a, 17, 18]. However, also in these cases, the abundance of integrative biomesogenic operation modes of native standards will be insufficiently met by scientific progress in mesogenic abstractions. Even more interesting than the beautiful phases of idealized helical entities of nucleic acids, for instance, seem in this connection, the individual appearances of "unspecific" singular phase textures of polydisperse evolutionary DNAs and RNAs [73, 75], which "live" the whole complexity of mesogenic life-pattern possibilities.

Carbohydrates [7 a, 17, 52 e, 60, 61], finally, so generally effective in phase stabilizations that even polyethyleneglycols, as strange artificial abstractions, promote classical nucleic acid phase expressions [52 t, 64–69], contribute with a spectrum of quite different approaches, under way in many laboratories, to a growing understanding of rapidly enlarging details, while so far only insufficiently coping with the urgent quest for generally attributable carbohydrate recognition codes [50, 60, 61].

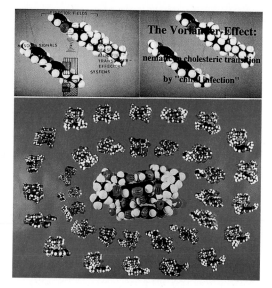

Figure 25. Low-molecular-weight effectors directing mesophase patterns of high-molecular-weight biomesogen organizations (left to right and top to bottom): mesogenic ligand-biopolymer interactions – Reinitzer's chiral cholesterylbenzoate [21, 22] and Kelker's MBBA [13f, g], elucidating Vorländer's "zirkulare Infektion" of nematic phases by chiral doping compounds [70], an early biomimetic thermotropic hormone model, simulating biomesogenic amplifications of molecular signals in biomesogenic surroundings [7a, 17, 18]; native and artificial small-molecule effectors of nucleic acid operation modes, modeling Vorländer's "chiral infection" within complex bio-mesogen systems and reminding of their intriguing initiatory roles for the redetection of the Vorländer-effect of switching mesophase design in artificial and native systems.

Just like the traditional and obviously difficult to overcome scattered views of proteins, nucleic acids, membranes and carbohydrates as kingdoms of their own in molecular biology [7a, 8c, 33p, q], the experimental findings in these new fields of liquid-crystal research constitute at present still a dispersed mosaic of details rather than the needed adequately complex and integrative picture [7a, 17, 18]. While liquid-crystal phase investigations remain preferably limited to helical aspects of only a few contributing natural structures and their accompanying artificial derivatives, general questions [72], such as microphase and domain tuning and regulation [7a, 17, 18], microstate distributions [72d] within interchanging organizations relating to entropy, information and complexity relationships [72d, e], complex energy/entropy compensations [33k, 72k], general field extensions to quasicrystals [72b], glass and plastic states [72e, f], as well as evolutionary relationships to spin-glasses [72d], intimate coherences in nucleic acid sequence patterns [72s–w] and molecular dynamics [77b–d, k, n] within supramolecular biomesogenic organizations and their developmental aspects [7a, 17, 77] still have to await unifying conceptions. Perhaps the previously offered "biomesogen" views [7a, 17, 18] can provide integrative guiding patterns for all (potential) molecular biological meso(phase) "builders" – ranging from their individual statics and dynamics, their amplifications into interactions of highly cooperative (transient) domain organizations to the integrative (domain-modulated) phase transition strategies and regulation modes of classical liquid-crystalline states (Fig. 26).

While so far the availability of computers has failed to extend individual molecular dynamics [77] into the developmental fields of supramolecular biomesogenic organizations, and phase/domain characterizations only succeed in the elucidation of uniform large molecular collectives [25, 51, 52], experiments following mesophase characteristics of self-organizational and self-reproductive processes down to the nanometer scale [73, 74] are receiving increasing interest and try to make up for the persisting difficulties of the lack of both far-reaching and profound theoretical interpretations [76].

Molecular structure amplifications and behavior-determining phase reductions meet

a)

b)

c)

d)

e)

f)

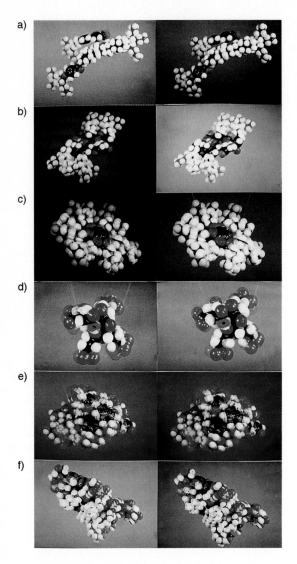

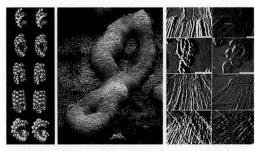

Figure 27. Visualization of biomesogenic nucleic-acid selforganizations by molecular-resolving micros-copies (top to bottom and left to right): nucleic acid strand patterns: B-DNA and A-RNA duplexes, DNA/RNA triplex, G-quadruplex, DNA/RNA-duplex/pro-tein-β-sheet complexations; STM and SFM-investi-gations of nucleic acid organizational behavior: DNA-plasmid [74 c]; RNA- and RNA/peptide-organiza-tions: $(U)_n \cdot (A)_n$-duplexes; $(U)_n \cdot (A)_n \cdot (U)_n$-tri-plexes; $(G)_n$-quadruplexes, $(L\text{-Lys})_n/(U)_n \cdot (A)_n$-com-plexes in 2D and 3D representation [7 a, 17, 18, 33 a, c, f, p, q, 75].

Figure 26. (Bio)mesogenic order–disorder patterns rather than geometric anisotropies as general precon-ditions for the appearance of mesophases: a) classical thermotropic cyanobiphenyl nematic [25 a, 51 e] and thermotropically designed steroid hormone cholester-ic [17 a, c, d, 26 f, 51 a, s, u]; b) protein-β-sheet simu-lative diacylhydrazine smectic [17 a, c, d, 26 f, 51 t]; c) hexaalkanoyloxybenzene discoid [28 a, 51 c] end-ing the century of "rod-like" dominance; d) lyotropic poly(L-glutamic acid) [51 d, 52 k, l] changing with an alkyl-disorder shell from lyotropic to thermotropic [58 e]; e) ds A-RNA lyotropic [52 n–w, 75]; f) am-photropic $(L\text{-Lys})_n$/RNA recognition complex [17, 18, 75].

within the creative fields of supramolecular biomesogenic organizations [17, 30]. Com-plex molecular recognition and information processing amplify the continuation of ad-equate phase/domain transition strategies into new qualities of intelligently develop-ing self-organizational and self-reproduc-tive systems that gain their transition to life and life-survival strategies by surprising entropy – information transformations far from thermal equilibrium [7 a, 17]. Suitable scientific approaches suffer from the restric-tions of methodologies that meet the re-quirements for solid- and liquid-phase in-vestigations but are insufficient for the treat-ment of mesophase characteristics and de-velopmental aspects of highly condensed systems, which play crucial roles in self-or-ganization, self-reproduction, information processing, and optimization. While a mul-titude of experimental findings in the fields of polarization microscopy, chiroptical in-vestigations, static and dynamic X-ray dif-

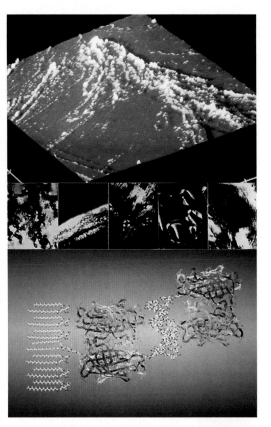

Figure 28. Complex biomesogenic organizations modeling evolutionary developments: (a) SFM visualization of polydisperse DNA [75]; (b) nucleic acid textures (left to right): chicken-DNA, $(U)_n \cdot (A)_n$ duplex $(U)_n \cdot (A)_n \cdot (U)_n$ triplex, $(G)_n \cdot (G)_n \cdot (G)_n \cdot (G)_n$ quadruplex, $(U)_n \cdot (A)_n \cdot (L\text{-}Lys)_5$ complex [75]; (c) building up of liquid–protein–nucleic acid multilayer systems as two-dimensional simulations of the grand evolutionary triad – at least in partial cooperation with a mediating water milieu [7 a, 17, 18, 33 a, c, p, q, 63 a, 75].

approaches and their scientific aims in direct nanometer-elucidations of the statics and dynamics of biomesogen nucleation patterns [74]. Adlayer arrangements provide direct visualization not only of supramolecular biomesogenic organizations, but also of their inherent dynamics (Fig. 27) [74, 75]. Dimension reductions of three-dimensional large scale bulk-phase repetition patterns into "two-dimensional mesophases" with their ability to survey multiple monolayer arrangements at the molecular (nanometer) level [73] display insights horizontally into the two-dimensional lateral extensions of phase relationships and vertically into the three-dimensional individualizations of creative nucleation centers of supramolecular biomesogenic developments (Fig. 28).

2.5 The Developed Biomesogenic Systems

Emergence of new qualities of the universe; life spatiotemporal coherences between nonclassical chaos and fundamental complexity achievements; transiently interchanging supramolecular organizations and their biomesogenic regulation; amplifiable "parts" permanently qualifying changing "wholes", developing "wholes" qualifying operational "parts", structure–phase dualities inherent in the developments of nucleo-protein systems; stabilization within the dynamics; morphogenetic programs and regained operational symmetries on the basis of fundamental asymmetries – scientific approaches to life patterns (Fig. 29) unravel a complex of problems, fitting for the complexity of life patterns themselves [1–8, 17, 18, 77–80].

After a century of mainly parallel and independent developments, artificial meso-

fraction techniques and nuclear magnetic resonance methods will provide – besides the often familiar and convenient texture patterns – a wealth of information on complex phase-domain organizations and their operation modes in leveling-off of statistic multitudes [51–72], the rapidly enlarging spectrum of molecular-resolution microscopies is receiving growing interest as to their

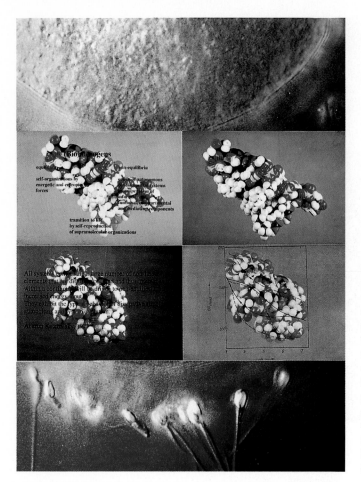

Figure 29. Biomesogens: (smaller pictures, top) transitions to life illustrated by the nucleation of nucleoprotein intimacies [7a, 17, 18]; (smaller pictures, bottom) replicative and informative nucleic acid triplexes [48a–q, t] as forerunners, displaying molecular hysteresis as basic mechanisms of memory imprints and oscillation and rhythm generation [7a, 17, 18, 33a, c, f, 48]; (top and bottom) the late followers' ontogenesis, retracing biomesogen phylogenesis in "organismic first intimacies" [16, 23].

gens and their biomesogenic native standards (Fig. 30) seem, nowadays, capable of a creative dialectic synthesis of the thus far dominant theses/antitheses tensions. The unique spatiotemporal coherences of (bio)mesogen systems promote self-consistent evolutionary views and afford intriguing perceptions of the grand process.

Even very simple and primitive mesogens differ from nonmesogenic species in that they bear some sort of rudimentary intelligence. At present, however, we appear to be faced with difficulties in treating this gleam of preintelligent behavior scientifically. Even our most advanced theoretical approaches prefer rather abstract and ethereal

shadows of real entities [76]. Changing from here to complex artificial or even native (bio)mesogen organizations amplifies the problems for quantitative treatments [7a, 17, 18, 33]. The constraints leading to only qualitative inferences – if not the temptation to speculate – are growing.

Although artificial mesogens have been developed in terms of thermotropic and lyotropic characteristics, biomesogens both blur and dialectically combine these extreme positions within the cooperative network hierarchies of their interdependent chiral complexity patterns (Fig. 26) [7a, 17, 18]. Thermotropics resemble lyotropics in that their flexible, more disordered seg-

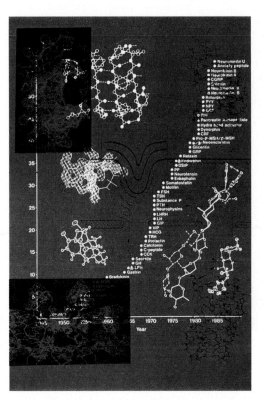

Figure 30. Anabolic and catabolic regulation hierarchies that channel biomesogenic information processing in a highly condensed phase [8e, 18, 52x, y].

ments serve the purpose of dynamization of the rigid, more ordered parts. Lyotropics, on the other hand, display, within their complex solvent – solute distributions, far more interactive coherencies than had originally been thought in classical views. While for lyotropics suitable solvents – especially water – provide a spectrum of entropy-driven organizational forces, thermotropics will profit by comparable entropy effects of order minimizations on the complex domain interfaces between rigid cores and flexible terminals. Within this picture, the solvent-like labilizing areas of interacting biopolymer organizations appear as special expressions of more general mobility characteris-

tics of mesogens, which are able to build up within their dynamic chiral order – disorder distributions transiently functional and transiently acting chiral order – disorder patterns. Within the dynamics of water-mediated protein, nucleic acid and membrane organizations, dynamic chirally instructed parts – submerging the individualities of their respective partners – exert functions of partial solubilizations for the mobilization of rather static chiral solute areas, the pre-intelligent handling of which is a prerequisite for some new properties as, for instance, recognition, functional catalysis, semi-permeable compartmentation, collective and cooperative operation modes, isothermal self-organizational behavior and self-reproduction of transiently acting chiral domain systems, information processing and optimizations.

Our original ideas about mesogens, which had been attracted by the unifying principles of huge artificial and preferentially achiral molecular ensembles and the overwhelming symmetries of their mesophase relationships [7a, 13, 14, 17, 18, 24 – 28, 51, 52], are thus redirected into the limited and methodologically extremely difficult areas of cooperatively processed chiral phase/domain systems, which govern with increasing complexity a more and more precisely tuned and refined, highly sophisticated instrumentary of modulated phase/domain transitions [7a, 17, 18, 69, 71, 72, 77] (Fig. 31). The chiral mesogen individualities of the grand supra-chirally organized amphiphilic patterns of life display within their molecular imprints the prerequisites for the projection of individual molecular facilities into the structural and functional amplifications of cooperative dynamic mesogen domain ensembles. The chiral order – disorder designed individual appears as a holographic image of the whole. By this, the classical views of interacting structural

a)

b)

c)

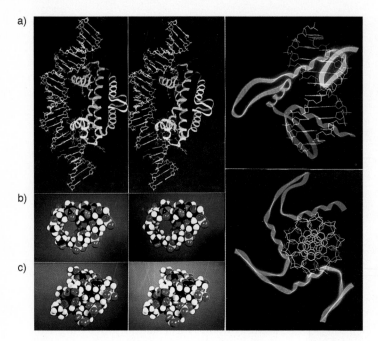

Figure 31. Biomesogen operation modes in nucleoprotein systems: (a) FIS/DNA recognition dynamics [71q]; DNA/protein unspecific (b) [38] and specific (c) [33 m, n] recognitions by β-sheet and α-helix fits to DNA minor and major grooves, respectively; d, e) Zn-finger protein handling DNA [71 f].

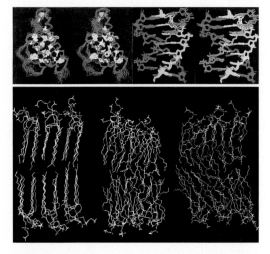

Figure 32. Biomesogen dynamics: (top left) interleukin-4 dynamics as derived from multidimensional NMR investigations [77 k], depicting among others labilized loops; (top right) the more rigid designs of nucleic and triplexes [77 l]; (bottom) dependence of membrane dynamics (MD) [77 m, n] on structural prerequisites and temperature range, MD snapshots in pico-second range; dilauroylphosphatidylethanolamine (95 ps), dioleylphophatidylethanolamine (120 ps), dioleylphosphatidylcholine (100 ps).

individuals submerge into the new qualities of transiently acting mosaics of mutual domain cooperativities, where chirally instructed stereoelectronic patterns of individual representatives of the grand triad, together with the "liquid polymers" of mediating water-clusters, anneal into the spatio-temporal coherencies of newly achieved biomesogenic domain organizations, their self-reproduction, and informational and functional optimizations. It is within this dual view of biomesogens that both the structural and the phase/domain aspects will contribute intriguingly to a consistent picture of life patterns and their operation modes, and it had been, moreover, a prerequisite to establishing self-organizational, self-reproductive, and morphogenetic processes under nearly isothermal conditions [7 a, 17, 18].

At least partially reflected in modern theories of nonclassical chaos [6, 7 a, 72 e, f], the developed chiral amphiphilic patterns,

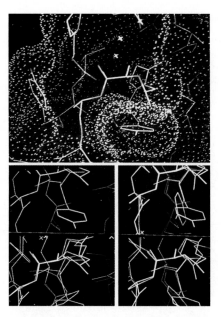

Figure 33. Detailed biomesogen dynamics treatment: rotational isomerizations of Tyr-35 of bovine pancreatic trypsin inhibitor mediated by structural changes in the surrounding biomesogenic protein matrix – overall view (top) and "film" in skeleton presentation (bottom) [77 b – d].

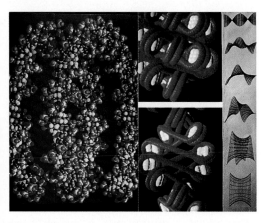

Figure 34. First insights into mesogenic chromatin regulations: superhelical DNA chromatin arrangement [29 g, h, 33 f]; winding up chromatin packages by Bonnet transformation operation-modes [77 h].

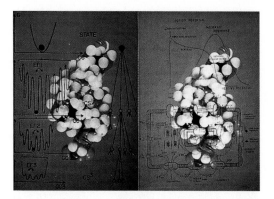

Figure 35. β-Casomorphin peptide messenger in addressing opioid and cardiovascular-cardiotonic (modulatory) subroutines of general psychoneuroimmunological programs, superimposed with organizational microstate [72 d] distributions and simplified regulation schemes of heart muscle contractions [18, 36, 52 y, z].

gen strategies, ranging from the localized motions of molecular segments [77 b – d] (Fig. 32, 33), via their coupling to collective processes (Fig. 34) in the surroundings [72 b – d, n], to the interdependences of complex regulations [18, 51 y, z] (Figs. 35, 36). Statics and dynamics of asymmetrically liberated and directed multi-solvent-solute systems and their sensitive phase and domain-transition strategies characterize our picture of today's biomesogen organizations. Synergetics [7 b] within complex solvent – solute subsystem hierarchies [7 a, 17, 18 v] and their mutual feedback found new aspects of chirally determined nonlinear dynamics with a number of preintelligent operation modes, such as reentrant phenomena [48 i], molecular hystereses and memory imprints, oscillations and rhythm generation [48 a – i, r, s], complex structure – motion linguistics and signal amplifications, self/nonself recognitions and discriminations, information processings and general optimization strategies [7 a, 17, 18] (Fig. 37).

amphiphilic both in terms of their order – disorder dialectics and the spatiotemporal coherencies of their nearly isothermal processings, are acting according to meso-

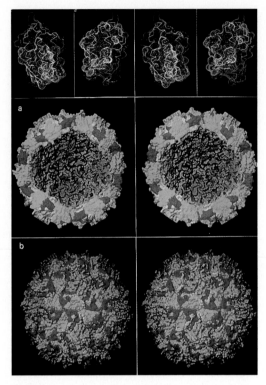

Figure 36. Biomesogen patterns of immunorelevant structures: (top, left to right) mobility sites of myohaemoerythrin, engaged in antigen–antibody recognitions [77 e] – molecular surface and α-carbon backbone color-coded: highly mobile, red and yellow, respectively; relatively well-ordered, blue; not studied, magenta; (center and bottom) polio virus exterior and interior surface patterns [71 a] reminiscent of the birefringent solutions of rodlike tobacco mosaic particles and their role in Onsager theory [37].

Figure 37. Chiral biomesogen dynamics in informational and functional processing (from top to bottom): complex structure–motion linguistics of a small molecule effector (β-casomorphin-5) in addressing neuroendocrine, cardiovascular-cardiotonic and immunomodulatory subroutines of general regulation programs [7 a, 17, 52 y, z]; cholesterol regulations in functional membranes [7 a, 54 b, c]; valinomycin interplays, mediating K^+ ions across membranes [7 a, 35]; biomesogenic signal amplification in the chromatin: steroid hormone efficiencies in domain regulations [7 a, 17, 18, 51 s]; molecular hystereses and memory imprints in polynucleotide triplexes [7 a, 17, 18, 33 a, c, f, p, q, 48 a–q, t]; intimacies of nucleic acids and proteins [7 a, 17, 18] along nucleation trajectories of the nucleoproteinic system.

An almost infinite number of examples characterize complex motions of the grand process that are stabilized just in their dynamics; dynamics that continue asymmetrically the spatial order–disorder distributions of biomesogenic life patterns into the spatiotemporal coherencies of dynamic order [5 b, 6 b, 7 a, 17, 18]. The contradictions between crystalline order and liquid disorder, which rendered Lehmann and Vorländer's "liquid crystals" [13, 14, 24] so extremely suspect to their contemporaries, represent and rule in fact the survival principles of the grand process. Biomesogens, which display within their chiral molecular imprints and designs not only the richness of experienced affinity patterns but also the capabilities of their play-educated dynamics, link together entropy and information within a delicate mesogen balance (Fig. 38). In this connection, Schrödinger's [29 c] question concerning the two so extremely different views of order appears to be dramatically relevant. We seem to be badly in need of new theoretical treatments of old terms that have changed their meaning somewhat on switching from statics to dynamics, from thermal equilibrium states to far from thermal equilibrium developments. What seems to be helpful are extensions of classical static descriptions to new self-organizing, self-stabilizing, and self-reproductive dynamic processes. The abstract and mathematically perfect order of an idealized crystal (Schrödinger's "dull wallpaper" in all its symmetries) and the strange dynamic order of life patterns (Schrödinger's "beautiful Raffael tapestry" – in all its asymmetries not completely matching the real process dynamics) are competitors for supreme roles. Covered by the general uncertainty principle – with the classical views of order and disorder as borderline cases – the new qualities and description capabilities of dynamic order unravel pattern developments from the coherence of life.

2.6   The Human Component

"Subject" and "relatedness" – developments via fermion – boson, particle – wave, structure – phase, and processing – information dualities into the preliminary final destinations of matter and consciousness; ontogenesis retracting (universe) phylogenesis,

quantum mechanics and general relativity theory longing for "theories of everything", self-consciousness and self-awareness, questions for quantum-selves within a creative universe – man at the interfaces to the universe, awakened by consciousness of himself [1 – 8, 17, 18, 76, 80]?

"Will there ever be such an achievement as a computer simulation of our central nervous system (CNS), a simulation of any given complexity, but nevertheless deterministic? Or is this CNS already a new quality, from the most highly developed computer as far away as a quantum mechanical approach from some one of Newton mechanics? Is Bohr's remark 'that the search for a value equivalent of thought resembles the measurement effect on a quantum object' more than only a formal analogy? Does our CNS work for the engrams of its information patterns with a hardly estimable diversity of the quantum fields of its stereo-electrically manipulated biopolyelectrolyte patterns? Is there more behind the hologram analogy of our brain storage? Does the whole giant information content of our brain equal the utmost complexity of collective quantum states of the macrostructures of the CNS? Is superconductivity of biopolymeric macrostructures a leading model? Are Josephson contacts perhaps relevant switches? Will the forecast of Kompaneyez as to the Bohr-analogy of thought being more suggestive of a giant continuous change of the collective states of our brain as a whole with its complex arrangements in the 'enlightened spot of our consciousness' one day become certain reality? Quantum theory so far has not approached such complex systems. There remain at present final speculative questions without any reasonably, competent answer" [7 a] – the early-1970s visions match supreme themes of our time from the suggestions for supreme facilities of future quantum computers

BIOMESOGENS

Amphiphilicity
Holographic design
Non-linear dynamics
Selforganizational forces
Self/non-self recognition and discrimination
Structure-motion linguistics

		Supramolecular		
Molecule	Macromolecule	Organization Domain	Microphase	Phase

Transient order-disorder patterns
Multi solvens-solute systems
Phase-domain transition strategies
Phase-domain cooperativity
Reentrant phenomena
Molecular hystereses, memory imprints,
oscillations and rhythm generations
Signal transduction and amplification
Information processings and preintelligent
operation modes
Spatio-temporal coherences

COMPLEXITY

Figure 38. Supramolecular biomesogenic organizations between "structure" and "phase" (left to right and top to bottom): developing biomesogenic textures: ammonium oleate – Lehmann's [14] first "scheinbar lebender Kristall" (apparently living crystal) (original photograph as a kind gift of Hans Kelker [13]), Lehmann's worm-like features [14], the Halle rediscovery of phase chemistry by a miscibility rule for promesogenic imidazole derivatives in the 1950s [26d–f], cholesteric streak texture of thermotropically designed steroid hormone derivatives [51t] (three images), textures of 146 bp DNA fragment (three images), cholesteric organization of Dinoflagellate chromosome [52q, 62], cholesteric oily streak and chevron textures of high-molecular-mass chicken-erythrocyte DNA [52q, 62], cholesteric oily streak and chevron textures of high-molecular-weight chicken-erythrocyte DNA [52n, o, s] (two images); dual structure–phase view of nucleic acid–protein interactions modeling supramolecular biomesogenic nucleation states of the nucleoprotein systems as visualized by RNA/(L-Lys)$_n$ intimacies [7a, 17, 18, 75]. These images have superimposed on them supramolecular domain organization between structure and phase: chiral secondary structure basic geometries [28d], continued with water-shell cover [53] and counterion clouds [69a], amplification of individual structures into order–disorder patterns of cholesteric phase arrangements with superhelical twist; suggestions for morphogenetic mesophase extensions, exemplified in mitosis of *Haemanthus* cells [8c]; supramolecular biomesogenic organizations as approached from "structure" and "phase" [7a, 17, 18].

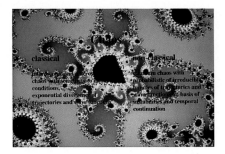

Figure 39. Classical and nonclassical approaches to chaos theories [5–7, 72e, f].

Figure 40. Order–disorder patterns in artificial mesogens and native biomesogens promoting spatial chiral order–disorder distributions into spatiotemporal coherences (top to bottom): amphotropic DNA/prealbumin complex [10], reminiscent in its order–disorder distributions of the simple achiral thermotropic n-alkoxybenzoic acid dimers [52 f]; periodic and chaotic oscillations of multiple oscillatory states in glycolysis [78] – all this superimposed on the details of the "very" event – modeling in all its interfacial relationships early evolutionary origins of the nucleoproteinic system and its inherent structure–phase dualities [7 a, 16, 17, 18].

[80 b], the importance of Bose–Einstein condensates up to the last questions of quantum-self-constituted self-consciousness fields as some sort of independent quality in the universe [7, 80 a].

An instability-induced grand transition from quantum vacuum to material existence might have created our universe. Energy and entropy gradients mediated duality developments and transitions between fermions and bosons, particles and waves, structure and phase into the process–information dualities of life patterns. At the interface to the universe, life patterns developed a preliminary finite duality between existent matter and self-consciousness fields. By this grand transition a facility of awareness beyond space and time originated that transformed its theses/antitheses tensions into creativity.

Monod's vagrant at the borders of "his" universe (Fig. 39) – realizing the extreme and unique exposition of his situation – becomes aware of the intriguing and inspiring motivations resulting from his borderline case. He begins to conceive of himself as an exciting creative periphery of matter and interfaces beyond, which adds to so far irrelevant evolutionary developments advancing criteria of conscience-controlled consciousness patterns feeding back to the grand process itself (Figs. 40, 41). He faces a future where his dreams of a universe will no longer fade away in senselessness, but seem increasingly enriched by blessed capabilities of self-realization within developing evolutionary fields. Descartes's proud "cogito ergo sum" is also – by descent and development, by creation and destination, by obligation and fairness – something like a first grand self-affirmation of our special space-time within the universe. A maternal developing universe awakened in his consciousness, recognizing by himself also itself. The light of self-consciousness born in the dark – science seems to cope with biblical motifs.

Figure 41. Mesogen patterns (bottom to top): cholesteric DNA [63 a, 75] superimposed on early universe computer simulations [12] and Lehmann's paintings of cholesteric phase [14]; "Γαια" mesogenic surface; human brain [79 a] with continued mesogenic patterns from bottom, reminiscent of mesomorphic relatedness of neuronal-activity patterns [80 a] to mesogenic operation modes and quite recent quests for quantum-computer capacities [80 b].

3   Outlook

Since that incomprehensible beginning, when the time-arrow of entropy marked the directionality of this outset, entropy seems

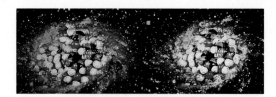

Figure 42. Evolutionary dynamic order patterns – from the maternal galaxy to its transition to life in self-replicational informational species [7 a, 17, 18].

not only to have endeavored to make linear uninformative equilibria states vanish in senseless disorder, but has, from the very beginning, developed extraordinary facilities to endow nonlinear systems far from thermal equilibrium with the rapidly enlarging capabilities of growing dynamic order (Fig. 42). It was the very molecular order–disorder design of their constituents that rendered those patterns sensitive to the amazing influences of entropy. Counteracting these rather "intelligent" molecules in its usual eagerness and need for disorder, entropy compensated for their lost homelands between order and disorder with the gift of cooperativity, self-organization, and some sort of newborn directed dynamic order within supramolecular ensembles and growing organizations. And, like Mephisto in Goethe's grand opus, who introduced himself as part of the power that always wants evil while always doing good, entropy enabled those beloved playmates to propagate in a dramatic outburst of directed dynamics over the deserts of order and the abyss of disorder as an endangered, but incomparably beautiful, grand mesogenic pattern, set out in a grand beginning for far unreachable (?) goals. Intimately interconnected with information and complexity within the framework of Boltzmann's H-theorem, entropy valued and promoted the informational contents of complexity states and opened, with the grand bifurcation of

Figure 43. Scalar-field explosion and developmental universe: quantum vacuum mediating within a grand fluctuation existence. Order–disorder coherences forwarding far from thermal equilibria chaos offers to levels of complexity that originate on basis of indeterministic quantum dualities consciousness patterns – sharing the whole [3, 7 a, 17, 18, 75].

the appearance of self-replicating systems, the ways for the development of what we now call life.

Referring to an independent "what is life" definition, proposed by Eschenmoser as "life is a catalyst that equilibrates its surroundings" within a common-term discussion at Maratea in 1993 [5 d], it seems perhaps not without a certain elegance to conceive of living systems just in their skillful utilization of energy gradients provided by the universe by their own attracting entropy gradients. Abilities that give rise to kingdoms of transitoriness, transientness and dynamics that establish supreme fundamental complexity patterns from creative chaos within their order–disorder coherences and life in the extreme relatedness of quantum-duality-based self-consciousness fields, a first recognition and perception of the grand process, a first creative quality beyond space and time (Fig. 43).

"A sun system, probably even more, that – embedded within the boundlessness of further intra- and intergalactic relationships – created evolutionary fields that became capable of information generation, process-

ing and optimization within selection-determined processes of self-organization and self-reproduction. A dynamic movement – seduced from the homing states of sterile rigid order hierarchies into the homelessness of dialectically stimulated action patterns, becoming increasingly capable of developing a simulative, and from there, within a multilevel orientating and interchanging process, also operational and creative consciousness of the dynamics of the grand process and its possible active promotion, a grand dialectically driven and enlarging consciousness patterns which escapes the static plains of persistence in dynamic actions between space and time. The statics of our ways so far, transformed, furthered and developed into the dynamics of the present and future – an admirable grand process amidst the finiteness of itself and its surroundings – set out in a billionfold lived and renewed forwarding effort of improvement and optimizations for far horizons: will that be our further way? A way that means for the transitoriness of our individualities, the finiteness of their endeavors the final goal, while the integrative vector of the grand process itself aims beyond into far inconceivabilities" [7 a].

Acknowledgements

I cordially thank Hans Kelker for stimulation and encouragement and for unforgettable hours in Halle. The extensive computer searches by Christian Bohley, Frank Klimaszewsky and Werner Loch concerning "living systems" and the molecular modeling contributions of Christian Bohley and Maren Koban are gratefully acknowledged.

4 References

[1] Transitions: a) I. Prigogine, I. Stengers, *Das Paradox der Zeit (Time, Chaos and the Quantum. Towards the Resolution of the Time Paradox)*, Piper, Munich **1993**; Science **1978**, *201*, 777; T. Petrosky, I. Prigogine, *Adv. Chem. Phys.* **1997**, *99*, 1; *Z. Naturforschung* **1997**, *52*, 37; b) C. F.

von Weizsäcker, *Die philosophische Interpretation der modernen Physik*, Nova Acta Leopold. **1986**, *NF/207*, 37/2; *Zeit und Wissen*, Hanser, Munich **1992**.

[2] Universe processings: a) J. S. Trefil, *The Moment of Creation*, Scribner's, New York **1983**; b) S. Weinberg, *Dreams of Final Theory*, Pantheon Books, New York **1993**; *The First Three Minutes*, Basic Books, New York **1977**; Sci. Am. **1994**, *271 (10)*, 22; c) S. W. Hawking, *A Brief History of Time*, Bantam, New York **1988**; *Black Holes and Baby Universes*, Bantam, New York **1993**; *Phys. Rev. D* **1995**, *52*, 5681; R. Bousso, S. W. Hawking, *Nucl. Phys. B* **1997**, *57*, 201; S. W. Hawking, R. Penrose, *Mercury* **1997**, *26*, 33.

[3] Versus processing of universes: A. Linde: *Elementarteilchen und inflationärer Kosmos*, Spektrum, Heidelberg **1993**; *Spektrum Wiss.* **1995**, *(1)*, 32; *Phys. Rev. D* **1997**, *56*, **1841**; A. Linde, D. Linde, A. Mezhlumian, *Phys. Rev. D* **1994**, *49*, 1783.

[4] The arrow of evolution: a) E. Jantsch, *Die Selbstorganisation des Universums*, Hanser, Munich **1992**; b) M. Gell-Mann, *The Quark and the Jaguar*, Freeman, New York **1994**; c) R. Penrose, *The Emperor's New Mind Concerning Computers, Minds, and the Law of Physics*, Oxford University Press, New York **1989**; *Computerdenken*, Spektrum, Heidelberg **1991**; *Shadows and the Mind*, Oxford University Press, New York **1994**; *Schatten des Geistes*, Spektrum Heidelberg **1995**; *Astrophys. Space Sci.* **1996**, *244*, 229; *General Relativity Gravitation* **1996**, *28*, 581; *Intern. Stud. Philosophy Sci.* **1997**, *11*, 7; S. W. Hawking, R. Penrose, C. Cutler, *Am. J. Phys.* **1997**, *65*, 676; d) F. J. Tipler, *The Physics of Immortality*, Doubleday, New York **1994**; e) C. E. Shannon, W. Weaver, *The Mathematical Theory of Communication*, University of Illinois Press, Urbana, Il. **1964**.

[5] Transitions to life: a) J. Monod, *L'Hazard et la nécessité*, Seuil, Paris **1970**; b) M. Eigen, *Stufen zum Leben*, Piper, Munich **1987**; *Cold Spring Harbor Symp. Quant. Biol.* **1987**, *LII*, 307; *Naturwissenschaften* **1971**, *58*, 465; *Trends Microbiol.* **1996**, *4*, 216; c) C. de Duve, *Blue-print for a Cell, The Nature and Origin of Life*, Patterson, Burlington **1991**; *Ursprung des Lebens*, Spektrum, Heidelberg **1994**; d) *Reproduction of Supramolecular Structure*, (Eds.: G. R. Fleischaker, S. Colonna, P. L. Luisi), *NATO ASI Ser. Vol. 496*, Kluwer, Dordrecht **1994**.

[6] The evolutionary field: a) R. Lewin, *Complexity – Life at the Edge of Chaos*, Macmillan, New York **1992**; *Die Komplexitätstheorie*, Hoffmann & Campe, Hamburg **1993**; b) F. Cramer: *Chaos and Order*, VCH, Weinheim **1993**; *Der Zeitbaum*, Insel, Franfurt a. M. **1993**; *Symphonie des Lebendigen*, Insel, Frankfurt a. M. **1996**.

[7] Coherences: a) S. Hoffmann, *Molekulare Matrizen (I Evolution, II Proteine, III Nucleinsäuren, IV Membranen)*, Akademie-Verlag, Berlin **1978**; b) H. Haken, *Advanced Synergetics*, Springer, Berlin **1983**; *Naturwissenschaften* **1981**, *68*, 293; *Rep. Progr. Phys.* **1989**, *52*, 513; c) J. C. Eccles, *Evolution of the Brain: Creation of the Self*, Routledge, London **1989**; d) D. Zohar, *The Quantum Self*, Quill/Morrow, New York **1990**; *Quantum Society*, Quill/Morrow, New York **1994**.

[8] From physics to genetics: a) W. Ebeling, R. Feistel, *Physik der Selbstorganisation und Evolution*, Akademie-Verlag, Berlin **1982**; H. Herzel, W. Ebeling, A. O. Schmitt, M. A. Jiménez Montaño, in *From Simplicity to Complexity in Chemistry – and Beyond* (Eds.: A. Müller, A. Dress, F. Vögtle), Vieweg, Braunschweig **1996**, p. 13; b) C. Tudge, *The Engineer in the Garden. Genes and Genetics. From the Idea of Heredity to the Creation of Life*, Cape, London **1993**; *Wir Herren der Schöpfung*, Spektrum, Heidelberg **1994**; c) B. Alberts, D. Bray, J. Lewis, M. Raff, K. Roberts, J. D. Watson, *Molecular Biology of the Cell*, Garland, New York **1983**.

[9] X-ray proof of Watson – Crick: R. E. Dickerson, in *Methods of Enzymology* (Eds.: D. M. J. Lilley, J. E. Dahlberg), Academic Press, New York **1992**, p. 67.

[10] C. C. F. Blake, *Endeavour* **1978**, *2*, 137.

[11] "Advanced" molecular modeling: P. K. Weiner, R. Langridge, J. M. Blaney, R. Schaefer, P. A. Kollman, *Proc. Natl. Acad. Sci. USA* **1982**, *79*, 3754.

[12] J. O. Burns, *Kosmologie und Teilchenphysik*, Spektrum, Heidelberg **1990**.

[13] Liquid crystals – history and facts: H. Kelker. a) *Mol. Cryst. Liq. Cryst.* **1973**, *21*, 1; b) in *Zehn Arbeiten über flüssige Kristalle*, 6. Flüssigkristallkonferenz sozialistischer Länder, Halle 1985 (Ed.: H. Sackmann), Wiss. Beitr. Martin-Luther-Univ. **1986**/52 (N 17), 193; c) *Naturwiss. Rundschau* **1986**, *39*, 239; d) *Mol. Cryst. Liq. Cryst.* **1988**, *165*, 1; e) *Liq. Cryst.* **1989**, *5*, 19; f) H. Kelker, R. Hatz, *Handbook of Liquid Crystals*, VCH, Weinheim **1980**; g) H. Kelker, B. Scheurle, *Angew. Chem.* **1969**, *81*, 903; *Angew. Chem. Int. Ed. Engl.* **1969**, *8*, 884; h) H. Kelker, B. Scheurle, J. Sabel, J. Jainz, H. Winterscheidt, *Mol. Cryst. Liq. Cryst.* **1971**, *12*, 113; i) H. Trebin, *Phys. Bl.* **1988**, *44*, 221; k) H. Falk, P. Laggner, *Österr. Chem. Ztg.* **1988**, *9*, 251; l) V. Vill, *Mol. Cryst. Liq. Cryst.* **1992**, *213*, 67; m) *Condens. Matter News* **1992**, *1*, 25; n) V. Vill, J. Thiem, *LIQCRYST – Datenbank flüssigkristalliner Verbindungen*, Springer, Heidelberg **1992**; o) R. Brauns, *Flüssige Kristalle und Lebewesen*, Schweizerbart, Stuttgart **1931**.

[14] Liquid crystals – the romantic period: O. Lehmann, a) *Molekularphysik,* Engelmann, Leipzig **1888/89**; b) *Plastizität von flüssigen Kristallen sowie ihre Umlagerungen,* Engelmann, Leipzig **1904**; c) Flüssige Kristalle und die Theorien des Lebens, 78. Versammlung deutscher Naturforscher in Stuttgart, **1906**; d) *Die scheinbar lebenden Kristalle,* Schreiber, Esslingen **1907**; e) *Die neue Welt der flüssigen Kristalle und deren Bedeutung für Chemie, Technik und Biologie,* Akademische Verlagsgesellschaft, Leipzig **1911**; f) *Die Lehre von den flüssigen Kristallen und ihre Beziehungen zu den Problemen der Biologie,* Bergmann, Wiesbaden **1918**; g) *Ergeb. Physiol.* **1918**, *16*, 255; h) *Flüssige Kristalle und ihr scheinbares Leben – Forschungsergebnisse dargestellt in einem Kinofilm,* Voss, Leipzig **1921**.

[15] Phase beauties in liquid crystals: T. E. Strzelecka, M. W. Davidson, R. L. Rill, *Nature* **1988**, *331*, 457.

[16] Continued in cellular beauties: P. Jaret, L. Nilsson, *Natl. Geograph. Soc. USA* **1986**, *169*, 702.

[17] Life sciences – biomesogen view: a) S. Hoffmann, W. Witkowski, in *Mesomorphic Order in Polymers and Polymerization in Liquid Crystalline Media* (Ed.: A. Blumstein), *Am. Chem. Soc. Symp. Ser.* **1978**, *74*, 178; *Polym. Prepr.* **1977**, *18*, 45; b) S. Hoffmann, in *Darwin Today* (Eds.: E. Geissler, W. Scheler) Akademie-Verlag, Berlin **1983**, 193; c) in *Liquid Crystals and Ordered Fluids* (Eds.: A. C. Griffin, J. F. Johnson, R. S. Porter), *Mol. Cryst. Liq. Cryst.* **1984**, *110*, 277; d) in *Polymeric Liquid Crystals* (Ed.: A. Blumstein), Plenum, New York **1985**, 432; *Polym. Prepr.* **1983**, *24*, 259; e) *Z. Chem.* **1987**, *27*, 395; f) *Wiss. Z. Univ. Halle* **1989**, *38/H4*, 3; *ibid.* **1992**, *41/H6, 37*, 51; g) in *Polymers and Biological Function* (Ed.: D. Braun), *Angew. Makromol. Chem.* **1989**, *166/167*, 81; h) see also: H. Kelker, in *Zehn Arbeiten über flüssige Kristalle* (Ed.: H. Sackmann), *Wiss. Beitr. MLU* **1986**/52, (N. 17), 193.

[18] Liquid crystals – molecular biological aspects: a) S. Hoffmann, W. Witkowski, in *Wirkungsmechanismen von Herbiciden und synthetischen Wachstumsregulatoren,* RGW-Symposium Halle 1972 (Eds.: A. Barth, F. Jacob, G. Feyerabend), Fischer, Jena **1972**/3, p. 291, 306; b) *Nucleic Acids Res.* **1975**, *SP 1*, 137; c) *Nucleic Acids Symp. Ser.* **1978**, *4*, 221; d) S. Hoffmann, W. Witkowski, S.-R. Waschke, A. Veckenstedt, *Nucleic Acids Symp. Ser.* **1981**, *9*, 239; e) S. Hoffmann, *Wiss. Z. Univ. Halle* **1983**, *32/H4*, 51; f) *Nucleic Acids Symp. Ser.* **1984**, *14*, 7; in *Plenary Lectures – Symp. Chem. Heterocycl. Comp. (VIIIth) and Nucleic Acids Comp. (VIth)* (Eds.: J. Beranek, A. Piskala), Czechoslovak. Acad. Sci., Prague **1985**, p. 45; g) in *Wirkstofforschung* *1980* (Ed.: H. Possin), Wissenschaftspubl. Martin-Luther-Univ., Halle **1981**/2, *11*, 35; h) S. Hoffmann, W. Witkowski, in *Wirkstofforschung '82* (Ed.: H. Possing), Wissenschaftspubl. Martin-Luther-Univ., Halle **1984**, p. 102; i) S. Hoffmann, in *Wirkstofforschung* (Eds.: A. Barth, H. Possin), Wiss. Beitr. Martin-Luther-Univ., Halle **1988**, 3(P32), 48; k) in *2nd Swedish – German Workshop on Modern Aspects of Chemistry and Biochemistry of Nucleic Acids* (Ed.: H. Seliger), *Nucleosides Nucleotides* **1989**, *7*, 555; l) in *Chirality – From Weak Bosons to the α-Helix* (Ed.: R. Janoschek), Springer, Heidelberg **1991**, p. 205; m) *Angew. Chem.* **1992**, *104*, 1032; *Angew. Chem. Int. Ed. Engl.* **1992**, *31*, 1013; n) in *Reproduction of Supramolecular Structure* (Eds.: G. R. Fleischaker, S. Colonna, P. L. Luisi), Kluwer, Dordrecht *NATO ASI Ser.* **1994**, *496*, 3.

[19] "Soft matter": P. G. de Gennes, *The Physics of Liquid Crystals,* Oxford University Press, Oxford **1984**; *Angew. Chem.* **1992**, *104*, 856; *Angew. Chem. Int. Ed. Engl.* **1992**, *31*, 842; P.-G. de Gennes, M. Hébert, R. Kant, *Macromol. Symp.* **1997**, *113*, 39; see also [24e].

[20] The "predecessors": a) W. Heintz, *Jahresber. Chem.* **1849**, 342 (Gießen 1850); **1852**, 506 (Gießen 1853); **1854**, 447 (Gießen 1855); b) *J. Prakt. Chem.* **1855**, *66*, 1; c) P. Duffy, *J. Chem. Soc.* **1852**, *5*, 197; d) R. Virchow, *Virchow's Archiv* **1854**, *6*, 571; e) M. Berthelot, *C. R. Acad. Sci.* **1855**, *41*, 452; f) *J. Prakt. Chem.* **1856**, *67*, 235; f) C. Mettenheimer, *Corr.-Blatt d. Verein. f. gem. Arbeit zur Förderung d. wiss. Heilkunde* **1857**, *24*, 331; g) L. Planer, *Liebigs Ann. Chem.* **1861**, *118*, 25; h) W. Loebisch, *Ber. Dtsch. Chem. Ges.* **1872**, *5*, 510.

[21] The beginning of the history of liquid crystals: F. Reinitzer, *Monatsh. Chem.* **1988**, *9*, 421.

[22] P. M. Knoll, *Fridericiana – Z. Univ. Karlsruhe* **1981**, 43.

[23] Last-century presentiments of inanimate – animate world coherences: E. Haeckel, a) *Kristallseelen,* Kröner, Leipzig **1917**; b) *Kunstformen der Natur,* Bibl. Inst. Leipzig **1899 – 1900**; c) F. Rinne, *Grenzfragen des Lebens,* Quelle & Meyer, Leipzig **1931**; d) *Ber. Naturforsch.* **1930**, *30*, 1; e) *Naturwissenschaften* **1930**, *18*, 837.

[24] Liquid crystals – beginning of the academic period: D. Vorländer, a) *Kristallin-flüssige Substanzen,* Enke, Stuttgart **1908**; b) *Chemische Kristallographie der Flüssigkeiten,* Akademische Verlagsgesellschaft, Leipzig **1924**; c) D. Vorländer, *Ber. Dtsch. Chem. Ges.* **1907**, *40*, 1970; *ibid.* **1908**, *41*, 2035; d) *ibid.* **1910**, *43*, 3120; e) *Z. Phys. Chem.* **1919**, *105*, 211; f) D. Vorländer, W. Selke, *ibid.* **1928**, *129*, 435; g) R. Schenck, *Kristallinische Flüssigkeiten und flüssige Kristalle,* Engelmann, Leipzig **1905**; h) *Ann. Phys.*

1902, *9*, 1053; i) *Z. Elektrochem.* **1905**, *50*, 951; k) F. Hoffmann, personal communication.

[25] Liquid-crystal views: a) G. W. Gray, *Molecular Structure and the Properties of Liquid Crystals,* Academic Press, New York **1962**; b) S. Chandrasekhar, *Liquid Crystals,* Cambridge University Press, New York **1977**; c) J. D. Litster, J. Birgeneau, *Phys. Today* **1982**, *35 (5)*, 52.

[26] Sleeping beauty's reawakening: a) F. Sauerwald, *Z. Metallkunde* **1954**, *45*, 257; b) H. Sackmann, *Z. Elektrochem.* **1955**, *59*, 880; c) H. Arnold, H. Sackmann, *ibid.* **1959**, *63*, 1171; *Z. Phys. Chem. (Leipzig)* **1960**, *213*, 137; d) S. Hoffmann, *Diplomathesis,* Halle University **1957**; e) G. Jaenecke, *Diplomathesis,* Halle University **1957**; f) S. Hoffmann, G. Jaenecke, W. Brandt, W. Kumpf, W. Weissflog, G. Brezesinski, *Z. Chem.* **1986**, *26*, 284; g) H. Schubert, *Wiss. Z. Univ. Halle* **1981**, *30*, 126.

[27] And animation: a) H. Sackmann, *Nova Acta Leopoldina* **1977**, *NF 47*, 137; in *Zehn Arbeiten über flüssige Kristalle* (Ed.: H. Sackmann), Wiss. Beitr. Martin-Luther-Univ. **1986/52**, (N 17), 193; b) D. Demus, H. Zaschke, *Flüssige Kristalle in Tabellen I und II,* Deutscher Verlag für Grundstoffindustrie, Leipzig **1974**, 1984; c) D. Demus, S. Diele, S. Grande, H. Sackmann, *Adv. Liq. Cryst.* **1983**, *6*, 1; D. Demus, L. Richter, *Textures of Liquid Crystals,* Deutscher Verlag für Grundstoffindustrie, Leipzig **1978**; d) H.-D. Dörfler, S. Diele, M. Diehl: *Colloid Polym. Sci.* **1984**, *262*, 139; e) H.-J. Deutscher, H.-M. Vorbrodt, H. Zaschke, *Z. Chem.* **1981**, *21*, 8; f) H. Zaschke, A. Isenberg, H.-M. Vorbrodt, in *Liquid Crystals and Ordered Fluids* (Ed.: A Griffin), Plenum, New York **1984**, 75; g) S. Hoffmann, *Z. Chem.* **1979**, *19, 241, ibid.* **1982**, *22*, 357; *ibid.* **1987**, *27*, 395.

[28] Strange discoid versus rod-like artificial classic and native alien: a) C. Destrade, H. Gasparoux, P. Foucher, N. H. Tinh, J. Malthete, J. Jacques, *J. Chim. Phys.* **1983**, *80*, 137; b) R. Blinc, *Ber. Bunsenges. Phys. Chem.* **1983**, *87*, 271; c) W. Schadt, M. Petrzila, P. R. Gerber, A. Villiger, *Mol. Cryst. Liq. Cryst.* **1985**, *122*, 241; d) R. Langridge, T. E. Ferrin, I. D. Kuntz, M. L. Conolly, *Science* **1981**, *211*, 661 – courtesy of Laurence H. Hurley.

[29] Structure and phase – routes to mesogenic nucleoprotein systems: a) F. Miescher, "On the chemical composition of pyocytes", in *Hoppe-Seyler's med. Untersuchungen* **1971**; in *Die histochemischen und physiologischen Arbeiten* (Ed.: F. C. W. Vogel), Leipzig **1897**; b) R. Altmann, *Arch. Anat. Phys. Phys. Abt.* **1889**; *Die Elementarorganismen,* Veit, Leipzig **1890**; c) E. Schroedinger, *What is Life?,* Cambridge University Press, New York **1944**; d) L. Pauling, R. B. Corey, H. R. Branson, *Proc. Natl. Acad. Sci. USA*

1951, *37*, 205, 735; L. Pauling, *The Nature of the Chemical Bond,* Cornell University Press, Ithaca, NY **1960**; e) E. Chargaff, *Experientia* **1950**, *6*, 201; *ibid.* **1970**, *26*, 810; *On some of the Biological Consequences of Base-Pairing in the Nucleic Acids in Development and Metabolic Control Mechanisms and Neoplasia,* Williams and Wilkins, Baltimore, MD **1965**; *Nature* **1974**, *248*, 776; f) J. D. Watson, F. H. C. Crick, *Nature* **1953**, *177*, 964; F. H. C. Crick, J. D. Watson, *Proc. R. Soc.* **1954**, *A233*, 80; g) R. A. Weinberg, *Sci. Am.* **1985**, *253*, 48; h) *Die Moleküle des Lebens* (Ed.: P. Sitte) Spektrum Wiss. **1988**; see also [5 d, 7 a, 17, 18].

[30] Life sciences – the supramolecular view: a) J. M. Lehn, *Science* **1985**, *227*, 849; *Angew. Chem.* **1988**, *100*, 91; *Angew. Chem. Int. Ed. Engl.* **1988**, *27*, 89; *ibid.* **1990**, *102*, 1347 (**1990**, *29*, 1304); *Chem. Biol.* **1994**, *1*, XVIII; *Supramolecular Chemistry,* VCH, Weinheim **1995**; b) R. Krämer, J.-M. Lehn, A. Delian, J. Fischer, *Angew. Chem.* **1993**, *105*, 764; *Angew. Chem. Int. Ed. Engl.* **1993**, *32*, 703; c) H. Ringsdorf, *Supramol. Sci.* **1994**, *1*, 5; – "Order and Mobility" between "Life Sciences" and "Advanced Materials"; d) H. Ringsdorf, B. Schlarb, J. Venzmer, *Angew. Chem.* **1988**, *100*, 117; *Angew. Chem. Int. Ed. Engl.* **1988**, *27*, 113; e) M. Ahlers, W. Müller, A. Reichert, H. Ringsdorf, J. Venzmer, *ibid.* **1990**, *102*, 1310 (**1990**, *29*, 1269); f) S. Denzinger, A. Dittrich, W. Paulus, H. Ringsdorf, in *From Simplicity to Complexity in Chemistry – and Beyond,* (Eds.: A. Müller, A. Dress, F. Vögtle), Vieweg, Braunschweig **1996**, p. 63; g) K. Fujita, S. Kimura, Y. Imanishi, E. Rump, H. Ringsdorf, *Adv. Biophys.* **1997**, *34*, 127; h) D. Philp, J. F. Stoddart, *Angew. Chem.* **1996**, *108*, 1242; *Angew. Chem. Int. Ed. Engl.* **1996**, *35*, 1154; Order–disorder coherences between chaos and complexity: see also [5 d, 7 a, 17, 18].

[31] Parity violation and chirality development: a) L. Pasteur, in *Leçons de chimie professées en 1860 par MM,* **1886**; L. Pasteur, A. A. T. Cahours, C. A. Wurtz, P. E. M. Berthelot, H. E. Sainte-claire Deville, R. F. Barral, J.-B. A. Dumas, *Sur la dissymétrie moléculaire,* Collection Epistème Paris Hachette Paris, **1861**; b) R. Janoschek, *Naturwiss. Rundschau* **1986**, *39*, 327; in *Chirality – From Weak Bosons to the α-Helix* (Ed.: R. Janoschek), Springer, Berlin **1991**, p. 18; c) H. Latal, *ibid.,* p. 1; d) S. Weinberg, *Science* **1980**, *210*, 1212; e) S. Mason, *Chem. Br.* **1985**, *21*, 538; *Trends Pharmacol Sci.* **1986**, (7), 20; *Nouv. J. Chim.* **1986**, *10*, 739; f) G. E. Tranter, *Nature* **1985**, *318*, 172; *Mol. Phys.* **1985**, *56*, 825; *Nachr. Chem. Techn. Lab.* **1986**, *34*, 866; *J. Theor. Biol.* **1986**, *119*, 469; g) D. K. Kondepudi, G. W. Nelson, *Nature* **1985**, *314*, 438; h) D. Z. Freedman, P. van Nieuwenhuizen, in *Teilchen, Felder und*

Symmetrien, Spektrum, Heidelberg **1988**, p. 170; in *Kosmologie und Teilchenphysik*, Spektrum Heidelberg **1990**, p. 120; i) M.-A. Bouchiat, L. Pottier, in *Elementare Materie, Vakuum und Felder*, Spektrum, Heidelberg **1988**, p. 130; k) K. Mainzer, *Chimia* **1988**, *42*, 161; l) *Chirality* (Eds.: I. W. Wainer, J. Caldwell, B. Testa), **1989**, p. 1; m) M. Quack, *Angew. Chem.* **1989**, *101*, 578; *Angew. Chem. Int. Ed. Engl.* **1989**, *28*, 959; n) R. A. Hegstrom, D. K. Kondepudi, *Sci. Am.* **1990**, *262*, 98; o) G. Jung, *Angew. Chem.* **1992**, *104*, 1484; *Angew. Chem. Int. Ed. Engl.* **1992**, *31*, 1457; see also [18 l–n].

[32] Evolutionary maternal inorganic matrices: a) M. Paecht-Horowitz, J. Berger, A. Katchalsky, *Nature* **1970**, *228*, 636; *Angew. Chem.* **1973**, *85*, 422; *Angew. Chem. Int. Ed. Engl.* **1973**, *12*, 349; b) A. G. Cairns-Smith, *Genetic Takeover and the Mineral Origins of Life*, Cambridge University Press, New York **1982**; c) H. G. von Schnering, R. Nesper, *Angew. Chem.* **1987**, *99*, 1087; *Angew. Chem. Int. Ed. Engl.* **1987**, *26*, 1059; d) R. J. P. Williams, *Biochem. Soc. Trans.* **1990**, *18*, 689; e) L. Addadi, S. Weiner, *Angew. Chem.* **1992**, *104*, 159; *Angew. Chem. Int. Ed. Engl.* **1992**, *31*, 153; f) S. J. Lippard, J. M. Beng, *Principles of Bioinorganic Chemistry*, University Science Books, Mill Valley, CA **1994**; *Bioanorganische Chemie*, Spektrum, Heidelberg **1994**; g) D. E. Devos, D. L. Vanoppen, X.-Y. Li, S. Libbrecht, Y. Bruynserade, P. P. Knops-Gerrits, P. A. Jacobs, *Chem. Eur. J.* **1995**, *1*, 144; h) A. Fironzi, D. Kumar, L. M. Bull, T. Bresier, P. Singer, R. Huo, S. A. Walker, J. A. Zasadzinski, C. Glinka, J. Nicol, D. Margolese, G. D. Stucky, B. T. Chmelka, *Science* **1995**, *267*, 1138.

[33] Constituents of nucleoprotein system: a) W. Guschlbauer, *Nucleic Acid Structure*, Springer, Berlin **1976**; in *Encyclopedia of Polymer Science and Engineering*, Wiley, New York **1988**, Vol. 12, p. 699; b) Y. Mitsui, Y. Takeda, *Adv. Biophys.* **1979**, *12*, 1; c) *Methods Enzymology* (Eds: D. M. J. Lilley, J. E. Dahlberg), Academic, San Diego, CA **1992**; d) R. W. Roberts, D. M. Crothers, *Science* **1992**, *258*, 1463; e) O. Kennard, W.-N. Hunter, *Angew. Chem.* **1991**, *103*, 1280; *Angew. Chem. Int. Ed. Engl.* **1991**, *30*, 1254; f) *Structural Biology I/II* (Eds.: R. H. Sarma, M. H. Sarma), Adenine Press, New York **1993** – with emphasis also on V. I. Ivanov's concluding remarks: "Simplicity Lost"; g) R. E. Dickerson, I. Geis, *The Structure and Action of Proteins 1971; Struktur und Funktion der Proteine*, VCH, Weinheim **1971**; h) R. E. Dickerson, M. L. Kopka, P. Pjura, *Chem. Scr.* **1986**, *26B*, 139; i) H. A. Scheraga, K.-C. Chou, G. Némethy, in *Conformations in Biology* (Eds.: R. Srinavasan, R. H. Sarma), Adenine Press, New York **1983**, p. 1; k) R. Jaenicke, *Prog. Biophys.*

Mol. Biol. **1987**, *49*, 117; *Angew. Chem.* **1984**, *96*, 385; *Angew. Chem. Int. Ed. Engl.* **1984**, *23*, 395; *Naturwissenschaften* **1988**, *75*, 604; l) M. Mutter, *Angew. Chem.* **1985**, *97*, 639; *Angew. Chem. Int. Ed. Engl.* **1985**, *24*, 639; M. Mutter, S. Vuilleumier, *ibid.* **1989**, *101*, 551 and **1989**, *28*, 535; m) W. F. Anderson, J. H. Ohlendorf, Y. Takeda, B. W. Matthews, *Nature* **1981**, *290*, 754; D. H. Ohlendorf, W. F. Anderson, Y. Takeda, B. W. Matthews, *J. Biomol. Struct. Dyn.* **1983**, *1*, 533; n) S.-H. Kim, in *Nucleic Acids Research: Future Developments* (Eds.: K. Mizoguchi, I. Watanabe, J. D. Watson), Academic, New York **1983**, p. 165; o) E. Katchalski-Katzir, *Makromol. Chem. Macromol. Symp.* **1988**, *19*, 1; p) L. Stryer, *Biochemistry*, Freeman, New York **1988**; *Biochemie,* Spektrum, Heidelberg **1990**; q) D. Voet, J. G. Voet, *Biochemistry,* Wiley, New York **1990**; *Biochemie*, VCH, Weinheim **1992**.

[34] Emergence of an "RNA world": a) S. Altman, *Angew. Chem.* **1990**, *102*, 735; *Angew. Chem. Int. Ed. Engl.* **1990**, *29*, 707; b) T. R. Cech, *ibid.* **1990**, *102*, 745 and **1990**, *29*, 716; *Spektrum Wiss.* **1987**, (*1*), 42; c) T. Inoue, F. Y. Sullivan, T. R. Cech, *J. Mol. Biol.* **1986**, *189*, 143; d) R. B. Waring, P. Towner, S. J. Minter, R. W. Davies, *Nature* **1986**, *321*, 133; e) P. A. Sharp, *Angew. Chem.* **1994**, *106*, 1292; *Angew. Chem. Int. Ed. Engl.* **1994**, *33*, 1229.

[35] Proteins' own ambitions: a) Yu. A. Ovchinnikov, V. T. Ivanov, *Tetrahedron* **1975**, *31*, 2177; b) Yu. A. Ovchinnikov, *Bioorganicheskaya chimya*, Prosvyeschenye, Moscow **1987**.

[36] Highlighting classical organic chemistry: a) E. Fischer, *Ber. Dtsch. Chem. Ges.* **1894**, *27*, 2985, 3189; b) E. Fischer, B. Helferich, *Liebigs Ann. Chem.* **1911**, *383*, 68; c) E. Fischer, K. Freudenberg, *Ber. Dtsch. Chem. Ges.* **1915**, *45*, 915; d) W. Langenbeck, *Lehrbuch der Organischen Chemie,* Steinkopf, Dresden **1954**; e) R. B. Woodward, in *Perspectives in Organic Chemistry* (Ed.: A. Todd), Interscience, New York **1956**, p. 155; f) A. Eschenmoser, *Angew. Chem.* **1988**, *100*, 5; *Angew. Chem. Int. Ed. Engl.* **1988**, *27*, 5; g) A. Eschenmoser, *ibid.* **1994**, *106*, 2455 and **1994**, *33*, 2363; h) F. W. Lichtenthaler, *ibid.* **1992**, *104*, 1577 and **1992**, *31*, 1541; *ibid.* **1994**, *106*, 2456 and **1994**, *33*, 2364; i) D. E. Koshland, *ibid.* 2468 and 2375; k) E. H. Fischer, *ibid.* **1993**, *105*, 1181 and **1993**, *32*, 1130.

[37] Complex "rod-like" virus: L. Onsager, *Ann. N. Y. Acad. Sci.* **1949**, *51*, 62Z.

[38] Mesogenic nucleation capabilities of nucleoprotein system: a) C. W. Carter, J. Kraut, *Proc. Natl. Acad. Sci. USA* **1974**, *71*, 283; b) G. M. Church, J. L. Sussman, S.-H. Kim, *ibid.* **1977**, *74*, 1458.

[39] Omne vivum e matribus – highly condensed states and the mesogenic matrix problem: a) A. Todd, in *Perspectives in Organic Chemistry* (Ed.:

A. Todd), Interscience, New York **1956**, p. 245;
b) G. Schramm, H. Grötsch, W. Pollmann, *Angew. Chem.* **1962**, *74*, 53; *Angew. Chem. Int. Ed. Engl.* **1962**, *1*, 1; c) J. H. Winter, *Angew. Chem.* **1966**, *78*, 887; *Angew. Chem. Int. Ed. Engl.* **1966**, *5*, 862; d) W. Kern, K. Kämmerer, *Chem. Ztg.* **1967**, *91*, 73; e) H. Kämmerer, *ibid.* **1972**, *96*, 7; f) S. Hoffmann, *Habilitationthesis* Halle University **1968**; *Z. Chem.* **1979**, *19*, 241; *ibid.* **1982**, *22*, 357; g) S. Hoffmann, H. Schubert, W. Witkowski, *ibid.* **1971**, *11*, 345, 465; *ibid.* **1974**, *14*, 154; h) D. Pfeiffer, L. Kutschabsky, S. Hoffmann, *Cryst. Struct. Commun.* **1982**, *11*, 1635; i) D. Scharfenberg-Pfeiffer, R.-G. Kretschmer, S. Hoffmann, *Cryst. Res. Technol.* **1988**, *23*, 881; k) G. Challa, Y. Y. Tan, *Pure Appl. Chem.* **1981**, *53*, 627; l) A. I. Scott, *Angew. Chem.* **1993**, *105*, 1281; *Angew. Chem. Int. Ed. Engl.* **1993**, *32*, 1223; see also [5 d, 7 a, 17, 181–n, 29, 30].

[40] Evolutionists: a) T. C. Lucretius, *De rerum natura*, Weidmann, Leipzig **1795**; b) A. I. Oparin, *Origin of Life*, Moskau **1924**; *The Origin of Life on Earth*, Academic, New York **1957**; c) J. B. S. Haldane, *The Course of Evolution*, Longman, New York **1932**; d) S. W. Fox, *Science* **1960**, *132*, 200; *The Origin of Prebiological Systems and of Their Molecular Matrices* (Ed.: S. W. Fox), Academic, New York **1965**; e) S. Spiegelman, *Q. Rev. Biophys.* **1971**, *2*, 213; f) H. Kuhn, *Angew. Chem.* **1972**, *84*, 838; *Angew. Chem. Int. Ed. Engl.* **1972**, *11*, 798; g) L. E. Orgel, *The Origins of Life on Earth*, Prentice Hall, Englewood Cliffs, NJ **1974**; h) M. Eigen, P. Schuster, *Naturwissenschaften* **1977**, *64*, 541; *ibid.* **1978**, *65*, 7; M. Eigen, *Chem. Scr.* **1986**, *26 B*, 13; i) J. J. Wolken, in: *Molecular Evolution and Protobiology* (Eds.: K. Matsuno, K. Dose, K. Harada, D. L. Rohlfing), Plenum, New York **1984**, 137; k) E. Szathmáry, in [5 d], p. 65; J. M. Smith, E. Szathmáry, *The Major Transitions in Evolution*, Freeman, Oxford **1995**; Evolution, Spektrum, Heidelberg **1996**; l) F. J. Varela, in [5d], p. 23; m) L. F. Orgel, *Sci. Am.* **1994**, *271 (10)*, 53; see also [1 – 8, 17, 18, 30].

[41] Approaches to directed artificial evolution: a) G. F. Joyce, *Cold Spring Harbor Symp Quant Biol* **1987**, *LII*, 41; *Nature* **1989**, *338*, 217; b) A. A. Beaudry, G. F. Joyce, *Science* **1992**, *257*, 635; c) G. F. Joyce, *Sci. Am.* **1992**, *269 (12)*, 49; d) R. R. Breaker, G. F. Joyce, *Proc. Natl. Acad. Sci. USA* **1994**, *91*, 6093; e) J. Tsang, G. F. Joyce, *Biochemistry* **1994**, *33*, 5966; f) R. R. Breaker, G. F. Joyce, *Chem. & Biol.* **1994**, *1*, 223; g) R. R. Breaker, G. F. Joyce, in [5d], 127; h) T. Li, Z. Zeng, V. A. Estevez, K. U. Baldenius, K. C. Nicolaou, G. F. Joyce, *J. Am. Chem. Soc.* **1994**, *116*, 3709; i) A. Plückthun, L. Ge, *Angew. Chem.* **1991**, *103*, 301; *Angew. Chem. Int. Ed. Engl.* **1991**, *30*, 296; k) G. Jung, A. G. Beck-

Sickinger, *ibid.* **1992**, *104*, 375 (**1992**, *31*, 367); l) D. P. Bartel, J. W. Szostak, *Science* **1992**, *261*, 1411; m) D. E. Huizenga, J. W. Szostak, *Biochemistry* **1995**, *34*, 656; n) J. W. Szostak, *Protein Engin.* **1995**, *8*, 3; *Chem. Rev.* **1997**, *97*, 347; o) A. J. Hager, J. W. Szostak, *Orig. Life* **1996**, *26*, 269; p) J. R. Lorsch, J. W. Szostak, *Acc. Chem. Res.* **1996**, *29*, 103; q) J. S. McCaskill, *Nachr. Chem. Techn. Lab.* **1995**, *43*, 199; r) G. von Kiedrowski, *Angew. Chem.* **1991**, *103*, 839; *Angew. Chem. Int. Ed. Engl.* **1991**, *30*, 892; s) P. Burgstaller, M. Famulok, *ibid.* **1995**, *107*, 1303 (**1995**, *34*, 1188); t) M. Egli, *ibid.* **1996**, *108*, 486, 2020 (**1996**, *35*, 432, 1894); *ibid.* **1997**, *109*, 494 (**1997**, *36*, 480); u) H. Seliger, R. Bader, M. Hinz, B. Rotte, A. Astriab, M. Markiewicz, W. T. Markiewicz, *Nucleosides Nucleotides* **1997**, *16*, 703; see also [5 – 7, 17].

[42] The enzyme-free approach in nucleic acid matrix reactions: a) L. E. Orgel, R. Lohrmann, *Acc. Chem. Res.* **1974**, *7*, 368; b) R. Lohrmann, L. E. Orgel, *J. Mol. Biol.* **1980**, *142*, 555; c) T. Inoue, L. E. Orgel, *Science* **1983**, *219*, 859; d) T. Inoue, G. F. Joyce, K. Grzeskowiak, L. E. Orgel, J. M. Brown, C. B. Reese, *J. Mol. Biol.* **1984**, *178*, 669; e) C. B. Chen, T. Inoue, L. E. Orgel, *ibid.* **1985**, *181*, 271; f) A. W. Schwartz, L. E. Orgel, *Science* **1985**, *228*, 585; g) O. L. Acevedo, L. E. Orgel, *Nature* **1986**, *321*, 790; h) L. E. Orgel, *Cold Spring Harbor Symp. Quant. Biol.* **1987**, *LII*, 9; *J. Theor. Biol.* **1986**, *123*, 127; *Folia Biol. (Praha)* **1983**, *29*, 65; *J. Mol. Biol.* **1995**, *181*, 271; *Orig. Life* **1996**, *26*, 261; i) J. P. Ferris, A. R. Hill Jr., R. Liu, L. E. Orgel, *Nature* **1996**, *381*, 59; k) B. C. F. Chu, L. E. Orgel, *Bioconj. Chem.* **1997**, *8*, 103; see also [43, 44].

[43] Self-replicative systems: a) L. E. Orgel, *Nature* **1992**, *358*, 203; b) S. Hoffmann, *Angew. Chem.* **1992**, *104*, 1032; *Angew. Chem. Int. Ed. Engl.* **1992**, *31*, 1013; in [5d], 3; c) J. Rebek, Jr., *Sci. Am.* **1994**, *271 (7)*, 34; in [5d], 75; d) D. Sievers, T. Achilles, J. Burmeister, S. Jordan, A. Terford, G. von Kiedrowski, in [5d], 45; see also: e) P. Strazewski, C. Tamm, *Angew. Chem.* **1990**, *102*, 37; *Angew. Chem. Int. Ed. Engl.* **1990**, *29*, 36; see also [5 – 7, 17, 18, 30].

[44] "Minimal models" of replication: a) G. von Kiedrowski, *Angew. Chem.* **1986**, *93*, 932; *Angew. Chem. Int. Ed. Engl.* **1986**, *25*, 932; in *40 Jahre Fonds der Chemischen Industrie 1950–90*, VCI, Frankfurt **1990**; *Bioorg. Chem. Frontiers* **1993**, *3*, 113; *Nature* **1996**, *381*, 20; b) W. S. Zielinski, L. E. Orgel, *Nature* **1987**, *327*, 346; *J. Mol. Evol.* **1989**, *29*, 281; c) G. von Kiedrowski, G. B. Wlotzka, J. Helbing, *Angew. Chem.* **1989**, *102*, 1259; *Angew. Chem. Int. Ed. Engl.* **1989**, *28*, 1235; d) G. von Kiedrowski, G. J. Helbing, B. Wlotzka, S. Jordan, M. Mathen, T. Achilles, D. Sievers, A. Terfort, B. C. Kahrs, *Nachr. Chem.*

Techn. Lab. **1992**, *40*, 578; e) A. G. Terfort, G. von Kiedrowski, *Angew. Chem.* **1992**, *104*, 626; *Angew. Chem. Int. Ed. Engl.* **1992**, *31*, 654; f) D. Sievers, G. von Kiedrowski, *Nature* **1994**, *369*, 221; g) Th. Achilles, G. von Kiedrowski, *Angew. Chem.* **1993**, *105*, 1225; *Angew. Chem. Int. Ed. Engl.* **1993**, *32*, 1198; h) B. G. Bag, G. von Kiedrowski, *Pure Appl. Chem.* **1996**, *68*, 2145; i) J. Burmeister, G. von Kiedrowski, A. D. Ellington, *Angew. Chem.* **1997**, *36*, 1321; *Angew. Chem. Int. Ed. Engl.* **1997**, *109*, 1379; see also [5–7, 18, 39, 42, 43].

[45] "Extrabiotic replication and self-assembly": a) J. Rebek Jr., *Angew. Chem.* **1990**, *102*, 261; *Angew. Chem. Int. Ed. Engl.* **1990**, *29*, 245; b) T. Tjivikua, P. Ballester, J. Rebek, Jr., *J. Am. Chem. Soc.* **1990**, *112*, 1249; c) T. K. Park, Q. Feng, J. Rebek Jr., *ibid.* **1992**, *114*, 4529; *Science* **1992**, *256*, 1179; d) M. Famulok, J. S. Nowick, J. Rebek Jr., *Act. Chem. Scand.* **1992**, *46*, 315; in [5 d], p. 75; e) R. J. Pieters, I. Huc, J. Rebek Jr., *Chem. Eur. J.* **1995**, *1*, 183; f) J. Kang, J. Rebek Jr., *Nature* **1996**, *382*, 239.

[46] 2-Pyridones as drastically abstracted nucleic acid analogs: a) S. Hoffmann, W. Witkowski, G. Borrmann, H. Schubert, *Z. Chem.* **1974**, *14*, 154; b) *ibid.* **1978**, *18*, 403; c) F. Persico, J. D. Wuest, *J. Org. Chem.* **1993**, *58*, 95.

[47] From supramolecular chirality selectors to elementorganic matrices: a) T. Gulik-Krzywicki, C. Fouquey, J.-M. Lehn, *Proc. Natl. Acad. Sci. USA* **1993**, *90*, 163; b) J.-M. Lehn, A. Rigault, *Angew. Chem.* **1988**, *100*, 1121; *Angew. Chem. Int. Ed. Engl.* **1988**, *27*, 1059; c) G. Bernardinelli, C. Piquet, A. F. Williams, *ibid.* **1992**, *104*, 1626 and **1992**, *31*, 1622; d) R. Krämer, J.-M. Lehn, A. DeCian, J. Fischer, *ibid.* **1993**, *105*, 764 and **1993**, *32*, 703; e) M. Kotera, J. M. Lehn, J. P. Vigneron, *J. Chem. Soc., Chem. Commun.* **1994**, 197; see also [1–8, 181–n].

[48] Triplexes between molecular hystereses and self-replication: a) R. A. Cox, A. S. Jones, G. E. Marsh, A. R. Peacocke, *Biochim. Biophys. Acta* **1956**, *21*, 576; b) E. Neumann, A. Katchalsky, *Ber. Bunsenges. Phys. Chem.* **1970**, *74*, 868; c) *Proc. Natl. Acad. Sci. USA* **1972**, *69*, 993; d) A. Katchalsky, E. Neumann, *Isr. J. Neuroscience* **1972**, *3*, 175; e) E. Neumann, *Angew. Chem.* **1973**, *85*, 430; *Angew. Chem. Int. Ed. Engl.* **1973**, *12*, 356; f) W. Guschlbauer, D. Thiele, M.-Th. Sarvechi, Ch. Marck, in *Dynamic Aspects of Conformation Changes in Biological Macromolecules* (Ed.: C. Sadron), Reidel, Dordrecht **1973**; g) S. Hoffmann, W. Witkowski, H.-H. Rüttinger, *Z. Chem.* **1974**, *14*, 438; h) S. Hoffmann, W. Witkowski, *ibid.* **1975**, *15*, 149; *ibid.* **1976**, *16*, 442; i) S. Micciancio, G. Vassallo, *Nuovo Cimento* **1982**, *1*, 121; k) R. W. Roberts, D. M. Crothers, *Science* **1992**, *258*, 1463;

l) N. T. Thuong, C. Hélène, *Angew. Chem.* **1993**, *105*, 697; *Angew. Chem. Int. Ed. Engl.* **1993**, *32*, 666; m) J. F. Mouscadet, S. Carteau, H. Goulaveic, F. Subra, C. Anclair, *J. Biol. Chem.* **1994**, *269*, 21 635; n) T. Li, K. C. Nicolaou, *Nature* **1994**, *369*, 218; o) I. Radnakrishnan, D. J. Patel, *Structure* **1994**, *2*, 17; p) V. B. Zhurkin, G. Raghunathan, N. B. Ulyanov, R. D. Camerini-Otero, R. L. Jernigan in [33 f], p. 43; q) C. Y. Sekharudu, N. Yathindra, M. Sundaralingam, in [33 f], p. 113; r) H. Träuble, H. Eibl, *Proc. Natl. Acad. Sci. USA* **1974**, *71*, 214; s) D. Nachmansohn, *ibid.* **1976**, *73*, 82; t) V. N. Soyfer, V. N. Polaman, *Triple-Helical Nucleic Acids,* Springer, Heidelberg **1996**, see also [17, 18, 52 o, s].

[49] "Minima vita" models: a) P. A. Backmann, P. Walde, P.-L. Luisi, J. Lang, *J. Am. Chem. Soc.* **1990**, *112*, 8200; *ibid.* **1991**, *113*, 8204; b) P. L. Luisi, F. J. Varela, *Orig. Life* **1990**, *19*, 633; c) P. A. Bachmann, P. L. Luisi, J. Lang, *Nature* **1992**, *357*, 57; d) P. L. Luisi, P. A. Vermont-Bachmann, M. Fresta, P. Walde, E. Wehrli, *J. Liposome Res.* **1993**, *3*, 631; e) P. L. Luisi, in [5 d], 179; f) P. L. Luisi, personal communication; g) T. Oberholzer, R. Wick, P. L. Luisi, C. K. Biebricher, *Biochem. Biophys. Res. Commun.* **1995**, *207*, 250; h) P. L. Luisi, *Adv. Chem. Phys.* **1996**, *92*, 425; *Orig. Life* **1996**, *26*, 272; i) P. L. Luisi, P. Walde, *ibid.* **1996**, *26*, 429; k) D. Berti, L. Franchi, P. Baglioni, P. L. Luisi, *Langmuir* **1997**, *13*, 3438; l) C. Cescato, P. Walde, P. L. Luisi, *ibid.* **1997**, *13*, 4480; see also [5–7, 40].

[50] Playing the game of evolution: a) A. Eschenmoser, *Angew. Chem.* **1988**, *100*, 5; *Angew. Chem. Int. Ed. Engl.* **1988**, *27*, 5; *Nachr. Chem. Techn. Lab.* **1991**, *39*, 795; *Nova Acta Leopold.* **1992**, *NF 67/281*, 201; *Chem. Biol.* **1994**, *1*, IV; *Orig. Life* **1994**, *24*, 389; *ibid.* **1996**, *26*, 235; *NATO ASI Series E* **1996**, *320*, 293; b) A. Eschenmoser, M. Dobler, *Helv. Chim. Acta* **1992**, *75*, 218; c) A. Eschenmoser, E. Loewenthal, *Chem. Soc. Rev.* **1992**, 1; d) J. Hunziker, H.-J. Roth, M. Böhringer, A. Giger, U. Diederichsen, M. Göbel, R. Krishnan, B. Jaun, C. Leumann, A. Eschenmoser, *Helv. Chim. Acta* **1993**, *76*, 259; e) S. Pitch, A. Eschenmoser, B. Gedulin, S. Hui, G. Arrhenius, *Orig. Life* **1995**, *25*, 297; f) A. Eschenmoser, M. V. Kisakürek, *Helv. Chim. Acta* **1996**, *79*, 1249; R. Micura, M. Bolli, N. Windhab, A. Eschenmoser, *Angew. Chem.* **1997**, *109*, 899; *Angew. Chem. Int. Ed. Engl.* **1997**, *36*, 870; g) W. K. Olson, *Proc. Natl. Acad. Sci. USA* **1977**, *74*, 1775; see also h) G. Quinkert, *Nachr. Chem. Tech. Lab.* **1991**, *39*, 788; see also [5–7].

[51] Liquid crystals – end of simplicity: a) A. Blumstein, L. Patel, *Mol. Cryst. Liq. Cryst.* **1978**, *48*, 151; A. Blumstein, S. Vilasagar, S. Ponrathnam, S. B. Clough, R. B. Blumstein, G. Maret, *J. Polym. Sci. Polym. Phys. Ed.* **1982**, *20*, 877;

Polymeric Liquid Crystals (Ed.: A. Blumstein) Plenum, New York **1985**; R. B. Blumstein, O. Thomas, M. M. Gauthier, J. Asrar, A. Blumstein, in *Polymeric Liquid Crystals* (Ed.: A. Blumstein), Plenum, New York, **1985**, p. 239; b) G. M. Brown, *Am. Sci.* **1972**, *60*, 64; *J. Chem. Ed.* **1983**, *60*, 900; G. M. Brown, P. P. Crooke, *Chem. Eng. News* **1983**, *60*, 24; c) S. Chandrasekhar, *Liquid Crystals,* Cambridge, University Press, New York **1977**; S. Chandrasekhar, B. K. Sadashiva, K. A. Suresh, N. V. Madhusudana, S. Kumar, R. Shashidhar, G. Venkatesh, *J. Phys.* **1979**, *40*, C-3; d) E. T. Samulski, A. V. Tobolsky, in *Liquid Crystals and Ordered Fluids* (Eds. J. F. Johnson, R. S. Porter), Plenum, New York **1970**, p. 111; *Liquid Crystalline Order in Polymers* (Ed.: E. T. Samulski), Academic Press, New York **1978**; E. T. Samulski, *Phys. Today* **1982**, *35 (5)*, 40; *Faraday Discuss. Chem. Soc.* **1985**, *79*, 7; E. T. Samulski, M. M. Gauthier, R. B. Blumstein, A. Blumstein, *Macromolecules* **1984**, *17*, 479; E. T. Samulski, in *Polymeric Liquid Crystals* (Ed.: A. Blumstein), Plenum, New York **1985**, p. 65; e) D. Demus, *Mol. Cryst. Liq. Cryst.* **1988**, *165*, 45; f) G. W. Gray, A. J. Leadbetter, *Phys. Bull.* **1977**, *28*, G. W. Gray, *Philos. Trans. R. Soc. London* **1983**, *A309*, 77; *Proc. R. Soc. London A* **1985**, *402*, 1; *Mol. Cryst. Liq. Cryst., Lett. Sect.* **1990**, *7*, 47; *Spec. Publ. – R. Soc. Chem.* **1991**, *88*, 203; *ibid.* **1991**, *96*, 288; *Mol. Cryst. Liq. Cryst.* **1991**, *204*, 91; G. W. Gray, J. W. Goodby, *Smectic Liquid Crystals,* Leonard Hill, Glasgow **1984**; g) A. C. Griffin, S. R. Vaidya, M. L. Steele, in *Polymeric Liquid Crystals* (Ed.: A. Blumstein), Plenum, New York **1985**, p. 1; h) H. Finkelmann, H. Benthak, G. Rehage, *J. Chim. Phys.* **1983**, *80*, 163; H. Finkelmann, G. Rehage, *Adv. Polym. Sci.* **1984**, *60*, 100; H. Finkelmann, *Angew. Chem.* **1987**, *99*, 840; *Angew. Chem. Int. Ed. Engl.* **1987**, *26*, 816; i) P. S. Pershan, *Structure of Liquid Crystal Phases,* World Scientific, Singapore **1988**; *Phys. Today* **1982**, *35 (5)*, 34; k) H. Bader, K. Dorn, B. Hupfer, H. Ringsdorf, *Adv. Polym. Sci.* **1985**, *64*, 1; l) *6. Flüssigkristallkonferenz sozialistischer Länder, Halle 1985* (Ed.: H. Sackmann), Wiss. Beitr. Martin-Luther-Univ. Halle **1985**/37 (N 14); m) H. Stegemeyer, *Angew. Chem., Adv. Mater.* **1988**, *100*, 1640; n) M. Ballauf, *Angew. Chem.* **1989**, *101*, 260; *Angew. Chem. Int. Ed. Engl.* **1989**, *28*, 253; o) A. M. Giraud-Godquin, P. M. Maitlis, *ibid.* **1991**, *103*, 370 and **1991**, *30*, 375; p) W. F. Brinkman, P. E. Cladis, *Phys. Today* **1982**, *35 (5)*, 48; I. D. Litster, R. J. Birgeneau, *Phys. Today* **1982**, *35 (5)*, 26; J. D. Brock, R. J. Birgeneau, J. D. Litster, A. Aharony, *Phys. Today* **1989**, *42 (7)*, 52; q) M. Kléman, *Rep. Prog. Phys.* **1989**, *52*, 515; r) C. A. Angell, *Science* **1995**, *267*, 1924; s) S. Hoffmann, W. Weissflog, W. Kumpf, W. Brandt, W.

Witkowski, H. Schubert, G. Brezesinski, H.-D. Dörfler, *Z. Chem.* **1979**, *19*, 62; H.-D. Dörfler, G. Brezesinski, S. Hoffmann, *Stud. Biophys.* **1980**, *80*, 59; S. Hoffmann, *Z. Chem.* **1985**, *25*, 330; *ibid.* **1989**, *29*, 173, 449; S. Hoffmann, W. Kumpf, *ibid.* **1986**, *26*, 252, 293, 392; S. Hoffmann, W. Kumpf, W. Brandt, G. Brezesinski, W. Weissflog, *ibid.* **1986**, *26*, 253; S. Hoffmann, W. Brandt, H.-M. Vorbrodt, H. Zaschke, *ibid.* **1988**, *28,* 100; t) H. Schubert, S. Hoffmann, J. Hauschild, I. Marx, *ibid.* **1977**, *17*, 414; S. Hoffmann, H. Schubert, A. Kolbe, H. Krause, Z. Palacz, C. Neitsch, M. Herrmann, *ibid.* **1978**, *18*, 93; H. Schubert, J. Hauschild, D. Demus, S. Hoffmann, *ibid.* **1978**, *18*, 256; D. Pfeiffer, R.-G. Kretschmer, S. Hoffmann, *Cryst. Res. Technol.* **1986**, *21*, 1321; u) S. Hoffmann, E. Günther, W.-V. Meister; W. Witkowski, *Wiss. Z. Univ. Halle* **1988**, *37/H1*, 32; v) M. Descamps, C. Descamps, *J. Phys. Lett.* **1984**, *45*, L-459; w) A. E. Elliot, E. J. Ambrose, *Discuss. Faraday Soc.* **1950**, *9*, 246; x) E. Sackmann, in *Biophysics* (Eds.: W. Hoppe, W. Lohmann, H. Markl, H. Ziegler), Springer, Berlin **1983**, p. 425; y) J. W. Goodby, *Mol. Cryst. Liq. Cryst.* **1984**, *110*, 205, 221; z) H. Hoffmann, G. Ebert, *Angew. Chem.* **1988**, *100*, 933; *Angew. Chem. Int. Ed. Engl.* **1988**, *27*, 902; see also [5–7, 17, 18, 30, 52].

[52] Mesomorphs and their order-disorder gradients: a) H. Eibl, *Angew. Chem.* **1984**, *96*, 247; *Angew. Chem. Int. Ed. Engl.* **1984**, *23*, 257; b) S. G. Kostromin, R. V. Talroze, V. P. Shibaev, N. A. Plate, *Macromol. Chem. Rapid Commun.* **1982**, *3*, 803; c) W. G. Miller, *Annu. Rev. Phys. Chem.* **1978**, *29*, 519; d) S. L. Kwolek, P. W. Morgan, J. R. Schaefgen, L. W. Gulrich, *Macromolecules* **1977**, *10*, 1390; e) D. L. Patel, R. D. Gilbert, *J. Polym. Sci., Polym. Phys. Ed.* **1982**, *20*, 877; f) S. Hoffmann, N. thi Hanh, K. Mandl, G. Brezesinski, E. Günther, *Z. Chem.* **1986**, *26*, 103; g) R. F. P. Bryan, P. Hartley, R. W. Miller, M.-S. Shen, *Mol. Cryst. Liq. Cryst.* **1980**, *62*, 281; h) D. W. Urry, *Biochim. Biophys. Acta* **1972**, *265*, 115; *Proc. Natl. Acad. Sci. USA* **1972**, *69*, 1610; i) S. Hoffmann, M. G. Klotz, E. Müller, in *6th Int. Conf. on Surface Active Compounds* (Ed.: F. Scheel), Akademie-Verlag, Berlin **1987**, p. 545; k) C. Robinson, J. C. Ward, R. B. Beevers, *Discuss. Faraday Soc.* **1958**, *25*, 29; l) N. Ise, T. Okubo, K. Yamamoto, H. Matsuoka, H. Kawai, T. Hashimoto, M. Fujimura, *J. Chem. Phys.* **1983**, *78*, 541; m) C. Piechocki, J. Simon, A. Skoulios, D. Guillon, P. Weber, *J. Am. Chem. Soc.* **1982**, *104*, 5245; n) I. Thondorf, O. Lichtenberger, S. Hoffmann, *Z. Chem.* **1990**, *30*, 171; o) W.-V. Meister, E. Birch-Hirschfeld, H. Reinert, H. Kahl, W. Witkowski, S. Hoffmann, *ibid.* **1990**, *30,* 322; p) S. Hoffmann, W. Witkowski, G. Borrmann, *ibid.* **1977**, *17*, 291; S. Hoffmann,

R. Skölziger, W. Witkowski, *ibid.* **1986**, *26*, 331, 398; W.-V. Meister, R. Skölziger, G. Luck, Ch. Radtke, D. Munsche, W. Witkowski, S. Hoffmann, *ibid.* **1990**, *30*, 95; q) F. Livolant, *Eur. J. Cell. Biol.* **1984**, *33*, 300; r) R. L. Rill, *Proc. Natl. Acad. Sci. USA* **1986**, *83*, 342; s) W.-V. Meister, A.-M. Ladhoff, S. I. Kargov, D. Burckhardt, G. Luck, S. Hoffmann, *Z. Chem.* **1990**, *30*, 213; t) Yu. M. Yevdokimov, V. I. Salyanov, A. T. Dembo, H. Berg, *Biomed. Biochim. Acta* **1983**, *42*, 855; u) E. Iizuka, *Polym. J.* **1978**, *10*, 235; 293; v) G. Damaschun, H. Damaschun, R. Misselwitz, D. Zirwer, J. J. Muller, I. A. Zalluskaya, V. I. Vorobev, T. L. Pyatigorskaya, *Stud. Biophys.* **1986**, *112*, 127; S. G. Skuridin, H. Damaschun, G. Damaschun, Yu. M. Yevdokimov, R. Misselwitz, *ibid.* **1986**, *112*, 139; w) S. Hoffmann, W. Witkowski, A.-M. Ladhoff, *Z. Chem.* **1976**, *16*, 227; x) J. Feigon, F. W. Smith, R. F. Macaya, P. Schultze, in [33 f], p. 127; y) S. Hoffmann, *Nucleic Acids Res. Symp. Ser.* **1987**, *18*, 229; *Z. Chem.* **1990**, *30*, 94; S. Hoffmann, C. Liebmann, P. Mentz, K. Neubert, A. Barth, *Wiss. Z. Univ. Halle* **1992**, *41/H6*, 37; S. Hoffmann, M. Koban, C. Liebmann, P. Mentz, K. Neubert, A. Barth: in *β-Casomorphins and Related Peptides: Recent Developments* (Eds.: V. Brantl, M. Teschemacher), VCH, Weinheim **1993**; z) S. Hoffmann, C. Liebmann, P. Mentz, K. Neubert, A. Barth, *Z. Chem.* **1989**, *29*, 210; see also [51].

[53] Mediating water: a) E. Clementi, G. Corongin, *J. Biol. Phys.* **1983**, *11*, 33; b) M. U. Palma, in *Structure and Dynamics of Nucleic Acids and Proteins* (Eds.: E. Clementi, R. H. Sarma), Adenine Press, New York **1983**, 125; c) W. Saenger, W. N. Hunter, O. Kennard, *Nature* **1986**, *324*, 385; d) W. Saenger, *Annu. Rev. Biophys. Chem.* **1987**, *16*, 93; e) N. Thantzi, J. M. Thoruton, J. M. Goodfellow, *J. Mol. Biol.* **1988**, *202*, 637; f) E. Westlof, *Annu. Rev. Biophys. Chem.* **1988**, *17*, 125; g) W. Stoffel, *Angew. Chem.* **1990**, *102*, 987; *Angew. Chem. Int. Ed. Engl.* **1990**, *29*, 958; h) C. G. Worley, R. W. Linton, E. T. Samulski, *Langmuir* **1995**, *11*, 3805; A. F. Terzis, P. T. Snee, E. T. Samulski, *Chem. Phys. Lett.* **1997**, *264*, 481; see also [7 a, 17].

[54] Membranes – from bilayers and their different modeling aspects to biomembrane complexity: a) P. B. Hitchcook, R. Mason, *J. Chem. Soc., Chem. Commun.* **1974**, 539; b) M. Inbach, E. Shinitzky, *Proc. Natl. Acad. Sci. USA* **1974**, *71*, 4229; c) E. Shinitzky, *Biomembranes 1–III*, VCH, Weinheim **1994**; d) H. Bader, K. Dorn, B. Hupfer, H. Ringsdorf, in *Polymer Membranes*, Springer, Heidelberg **1985**; *Adv. Polym. Sci.* **1985**, *64*, 1; e) P. Laggner, in *Topics in Current Chemistry*, Vol. 145, Springer, Berlin **1988**, p. 173; f) P. Laggner, M. Kriechbaum, A. Her-

metter, F. Paltauf, J. Hendrix, G. Rapp, *Prog. Colloid Polym. Sci.* **1989**, *79*, 33; g) T. Kunitake, *Angew. Chem.* **1992**, *104*, 692; *Angew. Chem. Int. Ed. Engl.* **1992**, *31*, 709; h) L. F. Tietze, K. Boege, V. Vill, *Chem. Ber.* **1994**, *127*, 1065; i) D. D. Lasic, *Angew. Chem.* **1994**, *106*, 1765; *Angew. Chem. Int. Ed. Engl.* **1994**, *33*, 1685; k) P. Huang, J. J. Perez, G. M. Loew, *J. Biomol. Struct. Dyn.* **1994**, *11*, 927; l) P. Huang, E. Bertaccini, G. M. Loew, *ibid.* **1995**, *12*, 725; see also [7 a, 17, 33, 51, 52].

[55] Membrane – carbohydrate systems: a) O. Lockhoff, *Angew. Chem.* **1991**, *103*, 1639; *Angew. Chem. Int. Ed. Engl.* **1991**, *30*, 1611; b) D. E. Devros, D. L. Vanoppen, X.-Y. Li, S. Libbrecht, Y. Bruynserade, P.-P. Knops-Gerrits, P. A. Jacobi, *Chem. Eur. J.* **1995**, *1*, 144; see also [7 a, 17, 33, 51, 52].

[56] Membrane – protein systems: a) H. R. Leuchtag, *J. Theor. Biol.* **1987**, *127*, 321, 341; b) H. Repke, C. Liebmann, *Membranrezeptoren und ihre Effektorsysteme*, Akademie-Verlag, Berlin **1987**, c) W. Stoffel, *Angew. Chem.* **1990**, *102*, 987; *Angew. Chem. Int. Ed. Engl.* **1990**, *29*, 958; d) K. Eichmann, *ibid.* **1993**, *105*, 56 and **1993**, *32*, 54; e) T. E. Ramsdall, P. R. Andrews, E. C. Nice, *FEBS Lett.* **1993**, *333*, 217; f) S.-E. Ryu, A. Truneh, R. W. Sweet, W. A. Hendrickson, *Structure* **1994**, *2*, 59; g) R. Smith, F. Separovic, T. J. Milne, A. Whittaker, F. M. Bennett, B. A. Cornell, A. Makriyannis, *J. Mol. Biol.* **1994**, *241*, 456; h) A. Koiv, P. Mustonen, P. K. J. Kinnunen, *Chem. Phys. Lipids* **1994**, *70*, 1; i) H. Chung, M. Caffrey, *Nature* **1994**, *368*, 224; k) A. P. Starling, J. M. East, A. G. Lee, *Biochemistry* **1995**, *34*, 3084; l) Y.-P. Zhang, R. N. A. H. Lewis, R. S. Hodges, R. N. McElhaney, *Biochemistry* **1995**, *34*, 2362; m) Y. Fang, D. G. Dalgleish, *Langmuir* **1995**, *11*, 75; see also [7 a, 17, 33, 51, 52].

[57] Membrane – nucleic acid systems: a) F. A. Manzoli, J. M. Muchmore, B. Bonara, S. Capitani, S. Bartoli, *Biochem. Biophys. Acta* **1974**, *340*, 1; b) H. Kitano, H. Ringsdorf, *Bull. Chem. Soc. Jpn.* **1985**, *58*, 2826; c) J.-P. Behr, *Tetrahedron Lett.* **1986**, *27*, 5861; d) G. Gellissen, *Biol. unserer Zeit* **1987**, *17*, 15; e) K. Shirahama, K. Takashima, N. Tarisawa, *Bull. Chem. Soc. Jpn.* **1987**, *60*, 43; f) Z. Khan, M. Ariatti, A. O. Manutrey, *Med. Sci. Rev.* **1987**, *15*, 1189; g) G. M. Palleos, J. Michas, *Liq. Cryst.* **1992**, *11*, 773; see also [7 a, 17, 33, 51, 52].

[58] Liquid crystalline peptide systems between lyotropy and even thermotropy: a) E. Iizuka, *Adv. Polym. Sci.* **1976**, *20*, 80; b) E. Iizuka, Y. Kondo, Y. Ukai, *Polym. J.* **1977**, *9*, 135; c) H. Toriumi, Y. Kusumi, I. Uematsu, Y. Uematsu, *ibid.* **1979**, *11*, 863; d) N. Ise, T. Okubo, K. Yamamoto, H. Matsuoka, H. Kawai, T. Hashimoto, M. Fuji-

mura, *J. Chem. Phys.* **1983**, *78*, 541; e) J. Wata-nabe, Y. Fukuda, R. Gehani, T. Uetmatsu, *Macro-molecules* **1984**, *17*, 1004; f) J. T. Yang, S. Ku-bota, in *Microdomains in Polymer Solution* (Ed.: P. Dubin), Plenum, New York **1985**, p. 311; g) M. Mutter, R. Gassmann, U. Buttkus, K.-H. Alt-mann, *Angew. Chem.* **1991**, *103*, 1504; *Angew. Chem. Int. Ed. Engl.* **1991**, *30*, 1514; h) T. Ueha-ra, H. Hirata, H. Okabayashi, K. Taga, Y. Yoshi-da, M. Kojima, *Colloid Polym. Sci.* **1994**, *272*, 692; i) A. Schreckenbach, P. Wünsche, *Polymer* **1994**, *35*, 5611; see also [7 a, 17, 33, 51, 52].

[59] Protein mesomorphs: a) G. H. Brown, R. J. Mishra, *J. Agr. Food Chem.* **1971**, *19*, 645; b) R. J. Hawkins, E. W. April, *Adv. Liq. Cryst.* **1983**, *6*, 243; c) R. E. Buxbaum, T. Dennerll, S. Weiss, S. R. Heidemann, *Science* **1987**, *235*, 1511; d) R. Huber, *Angew. Chem.* **1988**, *100*, 79; *Angew. Chem. Int. Ed. Engl.* **1988**, *27*, 79; e) D. Oesterhelt, C. Bräuchle, N. Hampp, *Q. Rev. Bi-ophys.* **1991**, *24*, 425; f) W. H. Daly, D. Poche, P. S. Russo, I. Negulescu, *Polym. Prepr.* **1992**, *33*, 188; g) J. J. Breen, G. W. Flynn, *J. Phys. Chem.* **1992**, *96*, 6825; h) D. W. Urry, *Angew. Chem.* **1993**, *105*, 859; *Angew. Chem. Int. Ed. Engl.* **1993**, *32*, 819; *Sci. Am.* **1995**, *272 (1)*, 44; i) C. La Rosa, D. Milardi, S. Fasone, D. Grasso, *Calorim. Anal. Therm.* **1993**, *24*, 233; k) A. Abe, in *Macromolecules 1992*, 34th IUPAC Int. Symp. on Macromolecules (Ed.: J. Kahovec), VSP, Utrecht **1993**, p. 221; l) W. H. Daly, I. I. Negulescu, R. M. Ottenbrite, E. C. Buruiana, *Polym. Prepr.* **1993**, *34*, 518; m) S. Kunugi, N. Tanaka, N. Itoh, in *New Functional Materials* (Ed.: T. Tsuruta), Elsevier, Amsterdam **1993**, Vol. B., p. 125; n) R. Furukawa, R. Kundra, M. Fechheimer, *Biochemistry* **1993**, *32*, 12 346; o) M. M. Giraud-Guille, *Microsc. Res. Tech.* **1994**, *27*, 420; p) Protein – protein mediators: H. Wiech, J. Buchner, R. Zimmermann, C. Jacob, *Nature* **1992**, *358*, 169; q) R. A. Stewart, D. M. Cyr, E. A. Craig, W. Neupert, *Trends Biol. Sci.* **1994**, 87; r) G. Fischer, *Angew. Chem.* **1994**, *106*, 1479; *Angew. Chem. Int. Ed. Engl.* **1994**, *33*, 1479; see also [7 a, 17, 33, 51, 52].

[60] The difficulties around carbohydrates and their mesomorphic ambitions: a) G. A. Jeffrey, S. Bhattacharjee, *Carbohydr. Res.* **1983**, *115*, 53; G. A. Jeffrey, *Acc. Chem. Res.* **1986**, *19*, 168; *J. Mol. Struct.* **1994**, *322*, 21; P. M. Matias, G. A. Jeffrey, *Carbohydr. Res.* **1986**, *153*, 217; b) K. Praefcke, P. Psaras, A. Eckert, *Liq. Cryst.* **1993**, *13*, 551; K. Praefcke, D. Blunk, *ibid.* **1993**, *14*, 1181; D. Blunk, K. Praefcke, G. Legler, *ibid.* **1994**, *17*, 841; R. Miethchen, H. Prade, J. Holz, K. Praefcke, D. Blunk, *Chem. Ber.* **1993**, *126*, 1707; K. Praefcke, D. Blunk, J. Hempel, *Mol. Cryst. Liq. Cryst. Sci. Technol. Sect. A* **1994**, *243*, 323; c) V. Vill, H.-W. Tunger, H. Stegemeyer, K.

Diekmann, *Tetrahedron* **1994**, *5*, 2443; P. Stan-gier, V. Vill, S. Rohde, U. Jeschke, J. Thiem, *Liq. Cryst.* **1994**, *17*, 589; S. Fischer, H. Fischer, S. Diele, G. Pelzl, K. Jankowski, R. R. Schmidt, V. Vill, *ibid.* **1994**, *17*, 855; H. Fischer, V. Vill, C. Vogel, U. Jeschke, *ibid.* **1993**, *15*, 733; V. Vill, H.-W. Tunger, A. Borwitzky, *Ferroelectrics* **1996**, *180*, 227; P. Sakya, J. M. Seddon, V. Vill, *Liq. Cryst.* **1997**, *23*, 409; d) J. F. Stoddart, *An-gew. Chem.* **1992**, *104*, 860; *Angew. Chem. Int. Ed. Engl.* **1992**, *31*, 846; e) M. Kunz, K. Ruck, *ibid.* **1993**, *105*, 355 (**1993**, *32*, 336); U. Spren-gard, G. Kretzschmar, E. Bartnik, Ch. Hüls, H. Kunz, *ibid.* **1995**, *107*, 1104 (**1995**, *34*, 990); f) G. Wenz, *ibid.* **1994**, *106*, 851 (**1994**, *34*, 803); g) P. Zugenmaier, *Cellul. Chem. Biochem. Ma-ter. Aspects* **1993**, 105; W. V. Dahlhoff, K. Riehl, P. Zugenmaier, *Liebigs Ann. Chem.* **1993**, *19*, 1063; N. Aust, C. Derleth, P. Zugenmaier, *Macromol. Chem. Phys.* **1997**, *198*, 1363; h) U. Hinrichs, G. Büttner, M. Steifa, Ch. Betzel, V. Zabel, B. Pfannenzütter, W. Saenger, *Science* **1987**, *238*, 205; i) J. H. M. Willison, R. M. Abeysekera, *J. Polym. Sci., Part C, Polym. Lett.* **1988**, *26*, 71; k) C. A. A. van Boeckel, M. Pe-titon, *Angew. Chem.* **1993**, *105*, 1741; *Angew. Chem. Int. Ed. Engl.* **1993**, *32*, 1671; l) C. Un-verzagt, *ibid.* **1993**, *105*, 1762 (**1993**, *32*, 1691); m) T. J. Martin, R. R. Schmidt, *Tetrahedron Lett.* **1993**, *34*, 6733; n) H. Fischer, A. Keller, J. A. Odele, *J. Appl. Polym. Sci.* **1994**, *54*, 1785; o) T. Ogawa, *Chem. Soc. Rev.* **1994**, 397; p) T. P. Trouard, D. A. Mannock, G. Lindblom, L. Eil-fors, M. Akiyama, R. N. McElhaney, *Biophys. J.* **1994**, *67*, 1101; q) A.-C. Eliasson, *Thermochim. Acta* **1994**, *246*, 343; r) C.-H. Wong, R. L. Hal-comb, Y. Ichikawa, T. Kajimoto, *Angew. Chem.* **1995**, *107*, 453, 569; *Angew. Chem. Int. Ed. Engl.* **1995**, *33*, 412, 576; s) X. Auvray, C. Petipas, R. Anthore, I. Rico-Lattes, A. Lattes, *Langmuir* **1995**, *11*, 433; t) D. E. Devros, D. L. Vanoppen, X.-Y. Li, S. Libbrecht, Y. Bruynserade, P.-P. Knops-Gerrits, P. A. Jacobi, *Chem. Eur. J.* **1995**, *1*, 144; see also [7 a, 17, 33 p, q, 51, 52 e].

[61] Carbohydrate – nucleic acid systems – the two sugar mates: a) T. Li, Z. Zeng, V. A. Estevez, K. U. Baldenius, K. C. Nicolaou, G. J. Joyce, *J. Am. Chem. Soc.* **1994**, *116*, 3709; b) K. C. Nico-laou, K. Ajito, H. Komatsu, B. M. Smith, F. Liun, M. G. Egan, H. Gomez-Paloma, *Angew. Chem.* **1995**, *107*, 614; *Angew. Chem. Int. Ed. Engl.* **1995**, *34*, 576; c) G. Pratviel, J. Bernadou, B. Mennier, *ibid.* **1995**, *107*, 819 (**1995**, *34*, 746); see also [7 a, 17, 33, 51, 52].

[62] Nucleic acid liquid crystals – from phase eluci-dation to chromatin regulation: a) F. Livolant, Y. Bouligand, *Chromosoma* **1978**, *68*, 21; b) F. Livolant, *Tissue* **1984**, *16*, 535; c) *Eur. J. Cell. Biol.* **1984**, *33*, 300; d) *J. Phys.* **1986**, *47*, 1813;

ibid. **1986**, *47*, 1605; *ibid.* **1987**, *48*, 1051; e) Y. Bouligand, F. Livolant, *ibid.* **1984**, *45*, 1899; f) F. Livolant, M. F. Maestre, *Biochemistry* **1988**, *27*, 3056; g) F. Livolant, Y. Bouligand, *Mol. Cryst. Liq. Cryst.* **1989**, *166*, 91; h) F. Livolant, A. M. Levelut, J. Doucet, J. P. Benoit, *Nature* **1989**, *339*, 724; i) F. Livolant, *J. Mol. Biol.* **1991**, *218*, 165; k) D. Durand, J. Doucet, F. Livolant, *J. Phys. (Paris) II* **1992**, *2*, 1769; l) A. Leforestier, F. Livolant, *ibid.* **1992**, *2*, 1853; m) A. Leforestier, F. Livolant, *Biophys. J.* **1993**, *65*, 56; n) F. Livolant, A. Leforestier, D. Durant, J. Doucet, *Lect. Notes Phys.* **1993**, *415*, 33; o) A. Leforestier, F. Livolant, *Liq. Cryst.* **1994**, *17*, 651; p) J.-L. Sikorav, J. Pelta, F. Livolant, *Biophys. J.* **1994**, *67*, 1387; q) J. Pelta, J. L. Sikorav, F. Livolant, D. Durand, J. Doucet, *J. Trace Microprobe Techn.* **1995**, *13*, 401; r) J. Pelta, D. Durand, J. Doucet, F. Livolant, *Biophys. J.* **1996**, *71*, 48; s) F. Livolant, A. Leforestier, *Progr. Polym. Sci.* **1996**, *21*, 1115; see also [7a, 17, 33, 51, 52q].

[63] The "unspecific" textures of "evolutionary" polydisperse DNAs: a) W.-V. Meister, A.-M. Ladhoff, S. I. Kargov, G. Burckhardt, G. Luck, S. Hoffmann, *Z. Chem.* **1990**, *30*, 213; b) A. Leforestier, F. Livolant, "Phase Transition in a Chiral Liquid Crystalline Polymer". Eur. Conf. Liq. Cryst. Sci. Technol., Conf. Abstr., Films **1993**, *1-19*, 24; see also [62].

[64] The beauties of nucleic acid textures – imaging phase behavior of adjusted "rod-likes": a) R. L. Rill, P. R. Hilliand, G. C. Levy, *J. Biol. Chem.* **1983**, *258*, 250; b) R. L. Rill, *Proc. Natl. Acad. Sci. USA* **1986**, *83*, 342; c) T. E. Strzelecka, R. L. Rill, *J. Am. Chem. Soc.* **1987**, *109*, 4513; d) T. E. Strzelecka, M. W. Davidson, R. L. Rill, *Nature* **1988**, *331*, 457; e) R. L. Rill, F. Livolant, H. C. Aldrich, M. W. Davidson, *Chromosoma* **1989**, *98*, 280; f) D. M. van Winkle, M. W. Davidson, R. L. Rill, *J. Chem. Phys.* **1992**, *97*, 5641; g) K. H. Hecker, R. L. Rill, *Anal. Biochem.* **1997**, *244*, 67; h) D. H. Van Winkle, A. Chatterjee, R. Link, R. L. Rill, *Phys. Rev. E* **1997**, *55*, 4354; see also [15, 51, 52r].

[65] Emergence of helical and suprahelical motifs from different nucleic acid composites: a) G. P. Spada, A. Carcuro, F. P. Colonna, A. Garbesi, G. Gottarelli, *Liq. Cryst.* **1988**, *3*, 651; b) G. P. Spada, P. Brigidi, G. Gottarelli, *J. Chem. Soc. Chem. Comm.* **1988**, 953; c) P. Mariani, C. Mazabard, A. Garbesi, G. P. Spada, *J. Am. Chem. Soc.* **1989**, *111*, 6369; d) S. Bonazzi, M. M. de Morais, G. Gottarelli, P. Mariani, G. P. Spada, *Angew. Chem.* **1993**, *105*, 251; *Angew. Chem. Int. Ed. Engl.* **1993**, *105*, 248; e) P. Mariani, M. M. De Morais, G. Gottarelli, G. P. Spada, H. Delacroix, L. Tondelli, *Liq. Cryst.* **1993**, *15*, 757; f) A. Garbesi, G. Gottarelli, P. Mariani, G. P.

Spada, *Pure Appl. Chem.* **1993**, *65*, 641; g) F. Ciuchi, G. Di Nicola, H. Franz, G. Gottarelli, P. Mariani, M. G. Ponzi Bossi, G. P. Spada, *J. Am. Chem. Soc.* **1994**, *116*, 7064; h) A. J. MacDermott, L. D. Barron, A. Brack, T. Buhse, A. F. Drake, R. Emery, G. Gottarelli, J. M. Greenberg, R. Haberle, *Planet. Space Sci.* **1996**, *44*, 1441; i) G. P. Spada, S. Bonzzi, A. Garbesi, S. Zanella, F. Ciuchi, P. Mariani, *Liq. Cryst.* **1997**, *22*, 341; k) F. M. H. deGroot, G. Gottarelli, S. Masiero, G. Proni, G. P. Spada, N. Dolci, *Angew. Chem.* **1997**, *109*, 990; *Angew. Chem. Int. Ed. Engl.* **1997**, *36*, 954; l) K. H. Hecker, R. L. Rill, *Analyt. Biochem.* **1997**, *244*, 67; m) D. H. Van Winkle, A. Chatterjee, R. Link, R. L. Rill, *Phys. Rev. E* **1997**, *55*, 4354; see also [51, 52].

[66] The detailed vision of helical liquid crystal motifs exemplified also in the case of nucleic acid systems: a) E. Iizuka, Y. Kondo, Y. Ukai, *Polymer J.* **1977**, *9*, 135; b) E. Iizuka, *ibid.* **1978**, *10*, 293; c) *ibid.* **1983**, *15*, 525; d) *Adv. Biophys.* **1988**, *24*, 1; e) E. Iizuka, J. T. Yang, in *Liquid Crystals and Ordered Fluids* (Eds.: J. F. Johnson, R. S. Porter), Plenum, New York **1984**, p. 197; see also [51, 52u].

[67] Liquid crystal behavior of nucleic acids accompanying molecular biological aspects: a) N. Akimenko, V. Kleinwächter, Yu. M. Yevdokimov, *FEBS Lett.* **1983**, *156*, 58; b) Yu. M. Yevdokimov, S. G. Skuridin, V. I. Salyanov, *Liq. Cryst.* **1988**, *3*, 1443; c) Yu. M. Yevdokimov, S. G. Skuridin; B. A. Chemukha, *Biotekhnologya* **1992**, *5*, 103; d) V. I. Salyanov, M. Palumbo, Yu. M. Yevdokimov, *Mol. Biol. (Moscow)* **1992**, *26*, 1036; e) V. I. Salyanov, P. L. Lavreutier, B. A. Chernukha, Yu. M. Yevdokimov, *ibid.* **1994**, *28*, 1283; f) D. N. Nikogosyan, Yu. A. Repeyev, D. Yu. Yakovlev, Y. I. Salyanov, S. G. Skuridin, Yu. M. Yevdokimov, *Photochem. Photobiol.* **1994**, *59*, 269; g) Yu. M. Yevdokimov, V. I. Salyanov, E. Gedig, F. Spener, *FEBS Lett.* **1996**, *392*, 269; h) Yu. M. Yevdokimov, V. I. Salyanov, S. V. Semenov, *Biosens. Bioelectron.* **1996**, *11*, 889; i) V. A. Belyakov, V. P. Orlov, S. V. Semenov, S. G. Skuridin, Yu. M. Yevdokimov, *Liq. Cryst.* **1996**, *20*, 777; see also [52t].

[68] Nucleic acid liquid crystals by dynamic approaches: a) R. Brandes, D. R. Kearns, *Biochemistry* **1986**, *25*, 5890; b) R. Brandes, R. R. Vold, D. R. Klarus, A. Rupprecht, *J. Mol. Biol.* **1988**, *202*, 321; c) R. Brandes, D. R. Kearns, *Biochemistry* **1986**, *25*, 5890; d) *J. Phys. Chem.* **1988**, *92*, 6836; e) D. R. Kearns, *CRC Crit. Rev.* **1988**, *15*, 237; f) V. L. Hsu, X. Jia, D. R. Kearns, *Toxicol. Lett.* **1996**, *82*, 577; see also [51, 52].

[69] Complexity of further nucleic acid mesomorph approaches: a) G. L. Seibel, U. C. Singh, P. A. Kollman, *Proc. Natl. Acad. Sci. USA* **1985**, *82*,

6537; b) S. P. Williams, B. D. Athey, L. J. Mugliar, S. Schapp, A. H. Gough, J. P. Langmore, *Biophys. J.* **1986**, *49*, 233; c) Th. Ackermann, *Angew. Chem.* **1989**, *101*, 1005; *Angew. Chem. Int. Ed. Engl.* **1989**, *28*, 981; d) A. Fernandez, *Naturwissenschaften* **1989**, *76*, 469; e) Z. Reich, S. Levin-Zaidman, S. B. Gutman, T. Arad, A. Minsky, *Biochemistry* **1994**, *33*, 4177; f) Z. Reich, E. J. Wachtel, A. Minsky, *Science* **1994**, *264*, 1460; g) Z. Reich, O. Schramm, V. Brumfeld, A. Minski, *J. Am. Chem. Soc.* **1996**, *118*, 6435; h) S. Levin-Zeitman, Z. Reich, E. J. Wachtel, A. Minski, *Biochemistry* **1996**, *35*, 2985; i) A. Kagemoto, T. Sumi, Y. Baba, *Thermochim. Acta* **1994**, *242*, 77; k) L. C. A. Groot, M. E. Kuil, J. C. Leyte, J. R. C. van der Maarel, *Liq. Cryst.* **1994**, *17*, 263; l) A. Koiv, P. Mustonen, P. K. J. Kinnunen, *Chem. Phys. Lipids* **1994**, *70*, 1; m) T. M. Bohanon, S. Denzinger, R. Fink, W. Paulus, H. Ringsdorf, M. Weck, *Angew. Chem.* **1995**, *107*, 102; *Angew. Chem. Int. Ed. Engl.* **1995**, *34*, 58; n) A. Chatterjee, D. H. van Winkle, *Phys. Rev. E.* **1994**, *49*, 1450; see also [7 a, 17, 18, 51, 52].

[70] Vorländer's "chiral infection" modeling hormonal amplification: a) D. Vorländer, F. Jaenecke, *Z. Phys. Chem. A* **1913**, *85*, 697; b) S. Hoffmann, *Z. Chem.* **1989**, *29*, 102; see also [7 a, 17, 18].

[71] Structure-phase problems around nucleic acid–protein interactions: a) J. H. Hogle, M. Chow, D. J. Filman, *Science* **1985**, *229*, 1358; b) D. L. Ollis, S. W. Weute, *Chem. Rev.* **1987**, *87*, 981; c) G. Dreyfuss, M. S. Swanson, S. Pinol-Roma, *Trends Biochem. Sci.* **1988**, *13*, 80; d) R. M. Evans, S. M. Hollenberg, *Cell* **1988**, *52*, 1; e) J. P. Leonetti, B. Rayner, M. Lemaitre, C. Gagnor, P. G. Milhand, J.-L. Imbach, B. Lebleu, *Gene* **1988**, *72*, 323; f) T. J. Gibson, J. P. M. Postma, R. S. Brown, P. Argos, *Protein Engl.* **1988**, *2*, 209; g) R. G. Brennan, B. W. Matthews, *J. Biol. Chem.* **1989**, *264*, 1903; h) R. Bandziulis, M. S. Swanson, G. Dreyfuss, *Genes Devel.* **1989**, *3*, 431; i) R. D. Klausner, J. B. Harford, *Science* **1989**, *246*, 870; k) T. A. Steitz, *Q. Rev. Biophys.* **1990**, *23*, 205; D. Wassaman, J. A. Steitz, *Nature* **1991**, *349*, 463; l) M. Ptashne, A. A. F. Gann, *ibid.* **1990**, *346*, 329; m) D. Latchman, *Gene Regulation, A Eukaryotic Perspective,* Hyman, London **1990**; n) N. P. Pavletich, C. O. Pabo, *Science* **1991**, *252*, 809; o) S. W. Ruby, J. Abelson, *Trends Genet.* **1991**, *7*, 79; p) C. Cuthrie, *Science* **1991**, *253*, 157; q) D. Kostrewa, J. Granzin, D. Stock, H. W. Choe, J. Lahban, W. Saenger, *J. Mol. Biol.* **1992**, *26*, 209; W. Saenger, C. Sandmann, K. Theis, E. B. Starikow, G. Kostrewa, J. Laban, J. Granzin, in *Nucleic Acids and Molecular Biology* (Eds.: F. Eckstein, D. M. J. Lilly), Springer, Berlin **1993**, 158; R. Giege, J. Drenth, A. Ducruix, A. McPherson,

W. Saenger, *Progr. Cryst. Growth Charact. Mater.* **1995**, *30*, 237; C. Sandmann, F. Cordes, W. Saenger, *Proteins: Struct., Funct., Genet.* **1996**, *25*, 486; W. Saenger, D. Kostrewa, J. Granzin, F. Cordes, C. Sandmann, C. Kisker, W. Hinrichs, in *From Simplicity to Complexity in Chemistry – and Beyond* (Eds.: A. Müller, A. Dress, F. Vögtle), Vieweg, Braunschweig **1996**, p. 51; r) D. G. Gorenstein, *Chem. Rev.* **1994**, *94*, 1315; s) M. Famulok, D. Faulhammer, *Angew. Chem.* **1994**, *106*, 1911; *Angew. Chem. Int. Ed. Engl.* **1994**, *33*, 1827; see also [7 a, 17, 18, 33, 51, 52].

[72] Selected problems around structure and phase – general questions: a) G. Solladié, R. G. Zimmermann, *Angew. Chem.* **1984**, *96*, 335; *Angew. Chem. Int. Ed. Engl.* **1984**, *23*, 348; b) S. Andersson, S. Lidin, M. Jacob, O. Terasaki, *ibid.* **1991**, *103*, 771 (**1991**, *30*, 754); c) A. Amann, W. Gans, *ibid.* **1989**, *101*, 277 (**1989**, *28*, 268); d) H. Frauenfelder, *Ann. NY Acad. Sci.* **1986**, *504*, 151; e) *Disordered Systems and Biological Models* (Ed.: L. Peliti), CIF Series, World Scientific, Singapore **1989**, 14; f) J. M. Ottino, F.-J. Muzzio, M. Tjahjadi, J. G. Framjione, S. C. Jana, H. A. Kusch, *Science* **1992**, *257*, 754; g) D. Seebach, *Angew. Chem.* **1990**, *102*, 1362; *Angew. Chem. Int. Ed. Engl.* **1990**, *29*, 1320; h) F. M. Stillinger, *Science* **1995**, *267*, 1935; i) B. Frick, D. Richter, *ibid.* **1995**, *267*, 1939; k) R. Jaenicke, *Angew. Chem.* **1984**, *96*, 385; *Angew. Chem. Int. Ed. Engl.* **1984**, *23*, 395; l) U. Sprengard, G. Kretzschmar, E. Bartnik, Ch. Hüls, H. Kunz, *ibid.* **1995**, *107*, 1104 (**1995**, *34*, 990); m) S. Friberg, *Naturwissenschaften* **1977**, *64*, 612; n) C. N. Paul, U. Heinemann, U. Mahn, W. Saenger, *Angew. Chem.* **1991**, *103*, 351; *Angew. Chem. Int. Ed. Engl.* **1991**, *30*, 343; o) C. A. Angell, *Science* **1995**, *267*, 1924; p) M. Quack, *Angew. Chem.* **1989**, *101*, 578; *Angew. Chem. Int. Ed. Engl.* **1989**, *28*, 959; q) A. Fernandez, *Naturwissenschaften* **1989**, *76*, 469; r) K. Shinoda, *Pure Appl. Chem.* **1988**, *60*, 1493; s) Nucleic acid frequence correlations and coherences?: C. A. Chatzidimitriou-Dreismann, *Adv. Chem. Phys.* **1991**, *80*, 201; *Helv. Chem. Acta* **1992**, *75*, 2252; *Nature* **1993**, *361*, 212; *Adv. Chem. Phys.* **1997**, *99*, 393; D. Lavhammar, C. A. Chatzidimitriou-Dreismann, *Nucleic Acids Res.* **1993**, *21*, 5167; E. J. Brändas, C. A. Chatzidimitriou-Dreismann, *Int. J. Quant. Chem.* **1995**, *53*, 95; C. A. Chatzidimitriou-Dreismann, R. M. F. Streffer, D. Larhammar, *Nucleic Acids Res.* **1996**, *24*, 1676; t) J. Maddox, *Nature* **1992**, *358*, 103; u) O. Kennard, W.-N. Hunter, *Angew. Chem.* **1991**, *103*, 1280; *Angew. Chem. Int. Ed. Engl.* **1991**, *30*, 1254; v) R. E. Dickerson, *J. Mol. Biol.* **1989**, *205*, 787; w) E.-L. Winnacker, *Angew. Chem.* **1992**, *104*, 1616; *Angew. Chem. Int.*

Ed. Engl. **1992**, *31*, 1578; x) Fractal motifs: H. Jürgens, H.-O. Peitgen, D. Saupe, *Chaos und Fractale,* Spektrum, Heidelberg **1989**; y) A. S. Borovik, A. Y. Grosberg, M. D. Frank-Kamenetski, *J. Biomol. Struct. Dyn.* **1994**, *12*, 656; z) A. Blumen, M. Schnörer, *Angew. Chem.* **1990**, *102*, 158; *Angew. Chem. Int. Ed. Engl.* **1990**, *29*, 126; z') V. Percec, J. Heck, G. Johansson, D. Tomazos, M. Kawasumi, P. Chu, G. Ungar, *Mol. Cryst. Liq. Cryst. Technol. Sect. A* **1994**, *254*, 137; see also [7 a, 17, 18, 19, 25, 33, 51, 52].

[73] Layers and adlayers approaching nanometer-scale–lipid-protein-systems: a) H. Ebato, J. N. Harron, W. Müller, Y. Okahata, H. Ringsdorf, P. Suci, *Angew. Chem.* **1992**, *104*, 1064; *Angew. Chem. Int. Ed. Engl.* **1992**, *31*, 1078; b) W. Müller, H. Ringsdorf, E. Rump, X. Zhang, L. Augermaier, W. Knoll, M. Lilly, J. Spinke, *Science* **1993**, *262*, 1706; c) D. Vaknin, K. Kjaer, H. Ringsdorf, R. Blankenburg, M. Piepenstock, A. Diederich, M. Lösche, *Langmuir* **1993**, *9*, 1171; d) M. Lösche, C. Erdelen, E. Rump, H. Ringsdorf, K. Kjaer, D. Vaknin, *Thin Solid Films* **1994**, *242*, 112; e) R. Naumann, A. Jonczyk, C. Hampel, H. Ringsdorf, W. Knoll, N. Bunjes, P. Graeber, *Bioelectrochem. Bioenerg.* **1997**, *42*, 241; – Lipid-nucleic acid-systems: f) R. C. Ebersole, J. A. Miller, J. R. Moran, M. D. Ward, *J. Am. Chem. Soc.* **1990**, *112*, 3239; g) Y. Okahata, Y. Matsunobu; K. Ijiro, M. Mukae, A. Murakami, K. Makino, *J. Am. Chem. Soc.* **1992**, *114*, 8300; h) Y. Okahata, K. Ijiro, Y. Matsuzaki, *Langmuir* **1993**, *9*, 19; i) G. Decher, *Nachr. Chem. Tech. Lab.* **1993**, *41*, 793; k) A. U. Niu, L. Shain, W. Ling, J. Werenka, D. Larson, E. Henderson, *SO Proc. SPIE Int. Soc. Opt. Eng., Proc. Absorb. DNA Sequencing Technol.* **1993**, 72; l) S. Yamaguchi, T. Shinomura, T. Tatsuma, N. Oyama, *Anal. Chem.* **1993**, *65*, 1925; – General aspects: m) B. M. Robinson, N. C. Seeman, *Protein Eng.* **1987**, *1*, 295; n) N. C. Seeman, Y. Zhang, J. Chen, *J. Vac. Sci. Technol. A* **1994**, *12*, 1895; o) J. Economy, *Angew. Chem.* **1990**, *102*, 1301; *Angew. Chem. Int. Ed. Engl.* **1990**, *29*, 1256; p) F. M. Menger, *ibid.* **1991**, *103*, 1104 (**1991**, *30*, 1086); q) K. Camman, U. Lemke, A. Rohrer, J. Sander, H. Wilken, B. Winter, *ibid.* **1991**, *103*, 519 (**1991**, *30*, 516); r) J. M. Wendorff, *ibid.* **1991**, *103*, 416 (**1991**, *30*, 405); s) D. D. Lasic, *ibid.* **1994**, *108*, 1765 (**1994**, *33*, 1685); t) J. Majewski, R. Edgar, R. Popovitz-Biro, K. Kjaer, W. G. Bouwman, J. Als-Noelsen, M. Lahav, L. Leiserowitz, *ibid.* **1995**, *107*, 707 (**1995**, *34*, 649); see also [7 a, 17, 18, 51, 52].

[74] Molecular resolving microscopies: a) E. Spiess, R. Lurz, *Methods Microbiol.* **1988**, *20*, 293; b) J. Frommer, *Angew. Chem.* **1992**, *104*, 1325; *Angew. Chem. Int. Ed. Engl.* **1992**, *31*, 1298; c) D. Jacquemain, S. G. Wolf, F. Leveiller, M.

Deutsch, K. Khaer, J. Als-Nielson, M. Lahav, L. Leiserowitz, *ibid.* **1992**, *104*, 134 (**1992**, *31*, 130); d) *Scanning Tunneling Microscopic and Related Methods* (Ed.: R. J. Brehns), Kluwer, Dordrecht **1993**, e) *Scanning Tunneling Microscopy* (Eds.: R. Wiesendanger, H.-J. Güntherodt), Springer, Berlin **1993**; f) E. Delain, E. Lecane, A. Barbin-Asbogant, A. Fourcade, *Microsc. Microanal. Microstruct.* **1994**, *5*, 329; g) T. Basche, *Angew. Chem.* **1994**, *106*, 1805; *Angew. Chem. Int. Ed. Engl.* **1994**, *33*, 1723; h) H. Rohrer, *Microsc. Microanal. Microstruct.* **1994**, *5*, 237; i) E. F. Aust, M. Savodny, S. Ito, W. Knoll, *Scanning* **1994**, *16*, 353; approaches to self-organizing systems: k) C. D. Frisbie, L. F. Rozsnyai, A. Noy, M. S. Wrighton, C. M. Lieber, *Science* **1964**, *264*, 2071; l) L. Häußling, B. Michel, H. Ringsdorf, H. Rohrer, *Angew. Chem.* **1991**, *103*, 568; *Angew. Chem.* **1991**, *103*, 568; *Angew. Chem. Int. Ed. Engl.* **1991**, *30*, 569; m) J. J. Breen, G. W. Flynn, *J. Phys. Chem.* **1992**, *96*, 6825; n) D. P. E. Smith, A. Bryant, C. F. Quate, J. P. Rabe, C. Gerber, J. D. Swallen, *Proc. Natl. Acad. Sci. USA* **1987**, *84*, 969; o) S. Buchholz, J. P. Rabe, *Angew. Chem.* **1992**, *104*, 88; *Angew. Chem. Int. Ed. Engl.* **1992**, *31*, 189; p) S. Lincotti, J. P. Rabe, *Supramol. Sci.* **1994**, *1*, 7; q) H. G. Hansma, M. Bezanilla, F. Zerchansern, M. Adrian, R. L. Sinsheimer, *Nucleic Acids Res.* **1993**, *21*, 505; r) T. E. Ramsdale, P. R. Andrews, E. C. Nice, *FEBS Lett.* **1993**, *333*, 217; s) B. Samori, C. Nigro, V. Armentanof, S. Cimieri, G. Zuccheri, C. Quagliariello, *Angew. Chem.* **1993**, *105*, 1482; *Angew. Chem. Int. Ed. Engl.* **1993**, *32*, 1461; t) T. Thundat, R. J. Warmack, D. P. Allison, K. B. Jacobson, *Scanning Microsc.* **1994**, *8*, 23; u) P. Pasero, C. Blettry, M. Marilley, B. Jordan, S. Granjeaud, M. Dayez, A. Humbert, *J. Vac. Sci. Technol. B* **1994**, *12*, 1521; v) G. Zuccheri, G. A. Ranieri, C. Nigro, B. Samori, *J. Vac. Sci. Technol. B* **1995**, *13*, 158; see also [5–7, 17, 18, 33, 51, 52].

[75] Molecular resolving microscopies – ways to self-organizing nucleoprotein systems: a) C. Bohley, W.-V. Meister, S. Lindau, F. Klimaszewsky, J. Barthel, S. Hoffmann, unpublished; b) E. Birch-Hirschfeld, J. van Esch, S. Hoffmann, I. Kuniharu, H. Ringsdorf, W. Saenger, C. Sandmann, U. Schilken, M. Strube, I. Thondorf, unpublished; c) R. Heinz, J. P. Rabe, W. V. Meister, S. Hoffmann, *Thin Solid Films* **1995**, *264*, 246; d) C. Bohley, D. Matern, W.-V. Meister, S. Kargov, S. Lindau, J. Barthel, S. Hoffmann, *Surface Interface Analysis* **1997**, *25*, 614; e) C. Bohley, T. Martini, G. Bischoff, S. Lindau, E. Birch-Hirschfeld, S. Kargov, W.-V. Meister, J. Barthel, S. Hoffmann, *Nucleosides Nucleotides* **1997**, *16 (10)*; f) C. M. Niemeyer, *Angew. Chem.* **1997**, *109*, 603; *Angew. Chem. Int. Ed. Engl.*

1997, *36*, 585; g) S. C. Tsang, Z. Guo, Y. K. Chen, M. L. H. Green, H. A. O. Hill, T. N. Hambley, P. J. Sadler, *ibid.* **1997**, *109*, 2292 (**1997**, *36*, 2198); see also [7 a, 17, 18, 51, 52].

[76] Leading theories in the constraints of reductionistic approaches: a) L. Onsager, *Ann. N. Y. Acad. Sci.* **1949**, *51*, 62Z; b) P. J. Flory, *Proc. R. Soc. London Ser. A* **1956**, *234*, 73; *Adv. Polym. Sci.* **1963**, *61*, 1; *J. Polym. Sci.* **1961**, *49*, 104; c) W. Maier, A. Saupe, *Z. Naturforsch. A* **1959**, *14*, 882; *ibid.* **1960**, *15*, 287; d) V. I. Ginzburg, *Phys. Solid State* **1960**, *2*, 184; e) L. D. Landau, E. M. Lifshitz, *Statistical Physics,* Pergamon, New York **1980**; f) I. W. Stewart, T. Carlsson, F. M. Leslie, *Ferroelectrics* **1993**, *148*, 41; g) F. M. Leslie, *Liq. Cryst.* **1993**, *14*, 121; h) B. Samori, S. Masiero, G. R. Luckhurst, S. K. Heeks, B. A. Timimi, P. Mariani, *Liq. Cryst.* **1993**, *15*, 217; i) G. R. Luckhurst, *Ber. Bunsen-Ges. Phys. Chem.* **1993**, *97*, 1169; G. R. Luckhurst, *Mol. Phys.* **1994**, *82*, 1063; k) S. M. Fan, G. R. Luckhurst, S. J. Picken, *J. Chem. Phys.* **1994**, *101*, 3255; l) A. P. J. Emerson, G. R. Luckhurst, S. G. Whatling, *Mol. Phys.* **1994**, *82*, 113; see also [17, 19, 25, 37, 51, 52].

[77] Stabilization within the dynamics: a) S. L. Kwolek, P. W. Morgan, J. R. Schaefgen, L. W. Gulrich, *Macromolecules* **1977**, *10*, 1390; b) J. A. McCammon, C. Y. Lee, S. H. Northrup, *J. Am. Chem. Soc.* **1983**, *105*, 2232; c) M. Karplus, J. A. McCammon, *Sci. Am.* **1986**, *254*, 42; d) J. A. McCammon, S. C. Harvey, *Dynamics of Proteins and Nucleic Acids,* Cambridge University Press, New York **1987**; e) J. A. Trainer, E. D. Getzoff, H. Alexander, R. A. Houghten, A. J. Olson, R. A. Lerner, *Nature* **1984**, *312*, 127;

f) J. M. Magle, M. Chow, D. J. Filman, *Science* **1985**, *229*, 1358; g) R. A. Shalaby, M. A. Lauffer, *Arch. Biochem. Biophys.* **1985**, *236*, 300; h) Z. Blum, S. Lidin, *Acta Chem. Scand.* **1988**, *342*, 417; i) W. F. van Gunsteren, M. J. C. Berendsen, *Angew. Chem.* **1990**, *102*, 1020; *Angew. Chem. Int. Ed. Engl.* **1990**, *29*, 992; k) H. Oschkinat, T. Müller, T. Dieckmann, *ibid.* **1994**, *106*, 284 (**1994**, *33*, 277); l) I. Radnakrishnan, D. J. Patel, *Structure* **1994**, *2*, 17; m) P. Huang, J. J. Perez, G. M. Loew, *J. Biomol. Struct. Dyn.* **1994**, *11*, 927; n) P. Huang, E. Bertaccini, G. H. Loew, *ibid.* **1995**, *12*, 725; see also [1–8, 17, 18, 51–76].

[78] Structure-phase up to morphogenesis: a) B. Hess, M. Markus, *Trends Biochem. Sci.* **1987**, *12*, 45; b) C. Nüsslein-Volhard, *Angew. Chem.* **1996**, *108*, 2316; *Angew. Chem. Int. Ed. Engl.* **1996**, *35*, 2176; c) E. Wieschaus, *ibid.* **1996**, *108*, 2330 (**1996**, *35*, 2118); see also [1–8, 17, 18, 51, 52].

[79] Human brain between deterministics and probabilistics: a) *Gehirn und Bewußtsein* (Ed.: W. Singer), Spektrum, Heidelberg **1994**; b) J. Travis, *Science* **1992**, *258*, 216; modeling by neuronal nets: c) D. Haarer, *Angew. Chem.* **1989**, *101*, 1576; *Angew. Chem. Int. Ed. Engl.* **1989**, *28*, 1544; d) G. Schneider, P. Wrede, *ibid.* **1993**, *105*, 1192 (**1993**, *32*, 1141); e) J. Gasteiger, J. Zupan, *ibid.* **1993**, *105*, 510 (**1993**, *32*, 503); see also [1–8, 17, 18].

[80] Human brain – the quantum view: a) K. J. Schulten, in *Signalwandlung und Informationsverarbeitung* (Ed.: W. Köhler), *Nova Acta Leopoldina* **1995**, *NF 294/72*, 9; b) A. Ekert, *Quantum processings of information, ibid.* **1995**, *294/72*, 14; see also [1–8, 17, 18].

Chapter IX
Cellulosic Liquid Crystals

Peter Zugenmaier

1 Introduction

Cellulose represents an important polymer, which is most abundant in nature, and serves as a renewable resource in many applications, e.g., fibers, films, paper, and as a composite with other polysaccharides and lignin in wood. Cellulose derivatives will also be used as films and fibers, food additives, thermoplastics, and construction materials, to name just a few. Cellulose and cellulose derivatives have played an important role in the development of the macromolecular concept. So far, little use has been made of the fact that cellulose represents a chiral material except, e.g., in a rare case as stationary material in liquid chromatography for the separation of chiral compounds. Nature ifself uses the chirality of cellulose occasionally, and twisted structures of cellulose molecules are found in cell walls.

Cellulose and some derivatives form liquid crystals (LC) and represent excellent materials for basic studies of this subject. A variety of different structures are formed, thermotropic and lyotropic LC phases, which exhibit some unusual behavior. Since chirality expresses itself on the configuration level of molecules as well as on the conformation level of helical structures of chain molecules, both elements will influence the twisting of the self-assembled supermolecular helicoidal structure formed in a mesophase. These supermolecular structures of chiral materials exhibit special optical properties as iridescent colors, and may be used for special coloring effects such as crustaceans provide with the help of chitin.

2 The Structure of Cellulosics

The structure and arrangement of cellulosic chains play an important role in the formation of liquid crystals. At present, neither the conformation of cellulosics nor the solvent bound to the chain in the case of a lyotropic mesophase are known for these liquid-crystalline systems. Nevertheless, these structural features form the basis for a discussion of structural and thermodynamic aspects. Information on cellulosics is available for the two borderline cases next to the LC state, i.e., for the solvent built-in solid state as well as for the pure solid state, obtained by X-ray, NMR, and potential energy analysis on one side, and for the semi-dilute state from light-scattering experiments on the other side. These data have to be evaluated for a discussion of possible structures and models in liquid-crystalline phases.

Cellulose consists of β-D-anhydroglucose-pyranose units linked by (1-4)-glucosidic bonds in contrast to amylose in which the units are linked by α-(1-4) glucosidic bonds (Fig. 1). Cellulose is considered to form somewhat stiff chain molecules, whereas

a)

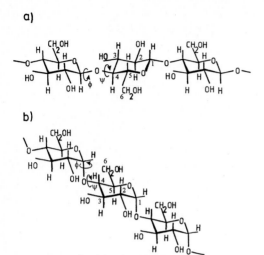

b)

Figure 1. Schematic representation of the configuration of (a) cellulose and (b) amylose with conventional atom labeling and the introduction of the torsion angles ϕ and ψ.

amylose has flexible ones. The solubility is quite different for these two polysaccharides. The monomer units are relatively stiff for cellulose, but more flexible for amylose, with some allowed rotations of the hydroxyl group at C6 leading to three distinct energy minima for both polysaccharides. The two dihedral or torsion angles ϕ and Ψ, involving the linkage of two adjacent residues, determine the chain conformation, which, especially for cellulosics, is quite limited by steric interactions between two adjacent residues and only allows extended chains as low energy systems. Nevertheless, sheet-like structures such as 2/1 helices or rod-like structures such as left-handed 3/2 or 8/5 helices are possible and found in cellulosics with a uniform distance of about 5 Å (0.5 nm) for one monomeric unit projected on the helix axis.

The three hydroxyl groups of a monomeric unit form a dense hydrogen-bonded network in solid cellulose and provide sites for substitution by ester, ether, or carbamate groups. Cellulose commonly exhibits a sheet-like structure such as a 2/1 helix in the solid state, and it was not until recently that the packing of natural cellulose was discovered to consist of two polymorphs of different ratio for the various kinds of natural cellulose. Strong interactions between cellulose and small molecules or ions, e.g., sodium hydroxide, may lead to a change in the cellulose chain conformation, and rod-like 3/2 helices in a solvent built-in crystalline state may be formed.

Cellulose derivatives pack as sheet-like structures, for example, cellulose triacetate or butyrate as well as some ethers, e.g., trimethyl cellulose. Regio-selective derivatives of these mentioned esters or ethers placed at C2 and C3 with various other groups at C6 may belong to the same conformation class, that is, their conformation is determined by the substituent at positions 2 and 3. Many cellulose derivatives crystallize as rod-like helices of various kinds, e.g., 3/2, 8/5, and these structures are also found in solvent built-in crystals. It should be mentioned that several cases are known where a change in conformation occurs if a polar solvent is built into the interstitial spaces between the polysaccharide chains. The sheet-like cellulose triacetate conformation converts to a rod-like 8/5 helix with nitromethane present.

Sophisticated experimental methods allow the development of models for polymers in dilute and semi-dilute solutions. Chain stiffness may be represented by the Kuhn segment lengths and determined in dilute solution. Models for cellulose and cellulose derivatives have recently been published whose main features are the irreversible aggregation of chains, if hydrogen bonding is possible even in dilute solutions. Trisubstituted cellulose derivatives or cellulose in hydrogen bond breaking solvents exist as molecular dispersed chains. However, at higher concentrations, in the semi-dilute regime, reversible cluster or association occur and, depending on the geometry of these particles, liquid crystals or gels are obtained. The association process is strongly influenced by the molecular mass, and above a certain length of the cellulosic chain, only gel formation is observed. Aggregation and association processes are comparable with micelles for amphiphilic molecules and lead to lyotropic liquid crystals upon increasing their concentration. Such a process of creating larger entities first, which then form lyotropic liquid crystals, has to be considered for cellulosics. The models on which theories on thermodynamics, flow behavior, and the structural features of a cholesteric phase depend have to take these novel ideas into account. In this respect, the term 'lyotropic' describes the feature of creating liquid crystals by forming clusters or other arrangements first, and in a further step overall order is introduced. 'Thermotropic' then means obtaining or changing the order through temperature.

3   The Chiral Nematic State

3.1   Methods for the Detection of Cellulosic Mesophases

Lyotropic cellulosics mostly exhibit chiral nematic phases, although columnar phases have also been observed. The molecules in the thermotropic state also form chiral nematic order, but it is sometimes possible to align them in such a way that a helicoidal structure of a chiral nematic is excluded. Upon relaxation they show banded textures. Overviews on lyotropic LC cellulosics are

provided by Gray and co-workers [1, 2], on thermotropics by Fukuda et al. [3], on polymer solvent interactions by Zugenmaier [4], and on LC cellulosics by Gilbert [5].

A first characterization of LC materials normally starts with the observation of textures in the polarizing microscope (see Fig. 2), followed by additional investigations evaluating other means of identifying the kind of phase present. Chiral nematics, also termed cholesteric phases, show unusual optical properties which are caused by a supermolecular structure by which the phase may be determined on applying spectroscopic means. These optical properties include a wave-guiding effect, selective reflection, optical rotation, and birefringence. Studies on the phase transition and the flow behavior add to our knowledge of chiral nematic phases. A survey on the structure and properties of cholesteric phases is given by Dierking and Stegemeyer [8], and theoretical models are discussed.

The wave-guiding regime occurs for a pitch $P \gg \lambda$ and was investigated by Maugin [10]. This effect plays an important role in display technology. The plane of polarization of light follows the twist of the nematic director defined within a nematic sheet. The wave-guiding regime has yet to be considered in the field of cellulosics.

At the selective reflection wavelength, a component of linearly polarized light, e.g., right-handed or left-handed circularly polarized light, is totally reflected without a phase reversal by a twisted helicoidal structure at a wavelength correlated to the pitch P. An oblique incident of light is presumed onto the substrate plane or nematic sheet (see Fig. 3)

$$P = \frac{\lambda_o}{\bar{n}} \tag{1}$$

where λ_o is the selective reflection wavelength, $\bar{n} = (n_{o,nem} + n_{e,nem})/2$, the mean re-

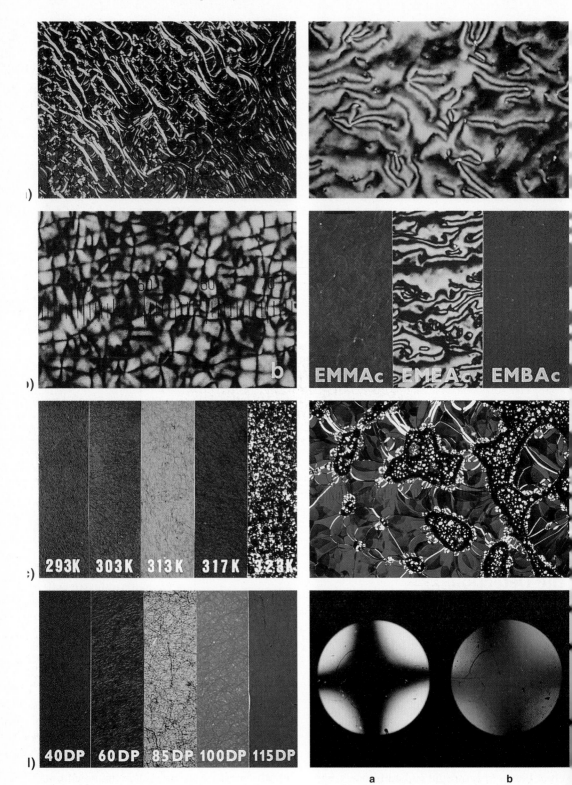

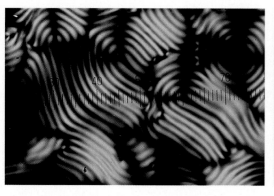 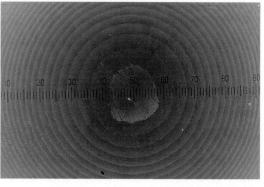

k)

Figure 2. Characteristic textures of cellulosics observed in the polarizing microscope at room temperature, if not otherwise noted (abbreviations: CTC cellulosetricarbanilate, cf. Table 4 for solvents): (a) Oily streaks of a planar oriented lyotropic LC solution of CTC/MAA ($c=0.7$ g/ml, 1 unit of scale 17 µm, adapted from [6]). (b) Parabolic focal conic texture of CTC/MAA ($c=0.85$ g/ml) found after annealing (1 unit of scale: 5.7 µm, adapted from [6]). (c) Planar and parabolic focal conic (at 323 K) textures of CTC/DEMM at increasing temperatures ($c=0.8$ g/ml, DP$=125$, adapted from [7]). (d) Planar textures of CTC/DEME dependent on the molecular mass (narrow fractions, denoted by the degree of polymerization DP; $c=0.8$ g/ml, adapted from [8]). (e) Schlieren texture of a nematic-like LC of CTC/EMAc+EDAc (1:1) (DP$=125$, $c=0.7$ g/ml, adapted from [7]). (f) Planar and schlieren texture of LC CTC caused by a change of solvent from EMMAc to EMEAc to EMBAc, exhibiting in sequel a left-handed, nematic-like, right-handed structure (DP$=80$, $c=0.8$ g/ml, adapted from [8]). (g) Biphasic region of LC CTC/DEME; red areas remaining planar texture, black areas isotropic phase. Such polarization microscopic textures may serve for establishing phase diagrams ($c=0.65$ g/ml, DP$=125$, adapted from [8]). (h) Conoscopic pictures of CTC/EME (a) without and (b) with λ plate to establish the positive birefringence of CTC/EME (adapted from [8]). (i) Fingerprint texture of CTC/EMEAc (DP$=125$, 1 unit of scale: 3 µm, adapted from [8]. (k) Cano rings of LC CTC/TRIME ($c=0.8$ g/ml, DP$=125$, 1 unit of scale: 17 µm, adapted from [8]).

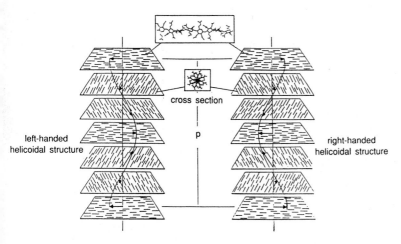

left-handed
helicoidal structure

cross section

p

right-handed
helicoidal structure

Figure 3. Schematic representation of the cholesteric liquid-crystalline structure of cellulosics; $P=\lambda_o/\bar{n}$ where P represents the pitch, λ_o the reflection wavelength, and $\bar{n}$ the mean refractive index of a sheet. $P>0$ for a right-handed twist, and $P<0$ for a left-handed twist.

fractive index of a nematic sheet; $n_{o,\text{nem}}$ and $n_{e,\text{nem}}$ being the ordinary and extraordinary refractive index of the considered sheet.

The width of the reflection band is given by $\Delta\lambda=\Delta n\,P$, where Δn is the birefringence

($\Delta n=n_e-n_o$; n_e being the extraordinary refractive index, n_o the ordinary one). Most cholesteric phases show a negative birefringence, except for a few compounds whose polarizability normal to the rod axes is larger than along the rod direction. Such a case is observed for cellulosetriphenyl carbamate or derivatives thereof; the phenyl ring is placed almost perpendicular to the cellulose backbone axis (see Fig. 4). The sign of the birefringence may also be determined by observation of a planar texture along the uniaxial optical axis, as found in cholesterics in the polarization microscope (conoscopic mode) with a λ plate (see Fig. 2h). This observation then leads, with some knowledge of the main direction of the polarizability, i.e., for rods mainly in the rod axis direction, to the determination of the chiral nematic phase. The relative difference of the two main polarizabilities $\Delta\alpha/\bar{\alpha}$ is available by an extrapolation method from Haller [12], which is commonly used for the determination of the order parameter S. The order parameter S was introduced by Hermans and Platzek [14] and for nematic phases by Maier and Saupe [30] by $S = 1/2 \langle 3\cos^2\vartheta - 1\rangle$, ϑ being the angle between the molecular axis and the director $\boldsymbol{n}$. S actually represents a mean value of the distribution for the long axes of the molecules in the mesophase and plays an important role in all statistical and thermodynamic considerations.

If $\lambda \neq \lambda_o$ the plane of polarization of incident light is rotated by an angle Θ and converted to elliptically polarized light. De Vries [13] applied Maxwell's equations to twisted sheets and derived for Θ, the optical rotatory power

$$\Theta = \frac{\pi \Delta n_{nem}^2}{4\lambda^2} P \frac{1}{1-(\lambda/\lambda_o)^2} \qquad (2)$$

Equations (1) and (2) contain the pitch P besides experimentally available quantities, i.e., the selective reflection wavelength λ_o, the optical rotation Θ, and the refractive indices n_o and n_e of a nematic sheet. The usually determined cholesteric birefringence can be converted to the nematic one by $\Delta n_{nem} = -2\Delta n_{ch}$. By fitting Eq. (2) to the actual measured data, the pitch is obtained and the validity of Eq. (2) tested. In the limit of $\lambda \ll \lambda_o$, Eq. (2) leads to

$$\Theta = \pi \frac{\Delta n_{nem}^2}{4\lambda^2} P \qquad (3)$$

This equation was used quite often to prove the existence of cholesteric phases in the absence of other methods, especially if the selective wavelength lies outside the visible range. A plot of Θ versus λ^2 should result in

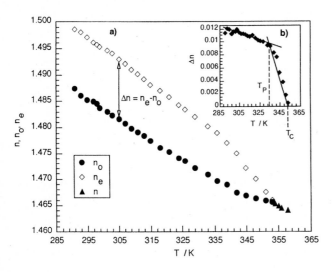

Figure 4. Refractive indices of a lyotropic cellulosic system (3Cl-CTC/EMMAc, concentration $c = 0.8$ g/ml) as a function of temperature, measured on samples exhibiting planar textures with an Abbé refractometer. n_o: ordinary refractive index ●; n_e: extraordinary refractive index ◇; n: refractive index in the isotropic phase ▲. The birefringence is defined by $\Delta n_{ch} = n_e - n_o$ and shown in the insert (b); $T_p \rightarrow T_c$ defines the biphasic region and may serve for establishing the phase diagram; T_c is the clearing temperature (adapted from [11]).

a straight line. De Vries' evaluation neglects the optical activity of the chiral molecules (conformation and configuration) and sets the sample size infinite, that is, the number of helicoidal repeats has to be large.

The Kramers–Kronig [15] transformation relates the optical rotation (ORD) to the circular dichroism (CD). This spectroscopic method is widely used for the determination of the pitch as well. Figure 5 shows examples for left- and right-handed helicoidal structures of cellulosics as they appear in ORD and CD measurements.

The selective reflection also affects the transmission of normal light, causing a diminution of intensity, and this loss of intensity can be measured by transmission spectroscopy (UV–VIS). A minimum in intensity corresponds to the selective wavelength and leads directly to the pitch of the helicoidal structure and, with the application of circularly polarized light, to the handedness of the structure (see Fig. 6).

Besides spectroscopic methods, the Grandjean–Cano [18] (Fig. 2k) and fingerprint techniques (Fig. 2i) represent a direct method for the determination of the pitch in cellulosic mesophases, especially for large sizes of the pitch outside the spectroscopic range. Cano placed a lens, Grandjean a wedge, on top of the LC sample in a polarization microscope. If the sample orients well, bright and dark rings or stripes in the case of the Grandjean wedge are observed.

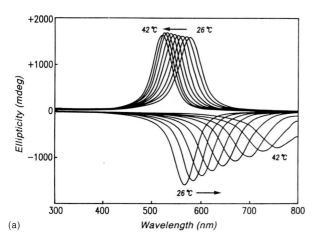

(a)

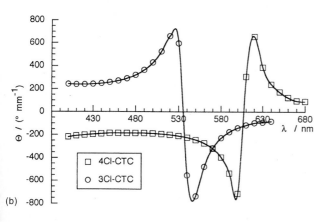

(b)

Figure 5. (a) CD and (b) ORD spectra of lyotropic mesophases of cellulosics with reversal of the handedness. (a) Temperature dependence of lyotropic (acetyl) ethyl cellulose (DS of ethyl groups 2.5) in chloroform; positive ellipticity for a left-handed twist (acetyl DS=0.06); negative ellipticity for a right-handed twist (acetyl DS=0.5); concentration ≈50% by weight (adapted from [2]). (b) Optical rotatory power Θ as a function of wavelength for the lyotropic mesophase systems cellulose-tri(4Cl-phenyl)carbamate (4Cl-CTC)/ DEMM □□ with left-handed twist corresponding to $\Theta < 0$ below the reflection wavelength λ_o and cellulose-tri(3Cl-phenyl)carbamate (3Cl-CTC)/ DEMM ○○ with right-handed twist corresponding to $\Theta > 0$ below λ_o (adapted from [11]).

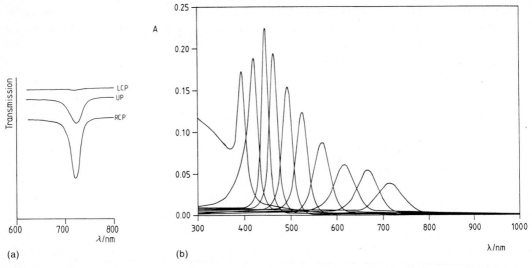

Figure 6. Transmission and absorbance spectra of CTC/solvent systems in the UV–VIS range. (a) Transmission for the right-handed supermolecular structure CTC/2-pentanone: left circularly polarized (LCP) light, right circularly polarized (RCP) light, normal light (unpolarized, UP); concentration $c = 0.6$ g/ml, temperature 16.7 °C, DP = 100 for CTC (adapted from [16]). (b) Absorbance spectra for left-handed twisted cholesteric CTC/MAA ($c = 1.0$ g/ml) at various temperatures; (from left to right: 10, 15, 20, etc., 55 °C) (adapted from [17]).

The diameters of the rings or the distances of the stripes are correlated with the pitch. For the Cano technique, the Eq. (4) is obtained by a geometric consideration (see Fig. 2k)

$$P = \frac{\Delta x^2}{r} \qquad (4)$$

where r is the radius of the lens, $\Delta x^2 = x_{i+1}^2 - x_i^2$, where x_i is the radius of the ith ring. The site of the ring and whether to use the edge or center can be checked by simple relations that have to be fulfilled regardless of the scale or magnification [19]. If the pitch is larger than several micrometers, the fingerprint method may lead to a good estimation. Here, the helical axes of the chiral nematic supermolecular structure are placed parallel to the substrate, which might be achieved by certain alignment layers coated on a glass surface or by certain procedures. Well-aligned samples then lead to the observation of dark and light

stripes in the polarizing microscope. The distance of these stripes amounts to $P/2$ ideally (see Fig. 2i). However, caution is advised, since the periodicity of the lines may be very sensitive to surface preparation and alignment.

All spectroscopic methods for an evaluation of the pitch lead to the size of the repeat distance of the supermolecular LC structure and the handedness of the twisting, although in UV–VIS spectroscopy, circularly polarized light has to be used. With some skill, the Cano technique may also result in the twisting sense of the helicoidal structure.

The formation of liquid-crystalline phases of cellulosics, especially of the lyotropic kind, has yet to be explained. Certainly the chain stiffness may have to be taken into account as one of the factors in question, but the solvent–polymer interaction may have to be considered as well. In the next section, models for the description of the pitch as a chiral property and models to describe the phase transition nematic (cholesteric)–isotropic by thermodynamics will be briefly considered. For further discussion the references or review articles should be consulted (see [9] or other chapters of this Handbook).

3.2 Models of Chiral Nematic Cellulosics

Several models are presently used to describe the chiral nematic structure and its temperature dependence for cellulosics. The following behavior has been observed for their supermolecular structure

(1) decreasing pitch with increasing temperature,
(2) increasing pitch with increasing temperature,
(3) inversion of the cholesteric handedness with temperature variation; this was found in two thermotropic oligomer cellulosics. This behavior is more common for lyotropic mesophases of cellulose derivatives where more variables are available, because the kind of solvent, the concentration, and various substituents, may all interact.

Various models proposed may not account for all these experimental facts. The Keating [20] and Böttcher [21] evaluation does not account for such a variety of behavior. Goossens [22] proposed the chiral nematic structure as the result of an anisotropic dispersion energy between chiral mesogens, and predicts for thermotropic LCs a pitch that is essentially independent of temperature. Lin-Liu et al. [23] developed a theory that accounts for all the above-stated temperature effects. The temperature dependence of the pitch is determined by the shape and position of the intermolecular potential as a function of the intermolecular twist

angle. Varichon et al. [24] extended these considerations by explicitly introducing the orientational entropy and concentration dependence of the pitch. This theory predicts a large diversity of behavior.

A statistical model was introduced by Kimura et al. [25] which included attractive and repulsive asymmetric intermolecular forces for rod-like molecules. The results are summarized as Eq. (5), which is widely used for the lyotropic chiral nematic state

$$P^{-1} = Q\left(\frac{T_n}{T} - 1\right) \tag{5}$$

where Q is a factor that includes the geometry of the rods and the polymer volume fraction. $Q > 0$ for a right-handed twist and $Q < 0$ for a left-handed one. T_n represents the compensation temperature for which the twisting power $P^{-1} = 0$ or a nematic-like structure appears. Equation (5) is represented in Fig. 7 and discussed later with experimental data. Osipov [26] also developed a statistical theory with steric and chiral interactions between molecules in solution and arrived at the following equation for cellulosics

$$P^{-1} = U(\chi - \lambda_s k_B T) \tag{6}$$

where U is a factor describing the concentration and physical properties, χ is related to the attraction, λ_s represents the steric repulsion between the molecules, and k_B is the Boltzmann constant. This relation fulfills all the experimental observations and leads to the twisting power $P^{-1} = 0$, if the attraction and steric repulsion cancel.

An empirical expression was also introduced with

$$P^{-1} = a\left(1 - \frac{T}{T_n}\right) \tag{7}$$

where a is related to the concentration by $a \propto c^m$ with $m = 2$ in the case of polypeptides.

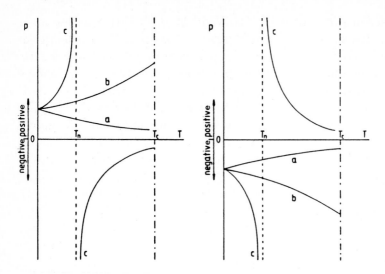

Figure 7. Schematic representation of pitch P versus T according to Kimura et al. [25] for $Q>0$ left (right-handed pitch) and $Q<0$ right (left-handed pitch); T_n is the compensation temperature, T_c is the clearing temperature. (a) Predominantly polar, (b) polar and steric, and (c) steric interactions (adapted from [17]).

In many studies m was found to lie between 1 and 3.

4   Mesophase Formation of Cellulosics

Rod-like macromolecules or semi-flexible chains such as cellulosics with a certain rigidity may form thermotropic and, in highly concentrated solution with suitable interactions, lyotropic liquid-crystalline phases.

Onsager [27] predicted a phase separation of a solution of rod-like molecules above a critical concentration that depends on the aspect ratio of the molecules. The interactions are of a steric nature, and higher order terms in the virial expansion employed in his considerations have been disregarded. Khokhlov and Semanov [28] expanded the Onsager approach from rods to worm-like chains, and it was shown that for stiff chains good agreement with experimental data was achieved. Cellulosics, however, represent semi-flexible chains with short persistence length or Kuhn segment

length l_K and a reasonable agreement was not obtained.

Flory [29] proposed a lattice theory that accommodates chain flexibility in liquid-crystalline polymers in a later version. The critical concentration for mesophase formation depends for freely jointed rods on the aspect ratio of the individual Kuhn segment, that is, instead of the contour length, the axial ratio of l_K is used for the calculation. The critical volume fraction of the polymer V_p^* is determined by the aspect ratio x, and Eq. (8) was derived as

$$V_p^* \approx \frac{8}{x}\left(1 - \frac{2}{x}\right) \qquad (8)$$

The phase separation of the solution is driven by the entropy of forming a liquid-crystalline phase at a high polymer concentration. However, Maier and Saupe [30] concluded that the formation of a nematic phase arises from orientation-dependent interaction of induced dipoles. From these considerations, it appears that both entropy and enthalpy contribute to the stability of mesophases, although in a different ratio for large and small molecules. Asymmetric attractions have been included in a later development of the Flory lattice theory by Warner and Flory [31].

In a further step, the Onsager and lattice model were adjusted to account for some shortcomings by the Kyoto group and applied for the description of the molecular mass dependence of thermotropic cellulosics [3].

5 Mesophases of Cellulose

Only a few solvents are known to dissolve cellulose completely, and solid cellulose decomposes before melting. Therefore, it is difficult to study the mesophase behavior of cellulose. Chanzy et al. [32] reported lyotropic mesophases of cellulose in a mixture of N-methyl-morpholine-N-oxide and water (20–50%), but were unable to determine the nature of the mesophase. Lyotropic cholesteric mesophase formation in highly concentrated mixtures of cellulose in trifluoroacetic acid + chlorinated-alkane solvent [33] and in ammonia/ammonium thiocyanate solutions [34] has been studied, and although poor textures were obtained in the polarizing microscope, high optical rotatory power has been measured in an optical rotation (ORD) experiment, which could be fitted to the de Vries equation [Eq. (3)] for selective reflection beyond the visible wavelength region and was taken as proof of a lyotropic chiral nematic phase.

Interestingly, the suspension of sulfuric acid treated micro-crystalline cellulose in water leads to a lyotropic LC cellulose state [35]. Well-formed textures and disclination characteristics of a chiral mesophase have been observed. The critical volume fraction for phase separation of salt-free suspensions has been as low as 0.03, with a relatively narrow biphasic region. From the dimensions of the crystallites present (200–300 nm long and approximately 7 nm wide), the anisotropic phase requires a concentration of more than 0.1 (volume fraction) according to Onsager's approach for predicting the biphasic region. The interaction between the crystallites is sensitive to the electrolyte content and is believed to be strongly influenced by charges on the rods. The pitch of these lyotropic structures lies in the 10–100 μm range.

Most of the investigations to obtain LC cellulose were undertaken to achieve high-performance films or fibers from anisotropic solutions. The development of stable cellulose LiCl/dimethylacetamide (DMAC) systems led to an attempt to produce anisotropic solutions [36, 37]. Evidence was found of mesophase formation at 10–15% by weight depending on the salt concentration, with some problems due to limited solubility at high concentration (>15%). Measurements of the persistence length of cellulose in a dilute solution of this system indicate that the cellulose chains are stiffer than those of cellulose derivatives [38], and therefore lower the critical concentration for

Table 1. Cellulose forming lyotropic mesophases.

System	Handedness
Cellulose/NMMNO (N-methyl-morpholine-N-oxide) [32, 40, 41]	
Cellulose/TFA + chlorinated alkanes (1,2-dichloroethane, CH_2Cl_2) [33]	right-handed
Cellulose/NH_3 + NH_4SCN [34]	left-handed [42]
Cellulose/LiCl-DMAc [36–38]	left-handed
Cellulose/H_2O (suspension of crystallites) [35]	

phase separation. The enhanced stiffness may be due to cellulose complexes and intra-chain hydrogen bonding and polyelectrolyte effects. Recent results from light-scattering investigations suggest that, in cases when hydrogen bonds between chains are involved, irreversible aggregates are formed [39]. This behavior may lead to a situation as discussed for the cellulose suspension, now with the aggregates serving as the structural units for mesophase formation rather than single, stiff chains as commonly expected.

Presently known lyotropic LC celluloses are collected in Table 1.

6   Lyotropic Liquid-Crystalline Cellulose Derivatives

Lyotropic LC cellulose derivatives are formed in highly concentrated solutions (~20–70 wt%). As general goals for studying these mesophases, the following incentives might serve.

(1) To get more insight into the polymer–solvent interaction, especially to investigate the correlation between chirality on a molecular level (chiral centers, chiral conformation) and the helical twisting power.

(2) To understand the cholesteric–isotropic phase transition regarding interactions and models of the molecules in these two phases. Clearly experiments with changes in temperature, concentration, substituents, solvents, and molecular mass will be very helpful in achieving these goals and may serve as tests for the proposed ideas and models.

(3) To produce new materials and developing better performance fibers and films by new processes.

6.1   Supermolecular Structure and Optical Properties

The optical properties of chiral nematic phases are closely related to their supermolecular structures, as stated by the considerations of de Vries. In particular, the planar textures exhibit beautiful colors correlated to the pitch P of the helicoidal structures by Eq. (1), if the selective reflection wavelength lies in the visible range, and many examples are shown in Fig. 2.

Cellulose derivatives exhibit Kuhn segment lengths in the range 10–100 nm in dilute solution and are regarded as semi-flexible polymers. In the highly concentrated state, these solutions should in general lead to chiral mesophases, as predicted by theory, and may exhibit iridescent colors. However, several exceptions are found. Cellulose derivatives with similar chain stiffness to those leading to cholesteric phases will form microgels or gels instead. From a crude point of view, lyotropic LC cellulose derivatives ought to have the 'right solubility'. This means, if they do not dissolve well, the high concentration necessary for forming a mesophase will not be obtained, and gels or some other structures will form instead. On the other hand, if they dissolve too well, the lyotropic liquid-crystalline state with the needed aggregates or clusters of a certain shape will be passed and gel formation occurs in the end. A balanced solubility might be a necessity.

Also, the shape of the entities in solution has to be of a certain geometry. Very large molecules might not be suitable and collapse into gels. Many of the lyotropic LC derivatives show, at low or semi-dilute concentrations, aggregation or association behavior, which will be reversible or irreversible, as investigated by light-scattering experiments. Models have been developed for the extensively studied, fully substituted cellulosetricarbanilate (CTC) for which molecularly dispersed chains are found in dilute solutions, which are very much extended for small molecules, and a random coil with some polymer–polymer contact for larger molecules [11] (Fig. 8a). In the semi-dilute solution range (Fig. 8b), reversible clusters appear, which according to a DSC study carry bound solvent [17].

Hydroxypropyl cellulose (HPC) is the most widely studied lyotropic LC cellulose derivative. Werbowyj and Gray discovered the lyotropic LC state for cellulosics of this derivative [43]. HPC exhibits poor solubility and problems occur with handling at a molecular dispersed level, even at low concentration. However, HPC has the advantage of being commercially available and is able to dissolve in water, hence rheological studies can be carried out with the large amount of materials needed. It has the dis-

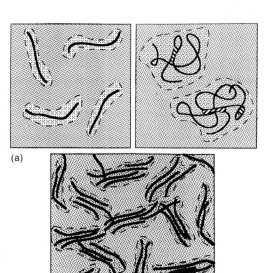

(a)

(b)

Figure 8. Schematic models for CTC in (a) dilute solution for small and large molecules, respectively, and (b) semi-dilute solution for small molecules. The dashed lines enclose domains with solvent shells (adapted from [11]).

advantage that a precise characterization of the substituents and their distribution along the chain is rather difficult to achieve. Precise measurements on the phase behavior, even for narrow molecular mass fractions, have been performed as well as on the supermolecular helicoidal structures of the chiral nematic phase. Fully trisubstituted cellulose derivatives do not have this disadvantage, and therefore, are more suitable for basic studies. A large amount of information has been gathered on lyotropic cellulose derivative mesophases (see Table 2) and is now available for the evaluation of various kinds of structures and will be discussed below.

6.2   Handedness

Cholesteric mesophases of cellulosics represent a special kind of materials inasmuch as left-handed and right-handed twisted supermolecular structures occur. HPC in water exhibits a right-handed helicoidal structure in contrast to cellulosetriacetate (CTA) in trifluoroacetic acid (TFA), whose structure is left-handed. According to simple modeling with springs [66], a left-handed twist of the wires, corresponding to a left-handed helix conformation, leads to a left-handed supermolecular twisted structure, a right-handed conformation to a right-handed cholesteric helix. This idea is confirmed by the theoretical considerations of Kimura et al. [25] and will be discussed in more detail in the section on the temperature dependence of the supermolecular pitch (Sec. 6.3). The observation of the twist sense of the supermolecular structures, then leads, if the model is accepted, directly to the handedness of the chain conformations. In all conformational studies for cellulosics in the solid state, only left-handed helices and 2/1 structures are established. The only pub-

Table 2. Cellulose derivatives forming lyotropic cholesteric mesophases.

System	Handedness
HPC (hydroxypropyl cellulose)/H_2O [43], CH_3OH, C_2H_5OH [44]	right-handed
HPC (hydroxypropyl cellulose)/$Cl_2CHCOOH$, CH_3COOH, DMAc [45–47]	
HPC (hydroxypropyl cellulose)/pyridine, 2-methoxyethanol, 1,4-dioxane [48]	
HPC (hydroxypropyl cellulose)/m-cresol [49]	
Celluloseacetatebutyrate/CH_3COOH, MEK, DMAc [45]	
CTA (cellulose triacetate)/dichloroacetic acid (DCA) [45]	
CTA/TFA (trifluoroacetic acid)+chlorinated alkanes [50]	(left-handed)
CTA/TFA [51]	left-handed
CTC (cellulosetriphenylcarbamate)/2-pentanone, methyl ethyl ketone [52]	right-handed
EC (ethyl cellulose)/glacial acetic acid [45]	left-handed [52]
EC (ethyl cellulose)/dichloro acetic acid [45]	right-handed [53]
EC (ethyl cellulose)/glacial acetic acid+dichloro acetic acid [53]	left- → right-handed
CDC (cellulosediphenylcarbamate)/pyridine (by → concentration) [52]	right- → left-handed
CTC/diethylene glycol monoethyl ether (DEME) [54, 55]	left-handed
CTC/DEME+2-pentanone [57]	left- → right-handed
CTC/methyl ethyl ketone+2-pentanone [52]	righ-handed
Cellulose acetate-trifluoro acetate/TFA [56]	left- → right-handed
Acetoxypropyl-aceto cellulose/acetic acid, dibutyl phthalate [75, 62]	left-, right-handed
3Cl-CTC/DEME or other solvents [57, 11]	right-handed
CTC+3Cl-CTC/TRIMM [58]	left- → right-handed
CTC+3Cl-CTC/DEMH [58, 4] (all compositions)	right-handed
CA (acetyl cellulose)/TFA, other solvents [76]	left-handed
(Acetyl)ethyl cellulose (AEC, $DS_{eth}=2.5$)/$CHCl_3$, CH_2Cl_2, DCA, aqueous phenol, acetic acid, m-cresol [60, 61], (by acetyl composition in m-cresol)	left- → right-handed
(Acetyl)ethyl cellulose (AEC, $DS_{eth}=2.5$)/$CHCl_2COOH$ [60]	right-handed
2-Ethyloxypropyl cellulose/acetonitrile, dioxane, methanol [59] (also thermotropic)	right-handed
Acetoacetoxypropyl cellulose/acetic acid [62]	left-handed
Phenylacetoxy cellulose (PAC; DS=1.9), 4-methoxyphenylacetoxy cellulose (4MPAC; DS=1.8)/CH_2Cl_2 [63] (also thermotropic)	(left-handed)
Trimethylsilyl cellulose (TMSC; DS=1.55), p-tolylacetoxy cellulose (TAC; DS=0.5–1.8)/CH_2Cl_2 [63] (also thermotropic)	(left-handed)
CTC/MAA (methyl acetoacetate) [17]	left-handed
CTC/glycols [55]	left- or right-handed
3-Methyl, 3-methoxy, 3-fluoro, 3-chloro, 3-CF_3-CTC/different solvents [11]	right-handed
4-Methyl, 4-methoxy, 4-chloro, 4-bromo-CTC/different solvents [11]	left-handed
3,4-Chloro-CTC/different solvents [11]	right-handed
Regio-selectively substituted CTC (in 2,3,6 position along the chain; meta position at the phenyl ring with Cl, methyl, H)/EMMAc [64]	left- or right-handed
6-O-trityl-2,3-O-hexyl cellulose/acetic acid, tetrahydrofuran [65]	left-handed
6-O-trityl-2,3-O-pentyl cellulose/tetrahydrofuran [65]	right-handed

lished right-handed helix for cellulose trini-trate might actually be a left-handed one, since the conformational energy is lower for the left-handed structure.

The problem that has to be solved then is: How might a right-handed supermolecular structure be formed whose bases have to be right-handed helical conformations? It can be simply visualized that, for example, a left-handed threefold helix will easily turn into a right-handed threefold one, if the dimer is considered as the building unit of the chain conformation. Only the pitch has to be doubled. Different positions of O6 at adjacent residues or solvent attached to the chain may provide such dimers and lead to a reversal of the twisting direction of the cellulosic chain.

The same cellulose derivatives, fully or not fully substituted, sometimes show in one solvent a right-handed, and in another solvent a left-handed supermolecular structure. In this case a mixture of the solvents with a certain composition then leads to a so-called compensated cholesteric structure, which is nematic-like, that is, the twisting power $P^{-1}=0$ or the pitch is infinite (see Fig. 9). This structure resembles a nematic phase, not only from a structural modeling point of view, but rather nematic-like schlieren textures are observed (see Fig. 2e). A twist inversion was also observed for cellulose derivatives with two different side groups statistically distributed along the chain at a certain composition. Ethyl cellulose (degree of substitution $DS=2.5$) was acetylated and inversion occurred at an acetyl DS as low as 0.19 for (acetyl)ethyl cellulose in m-cresol (Fig. 5a). The ethyl-acetyl composition may change when other solvents are used. Another example is provided by a fully trisubstituted carbanilate. The distribution of 3-chlorophenyl- and phenyl-carbamate groups at about a 1:1 ratio random along the cellulose chain leads to a compensated

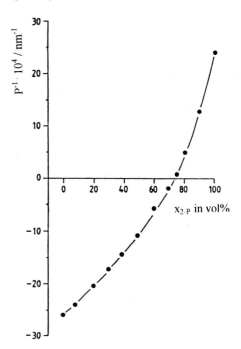

Figure 9. Twisting power P^{-1} of a mixed solvent system: CTC/DEME + 2-pentanone ($c=0.8$ g/ml, $T=298$ K; spectroscopic method, Cano technique); $x_{2\text{-}P}$: volume fraction of 2-pentanone (adapted from [7]).

structure in triethylene glycol monomethyl ether (TRIMM) (Fig. 10; the abbreviations of solvents are summarized in Table 4). Cellulose tri(3-chlorophenyl)carbanilate in the same solvent exhibits a right-handed structure; CTC itself is left-handed. On the other side, both pure trisubstituted compounds in DEMH show structures with right-handed twist sense. A variation of the two side groups along the chain does not change the twist sense of the structures at all. However, it should also be noted that the twisting power as a function of composition cannot be represented by a single straight line connecting the two initial twisting powers of the pure compounds (see Fig. 10).

Recently, regio-selectively substituted tricarbanilates were investigated and a strong influence of the twisting power found

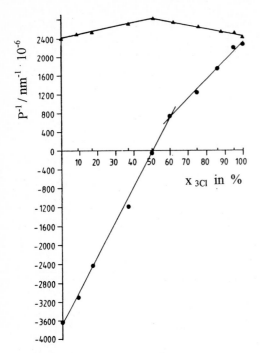

Figure 10. Twisting power P^{-1} as a function of composition for a copolymeric CTC–3Cl-CTC/solvent system; solvent TRIMM ● (0.8 g/ml, room temperature), and DEMH ▲ (0.7 g/ml, room temperature) (adapted from [4]).

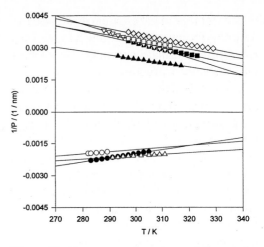

Figure 11. Temperature dependence of the twisting power P^{-1} of various right- and left-handed cholesteric structures of cellulose derivatives regio-selectively substituted in EMMAc. HHH ($c=0.9$ g/ml) ●, CCC (0.7 g/ml) ■, CCH (0.7 g/ml) ▲, MMC (0.7 g/ml) ▽, CCM (0.8 g/ml) ◇, MMM (0.7 g/ml) □, HHF (0.9 g/ml) △, HHF (0.8 g/ml) ○ (for abbreviations, see Table 3; adapted from [64]).

for the substituents dependent on their placement in the 2, 3, 6 position at the anhydroglucose residue (Fig. 11). The twisting power P^{-1} of a trisubstituted LC cellulose derivative/solvent system predominantly depends on the substituent in the 2 and 3 position of the anhydroglucose unit. The 6 position is of minor importance. These conclusions are also supported by investigations on 2,3-di-O-alkylcellulose, 6-O-alkylcellulose [67] and 6-O-trityl-2,3-

Table 3. Overview of cellulose derivatives and their corresponding polymer code for the position of substitution 2, 3, 6 at the anhydroglucose unit with phenyl urethane by fluoro (F) in the para, and by chloro (C), methyl (M), and hydrogen (H) in the meta site.

Cellulose derivative	Polymer code 2, 3, 6
Cellulosetris(3-chlorocarbanilate)	CCC
Cellulose-2,3-bis-(3-chlorocarbanilate)-6-(carbanilate)	CCH
Cellulose-2,6-bis-(3-chlorocarbanilate)-3-(carbanilate)	CHC
Cellulose-2,3-bis-(carbanilate)-6-(3-chlorocarbanilate)	HHC
Cellulosetricarbanilate	HHH
Cellulose-2,3-bis-(carbanilate)-6-(4-fluorocarbanilate)	HHF
Cellulose-2,3-bis(4-fluorocarbanilate)-6-(carbanilate)	FFH
Cellulose-2,3-bis-(3-chlorocarbanilate)-6-(3-methylcarbanilate)	CCM
Cellulose-2,3-bis-(3-methylcarbanilate)-6-(3-chlorocarbanilate)	MMC
Cellulosetris(3-methylcarbanilate)	MMM

O-alkyl cellulose [65], which were found to be sensitive to the kind of alkyl substituents. Experiments carried out with various phenyl carbamates suggest that P^{-1} is also strongly influenced by the substituent at the phenyl ring, mostly by the position of the substituent, but less by the kind of substituent attached. Almost the same twisting power is observed for a chlorine or a methyl group in the meta position at the phenyl ring. This is also true for the same substituents in the para position in the same solvent. However, a twist inversion from right- to left-handed structures occurs by going from the cellulosetri(3-chlorophenyl)carbamate to cellulosetri(4-chlorophenyl)carbamate in several glycols as the solvent (see Fig. 5b). Because of this reversal of the twist, a balanced regio-selective trisubstitution along the chain leads to a compensated lyotropic derivative/solvent system.

The handedness of the twist of the supermolecular structure may also be changed by a variation of the side group length of a diethylene glycol solvent (Fig. 12a, cf. also Fig. 2f). In this case a strong linear dependence of the clearing temperature T_c is also observed, which is considerably lower for a longer side group. The compensation of the chiral structure does not specifically influence the phase transition temperature (Fig. 12b). These findings clearly demonstrate that a fine balance exists between the solvent and the polymer chain, and that the chirality of the supermolecular structure is strongly influenced by this interaction, even with a nonchiral solvent involved. It is noteworthy that the Kuhn segment length l_K for the right-handed system 3Cl-CTC/EMMAc (for the abbreviations of the solvents, see Table 4) and the left-handed system 4Cl-CTC/EMMAc differs by almost a factor two, i.e., 37.6 nm versus 20.2 nm. On the other hand, the left-handed system 3Cl-CTC/EMMAc is comparable in chain stiffness with the right-handed 3,4Cl-CTC/

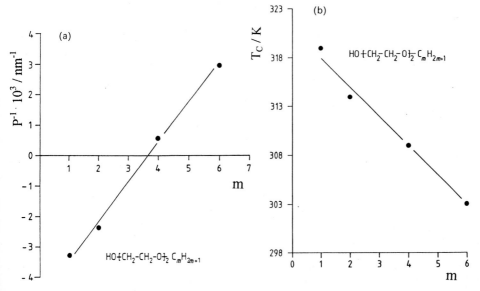

Figure 12. (a) Solvent dependent twisting power P^{-1} at $T = 298$ K, and (b) clearing temperature T_c for CTC/solvent systems ($c = 0.8$ g/ml, DP = 90) with the variation of the length of the alkyl side group of the solvent (adapted from [4]).

EMMAc with $l_K = 24.8$ nm [11]. In contrast, if commercially available methyl cellulose (DS = 1.6–1.7) is substituted to DS = 3 with the appropriate phenylcarbamate groups (3Cl, 4Cl), these derivatives exhibit no mesophase in dioxane, in fact only gels are formed, although the Kuhn segment length is comparable with those of 3Cl-CTC (in dioxane this is 49.5 nm) with 32.8 nm (3Cl) and 26.8 nm (4Cl) [11].

It should be emphasized that the compensation of the twist and changes in the pitch for statistically trisubstituted LC derivative/solvent system, e.g., 3Cl-CTC–CTC/TRIMM, occur when no full turn of the conformational helix with one kind of substituent group most probably exists. At present, a suitable explanation cannot be provided for this behavior of lyotropic LC cellulose derivatives.

cial case for lyotropic systems that show a compensation of the twist, which means that the pitch tends towards infinity, is also included in this representation. A strong negative or positive gradient is then observed in this region. The compensation of the twist leads to a nematic-like structure, which is confirmed by texture observations, and schlieren textures appear in the polarization microscope (see Fig. 2e).

The temperature dependence of the pitch can be described with Kimura and coworkers' equation [Eq. (5)], and some general conclusions derived. For most systems, polar and steric interactions have to be considered, as for the left-handed twisted structure of CTC/DEME (Fig. 13) or for the right-handed one of 3Cl-CTC-DEME (Fig. 14a). Here a break in the pitch versus the temperature curves can be detected, which

6.3   Temperature Dependence of the Pitch

The pitch of cellulosic mesophases can be determined by the methods introduced earlier, by spectroscopic means, or directly with the Grandjean–Cano technique. Its variation with temperature then leads to the temperature gradient. Most cellulose derivative/solvent systems exhibit a positive temperature gradient for values of pitch that according to Kimura and co-workers are caused by polar and steric effects (see Fig. 7). Lyotropic mesophases of ethyl cellulose/chloroform show a negative temperature gradient, in contrast to LC ethyl cellulose/dichloroacetic acid with a positive one. All possibilities are found for CTC mesophases, left- and right-handed twists with positive and negative temperature gradients (see Table 4, cf. also Fig. 2c), and can be accounted for by the graphs in Fig. 7. The spe-

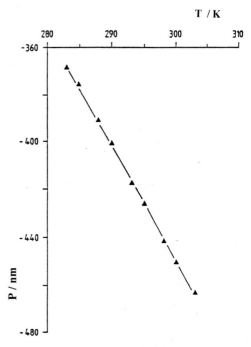

Figure 13. Pitch P as a function of temperature T for the left-handed LC system CTC/DEME ($c = 0.8$ g/ml) (adapted from [4]).

Table 4. Overview of the temperature, concentration gradients, and handedness of the pitch for various solvents [a,b].

	P	∂P/∂T	∂P/∂c		P	∂P/∂T		∂P/∂c		P	∂P/∂T	∂P/∂c
EMM	●	▲		DEMM	●	▲		■	TRIMM	●	▲	□
				DEME	●	▲	△	■	TRIME	●	▲	■
EMB	○		□	DEMB	○							
				DEMH	○	▲						
EDM	●	△	□	DEDM	●	▲		□	TRIDM	●	▲	□
EMMAc	●	▲	□									
EMEAc	●		□	DEMEAc	●							
EMBAc	○	▲	■	DEMBAc	○	△		■				
EDAc	●	△	□									
				DEMCl	●	▲		□	TRIMCl	●	▲	□
DIOX	●	▲	□	cHEX	●	▲		□	AcBL	●	▲	□

$P < 0$:	●	$\partial P/\partial T < 0$:	▲	$\partial P/\partial c < 0$:	■
$P > 0$:	○	$\partial P/\partial T > 0$:	△	$\partial P/\partial c > 0$:	□
		$\partial P/\partial T \approx 0$:	▲		

[a] Adapted from [4].
[b] List of solvents and their abbreviations:

Mono- and disubstituted glycols with alkyl groups
Ethylene glycol monomethyl ether	EMM
Diethylene glycol monomethyl ether	DEMM
Triethylene glycol monomethyl ether	TRIMM
Ethylene glycol monoethyl ether	EME
Diethylene glycol monoethyl ether	DEME
Triethylene glycol monoethyl ether	TRIME
Ethylene glycol monobutyl ether	EMB
Diethylene glycol monobutyl ether	DEMB
Diethylene glycol monohexyl ether	DEMH
Ethylene glycol dimethyl ether	EDM
Diethylene glycol dimethyl ether	DEDM
Triethylene glycol dimethyl ether	TRIDM

Substituted glycols with alkyl/acetate groups:
Ethylene glycol monomethyl ether acetate	EMMAc
Ethylene glycol monomethyl ether acetate	EMEAc
Ethylene glycol monobutyl ether acetate	EMBAc
Diethylene glycol monoethyl ether acetate	DEMEAc
Diethylene glycol monobutyl ether acetate	DEMBAc

Substituted glycols with acetate groups:
Ethylene glycol diacetate	EDAc
Ethylene glycol monoacetate	EMAc

Glycol monochlorohydrins:
Diethylene glycol monochlorohydrin	DEMCl
Triethylene glycol monochlorohydrin	TRIMCl

Cyclic systems and others:
1,3-Dioxolane	DIOX
2-Acetyl-γ-butyrolactone	AcBL
Cyclohexanone	xHEX
Ethylmethyl ketone	MEK
Methylpropyl ketone	MPK
Methyl acetoacetate	MAA

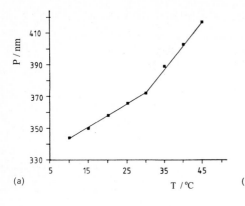

(a)

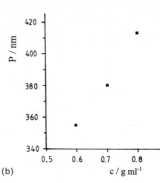

(b)

Figure 14. Pitch P for the right-handed system 3Cl-CTC/DEME as a function of (a) temperature ($c=0.8$ g/ml) and (b) concentration ($T=25\,°C$) (adapted from [4]).

might be due to a conformational change of the cellulosic chain. Different polymer–solvent interactions might be caused by another conformation. Predominant polar interactions are rare, as are twist inversions with changing temperature gradients. (Acetyl)-ethyl cellulose represents an example with positive and negative temperature gradients of the pitch (Fig. 15). The fully trisubstituted derivative ($DS_{eth}=2.5$ fixed, $DS_{ac}=0.5$) exhibits predominantly steric interactions with a positive slope in the diagram. This situation changes when enough free hydroxyls are present for the not fully substituted derivatives, and polar interactions are introduced. The turning point occurs in the vicinity of the twist inversion at $DS_{ac}=0.19$ as predicted by Eq. (5).

6.4   Concentration Dependence of the Pitch

Theoretical considerations are rare for the concentration dependence of LC cellulosic/solvent systems and the same is true for experimental studies. In many instances it is difficult to keep a constant concentration over a longer time during which experiments are carried out. If a functionality $P=kc^{-m}$ is assumed (k: a constant, c: concentration, m: a variable that might depend on the stiffness of the molecules as well as on the solvent, the temperature, and the molecular mass), the value of m is 2 for somewhat stiff LC poly-γ-benzyl-L-glutamate in dioxane and 3 for semi-flexible LC HPC in water. A detailed study of LC CTC/MAA [17] led to a value of $m=2.4$ for lower concentrations between 38 and 44 wt%, independent of the temperature. At a higher concentration of 44–50 wt%, a value of $m=1.1$ at 20 °C was determined, which increases

Figure 15. The value of the pitch P as a function of temperature for LC (acetyl)ethyl cellulose/chloroform (45 wt%, DS of ethyl groups 2.5). The filled and unfilled symbols represent the samples with acetyl DS below and over the compensation of the twist (0.19), respectively. Acetyl DS$=0.05$ □, 0.06 ○, 0.08 △, 0.30 ■, 0.42 ●, and 0.50 ▲ (adapted from [1]).

to 2.3 at 35 °C with temperature. Above 35 °C, the exponent m remains constant, indicating the limiting value of the system considered. The smaller value at a higher concentration might be interpreted in terms of a stronger twisting of the nematic sheets being hindered. Since the limiting value for m is between the ones observed for poly-γ-benzyl-L-glutamate and HPC, it can be concluded that CTC exhibits a stiffer chain conformation than HPC. Actually, the persistence length of HPC in dimethyl acetamide is listed as 6.5 nm [69], and for CTC in dioxane as 11 nm [70]. These data confirm the conclusions drawn.

Generally, a decrease in the pitch with increasing concentration occurs for most lyotropic LC cellulosics. However, in some cases, such as CTC/ketone systems or CTC in some glycols, an increase in pitch is observed (see Fig. 14b), and in rare cases a sharp change has been detected in the concentration gradient of the pitch, quite similar to the temperature gradient.

6.5 Molecular Mass Dependence of the Pitch

An increase of the pitch with molecular mass has been observed for LC cellulose acetate in trifluoroacetic acid and for CTC in DEME (cf. also Fig. 2d). The pitch rapidly changes at low molecular masses and remains almost constant above a DP of 150 for the left-handed CTC/DEME system [54]. At very high molecular masses, the LC state was not obtained, gels formed instead. A similar leveling off was observed for the clearing temperature T_c (Fig. 16). For the pitch P as well as for the clearing temperature T_c, an analytical expression was derived with fitted parameters as a function of the molecular mass M_w [4]. This observa-

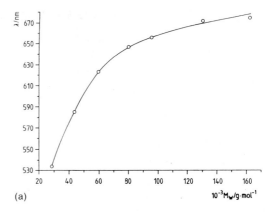

(a)

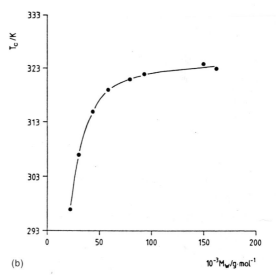

(b)

Figure 16. Molecular mass M_w dependence of (a) the reflection wavelength λ_o and (b) the clearing temperature T_c for narrow fractions of the left-handed system CTC/DEME (0.8 g/ml, $T = 293$ K, method UV–VIS and ORD) (adapted from [54]).

tion is opposite to the findings for LC hydroxypropyl cellulose HPC and acetoxypropyl cellulose APC. In these systems, hydrogen bonding may lead in dilute solution to the formation of irreversible aggregate, as discussed for several cellulosic systems [39], which might result in a different behavior when LC systems are formed.

It was demonstrated that the molecular mass of CTC [54], a derivative of cellulose, which is synthesized without degradation, can be determined by using the graphs in Fig. 16 as well as the molecular mass of cellulose itself. The experimental knowledge of the selective wavelength λ_o or the clearing temperature T_c leads directly to the molecular mass M_w. A rough estimation of the selective color observed in the polarizing microscope already gives good results (see Fig. 2d).

6.6   Solvent Effects

The interaction of nonchiral solvent with the chiral cellulosic chain plays a decisive role in the formation of lyotropic liquid crystals. Depending on this interaction, right- or left-handed supermolecular structures are formed, or no twisting at all may occur; this was extensively studied for a variety of CTC derivatives and solvent systems [11, 55]. NMR and IR experiments suggest that the acidity of the N–H group of the carbamate changes depending on the substituent at the phenyl ring in the meta, para, or ortho position. Likewise, steric hindrance may predominantly occur for certain positions. If these findings are accepted, dipolar interactions are different for protic or aprotic solvents and changes in the LC helicoidal structures are to be expected. Experiments have been carried out to determine the free and bound solvent of CTC dissolved in MAA [17]. Only in the dilute state is all the solvent free. Even in the semi-dilute state most of the solvent is bound, and in the liquid-crystalline phase almost all solvent seems to be bound. On the other hand, light-scattering experiments [11] also clearly indicate that in the semi-dilute state of various CTC in EMMAc and in other glycols clusters are formed, which are of different geometry for small and large molecules and

are capable of binding solvent. These clusters have to be regarded as the basic units of the lyotropic LC state. X-ray diffraction of lyotropic LC cellulosics confirms the cluster formation, since a diffraction ring with a d-spacing of the diameter of the conformational helix of CTC was detected. Such a Bragg reflection can only be observed if packing of the chains is present. A similar diameter of the cellulosic chain was also confirmed by electron diffraction of single crystals of CTC with solvent built in.

The effect of a particular solvent on the chirality of a lyotropic phase can be altered by the mixing of solvents and additivity of the chiral quantity as the twisting power P^{-1} or the pitch is observed in a first approximation. This may even lead to a compensation of the twist, as represented in Fig. 9. Another example is provided in Fig. 17, where CTC was dissolved in a mixture of ketones and the right-handed twist preserved.

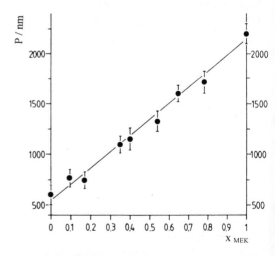

Figure 17. Pitch P of a mixed solvent system: CTC/ MEK + 2-pentanone ($c = 0.6$ g/ml, room temperature, method: Cano rings); x_{MEK}: volume fraction of MEK (methyl ethyl ketone) (adapted from [16]).

6.7   Order Parameter

The order parameters S represent a basic quantity for a thermodynamic description of liquid crystals, and has been determined with an extrapolation method of Haller [12] for several CTC liquid-crystalline systems. A value of S near 0.5 was found at the beginning of the anisotropic–biphasic region (see Fig. 2g), independent of the concentration and molecular mass distribution, and S decreases rapidly in the biphasic region [17]. In the anisotropic region, the order parameter increases slightly with decreasing temperature and reaches approximately $S=0.7–0.8$ for the systems studied. As an example, the dependence of the order parameter S on temperature is shown for the right-handed supermolecular structure of 3Cl-CTC/DEME in Fig. 18. The mean field theory of Maier and Saupe [30] states for a one component system that S has a value of 0.4289 at the first order phase transition at T_c. The order parameter S increases with decreasing temperature T to about 0.65 at a $T^*=T/T_c$ of about 0.9. Although some experimental deviations from the expected value at the phase transition temperature T_c are normally observed for nematic–isotrop-ic transitions, the order parameter shows small deviations at lower temperatures. In this respect, the order parameter of $0.7–0.8$ for the cellulosic systems appears somewhat high at lower temperatures ($T^*=0.9$), especially when considering synthetic polymer systems for which the curves of the order parameter lie far below the ones for small rod-like molecules.

6.8   Phase Behavior

Although several theories for phase transitions exist for stiff and semi-flexible chains, experimental investigations are rare. The published data are mostly concerned with the critical concentration for mesophase formation at a certain temperature. It was found that the critical concentration varies with the nature of the solvent, but is not affected by the degree of polymerization, as predicted by Flory's theory and verified for HPC in dimethylacetamide and acetoxypropyl cellulose (APC) in dibutyl phthalate with polydisperse molecular mass. However, for cellulose acetate in trifluoroacetic acid and cellulose in N-methyl-morpholine-N-oxide, a dependence was found on the degree of polymerization. Careful studies of fractionated HPC in water confirm a molecular mass dependence. A curious intermediate region was detected (Fig. 19) with two transitions which are not due to changes in the nature of the phases but to sharp changes in their composition [72]. It was suggested that the changes occur in the isotropic phase, and the isotropic–anisotropic transition is driven by entropic factors. This phase behavior might be a consequence of the tendency for an aqueous system to demix on heating, which is different from most lyotropic cellulose derivatives.

Studies of phase diagrams of cellulose-tricarbanilate in MAA [17] and EMMAc

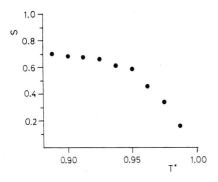

Figure 18. Order parameter S as a function of the reduced temperature $T^* = T/T_c$ for 3Cl-CTC/DEME; $c = 0.7$ g/ml (adapted from [71]).

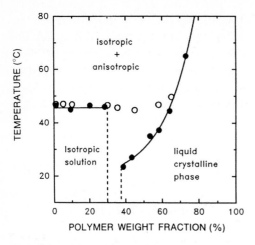

Figure 19. Phase diagram of a narrow fraction of HPC (molar substitution $MS = 5.0$, $M_w = 209 \times 10^3$ g/mol) in H_2O. Onset of phase separation by turbidimetry ●, transition temperatures with scanning Calvet calorimeter ○ (adapted from [2] and [72]).

[11] as solvents show basically very little influence of the molecular mass or the molecular mass distribution above a certain degree of polymerization (Fig. 20); a narrow biphasic region, almost independent of the molecular mass distribution, and a bending

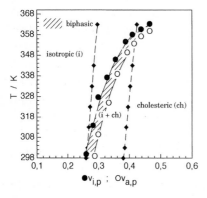

Figure 20. Phase diagram for the right-handed system 3Cl-CTC/EMMAc. Experimentally determined phase boundaries ● ○; theory of Warner and Flory with anisotropic dispersion interaction and an aspect ratio of $x = 20$ ◆ ◆. V_p^i, V_p^a volume fraction of the polymer in the isotropic and anisotropic phase at equilibrium, respectively (adapted from [11]).

of the curves at higher temperatures are apparent. This bending might be due to a change in the Kuhn segment length at elevated temperatures. Best agreement with theoretical considerations is obtained with the Warner–Flory lattice model with anisotropic interactions. However, the small biphasic region cannot be explained. The Kuhn segment length for 3Cl-CTC in EMMAc was determined as 37 nm at room temperature, and the diameter of CTC amounts to approximately 18 Å (1.8 nm) from diffraction studies. Therefore an aspect ratio of $x = 20$ seems appropriate.

An interesting phenomenon occurs when two quite similar cellulose derivatives, CTC and 3Cl-CTC, are mixed in one solvent, TRIMM. Complete mixing occurs at the borders of the polymer compositions, but in the vicinity of a 1:1 ratio demixing takes place and can be studied by measuring the pitch of the liquid-crystalline phase (Fig. 21). A single pitch is observed at the borders and two pitches near the center of the graph. This finding is confirmed by observations in the polarizing microscope, where one selective color appears at each border, red and blue, respectively, and both colors are seen in various domains for a 1:1 mixture. ORD measurements, as depicted in Fig. 22, can also be regarded as proof of the demixing at a 1:1 polymer ratio.

According to recent investigations, cluster formation and bound solvent have to be considered, which will influence the phase behavior considerably. Also, the interpretation of careful measurements has to take into account the fact that a high molar mass fraction mixed into the sample will not turn into a liquid crystal, and a too low molar mass fraction may have an influence on the clearing temperature, as shown in Fig. 16.

At higher concentrations, a further anisotropic phase termed the columnar phase has been observed for some lyotropic LC cellu-

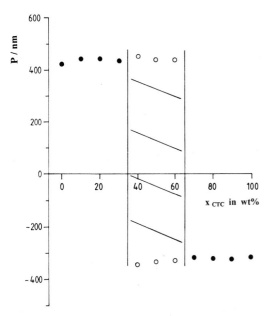

Figure 21. Pitch P of the ternary system CTC + 3Cl-CTC/TRIMM as a function of the composition of CTC/3Cl-CTC ($c = 0.8$ g/ml; polymer mixture/solvent; $T = 293$ K); x_{CTC}: weight fraction of CTC (adapted from [73]).

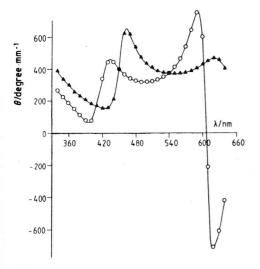

Figure 22. Optical rotatory power Θ as a function of wavelength for the ternary system CTC + 3Cl-CTC/TRIMM with 50 wt% CTC ▲ and two adjacent thin samples of CTC/TRIMM and 3Cl-CTC/TRIMM ○ (adapted from [71].

losics. A phase transition from the chiral nematic phase to the columnar one was confirmed by DSC measurements [4]. X-ray fiber diffraction led to a hexagonal packing, and an orientation occurs along the cellulosic chain axis when the samples are stretched.

7 Thermotropic Liquid Crystals of Cellulosics

Cellulose and oligocellulose derivatives have been studied and a variety of thermotropic LC phases established, i.e., cholesteric and columnar structures for cellulose derivatives, discotic columnar and smectic-type for the oligomers, depending on the side group and the main-chain lengths. A list of compounds exhibiting thermotropic mesophases is presented in Table 5.

In particular, derivatives of the easily available HPC were used in early investigations, although experimental data are sometimes difficult to obtain and to interpret because of high clearing temperatures causing degradation and irreversible changes during the measurements. Later, fully substituted derivatives, e.g., trialkyl cellulose, cellulose trialkanote, and trialkylester of (tri-O-carboxymethyl) cellulose, were synthesized for a comparison of their thermal behavior and theoretical studies. In a further development, regio-selectively substituted derivatives as 6-O-α-(1-methylnaphthalene)-2,3-O-pentyl cellulose [68] were introduced for chirooptical investigations.

2-Ethoxypropyl cellulose [59], an ethyl ether of HPC forms excellent thermotropic and lyotropic mesophases, the lyotropic ones with acetonitrile, dioxane, and methanol. Both thermotropic and lyotropic systems exhibit cholesteric phases with a right-handed helicoidal supermolecular structure,

Table 5. Cellulose derivatives forming thermotropic LC mesophases.

Compound	Remarks
2-Acetoxypropyl cellulose (APC) [74, 75]	thin films
2-Hydroxypropyl cellulose (HPC) [76]	
Trifluoroacetate ester of HPC [77]	
Propanoate ester of HPC [78]	
Benzoate ester of HPC [79]	
2-Ethoxypropyl cellulose [59]	right-handed
Acetoacetoxypropyl cellulose [62]	
Trifluoroacetoxypropyl cellulose [77, 80]	
Phenylacetate and 3-phenylpropionate of HPC [57]	banded texture
Phenylacetoxy, 4-methoxyphenylacetoxy, *p*-tolylacetoxy cellulose, trimethylsilyl cellulose [63]	leading to cellulose
Trialkyl cellulose (TALC) [81]	
Cellulose trialkanoate (CTAL) [82]	
Trialkyl ester of (tri-*O*-carboxymethyl) cellulose (CMC) [83]	
Oligocellulose derivatives [84, 3] tri-*O*-(2-methoxyethoxy)ethyl cellulose, tri-*O*-heptyl cellulose (DP = 11)	right- → left-handed
6-*O*-α-(1-methylnaphthalene)-2,3-*O*-pentyl cellulose [68]	right-handed, biphasic

and exhibit excellent ORD and CD spectra (Fig. 23) in the thermotropic state. The pitch dependence on temperature for thermotropic mesophases follows Kimura et al.'s [25] predictions, although it should be emphasized that the empirical relationship

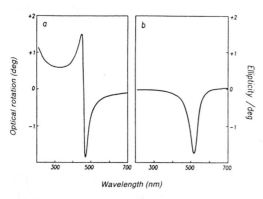

Figure 23. Spectra by ORD (a) at 140 °C and CD, (b) at 152 °C of a thin layer of ethoxypropyl cellulose that forms a thermotropic right-handed helicoidal structure (adapted from [2]).

[Eq. (7)] established for lyotropic LC poly(alkyl) glutamates holds also for different esters of HPC [75]. In addition, these derivatives exhibit a coinciding clearing and compensation temperature, T_c and T_n, which might just happen accidentally. However, 2-ethoxypropyl cellulose shows a finite pitch at the clearing temperature, as is generally observed for lyotropic cellulose derivatives. X-ray investigations on thermotropic (acetoxypropyl) cellulose [75] cast as a film or on 3-phenylpropionate of HPC [57], stretched and exhibiting a banded texture upon relaxation, show a reflection at a *d*-spacing correlated with the diameter of the helix chain conformation [16 Å (1.6 nm)]. For the latter derivative, a meridional reflection of the well-oriented fiber pattern was clearly visible, corresponding to a fiber repeat of 5.02 Å (0.502 nm), which is the rise per glucose unit in cellulosics. From these data on the stretched sample, cluster formation in the thermotropic LC phase has been concluded for well-aligned chains (probably a 3/2 helix).

A thorough investigation on thermotropic LC cellulosics has been carried out by the Kyoto group (see [3]) on the thermal behavior and phase transitions of fully substituted trialkyl cellulose (TALC), cellulose trialkanate (CTAL), and the trialkyl ester of (tri-*O*-carboxymethyl) cellulose (CMC) with a variation of side-group lengths. Besides regular cholesteric phases, a hexagonal columnar phase of high viscosity and low birefringence for CTAL was observed (see Fig. 24). The clearing temperature depends differently on the lengths of the side groups for ethers and esters, as does the entropy change ΔS_c, supporting the idea that different structured phases are present for the ester and ether compounds.

The dependence of the clearing temperature T_c of trialkyl cellulose on the side-group lengths can be predicted if a reference sample with known persistence length q of the cellulosic chain and diameter D is available at T_c. The following relationship for the segmental axial ratio holds

$$x = \frac{q}{D} = \frac{q_r}{D_r} \tag{9}$$

where r denotes reference. The temperature dependence of q can be taken from intrinsic viscosity and light-scattering studies; $q = 432 \exp(-5.0 \times 10^{-3} T)$ (Å) for triheptyl cellulose (300 K $< T <$ 370 K). D may be estimated. Figure 25 shows the comparison between this estimation and experimental data.

The importance of the chain length (DP) dependence of the thermotropic systems is stressed by an investigation on cellulose tridecanoate (CTD). Narrow fractions have been used and the clearing (T_c) and melting (T_m) temperatures studied as a function of DP (see Fig. 26). T_m appears almost independent of DP, whereas T_c drops for small DP. For DP < 5, a discotic columnar phase was established, while for longer chains a hexagonal columnar structure was present. The molecular axis thus changes in the columns: perpendicular for small DP and parallel for large DP. This change occurs at a DP of about 10, but cannot be observed because the melting temperature here lies above T_c.

An interesting phenomenon occurs for the oligomeric tri(2-methoxyethoxy) ethyl cellulose (TMEC) [85, 86] and triheptyl cel-

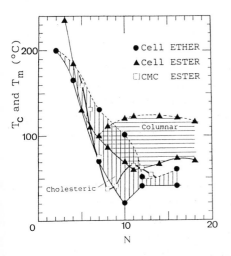

Figure 24. Dependence of side-group length N on T_c and T_m for thermotropic cellulose derivatives: trialkyl cellulose ●, cellulose trialkonates ▲, trialkylesters of CMC □ (no mesophase) (adapted from [3]).

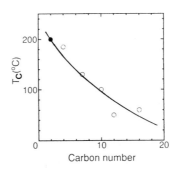

Figure 25. Clearing temperature T_c of trialkyl cellulose as a function of carbon number of the side group. Prediction: solid curve (adapted from [3]).

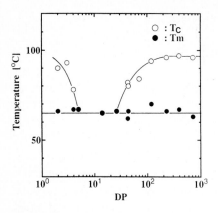

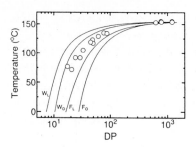

Figure 28. Molecular mass (DP) dependence of the clearing temperature T_c for narrow fractions of triheptyl cellulose (THC). Solid line: predictions by various models (adapted from [3]).

Figure 26. Clearing (T_c) and melting (T_m) temperatures of fully decanoated cellulose as a function of DP (adapted from [3]).

lulose (THC) [86], where the rare case of a temperature-induced inversion of the twist is observed, represented by the twisting power $P^{-1} = 0$ in Fig. 27. The compensation temperature T_n increases with chain length for THC, as does the clearing temperature T_c. At high DP, however, T_c levels off (Fig. 28), as observed for the lyotropic LC CTC/DEME system. This behavior can qualitatively be predicted by an Onsager-type model or a lattice type model with freely jointed chains or worm-like chains (Fig. 28) (see [3]).

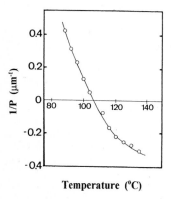

Figure 27. Twisting power P^{-1} of an oligomeric tri(2-methoxyethoxy) ethyl cellulose (TMEC) (adapted from [3]).

8   Concluding Remarks

A basic understanding of the structure and behavior of liquid-crystalline cellulosics has yet to evolve. From a conceptual point of view, the chirality of the cellulosic chain is most sensitively expressed in the supermolecular structure of the cholesteric phase, which may be described by the twisting power P^{-1} or the pitch. At present, no information is available about domains or domain sizes (correlation lengths) of supermolecular structures. The chirality in the columnar phases has not been addressed at all. The principal problem, i.e., how does chirality on a molecular or conformational level promote chirality on the supermolecular level, has not been solved. If this correlation were known, it would enable the determination of the conformation of cellulosic chains in the mesomorphic phase and the development of models for the polymer–solvent interactions for lyotropic systems. On the other hand, direct probing of this interaction would provide a big leap towards an understanding of lyotropic phases.

Turning to the phase diagram: The free energy that determines the phase behavior of a system contains chiral and nonchiral quantities. Several models that are used for the description of cholesteric phases do not contain chiral properties. On the other hand, the basic building units in the application of theories concerning lyotropic phases are still not clear. Although a big impact in progress for an understanding of the cholesteric phases would result from statistical and thermodynamic considerations, the fine details have to be introduced by chirality evaluation through experimental data as a first step.

Further important data for an understanding of liquid-crystalline phases of celluloscis can be provided by a number of investigations, such as flow behavior, induced circular dichroism, and many more. A collection of literature and details is given in [2]. We have limited ourselves to a discussion of basic chirality and phase behavior.

9 References

[1] J.-X. Guo, D. G. Gray in *Cellulosic Polymers* (Ed.: R. D. Gilbert), Hanser, Munich **1994**, p. 25.

[2] D. G. Gray, B. R. Harkness in *Liquid Crystalline and Mesomorphic Polymers* (Eds.: V. P. Shibaev, L. Lam), Springer, New York **1994**, p. 298.

[3] T. Fukuda, A. Takada, T. Miyamoto in *Cellulosic Polymers* (Ed.: R. D. Gilbert), Hanser, Munich **1994**, p. 47.

[4] P. Zugenmaier in *Cellulosic Polymers* (Ed.: R. D. Gilbert), Hanser, Munich **1994**, p. 71.

[5] R. D. Gilbert in *Agricultural and Synthetic Polymers, ACS Symp. Ser. 433* (Eds.: J. E. Glass, G. Swift), American Chemical Society, Washington, DC **1990**, p. 259; R. D. Gilbert in *Polymeric Materials Encyclopedia* (Ed.: J. C. Salamone), CRC, Boca Raton, FL **1996**, p. 118.

[6] P. Haurand, Dissertation, TU Clausthal, Clausthal-Zellerfeld, Germany **1990**.

[7] M. Siekmeyer, P. Zugenmaier in *Cellulose: Structural and Functional Aspects* (Eds.: J. F. Kennedy, G. O. Phillips, P. A. Williams), Ellis Horwood, Chichester **1989**, p. 347.

[8] M. Siekmeyer, Dissertation, TU Clausthal, Clausthal-Zellerfeld, Germany **1989**.

[9] I. Dierking, H. Stegemeyer in *Chiral Liquid Crystals* (Eds.: L. Komitov, S. T. Lagerwall, B. Stebler), World Scientific Publishing, Singapore, unpublished.

[10] M. C. Maugin, *Bull. Soc. Fr. Minéral. Cristallogr.* **1911**, *34*, 71.

[11] E. Klohr, Dissertation, TU Clausthal, Clausthal-Zellerfeld, Germany **1995**.

[12] I. Haller, *Progr. Solid State Chem.* **1975**, *10*, 103.

[13] H. De Vries, *Acta Cryst.* **1951**, *4*, 219.

[14] P. H. Hermans, P. Platzek, *Kolloid Z.* **1939**, *88*, 68.

[15] Kramers–Kronig relationship, see H. Ibach, H. Lüth, *Festkörperphysik*, Springer, Berlin **1988**, p. 251 (and other texts).

[16] P. Zugenmaier in *Cellulose: Structure, Modification and Hydrolysis* (Eds.: R. A. Young, R. M. Rowell), Wiley, New York **1986**, p. 221.

[17] P. Haurand, P. Zugenmaier, *Polymer* **1991**, *32*, 3026.

[18] F. Grandjean, *C. R. Acad. Sci. Fr.* **1921**, *72*, 91; R. Cano, *Bull. Soc. Fr. Minéral. Cristallogr.* **1968**, *91*, 20.

[19] U. Vogt, P. Zugenmaier, *Makromol. Chem., Rapid Commun.* **1983**, *4*, 759.

[20] P. N. Keating, *Mol. Cryst. Liq. Cryst.* **1969**, *8*, 315.

[21] B. Böttcher, *Chemiker-Zeitung* **1972**, *96*, 214.

[22] W. J. A. Goossens, *Mol. Cryst. Liq. Cryst.* **1971**, *12*, 237.

[23] Y. R. Lin-Liu, Y. M. Shih, C.-W. Woo, H. T. Tan, *Phys. Rev. A* **1976**, *14*, 445; Y. R. Lin-Liu, Y. M. Shih, C.-W. Woo, *Phys. Rev. A* **1977**, *15*, 2550.

[24] L. Varichon, A. Ten Bosch, P. Sixou, *Liq. Cryst.* **1991**, *9*, 701.

[25] H. Kimura, M. Hosina, H. Nakano, *J. Phys. (France)* **1979**, *30*, C3-174; *J. Phys. (Jpn.)* **1982**, *51*, 1584.

[26] M. A. Osipov, *Chem. Phys.* **1985**, *96*, 259; *Nuovo Cimento* **1988**, *10D*, 1249.

[27] L. Onsager, *Ann. N. Y. Acad. Sci.* **1949**, *51*, 627.

[28] A. R. Khokhlov, A. N. Semanov, *Physica* **1981**, *108A*, 546; **1982**, *112A*, 605.

[29] P. J. Flory, *Proc. R. Soc. London Ser. A* **1956**, *234*, 73; P. J. Flory, G. Ronca, *Mol. Cryst. Liq. Cryst.* **1979**, *54*, 289 and 311.

[30] W. Maier, A. Saupe, *Z. Naturforsch.* **1958**, *13a*, 564; **1959**, *14a*, 882; **1960**, *15a*, 287.

[31] M. Warner, P. J. Flory, *J. Chem. Phys.* **1980**, *73*, 6327.

[32] H. Chanzy, A. Péguy, Y. S. Chaunis, P. Monzie, *J. Polym. Sci., Polym. Phys. Ed.* **1980**, *18*, 1137.

[33] D. L. Patel, R. D. Gilbert, *J. Polym. Sci., Polym. Phys. Ed.* **1981**, *19*, 1231.

[34] K.-S. Yang, M. H. Theil, J. A. Cuculo in *Polymer Association Structures, ACS Symposium Series 384* (Ed.: M. A. El-Nokaly), American Chemical Society, Washington, DC **1989**, p. 156.

[35] J.-F. Revol, H. Bradford, J. Giasson, R. H. Marchessault, D. G. Gray, *Int. J. Biol. Macromol.* **1992**, *14*, 170.

[36] E. Bianchi, A. Ciferri, G. Conio, A. Tealdi, *J. Polym. Sci., Polym. Phys. Ed.* **1989**, *27*, 1477.

[37] C. L. McCormick, P. A. Callais, B. H. Hutchinson, *Macromolecules* **1985**, *18*, 2394.

[38] M. Terbojevich, A. Cosani, G. Conio, A. Ciferri, E. Bianchi, *Macromolecules* **1985**, *18*, 640; E. Bianchi, A. Ciferri, G. Conio, A. Cosani, M. Terbojevich, *Macromolecules* **1985**, *18*, 646.

[39] W. Burchard, *TRIP* **1993**, *1*, 192.

[40] P. Navard, J. Haudin, *Br. Polym. J.* **1980**, *12*, 174.

[41] I. Quenin, H. Chanzy, M. Paillet, A. Péguy in *Integration of Fundamental Polymer Science and Technology* (Eds.: L. A. Kleintjens, P. J. Lemstra), Elsevier Appl. Science, London **1986**, p. 593.

[42] Y.-S. Chen, J. A. Cuculo, *J. Polym. Sci., Polym. Chem. Ed.* **1986**, *24*, 2075.

[43] R. S. Werbowyj, D. G. Gray, *Mol. Cryst. Liq. Cryst. (Lett.)* **1976**, *34*, 97.

[44] R. S. Werbowyj, D. G. Gray, *Macromolecules* **1980**, *13*, 69.

[45] J. Bheda, J. F. Fellers, J. L. White, *Colloid Polym. Sci.* **1980**, *258*, 1335.

[46] Y. Onogi, J. L. White, J. F. Fellers, *J. Non-Newt. Fluid. Mech.* **1980**, *7*, 121.

[47] Y. Onogi, J. L. White, J. F. Fellers, *J. Polym. Sci., Polym. Phys. Ed.* **1980**, *18*, 663.

[48] T. Tsutsui, R. Tanaka, *Polym. J.* **1980**, *12*, 473.

[49] S. Suto, W. Nishibori, K. Kudo, M. Karasawa, *J. Appl. Polym. Sci.* **1989**, *37*, 737.

[50] D. L. Patel, R. D. Gilbert, *J. Polym. Sci., Polym. Phys. Ed.* **1981**, *19*, 1449.

[51] G. H. Meeten, P. Navard, *Polymer* **1982**, *23*, 1727.

[52] U. Vogt, P. Zugenmaier, *Ber. Bunsenges. Phys. Chem.* **1985**, *89*, 1217.

[53] P. Zugenmaier, P. Haurand, *Carbohydr. Res.* **1987**, *160*, 369.

[54] M. Siekmeyer, P. Zugenmaier, *Makromol. Chem., Rapid Commun.* **1987**, *8*, 511.

[55] M. Siekmeyer, P. Zugenmaier, *Makromol. Chem.* **1990**, *191*, 1177.

[56] A. M. Ritcey, K. R. Holme, D. G. Gray, *Macromolecules* **1988**, *21*, 2914.

[57] H. Steinmeier, P. Zugenmaier, *Carbohydr. Res.* **1988**, *173*, 75.

[58] A. San-Torcuato, *Diploma Thesis*, Institut für Physikalische Chemie der TU Clausthal, Clausthal-Zellerfeld, Germany **1989**; P. Zugenmaier, *Das Papier* **1989**, *43*, 58.

[59] A. M. Ritcey, D. G. Gray, *Macromolecules* **1988**, *21*, 1251.

[60] J.-X. Guo, D. G. Gray, *Macromolecules* **1989**, *22*, 2082.

[61] J.-X. Guo, D. G. Gray, *Macromolecules* **1989**, *22*, 2086.

[62] W. P. Pawlowski, R. D. Gilbert, R. E. Fornes, S. T. Purrington, *J. Polym. Sci., Polym. Phys. Ed.* **1987**, *25*, 2293.

[63] W. P. Pawlowski, R. D. Gilbert, R. E. Fornes, S. T. Purrington, *J. Polym. Sci., Polymer Phys. Ed.* **1988**, *26*, 1101.

[64] P. Zugenmaier, C. Derleth in *Recent Advances in Cellulose Derivatives, ACS Symposium Series* (Eds.: T. Heinze, W. Glasser), American Chemical Society, Washington, DC, unpublished.

[65] B. R. Harkness, D. G. Gray, *Can. J. Chem.* **1990**, *68*, 1135.

[66] N. H. Hartshone, *The Microscopy of Liquid Crystals*, Microscope Publications, London, England **1974**, p. 80.

[67] T. Kondo, T. Miyamoto, *Polymer*, submitted.

[68] B. R. Harkness, D. G. Gray, *Macromolecules* **1991**, *24*, 1800.

[69] G. Conio, E. Bianchi, A. Ciferri, A. Tealdi, M. A. Aden, *Macromolecules* **1983**, *16*, 1264.

[70] A. K. Gupta, J. P. Cotton, E. Marchal, W. Burchard, H. Benoît, *Polymer* **1976**, *17*, 363.

[71] H. Steinmeier, Dissertation, TU Clausthal, Clausthal-Zellerfeld, Germany **1988**.

[72] L. Robitaille, N. Turcotte, S. Fortin, G. Charlet, *Macromolecules* **1991**, *24*, 2413.

[73] P. Zugenmaier, *Das Papier* **1988**, *42*, 673.

[74] S. L. Tseng, A. Valente, D. G. Gray, *Macromolecules* **1981**, *14*, 715.

[75] G. V. Laivins, D. G. Gray, *Polymer* **1985**, *26*, 1435.

[76] K. Shimamura, J. L. White, J. F. Fellers, *J. Appl. Polym. Sci.* **1981**, *26*, 2165; S. Dayan, P. Maissa, M. J. Vellutini, P. Sixou, *J. Polym. Sci., Polym. Lett. Ed.* **1982**, *20*, 33.

[77] S. M. Aharoni, *J. Polym. Sci., Polym. Lett. Ed.* **1981**, *19*, 495.

[78] S. L. Tseng, G. V. Laivins, D. G. Gray, *Macromolecules* **1982**, *15*, 1262.

[79] S. N. Bhadani, D. G. Gray, *Makromol. Chem., Rapid Commun.* **1982**, *3*, 449. S. N. Bhadani, S.-L. Tseng, D. G. Gray, *Makromol. Chem.* **1983**, *184*, 1727.

[80] P. Navard, A. E. Zachariades, *J. Polym. Sci., Polym. Phys. Ed.* **1987**, *25*, 1089.

[81] T. Yamagishi, Ph. D. Thesis, Kyoto University, Japan **1989**.

[82] T. Yamagishi, T. Fukuda, T. Miyamoto, Y. Yakoh, Y. Takashina, J. Watanabe, *Liq. Cryst.* **1991**, *10*, 467.

[83] T. Fukuda, M. Sugiura, A. Takada, T. Sato, T. Miyamoto, *Bull. Inst. Chem. Res.*, Kyoto University **1991**, *69*, 211.

[84] T. Itoh, A. Takada, T. Fukuda, T. Miyamoto, Y. Yakoh, J. Watanabe, *Liq. Cryst.* **1991**, *9*, 221.

[85] T. Yamagishi, T. Fukuda, T. Miyamoto, T. Ischizuka, J. Watanabe, *Liq. Cryst.* **1990**, *7*, 155.

[86] T. Yamagishi, T. Fukuda, T. Miyamoto, J. Watanabe in *Cellulose: Structural and Functional Aspects* (Eds.: J. F. Kennedy, G. O. Phillips, P. A. Williams), Ellis Horwood, Chichester **1989**, p. 391.

Index Vol. 1–3